Handbook of Science and Technology Studies

HANDBOOK OF SCIENCE AND TECHNOLOGY STUDIES

Sponsored by the Society for Social Studies of Science

HANDBOOK OF SCIENCE AND TECHNOLOGY STUDIES

SHEILA JASANOFF
GERALD E. MARKLE
JAMES C. PETERSEN
TREVOR PINCH
EDITORS

PUBLISHED IN COOPERATION WITH THE
SOCIETY FOR SOCIAL STUDIES OF SCIENCE

SAGE Publications
International Educational and Professional Publisher
Thousand Oaks London New Delhi

For information address:

SAGE Publications, Inc.
2455 Teller Road
Thousand Oaks, California 91320

SAGE Publications Ltd.
6 Bonhill Street
London EC2A 4PU
United Kingdom

SAGE Publications India Pvt. Ltd.
M-32 Market
Greater Kailash I
New Delhi 110 048 India

Library of Congress Cataloging-in-Publication Data

Main entry under title:

Handbook of science and technology studies / editors, Sheila Jasanoff
. . . [et al.].
p. cm.
Published in cooperation with the Society for Social Studies of Science.
Includes bibliographical references and index.
ISBN 978-0-7619-2498-2
1. Science. 2. Technology. I. Jasanoff, Sheila. II. Society for Social Studies of Science.
Q158.5.H36 1994
306.4'5—dc20 94-16787

95 96 97 98 99 10 9 8 7 6 5 4 3 2 1

Sage Production Editor: Astrid Virding

Contents

Foreword

THE Society for Social Studies of Science (4S) was founded to "promote research, learning, and understanding in the social analysis of science" with membership open to all interested in the "social and policy aspects of science." From the outset, the 4S has had two notable characteristics: multidisciplinarity and internationalism in its membership and their contributions. Thus in 1988, when the 4S proposed sponsorship of a handbook on science and technology studies in collaboration with Sage Publications, we sought to promote a volume with breadth in contributions, by national region, discipline, and theoretical and methodological perspectives. The handbook project, and the volume now produced, were designed to reflect, and we hope sustain, the vitality of the 4S and the contributions of those working in the study of science and technology.

Gratitude for the project is extended throughout the 4S. For their long and committed work on the volume, appreciation goes to the editors. The authors were fundamental to the volume, as were the reviewers, who read and refereed each contribution, and we thank them. The 4S Handbook Committee served as an editorial and advisory board for the project. Appreciation goes to each member for his or her dedication and responsiveness. In addition, over the course of the project, four 4S presidents were involved: Presidents Arie Rip, Harriet Zuckerman, Harry Collins, and Sal Restivo lent their wisdom, judgment, and experience to the project's inception, development, and completion. Finally, for his colleagueship, good humor, and reliable good sense, we are grateful to Mitch Allen of Sage Publications.

We believe that this *Handbook* captures the energy and directions in the field, and it is our hope that the volume has a role in generating continuing research and learning in science and technology studies.

Mary Frank Fox
Handbook Committee

Introduction

EDITING a handbook, especially for a still emerging field such as science and technology studies (STS), is like constructing a map of a half-seen world. For the editors, shaping this volume has been as much an act of imaginative risk taking as of diligent codification. What is the form of this world that we call STS? What are its divisions and boundaries? How might it be split into continents, let alone lesser domains, in ways that are fair to all its inhabitants? What, to begin with, does STS stand for? Is it "science, technology, and society" as in the last handbook for the field, or is it time now to adopt the newer guise of S&TS—"science and technology studies"? In asking these questions, the editors found that they had to reenvision their own role: It would not be possible to act as "neutral" gazetteers of already charted territory; any map making for STS would necessarily entail statecraft as well as politics.

Like all good cartographers, we began by drawing the meridians and parallels, seeking to divide the spaces as even-handedly as possible. Our national and disciplinary backgrounds served us well in this enterprise. Among ourselves, we could claim reasonable familiarity with most recent strands of sociological, historical, political, and legal studies of STS. Building on these perspectives, we drafted a "proposal" that was approved by the Society for Social Studies of Science (4S), the sponsor of this *Handbook*. Even at this early stage, however, we conceived of the project as something more than the traditional, treatiselike handbook that would clinically describe the world of STS. The field, in our view, had not yet achieved the hoary respectability that merits such dispassionate, and unimaginative, treatment.

To be sure, we wanted to compile scholarly assessments of the literature that could be presented to neophyte graduate students as the state of the art in STS. We also wanted definitive road maps of the terrain—careful summaries of work done in the 1970s and 1980s that would set new STS researchers on course for the 1990s. Equally, we wanted the book to project the field's broad interdisciplinary and international outlook. We hoped that no single depiction—especially that of one disciplinary specialty or national tradition—would dominate. But we wanted, above all, to capture for readers who come fresh to STS a little of the excitement and unpredictability that have drawn scholars from such a diversity of backgrounds to claim STS as their primary intellectual home.

We conceived of this *Handbook*, then, as presenting an unconventional but arresting atlas of the field at a particular moment in its history. Not all the chapters would be of equal length and density or identical in theoretical and methodological orientation. Where appropriate, the book would present different "takes" on the same issues and several chapters would deal with them in a lively, theoretically informed fashion. Included in our first proposal therefore were empirical topics such as the human genome project and computers, which seemed to be turning into focal points for some of the most original work in STS. We are happy to report that this aspect of our proposal survived the politics of map making. Chapters by Harry Collins on artificial intelligence and science studies, by Paul Edwards on computers, and by Stephen Hilgartner on the human genome project all attest to the richness of STS's engagement with these novel areas of science and technology.

We also encouraged authors addressing more established subjects to challenge and stretch the standard definition of a handbook chapter. Thus Wiebe Bijker metaphorically undergirds his essay on technology studies with a case study of the constructions that keep the Netherlands afloat—the technological system of drains, dikes, and polders. Similarly, Brian Martin and Evelleen Richards draw on their own researches on the vitamin C-cancer and fluoridation debates to support their general argument about scientific controversies. Malcolm Ashmore, Greg Myers, and Jonathan Potter adopt a consciously self-referential strategy in reviewing work on the discourse and rhetoric of science: By writing in an unconventional diary form, they exemplify the very experiments in "new literary forms" that their chapter surveys. Evelyn Fox Keller fittingly couches her review of gender and science in the style of personal reflection.

Although the initial proposal successfully defined the oceans and continents, we felt that the lesser territorial markers—mountains, rivers, and lakes —should be supplied by our colleagues in STS. As editors, we were most concerned about the risk of creating black holes, but we consoled ourselves with the thought that this was mixing metaphors.

To fill out the details of the map, we advertised widely in professional STS journals, newsletters, and the like, and we solicited contributions from lists of established scholars in the field. The process of solicitation provided our second major lesson in political geography. As some 160 prospective authors responded with chapter outlines, some imagined countries disappeared and even a continent was threatened. Other countries were balkanized as authors vigorously asserted a more refined geographic vision and sensibility than the editors possessed. Some challenging proposals arrived unsolicited as word of the handbook project spread. There seemed to be no shortage of authors who wanted to place their personal stamp on the contours of the field.

In the negotiations between authors and editors, the history of science and technology, which has played a pivotal role in recent science studies, disappeared from the map as a discrete entity. Although there is no single chapter reviewing historical studies in STS, it is not far-fetched to say that historical methods have left their imprint on the field as a whole, contributing greatly to the convergence of "science, technology, and society" with "science and technology studies." To a lesser extent, a similar statement could be made about the philosophy of science. Perhaps more surprising to some readers, the area of stratification, once the heartland of sociology of science, attracted little notice from potential contributors. Where the gaps seemed important enough, as in this particular case, we encouraged other authors to cover the ground. Thus Mary Frank Fox in her chapter on women in science draws heavily on the stratification literature.

While some familiar territories were being annexed to others, new nation-states were also arising. Proposals came flooding in for chapters on topics such as rhetoric of science, gender and science, science policy and politics, and various forms of technology studies. The field, it seemed, was intent on defining itself in ways not initially contemplated. We decided to accept this movement toward self-definition. Rather than continue the search for authors to occupy every vacant slot in the proposal, we decided to redraw the boundaries so as to include more of the topics that authors did wish to address. As a result, STS theory, for example, is examined both through the lens of a French actant-network theorist, Michel Callon, and through the lens of a North American critical sociologist, Sal Restivo. The areas of gender and science, science and politics, and technology studies were additional beneficiaries of our restructuring.

Supplementing the authors and editors, a multitude of referees and the supervisory 4S Handbook Committee offered their own cartographic critiques. A few mountains were moved, some rivers burst their banks, a few were dammed, and, of course, innumerable roadways were rerouted. The editors hope that the result of all this collective effort is a more interesting and comprehensive, if not always more coherent, guide to the field. That our original vision was refined and even partly displaced is no doubt a positive outcome. The landscape now abounds with inviting streams and eddies. There are to be sure a few gorges that are weakly bridged and must be crossed with care; readers will not find in this volume, for example, the ultimate resolution of the realism-relativism debate between philosophers and sociologists of science. In the final instance, however, the picture that emerges from the following pages is, as we had intended, a composite of many individual visions and much communal debate and reflection.

Some lacunae inevitably remain. Most notable is the absence of distinct contributions on the economics of science and technology (although both Bijker and Callon mention such work in passing), on the psychology of science (except insofar as it is touched upon in Wynne's treatment of mental models), on law and science (possibly because there is not yet a sufficient volume of work to survey), and on science and race. The last topic was one that we, as editors, specifically debated. In the end we settled for a policy of avoiding marginalization by requesting that issues of race be raised in several of the chapters. Finally, what of the temptations of postmodernity, not to mention cyborgs and hyperspace? We assure the reader that these are present in the shadows, lurking in the hypertext surrounding our own map. Whether they will consume STS in the next millennium we leave to the next handbook to decide.

David Edge's introductory essay—Chapter 1, "Reinventing the Wheel"—usefully reminds us of the distance that the field of S&TS has traveled since its inception and the important ways in which the obstacles it confronts remain unchanged. As the founder of the Science Studies Unit at Edinburgh and the longtime editor of *Social Studies of Science*, Edge is particularly well placed to provide this historical overview. He traces S&TS from the mid-1960s, when it was still possible to think that there was "no subject" corresponding to these initials, up to the present moment, when a shifting, complex, kaleidoscopic array of research projects are continually grouping and regrouping themselves under the banner of science and technology studies. Edge ends his chapter on a note of caution, however, as he observes how diehard, positivistic notions of science keep reasserting themselves in the public domain despite the best efforts of S&TS scholarship. Finding ways to combat this phenomenon is the major challenge that his chapter holds out to readers of this volume.

A project as ambitious as this one could hardly have failed to be (as it assuredly was in turn) arduous, contentious, frustrating, time consuming, and exhausting. That it has also at times been energizing and fun speaks well not only for the intellectual vitality of STS but for the strength of many of its internal support systems. Thanks are due in particular to all the authors for their remarkable patience and perseverance; to Mary Frank Fox for gently mediating among the often vociferous parties to the project; to Harry Collins and Harriet Zuckerman for lending support as presidents of 4S; to the members of the handbook committee who offered both overt and covert encouragement; to the referees who valiantly reviewed the sometimes unorthodox contributions to the literature; to our secretarial staffs at Cornell and Western Michigan for facilitating complex editorial correspondence; and, not least, to Mitch Allen of Sage Publications for bearing with what must many times have seemed like terminal uncertainty and indecisiveness.

Both editors and the Society owe a special debt of gratitude to the following individuals who provided invaluable critical comments on one or more draft chapters: Harvey Brooks, Frederick Buttel, Cynthia Cockburn, Steven Cole, Peter Dear, Sharon Dunwoody, Michael Gorman, Herbert Gottweis, Donna Haraway, Lowell Hargens, Dale Jamieson, Michael Lynch, Allan Mazur, Peter Meiksins, Judith Perrolle, Andrew Pickering, Judith Reppy, Aire Rip, Steven Shapin, Wesley Shrum, Stephen Turner, Ron Westrum, Peter Whalley, Rick Worthington, Steven Yearly. In coordinating the refereeing of chapters, as in other aspects of producing this volume, the editors attempted to maintain common standards and common vision of the enterprise, as reflected in this jointly written introduction.

A good map should give the reader a sense of familiarity with the unknown, an understanding of previously unsuspected spatial relationships, and an educated eye for the appealing details of the landscape. We hope that ours will meet some of these needs for the once and future traveler in STS. In this spirit of expectation tempered by modesty, we offer this volume to the ever-expanding community that it seeks to represent.

PART I
OVERVIEW

1

□

Reinventing the Wheel

DAVID EDGE

IN the early morning of March 1, 1966, I arrived at Waverley Station in Edinburgh, on the night train from London, to start the Science Studies Unit at Edinburgh University. Later that day, I was shown my bare office: no phone, no books, no bibliographical resources, no files, no staff—indeed, it was tempting to think, *no subject*! To recall it now is to recall a spirit of initiative and intellectual expansion that seems to have vanished from the British academic scene—and, I guess, in the light of the financial constraints associated with a worldwide recession, from the academic scenes of many other industrialized states. Now, 26 years later, the spirit on far too many campuses is more one of intellectual contraction, and the initiative has passed to politicians who, paradoxically, in seeking low-cost expansion in student numbers, are forcing hard-pressed academics to "play safe"—to aim at the mass production of a standard product. In such circumstances, STS can easily be made to seem to be an optional extra, a fancy gloss. Is this the end of an era, or just a passing phase? Has the heady sense of interdisciplinary adventure, of the seductive combination of academic priority and practical urgency, disappeared for good?

AUTHOR'S NOTE: I am most grateful to Sheila Jasanoff for many useful comments and detailed suggestions on this draft, and for creatively mediating the criticisms of a justifiably harsh referee.

Of course not: In some ways we are where we always were, and always will be. For one thing, the illusion that I flirted with in my empty office—that there was "no subject"—was clearly unrealistic, if not wilfully misinformed. At that time, several strong disciplinary traditions were developing their own lively analyses of science in its social context—of science as a social phenomenon. In the mid-1960s, contingent circumstances encouraged these streams to converge, consciously and explicitly. From their diverse confluences, the "subject" of STS emerged. This book commemorates that achievement and points us again to the future. For, while it may be true that, in some countries, contraction is the keynote, optimistic expansion rules elsewhere.[1]

What were these disciplinary streams? They were to be found, especially, in a number of well-established specialties in history and philosophy (increasingly interacting with emerging themes in the history and philosophy of science, to their mutual benefit), in sociology, in anthropology, and (less strikingly) in economics and the political and legal sciences; the list is not complete.[2] These quasi-independent strands still live their own lives, and scholars within them pursue their own careers. And the range of contexts in which this confluence took place is similarly wide: Some scholars had "pure" epistemological and sociological aims; others were teaching students in the sciences and engineering; some were active as policy advisers. Local contingencies set local targets and determined the form of the emerging interdisciplinary dialogue. Since its inception, STS has been marked by its inherent diversity; a thousand flowers have bloomed, and some have withered too. Yet, as this book bears witness, the movement has constructed a new picture of science and technology, with some coherence and great imaginative and practical power.

Maintaining that vision, and realizing its potential, is an ongoing task. I want to argue that the original aims of the STS pioneers are still very much alive and relevant; that the intellectual and diplomatic tasks they entail still offer exciting challenges; but that, for all the progress of the past quarter century, summarized so cogently in the chapters to follow, the insights we have gained still seem, too often, to be almost wantonly disregarded. The potential audiences for our "messages" are stubbornly unreceptive. Sadly, I see others (mainly outside the STS community) refinding the old sense of practical urgency and unease at the social status and role of science and technology, and then, in attempting to relieve that unease, "reinventing the wheel"—in blissful ignorance of the fact that a solid body of scholarship now exists that suggests that the simple "wheel" cannot bear the strain they wish to put upon it. Perhaps the next phase in the development of STS must be a more urgent concern for *communication* and *translation*: for "making real" its true potential.

For the "wheel" that is continually rediscovered and reapplied is the "received view" of science and technology as asocial, impersonal activities—a positivistic, even mechanistic, picture of an endeavor that defines its own logic and momentum, its own values and goals, and legitimates its progress by appealing to the assumption that the authority of nature is independent of, and prior to, the authority of society. The public discussion of science is cast in these terms. There is talk of "the impact of science on society" and of the need for greater "public understanding" of it; of a "linear" model of (applied) science giving rise to technology;[3] of the need for a *moral* response, to impose "values" on an apparently autonomous and uncontrolled process. Sometimes even members of the STS community lapse into such a rationalistic, instrumental, and self-contained view of science. However, in doing so, they swim against the scholarly tide in STS. For, in contrast with this positivist tendency (indeed, often in conscious reaction against it), STS holds out the "new" view of science and technology as essentially and irredeemably *human* (and hence social) enterprises—both in the context that nourishes, supports, and directs them and in their inner character. And this is a triumphant, positive humanism: not the miserable confession that "scientists are only human" because you can catch them making mistakes, getting angry, being secretive and fraudulent. What STS scholarship has gained is a view of science and technology as human *achievements*, hewn precariously out of the recalcitrant ambiguity of nature, the random contingencies of history, and the exacting disciplines of social negotiation by humankind's imagination, ingenuity, and wit. It is a rich and detailed tapestry, and these pages contain its elements; but it is a prospect that fades if it is neglected or not mobilized, a challenge that can deter the naive intruder who stumbles, uninformed, upon the problems it addresses. Even true believers can lapse from its exacting disciplines. The scholarly achievement exits, yet somehow the world's practical problems stubbornly resist its advances. I suppose that I have myself reinvented an old French "wheel"—*plus ça change, plus c'est la même chose*. In some ways, we have come a long way; in others, we are almost where we started. And I want to try to suggest why this is so.

THE MOOD OF THE MID-1960s

Let me take you back to the mid-1960s, for it is important to realize something of the context (or, rather, the variety of contexts) in which STS emerged.[4] The sheer growth of STS scholarship since then has been astonishing;[5] but, in many ways, the challenges the field faces today are exactly the same. The voices from the past have a very contemporary ring. I want to highlight three strands in the thinking that led to the formation of the early interdisciplinary STS groups and shaped the main themes of their work: I am sure that readers in many countries will recognize the rhetoric.

Research on Science as a Social System

It is impossible to remember the 1960s without recalling the intellectual influence of Derek de Solla Price, of his seminal book *Little Science, Big Science* (1963), of his claims for the exponential growth of science (and its consequences), and, in particular, of his famous and oft-repeated quotation:

> It is clear that we cannot go up another two orders of magnitude as we have climbed the last five. If we did, we should have two scientists for every man, woman, child, and dog in the population, and we should spend on them twice as much money as we had. Scientific doomsday is therefore less than a century distant. (Price, 1963, p. 19)[6]

Such claims led to vigorous debate. One point at issue (raising strong echoes of the "is/ought" problem, and not just to a Scottish fan of the work of David Hume) concerned whether or not one should plan for the "leveling off" of expenditure on science. (Derek Price himself contributed to this debate; see Price, 1971.) One influential voice in Britain in favor of such a move was that of Lord Bowden, a technologist who was then a minister of state for education and science in the Labour administration. Bowden thought that we should carefully plot curves of rising expenditure in all scientific specialties and strive to limit their financial ambitions. In a striking exchange in the pages of the *New Scientist*, he was challenged by J. D. Bernal, who thought Bowden's approach was folly. To him, the obvious move was to find those specialties for which the curves were rising fastest—and then, far from *limiting* the expenditure, to *increase* it as fast as possible, to "send the curve right through the roof" (Bernal, 1965; Bowden, 1965a, 1965b, 1965c)! Essentially the same dilemma recurs in recent debates; see, for example, Ziman (1989).

Exchanges of this kind were repeated worldwide and were part of a climate of discussion that began to stress the urgent practical need for research *on science*. We needed, it was argued, to understand its extension and development, its relationship to technology and economic growth—above all, perhaps, we needed to work out how to "get returns on the money we spent on science" before its relentless exponential growth rendered us all penniless.[7] In other words, people began to articulate a need for that kind of knowledge about science that would underpin rational policy decisions on its finance and development. There were calls for a "science of science," and for attempts to devise explicit and rational "science policy" (several items listed in note 2 reflect this missionary zeal). In 1965 the Science of Science Foundation was established in London.[8] Its director, then as now, was Maurice Goldsmith. He neatly expressed the underlying instrumentalism that led to the SSF's foundation when he claimed that "science is a cow that we do not yet know how to milk."[9] This crude metaphor rang many a bell in government offices around the world: Wherever centralized bureaucracies were struggling with decisions on investments in science and technology (and that meant everywhere in the industrialized world), it offered the hope of expert guidance. Science of science found a particularly sympathetic audience in the east European and Soviet countries and in China. Its quantitative drive has since developed into the sophisticated techniques of bibliometrics and "citation studies." This strand within STS has tended to maintain a rationalistic, uncritical view of science (see below).

In parallel with this development, and largely independent of (indeed often hostile to) it, another major thrust of STS scholarship approached the analysis of "science as a social system" from quite a different direction. One disciplinary tradition that had already addressed this issue, and that had to its credit achievements predating the birth of STS, was the work in the sociology of science associated particularly with the name of Robert K. Merton (1973b). But, as a number of critics had early pointed out, this tradition took for granted the essentially positivistic view of science that was also implied in the drive for a rational science policy. Drawing from work in history, sociology, philosophy, anthropology, cognitive psychology, and linguistics, the 1970s saw the emergence of a radical, relativistic new "sociology of scientific knowledge" (SSK). This research had an academic, humanistic aim: to develop an empirically informed view of the social nature of scientific knowledge. It drew much of its initial inspiration from the work of Thomas Kuhn (1962/1970) and, to a lesser extent, from J. D. Bernal (1939, 1954) and Michael Polanyi (1958, 1967). It was largely pioneered in Britain, among scholars working in an educational context, but it has since been taken up, developed, and applied much more widely. As it is one of the main achievements of STS scholarship, its results occupy many of the later pages in this volume. I hope I can therefore be excused from having to summarize it here.[10] The scholarly thrust that it represents has helped to shape and mold many subsequent fruitful innovations: micro-scale ethnographic studies of laboratory practice; the analysis of scientific rhetoric and technical discourse; new approaches in the sociology of technology, such as the emphasis on "sociotechnical systems"; and the often paradoxical subtleties of "actor-network" theory. All this can be seen as stemming from SSK's initial urgings.

Educational Considerations

Another influential context, especially in Europe, was a long-standing dispute over the proper principles for the education of scientists. The early 1960s saw the so-called two cultures debate, sparked by C. P. Snow's famous Rede Lecture at Cambridge (1959; see also his 1971). Not surprisingly, this discussion, like the parallel groping toward a basis for science policy, made assumptions about the unique status of science. It was common to hear talk of a "scientific culture," objective and hard-nosed, contrasted with a "humane culture," which consisted (presumably) of everything else, and was characterized by a regard for subjective feelings and values. Liam Hudson (1962, 1963, 1966), a British educational psychologist, had popularized talk of "Convergers" and "Divergers"—the former of channeled imagination (and tending to choose to study science); the latter exhibiting all the creative virtues (and tending to go elsewhere). Toward the end of the decade, concerns of this kind had given rise, in many countries, to a consideration of the problems raised by the "swing away from science in the schools" and the subsequent reluctance of science graduates to seek industrial employment. How best could we restructure the education of our technical experts so that our investment in their training brought tangible social rewards? There were also voices (many with a Dutch accent) urging that science education should prepare scientists to act with "social responsibility" (see Rip & Boeker, 1974; for an American voice, Brown, 1971, and items in note 12 below).

Taken together, these public discussions led to widespread talk of the necessity to *reform* science education—to "liberalize" it, to "make it more human." It was held both that the alleged "inhuman" face of science discouraged its study among schoolchildren and that this same "depersonalization" led to the production of graduates in the sciences and technology who were ill-suited to the roles they would eventually play in industry and society. Science education, it was commonly argued, was too "narrow," too "specialized"; it was *incomplete*. We needed not only specialists but also "generalists."

Such views led to proposals to incorporate into science education courses on "Contextual Studies" as a clear social need. An influential British report quoted A. B. Pippard, the distinguished physicist, to the effect that science education in the universities had "become fossilized to the point where it was almost impossible for a student to read science as an instrument of general education" (U.K. Committee on Manpower Resources, 1968, Annex E, p. 106). Many believed that it was both necessary and desirable to broaden science education; there should be an attempt "to acquaint the student in some depth with at least a few facets of the working of society, primarily (but not exclusively) in relation to industry, government and education," looking forward to the day "when all science, technology and engineering students gain some understanding of the society in which they will work. Only in this way can our educational system produce a well-balanced output of such graduates" (U.K. Committee on Manpower Resources, 1968, Annex E, pp. 76, 77). The rhetoric was repeated in many countries and, by the end of the decade, innovations on these lines were established. In the Netherlands, for example, some induction in the principles of "social responsibility" was incorporated, by law, in all technical training; and in Britain the newly formed Council for National Academic Awards (CNAA) had formally enshrined a similar requirement in the syllabuses of all its degrees, so creating the "audience" for the U.K. STS Association. (Sadly, and conversely, the decline of the CNAA's influence on degree content in the late 1980s precipitated the sudden end of the STSA.) One major result of these educational innovations was that serious attention was paid to "interdisciplinary" collaboration, in both teaching and (eventually) research; this feature has profoundly influenced the course of STS studies (for a British example of this interest in "interdisciplinarity," see Nuffield Foundation, 1975). And the pedagogical context created by such moves gave considerable scope for the incorporation of STS scholarship (especially the humanistic insights of SSK and its derivatives) into the syllabuses of budding scientists and technologists.

To my mind, it is no accident that such "critical" scholarship evolved hand in hand with developments in the training of technical experts, in the nurturing, humane tradition that characterizes the European ideal in higher education. As Clive Morphet, who was an influential member of the old U.K. STSA, has argued (in an unpublished conference paper):

> The justification for the inclusion of these non-vocational elements has . . . always involved at its core the notion that courses in science and technology are essentially illiberal, being concerned with facts, objectivity, abstraction, specialization, manipulation, etc., and that such a diet is unwholesome and should be "balanced" or "complemented" with a course or courses which expose students to more subjective ways of knowing and doing. . . . A cruder expression of the same sentiments will describe science and technology students [and, presumably, their teachers] as uncultured. (Quoted in Edge, 1988a, p. 18)

Even where all parties share an essentially positivist view of science, this claim could engender not only opposition but active hostility. Those of us who have experienced it know that this is one feature that has not changed over time! But it must be said that, rereading these pleas from the 1960s, I am struck again by how relevant their challenge still is. (For a recent example that repeats old rhetoric, see Russell, 1992.) Indeed, sharpened by subsequent debates about the need to recruit more women to science and technology, and their status within those professions, and by feminist critiques of science, the message has been amplified.[11] The positivist terms in which the "two cultures" debate was set tended to separate science from the humane domain. Ironically, out of a (misguided) attempt to make science social and "civilized" (as if it were part of neither society nor civilization) has arisen a richer STS criticism. And yet the earlier rhetoric is constantly reemerging. The British, self-flagellating as ever, keep the flag flying: A recent U.K. report, arguing (yet again) for "broader" science education, states that

> the typical science student is thought of as someone who is male, boring but clever, who has difficulty in writing essays and dealing with the uncertainties of open-ended issues. He has usually become locked into the science stream at an early age and does not want, or have, any other choice. He copes well with, even enjoys, the content laden curriculum and atmosphere of revealed and unambiguous truth. (U.K. Cabinet Office, 1991, p. 9)

This is the challenge of the 1960s restated with, again, an official endorsement. It promises, again, moves toward the humanization of science education. Is this merely another false dawn? Will the promises be forgotten when the price becomes clearer? For the richer STS criticism I referred to brings a radicalism that institutionalized science finds hard to bear: The yearning unhappiness with the unreformed education of technical experts lingers, but its humanization may be too costly. The wheel is then reinvented. Meanwhile, from Britain, where we have seen the death of our STSA, I gaze over the Atlantic at the recent rise of the (U.S.) National Association for Science, Technology and Society (NASTS): Is this a similar zeal bound for a similar end? If not, what is it about the political culture of Britain and North America that makes the difference?

The Democratic Impulse

While it can be argued that these two strands—moves toward a rational basis for science policy, and a reform of science education—were dominant in Britain and Europe, and their inner complexity and contradictions were beginning to unfold, a third, quasi-independent theme was also emerging. Under the influence of the Vietnam war, and coincident with the rise of civil rights, feminist, and environmental movements, an urge arose to explore the possibilities of the democratization of science and technology. The horrors of war, and the unleashed power of military technology, uncovered latent resentments in many democratic countries over the power and influence of expert élites. Perhaps predictably, the American distrust of centralized power muted any campaigns for nationwide curricular reform. Initiatives, such as those arising from Conant's pioneer experiments in teaching the history of science at Harvard (Conant, 1948), tended to be isolated. But, as Barbara Beigun Kaplan (1991) has recently argued:

> The 1960s . . . was a time of particularly heightened awareness of the need for educational institutions to become more socially responsible. This effort was spurred by social reform movements outside of academia, such as civil rights, environmental accountability, and the anti-war movement. It was during this period that . . . STS gained a foothold on American college campuses. The social protest movements . . . spawned the initial interest in these subjects and thus provided justification for their study within colleges and universities. . . . STS, as a subject of study, shares [with women's studies] this unique situation of having its origins in a social movement outside of academia. (pp. 1-2)

Both in the United States and elsewhere, these complex social and educational forces, allied with the wish of centralized bureaucracies for firmer science policy guidelines, have led to moves to enfranchise the technically ignorant public and to both empower and inform them appropriately; and this, in turn, has given further impetus to STS activities. One of the pioneers of American STS education, Stephen Cutcliffe (1989b), confirms that this remains a central motive when he writes that "we are now beginning to move into a phase where STS may help to shape public response and involvement in decision making regarding scientific and technical change, both by providing an awareness of the public's intimate involvement, and by offering suggestions regarding the specific role that it has to play" (p. 23).

Over the same period, the rise of such organizations as "Science for the People" has added to this mix the critical intellectual spice of a thoroughly *political* sociology of science.[12] And this, in turn, has influenced that strand of STS scholarship that might fairly be called "controversy and risk studies."[13] Exemplified by such work as Dorothy Nelkin's, and now firmly entrenched worldwide (in addition to the United States, one thinks of Holland, Sweden, and Australia, for example), this theme has engaged social scientists in interdisciplinary analysis of technical disputes, contributing a distinctive element to the demystifying urge within STS. Indeed, many of those who have become committed to the educational goals of STS, both within and outside the scientific community, have shared the (at least relatively) radical political aims that are characteristic of researchers in these aspects of the political sociology of science and technology—a fact that has tended to complicate the successful institutionalization of these goals in routine pedagogical practice.

GROWTH, DIFFERENTIATION, AND "CREATIVE TENSIONS"

However restrictive the current climate may be, the past 25 years have seen a steady growth and differentiation of the STS community. A glance at the variety of journals in the field shows how diverse it is.[14] Given this flourishing variety, it would be tempting to try to depict the STS movement as a benignly homogeneous "broad church." But this would be misleading: Internal tensions, at both the "intellectual" and the "political" levels, have been a feature of its brief history to date, and have often generated heated conflict. It seems to me that much of the intellectual vigor of the field rests on just that fact. Early examples of such tensions were the criticisms of the functionalist "Mertonian" emphasis on the explanatory role of norms and values (see, e.g., Barnes & Dolby, 1970; Mulkay, 1969, 1976; Rothman, 1972); disputes over the range and scope of quantitative and citation methods (Edge, 1977, 1979; Gilbert & Woolgar, 1974); and the mutual suspicion and lack of interaction between groups with "academic" and "policy" aims—a division recently reformulated as a "Low Church/High Church" distinction (Fuller, 1992; Ilerbaig, 1992). But the reader may judge the justice of my claim against the contents of this volume, for all the contemporary variety of opinion, posture, and conceptualization are exhibited within. Essentially, I see this vigor as stemming from an old philosophical conundrum to which I have already referred—namely, the relationship between fact and value. And I pin my hopes for the survival of the field to the fruitful continuation of this creative tension. Let me now try to make sense of these assertions.

"Technocratic" or "Critical"

The educational concerns I have described above have molded much STS scholarship; not only has that scholarship been exploited to provide material suitable for STS courses (for both technical and general students), but it has also shaped studies intended to illuminate and inform democratic involvement in "participatory technology" (Carroll, 1971; Nelkin & Pollak, 1979). History, philosophy, and the more "critical" elements of sociology and political science were swept up into this aspect of STS studies and have since developed a quasi-autonomous momentum. I repeat that I do not consider it to be a coincidence that a rigorous approach to the sociology of scientific knowledge (SSK), often apparently "internalist" in its emphasis on the *minutiae* of scientific belief and practice, arose out of the European educational context, with its more focused, specialized emphases in undergraduate training; in other settings (especially in North America, where the "received view" of science is less strongly embedded in political and legal institutions), an essentially compatible, "critical" approach to the nature of scientific expertise arose in the more "external" context of wider studies in political science and legal practice.

On the other hand, as we have seen, the concern to derive a "sound" basis for rational policy decisions on investments in science and technology, and for training appropriate personnel to oversee this activity, led to the parallel development of a much less reflective body of STS scholarship. The aim of this putative "science of science" was to provide the "objective, value-free" foundations for such decisions and thus to give them a scientistic credibility. Although this particular vision has somewhat faded over the years, it has inspired much of enduring worth: Quantitative approaches in STS studies, citation and co-citation methods, and many helpful analyses deriving from the policy sciences—all stem from this concern. (The journal *Scientometrics*, founded with Derek Price's enthusiastic support, has logged many of these developments exhaustively; see also Woolgar, 1991a.)

But these two streams in STS still do not flow together. Now, 26 years later, we find both concerns—the educational and the political—not only alive and kicking but up to much the same old rhetorical tricks. The two strains of thought I have been delineating—the reflective and the rationalistic—duck and weave as circumstances demand. As my earlier quotations have, I hope, illustrated, the reform of science education, and calls for appropriate changes in the training of technical experts to ensure their "social responsibility," are still heard in the land, and sung to much the same old unreflective tune;[15] and recurrent problems of science and technology policy suggest research agendas that aim to resolve endemic uncertainties by means of scientistic nostrums (the recent enthusiasm for "risk assessment" is a case in point).[16] These two imperatives generate their own creative tension as the strains of thought embedded in them interact.

Perhaps I can best spell out the character of this creative tension by using an educational analogy.[17] How can we best prepare our future scientists and technologists to enter society? How can we best educate them to meet the complex social and ethical problems they will necessarily encounter in the jobs they eventually enter?

There are, broadly, two kinds of response to this kind of question. The first, which I will call "technocratic," runs as follows. First, conduct a survey of these young graduates in their various employments and practices to see what skills and techniques they have to deploy to survive. Second, deduce which of these skills and techniques are not present in the school and college courses they have previously attended. Third, devise model syllabuses that will include these skills and techniques, and so will remedy these revealed initial defects in their repertoires. And, finally, armed with these urgent and well-researched educational innovations, convince the school and college authorities to adopt them. Many "broadening courses" for science students are derived in essentially this way: If they don't know about cost-benefit analysis, and they will eventually need to know about it, then teach them it. This "technocratic" approach does not question whether or not people *should* adopt the techniques and play the roles that they do. In other words, it does not question the credibility or status of technical expertise or the ultimate justification for that social role; its concern is, essentially, to allow the status quo to operate more smoothly.

The second response I will call the "critical" or "self-awareness" approach. This starts from the assumption that what young scientists most need (as, indeed, do we all, desperately) is some way of making sense of the institutional forces and pressures to which they will be subject, and of the existence of which they are initially unaware. They need, in other words, some critical understanding of the nature of the situation into which they are being propelled, so that they can work out what is happening to them before it is too late and so allow them (at least in principle) to retain some control over their fate. You cannot "resist temptation" unless you are able to recognize it for what it is. You cannot make rational choices of your course of future action without an understanding of this kind, and also a reflexive self-awareness of your own strengths and weaknesses, skills and aptitudes, inclinations and preferences, values and feelings, and so on. You cannot "exert willpower" without such preconditions.

Applying this principle to science education, the "critical" approach implies teaching material about science, its institutionalization and social structure, its values and practices, so as to stress *ideas* about its social nature and its relationship with other social institutions; these ideas are taught in a reflexive manner (that is, the students are asked to think about their own experience and to articulate their own reactions to the problems revealed by that reflection) and are then available, on starting a career, to guide those perceptions on which a graduate's autonomy and freedom of choice depend. In other words, the educational goals of teaching in this mode are essentially *cognitive*; the aim is an understanding of the social roles that the graduate might eventually assume, and of the institutional pressures implied by those roles, linked with the self-awareness that will activate those roles creatively. The status and credibility of technical expertise, and the justification of those social roles, are *not* taken for granted.

This critical goal is both subtle and profound, and it extends well beyond the strictly educational confines in which I have discussed it here, into much wider issues of practice and policy for science. It challenges many deeply embedded received opinions and beliefs about science and technology—it is thus potentially subversive. It is not politically committed, which earns it the opprobrium of those more radically inclined—but it is at least not politically "blind," which is an accusation to which the "technocratic" approach is wide open. Critical goals, often cast explicitly in the language of epistemology, have been one major motor of STS scholarship. The so-called relativist tradition is now so well established that its "policy" implications are being pressed (I will say more on this point later).[18] But the picture it gives of the nature of scientific activity, and hence of the status of scientific knowledge and expertise, departs strikingly from the traditional ideologies of science. A flexibility and endemic uncertainty, always problematic and never resting, characterizes the "new view" of science.

This critical perspective challenges "the atmosphere of revealed and unambiguous truth" that surrounds science, and thus draws fire not only from the politically committed but also from an influential body of active scientists, who denounce it as a false picture, likely to decrease public confidence in (and hence public support for) scientific activities (witness the extraordinary attack on the philosophy of science by Theocharis & Psimipoulis, 1987, and the subsequent correspondence; and criticism such as that mounted by Wolpert, 1992). On the other hand, many STS scholars argue that, unless this critical image of science is more widely diffused, any realistic appraisal of its proper role in society is doomed (see the first volume, 1992, of the journal *Public Understanding of Science*, and such work as H. Collins, 1987, 1988; Jasanoff, 1986, 1990a; Silverstone, 1985; Wynne, 1982, 1993); and the diffusion has begun (e.g., Barnes, 1985; Collins & Pinch, 1993). Here lies not only one focus of diplomatic difficulty for STS scholarship but a potentially fruitful line of further STS research. We need a more detailed understanding not only of the topography of the public's image of science but also of how (and to what extent) that image can be manipulated by those in whose interest it is to do so; and, in that inquiry, we will have to deploy the full range of techniques and analysis now available to STS scholars—both qualitative and quantitative methods, political and textual (rhetorical and discourse) analyses, historical and contemporary studies, "critical" and "political" goals. This process has, of course, already started (see Edge, 1990, esp. pp. 223-225); but it still has a very long way to go.

TOWARD A CREATIVE RECONCILIATION?

The more distant history of STS studies has been marked, perhaps too often and at times too dramatically, by conflict and a lack of communication that has appeared almost wilful. Co-citationists and their critics, strong programmers and reflexive discourse analysts, ethnomethodologists and policy analysts, have all been guilty of such standoffishness. But now the mood seems to be changing—and not just within the STS community itself. The early relativist Kuhnians, eager to point out to the poor functionalist Mertonians the obvious errors of their ways, struck out into the inner country, so as to bring the substantive content of science into the proper domain of a full-blown sociology of scientific knowledge. We are all immensely the richer for their pioneering endeavors: Some of their early dispatches from the interior were perhaps too lucid for comfort, but other smoke signals and talking drums seemed, at times, to be esoterically coded for private consumption. Historians and philosophers have now got their message—the debates that their findings provoke are truly interdisciplinary. Political scientists (and scientists themselves) have proved more resistant but are increasingly joining in the game. Now, and this volume testifies to this, there are more general signs of a coming *rapprochement* (and see S. Cole, 1992). Perhaps, after all, the old wheel can bear at least *some* strain.

Normative and Reflexive: Fact and Value Revisited

In a number of places around the world, the "critical" are now talking to the "technocratic." I could cite many examples of this move, but will mention just two. In the United States, the National Science Foundation has, for many years, maintained a program of support for activities in the field of "ethics and values." This has survived many metamorphoses but has recently come to embrace aspects of policy research and has joined forces with the history and philosophy of science. Under its auspices, much fruitful cross-boundary dialogue has been encouraged. And some years ago, in Britain, John Ziman and Peter Healey founded the "Science Policy Support Group." The object of this enterprise was not just to try to ensure a healthy flow of reliable finance to the body of U.K. STS scholars but also to explore to what extent the critical insights they had established could be usefully related to technocratic (or, at least, policy) goals. The dialogue that ensued was patchy; some of the seed fell on stony ground. But in some areas (in particular, military science, the public understanding of science, and university/industry relations),[19] the partnership has been fruitful. One result of dialogue of this kind has been to raise sharply the old "is/ought" problem in the form of a question: "Given that critical STS scholarship paints a distinctive and fresh picture of science—a new "*is*"—what are the policy implications (if any)—the new "*ought*"—that follow?" As Brian Wynne (1992a), reflecting on a study of the formulation of Canadian biotechnology policy, comments:

> What is such analysis attempting to offer, or achieve? It would be pretentious to claim that this kind of analysis makes, or should aim for, any direct practical

> influence on the world. Nevertheless the hope might still be admitted that it did at least offer others some of the necessary intellectual resources for such struggles, whether or not they are taken up, or meet with success if they are. (p. 579)

Is there any clear sense of how the new *description* entails any new and different *prescription*? The situation seems paradoxical: The intensity with which some people reject the new view suggests that it must have strong implications for action (and that these implications are perceived as threatening), but spelling out in precise detail what those implications might be proves to be an elusive task.

It might be thought that such a question would simply lead to a sterile replay of the kind of discussion familiar to students of Moral Philosophy 1, armed with Hume et al. and knowing only too well that there are "standard responses" to every "standard objection"—that any disputant can counter any adversary with a bag of rhetorical tricks. But the dialogue within STS is much more promising. For one thing, the body of scholars who have developed the sociology of scientific knowledge—and who thus form a major part of the STS "mainstream"—have always, and obviously, held differing views on important issues. The idea that we are seeing an entirely consistent and homogeneous "new picture" of science is clearly untrue. So, it would appear, internal differences of "value" can negate any apparent consensus on "fact." And those SSK scholars who have (scientistically?) maintained that they are advancing an essentially objective, empirical ("factual") account of scientific practice have, for some time, been open to attack on that score—thus the criticisms voiced, over 10 years ago, of the explanatory use of the notion of "interests," which stressed the role of *reflexivity* in assessing the status of any putative explanation (see Woolgar, 1981). Now, much more generally, we are seeing a move toward a more explicitly *normative* sociology of science and the reemergence of an emphasis on broader social philosophy (see, for example, Cutcliffe, 1990; Radder, 1992; Sismondo, 1993; Winner, 1993; Woodhouse, 1991). Essentially, what is at the heart of these discussions is a dispute over whether or not *any* scholarly "picture" can claim to "stand outside" the political arena and offer reliable guidance from a "detached" stance. What determines "closure" on the "best" available picture—and then on its policy implications? The reflexivity that SSK has itself spawned comes home to roost here. The essential inseparability of "facts" and "values" (which, in the mainstream corpus of STS literature, has been used, for example, to abort the idea of a "Science Court"; for a discussion, see Casper & Wellstone, 1978, 1981; Mazur, 1977; Nelkin, 1977b) has to be confronted directly within the confines of STS research. This is, I believe, likely to be a very fruitful focus of STS debate for the foreseeable future. It will not be an introspective, navel-gazing exercise, for the disputants are engaged in both empirical research and political activity, which "roots" their scholarship.

This pragmatic "rooting" of STS research may also help to build bridges between the "technocratic" and the "critical" within the STS community; and, in attempting to cross those bridges, some tentative weight might be carried, at times, by the old, discredited wheel. Another hopeful aspect of the current scene—again fully illustrated in this volume—is that, whereas some years ago no "critical" scholar would be seen dead saying a good word for the "technocratic" tyranny of quantitative research methods, we are seeing the development of new, computer-based methods by scholars whose "critical" credentials are impeccable. The emergence of "actor-network" theories, and "sociotechnical graphs" to fill them out, is a case in point (see Latour, Mauguin, & Teil, 1992; J. K. Scott, 1992, and, for another example of a sympathetic critical scholar, H. Collins, 1992b). And among the "traditional" citation-based STS scholars themselves, there is a more sensitive awareness of the need to exercise reflexive restraints. This is not to say that the entire community now stands redeemed—"momentum" is not so easily mocked! But, where stand-off and silence once reigned, now a conversation is beginning.

THE REMORSELESS WHEEL

This volume unfolds for you, in all the exciting richness that recent scholarship has achieved, the "new view" of science and technology—and of society itself—that is now available in the STS literature. But, in my opening remarks, I referred to the recurrent tendency to "reinvent the wheel." It seems that the "old," positivistic image of science, as an abstract, timeless search for irrefutable facts—ending the pain of uncertainty, the burden of dilemma and choice, separable from "society," and leading inexorably to technical innovations for the good of all—exhibits an apparently puzzling tenacity. But it should not surprise us that this is so. The old ideology underpins many of our most respected social institutions, and to doubt *their* credibility would be to lose one's social nerve. So long as the state remains the main provider of research finance,[20] scientists themselves will hold fast to the old picture, as it has earned them, in times past, access to the political arena and to the resources that they need. It is therefore difficult for them to drop the claim that they bring only a consensual certainty to bear when they find "facts" or offer advice, difficult to come to terms with any public admission that they, too, have to reconcile observations and values in reaching their judgments.

Perhaps more profoundly, studies of the relationship between scientific and legal rationality have stressed how far our legal institutions rest on a crudely positivistic view of scientific truth (indeed, of "truth" in general)—and have shown the complex variation of this relationship both within and between legal and cultural traditions. In some cultures (British, for instance), the legal mind finds great difficulty in the idea that two scientific witnesses can, in completely good faith, and with access to identical evidence, come to differing judgments and give contrary testimony (Wynne, 1982, 1989a); but in others (America springs to mind), trial lawyers thrive on a completely relativistic view of science, to the distress of the expert witnesses they cross-examine and of many of the judges who preside.[21] The challenge by critical STS scholarship to the old set of beliefs could thus erode the credibility and validity of some fraction of the legal profession and of their institutional forms and practices, in some countries and cultures. Given the centrality of legal institutions in all cultures, such conflicts have profound political implications. And this is, I believe, also true of educational institutions. Through the vicissitudes of historical contingency, the old image of science has become firmly embedded in the educational subconscious of modern industrialized societies. This is precisely why STS insights are (as I put it earlier) "wantonly neglected" and STS courses are constantly under threat when times are hard. Inasmuch as STS scholarship can successfully elucidate why the present structure of science education seems so easily to foster "illiberal" attitudes and beliefs, and then effect the relevant restructuring, the need for STS courses in technical education may diminish. But the political shifts entailed by such a prospect make it unlikely, in the foreseeable future. For most of us, the critical perspective offered by STS will be a permanent feature of all liberal education, for as long as we remain active.

No one can deny that the subject matter of STS studies is of central concern to humankind. STS analysis points to all the "higher" aspects of human endeavor—truth and power and justice and equity and democracy—and asks how these can be conserved and consolidated in modern society, so that the immense possibilities of scientific knowledge and technological innovation can be harnessed (in Bacon's words) "for the relief of man's estate." As Langdon Winner (1986a) has recently written, STS "occupies an enormous void, one created by a society that long ago committed itself to forge ahead full bore with scientific and technological advance, but never to forge ahead in developing the critical self-reflection such change seems to require" (p. 7).

But, as STS scholars know well, prophets are not always appreciated in their own countries; to analyze such a central social institution as science—the source of accredited knowledge—is to wield not a plowshare but a sword. And those who live by it should be prepared to die by it. For the conventional ideologies and myths that STS scholars challenge are the very myths that justify and legitimate current practices in the scientific, educational, legal, and many other institutions. To displace them would thus be to tamper with the legitimation of powerful vested interests, not only in science but in society at large. Ideas about the nature of science are but one aspect of modern political rhetoric, and changes in the exercise of power and authority in society—in other words, *political changes*—are a necessary precondition for their effective reform. This insight seems to me to lie behind many of the claims that science is, by its very nature, "impersonal." Traditionally, it is (at least tacitly) assumed that the authority of nature is independent of social authority, and transcends it. Ultimately, the alternative view turns this conventional wisdom on its head (see Collins & Shapin, 1983; Shapin, 1982, 1988a). Here, indeed, is a radical challenge to both thought and action. It will need a "wheel" that can really bear the strain. And it is to the kind of studies to which this volume is dedicated that readers wishing to find, invent, and manipulate this new wheel should turn.

NOTES

1. Where attempts are being made to expand participation in higher education at low cost, STS appears to be under pressure: but where participation levels have traditionally been high, and resources relatively lavish (as, for instance, in North America), innovation is still possible. I have in mind, for example, that we have recently seen new or expanded programs at UCSD, Cornell, and MIT; and there are many flourishing groups throughout the world that are, at the very least, happily treading water. Yet the pressures are a fact of life in the profession, and career prospects are tight. The 1960s and 1970s were, indeed, times of unusual optimism and expansion.

2. It would be quite impossible to do justice to all the quasi-independent disciplinary strands, on their own. However, the following (incomplete) list of influential works, in order of publication date, should give the reader some idea of the nature and scope of the kind of "focused confluence" that occurred during the 1960s: Cardwell (1957); Marcson (1960); Schoeck and Wiggins (1960); B. Barber (1962); Barber and Hirsch (1962); Kornhauser (1962); Rogers (1962); Coler (1963); Taylor and Barron (1963); Bright (1964); Haber (1964); Gilpin and Wright (1964); Ginzberg (1964); Orth, Bailey, and Wolek (1964); Armytage (1965); Boguslaw (1965); N. Kaplan (1965); Lapp (1965); Penick, Pursell, Sherwood, and Swain (1965); A. K. Smith (1965); Tybout (1965); Warner, Morse, and Eichner (1965); J. G. Burke (1966); Hagstrom (1966); Lakoff (1966); Morison (1966); Morse and Warner (1966); Pelz and Andrews (1966); Schmookler (1966); Storer (1966); Van Tassel and Hall (1966); Greenberg (1967); Kranzberg and Pursell (1967); Nelson, Peck, and Kalachek (1967); D. K. Price (1967); Schön (1967); Skolnikoff (1967); Williams (1967); Danhof (1968); Hirsch (1968); Nelson (1968); Orlans (1968); J. Watson (1968); Ziman (1968); Gruber and Marquis (1969); Haberer (1969); Jewkes, Sawers, and Sillerman (1969); Landes (1969); Rose and Rose (1969); and Lakatos and Musgrave (1970).

3. Although STS scholars abandoned the "linear model" of technical innovation decades ago, it still retains its force. When the chairman of the U.K. Economic & Social Research Council, in a recent prestigious public lecture, came out of the closet and confessed his lack of faith in the model, it was represented as a significant shift (see, Newby, 1992). And Sir John Fairclough, chairman of the U.K. Engineering Council and of the Centre for the Exploitation of Science & Technology, and from 1986 to 1990 chief scientific adviser to the Cabinet Office, recently declared: "I have come to appreciate that the widespread belief in the 'linear model' has not helped but

positively hindered the development of effective policies" (Fairclough, 1992, p. 17). Yet the model continues to exert its tenacious influence on everyday thinking, and hence on policy initiatives. Only a week after Fairclough's statement, in the same journal, in a prominently displayed letter to the editor titled "Routes to Exploiting All Research" (p. 15), an academic electronic engineer argued that "all [pure and applied] research proposals should identify one or more 'practical' problems that the research may help to solve, with some explanation of the connection between the research and the problems that have been identified, and an indication of likely time scales. This . . . should help to plot the route to exploitation for all kinds of research." A year later, the British Government is implementing a White Paper on the organization of science and technology that seems to enshrine just such a principle (disguised as "foresight") at the very heart of routine practice in science and technology policy. Whether this avoids Fairclough's condemnation remains to be seen.

4. I use mainly British illustrations, because that is where my experience has been located, but I hope to be surveying and expressing very general trends and features. For the sources of much of my information of developments worldwide, see the following Country Reports & Commentaries from *Science Studies* and *Social Studies of Science*: for France, Bowker and Latour (1987) and Comment by Freudenthal (1990); for Sweden, Elzinga (1980); for the former Soviet Union, Mirsky (1972), Hoffmann (1979), and Levin (1984); for Bulgaria, Kachaunov and Simeonova (1979); for Japan, Low (1989); for Italy, Pancaldi (1980); for Germany, Pfetsch (1979) and Spiegel-Rösing (1973); for the Netherlands, Rip and Boeker (1974); for Australia, Ronayne (1978); for the United States, Teich and Gold (1986); for Latin America, Vessuri (1987); and for Poland, Walentynowicz (1975).

5. Twenty years ago, when Barry Barnes prepared a Penguin Reader in the Sociology of Science, he had great difficulty finding enough good material to fill its pages; ten years later, when Barry and I prepared a similar volume for the Open University Press, our problem was how to eliminate material from a flood of possible readings (and our Bibliography ran to 377 items, covering 36 pages)! See Barnes (1972) and Barnes and Edge (1982).

6. I have been unable to trace the location of a much-reprinted shorter version: "By 2020, every man, woman, child and dog in the United States will be a scientist, and we will spend more on science than the whole GNP."

7. In America, Edwin Mansfield, Zvi Griliches, and others had pioneered such studies (see Griliches, 1958; Mansfield, 1965). In the United Kingdom, the newly convened Council for Scientific Policy commissioned a number of influential studies; see, for example, Cohen and Ivens (1967) and Byatt and Cohen (1969). For further evidence of early attention to this topic, see U.S. National Science Foundation (1971). See also several items listed in note 2. This concern is still very much alive today; see, for example, Rosenberg (1990), Pavitt (1991), and K. Smith (1991).

8. The origins of the Science of Science Foundation (now the International Science Policy Foundation and the publishers of *Science and Public Policy*) can be traced to November 1963, when Alan MacKay and Maurice Goldsmith planned a collection of essays, inspired by Bernal's *The Social Function of Science* (1939); this was eventually published under their joint editorship as *The Science of Science* (1964/1966). For the formation of the SSF, see Goldsmith (1965). In Paris, both the Organization for Economic Cooperation and Development (OECD), guided by Alexander King and Jean-Jacques Solomon, and the U.N. Educational, Scientific and Cultural Organization (UNESCO) coordinated these initiatives at the international level; see, for example, the three OECD reports, *The Research and Development Effort* (1965), *Government and Allocation of Resources to Science* (1966a), and *Fundamental Research and the Policies of Governments* (1966b) as well as Freeman (1969).

9. I cannot trace this quotation in print: I have it in my notes on a seminar addressed by Goldsmith in London in 1965.

10. And, as the person ultimately responsible for the Edinburgh group from which emerged the "Strong Programme" in SSK, I must plead to be doubly excused! For recent, contrasting views on this particular development, see Pickering (1992a, esp. pp. 1-4), M. Nye (1992, esp. pp. 233-235), and Kuhn (1992, passim). For another perspective, see Michel Callon (this volume).

11. For reviews, see Malcolm, Hall, and Brown (1976); U.S. National Science Foundation (1977); Høyrup (1978, 1987); Kelly (1979, 1987); J. Harding (1983); Bruer (1984); *Technology Review* (1984); and McIlwee and Robinson (1992). For a critical review of some feminist writing, see Grove (1989). Additional critical readings of the feminist literature are provided by Mary Frank Fox, Evelyn Fox Keller, and Judith Wajcman (in this volume).

12. See, for example, Henig (1979). Societies "for Social Responsibility in Science" sprang up in other countries; in Britain, the BSSRS published *Science for People*. Steven and Hilary Rose pioneered Marxist analysis; see their *Science and Society* (1969) and the two volumes they coedited, with the common title *Ideology of/in the Natural Sciences—The Radicalisation of Science* (1976a) and *The Political Economy of Science* (1976b). Robert M. Young also promoted Marxist studies; he was instrumental in founding the *Radical Science Journal*. In America, *Science for the People* and *Science and Nature* pursued similar paths.

13. For "controversy," see, for example, Nelkin (1979/1984/1991) and Engelhardt and Caplan (1987). For "risk," see Nelkin (1985), Douglas and Wildavsky (1984), and Douglas (1986b). For examples of interdisciplinary sophistication in this area, see Jasanoff (1986, 1990a) and Richards (1991).

14. Many STS scholars publish significant contributions in "mainstream" disciplinary journals—notably in history, philosophy, political science, and economics. This not only is a wise career strategy but also helps to ensure the quality of interdisciplinary dialogue in the field. There is also a rich interchange with the many journals now established in the history and philosophy of science and technology: *Isis, Osiris, Technology and Culture, Science in Context, Studies in History and Philosophy of Science, History of Science, Philosophy of Science, BJPS, BJHS*, and a wide range of more specialized and regional publications. Within STS itself, the variety is bewildering. Some "mainstream" scholarly journals are now well established: *Minerva, Social Studies of Science* (founded in 1971 as *Science Studies*), and *Research Policy* are examples, and *Science, Technology, & Human Values* (now the academic house journal of the Society for Social Studies of Science [4S]) has entered this list. Over the years, these pioneers have been joined by other, similarly scholarly, publications: Some, such as *Knowledge* and *Technology in Society*, address American traditions in political science; others, such as *Scientometrics*, develop methodologies; others, such as the Australian *Metascience* and the Scandinavian *Science Studies*, have emerged to cater to regional groups; others, such as *Public Understanding of Science*, to develop research on perceived social problems (in the case cited, a problem influentially raised *by scientists* in the face of recent public setbacks; for an illuminating comment, see Wynne, 1992b); yet others, such as *Radical Science Journal*, pursue distinctive theoretical and political analyses. A cluster of newsletters and bulletins also serve particular groups. For those interested in STS pedagogy, there are, for example, *Science, Technology & Society* (Lehigh University) and the *Bulletin of Science, Technology & Society* (NASTS); for bibliographic surveys, *Science and Technology Policy* (British Library); for society memberships, *Technoscience* (4S) and the *European Association for the Study of Science & Technology [EASST] Newsletter* (plus many local and national variants). And, along with all this, comes a penumbra of more "popular" publications dealing with issues in the STS field, but usually from a less analytical stance (and with, perhaps, more openly ideological and social interests); examples here are *Science for [the] People, Science as Culture, Issues in Science and Technology* (U.S. National Academy of Sciences), *Science and Public Affairs* (U.K. Royal Society), and the well-established monthly, *Technology Review* (MIT). All these (and many more) offer readers a rich mixture of topics and information, analytical perspectives and methodologies, political and social interests, and academic "levels."

15. As just one illustration of this, see my review of the volume published in 1986 by Pergamon for ICSU, in a prestigious series, *Science & Technology Education & Future Human Needs*, and titled *Ethics & Social Responsibility in Science Education* (Edge, 1988b).

16. The literature on this topic is voluminous. For a recent popular survey, see the British Medical Association (1987). Other useful references include Lowrance (1976), Council for Science and Society (1977), Hohenemser, Kasperson, and Kates (1977), Conrad (1980), Holdren (1982), Rosenfeld (1984), *Science* (1987, esp. the contribution by Wilson and Crouch, pp. 267-270), Freudenburg (1988), *Daedalus* (1990), and Drake and Wildavsky (1990).

17. Some things have indeed changed in the past 25 years! This section was originally cast in terms of an analogy from an unpublished paper given by J. R. Ravetz at a meeting of the British Association at Leeds in 1967: "The average scientist enters his [*sic*] career with no more preparation for the Social and Ethical problems he [*sic*] will encounter than an Irish girl arriving at Euston Station." However, I was advised (wisely) that this analogy "would not fly in America"; for its flight in Germany, see Edge (1988a, pp. 22-23).

18. For policy studies, see Wynne (1982) and Jasanoff (1986, 1990a); for media studies, see Silverstone (1985) and H. Collins (1987, 1988); for discussions of science education, see Edge (1974, 1975, 1985) and Collins and Shapin (1983) and, for technology, MacKenzie and Wajcman (1985).

19. For military science, see Gummett (1991) and MacKenzie (1990a); for "public understanding," see H. Collins (1987, 1988) and (for example) Wynne (1991b, 1993); for university/industry relations, see Webster and Etzkowitz (1991).

20. This situation is now changing; see the chapters by Elzinga and Jamison, and by Etzkowitz and Webster, later in this volume.

21. For an American example of an extreme positivist view, see Peter W. Huber (1991), *Galileo's Revenge*; see also Sheila Jasanoff (1992b), "What Judges Should Know About the Sociology of Science."

PART II
THEORY AND METHODS

The four pieces in this part are symbolic of the breadth of theoretical and methodological diversity that continues to drive science and technology studies. Although no single theoretical or methodological assumption dominates S&TS, a number of common themes recur throughout this part, and indeed throughout this *Handbook*.

The rich terrain of science studies is exemplified by Michel Callon's ambitious attempt to outline four models of scientific development that can be used to locate most of the central theoretical work in the field: (a) science as rational knowledge, (b) science as competition, (c) science as sociocultural practice, and (d) science as extended translation. A refreshing feature of this approach is that it produces some odd bedfellows. Thus one finds under the competition model authors as diverse as Bruno Latour (early vintage Latour!), Warren Hagstrom, Karl Popper, and Pierre Bourdieu. Despite the success of Callon's overarching schema, there are some approaches that, by his own admission, do not seem to fit: most notably, work on gender and on discourse analysis, new literary forms, and reflexivity. Happily, such topics are covered extensively in other chapters in this volume (for gender, see Evelyn Fox Keller, Mary Frank Fox, and Judy Wajcman; for discourse analysis, new literary forms and reflexivity, see Ashmore, Myers, and Potter).

One reason Callon has difficulty integrating this work can be seen in a useful distinction drawn by Gary Bowden between S&TS approaches that focus upon broad analytical themes (e.g., the problem of representation) and those that focus upon more narrowly defined specific topics (e.g., the problem of fraud in science). Analytical issues such as gender (a theme with profound political implications) and discourse analysis, reflexivity, and new literary forms are raised throughout the natural sciences, social sciences, and humanities. Although such themes may gain a particular salience in S&TS, Bowden argues that they cut across a wide variety of subject matters. In building S&TS solely upon such themes, there is a danger that we will lose what is unique about science and technology. Bowden finds this uniqueness in the authority of science and technology as forms of culture. It is this authority that S&TS challenges and that should provide a necessary disciplinary identity.

The continuing neglect of gender and science is precisely the topic that Evelyn Fox Keller takes up in her personal reflections. The driving force that has brought gender to the center of the S&TS agenda has been much more the growing influence of feminism and feminist scholars than the internal disciplinary workings of the philosophy, sociology, or history of science. In outlining future directions, Keller calls for the disaggregation of "gender and science" into studies of (a) women in science, (b) scientific constructions of sexual difference, and (c) the uses of gender in scientific constructions of subjects and objects that lie beneath and beyond the human skin (or skeleton). It is this third area of study—of "gender in science"—that Keller reviews in her chapter. She concludes by pleading for better integration of the strengths of feminist theory and of other approaches in science studies.

It is clear that the doctrine of social constructivism has left an indelible print on S&TS. However, the chapters in this part provide a salutary warning against any complacency that social constructivists might feel about the steady ascendancy of their approach over the last decade.

Sal Restivo takes social constructivists to task for not being thoroughgoing enough when it comes to issues concerning the sociology of mind and artificial intelligence (AI) (a challenge that Harry Collins and Paul Edwards take up in their separate ways in later chapters). Restivo finds the vestiges of Mertonianism still present in the field and especially in the arena of science policy (though this assumption is challenged in the chapters by Susan Cozzens and Edward Woodhouse and by Aant Elzinga and Andrew Jamison). Is social constructivism, moreover, capable of providing the radical activist agenda desired by some? Drawing from the classical sociological tradition, Restivo points to various paths within the field that do not always fit with today's constructivist interests but that, nevertheless, provide a means whereby sociology of science might avoid the pitfalls of becoming dependent upon the traditional ideology of the natural sciences.

These four pieces can be taken as symptomatic of the diversity and fragmentation that is always attendant upon an interdisciplinary enterprise such as S&TS. Indeed, the forces calling for diversity are perhaps even stronger today than when these pieces were first commissioned. The growing influence of "cultural" studies and postmodernity within S&TS suggests the appearance of yet another seemingly crosscutting theoretical agenda (see, e.g., Rouse, 1993). But a different, less pessimistic, reading is also possible. These chapters provide evidence of the continuing responsibility that scholars in S&TS feel toward synthesizing the different trends in the field. Whether in Callon's overarching schema of four models, in Bowden's interdisciplinary vision, in Keller's call for integrating feminist theory and science studies, or in Restivo's call for a more thoroughgoing sociology of science, one finds an articulate engagement with the search for a powerful, coherent theoretical vision of the field. We may not yet have reached that goal—indeed, we may never get there—but these chapters display both the magnitude of the task and the most fruitful paths to explore. They outline the theoretical agenda for much of the rest of this *Handbook.*

2

□

Four Models for the Dynamics of Science

MICHEL CALLON

"WE must explain why science—our surest example of sound knowledge—progresses as it does, and we must first find out how in fact it does progress" (Kuhn, 1970, p. 20). Many answers have been proposed to these two questions. In choosing to organize this chapter in terms of different models of scientific development, I have deliberately sought to emphasize the collective character of work in science studies. My aim is to avoid the repetitive and controversial step of taking a few selected books by a number of great authors—the science studies canon—as the point of departure. To be sure, my way of presenting the arguments has its drawbacks. For instance, the debates that have driven the field as it has grown do not come into focus. However, the theoretical structure of arguments and choices is made clearer, as is the fact that analysts are always struggling with a series of different dimensions. It is thus impossible to give a definition of, for example, the nature of scientific activity, without at the same time suggesting a certain interpretation of the overall dynamics of development and establishing the identity of the

AUTHOR'S NOTE: I received a lot of stimulating comments while this chapter was in preparation. In addition to two anonymous referees, I thank François Jacq, Bruno Latour, Dominique Pestre, Trevor Pinch, Vivian Walsh, Yuval Yonay, and all my colleagues from the Centre de sociologie de l'innovation. Without John Law's support and help, this chapter probably would never have been completed. This version was completed while I was a member at the Institute for Advanced Study in Princeton, which I thank for its support and hospitality.

actors involved. Even the most philosophical works imply a conception of the social organization of science, and reciprocally the purest sociological analyses assume views of the nature of scientific knowledge.

Also my approach draws attention to the overall coherence of what may appear to be different approaches to STS. It turns out that, once a decision has been made about the character of scientific findings, certain consequences for the description of the institutions and dynamics of science necessarily follow. Though it is true that authors often escape from the logic of a single model by combining several together, the use of the models reveals the way in which authors from different schools and disciplines sometimes in reality share a common framework of assumptions.

I have distinguished four models, each of which emphasizes a central issue. The first is that of science as rational knowledge where the object is to highlight what distinguishes science from other forms of knowledge. The second is that of science as a competitive enterprise where the main concern is the organizational forms that science takes. The third is the sociocultural model and particularly the practices and tacit skills that it brings into play. The fourth model, that of extended translation, attempts to show how the robustness of scientific statements is produced and simultaneously how the statements' space of circulation is created.

Each model is characterized by its answers to six questions that lay out the social and cognitive dimensions of scientific development. Though the list of questions might considered fragmented, from a practical point of view the schema appears to work. The questions are these: (a) What does scientific production consist of? (b) Who are the actors and what competence do they have? (c) How does one define the underlying dynamic of scientific development? (d) How is agreement obtained? (e) What forms of social organization (internal or external) are assumed? (f) How are the overall dynamics of science described?

MODEL 1: SCIENCE AS RATIONAL KNOWLEDGE

This model seeks to clarify what distinguishes science from other human activities. It focuses on scientific discourse and explores the links it establishes with the reality of which it speaks.

Nature of Scientific Production

The outcome of research activity consists of statements and networks of statements. The classification of these statements and the characterization of their relations is a central issue.

The most common classification is one that opposes observational statements (or empirical ones) and theoretical statements. This distinction accounts for the dual dimension of science: experiments and data collection, and also conjectures and generalizations.

If one takes the following statements:

a. Any electron placed in an electric field is subject to a force proportional to its charge.
b. In the circuit C situated in this laboratory, the intensity of the current is 50 amperes.
c. The needle on the ammeter placed in the circuit C points to the figure 100.

These three statements are independent and their vocabulary is different in each case. In statement a, the entities referred to are not directly observable by human beings by means of the five senses alone—nobody has ever seen an electron and even less so an electric field—which is why these notions are said to be abstract. In statement b, the repertoire is similarly abstract, although certain entities—for instance, "the circuit C" or "this laboratory"—are directly observable. It is with statement c that we enter into the realm of the senses. The figure 100 may be seen, as may the ammeter itself. The fact that the needle points to the figure 100 may be agreed upon after visual inspection only.

How does one go from statement a to statement c? The notion of translation may be used to describe these moves (strictly speaking, it should be called "limited translation" so as to maintain the distinction between this and the extended notion of translation in Model 4). The translations that permit statements a, b, and c to be related are far from obvious. Several means for creating these translations have been suggested, and all take the form of a sort of abstract calculation: for example, correspondence rules, coordinated definitions, dictionaries, or elaboration of an interpretative system.[1] In all cases it is generally recognized that it is not possible to move from one kind of statement to another by means of logic alone (Grünbaum & Salmon, 1988). Whatever the particular strategy, it leads to the creation of a third family of statements that associate terms from both observational and theoretical statements and consequently act as translation operators.[2]

With the proliferation of intermediate statements, the distinction between theoretical and observational statements is by no means clear. A first position, which may be termed *reductionist,* is to minimize the distance between both types of statements. It covers two extreme forms: (a) Theoretical statements are derived from observational statements (positivism and logical empiricism); such a doctrine can be mobilized either to provide a criterion of validity (the so-called inductionist theory) or to establish demarcation criteria between statements that have meaning and those that do not;[3] (b) observational statements are shaped by theoretical considerations without which they have no meaning:[4] This is the so-called theory-ladenness of observations.

A second position is a refusal to establish hierarchical links between theoretical and observational statements. Thus, though there are indeed connections, in the first instance it is assumed that the different categories of statements are relatively independent from one another. Under these circumstances, it is possible to test empirical predictions derived from theoretical statements or to decide whether one theory explains a set of observations better than another.

In this model, knowledge production is basically reduced to the production of statements between which translation relationships can be established. Translation is confined to its linguistic meaning—translation is not an exit from the universe of statements. This explains the natural drift toward philosophical and ontological questions. How can one avoid discussing what is "represented" by the statements and the essences they bring into play?[5]

Actors

The relevant actors are essentially the researchers but reduced to the role of statement utterers. Technicians with their skills, disseminators of knowledge, manufacturers of instruments, teachers, and pieces of experimental apparatus are totally absent; society is rarified, reduced to its simplest expression. A consequence of this work of purification is the attribution of wide-ranging competence to the (rare) actors involved. The more narrowly the group is circumscribed, the more the task entrusted to its survivors is complex.

The competencies assumed by the researchers are sensory and cognitive. The scientist must be capable of articulating statements that integrate her observations. She is thus dependent on her five senses, and particularly on sight (observation is always mentioned). The scientist must also be capable of imagining statements that are not directly linked to observation and of introducing translations between them. Her ability to produce metaphors and analogies is emphasized by authors like Holton (1973; Hesse, 1974). Others insist on aesthetic sensitivity; certain theories or reasoning seduce by their simplicity, their elegance, or their beauty.

To these cognitive competencies is added what we might call the rational dimension. The notion of rational activity rests on the capacity to make justifiable decisions. The rules for such justification have been described in many different, and indeed contradictory, ways. They may have to do with the promise of a given theory, its generality, its robustness, the extent to which it fits experimental data, its ability to resist rigorous testing, or its simplicity. However, such judgments don't take place in a vacuum but depend on existing theories and statements. In talking of this accumulation of objective knowledge—knowledge distributed in books, articles, libraries, or the memories of computers—Popper (1972) talks of a "third world." One could say, to paraphrase Boltanski and Thévenot (1991), that the particular competence of the scientist lies in the ability to justify why one statement is to be preferred rather than another in a specific set of circumstances.

In this model, the scientist is a monstrous being—one that incorporates a range of diverse competencies, normally thought to be distributed between different members of society.

Underlying Dynamic

Why does science advance? Or to formulate it in terms of the model, why do scientists tirelessly add new statements to existing ones? And why do they transform, amend, and invalidate them?

First, the critical and ongoing work on statements assumes that the scientist is endowed with a solid moral commitment. This is not so much to keep her from the temptation of fraud, for the debate between specialists is sufficient to eliminate fraud—honesty results from mutual scrutiny. Rather, it is to encourage her to produce ever more statements, which she must be prepared to test and possibly to abandon. The scientist is caught in a double bind; on the one hand, she must devise and produce an increasing number of statements but, on the other hand, she must submit these to the ruthless constraint of selection.

Second, elements of an answer can be found in the institution of science; this is where the complementarity of this model with Model 2, of science as a competitive enterprise, is most obvious. The reward system of science is essential, for even the scientist with the most acute moral sense will not strive to produce new statements without encouragement. Scientific institutions act to channel the force driving scientists—whether a passion for truth, the desire to participate in the collective enterprise of knowledge, the wish to control nature, or the relentless search to resolve problems or contradictions. The scientist is but an operator by whom statements are brought into existence and confronted with one another. The model brings into play a sort of Darwinism extended to statements.

Agreement

In this model, agreement covers statements of all kinds—for instance, theoretical or observational—and also the constructions and arrangements they engender.

Agreement is first to be explained by the fact that actors share similar competencies. They are able to agree on strict observational statements like these: The thread breaks or the curve peaks (which of course implies existing agreement on what a thread or a peak is). They are also capable of inference (a logical contradiction is obvious to all), able to establish the extent to which a statement is general[6] and/or make decisions based on what at the time is considered to be good reason.

Beyond shared competencies, what plays an essential role in the construction of agreement is the existence of a field of discussion where statements can be confronted. We thus find another way of defining rational activity that dates back to ancient Greece at least. Vernant (1990) suggests that science is but the continuation of political debate in a different arena—its transposition from the social to the cosmos. Reason comes into being where arguments occur. And because it is centered on statements, Model 1 treats very seriously the existence of the space of debate in which statements are expressed and their robustness tested. Discussion between scientists takes place in colloquia, in journals, or more informally between colleagues around the laboratory bench before publication or presentation. It also takes the more subtle form of interior dialogue when a scientist debates with herself to anticipate objections and simulate probable debate. But this self-imposed rigor is similar to that exerted by her colleagues—or that which she exerts on them in turn. Private and public space, or formal relations and informal ones, are not opposed to one another. If indeed there is a boundary, it is the one between the errors that are kept to oneself or one's immediate colleagues, and so may be repaired without damage, and those that can only be admitted to by losing part of one's credibility. This is more a question of reputation than a difference between modes of argumentation and discussion.[7]

But, to take the vocabulary used earlier, how are the translations that turn observational statements into theoretical (and vice versa) made robust enough to survive in the debate? There are many answers to these questions. Some have to do with tests and their interpretation. To be convincing and to overcome the subjectivity of a statement means subjecting its validity to experimental checks and to the criticism of colleagues who can verify that the tests are meaningful and correctly interpreted. Others suggest that theoretical statements need to be presumptively true about the world. Conformity with experience, which is always problematic, is less important than, for instance, the ability to make verifiable predictions. Yet others adopt a pragmatic view: the robustness of a translation is measured by applying criteria or rules that have been developed over the course of centuries. Examples of such standards include the extent to which a statement is general, the economy of the entities that it mobilizes, its ability to resist demanding tests, and its fertility in leading to unexpected applications. Finally, others consider criteria as pure conventions that convince only those who are already committed to them.

Whatever the solution adopted, the existence of explicit and shared standards is admitted, whether they are hypothetical or categorical or are reasonable conventions.

Social Organization

This model imposes severe constraints on the social organization of scientific work. The paradox is that the more one insists on the cognitive and discursive dimension of knowledge, the more one increases the demands on social organization. Those who utter statements can only undertake their work—that of discussing, testing, experimenting, selecting, falsifying, and so on—if they are protected by society as a whole and by particular institutions.

Without the public space of (free) discussion, science degenerates into beliefs stamped with subjectivity. Science is synonymous with democracy or, to use Popper's (1945) expression, open society. In an open society, institutions are the revisable creations of human activity; the critical mind knows no limits—gods, Caesars, or tribunes. Questioning is permanently renewed and no state of rest is satisfactory. The individual is privileged because she both introduces and judges novelty. There is an analogous concern in the writing of Habermas (1987) with pressing science into a space for public discussion and communication.

But it is not enough that science should be immersed in an open society; powerful institutions that guarantee the smooth functioning of critical debate are also necessary. What will be said in the presentation of Model 2 applies here without restriction; with respect to social organization, Model 2 can be considered a natural complement to Model 1.

Dynamics of the Whole

The development of science is expressed in the proliferation of statements that are the result of a dialogue between man and nature. A silent man, faced by an equally silent nature, could neither accumulate statements nor produce revisable knowledge. So a scientist does not simply read the great book of Nature; she transcribes it, translating it into statements inscribed in linguistically shaped argument. Putting the universe into words is the essential task of scientific knowledge. Science is thus developed in the form of a dual dialogue, first between scientists and Nature (observational statements and theoretical statements) and second between scientists themselves. These two dialogues are interdependent; they take the form of a triangle in which one of the protagonists (Nature) is content to reply in a cryptic way to the questions it is posed. As in any confrontation, contradictions and incomprehension constantly develop. What exactly is nature's message? How should experiments be conceived and the results interpreted? What theoretical statements can be put forward? What do they explain? These gaps and divergences restart the machine of knowledge.

Such a vision necessarily implies the notion if not of progress then at least of progression. There are statements, forever more statements that hold Nature as close as possible and ply it with ever more awkward and precise questions. So investigation has no end, yet it continues! Nature's reality is asserted and the statements produced are seen as an increasing theoretical approximation or a better experimental description. Or, alternatively, one may not express any opinion about this reality and simply concentrate on the endless production of more and more robust or reliable statements.[8] Whether the statements that are produced tend toward the truth, relate together ever-increasing numbers of empirical observations, or increase our ability to control and manipulate the world, the tragic beauty of Model 1 is that it is scientists and scientists alone who have to choose which statements to preserve and which to discard.

MODEL 2: COMPETITION

There are numerous variations of this model, but they all share the two fundamental tenets: (a) Science produces theoretical statements whose validity depends on the implementation of appropriate methods. (b) The evaluation of knowledge—an evaluation that leads to its certification—is the result of a process of competition or, more generally, of a struggle usually described with categories borrowed from economics or sociobiology.

Nature of Scientific Production

In this model nothing is said about the content of scientific work. It is simply assumed that the research scientist develops knowledge that is submitted to the judgment of colleagues. Knowledge is generally transmitted in the form of publications that are disseminated without any particular restriction.

These publications are in principle intelligible to specialists in the field. One can make use of the notion of information to speak of their contents. This knowledge or information is characterized by its novelty, its originality, or perhaps its degree of generality. An evaluation of its utility, as perceived by others—scientists or nonscientists—is also possible. This model does not exclude the existence of tacit skills, but this is alluded to without being turned into a specific component.

Actors

For this model, the actors in the production of scientific knowledge are the research scientists themselves. A distinction is made between the world of scientists (the specialists) and that of the layperson. Technicians are reduced to an instrumental role on the same level as the experimental apparatus. Science is first and foremost an intellectual adventure, and its practical and technical dimensions are eclipsed.

Research scientists are social beings whose individual competencies are not defined or analyzed as such. Their membership in a discipline or specialty determines their aims and ambitions, together with their theoretical and experimental choices. Therefore the rationality of scientific activity results from the interaction between scientists, and in particular from their competition, not from any particular inherent predisposition that distinguishes them from other human beings.

The motivations behind scientists' actions are not theirs alone. Various suggestions have been made by different authors. Merton (1973b) insists on the role of norms that define permissible behavior and on reward systems that institutionalize the production of knowledge. Bourdieu (1975a) sees scientists as agents guided by their habitus who develop strategies for positioning in a field structured by the interlacing of these different strategies. Hagstrom (1966) conceives of scientists as striving to maintain the confidence of their colleagues.

Scientists therefore have a dual role. This resembles a Darwinian struggle in which they are both judges and litigants. Every researcher judges his or her colleagues (Is the knowledge new and robust? Is the information useful?) but is similarly judged by them in the same way.

Underlying Dynamic

What are the mechanisms responsible for this organized and collective search for knowledge? Why are scientists led to produce more and more knowledge?

The answers provided by this model draw their inspiration from different versions of economic theory. One may first, like Hagstrom, conceive of an exchange economy. The scientist who is evaluated positively by her colleagues receives recognition, and this, in turn, bolsters confidence in her. This is a gift economy.

The model may be that of neoclassical economics. Here the scientist is comparable with an entrepreneur. The product she offers her colleagues is knowledge, which the latter evaluate as a function of its utility and quality. This evaluation is measured in the form of symbolic awards (see the section

"Social Organization," below). Each scientist is supposed to maximize her personal profit, that is, the recognition granted her. A climate of competition is thus created, which, as in the neoclassical market, channels individual passions and selfish interests into a collective, rational, and moral enterprise (Ben-David, 1991; J. Cole, 1973; Hull, 1988; Merton, 1973b).

Or the model may be that of a capitalist economy as described by Marxists. Scientists are not so much interested in recognition per se as in the possibility of obtaining ever more of it—the object is one of accumulation or circulation. Here one encounters Bourdieu's (see 1975a) works and Latour's first analyses (Latour & Woolgar, 1979). Researchers have no choice: If they want to survive among their colleagues, they have to accumulate credit or credibility, which constitutes their capital. Without capital they cannot obtain support for new programs. On the other hand, the more capital they have, the more they are able to carry out research, the results of which would increase their initial endowment. Scientists are thus caught up in a logic of success.

One of the features of the economic metaphor is that the psychological motivations or aims of scientists are not important. Competition coordinates individual behavior. So this is science without a knowing subject—a prospect dear to philosophers as different as Popper (1972) and Althusser (1974). The fact that some authors present a Darwinian metaphor, whereas others are committed to one drawn from economics, makes no fundamental difference to the interpretation proposed here.

Agreement

In this model the production of agreement implies what could be called free discussion between scientists. Science is caught in a double movement of openness and closure. Openness guarantees that all points of view can make themselves heard, and closure signifies that reaching agreement is the objective assigned to these discussions.

The openness of debate must not go beyond the scientific community. In this model researchers monopolize discussion and any interference from outside is a potential source of disorder. Not all exchange with the sociopolitical environment is excluded. External requests may be formulated, and preoccupations and convictions may be brought to the attention of the scientific community. The model tolerates, for example, the idea that industry or political decision makers ask questions and orient programs (Merton, 1938/1970). It also admits that metaphysical concerns and philosophical convictions motivate researchers. Thus Freudenthal (1986) has linked Newtonian physics to Hobbes's political philosophy. Yet such influences do not (or should not) go right to the heart of scientific activity. They contribute to formulating problems or putting them in a hierarchy. For this reason they play an important role in creating the preconditions for scientific agreement. This agreement, to be complete, must also relate to the technical matters that are immune to outside influence. Agreement cannot be determined therefore by positions of power or arguments from authority. There is an irreducible internal core that is the responsibility of the scientific community. The scientist can only be convinced by statements that in the last analysis draw on method for their robustness. What counts as an acceptable method could vary over time but is considered taken for granted during any given period.

Social Organization

Organization is one of the central variables of the model. Indeed, the viability of the scientific enterprise rests upon an organization that strictly separates the inside from the outside.

Internal Organization

The incentive system plays a vital role, driving scientists to produce knowledge. It is based on a double trigger mechanism. "Discoveries" (or, more broadly, "contributions") are identified and attributed to certain scientists, who are rewarded according to the quality of these contributions.

In this model, what counts as a discovery is the outcome of a social process. In the incessant flow of scientific production, how does one isolate identifiable units of knowledge that are more or less independent of one another? And how does one then decide on the origin of each of these elementary contributions? There is no universal answer. The delineation of contributions as well as their imputation (Gaston, 1973; Merton, 1973b) often gives rise to controversies and to reevaluation (Brannigan, 1981; Woolgar, 1976).

The identification of discoveries and their "authors" could not be stabilized without material devices and rules that codify the formulation of knowledge and its transmission. Thus the scientific article in its present form makes it possible to delimit a piece of information precisely, to organize its dissemination, to identify the authors who produced it, to date their contributions, and to mention what has been borrowed from other authors by means of quotations and citations (D. J. S. Price, 1967).[9]

The functioning of the reward system depends upon the identification and attribution of contributions. Their importance, their quality, and their originality are all evaluated simultaneously. And the possible forms of reward are varied, so that they may be adapted to the supposed importance of the contribution: promotion, prizes that vary from the most modest to the most prestigious, election to an academy, and eponymy (giving the name of the scientist to a result that has been attributed to him, as in the case of Ohm's law). The researchers—the actors in scientific development—are thus encouraged to contribute to the advancement of knowledge.

The model suggests that these rewards are symbolic in character; evaluation is not directly related with possible economic gain. Science is a public good.[10] Statements like "the structure of DNA is a double helix" are nonrival and nonappropriable. That Mr. Jones mobilizes the statement for his own activities does not diminish its usefulness to Mr. Brown and does not prevent the latter from doing the same thing. So with these public goods market mechanisms don't work efficiently. And this is why conventional economic incentives (valuing goods on the market) are replaced by another reward system—one that urges researchers to produce knowledge that they make public. Thus publication enables contributions (discoveries) to be identified and imputed, but it also ensures that these public goods are disseminated—which explains why publication is considered to be a cornerstone of science.

Another essential issue is the maintenance of free access to discussion. The social organization should encourage scientists to produce knowledge but must also favor open debate: seminars, colloquia, or the right to reply in journals. This ensures that any scientist wishing to participate is able to do so. Free access is a basic principle that is inscribed in the norms of science and its institutional forms.

The model stresses the role of individuals (the researchers). Yet scientific activity is more and more a matter of teamwork. So how is the emergence of joint research sites like laboratories explained? And what is their role? In this model the question is as problematic as the existence of firms was to economics (Coase, 1937). Even if some authors examine the organizational structures, performance, and strategies of laboratories (Whitley, 1984), in this model the laboratory is an anomaly. Its existence can simply be seen as a consequence of technical constraints: the management of large-scale equipment or experimental work that depends on the division of labor to secure economies of scale.

Relations With the Environment

Model 2 explores the relationship between science and its environment but does so by establishing a clear boundary between inside and outside. When this boundary is crossed, the norms, rules of the game, incentives, and types of resources break down. The notion of a scientific institution, with its own goals, values, and norms (Merton, 1973b), together with the notion of a scientific field (Bourdieu, 1975a), mark the existence of territory. Numerous historical analyses have shown how this social space governed by its own laws has become autonomous and how the role of the professional scientist has gradually emerged and been consolidated (Ben-David, 1971).

The existence of autonomy does not exclude exchange and influence with the outside world. For instance, Bourdieu conceives of two markets: a restricted market, limited to specialists, where scientific theories are debated, and a general market that transmits the products, thus stabilized, to the external actors interested in them—firms, state agencies, and the educational system (Bourdieu, 1971). Between the two markets there are mechanisms of control. The value of a product (a theory) on the external market depends partly on the value it is given by the internal market (and vice versa). Again David, Mowery, and Steinmueller (1982), adopting the economist's point of view, consider science to produce nonappropriable information that is reused by economic agents. The latter, in turn, produce the specific (and appropriable) information that they need, in a more predictable and less costly manner. And yet again A. Rip (1988) proposes a generalization of the cycles of credibility by introducing the "fundability" of research projects, linking the logic of scientific development to that of the politico-economic actors.

The duality of organizational forms is crucial in this model. The border between the internal and external is essential to science and protects its core yet must be sufficiently permeable to transmit the influences that nourish science and ensure its social utility.[11] The organizations that link science to its environment (industrial research centers and state agencies for encouraging research) play a crucial role in managing these exchanges in a proper way (Barnes, 1971; Cotgrove & Box, 1970; Kornhauser, 1962; Marcson, 1960).

Dynamics of the Whole

This model depends on a regular process of growth against which it is possible to explain historical "accidents" and decreasing returns. This growth is explained by the fact that scientists work in those areas of research where the anticipated symbolic profits are likely to be highest because the problems being tackled are considered important, and where there are still many areas of ignorance. Accordingly, everything that fosters mobility also tends to favor overall growth, whereas anything that impedes it tends to reduce the productivity of science (Ben-David, 1971, 1991; Mulkay, 1972). If free debate is hindered, if the incentive system malfunctions, and if positions of monopoly come into being, then that productivity may decrease. Here again one finds arguments that are fairly close to those used in the analysis of economic growth.

Moreover, if society does not guarantee the boundaries, and if it does not support the internal organization of science and its rules, then research as a whole will break down, as when the Nazis advocated a racialized and nationalized science or when the Soviet Communist party rejected Mendelian genetics. So the model goes beyond mere considerations of production. Science produces knowledge, but the institution that supports it has an essential function, that of enabling rational knowledge to develop. When the dynamics of science are hindered, reason is affected.

MODEL 3: SCIENCE AS SOCIOCULTURAL PRACTICE

This model says that science does not really differ from other activities and the certainties it leads to do not enjoy any particular privilege—an argument based on the fact that science is much more than the simple translation of statements. The third model suggests that science must be considered to be a practice whose cultural and social components are as important as the constraints that arise from the order of discourse.

Nature of Scientific Production

Model 1 was content to limit its investigation to statements and assumed that these were transparent, their meanings lying simply in the system of statements. Yet, as the pragmatists of language have taught us, a statement has no meaning without a context. Model 3 adopts this position and emphasizes the importance of nonpropositional elements (tacit skills) in the production of knowledge.

The contribution of authors such as Kuhn (1962) and Wittgenstein (1953) is essential. The notions of rules and how they may be followed, language games, forms of life, and learning by examples underline the importance of tacit knowledge—a notion developed by Polanyi (1958) to account for the transmission of noncodified information. Certain knowledge—for example, knowledge linked to the functioning of instruments or the interpretation of data supplied by these instruments—cannot be expressed in the form of explicit statements. In this view science is an adventure that depends on local know-how, on specific tricks of the trade, and on rules that cannot easily be transposed. Formal statements can only travel and be understood if their instrumental environment and the knowledge incorporated in human beings is the same. This theme was developed brilliantly by authors such as Fleck (1935) and then J. Ravetz (1971): "In every one of its aspects, scientific inquiry is a craft activity depending on a body of knowledge which is informal and partly tacit" (p. 103). Collins enriched this argument considerably in several studies. For instance, in his study of the construction of the TEA laser, he showed that the diffusion of knowledge could not be reduced to the mere transmission of information: "The major point is that the transmission of skills is not done through the medium of written words" (H. Collins, 1974).

Collins thus distinguished the algorithmic and the enculturation models. In the former, science consists of the production of codified transparent information; in the latter, tacit skills and learning are important—a scientific statement is always opaque, its meaning reducible neither to what it states nor to what is said by the system of statements to which it belongs. The distinction between algorithmic and enculturational models becomes essential when the question of replication of experiments is considered. The reproduction of an experiment always implies close interaction between scientists and experimental arrangements; an entire culture is transmitted with this know-how, these ways of seeing and interpreting, these observational statements.[12] As Collins (1974) says: "Only those scientists who spent some time in the laboratory where the success has been achieved prove capable of successfully building their own version of the laser." The affirmation of the enculturation model has a general implication: that practices incorporated in human beings (those who manipulate and interpret) are intertwined with experimental apparatus, protocols, and observational or theoretical statements. To extract statements from this whole and to transform them into a privileged object of scientific production is to take them out of their context and strip them of their meaning.

Actors

The actors involved in the dynamics of developing scientific knowledge are not limited to experimentalists and theoreticians. In a highly suggestive article, Collins and Pinch (1979) introduce a distinction between what they call the constitutive and the contingent forums. They show how groups outside the scientific community may be mobilized in the production of knowledge. The list of these groups depends on the particular situations under study: the manufacturers and distributors, the media, state agencies, firms with their engineers, or even external pressure groups (philosophers, ethical committees, and so on)—any or all of these may participate. The border between insiders and outsiders fluctuates and is negotiable. But what is analytically important is to explore the mechanisms by which constraints, demands, and interests outside the circle of researchers influence scientific knowledge. In an exemplary work devoted to the Great Devonian Controversy in geology, Rudwick (1985) follows the different actors who were directly or indirectly interested in the debate during the 1830s about the existence of a geological stratum (the Devonian). He gives real depth to all these characters, reconstituting the network of relations and locating them in the institutional frameworks of the period. Wise's (Wise & Smith, 1988) work on Lord Kelvin, Schaffer's (1991) on astronomers, MacKenzie's

(1981) on the emergence of statistics, and Pestre's (1990) on Neel are other examples of such analyses.

Attention is also paid to those who work in laboratories. In Models 1 and 2, technicians are present everywhere, but in the form of transparent shadows. They carry out experiments, collect samples, and determine measurements; yet their work has no influence on the content of knowledge and they have the same status as instruments. The sociocultural model repairs this omission. Just as it emphasizes experimental work, it also brings into play those who carry out the experiments and prepare the samples. Shapin (1989), in a highly instructive article, has greatly contributed to this rehabilitation. K. Knorr (Knorr Cetina, in press) stresses the particular role of Ph.D. students in laboratory life.

To be sure, the researchers are not forgotten. Their competences are diversified and include the capacity not only to formulate and interpret coded statements and algorithms but also to elaborate and control tacit skills or the rules of the art. The researchers (technicians must be included in this category) manipulate, decipher, inspect, tinker with, interpret, and reason (Knorr, 1981; Latour & Woolgar, 1979; Lynch, 1985a). They are furthermore capable of learning and memorizing. The notion of learning, although central to this model, has largely been left unexamined. Several different approaches to learning exist in literature. Bayesian analyses insist on the probabilistic character of knowledge and on the role of experiments in strengthening or transforming subjective probabilities (Hesse, 1974); others refer to Piagetian theories or those of artificial intelligence (Mey, 1982) or to Gestalt psychology (Kuhn, 1962). This offers a wide field for research. Whatever the theoretical stance adopted, the underlying hypotheses are clear; the learning capacity of actors endows them with both historical depth (they guarantee a certain continuity of knowledge) and a (permanent) faculty for invention, that is, for redefining routines and rules for coordinating action, which enables one to understand why science is not limited to repetition.

The stress laid on tacit skills and learning mechanisms leads to the social group. Interaction is only developed within the framework of a shared culture and scientific activity is no exception. This hypothesis has its source in the notion of a paradigm proposed by Kuhn, who refers on the one hand to the group and on the other to the scientific competence and production of each of its members. For Collins it is the core set that is the fundamental actor responsible for the production and transmission of knowledge. It groups together researchers who share the same problems and culture. Collins also refers to Granovetter (1973) to suggest that a researcher's impact is greatest if she enters into unusual or atypical social relationships (see Mulkay, 1972). Schaffer (1991) adopts an analogous point of view: "The coordination between these two networks was crucial, because it showed that observatory managers and experimental astronomers might collaboratively extend their control beyond the boundaries of celestial mechanics" (p. 6). Scientific groups are structured like social networks—they can become denser, close in on themselves, fragment, or merge (Crane, 1972; Mullins, 1972). The dynamics of these networks depend on the strategies of relationship building followed by their members, and each transformation of the social network implies a cultural transformation.

In extending the field of analysis by analyzing all the social groups that intervene in the process of creating knowledge (the "constituency of interest"), the defenders of Model 3 give the description a distinctively sociological flavor, without sinking into reductionism. For the first time, sociology treats the contents of science with the same degree of depth and the same concern for detail as any other human activity.

Underlying Dynamic

To account for the dynamics of scientific activity, there is no need to invent new sociological explanations. Barnes (1977) provides the clearest and most systematic presentation of this point of view. Inspired by the Marxist tradition, of which we may also find traces in Habermas's work, he writes: "Knowledge grows under the impulse of two great interests—an overt interest in prediction, manipulation and control, an overt interest in rationalization and persuasion" (Barnes, 1977, p. 38). Thus, in the phrenological controversy studied by Shapin (1979), we find a mixture of sociopolitical and cognitive interests. The endeavor to clarify the possible existence of the frontal sinuses is as much to score points in the class struggle in Edinburgh as to learn anything about the brain. These two families of interests are to be found in all societies; if certain ones like our own have developed science, it is for contingent historical reasons. Interests linked to prediction and control have been intensified and then inscribed in specific institutions.

More generally in Model 3, the explanation of the underlying scientific dynamics depends on the particular sociological models used. We have just evoked Barnes's macrosociology but there are more microsociological possibilities. In Pickering's recent texts, we find an explanation that makes no distinction between a scientist and any other goal-oriented social actor: "Doing science is real work" (Pickering, 1990). Science is a practice and is analyzed like all practices; a researcher has resources, tries to reach her goals, and seeks to create coherence between the disparate and sometimes intractable elements that make up her environment (instruments, theoretical, and experimental models), some of which resist all reorganization. K. Knorr

(1992b), relying upon Merleau Ponty's philosophy, gives an illuminating description of what she calls the "epistemic cultures" of high-energy physics and molecular biology. She stresses the disunity of scientific practices that depend on "their orientation toward and treatment of signs, on their relations to themselves, on the forms of alignments they institute between subjects and natural objects, on their general approach to capturing and engaging truth effects in inquiry" (p. 3). And there are other possibilities, including ethnomethodology (Lynch, 1985a), symbolic interactionism (Clarke & Gerson, 1990; Fujimura, 1992a; Star, 1989a), or cultural anthropology (Hess, 1992; Traweek, 1988). All these studies rest on the same assumption: that science is a human activity, one that is specific but that does not merit a change of analytical instruments. Possible explanations for the development of science are as numerous as sociological theories!

Agreement

Agreement between scientists must be explained in the same terms as consensus between social actors anywhere. The principles of Bloor's (1976) "Strong Programme" are the methodological translation of this hypothesis. Because nothing distinguishes science from other human activities, and because scientists are like other social actors, agreement, disagreement, success, and failure need not be explained in different terms.

This argument can be illustrated by the exemplary work of Collins on gravitational waves. As Golinski (1990b) puts it, for Collins,

> experiment is potentially open ended. At no point, in his view, does nature force a particular interpretation upon experimenters. . . . The evidence is always too much to fit within interpretive scheme and too little to determine the choice between any number of possible alternative schemes. . . . Controversy can be continued as long as a critic can find the resources to sustain these objections. . . . Sufficient differences between two versions of an experiment could always be found by a critic who wished to deny that a proper replication had been achieved. (p. 494)

Collins (1985) calls this type of dispute the "experimenter's regress." What remains to be explained is why protagonists with different interests, know-how, and practices end up considering that the debate is closed.

The answers given by the sociocultural model tend to fall into several classes. First, there are fairly traditional macrosociological explanations. Because agreement never rests on indisputable evidence, its construction depends solely on the state of social forces and particularly those outside the scientific community, or the group of researchers involved in the debate. The Edinburgh School (Barnes, Bloor, Shapin, and MacKenzie) carried out many case studies in which the influence of political, economic, or cultural interests created a balance of power favorable to a particular outcome. This approach can sometimes appear determinist and mechanistic. Thus it is occasionally claimed that identifiable external groups or social classes add their own weight to that of the scientists with whom they agree. Alternatively, scientists may be left to choose their allies themselves—in this case such alliances do not pervert science because nature is sufficiently ambiguous and experiments are sufficiently complex to support the different opinions and judgments. As there are never uncontroversially good reasons to choose one theory rather than another, there is room for sociological explanation without endangering the autonomy of scientific work (Barnes & Shapin, 1979; Wallis, 1979).

The alternative approach uses notions such as confidence. For instance, in his work on solar neutrinos, Pinch (1986) elegantly shows the importance of the creation of a climate of confidence throughout the design and conduct of experiments. By associating the representatives of several disciplines, by taking into account objections as they are raised, the project becomes a collective enterprise based on relations of reciprocity (exchange of information and so on); the agreement on the results is the fruit of this growing confidence. The nature and extent of relations formed during the conception and realization of experiments and during the elaboration of theories largely determine the likelihood of agreement as opposed to a continual experimental regress. The relationship between this type of analysis and developments in game theory (Axelrod, 1984; Kreps & Wilson, 1982) has been little explored.

Agreement may be facilitated by operations on the instruments themselves. Collins's research has shown that the difficulty with replication is largely ascribed to differences between pieces of experimental apparatus. As Collins suggests, standardization and the calibration of instruments reduce the likelihood of divergence and favor agreement. If this calibration is not achieved, one returns to the situation so well described by Schaffer (1989) with respect to Newton's experiments on the refraction of light in prisms:

> Newton's "law" did not compel experimenters such as Rizetti: "it could be a pretty situation," the Italian exclaimed, "that in places where experiment is in favor of the law, the prisms for doing it work well, yet in places where it is not in favor, the prisms for doing it work badly." For such critics, Newton's prisms never became transparent devices of experimental philosophy. (p. 100)

This transparency of instruments, created by scientists, which became important in the second half of the nineteenth century, "let nature speak for itself" (Daston & Galison, in press), led to the black-boxing of experimental methods and to their standardization (Latour, 1987). Of course, this agreement in turn depends on collaboration and compromise, which must again be explored. Once it is achieved, however, it is inscribed in the calibrated instruments and provides a solid basis for new agreements. This may prompt a distinction between passive instruments (Fleck, 1935), which are not reconsidered, and active instruments, which evolve and become controversial. Passive instruments form the common ground on which arguments and counterarguments can be deployed. They furnish a common measure. G. Bachelard's (1934) notion of "phénoménotechnique" greatly contributed, in his time, to emphasizing the importance of agreement sealed by instruments. Because the instruments come to embody the theories they are used to support, disagreement is made more difficult (Latour & Woolgar, 1979); the refutation of a statement implies the refutation of the instruments and their calibration. The theories are "hardwired," to use Galison's (1987) nice expression.

Finally, the sociocultural model allows the use of all available means. The possible mechanism for "closure" and the possible studies of these mechanisms are endless.

Social Organization

The sociocultural model is, paradoxically, only moderately interested in questions of organization and institutional forms. This observation applies as much to the internal organization of scientific activity as to its relations with the sociopolitical environment.

The notion of rules is probably one of the best suited to account for a social organization capable of managing scientific practice in its entirety. Rules are both implicit and explicit; they are not outside action but are interpreted, elaborated, and transformed within action. Again, they are both social and technical, ensuring a minimum of coherence and making anticipation and discussion possible. They are compatible with the proliferation of social groups and the diversity of their identities. Rules, which are more or less local and specific, form the fixed point around which relationships of power and influence can be developed. Sociological and economic work on the appearance of rules or conventions could be usefully mobilized to enrich the sociocultural model (Bloor, 1992; Favereau, in press; Lynch, 1992; Vries, 1992).

Emphasizing the role of learning, the model consequently stresses the importance of skills transmission and training. This produces relationships of dependence between masters and disciples, and also within laboratories between different actors with different types of skill. Shapin (1989) gives a good illustration of this type of analysis by highlighting the crucial position held by technicians in early laboratories. This sociology reintegrates more traditional considerations of power and domination into the world of science (Schaffer, 1988).

Finally, Model 3 considers boundaries between science and its environments as constructed by actors themselves in various hybrid settings. Jasanoff's (1990a) study of regulatory science, Abir-Am's (1982) investigation of Rockefeller Foundation policy in molecular biology, Dubinskas's (1988) work on high-technology organizations, Wynne's (1992c) analysis of the entanglement of science and policymaking in environmental issues are a few examples of this growing field of analysis.

Dynamics of the Whole

The sociocultural model challenges the idea of continuity in the development of scientific knowledge.

If science does not progress in a linear way, it is because it is involved in social relations that have their own logic. Barnes's notion of interest is very useful from this point of view. Scientific knowledge can be seen always as a response to one kind of interest, that of prediction and control, but its contents are organized and structured according to different and changing social configurations. Knowledge is marked by the conditions of its production; Kuhn's approach is exemplary in its insistence on the incommensurability of skills and paradigms. The historicity of science is expressed in the problems it asks itself and can be seen as a function of global history.

A more subtle analysis is also possible. Collins, for example, notes that the diffusion of knowledge cannot take place without transposition and adaptation to local circumstances. No replication has ever resembled the experiment that inspired it, even when the instruments have been perfectly calibrated and procedures highly standardized. Transfer involves loss and creation, elimination and addition. This view leads to the original interpretation of Kuhn's work by Masterman, which links paradigms and "Wittgensteinian" family resemblances. The argument is that any new instantiation of the paradigm creates a discrepancy from the original exemplar. The distance from the original grows from one instantiation to the next, and the paradigm ends up betraying itself (Masterman, 1970). The dynamics of science are born of these successive discrepancies—discrepancies that are nothing more and nothing less than the research process itself. It is because diffusion inherently involves transformation and transposition that science is forever developing. This leads to a conception in which the dynamics of science create "a genuinely historical process; facts and phenomena, concepts and theories as well as the instruments and institutions of science are bound to the wheel of what happened" (Pickering, 1990).

MODEL 4: EXTENDED TRANSLATION

We have seen in Model 1 that the notion of translation can be used to explain the establishment of links between different statements. Model 4 develops this definition beyond the domain of codified knowledge. *Translation* refers here to all the operations that link technical devices, statements, and human beings. The notion of translation leads to that of *translation networks,* which refers to both a process (that of translations that are joined together) and a result (the temporary achievement of stabilized relations). This model seeks to explain the proliferation of scientific statements and their broadening sphere of circulation. Finally, it calls for a deep reformulation of social theory.

The Nature of Scientific Production

Manufacturing Statements

Like Model 1, the extended translation model assumes that the prime objective of scientific activity is to produce statements. But, like Model 3, it stresses the process of production and the role of nonpropositional elements in this process. Take the following two statements: (a) "The structure of DNA is a double helix." (b) "The facade of the pension where Father Goriot lived had been covered with a layer of poor quality pink paint which bad weather had caused to crack." The difference between statements a and b does not lie in the statements themselves but in the extent to which the reader is able to work his or her way up the chain of elements that support the statements. Statement a refers to other statements, other objects, and other time spaces, which it sums up and condenses, and to which it gives access. The second statement refers to nothing other than texts and the inescapable fiction of the world of the novel. The notion of a translation chain describes the series of displacements and equivalences necessary to produce a particular type of statement.

Insofar as science is concerned, translation chains combine heterogeneous elements of which the most important are statements, technical devices, and the tacit skills that can rightly be called embodied skills. To understand how relationships can be built between these different elements, one must first introduce Latour's notion of inscription, which refers to all written marks (Latour, 1987; Latour & Woolgar, 1979). Inscriptions include graphic displays, laboratory notebooks, tables of data, brief reports, lengthier and more public articles and books. The notion of inscription points to the importance of writing and to its diversity. Thus the division between instruments (i.e., experiments) and statements (i.e., observations), implied by the preceding models, is replaced by a range of inscriptions, from the crudest marks to the most explicit and carefully crafted statements. From marks to diagram, from table to graph, from graph to statement, and from statement to statement—each is a translation.

Translation chain:

→ instrument → marks → diagrams → tables → curves → observational statement 1 → theoretico-observational statement 2 → theoretical statement 3 → and so on

Writing devices are important in all scientific fields and beyond. For example, Foucault (1975) analyzes the hospital as a device that places the individual in a "network of writing."[13] As entities are translated, resistance encountered, and answers gathered, the devices progressively take on form and materiality. Although the task of writing is general, experience shows that a charmed quark, a suffering body, a replicated gene, a humiliated social group, or a geological stratum and its fossils cannot be written in the same way.[14]

Science is a vast enterprise of writing, but to move from an inscription to a statement, and from one statement to another, requires embodied skills and/or technical devices. Without them the manufacture of knowledge (Knorr, 1981) would be unproductive. Thus it is the constant interaction between inscriptions, technical devices, and embodied skills that leads to the development of statements. These interactions may be observed in the composition of experiments (Hacking, 1983), in the interpretations of inscriptions (Amann & Knorr Cetina, 1988a, 1988b; Lynch, Livingstone, & Garfinkel, 1983; Pinch, 1985), in conversations between scientists, or between scientists and technicians, and the writing and the rewriting of articles or reports (Myers, 1990a). All these interactions are translations, and they all contribute to the production of statements—a process that Law (1986b) calls heterogeneous engineering. Ethnographic research has described many of them, and graphic methods such as those developed by Fujimura (n.d.) or Gooding (1992) make these easier to depict.

Taking Statements Out of Laboratories

Scientific activity is not simply a matter of manufacturing statements; often (if not always) it seeks to take statements out of the laboratory. But this challenges the conventional distinction between the content of knowledge and the context of production. The notion of translation makes it possible to understand how context and content are simultaneously reconfigured.

Translation leads to the identification and shaping of allies and to seeking their support. It means establishing an equivalence between, say, the biochemical study of an obscure polymer and its absorption by certain body organs and many other agents in society, for example, the groups and institutions that support the struggle against cancer, the field of biochemistry interested in such a polymer, or the pharmaceutical industry and the medical profession (Law, 1986a). A team of biochemists can define other actors and suggest the following translation: We want what you want, so ally yourselves with us by endorsing our research and you will have a greater chance of obtaining what you want (Callon, 1980b). Such translations are always tentative and in certain cases postulate completely new actors, which are then brought into existence. The translations might be inscribed in texts stating explicitly the contribution of the projected work, in material substances, or in skills and instruments. These translations might require huge investments. They link closely the definition of very technical problems with the constitution of a space of circulation for the knowledge that is produced.

Translation Networks

The notion of translation network refers to a compound reality in which inscriptions (and, in particular, statements), technical devices, and human actors (including researchers, technicians, industrialists, firms, charitable organizations, and politicians) are brought together and interact with each other. The networks vary in length and complexity. Some only rarely leave the laboratories or their communities of specialists and act primarily via instruments and statements. Others stabilize some of these entities and mobilize them to multiply connections with nonspecialists. Wise (1988), for instance, describes how machines act as material and durable mediators between engineering, industry, and the esoteric concerns of particular domains of research. Yet other networks are active on both fronts and enter into a dynamic of expansion, where each translation within the laboratory leads the network outside to be lengthened. In all cases it can be said of scientific activity that it establishes translation networks.

Inversion of Translations

When a network is established, scientists talk not only on behalf of electrons or DNA, which they translate in their laboratories, but also for the countless external actors they have interested and that have become the context for their actions. Their ability to act as legitimate spokespersons is due to the series of representations that have been set up. This led L. Star to propose the notion of *re-representation* (Star & Griesemer, 1989). For translation is also representation. In the system produced by Galileo to translate gravitational forces, there is a succession of representations: The clepsydra represents time; the angle of incline re-represents the difference in drop; the table re-re-represents the sphere's course; the curve (re)4-presents the table; and the mathematical formula (re)5-presents the curve. As with elections, one can talk of representation to the nth degree. But actors attracted to scientific work are also re-presented. Biochemists seek to re-present chemotherapy and the fight against cancer. The argument is that the scientist's particular strength is that she is able to accumulate both types of representation: to re-present herself as spokesperson of both nature and society.

This analysis sheds new light on the standard problem of reference. Thus the statement "the structure of DNA is a double helix" is the last link in a chain that, from translation to translation, refers to other inscriptions, embodied skills, and technical devices. Statements do not talk of an outside reality; they are simply one location point in a long and teeming network. There is no one "reference" but an entanglement of "microreferences": The statement refers to a table that refers to a trace; the trace refers to a technical device, and its interpretation refers to embodied skills. So it is only when attention is focused on the final statement that the translation chain is split, that one can talk of out-thereness. Then one has inversion: The pulsar is said to be the cause of the statement while it is present at each point of the chain of translation but in various forms including statements (Latour & Woolgar, 1979; Woolgar, 1988b). Similarly, context cannot be dissociated from scientific content unless we put the translations that define it between brackets. So the notion of translation is preferable to that of reference, even though the etymology is close. This is because, when it is said that a statement translates DNA, or that biochemists translate chemotherapists' projects, no hypotheses about reality or correspondence are made. One is instead reminded that reference is nothing more than an effect of a translation chain, and that its robustness depends entirely upon the latter.

Actors

The extended translation model substitutes the notion of an actor with that of an actant (a notion borrowed from semiotics; Latour, 1987, 1988). *Actant* refers to any entity endowed with the ability to act. This attribution may be produced by a statement (the statement "somostatine inhibits the release of the growth hormone" attributes the property of inhibition of the actant growth hormone to the actant somostatine), by a technical artifact (a chromatographer gives gases the ability to diffuse in a column having elements that are themselves defined as obstacles to this progression; it also implies a researcher inspecting the signs of diffusion as well as other technical artifacts required in its functioning), or by a human being who creates statements and constructs artifacts.

The notion of an actant is particularly important in the study of scientific activity. This is because the latter permanently modifies the list of entities making up the natural and social world. Out of laboratories come quarks, enzymes, and proteins, all new actants that did not exist before being brought into play by statements, tables, machines, or embodied skills. But within laboratories, social groups interested in scientific production are also being formed—groups that make up the famous social context. Before Einstein wrote to Roosevelt, politicians could not want the atom bomb; afterward, they wanted it very much. The actant "Roosevelt-who-wants-the-atom-bomb to combat the powers menacing the free world" is no less a laboratory creation than "somostatine-which-inhibits-the-growth-hormone." This, then, is the attraction of the notion of actant. It is sufficiently supple to account for the proliferation of entities that all contribute to scientific production: electrons and chromatographers, the president of the United States and Einstein, physicians with their assistants, the cancer research campaign, electron microscopes and their manufacturers—all are actants.

The list of actants and their definitions are liable to change, and these changes often give rise to debate. If it is claimed in another laboratory that somostatine also exists in the pancreas and does not inhibit the growth hormone but inhibits the production of insulin, then somostatine's definition changes (Latour, 1987). The very identity of the actant somostatine is transformed, even if its name stays the same. But this is also the case for Roosevelt if he is convinced of the impracticability of the Manhattan project. Actants may more or less successfully resist definition imposed on them and act differently. Then their identity depends on the state of the network and the translations under way, that is, on the history in which they are participating. Society and nature fluctuate like the networks that order them (Callon, 1986b, 1989; Latour, 1987, 1991a)—existence precedes essence. The latter has variable geometry, changing as time passes. And this is why the model rejects broad divisions, both between nature and society and between human and nonhuman. It does not challenge the existence of differences. On the contrary, it multiplies them by allowing the observer to register them all and follow them as they change. The analysis of science is a wonderful laboratory. It is a place where one may study social links in the making.

Underlying Dynamic

The extended translation model gives a broad definition of action. An actant may be a pharmaceutical firm that aims at developing anticancer drugs, a political party that supports cruise missiles, a technician working on a mass spectrometer, a researcher interpreting data charts, or an electron that does not interact with a flow of protons. All these actants are brought into play, mobilized in statements, instruments, or embodied skills. Each new translation may modify, transform, contradict, or alternatively strengthen former translations. Each, that is, may modify or stabilize the actants' universe. To translate is to describe, to organize a whole world filled with entities (actants) whose identities and interactions are thereby defined. In this model the notion of action disappears in favor of that of translation. What, then, is the explanation of scientific change?

To translate a device into an inscription, an inscription into a statement, or a statement into embodied skills, is to create a discrepancy, a betrayal. In short, equivalence is the exception. It is only obtained with difficulty and at great expense. Divergence between translations and the proliferation of entities is the rule, not the exception. The chromatograph traces a curve, the technician draws up charts, the scientist goes from one statement to another, her competence is reinscribed in an experimental device that produces new marks, and so on. Every new translation produces a discrepancy in relation to previous translations, which it then threatens.

So why are there so many proliferating translations? One doesn't have to imagine that actors are steeped in power, trying to impose their equivalences at all costs (though this is not impossible). The notion of action is distributed to all the actants. It is enough to imagine that even the most modest actant, the humblest electron microscope, the most docile technician, and the least imaginative researcher, all produce slightly differing translations. The proliferation of discrepancies lies in these small betrayals. The universe of translation is polytheist. History is an accumulation of such betrayals, and, as the sciences are nothing more than a set of extended translations, their dynamics are no different. This is another way of saying that uncertainty lies at the heart of scientific production. But it is also a way of saying that nature is neither more nor less active or malleable than society.

Agreement

The extended translation model does not talk of assent and dissent. Rather, it speaks more generally of alignment or dispersion of translation networks.

To speak of consensus as in Models 1 and 2, or of closure of debate as in Model 3, is to privilege the discursive dimension of science. By contrast, the translation model, even if it emphasizes the production of statements, assigns great importance to the hidden side of debates—to all that is not discussed but the presence of which allows dialogue to be established.[15] All controversies, even the most fierce and relentless, depend upon a tacit agreement about what is important and what is not. Collins himself shows this in his study of gravitational waves. The discussion between Weber and his colleagues, the exchange of arguments and counterarguments, would have been impossible without deeper agreement about the meaning of Einsteinian theories, the capabilities of computers, the character of mathematical tools, or the nature of torsional moment. For there to be disagreement on the interpretation of a recording, a whole invisible infrastructure of embodied skills, of known and recognized technical artifacts, is needed. Its existence makes discussion possible.

In Model 4 the meaning of a statement—the possibility it has of being taken up or discussed—depends on the chain of translation in which it is located. The explanation of a statement's force—its ability to convince—is no different than the explanation of its meaning. Again, it depends on translation chains and the references these create. Force, then, is a function of the robustness of chains and more particularly of the morphology of the networks they constitute. An isolated statement has no more force than it has meaning. It follows, then, that networks with differentiated elements, which have translated one another, are most forceful. And so are those with many intertwinings. This is because any attempt to question the network is rapidly confronted with a dense network of translations that all support one another. The translation network and the heterogeneity of its components (technical devices, statements, inscriptions, embodied skills, social groups outside laboratories) explain the robustness of arguments.

Such an interpretation is to be found in the work of Pickering or Hacking, though they are concerned mainly with laboratory translations. Pickering distinguishes three categories of elements: models of phenomena, experimental procedures, and interpretative models. The chains of translation are stabilized (another method for defining robustness) when these three subsets are given coherence, that is, when "the interpretation model affecte(d) a smooth translation between the material procedure and one of the two contending phenomenal models" (Pickering, 1990). It is their assembly and the translations that make them converge—Pickering calls this the mangle of practice (Pickering, in press)—that lead to robustness and stability. Hacking is concerned with the way in which "the laboratory sciences tend to produce a sort of self-vindicating structure that keeps them stable" (Hacking, 1992). This leads him to explore the interactions between a series of heterogeneous elements that strengthen one another—elements that he regroups into three large families of ideas, things, and marks[16] (see also Ackerman, 1985). Such groups and the iterative process giving significance and force to statements by making them coherent are another, novel, way of defining learning.

The robustness of networks depends on the alignment and interlacing of translations created in laboratories. But it extends far beyond these factors. Fujimura (1992a), for example, highlights the multiplicity of links that contribute to creating long and robust networks. As elegantly summarized by Pickering (1992b):

> Her examples include the cells that circulate between the operating room and medical and basic researchers, the recombinant-DNA techniques that flow between the different laboratories that constitute the various fields of technical practice, the computerized data bases that transport findings from one social world to the next, . . . and the oncogene theory that serves to organize conceptual, social and material relations between all the social worlds involved. (p. 13)

This leads the translation model to propose a local definition of the universal. According to this definition, statements, experimental devices, and incorporated know-how go no further than the translation networks they compose and in which they circulate. So the universality of science lies in the extension and extent of these networks. Model 4 thus accounts for the character of science stressed by Model 1: universality, capitalization, closure of dissent.

Social Organization

In the translation model, organization is seen from two different perspectives—either from the standpoint of the overall dynamics of networks or in terms of their internal management.

The creation and development of networks depend on a set of conditions that either facilitate or hinder the deployment of translations. Sometimes translations and devices in which they are inscribed may trigger opposition that they do not have the strength to overcome. Can any recording device be used to make an embryo write? Can a human being suffer so that the limits of his resistance may be studied? Is research on bacteriological warfare acceptable? Obviously, seeking answers to these questions is sometimes thought to be illegitimate. And the limits, in principle always revisable, are embodied in the protests, rules, or technical devices that together restrict the field of tolerated translations.

Other obstacles to the proliferation of translations lie in the more or less explicit arrangements that define the circulation of statements, instruments, and embodied skills or that distribute property rights (Cambrosio, Keating, & MacKenzie, 1990). Thus rules of confidentiality may hinder the ramification of networks, while exclusive rights to certain results limit the possibility of connection (as, for example, in the case of patents that could protect the identification of human genes). Finally, the mechanisms for designating the legitimate spokesmen (the actants authorized to speak on behalf of the networks) also influence the character of possible translations. This applies, for example, to the evaluation procedures of researchers, to the composition of commissions responsible for defining research programs, and to the conditions for exercising expertise.

Who is authorized to make whom talk? Who may ally herself with whom? Who speaks on whose behalf? The answers to these three questions define the space for the development of translation networks.

The model is also concerned with the internal management of networks and organizational forms in which they are embodied. The extension of networks and the diversity of their translations mean that the organization of interaction between their heterogeneous elements is an important strategic matter. New analytical tools are needed to study the distribution and links between instruments, statements, and embodied skills and, more generally, all the mobilized actants. Both the contents and the modes of circulation of what is produced depend on the dynamics of these interactions. Some recent studies (which are still small in number: Cambrosio & Keating, 1992; Cambrosio, Keating, & MacKenzie, 1990; Knorr Cetina, in press; Law, 1993; Vinck, Kahane, Larédo, & Meyer, 1993) highlight the variety of configurations and emphasize the increasing importance of networks of laboratories that are linked with firms, state agencies, or hospitals. The study of their organization and, notably, of their multiple forms of coordination (market, hierarchy, trust, technique, and so on) are of particular importance for the extended translation model.

Dynamics of the Whole

The notion of translation network suggests that it is not only the distinction between nature and society that is outdated, but that the conventional opposition between macro- and microanalysis (between global change and local action) is inappropriate.

In the past the opposition between society and nature was used to distinguish a world of passive entities from a world of human actors capable of imagination, invention, and expression. Translation networks establish a continuum between these two extremes—extremes that in practice are never reached. If one still wants to talk of nature and society, it is better to say that translation networks weave a socionature, an in-between that is inhabited by actants whose competence and identities vary along with the translations transforming them. Both passive beings and genuine actors are found there, but the dividing line is not laid down. The history of science is mixed up with the history of these socionatures, which are as varied and come in as many forms as the networks that shelter them.

Size and structural effects are properties of networks. Three concepts make it possible to describe the tension between local action and global change:

irreversibility, lengthening, and variety (Callon, 1991, 1992; Callon, Law, & Rip, 1986; Latour, 1991b; Law, 1991a).

A network becomes *irreversible* to the extent that its translations are consolidated, making further translations foreseeable and inevitable. Under such circumstances, embodied skills, experimental devices, and systems of statements become increasingly interdependent and complementary. The collective learning that takes place makes accumulation possible. A development ends up by following a perfectly determined sociotechnical path that progressively reduces the room for manoeuver of the actants involved. Other developments and other configurations are always possible in which the reversibility of networks are maintained and the translations remain open.

A translation network is *lengthened* to the extent that it enrolls an increasing number of diverse actants. These may come inside or outside the laboratories for what is important is the number of entities that are associated. The lengthening of a network is generally accompanied by "black-boxing" in which entire chains of translation are folded up and embodied in sentences, technical devices, substances, or skills. Indeed, this process of black-boxing lies at the heart of scientific dynamics (Latour, 1987). In this way, preceding extended networks are punctualized in a new actant; they are maintained, but in an easily manipulable and durable form. Furthermore, they contribute to the production of ever more statements, themselves doomed to pursue their existence silently in the bodies or machines that ensure the enterprise's continuity.

A translation network creates its own coherence. Where there are many diverse and disconnected networks, there are many translations. Conversely, when networks are strongly interconnected to form a system, the level of *diversity* is low. This level is obviously a product of history. But there are two elements that are particularly important in maintaining some degree of diversity. First, certain actors (e.g., state authorities) encourage the proliferation of translation networks. Second, the existence of boundary objects (Star & Griesemer, 1989) or mediators (Wise, 1988) may enable translation networks to coexist peacefully and may mean that one does not necessarily eliminate the others. Such boundary objects or mediators serve to link disjointed translation networks, which thus join together without necessarily fusing them into one. They are sufficiently ambiguous (polysemous in the case of notions and statements, multifunctional in the case of technical devices, complex in the case of embodied skills[17]) to serve as points of departure for divergent translation chains, to which they serve as gateways. Sometimes the weak links formed by boundary objects may strengthen, in which case fusion follows; connections multiply and the same statements, competencies, and technical devices circulate freely between the different points of the new network.

The model of extended translation does not oppose local and global nor does it negate agency and passive behaviors. Rather, it describes the dynamics of networks of different lengths, degrees of irreversibility, diversity, and interconnectedness. This double challenge to the opposition between micro and macro and the distinction between nature and society is to be found in the debate on the environment. Is global change linked to perfecting the design of the catalytic exhaust pipe? Does the feasibility of society depend on the creation of bacteria that have been programmed to destroy themselves? These questions are new because they blur the distinction between science and politics, and the one between human and nonhuman. Here it is clear that translation networks become both the protagonists and the subjects of debate. So it is that the eternal problem of political philosophy is posed: Who has the right to speak on behalf of whom? But the terms in which the question are answered are new. Unlike Model 3, it is not an application of politics to science, but science is now the source of new ideas and concepts of political philosophy.

CONCLUSION

The models presented here have allowed us to regroup scattered works into four coherent units. The argument suggests that, to understand the dynamics of science and its growth, we need to explore both its contents and its organization. The way each model grants priority to different questions, and the way in which it tackles other issues, depends on the way in which it treats these questions. We have explored this coherence, which is sometimes not visible when authors are considered individually.

It would be unfair not to mention that these models fail to capture some of the most promising new developments in social studies of science and technology. For example, work on discourse analysis (Chapter 15) attempts to grasp the irritating problem of reflexivity and to imagine new literary forms (Chapter 4) and last but not least politically crucial work on gender (Chapters 9 and 10) cannot be expected to be integrated with this presentation. That these works do not fit in with the four models is probably a consequence of focusing mostly on such general issues as the cultural and political place of scientific knowledge in modern societies rather than on the specific dynamics of science.

The schema adopted dramatizes difference rather than convergence. Yet convergences do exist to some extent. For instance, Models 4 and 1 share the notion of translation. Models 3 and 4 stress the role of instruments and incorporated skills in the dynamics of science; they also recognize the importance of entrenched networks of notions and/or technical standards for taming controversies. And to be sure, there are other points in common. But more important, in my view, each model has enriched the preceding one. The question of the soundness of such a progressiveness remains open. The reader should understand that giving a rhetorically plausible answer is difficult for the author! Each model's strength obviously hinges upon the number of allies it is able to enlist for its backing and defense. Model 4 is devised to satisfy simultaneously those who are obsessed with the need to explain a statement's robustness, those who see science as a competition between knowledge claims, and finally those who consider science to be a heterogeneous socio-cultural practice. How successful such an attempt will be depends on the reader and not on the author.

Be that as it may, future research could be undertaken in two directions:

1. Each model is strong in certain areas and weak in others. For example, in Model 2, the borrowings from economics are limited to the most general theories and, admittedly, the oldest. The concepts of industrial economy are not used. Notions such as barriers to entry, differentiated return on investments, imperfect competition, diversification or differentiation strategies would certainly enrich the analysis. More generally, the historical investigation of the emergence and the evolution of so-called scientific institutions deserves to be carefully scrutinized. Those who are committed to Model 1 might wish to explore how some new criteria for assessing the robustness of statements emerge and are accepted. The supporters of Model 3 might deepen their research on the establishment of agreement and develop a more articulated cultural history of scientific practices, paying attention to the boundary construction between science and its environment (Chapter 18). And those of Model 4 have at present little to say about the organizational forms accompanying or hindering translation networks.
2. The models presented in this chapter have much to say about the relationships between statements, technical artifacts, and embodied skills as well as the substitution or complementarity to which these give rise. But there is little work on the links between either the translation networks of science on one hand and technology (Bijker & Pinch, 1987) and economics on the other. Such investigations might show how networks develop in which statements, technical devices, money, embodied skills, confidence, and commands all circulate. If this is done, then a link will be built with neighboring disciplines, in particular with the economics of technical change whose recent results show a remarkable convergence with those of the sociology of science and technology. Such, at any rate, is an exciting possibility.

NOTES

1. For a clear presentation, see P. Jacob (1981).

2. Two opposite arguments on the conditions of possibility of such translations are proposed by D. Davidson (1984) and Quine (1969).

3. As in Carnap's radical theory, which rejects any meaning for statements that cannot be directly related to observation: This permits one to say that the correctly constructed statement "Caesar is a prime number" has no meaning. This point of view is also defended by Wittgenstein in the Tractatus (Wittgenstein, 1921).

4. This position was vigorously defended by Bachelard (1934) and taken up again by Hanson (1965). In general it corresponds to philosophical realism in which it is argued that theoretical progress increases access to natural reality. Curiously, however, philosophical relativists such as Quine insist on the theoretical (and hence arbitrary) character of all observation statements.

5. Here one finds all the debates between realism, pragmatism, positivism, and relativism. Realists insist that statements increasingly approximate reality (Putnam, 1978); positivists argue that the accumulation of observation statements extends and increases the precision of our knowledge (Carnap, 1955). For pragmatists, science is treated as a reliable tool that makes it possible to act on and control nature (Laudan, 1990). And relativists insist that statements teach us nothing about reality "out there" (Feyerabend, 1975; Quine, 1953).

6. For example, the statement "gravitational force is an inverse function of a power of distance" is less demanding than the statement "gravitational force is an inverse function of the square of distance."

7. The sociology of Goffman can be usefully applied to explain how actors try to avoid losing face (and try to keep their opponents from losing it too). See Wynne (1979) on this point.

8. For an appealing presentation, see Laudan (1990).

9. The article is only one of the ways of identifying and imputing discoveries; the laboratory notebook where experiments and their dates are written down is also important as is the strict separation between technicians and researchers. These elements are held together by norms, rules, or organizations. There are no journals without publishing houses, just as there are no articles without referees; there is no agreement about the date of experiments without commissions of experts who establish a chronology. The separation between technicians and researchers, essential if discoveries are not to be imputed to teams, is maintained by a system of diplomas, recruitment procedures, a strict hierarchy of occupational roles, and so on. A precise history of this system remains mostly to be written.

10. Goods are said to be public when their use by one person does not exclude their use by another person. According to economists, this property is an intrinsic characteristic of certain goods.

11. A detailed presentation and genealogy of this "eclectic" position is given by Shapin (1992a).

12. Numerous empirical studies have supported this hypothesis. Cambrosio (1988), in an article neatly titled "Going Monoclonal," is outstanding, for one sees everything that should be learned on the job, all the details that matter but that never appear in the texts or spoken word though they are essential for producing monoclonal antibodies by the world's best scientist. Cambrosio even notes amusingly that the tacit part of the practices is so important that certain superstitions are developed to account for the success or failure of an experiment that cannot be explained by explicit knowledge.

13. "The examination which places individuals in a field of surveillance also situates them in a network of writing; it involves them in a whole series of documents which capture and bind them. The examination procedures were immediately accompanied by a system of thorough registration and documentary accumulation" (Foucault, 1975, p. 191).

14. For a detailed analysis of the variety of the devices of translations in physics and molecular biology, see Knorr Cetina (in press).

15. Paradoxically, Model 3, which emphasizes the role of tacit knowledge, focuses the analysis of closure on explicit controversies.

16. Each of these three groups includes five different items. We find in them statements about anything from general theories to the modeling of instruments; the material elements mobilized including the targets (processed natural substances, laboratory animals, samples, and so on) subjected to checks but also tools and other generators of data; inscriptions (or "marks") produced by the generators of data and on which operations are carried out (evaluation, reduction, analysis, and interpretation). It is from the converging of these items, which is always difficult and constitutes the thread of science, that the robustness of knowledge and its stability results.

17. A statement may be a boundary object, as, for instance, in the case of the Lorentz equation that established a link between Newtonian and Einsteinian mechanics. Again, instruments are often powerful mediators as well as human beings (Downey, in press-b).

3

□

Coming of Age in STS

Some Methodological Musings

GARY BOWDEN

MARGARET Mead's *Coming of Age in Samoa* examines the link between formative childhood experiences and adult personality traits. This chapter argues that STS, as a scholarly enterprise, has come of age. Having passed through its formative infant experiences, the field is ready to adopt an adult personality. Unfortunately, both past and current STS research practice displays methodological confusion more symptomatic of an adolescent identity crisis than of a stable adult personality.[1]

This chapter sorts through the pandemonium of competing methodologies and charts a course toward a stable identity for the field. The first section provides a diagnosis. It argues (a) that there exist three distinct methods of explanation within STS, which are topic focused, issue focused, and combined focus, (b) that each method implies a different vision of the field and hence (c) that the selection of an appropriate method is intimately tied to the definition of the field. The second section focuses upon the formative experiences that defined STS's identity—the emergence of contextualist approaches to technology and to scientific knowledge. The third section turns

AUTHOR'S NOTE: This research was supported in part by SSHRC grant no. 499-89-0037. The comments of the reviewers and, in particular, of R. Steven Turner greatly improved this chapter.

from diagnosis to prescription. Using the identity defined in the second section as a ruler to measure the appropriateness of method, it is argued (a) that issue focused methods are incompatible with STS, (b) that the case justifying a combined focus approach to method within STS has yet to be made, and (c) that topic focused methods, though currently lacking the desired level of integrative coherence, display the most promise for STS.

METHOD AND THE VISION OF STS

Two distinct usages of the concept "method" appear in the STS literature. The most common usage focuses upon various strategies for data collection and analysis. This usage conceives the researcher as possessing a variety of tools. The researcher's chore is to select the tools appropriate for the chosen problem. In other words, this usage defines good research in terms of the appropriateness of the tools and their use rather than mandating the use of a particular tool. The second conception focuses less on mechanisms for collecting data and more on the method of explanation for data that have been collected. Bloor's (1976, pp. 4-5) specification of the four characteristics of a strong program explanation and Latour's (1987, p. 258) listing of seven "rules of method" are the most explicit examples of this conception within STS.

Methods of explanation are broader in scope and more prescriptive in tone. For example, only one of Latour's rules (the injunction to study science in action) bears directly upon considerations about data collection. Most of the rules deal with the form of acceptable explanations (e.g., don't adduce nature or society as explanatory factors). According to Latour (1987), his "rules of method are a package that do not seem to be easily negotiable. . . . With them it is more a question of all or nothing" (p. 17). Thus failure to follow the prescriptive rules results in explanations that are inappropriate at best and more likely (at least in the view of the method's proponents) erroneous.

Three factors underpin this chapter's focus upon methods of explanation. First, there exist a wide variety of disciplinary-based texts that competently treat issues of data collection and analysis. While the application of these techniques to the topic of science and technology certainly raises issues worthy of discussion, the diversity of the techniques used—from traditional approaches like participant observation, analysis of historical documents, or the comparison of official statistics to more recent developments such as textual analysis and the use of computer simulations—precludes an adequate treatment in this limited space.[2]

Second, while STS has been quite innovative in terms of methods of explanation, STS research, except in the area of bibliometrics, primarily uses traditional disciplinary approaches to data collection and analysis. In general, the strength of STS comes not from innovations in methods of collecting and analyzing data but from the ability of an inclusive approach to open up the deficiencies in understanding that flow from the ways that different disciplines collect data. The internalist data collection of laboratory studies, for example, focuses upon the local production of knowledge at the lab bench. This view, however, is challenged by the external focuses of politics and law in which standards other than those of research science are used to judge the acceptability of scientific findings (Jasanoff, 1990a; Salter, 1988). As another example, the armchair epistemology of traditional philosophy is challenged by the detailed empirical evidence provided by history and the sociology of scientific knowledge (SSK).

While this inclusive approach to data and the questions that data suggest has been a major engine in STS's theoretical development, the impact of methodological innovation is not necessarily uniform. Laboratory studies, for example, came as a shock to sociologists used to studying the institution of science and its verbal characterizations. A whole new area, the practice of science, was opened to sociological scrutiny. Ironically, the same methodological approach may be running a startlingly different course in history. Unlike sociologists, historians have a long tradition of studying the practice of (dead) scientists—through the examination of laboratory notebooks, correspondence, experimental apparatus—and a long extant tradition for explaining the results of such investigations—the internalist history of science. In short, the emphasis upon lab studies within history runs the danger of providing a methodological warrant to particularizing, individualizing, excessively micro-level studies that may inadvertently revive certain elements of historicism.[3] Thus, where the emphasis upon lab studies facilitated theory building in sociology, the same emphasis may actually be working to hinder theory development within the history of science.

This brings us to the third, and most important, reason for focusing upon methods of explanation over those of data collection. As noted previously, choices about the explanatory structure frequently carry implications for the types of data one uses. Thus choices about the method of explanation take analytical precedence over the details of data collection. Stated another way, it makes little sense to make choices about how you are going to get somewhere until you know your destination. But, as argued below, choice of destination involves selecting a particular disciplinary vision for STS.

The following paragraphs distinguish between three methods of explanation—topic focus, issue focus, and combined focus—and argue that each type of method is linked to a different disciplinary vision for STS. Durkheim's (1895/1938) *Rules of Sociological Method* provides a classic illustration of such a linkage. In this work Durkheim defines a discipline (sociology) through

TABLE 3.1 The Relationship Between Methodological Type and Vision of the Field

Focus of Method	*Topic*	*Analytic Issue*	*Combined*
Specifies topic?	Yes	No	Yes
Specifies explanatory form?	No	Yes	Yes
Corresponding disciplinary vision	Multidisciplinary	Transdisciplinary	Interdisciplinary

reference to a subject matter (social facts) and a method of explaining that subject matter (through reference to other social facts). Durkheim's method illustrates a combined focus because it *combines* stipulations about the characteristics of both the dependent variable (i.e., the thing to be explained) and the independent variable (i.e., the type of explanatory factor adduced). Topic focused methods specify only the *substantive topic* (the dependent variable) while issue focused methods specify the type of explanatory factor (or structure) that is appropriate given the *focus upon a particular analytic issue*. The major differences between these three approaches to method are summarized in Table 3.1 and described in the following three subsections.

Topic Focused Method

The vast majority of research within STS takes a topic focus approach to method: The researcher uses the methods and techniques of his or her particular academic discipline to study some aspect of science or technology. Merton's (1973a) work on the institutional structure of science provides a classic illustration of this approach. In methodological terms, Merton advocated an approach that drew its analytic resources (i.e., theory, concepts, method) from sociology. Science was interesting not because it requires different analytic resources to understand it but because the normative structure manifested itself in a distinctive way within the context of science. In general, topic focused research embodies the following characteristics: (a) a primary focus upon a particular substantive topic, that is, an aspect of science or technology; (b) considerable attention (either explicit or implicit) to data collection and analysis; and (c) little if any attention (either explicit or implicit) to methods of explanation. The latter issues go unexplored because topic focused research generally involves a taken-for-granted explanatory structure derived from the researcher's discipline.

The topic focused approach implies a multidisciplinary conception of the field; it views the field as held together by the topic, the study of science and technology, rather than any single method that is *unique* to STS. According to this conception, there exist many disciplines (e.g., political science, history, philosophy, economics), each of which has a particular way or several ways of examining the topic. Thus one can speak of the sociology *of* science or the economics *of* research and development. In these cases, theoretical concepts and methods of analysis appropriate to the given discipline are focused upon some aspect of science or technology. This view of STS holds (a) that science and technology can be effectively studied through the perceptual glasses provided by the various existing disciplines but (b) the understanding gained through exclusive reliance upon the perceptual glasses of any single discipline is necessarily incomplete. Thus the implied vision of the field is multidisciplinary in the sense that one must study the topic from the viewpoint of several different disciplines to gain a full understanding of the phenomenon.

Combined Focus Methods

Combined methods, like the topic focused approach, embody the desire to understand a particular substantive topic. Unlike the topic focused approach, however, combined methods presume that characteristics of the subject matter dictate certain choices about method. Thus, for Durkheim, the emergent properties that characterize social facts could only be explained through reference to other variables with similar emergent properties. Within STS, the social constructivist view of scientific knowledge, for example, entails certain methodological strictures (e.g., the symmetry thesis). In short, the combined focus approach to method, unlike the topic focus approach, holds that characteristics of the subject matter require a method that is unique to that topic. This is obviously a strong claim in that it forces the researcher to specify which characteristic(s) of the subject matter makes methodological innovation necessary and to link those characteristics to specific methodological strictures.

In contrast to the multidisciplinary vision of STS associated with the topic focused approach, combined methods imply an interdisciplinary vision of the field. In other words, combined methods presume a common culture of investigation (i.e., concepts, method, epistemology) shared by all participants. This contrasts sharply with the multiple cultures of investigation evident in a multidisciplinary field (Bauer, 1990). A second fundamental difference involves the degree of integration. Multidisciplinary approaches are frequently little more than several disciplinary perspectives held together by a staple. Interdisciplinary approaches, in contrast, involve a coherently integrated package of analytic resources, frequently including concepts other than those used by existing disciplines.

Analytic Issue Focused Methods

Methods involving an analytic issue focus give preeminence to a particular theoretical issue and then argue that certain methodological practices are necessary to deal with the implications of that issue. The "new literary forms" style of analysis (Woolgar, 1988a), for example, has emerged to cope with the vicissitudes of the problem of reflexivity. Similarly, the exclusive focus on scientific discourse advocated by Gilbert and Mulkay (1984) follows from an unwillingness to invest any specific account with the privileged status of authenticity. Unlike topic and combined focused methods, analytic issue-based methods are not intimately linked to the particulars of a specific *substantive* topic. The problem of reflexivity, for example, is not peculiar to social studies of science and technology. It is a much more fundamental problem that besets all social analysis.[4] In short, analytic issue-based methods derive their methodological strictures from attempts to resolve analytic issues exemplified in, but neither emergent from nor limited to, science and technology.

The methods that emerge from attempts to resolve general analytic issues are transdisciplinary in nature. They address generic research issues within the social sciences and humanities, and hence the methodological strictures that result are independent of particular substantive topics. Stated another way, the methodological strictures apply across a wide variety of subject matters. The most obvious current example of transdisciplinary analysis is postmodernism.

Having created this taxonomy of methods, an obvious question arises: Which type of method is appropriate for STS? Questions of appropriate methodology, like questions of appropriate technology, presume particular visions. The appropriate technology movement begins with a particular vision of society (i.e., controlled, sustainable growth; democratic participation; small is beautiful; and so on). Technologies become "appropriate" or "inappropriate" depending upon whether they facilitate such a society. Similarly, selection of the appropriate method for STS depends upon a particular vision of the field. In other words, selection of an appropriate methodology presumes a definition of STS.

DEFINING STS

But what exactly is STS? Like Mead, I believe that the identity of the adult (i.e., the definition of the field) is an outcome of its formative experiences. Prior to the 1960s, social science and humanities research on S&T consisted primarily of historical, philosophical, and, to a lesser extent, sociological research, which treated S&T as autonomous entities separated from their social context. Philosophers studied the logic of the scientific method, Whig historians documented the natural evolution of ideas and technological artifacts, while sociologists gazed at the institutional structure of science and its pattern of communication and reward.

The first break with this approach occurred in the mid-1960s, primarily among historians of technology and a variety of individuals interested in engineering education. Building on the works of Ellul (1964), Mumford (1967), and others, these thinkers eschewed the notions of autonomous technology and of technology as a neutral tool and replaced it with a view of technology as a complex enterprise that takes place in specific contexts shaped by and, in turn shaping, human values (Cutcliffe, 1989a). This development—institutionalized in such journals as *Technology and Culture* and *Science, Technology, & Human Values* and in the STS programs at Cornell, Penn State, MIT, and Lehigh—generated an enormous literature on ethics and values in relation to technology. Heavily influenced by Snow's (1959) "two cultures" metaphor, these individuals attempted to substitute a humanistic discourse about technology for the standard engineering discourse involving problem definition and solution. Take, for example, the engineer who "solves" the local sewage problem by constructing a pipeline that dumps the waste into the ocean. Engineering discourse sees this as a solution because the "problem" has been defined in a bounded manner (pollution within the city limits) that extracts it from the wider context and establishes the criteria that a solution must meet, that is, getting rid of the smell, groundwater pollution, and so on that upset the city residents. In contrast to this discourse of problem definition and solution, these scholars wished to substitute a discourse involving problem understanding and choice. When the issue is understood within context, the engineer has not solved the problem but has made the choice to pollute the ocean in order to improve living conditions within the city. In short, this viewpoint attempted to replace the authority of technical solutions with human values and the related issues of ethics and political choice.

In contrast to traditional history, which had focused primarily upon the development of particular scientific ideas or technological artifacts, these scholars placed a greater emphasis upon the social impacts of science and technology. One significant consequence of this focus was an emphasis upon science and technology policy. A considerable effort was expended in attempts to plan science and technology, anticipate potential social impacts, and rationalize for the future. By the 1980s, however, with the shift in political culture and the emerging emphasis upon free-market models of innovation and diffusion, the effort spent on such enterprises had declined significantly.

The second major break from traditional disciplinary views occurred in the mid-1970s when developments in the philosophy and history of science opened up the content of scientific knowledge to sociological scrutiny (H. Collins, 1983; Shapin, 1982). Institutionalized in journals such as *Social Studies of Science* and in the Science Studies departments at Edinburgh and Bath, this approach argued for an empirical examination of the social bases of scientific knowledge. Though it developed in relative isolation from the first strand, SSK has been animated by a goal similar to that of the first strand—to undermine the authority of science by destroying the epistemological privilege that has traditionally justified that authority. More academic and theoretical than the first strand, the society in science strand has gone through a number of significant transformations. Most significant among these is the growing retreat from extreme social relativism and the recognition that we live in a material world, that is, that the natural world places constraints upon the construction of scientific knowledge claims and technological artifacts.[5]

A third major event, the turn to technology within science studies (e.g., Bijker, Hughes, & Pinch, 1987), occurred in the late 1980s. In many ways, this event is more symbolic than substantive. Where the first two developments had involved profound conceptual reformations—the contextualization of science and technology and the display, in great detail, of the manner in which context affected both the creation of scientific knowledge and the impact of S&T upon society—the third primarily extended the insights of the second development to technology. By shifting the gaze to technology, however, this event brought the two communities, which had existed in relative scholarly autonomy, into much closer communication. The transformation of *Science, Technology, & Human Values* into the official journal of the Society for the Social Study of Science illustrates the institutionalization of this connection. Since that time a small but growing body of literature has begun to link the two strands.[6]

To summarize, the formative experiences of STS's youth provide a clear sense of a field organized around the varying historical and cross-cultural manifestations of the relationship between social context and the processes, cultures, and institutions involved with understanding, manipulating, and using nature. In addition, there exists an ideological component: a sense of tension between the authority vested in the expertise of scientists and engineers and the democratic values of the countries in which STS emerged, particularly in relation to those areas where the impacts of science and technology can adversely affect the public. Though there has not emerged any consensus about the specific analytic concepts relevant for understanding the topic, there is recognition that certain disciplines may provide necessary concepts and that any single discipline's concepts alone will not be sufficient. In addition, a variety of methodological principles have been widely accepted. From history and anthropology, STS has recognized that the manifestation of abstract analytic categories is neither temporally nor cross-culturally invariant. In addition, there is general recognition of the problems associated with Whig history and of the resulting necessity to distinguish between the concepts of the actor and those of the analyst. The tension between the descriptive, narrative tradition within history and a more theoretical, explanation-oriented emphasis within the social sciences has led to widespread acceptance of explanation at the level of thick description.[7] Similarly, the confrontation between philosophy and SSK resulted in (a) an acceptance of relativism as *methodologically* justified and (b) the view that normative statements need empirical, rather than logical, grounding.[8]

LINKING METHOD WITH VISION

STS and Multidisciplinarity

The description above defines the field by reference to the practices of its members, not by the programmatic statements of individual authors. Based on current practice, then, STS is a multidisciplinary field. The key factor holding the various STS research agendas together is a topical interest in science and technology, a topic that generally falls outside the mainstream interest of social scientists and humanists. This is not to suggest that STS is a generic field that subsumes all of the history, philosophy, economics, sociology, and so on of science and technology. It is, instead, an amalgamation of contextualist approaches that have their roots in each of these disciplines. As such, there exist a variety of humanities and social science perspectives toward science and technology that fall outside the domain of STS. Traditional internalist history of science and philosophical approaches that treat technology as autonomous are only two examples.

While it is safe to describe the vast majority of current STS research practice as multidisciplinary in orientation, one should note that few scholars promote or actively defend this vision of the field. One reason for this is that most STS research relies upon disciplinary-based master narratives.[9] In other words, few scholars have developed explanatory frameworks that successfully display how the myriad social, political, economic, technical, and so on factors come to be bound up in a seamless web. One well-known exception to this is Hughes (1983), who treats the seamless web as an outcome of the construction of a social system. Another less well-known, but equally useful, model has been articulated by Gingras (in press). At the most general level, Gingras builds a model of scientific change based upon a dynamic of communicative action inside a structured field—a model that he applies to science (Gingras, 1991), to technology (Gingras & Trépanier, 1993), and, reflexively, to SSK. While the details of Gingras's approach are not important here, two points should be mentioned. First, the framework reads very much like traditional, middle-of-the-road, social science. Second, the framework puts forward a structure that anticipates communication, argument, and the resultant modification/redefinition of both the topic of study and the appropriate analytic resources. In other words, the framework anticipates that future debate between STS and other fields will transform STS in the same way that, for example, ethnomethodological studies of work transformed SSK (Pickering, 1992a). As such, the framework makes no attempt to specify a list of explanatory resources but adopts an inclusive orientation toward the utility of concepts that emerge from the process of critical argumentation between scholars from different disciplines.

In contrast to Hughes and Gingras, a significant number of scholars actively criticize multidisciplinarity. Prominent objections come from two distinct directions. On the one hand are individuals with a transdisciplinary orientation who constantly nip at the heels of other researchers by showing how they have not properly addressed this or that analytic issue. On the other hand are those who aspire to an interdisciplinary STS. A number of factors underpin this desire. For many scholars it is the perceived inadequacies of the multidisciplinary approach rather than a commitment to an interdisciplinary approach per se. Frustrated with the limitations of disciplinary narratives, these individuals aspire to a total history, rather than an economic, political, or social history, of S&T. Unsatisfied with integrative multidisciplinary approaches, they decry reliance upon traditional disciplinary concepts and promote the development of nondisciplinary language. Another motive exists among individuals who desire to make STS less constructivist and more constructive in its orientation. They perceive STS to be unable to affect policy to the extent they would like because STS, unlike economics and other policy-relevant disciplines, is riddled with disciplinary factionalism that seriously undermines efforts to influence policy.

Analytic Issues and STS

While it may not be possible to define a vision of the field that all practitioners find acceptable, it is much easier to identify what STS is not. It is my contention that analytic issue-based methods should not be allowed to define the field. Stated more bluntly, these methods fall outside the arena generally recognized as STS. An appreciation of this point can be gained by drawing the distinction between social studies *of* science and social studies *in* science. In social studies *of* science, the focus is upon a substantive topic:

science and technology. In this case the analytic resources of the social sciences are applied to science and technology to understand something about that topic. Social studies *in* science concerns itself with doing "methodologically proper" social science. What counts as methodologically proper depends on the particular analytic issue being raised (e.g., reflexivity) and the methodologically proper resolution for dealing with that issue (e.g., new literary forms). Research is done on science and technology not to understand something about the topic itself but to exemplify the operation of methodologically correct social science within a particular domain. As such, it fails to meet a seemingly trivial criteria, which is that it be about S&T.

The above paragraph draws a boundary around STS that excludes several prominent approaches to STS, approaches that have served as strong critiques of alternate methodologies. The rationale for excluding these approaches flows from the vision of the field that they imply, not from the methods themselves. In other words, there is nothing inherently wrong with such approaches. The problem occurs when they claim to tell us something novel about S&T or, more perniciously, when they attempt to prescribe themselves as the *only* way to examine science and technology. The recognition that such methods should not dictate our approach to science and technology does not imply that we can dismiss the issues that they raise. These methods attract proponents because they address significant analytic concerns. The problem of reflexivity, for example, does not disappear once you dismiss new literary forms as the appropriate method for discussing STS. Rather than being sucked into the vortex of the reflexivist whirlpool, one needs a principled reason, similar to that provided by Collins and Yearley (1992a), as to why one is reflexive to a certain point and stops being reflexive beyond that point. Stated another way, the *criteria* that Collins and Yearley establish for stopping the relativist regress is the proper one for a field centered upon the desire to understand science and technology: Does being reflexive yet again tell me anything more about S&T? This does not imply, however, that their *application* of that criteria is the only, or even the best, response to the issue.

An Interdisciplinary STS?

If analytic methods fall outside acceptable approaches to STS, and if topic focused methods, despite their prevailing usage, find little explicit support, we are led to an examination of the combined methods that characterize interdisciplinary research. A key feature of combined methods, it will be remembered, is the notion that one or more unique characteristics of the subject matter dictate the appropriate type of explanatory resource. Thus combined methods can't merely be applied, they must be justified by explicitly stipulating the characteristic of the topic that necessitates methodological innovation.

Let us begin, then, by asking about uniqueness. What is there about S&T that *requires* a discrete and specialized form of scholarship? But merely posing this question opens a Pandora's box of problems for individuals committed to an interdisciplinary STS. The tacit premise of a name like "science and technology studies" is that one can conceive of S&T as discrete bodies of knowledge, skills, practices, or whatever that form an object of investigation. But such a view is clearly nonsense. On one hand, the issues of specialization and historical contingency undermine this claim. Once one adopts an empirical, as opposed to an a priori, approach to S&T, it becomes clear that the science of, for example, plasma physics has little in common with the science of conservation biology. This is a large part of the reason that historians have found the abstract concepts of science and technology so useless. On the other hand, if one construes S&T as a rubric encompassing all manner of inquiry about the natural world, then one is left wondering why areas such as medicine have generally been excluded from STS scholarship. But these objections pale in comparison to another. As discussed above, the predominant thread holding the fabric of STS together is the notion that S&T must be viewed *in context*. And this focus is clearly incompatible with treating S&T as abstract categories lacking historical anchors. But shifting one's gaze to the uniqueness of S&T in context, as suggested by the second verbal rendering of STS (science, technology, and society), does little to solve the problem. The triumph of contextualism has been to extend the boundaries of "S&T" further and further into the sociohistorical context. As such, these efforts have successfully broken down traditional barriers that marginalized the study of S&T within the social sciences and humanities and shown that foregrounding S&T is fundamental to understanding many traditional disciplinary topics. Thus it is not without some irony that one must view attempts to consolidate these gains by isolating STS as an independent disciplinary entity distinct from the rest of the social sciences and humanities. Indeed, both Mumford, one of contextualism's founding fathers, and Latour, one of its most prominent contemporary voices, realize that the path leads not to a new specialty but to a synthesizing enterprise that washes away the barriers separating STS from the rest of the social sciences and humanities.

This said, it must be noted that the methodological rationale behind current efforts at synthesis (such as actor-network theory) are seriously flawed. These individuals argue that science and technology are inextricably linked to their context in a manner that precludes separation. Categories are seen as indistinct, thus invalidating traditional distinctions between animate and inanimate objects, science and technology, nature and society, and so on because the social runs throughout the technical and thus cannot be separated from it. In a similar vein, Pickering (1990) argues that the unity and integrity of scientific practice cut across present disciplinary boundaries, and hence the use of existing disciplinary concepts and categories is likely to result in a serious misunderstanding of what science is like. Consequently, while Pickering is waiting for the Godot of a nondisciplinary language, actor-network theory has devised a whole new jargon, complete with dictionaries in both English and French, to avoid words with undesirable disciplinary and/or reductionist connotations.[10]

The flaw in this approach—that is, the presumption that the interpenetration of the natural and the social can only be understood through a language that denies traditional categories—can be seen by contrasting actor-network theory with the multidisciplinary orientation of Hughes (1983). He explicitly distinguishes between various entities (e.g., banks, engineering firms, political parties, animate and inanimate objects) and then proceeds to display how these various historically contingent entities are linked together in a manner that transforms them into a seamless web. Gingras (in press) likens this to the baking of a cake: The cook begins with a variety of heterogeneous ingredients (eggs, flour, water, and so on), which, combined in the proper proportions and manner, result in the creation of a homogeneous cake. Thus, for individuals such as Hughes, the difficulty in distinguishing between the technical and the social has a purely methodological basis. When one focuses upon the resulting system (the cake), it is difficult to distinguish one ingredient from the other. When one focuses upon the process of system creation, however, such distinctions become not only possible but necessary.

Equally as problematic is the logic that lies behind the search for new words. It simply does not follow from the fact that disciplinary concepts tend to be used within disciplinary master narratives that such concepts *necessarily entail* disciplinary master narratives.[11] This view presumes that traditional categories will always be reified and that scholars are incapable of remembering that such categories are analytical conventions that do not (and cannot) capture the totality of any phenomenon.

If, as I have attempted to show, the case for uniqueness has not yet been made and the attempts to develop a nondisciplinary language are misdirected, it does not follow that such a case *cannot* be made. If STS aspires to interdisciplinary status, one way to move toward that goal would be to look for parallels between STS and various successful interdisciplinary focuses. Following this path leads Bauer (1990) to conclude that STS does not fit the profile. His notion of successful interdisciplinary efforts (e.g., new disciplines such as biochemistry that emerge between the boundaries of existing disciplines), however, overlooks interdisciplinary projects that integrate concepts from a variety of disciplines into a coherent whole (e.g., the feminist epistemology approach within women's studies).

One key element in the coherence of feminist epistemology is its focus upon a large and important question: What is unique about women in society? The answer, of course, is found in the subordinate status of women throughout the world and the mechanisms that create and perpetuate that status. The equivalent STS question would be this: What is unique about science and technology in culture?[12] Where traditional disciplines answered this question by segregating nature from society, most recent approaches within STS pointedly deny the legitimacy of the question. These approaches argue that S&T are only superficially unique and distinct cultural forms; if we look closely enough, they can be reduced to politics, interests, career strategies, areas on the grid, and so on.

Methodologically, the recent approaches suffer from two problems. First, they substitute one type of segregation for another. Where positivists held that a single, nature-based approach (the scientific method) was appropriate for understanding both nature and society, recent theorists argue that a single, socially based approach (social constructivism) explains both the natural and the social world. This is the conceptual equivalent of the feminist argument that you counter one type of sexist perspective (male centered) by substituting a second type of sexist perspective (female centered).[13] Second as Eichler (1986) notes, while such substitution may be a necessary antidote, it should not be confused with the ultimate goal—the creation of a nonsexist paradigm. Analogously, the substitution of one type of segregationist approach toward S&T for another should not be confused with the ultimate goal—development of a perspective that truly integrates nature and society. And, as I have attempted to show, Hughes and Latour have outlined two fundamentally different approaches to the desegregation of nature and society. Hughes, like the U.S. Supreme Court, favors integration as a method for ending segregation. He sees the distinction between nature (blacks) and society (whites) as a legitimate one, as is the process of busing, which results in integrated schools. Latour, on the other hand, aims to desegregate through the imposition of an Orwellian newspeak that obliterates any offensive distinctions based on skin color.

If most current approaches avoid the question of cultural uniqueness, what remedies are available? Again feminist epistemology provides a useful hint; focus upon the issue of status within society. The unique thing about S&T is their dominant status and the mechanisms that create and perpetuate that status. Science, for whatever reason, is seen as producing "objective" knowledge, a characteristic that gives scientific knowledge a greater claim on legitimacy than that possessed by other forms of knowledge. Equally as significant, the institution of science has an unrivaled institutional coherence that allows it to be transported from one culture to another virtually intact. Unlike the introduction of Catholicism into Mexico, which resulted in a bizarre amalgamation of traditional and Catholic religious beliefs and practices, the introduction of science into Mexico resulted in scientific beliefs and practices that are largely indistinguishable from those in Britain or the United States. Significantly, focusing upon the authority of S&T also brings together the two strands of *STS* around a shared goal. However, this road, exemplified by Campbell's (1988) call for a comparative study of belief systems, leads beyond the boundaries normally associated with STS.

CONCLUSION

STS has come of age. No longer is it a child searching for guidance from other adult disciplines. STS possesses all of the traditional characteristics of a discipline with professional autonomy. No longer is it an adolescent experiencing rapid growth and dramatic transformation, embroiled in generational rebellion, and believing its accomplishments will single-handedly reshape the academic world. Gone are the heady days of the growth of SSK when novel theoretical approaches proliferated like weeds. STS has reached adulthood.

In one significant manner, however, STS still hearkens back to its recent youth. One clear mark of adulthood is the emergence of a stable identity. STS, caught in the middle of discussions involving three distinct approaches to method, has yet to establish such a stable identity. Such an identity will not emerge until discussions about method are informed by a conscious consideration of the link between method and disciplinary identity.

NOTES

1. For comments decrying such methodological confusion, see Spiegel-Rösing and Price (1977) or Hicks and Callebaut (1989, pp. 1-2).

2. One such issue is the inordinate reliance upon methodological convention. There exist a number of areas within STS where relatively large bodies of scholarship rest upon highly conventionalized approaches to data collection and analysis. Research on controversies (both scientific and technical), for example, tends to use the case study approach. Similarly, studies of communication patterns frequently use citation and co-citation analysis; examinations of the role of science and technology in economic development often employ cross-national comparisons; and so on. Frequently these conventions emerged because use of the particular technique highlights a key aspect of the substantive topic. Citation practices, for example, illuminate an aspect of scientific communication. The key point is that such linkages are an artifact of convention rather than methodological necessity. While the conventions may have emerged for good methodological reasons, strict adherence to them imposes artificial limitations.

3. See, for example, Holmes's (1992) concept of "investigative pathway" and the manner in which it returns biography to a central methodological position in the history of science.

4. For a discussion of the pervasiveness of reflexivity, see Giddens (1976). Like reflexivity, the issue of whether or not any particular discursive account can be treated as authoritative emerges in a wide range of areas outside science studies. While it is true that the theoretical point made by Gilbert and Mulkay (1984) carries a more ironic twist when made about science (because of the tremendous authority conventionally invested in scientific knowledge), this is not the same thing as claiming that the problem is specific only to science.

5. Rudwick (1985) was one of the first to distinguish between social "shaping" and social determinism of knowledge claims. Since that time, several theorists have moved in that direction. See, for example, Pickering (1990) or Callon and Latour (1992).

6. See, for example, Morone and Woodhouse (1986) or the books reviewed in Woodhouse (1991).

7. See, for example, Abrams (1980) or the dialogue between Buchanan (1991), Law (1991b), and Scranton (1991).

8. For an entertaining discussion of the philosophical issues surrounding the relativism debate, see Laudan (1990). For a defense of methodological relativism, see Collins and Yearley (1992a). For a focus upon the philosophers' concern with normative pronouncements within a social epistemology, see Fuller (1988).

9. For a definition and discussion of such narratives, see Pickering (1990).

10. See, for example, Callon, Law, and Rip (1986).

11. Indeed, in many cases, these master narratives are straw women. MacKenzie (1981), for example, frankly admits that the explanatory role of interests is of much greater importance for understanding one side of the debate than it is for understanding the other—a fact conveniently forgotten by those who characterize the work as identifying interests as *the* explanatory factor.

12. Many regional and ethnic studies programs (e.g., Latin American Studies, Black Studies) describe, in a multidisciplinary manner, the distinctive cultural attributes of the particular region or group. With few exceptions (e.g., the focus on American exceptionalism in American Studies), these areas have not generated explanations involving combined methods.

13. For a spirited defense of this view, see the discussion of alternation in Collins and Yearley (1992a, 1992b).

4

□

The Origin, History, and Politics of the Subject Called "Gender and Science"

A First Person Account

EVELYN FOX KELLER

MARY Poovey (1989) introduced the term *border cases* to denote historical phenomena that, by virtue of their location on the "border between two defining alternatives," constitute privileged sites for examining the ideological work of gender; "border cases," she writes, "mark the limits of ideological certainty" (p. 12). Peter Galison has introduced a closely kindred notion of "trading zones" to call attention to the extensive traffic across borders (in his case, between experiment, instruments, and theory) and hence to the impossibility of clear demarcations between these different domains (Galison, 1990). Both notions might be invoked to highlight the cultural and historical specificity of disciplinary perspectives in general and, at the same time, to problematize the very concept of disciplinary boundaries. "Gender and science," I suggest, is a border case par excellence. It sits not on one border but on multiple borders—indeed, on the borders between feminist theory and all the scientific and metascientific disciplines. It is also a trading zone, a domain

AUTHOR'S NOTE: Adapted from a paper presented at the History of Science Society meetings in Madison, October 1991. I am grateful to the Center for Advanced Studies in the Behavioral Sciences at Stanford for support during the year 1991-1992 when this was written.

of cross talk, exchange, and struggle. By its very existence, it calls into question the borders of all these disciplines.

As far as I can tell, the term itself first made its appearance in an article by that title published in 1978. The term was chosen to demand rather than to designate an inquiry, and the article—as it happened—was published not in a history of science journal but in a psychoanalytic journal. An extended quote will make the point:

> The historically pervasive association between masculine and objective, more specifically between masculine and scientific, is a topic that academic critics resist taking seriously. Why is that? Is it not odd that an association so familiar and so deeply entrenched is a topic only for informal discourse, literary allusion, and popular criticism? How is it that formal criticism in the philosophy and sociology of science has failed to see here a topic requiring analysis? The virtual silence of at least the nonfeminist academic community on this subject suggests that the association of masculinity with scientific thought has the status of a myth which either cannot or should not be examined seriously. It has simultaneously the air of being "self-evident" and "nonsensical"—the former by virtue of existing in the realm of common knowledge (that is, everyone knows it), and the latter by virtue of lying outside the realm of formal knowledge, indeed conflicting with our image of science as emotionally and sexually neutral. . . .
>
> The survival of mythlike beliefs in our thinking about science . . . ought, it would seem, to invite our curiosity and demand investigation. Unexamined myths, wherever they survive, have a subterranean potency; they affect our thinking in ways we are not aware of, and to the extent that we lack awareness, our capacity to resist their influence is undermined. The presence of the mythical in science seems particularly inappropriate. What is it doing there? From where does it come? And how does it influence our conceptions of science, of objectivity, or, for that matter, of gender? (Keller, 1978, pp. 187-188)

Although I was technically the author of this article, I did not then, and I do not now, regard the ideas behind it as originating with me. In the introduction to the book this eventually turned into, I tried to emphasize the ways in which these questions derived from the logic of a collective endeavor we had just begun to call "feminist theory," and in one of the early drafts I even tried to spell out the sense in which they were not mine at all but "ours." Feminists have recently become as suspicious of the first person plural as they had earlier been of the impersonal pronoun, so I need to try to be quite clear about my notion of a collective "we." Who, and what, did I have in mind? Certainly not historians or sociologists of science—at that time, I'm not sure I even knew any. Rather, what I had in mind was a very local collectivity of mostly white, middle-class women academics who had actively participated in what we now call "the women's movement" (recall that 1975 was the designated

"International Year of Women"), who called themselves feminists, who were involved in "consciousness-raising" groups, and who had begun to deploy their heightened consciousness in radical theoretical critiques of the disciplines (and the worlds) from which they had come. By the mid-1970s, works in "feminist theory"—the name we gave to this collective endeavor—had begun to appear in anthropology (e.g., Ortner, 1974; Reiter, 1975; Rosaldo & Lamphere, 1974; Rubin, 1975), history (Kelly-Godol, 1976), sociology (Chodorow, 1974; D. Smith, 1974), literature (Gilbert & Gubar, 1979; Millett, 1970; Showalter, 1970), psychology (Dinnerstein, 1977; Mitchell, 1975), and psychoanalysis (J. Miller, 1976)—though not yet in any discipline relating to the natural sciences.[1]

The first step had been to appropriate the term *gender* and distinguish it from *sex,* as a way of underscoring and elaborating Simone de Beauvoir's dictum that "one is not born a woman." In a classic and self-conscious deployment of naming as a form of political action, they/we redefined *gender* to demarcate the social and political, hence variable, meanings of *masculinity* and *femininity* from the biological or fixed category of male and female. The function of this redefinition was to redirect attention away *from* the meaning of sexual difference and *to* the question of how such meanings are deployed. To quote Donna Haraway (1991a), "Gender is a concept developed in order to contest the naturalization of sexual difference" (p. 131). Very quickly, feminists began to see, and as quickly to exploit, the analytic power of this distinction for exploring the force of gender and gender norms not only in the making of men and women but also as silent organizers of the cognitive and discursive maps of the social and natural worlds that we, as humans, simultaneously inhabit and construct—even of those worlds that women rarely enter—even, that is, of the natural sciences.[2]

In other words, it was just a matter of time before feminists who had been involved in these conversations, who had been reading these papers, and who knew something about the natural sciences would take on, as they say, the "hard" case. I may have been the first to use the term *gender and science,* but I was hardly alone in recognizing the kind of questions that had been brought into view and, once in view, that had come to demand analysis. By the late 1970s, a generation of feminists coming from a range of different disciplines, bringing with them varying senses of a collective "we," had come to take due note of the traditional naming of the scientific mind as "masculine" and the collateral naming of nature as "feminine" and, accordingly, to require examination of the meaning and consequences of these historical connotations.[3] We hoped, by that route, at one and the same time to undermine these traditional dichotomies and to pave the way for a restoration or relegitimation of just those values that had been excluded from science by virtue of being labeled "feminine." All of them/us were variously fueled, and to various degrees, by what Donna Haraway (1991a) has come to call "paranoid fantasies and academic resentments" (p. 183), by what I might call "utopian aspirations," as well as by more straightforward recognitions of intellectual (and soon, even academic) opportunities. For some, a suspicion that " 'objectivity' might be a code word for 'domination' " went hand in hand with the fantasy that we had hold of a lever with which we could not only liberate women but also turn our disciplines upside down—perhaps, even change the world. (Some of us were more humble—I, for example, merely thought of changing science.) In other words, it was a pretty heady time.

The first book-length response came from Carolyn Merchant, writing as a historian of science, as a Marxist, as a feminist, and as a committed environmentalist. In *The Death of Nature* (1980), she focuses squarely on the significance of the metaphor of nature as woman—for science, for capitalism, and for women—in the displacement of organicist by mechanist worldviews.[4] By arguing that this displacement implied, at least to the users of that language, a symbolic act of violence both against nature-as-woman and against woman-as-nature, her work played a major role in mobilizing, at least briefly, a coalition between feminists and environmentalists.

By the time my own book on "gender and science" came out (1985), I too had discovered history, and sought to tie such shifts in metaphors of nature (and simultaneously of mind and knowledge) to changing conceptions of individuality, selfhood, and masculinity—changes that were themselves neither epiphenomenal nor causal but deeply enmeshed in the social, economic, and political changes of the time. Especially, I attempted to argue for the confluence of new definitions of masculinity and new conceptions of what constituted a "proper" and epistemologically "productive" relation between "mind" and "nature." I sought, in a word, to locate the popular equation between "masculine" and "objective" in a particular historical transition. But to understand the relationship between "objectivity" and domination, I returned to the psychodynamics of individual development, adding to my earlier foray an analysis of the relations between love, sex, and power.

By the early 1980s, there was already quite a bit of feminist literature about science, written out of widely varying conceptions of feminism. In addition to the work emerging directly out of feminist theory, there was a growing literature on the history of women in science that had come more directly out of the history of science (most notably, Margaret Rossiter's book on "women in American science," 1982, and possibly even my own on McClintock —Keller, 1983); there was also a body of work, mostly by biologists, devoted to a critical examination of scientific constructions of "woman" (e.g., Bleier, 1984; Fausto-Sterling, 1989; Hubbard, Henefin, & Fried, 1982; Hubbard & Lowe, 1979). If one grouped these different literatures together, one could begin to offer courses under the rubric "gender and science"—provided, that is, one expanded the meaning of the term *gender* to include women and sex. Such an expansion and coalition seemed reasonable enough at the time, especially when other problems were more pressing: For example, where would one put such a course, composed of so many different disciplinary and intellectual agendas? And even, as became increasingly clear over the 1980s, of different political agendas? I was fortunate; I had access to an STS program that was hospitable; others tried women's studies; eventually, some even tried history of science programs. By the late 1980s, the catch-all label of "gender and science" had caught on, increasingly invoked in this country and abroad, to denote the rapidly growing interest in science studies in just about everything having to do with women, sex, or gender.[5] But the more popular the label became, the more evident became its internal tensions. Especially problematic was the slippage it invited between "women" and "gender."[6]

Even at the very beginning, this slippage was a source of discomfort for feminists. For one thing, and most trivially, the equation of *women* with *gender* is a logical error—in fact, as Donna Haraway (1991a, p. 243) points out, exactly the same kind of error involved in equating *race* with people of color. Whatever the term is taken to mean, when we use it to apply to people, strictly speaking, we mean it to apply to at least most if not all people. (My own interest in gender and science, for instance, focused neither on women nor on "femininity" but on men and conceptions of "masculinity.") What, then, invites the elision, and why would feminists permit such an elision? One answer to the first question is immediately made clear by the analogy with race: Women are culturally and historically marked by their sex or gender in a way that men are not, much as people of color are marked by their race. But the principal reasons that feminists at least initially tolerated and to some extent even supported the slippage between women and gender were twofold: First, the primary concern with which most of us had begun was with the force of gender on women's lives; second, even when our concerns moved outward, as they conspicuously did in feminist theory, that slippage was endured out of simple expedience. More recently, however, in most parts of the academy, it has widely come to be seen as necessary to mark the distinction between women and gender explicitly (witness, for example, the renaming of a number of feminist research programs at institutions such as Stanford and UC Berkeley as "Program for the Study of Women and Gender").[7] I want to suggest that this distinction has now become acutely necessary in "gender and science."

A decade ago, a loose coalition of works on women, sex, gender, and science could be held together by the common denominator of feminism—that is, by a commitment to the betterment of women's lives. By comparison, differences in other commitments (e.g., differences in disciplinary, theoretical, or political perspectives) might well pale—as long, that is, as that common denominator remained primary, as long as the women whose lives we sought to improve seemed to have common needs, and as long as we continued to see ourselves as a subversive force operating from the margins and interstices of academic life. In the intervening years, however—in large part because of the very successes of contemporary feminism—all these differences became both more visible and more pressing, perhaps especially as we, and our concerns, began to move into established academic and disciplinary niches. Today, it has become conspicuously evident that not all women have the same interests or needs; so has it become evident that not all scholars who call themselves "feminist" have common or even reconcilable theoretical and disciplinary agendas.

These remarks surely pertain to feminist scholarship in general, but I would like to try to spell out their particular relevance for the present *status* of the catch-all "gender and science" in the U.S. academy, and especially in science studies. For example, insofar as our interest in the history of women in science was initially motivated by a protest against a history of exclusion, and a political quest for equity, the dramatic changes that have occurred in the participation of at least some women in science over these 15 years need to be noted. I do not mean to suggest that women have achieved equity in the sciences[8] but that, where 15 years ago gender appeared as the principal axis of exclusion, today a glance at the racial and ethnic profile of the women who have entered scientific professions over these years suggests that it no longer does. Furthermore, from the perspective of those who have broken through the gender barrier and have now forged alliances within the scientific enterprise, it is not at all obvious how a continued focus on gender, especially given its critical force, might be in their interest. These, I think, are some of the issues that Londa Schiebinger (1989) had in mind when she stressed the importance of context in what she somewhat elliptically calls "arguments for and against gender differences" (p. 273); they are also the issues contributing to the increasing difficulty experienced by so many of those attempting to teach the disparate subjects of women, sex, and gender in science together under one rubric.[9]

But if the primacy of gender has receded as an occupational barrier in the sciences, and its utility as a critical wedge has been blunted by occupational success, recognition of both its cultural and its analytic importance has in other areas only increased—attesting, once again, to the successes of contemporary feminism. It is alas true that feminist theory has not proven itself powerful enough to change the world, as I once hoped it might. It has, however, radically, and I think irrevocably, changed the landscape of a number of academic disciplines—some more so than others: notably, for example, not the natural sciences or, at least not yet, in the philosophy of science; hardly at all in the social studies of science; and only in some areas of the history of science.

The recalcitrance of the natural sciences per se may not be surprising, but the relative unresponsiveness of the history, sociology, and philosophy of science is. For feminist theory to more fully realize its analytic promise in the various disciplines associated with the analysis of science, clearly historians, philosophers, and sociologists of science need to start reading its literature. Brief reviews such as the one I am attempting here cannot suffice as an introduction to the distinctions, the nuances, or the analytic groundwork that those familiar with that literature already take for granted. But also, I suggest, we need a new taxonomy; "gender and science" needs to be disaggregated into its component parts.

Schematically, these might be described as studies of (a) women in science, (b) scientific constructions of sexual difference, and (c) the uses of gender in scientific constructions of subjects and objects that lie both beneath and beyond the human skin (or skeleton). Each of these subjects has by now accumulated a rich literature in its own right and requires its own reconfiguration into new kinds of "trading zones." In the remainder of this chapter, I will focus on the third of these, what might be called "gender *in* science"—trading not between assorted analyses of women, sex, and gender in science but between multidisciplinary studies of gender, language, and culture in the production of science. For such studies of gender *in* science, I want to make a particular plea for a consolidated, two-way, effort toward integration of a number of analytic perspectives that are currently (mis)perceived as disjoint.

In proposing a disaggregation of what has come to be known as "gender and science," my aim is not to prescribe one particular position in any of its internal debates, or even to advocate the primacy of any particular political or intellectual agenda; rather, it is to provide room for all the different concerns they separately raise. I especially want to sidestep the question of whether one is "for or against gender differences" and to allow for the possibility of being either both or neither. The principal point I wish to emphasize is that the role of gender ideology is but one aspect of the constitutive role of language, culture, and ideology in the construction of science, and hence, though the roots of such analyses have been and must continue to lie in feminist theory, I take their place in science studies to be just one part of that more general inquiry. I suggest that work in this area not only has raised novel kinds of questions for historians, philosophers, and sociologists but also offers some novel models of and sites for analysis that might even be of use to those who are not women, who may not even be gendered, and who don't necessarily think of themselves as feminists.

To illustrate this claim, I will select some representative examples of the kinds of questions feminists have raised about the implications of a gendered vocabulary in scientific discourse, proceeding from those that have relatively straightforward implications for the reading of scientific texts to those with rather more indirect implications. In all of these examples, metaphors of gender can be seen to work, as social images in science invariably do, in two directions: They import social expectations into our representations of nature and, by so doing, they simultaneously serve to reify (or naturalize) cultural beliefs and practices. Although the dynamics of these two processes are almost surely inextricable, many feminists focus on the latter, emphasizing their effects (usually negative) on women; here my focus will be on the former, on their influence on the course of scientific research.[10]

Starting at the place where concerns about women, sex, and gender are most likely to intersect, the first questions that come into view arise in analyses of past and current work in the biology of reproduction and development. A basic form common to many of these analyses revolves around the identification of synecdochic (or part for whole) errors of the following sort: (a) The world of human bodies is divided into two kinds, male and female (i.e., by sex); (b) additional (extraphysical) properties are culturally attributed to those bodies (e.g., active/passive, independent/dependent, primary/secondary: read *gender*); and (c) the same properties that have been ascribed to the whole are then attributed to the subcategories of, or processes associated with, these bodies.[11]

The most conspicuous examples are no doubt to be found in the history of theories of generation. Nancy Tuana (1987), for example, has sought to augment the existing literature on reproductive theories from Aristotle to the preformationists by focusing on the imposition of prevailing views of women (i.e., as passive, weak, and generally inferior) onto their roles in reproduction; Laqueur's more recent and more probing analysis *Making Sex* (1990) adds substantially to such an effort. And some authors have undertaken corresponding analyses of contemporary discussions of fertilization; for instance, Scott Gilbert and his students (in Tuana, 1987) have followed the language of courtship rituals in standard treatments of fertilization in twentieth-century textbooks. Emily Martin (1991) has continued this effort by tracking the "importation of cultural ideas about passive females and heroic males into the 'personalities' of gametes" (p. 500) in the most recent technical literature. The argument goes as follows.

Conventionally, the sperm cell has been depicted as "active," "forceful," and "self-propelled," enabling it to "burrow through the egg coat" and "penetrate" the egg, to which it "delivers" its genes, and "activate[s] the developmental program." By contrast, the egg cell "is transported," "swept," or merely "drifts" along the fallopian tube until it is "assaulted," "penetrated," and fertilized by the sperm (Martin, 1991, pp. 489-490). The technical details that elaborate this picture have, until the last few years, been remarkably consistent: They provide chemical and mechanical accounts for the motility of the sperm, their adhesion to the cell membrane, and their ability to effect membrane fusion. The activity of the egg, assumed nonexistent, requires no mechanism. Only recently has this picture shifted and, with that shift, so too has shifted our technical understanding of the molecular dynamics of fertilization. In an early and self-conscious marking of this shift, two researchers in the field, Gerald and Helen Schatten, wrote in 1983,

> The classic account, current for centuries, has emphasized the sperm's performance and relegated to the egg the supporting role of Sleeping Beauty. . . . The egg is central to this drama, to be sure, but it is as passive a character as the Grimm brothers' princess. Now, it is becoming clear that the egg is not merely a large yolk-filled sphere into which the sperm burrows to endow new life. Rather, recent research suggest the almost heretical view that sperm and egg are mutually active partners. (p. 29)

And, indeed, the most current research on the subject routinely emphasizes the activity of the egg cell in producing the proteins or molecules necessary for adhesion and penetration. At least nominal equity (and who, in 1991, could ask for anything more?) seems even to have reached the most recent edition of *The Molecular Biology of the Cell*, where *fertilization* is defined as the process by which egg and sperm "find each other and fuse" (Alberts et al., 1990, p. 868).

For many, this recounting may leave more of a question than an answer, but the question is a critical one: What *is* the relation between the shift in metaphor in these accounts, the development of new technical procedures for representing the mechanisms of fertilization, and the concurrent embrace of at least nominal gender equity in the culture at large? If nothing else, tracking the metaphors of gender in this literature has provided us with an ideal site in which, with more extensive analysis, we can better appreciate and perhaps even sort out the complex lines of influence and interactions between cultural norms, metaphor, and technical development.[12]

Similar reviews could be provided of biological accounts of sex determination (e.g., Anne Fausto-Sterling, 1989, has explored the language of presence and absence in recent discussions of the sex gene), or of the relationship between cytoplasm and nucleus over the past 100 years, in which, far from coincidentally, the cytoplasm has at least tacitly been routinely figured as female, and the nucleus as male. Jan Sapp (1987) has given us an excellent account of the history of cytoplasmic inheritance, but he neglected to note the significant marks of gender in this history (some of which have been noted by Scott Gilbert and his students, in Tuana, 1987). Indeed, it is possible to tell a history of cytoplasm and nucleus that is remarkably parallel to that of the egg and sperm, in which recent support both of cytoplasmic inheritance and of the importance of the role of cytoplasmic determinants in development also parallels the rise in the 1980s of an ideology of gender equality. (And there are even some rumblings in the most recent discussion of sex genes.) I group these examples together both because of the similarity of their structure and the simplicity of their morals. Other examples, in which readings of gender are less closely tied to readings of biological sex, are correspondingly less straightforward both in their structure and in their implications. I discuss some of these next.

One general area of the history of science that has attracted the interest of a number of feminists has been that of the relation between genetics and developmental biology, and, in many of these discussions, an undercurrent of resistance to genetic determinism and a corresponding championship of the organizing models of developmental biology (or embryology) can clearly be seen (see, e.g., Birke & Silverton, 1984; Bleier, 1984; Gilbert, 1979; Hubbard & Lowe, 1979; Keller, 1983). Similar (or related) preferences can be seen in analyses of brain and behavior science (see, esp., Longino, 1989; Longino & Doell, 1983), of mechanism and organicism (e.g., Merchant, 1980), and even in certain analyses of models in physics (e.g., Kellert, 1993). The common denominator in these discussions might be described as a preference for interactionist, contextual, or global models over linear, causal, or "master molecule" theories.[13] The question is this: What does gender have to do either with these concerns or with these preferences?

One link to gender that may be of particular relevance to discussions of genetics and development can be traced to the tacit coding, already suggested, of the cytoplasm (and, more generally, of the body) as female and the corresponding coding of the nucleus or gene as male. Just as in the story about egg and sperm, such codings carry with them traces of social relations between male and female, inviting the suspicion that, even when merely implicit, they have been silently working to support correspondingly hierarchical structures of control in biological debates. To date, only fragmentary evidence has been brought to bear to support this suspicion, but sufficient evidence, I think, to demonstrate the necessity of incorporating an attentiveness to gender markings in future, more detailed, investigations of these subjects.[14]

But feminist concerns about "master molecule" theories do not necessarily depend on the allocation of gender labels to the constituent elements of debate. For some critics, the concern is more explicitly political, based on the fear that such hierarchical structures in biology are themselves rationales for existing social hierarchies, or that they reflect values supporting not only social but also individual constraint. Out of such concerns, Helen Longino (1989) has developed a sophisticated philosophical analysis of the ways in which political and social values inevitably enter into theory choice. Other scholars (e.g., Keller, 1985) have employed yet a different kind of argument, seeing in these hierarchical models the expression of a mind-set predicated on control and domination, unconsciously projecting its own sense of self-and-other onto representations of processes operating in the natural world.

Today, I find this argument by projection to be unduly limited—above all by its failure to take into account the particular kinds of material consequences that models or metaphors of domination have and, accordingly, the particular kinds of material ambitions such models support. Master Molecule theories are not only psychologically satisfying, they are also remarkably productive—productive, that is, in relation to particular kinds of aims (see, e.g., Keller, 1990). But even with this elaboration, feminist arguments for the use of gendered metaphors to (a) legitimate an agenda for scientific knowledge aimed at the domination of nature and (b) facilitate the projection of that agenda onto explanatory models for natural phenomena are often read in caricature, to approximate a kind of conspiracy theory. For just this reason, it is necessary to clearly emphasize what is not being claimed in these analyses. Of particular importance is the fact that no causal claims are being made, either for gender (hardly the only source of productive metaphors for science) or for language more generally. Metaphors—of gender or anything else—clearly do not, by themselves, drive the production of scientific knowledge; nor is language, by itself, capable of conjuring up material effects. But language does guide the human activities necessary to the construction of material effects. The very large question that now demands exploration is then: How? Feminists have accomplished the no mean feat of making visible the possibility of alternative mind-sets and metaphors. Now these efforts need to join with others to develop an analysis of the ways in which metaphors work to bridge the gulf between representing and intervening, of how they help to organize and define research trajectories (see, e.g., Keller, 1992, esp. chap. 4).

On the simplest and most obvious level, language gives us instruments of perception that conceptually magnify—in effect, create—precisely those similarities and differences with which metaphors begin. Nancy Stepan (1986), for instance, has traced the ways in which nineteenth-century metaphors of race and gender worked in the scientific construction of likenesses between women and African men, both set apart as if a separate species from white European males.

But language also guides the construction of instruments that not only conceptually but concretely bring new facts and objects into a user's purview. The metaphor of the "mind's eye" provided powerful motivation for the development of microscopes and telescopes, much as metaphors of information and Master Molecules did for the development of sequenators, Polymer

Chain Reaction techniques, and other technologies of genetic engineering. Both visually and technically, these instruments could be employed to expand the horizons of the models and metaphors that had shaped them—in the process, enlarging the domains for certain kinds of intervention and obscuring those of others. Genetic technology, for example, has actually brought genes closer to view and hence closer to hand. Inevitably, it simultaneously works to make other biological (and, of course, social) dynamics both less visible and less accessible.

Such studies have been well assisted by the linguistic sophistication Gillian Beer and Nancy Stepan have brought to our understanding of the conceptual dynamics of gendered metaphors, and by the philosophical sophistication Mary Hesse, Nancy Cartwright, and Ian Hacking have brought to our understanding of how science "works." These, together with the extraordinarily rich work on the politics of representation in the social studies of science, suggest the area of language and science as one of the most interesting new frontiers in science studies today.

More generally, and of particular importance for historians of science, there is a need for a deep and thoroughgoing contextualization of the analyses that were initially undertaken by feminists in such broad strokes. It may be, for instance, that it was feminist scholars who initially raised the question of cultural constructions of "objectivity" as a central issue for investigation, but it has taken mainstream historians of science to turn this question into a significant historiographic pursuit (I am thinking, for example, of the work of Lorraine Daston, Peter Dear, Peter Galison, and Ted Porter, some of which is collected in Megill, 1992). Now, surely, it is time for an integration of gender questions into this admirably careful and context-sensitive historiographic work.

Still, *context* is a big word, and it points in many different directions, a fact that the generic notion of historical specificity, even with the addition of gender, does not always capture. It may, for instance, fail to do justice to the need to attend to specificities of local disciplinary or even subdisciplinary interests, or to the specificities of local social (e.g., national, ethnic, and racial) interests. Donna Haraway offers us some useful ways of talking here that, for many, resonate simultaneously with current political and intellectual priorities: She defines feminist objectivity as *situated knowledges* and stresses the need for "partial perspectives." And, indeed, it is with Haraway's radical though controversial deployment of the method of partial perspectives that I will end this all too cursory review.

In *Primate Visions*, Haraway (1989) attempts, in contrast to a subject-rooted approach, a subject-free unpacking of how "love, power, and science [are] entwined in the constructions of nature in the late twentieth century" (p. 1), and she pursues this aim through an insistent scrutiny of the politics of narratives. As Gregg Mitman (1991) writes, "Her craft is the art of storytelling; her model for the construction of scientific knowledge is 'contested narrative fields.' Science, and in this instance primatology, is a story about nature, a tale circumscribed by its narrator, but one constantly evolving as new storytellers enter" (p. 164).

Haraway vigorously eschews the idea that gender can be understood independently of the politics of race and class, and her subject is an ideal one for making this case. As she makes abundantly clear, constructions of "nature" in twentieth-century primatology, like constructions of gender, are profoundly implicated in twentieth-century politics of race and colonialism. The fact that the (mostly white) women who entered the field in the 1970s and 1980s took the lead in restructuring traditional narratives does not, for Haraway, provide support for the idea of "a feminist science" (and with this I agree) but instead demonstrates once again the dependence of scientific narratives on their authors' historical "positioning in particular cognitive and political structures of science, race, and gender" (p. 303).[15] Haraway's very method precludes telling or even hoping for one coherent story, and many readers may be left feeling a bit too destabilized. But her reach toward a postmodern historiography has not only provided new models for working with "gender" in science, it also suggests new models for any politically oriented analysis of science.

No doubt, my choice of examples is idiosyncratic and has omitted much work on gender in science that has had an enormous impact on other disciplines.[16] It has even omitted much work that has had a significant impact on these disciplines. (Here, I think especially of the work of Sharon Traweek, Sandra Harding, Londa Schiebinger, and Rayna Rapp.) But I hope that the moral of my account will nonetheless be clear: For the future of these efforts in science studies, I now would like to add, complementing Haraway's emphasis on fractures and partiality, a plea for integration. Fifteen years ago, it took an organized effort on the part of feminists to rouse the attention of historians, philosophers, and sociologists to the marks and significance of gender. Working from the strengths of their political consciousness of gender and their irreverence for received boundaries operating both within and between disciplines, feminist theorists have brought home powerful lessons about the cognitive and institutional politics of gender that no one in science studies can now ignore. But to move the analysis of gender in science forward in these disciplines, I suggest that the strengths of feminist theory need now to be integrated with the strengths of other kinds of scholarship in these areas. There needs to be a lot more two-way traffic in this trading zone if it is to do the work it is capable of doing.

NOTES

1. There was, of course, also an important literature on "theory" emerging at the same time from collectivities of black women (see, e.g., B. Smith, 1983), and the omission of this literature from the list above is symptomatic of a crucial blind spot of white feminist theorists in that period, namely, the failure to address—or even, for the most part, to consider—the critical importance of race and ethnicity in the construction of gender.

2. The effect was explosive: *The Social Science Citation Index* lists 400 entries under "gender" (including "gendered," "genders," and so on) between 1971 and 1975; 1,380 between 1976 and 1980; 4,450 between 1981 and 1985; and 12,618 between 1986 and 1990. In some quarters, this effort was widely misunderstood as an attempt to reinscribe sexual difference. For the most part, however, the point was quite the opposite; it was an attempt to undermine (or deconstruct) the divisive and exclusionary work that gender was able to perform by virtue of being a silent signifier.

3. I might mention, for instance, Elizabeth Fee, Donna Haraway, Sandra Harding, Hilde Hein, Helen Longino, Carolyn Merchant, and Susan Leigh Star.

4. Brian Easlea's *Witch Hunting, Magic, and the New Philosophy*, on closely related issues, was published in England in the same year (1980).

5. Between 1976 and 1980, the *Social Science Citation Index* lists 1 entry under "gender and science"; between 1981 and 1985, 22 entries; and between 1986 and 1990, 51 entries. Over the same period, the use of this label as a title for courses, bibliographies, and conference proceedings showed a parallel increase.

6. If, for example, one attempts to sort through recent bibliographies bearing the titles either "Women and Science" or "Gender and Science" in an effort to distinguish these various literatures—on women in science, on scientific constructions of women, or on the role of gender ideology in science—the difficulty of nomenclature quickly becomes apparent. For example, in *The History of Women and Science, Health, and Technology: A Bibliographic Guide to the Professions and the Disciplines,* compiled in 1987 by Susan E. Searing and Rima D. Apple, all literature on the presence and influence of gender ideology in the natural sciences is included under subcategory I.D.2, titled "Science in Women's Lives." By contrast, the May 1988 newsletter of the Commission on the History of Women in Science, Technology and Medicine—intended to supplement the 1987 bibliography on "women of science, technology, and medicine" not by expanding its subject but by providing annotations to the literature previously cited—is titled *History of Gender and Science in New Books.*

7. The fierce debates around the naming of the research center at Berkeley (should it be called research on women or research on gender?)—that is, the debates that resulted in this linguistically awkward compromise—bear ample testimony to the importance of the distinction.

8. Betty Vetter's data indicate not only that equity has not yet been achieved for women but also that the goal of equity for women in science has actually begun to recede (see the most recent editions of *Professional Women and Minorities: A Manpower Data Resource Service*).

9. Such difficulty was widely attested to at the 1990 HSS meetings in the session on syllabi for courses in "gender and science."

10. For more conventional reviews of the literature in this area, see, for example, Haraway (1991a, esp. "Situated Knowledges"), S. Harding (1991), Keller (1990), Longino (1992), Nelson (1991). For comprehensive bibliographies, see Wylie et al. (1990) and Searing and Apple (1990).

11. Often, though not necessarily, such analyses have been undertaken from the vantage point of present, that is, superior, knowledge.

12. From the perspective of such a task, it is not necessary to embrace the most recent version as correct, either scientifically or politically (witness, e.g., the work of Churchill, 1979, and Farley, 1982, on nineteenth-century debates about sexual reproduction)—it is merely necessary

to register the extent to which gender ideologies are implicated in the construction of (at least some) scientific stories.

13. Actually, the term *master-molecule* was originally invoked by David Nanney (1957) in protest over the conception of genes as "dictatorial elements in the cellular economy" (p. 140) and only later appropriated by feminists for a larger critique. In the late 1980s, it was reappropriated by the NSF in the title of a glossy circular ("DNA: The Master Molecule of Life"), prepared in celebration of the successes of DNA.

14. Since the writing of this chapter, I have presented a significantly developed version of this argument at the International History of Science Summer School at Berkeley in 1992.

15. In different ways, Sandra Harding has also sought vigorously to challenge the Eurocentric hegemony evident in both science and science studies; in her most recent book, Harding (1991) has specifically sought to retrieve "feminist standpoint theory" from a "the perspective of 'global feminisms'—feminisms capable of speaking out of particular historical concerns other than the local Western ones that continue to distort so much of Western intellectual life" (p. viii).

16. For example, and especially close to my own heart, I have omitted discussion of the impact that feminist analyses of the personal and subjective dimensions of science have had on other disciplines, as well as their potential value for the history of science—were it not, that is, for the intellectual and political climate that currently prevails. Of interest, it was the gender coding of these areas that led us to examine their exclusion in the first place; might, I wonder, that same gender coding still be operative?

5

□

The Theory Landscape in Science Studies

Sociological Traditions

SAL RESTIVO

THE theoretical terrain in contemporary science studies is a diversified ecology. There are Weberian, Marxian, and Durkheimian niches. The renewed interest in Nietzsche across the intellectual world is showing up in science studies too. In theoretical studies of mathematics and mind, Durkheim, G. H. Mead, and O. Spengler are not merely ancestors and influentials, but leaders. And while the science studies movement long ago broke up the Mertonian hegemony in the sociology of science, the Mertonian niche is not only still viable but shows signs of regaining lost power. The Kuhnian paradigm has weathered a variety of criticisms and remains influential.

Nature, apparently vanquished by some of the early "relativists," seems to be making a comeback as an explanatory factor. The interdisciplinary nature of science and technology studies (STS) has not put an end to disciplinary competition, especially between sociology and philosophy. The constructivist paradigm that is the very core of STS, according to some theorists, is considered moribund, dead, or a grave error by others. Feminist theories of science continue to grow. But in spite of their significance for deepening our understanding of gender and knowledge, their development and influence during the 1980s occurred for the most part outside the professional core of the 4S community. Meanwhile, actor-network theories have become influential closer to the center of that science studies community.

My objective in this chapter is to focus on sociological traditions in the science studies theory terrain, and in particular on traditions that trace their roots more or less directly to classical sociological theory. I will also survey areas that show promise of blossoming and directing the future of science studies. Some areas that might normally fall within the scope of this sort of survey, including feminist and actor-network theories, are not covered here or are given only brief attention because they are covered in other chapters.

LIVING SOCIOLOGICAL TRADITIONS

The work of classical sociological theorists continues to provide new insights as contemporary theorists rediscover forgotten or neglected concepts or as well-known ideas take on new significance in the light of contemporary developments. Some of these classical insights guide the contributions to *Theories of Science in Society* (Cozzens & Gieryn, 1990). Spengler's (1926) long neglected work on mathematics and culture was resurrected by David Bloor (1976) in the mid-1970s and influenced the development of the sociology of mathematics in the 1980s (Collins & Restivo, 1983; Restivo, 1983). And Durkheim's (1961, p. 485) neglected remarks on the sociology of logical concepts, along with Nietzsche's (e.g., 1974, p. 298) widely unappreciated ideas on consciousness and thinking, have influenced social studies of mathematics and mind in the late 1980s and early 1990s (see, e.g., Restivo, 1990).

The modern roots of theoretical science studies can be traced to a variety of intellectual currents in early and mid-twentieth century scholarship: from Duhem-Quine and the empiricist turn in philosophy to Popper, the Vienna circle, and numerous positivist agendas; from the emergence of the sociology of knowledge in the works of Scheler and Mannheim to Fleck's social studies of medicine; from the emergence of the history and sociology of science in the works of George Sarton, Robert Merton, and Boris Hessen to the works of Ogburn, Gilfillan, and others on inventions and technology; from Wittgenstein to Mary Hesse in the philosophy of science; from the Grand Theories of Foucault, Gadamer, Derrida, Habermas, the Annales Historians, and Althusser to Mary Douglas's anthropology of knowledge and perception; from the reaction to positivism to the reaction to Hitlerism; from Veblen to Ellul; and at the junction of the modern and contemporary periods, from the Popper-Kuhn debates to the critiques of science that emerged in the 1960s.

The Mertonian paradigm (Merton, 1973b) in the sociology of science crystallized in the late 1960s and early 1970s, even as the STS movement was emerging. Merton's earliest work focused on the institutionalization of modern science in the West. His later work, and the work of his early students, described the structure and functioning of modern science as a social system. There seems to be a widespread sense in science studies that the Mertonian paradigm has been vanquished and relegated to the museum if not the attic of the field. There are two basic reasons that the Mertonian paradigm deserves attention in a contemporary science studies handbook: (a) The Mertonian paradigm is still the main source of ideas, perspectives, and advice on science policy in the United States, and elsewhere, especially where the scientometrics movement is influential (see Shapin, 1993, p. 839); (b) the Mertonian perspective continues to manifest itself in subtle and not so subtle ways—in, for example, the sort of resistance to the more profound implications of the sociology of scientific knowledge for a sociology of mind, thinking, and consciousness found in contemporary social studies of artificial intelligence and in self-organization theories of science (see notes 9 and 10). The continuing debates about social constructivism are paradoxically another sign of Mertonian revisionism. They reflect a resistance to sociology as the fundamental framework of science studies, but, more significantly, they reflect resistance to a radically post-social system of science analysis that makes not only knowledge but mind and consciousness social phenomena through and through. This is a Mertonian limitation that protects whatever is left of the autonomy of scientific knowledge. In addition to H. Collins (1990/1992), Hagendijk (1990), and Krohn, Kuppers, and Nowotny (1990), see the curious (for reasons detailed below) debate centered on Slezak (1989).

The Merton Thesis: The Hypothesis That Wouldn't Die

The Merton thesis is that, in the very earliest stages of its institutionalization, the legitimation of Western science was in substantial part an unintended consequence of the values and practices of ascetic Protestantism. Max Weber had already noted the propensity of ascetic Protestantism for mathematically rationalized empiricism. But he did not pursue this observation (Weber, 1904-1905/1958, p. 249, notes 1, 5).

In his most recent defense of the Merton thesis, Merton (1984, pp. 1095-1097) examines the thesis at three levels of theoretical abstraction.[1] The sociohistorical version is that ascetic Protestantism helped to "motivate and canalize the activities of men in the direction of experimental science." The middle-range hypothesis is that the development of science, like the development of any institution, had to be supported by group values. At the most general and abstract level, the hypothesis is that the interests, motivations, and behaviors in any given institutional sphere—such as religion or economy—are interdependent with the interests, motivations, and behavior in other institutional spheres—such as science. No matter how distinct and autonomous institutional spheres seem to be, they are linked through the multiple statuses and roles of given individuals.

The third formulation of the Merton thesis tends to undermine the simple "reciprocal influence" assumption that has generated decades of controversy over the relative validity of the Puritanism-science sequence versus the science-Puritanism sequence. That formulation gives greater priority to a structural and systemic argument as opposed to utilitarian hypotheses and hypotheses about the values and roles of individuals in "evolving" science (Ben-David, 1971, p. 31; Merton, 1970, pp. xix). Karp and Restivo (1974) constructed an alternative to reciprocal influence, utilitarian, and roles and values hypotheses. They claimed instead that modern science, Protestantism, and modern capitalism were linked parallel responses, in different institutional spheres, to underlying ecological and organizational conditions. The institutionalization of modern science was, they argued, the precondition for the scientific revolution. And that process was possible only in an ecological environment that facilitated an exchange-based economy and relatively decentralized political and military organization.[2]

Important contributions to the sociological theory of the scientific revolution were made in the 1980s by Carolyn Merchant (1980) and Margaret Jacob (1988). Merchant showed that there were alternative paths to the scientific revolution, and that the one forced by social and cultural circumstances was one that was oppressive to women, minorities, and laborers and exploitative of human and environmental resources. And Jacob gave the details of the intimate relationships between knowledge and the centers of political, commercial, and military power in the scientific revolution.[3] In spite of these developments, the idea that the scientific revolution was an organizational and institutional revolution rather than a purely intellectual one is still not widely appreciated.

The Social System of Science

Even as early as his dissertation, Merton was concerned with science as an autonomous institution, and more abstractly as a social system. This aspect of Merton's work has been described at length by Merton (1973b) and in surveys and reviews (see, e.g., Zuckerman, 1988b). Here I would like to point out certain features of the Mertonian paradigm that help to explain its continuing influence.

Merton is unusually sensitive to possible dysfunctions in the social structure of science. But his theories tend to block the realization of these dysfunctions. For example, Merton (1973b) claims that "joy in discovery and the quest for recognition by scientific peers are stamped out of the same psychological coin" (pp. 340, 401). This allows him to argue that departures from the norms of science can be "normal" and not at all detrimental to science.

Similarly, Merton recognized that the reward system could "get out of hand and defeat its original purposes" of reinforcing and perpetuating the emphasis on originality. But he has steadfastly held to the claim that only a few scientists "try to gain reputations by means that will lose them repute" (Merton, 1973b, pp. 300-302, 321).

Profound changes in science are allowed in principle by the Mertonian paradigm (Merton, 1973b, p. 329), but not changes in the ethos of science. Mertonian theory assumes *that,* as much as it explains why, once institutionalized, science cannot easily be moved off its moorings. In general, Merton tends to assume that the social system of science works *as a system,* that deviance is idiosyncratic, and that the social structure of science is (although dynamic and mutable in principle) fundamentally stable. Thus Merton (like many students of science) tends to formulate his claims about science in terms of what I call the grammar of the ever-present tense. This widespread tendency reflects the pervasiveness of and commitment to the hegemonic ideology of modern science.

The central dogma of the Mertonian paradigm is that the autonomy of science somehow makes scientific knowledge independent of social influences (Merton, 1973b, p. 209). But Merton (1973b, p. 204) is not a rigorous internalist. Society, he theorizes, has to have a certain shape or form to nourish the "immanent development of science." The idea of a *sociology* of "immanent development" became the great promise of post-Mertonian science studies.

Merton (1975, p. 52) has endorsed general theory as a goal in sociological inquiry. But he is known as the champion of a disciplined pluralism he called "theories of the middle range." This is an orientation to making "modest theoretical consolidations toward the ultimate and still very remote ideal of a unified comprehensive theory" (Merton, 1975, pp. 29, 52). Merton does not, however, indicate or suggest a path out of middle-rangeism toward a general theory.[4]

The Kuhnian Paradigm: A Mertonian Footnote

T. S. Kuhn (1962/1970) has moved in the footsteps of Merton as a great claims-maker about science in the ever-present tense. By the early 1980s, "T. S. Kuhn" had become a cultural resource more or less detached from T. S. Kuhn, his writings, and the social contexts of his arguments. "Kuhn" has served the interests of left, right, and center across the entire spectrum of intellectual discourse. It is no wonder that Kuhn has "regretfully" concluded that the success of *The Structure of Scientific Revolutions* rests on the fact that it is a text for all intellectual seasons and persuasions.

In a sense, Kuhn's discussion of social factors in scientific change in the early 1960s was a significant departure from positivistic and idealistic histories and philosophies of science. He argues that there are parallels between scientific and political revolutions. Both are inaugurated when a "narrow subdivision" of the given community has a "sense of malfunction" in the system. This is the precursor of "crisis" and a revolutionary period. This discussion of scientific change in terms of a *political* model (or, perhaps more important for understanding the limitations of Kuhn's approach, political *metaphor*), with its conception of *revolutionary* science and its emphasis on the *persuasive* aspects of scientific discourse, persuaded many students of science that Kuhn's account differs substantially from the account given by Merton. But neither the works of Merton and Kuhn, nor their own testimonies, nor the course of science studies itself in recent years, confirm the widespread and tenacious rumor of a Kuhnian revolution in science studies.

In the final analysis, the Mertonian and Kuhnian theories are functionalist and prescriptive accounts of science. Kuhn's theory has the added problem that it was originally conceived and carried out as a contribution to internalist history of science.[5]

Marxism is the most obvious traditional alternative to the Mertonian-Kuhnian paradigm. As we will see, it too has been restricted in science studies under the influence of the hegemonic ideology of modern science.

Marxist Science Studies

Karl Marx and Fredrich Engels set up the basic framework for the Marxist analysis of science (and technology) as social relations. Marx (1956) himself had a profound insight into the social nature of science (and of self and mind), expressed in the following words:

> [Even] when I am active *scientifically*, etc.,—when I am engaged in activity which I can seldom perform in direct community with others—then I am social, because I am active as a man. Not only is the material of my activity given to me as a social product (as is even the language in which the thinker is active): my own existence is social activity, and therefore that which I make of myself, I make of myself for society and with the consciousness of myself as a social being. (p. 102)

With Engels, he argued that even the simplest sensuous certainties are products of "social development, industry and commercial intercourse" (Marx & Engels, 1947, p. 35). In general, Marxist theory treats ideas as products of material and social conditions. Because the means of production include the means of mental production, the ruling class controls (or exerts control over) the production of ideas.

There is some ambivalence within the Marxist tradition concerning science as social relations and science as the preferred mode of inquiry and analysis (Aronowitz, 1988a). Marx (1956, pp. 110-111, 1973, pp. 699 ff.) himself conceived *modern* science as a bourgeois, alienated mode of inquiry. He associated the social transformation from capitalism to communism with a correlated transformation of science—the negation of science-as-it-is and the emergence of "human science," dealienated, unitary (not "Unified"), holistic, and global. At the same time, Marx argued for and developed a science of society. The only science he examined in any detail was mathematics. He did not study mathematics from the perspective of a social theorist but as a student of the conceptual foundations of calculus.

In his early work, the Marxist theorist Louis Althusser argued for a strict separation between science and ideology. This made explicit, and polarized, the ambivalence about science in Marxism. This problem is recalcitrant, but its solution depends on recognizing clearly that science *is* social relations. Modern science is a social institution of modern capitalist industrial technological society. This mode of science must be distinguished from the elements of "good inquiry" it may contain.

The materialist imperative in Marxism has made Marxists prominent innovators in the social study of scientific and mathematical knowledge, an area off-limits to Mertonians and Kuhnians. Marxist sociology (historical materialism) of science explores the relationship between the mode(s) of production in a social formation and the social organization of science as a form of work, the relationship(s) between science and the other human activities and products, and the effects of the mode(s) of production (first-order effects) and of the social organization of scientific work (second-order effects) on scientific ideas. Scientific and mathematical knowledge are not things "out there" in an eternal, universal Platonic realm that can be "discovered" in one revelatory way or another. Neither are they products of "pure" mental activity or of "geniuses" who create them out of thin air. In any given social formation, the prevailing mode of knowing grows out of practical activities (it is literally manufactured) and corresponds to the prevailing mode(s) of production and dominant social interests. Knowledge is not a simple "reflex" of "economic base." It reflects mediations among the various activities and products in a social formation. The more complex the social formation is, the more complex are the mediations and the causal chains.

A strong interpretation of historical materialism would lead to the conclusion that there is no "external reality" in and of itself. All knowledge is mediated by or coevolves with social practices, culture, and history. The best representative of this position, Oswald Spengler, is outside the Marxist tradition. In his pioneering analysis of mathematics, Spengler (1926) argues that *"there is not, and cannot be, number as such."* "There are several number-worlds as there are several cultures"; and "there is no mathematic, but only mathematics" (pp. 59-60).

The most ambitious project in the Marxist analysis of "pure science" is Alfred Sohn-Rethel's (1978) study, *Intellectual and Manual Labor.* Sohn-Rethel's basic theory is that (a) the original source of abstraction is commodity exchange; (b) this original abstraction contains the formal elements of conceptual thought articulated by the Greeks; and (c) the Greeks derived ideal abstraction from the real abstraction operative in exchange. The "pure reason" of classical Greece reflects the social relations of a *money economy. Commodity production* is the source of the idea that nature is an independent object world. And the process of *exchange* (which excludes use and is therefore abstract) is the source of the abstraction inherent in commodity production and therefore of the abstract intellectual categories that permeate commodity-producing societies. Already during the era of classical social theory, George Simmel (1900) had argued that the theoretical counterpart of the money economy is the exact mathematical interpretation of the cosmos. And of course Marx's writings on the nature of commodities are central to Sohn-Rethel's argument. Marx, remarking on the queerness of the commodity, described it as metaphysical and mysterious.

Sohn-Rethel's thesis is based in part on the assumption that commodity exchange and Euclidian abstractions arose at about the same time (actually over a 300-year period, 600-300 BCE). It is not, however, entirely clear why exchange and coinage should be the source of abstraction, Marx's remarks on exchange value notwithstanding, anymore than the elaborate (and coinless) credit systems (e.g., commodity-based transferable credit), "managed" legal tender currencies without metallic standards, or "substituted currencies" found in some of the early civilizations (including, for example, Babylonia).

From a Marxist perspective, it is clear that the fetishism attached to the products of labor under conditions of commodity exchange is reflected in the fetishism that attaches itself to the products of thought. Abstractions, fetishized under conditions of alienation and class struggle, become "pure" ideas. But this can be interpreted as a rationale for proposing the social rituals at the base of *religion* as the original roots of abstraction. In general, it appears that the capacities for abstraction and abstract theory reached a very high level too early to be accounted for in terms of the exchange abstraction thesis. This does not undermine the situational validity of Sohn-Rethel's theory.

Despite his well-thought-out criticism of *a priorism,* Sohn-Rethel tends to separate "imagination" from its material foundations and treats the economic "market" so abstractly that it is eliminated as a material basis for consciousness. In the end, he uses the exchange abstraction to defend the objectivity of "exact science." He manifests the ambivalence about science that is pervasive in Marxism and across the entire field of science and technology studies when he criticizes scientific Marxists who believe in timeless standards but then argues that Greek abstractions are universal yet "timebound": This contradiction, though, reflects the slow evolution of a line of inquiry in STS that has led in the direction of a sociology of objectivity (Restivo, 1983, pp. 148 ff.).

David Dickson has contributed to the sociology of mathematical knowledge in an important but widely ignored Marxist analysis of the "metaphysical shift" that gave primacy to mathematics in seventeenth-century Western Europe. The problems he addresses are these: (a) How did the mathematical reduction of an object's essential qualities coincide with the ideological requirements of seventeenth-century society? (b) How were these requirements built into the actual mathematics this society evolved?

In brief, Dickson's theory is that commodities are fetishized to make them compatible with capitalist social relations. The various aspects of the labor process must be reduced to quantified form before they can be dealt with in the capitalist process. There is, Dickson (1979) argues, "a formal correspondence between the way in which science subordinates the material world to the relations between mathematical symbols, and the way that capital subordinates the labor process to the relations between 'fetishized' characteristics" (pp. 22-24). At least from the seventeenth century on, these two processes reinforce and express one another. There is, Dickson claims, a "direct analogy" between the interchange the calculus permits between "process and product" and the interchange between process and product required by capital in its effort to articulate and control the links between the process of labor and the commodity. Just as earlier the merchants' needs in calculating commodity transactions had been met by the development of algebra, the need to control the labor process in capitalist society was met by the development of a mathematics of process. The historical context of this development is described in general terms for the period from pre-Alexandrian Greece to the Europe of Newton and Leibniz in Restivo (1983, pp. 239-266).

Marxism is at the root of most of what is innovative about science studies in relationship to traditional history, philosophy, and sociology of science.[6] This is especially the case for conflict theory, interest theory, and social constructivism. These theoretical orientations are thus the focus of the rest of this chapter. The theoretical power and explanatory potential of these theories rest precisely in their sociological content. That content goes beyond the works of Marx and encompasses the contributions of Durkheim, Weber, and George Mead. It is precisely this sociological content that has been underemphasized in science studies. Contemporary science studies researchers, even those who claim sociological competencies and commitments, tend to prefer the discursive strategies of traditional philosophy and natural science to those of sociology. I shall return to this claim toward the conclusion of this chapter.

Conflict Theory

The conflict theory of science, grounded in the theories of Marx, Weber, and Durkheim, is designed to explain the conditions under which particular forms of ideas and concepts are formulated, communicated, and accepted as true. The crux of the answer to the question, "What is science?" is located in intragroup communication about what is going to be accepted as "scientific knowledge."

Conflict theory builds an explanation of science on the established foundation of stratification and organization theory in general sociology. One major explanatory objective is to state the conditions under which scientists succeed or fail in their competition for attention. The history of science can be formulated in terms of the successes and failures of practitioners during their careers (this requires including more than just a few of the most prominent scientists in the analytic net; R. Collins, 1975, p. 481).

From the perspective of conflict theory, scientific activity (like intellectual activity in general) is analyzed in terms of four basic types of social roles: political, practical, leisure-entertainment, and teaching. The main activity of scientists in political roles is "defending the legitimacy of their organization and attacking the legitimacy of competitors" (Collins, 1975, p. 482). In practical roles, scientists "work to achieve some practical result for a customer, client, or boss." There are two main types of leisure-entertainment roles: In one, scientists who belong to a leisure class carry out scientific work for their own amusement; in the other, they are paid for entertaining patrons or a mass market. Teachers, finally, are involved in communicating scientific knowledge to specialized, full-time students. This always involves some degree of accumulating, assessing, and reorganizing the contributions of their predecessors. Because these roles are associated with occupational and status communities (and more generally with—following Weber's classification—class, power, and status communities), we can expect that scientists in different roles and in different communities will be different, have more or less predictable goals, outlooks, and ideas, and practice more or less predictable styles of work.

Teaching institutions have played a crucial role in the development of science. Scientific activity based on practical, political, or leisure-entertainment roles is relatively ephemeral (Ben-David, 1971, pp. 71-87). The European Scientific Revolution was concretized and sustained by the rise of autonomous teaching institutions, notably in France and Germany during the eighteenth and nineteenth centuries.

The conditions for the development of a relatively autonomous and generationally continuous science include a sizable and relatively autonomous educational system. These conditions allow for the formation of a community of teachers and students with their own ideals relatively independent of external control. Thus the size, autonomy, and—additionally—the degree of internal differentiation of educational systems are key factors in the emergence of scientific "golden ages" and traditions. The sustained development of these factors helps to explain why the Scientific Revolution occurred in Western Europe and not elsewhere.

Stagnation in scientific activity occurs when the autonomy of educational systems is degraded by external political, religious, or economic interests. Productive scientific activity depends on teachers and researchers. Degradation is facilitated when students carry into the schools the practical, ideological, and status goals and interests of the "outside" world.

Sciences are organizations: They consist of communication networks; there are sets of positions that are linked together in patterned ways; and there are relatively stable forms of influence and control. Organizations are characterized by different degrees of unity or disunity. The degree of unity or disunity is determined by how the resources for control are distributed.

Information, validation and recognition, and material resources are the bases of power in the division of labor in science. Validation of scientists' contributions and peer recognition integrate the individual scientist with his or her community and simultaneously give the community power over individual scientists. Material resources that sustain the community are provided, in general, by the wider social system (these are the "external conditions for science"; Collins, 1975, p. 495). Teaching positions are very important. They are the source of material resources; they provide channels of information and offer the initial forms of validation in the early years of scientific careers (Collins, 1975, pp. 475-496).[7]

In one of his most recent contributions to the conflict theory of science, Randall Collins (1989) has outlined a theory of intellectual change based on his ongoing study of the social causes of philosophies. Collins and Restivo (1983) have also proposed a novel way of looking at scandals in the history of science, in particular, scandals in the history of mathematics. The basic idea is that major scandals indicate shifts in the social organization of competition and production in science.

Interest Theory

The thesis that social interests determine (strongly or weakly, to one degree or another) ideas is not an innovation of modern students of STS. It is a centerpiece of Marxist thought and of the classical sociology of knowledge. But the modern STS movement has from its beginnings in the late 1960s and early 1970s included efforts to show that scientific knowledge is fueled by social interests. The "interests model" is widely associated with the writings of Barry Barnes, David Bloor, and their Edinburgh colleagues. Barnes (1983), for example, argues that behavior must be understood "in terms of its point, i.e., by the use of such notions as 'goal' or 'interest' "; and that "in general, the dynamics of institutions must be understood by reference to interests" (p. 31).

Knorr Cetina (1983), who favors a "constructionist" approach as opposed to an "interests" model, nonetheless notes that knowledge of the interests that inform scientists' theoretical preferences can supplement a constructionist analysis by showing "why it is likely that particular individuals hold particular beliefs" (p. 117). Conversely, studies of how knowledge is constructed in science may inform interest explanations.

Interest theory is viewed by some critics as a process of imputation, and they deny the possibility of imputing interests to social groups because we have no independent sources of knowledge about the groups, society, or even human beings (Latour, 1983, p. 144, referring to Woolgar, 1981). Latour distances himself from the disputes between interests theorists and their critics but does not banish interests from science studies. His move is to argue that interests are a consequence and not a cause of the scientists' efforts to translate what others want or what the scientist makes them want (Latour, 1983, pp. 144 ff.). This concept of "capturing" interests means that, "by pushing the explicit interests" of audiences and allies, the scientist furthers his or her own interests (Latour, 1987, p. 110). This idea informs the contribution of Gieryn and Figert (1990, pp. 67-96) to the "science-*in*-society" theoretical program (Cozzens & Gieryn, 1990).

Few interest theorists have tried to give a clear definition of what they mean by *interests*. This has led to some confusion among advocates and critics alike. In an effort to clarify what interests are, and to provide critics with something more substantive to attack, Restivo (1983) proposed the following:

> Ideas (and predicates, classifications, and representations) re-present social practices and social interests (Knorr-Cetina and Mulkay, 1983). Social interests are material or symbolic resources thought to be relevant to group survival and necessary for gaining, sustaining, or advancing advantages in relative power, privilege, and

> prestige. Attributed interests are social interests thought to be relevant to and necessary for a group's survival and relative power by outsiders, and may be more or less congruent with insider views. Interest attribution is itself a form of social interest. (pp. 124-125)

Social interests are expressed in the claims individuals make on cultural resources on behalf of the groups they represent (more or less explicitly), are members of, or aspire to membership in or association with. Social interests are activated in competitive environments. In cooperative settings, interests fuse with goals.

The point here was to show why interest attribution is not a simple and simplistic matter of "imputation" but, when carried out appropriately, an act of theory.

The Social Construction Conjecture

The idea of social construction is fundamental to sociological analysis (Berger & Luckmann, 1967; R. Collins, 1988, pp. 264-300). It has its classical roots in the works of Marx, and especially of Durkheim and his followers. Its application in modern science studies has drawn attention to the moment-to-moment activities of scientists as they go about producing and reproducing scientific culture. This is the significance of the social construction conjecture, and not its alleged relativistic implications.[8]

The social construction conjecture has recently been challenged by claims about "nonsocial" computer programs (e.g., BACON; Slezak, 1989). Because we know that computers and computer programs—like all artifacts—are cultural products and embody social practices, and because we also know that discoveries are never simply (or simple) matters of induction, it is curious that this kind of challenge can have a voice in science studies.

Science is social relations and social practice. The social group, not the individual, is the locus of knowledge. And neither machines (including computers and computer programs) nor humans are freestanding, autonomous, independent brains. Eventually, if artificial intelligence researchers are going to construct "intelligent machines," the machines are going to have to be ones that "share our culture."[9] So why should science studies researchers pay attention to alleged refutations of the "Strong Programme" or resist the social construction conjecture?

There is, in science studies as elsewhere in the intellectual world, a growing discomfort with the profound implications of sociology (and anthropology). Contemporary sociologists argue about the end of sociology, and a general clamor about the "death of the social" can be heard on the postmodern horizon. Somehow, these movements are tied to the growing visibility of a circle of science studies researchers who are more comfortable around natural scientists than they are around social scientists. These researchers also tend to have their natural scientist colleagues' penchant for philosophy. Valuable insights on the sociology of scientific facts thus take on, in the end, a sort of philosophical cast. All of this may help to explain why natural scientists and philosophical social scientists who do not claim too much for social sciences may be given more credibility in science studies than sociologists who claim that their goals are causal, predictive (when and where relevant), and explanatory. Causal theories, predictions, and explanations in sociology can in fact reflect the complexities of social life rather than mimic some sort of billiard balls physics model of science.

Hagendijk (1990) has recently criticized a version of constructivism that he attributes to Latour. His criticism is based on a view of science as part of a culturally created way of establishing facts that he believes constructivism undermines. He draws on Anthony Giddens's *structuration theory* as an alternative to constructivism. Here, science is considered to be generated by rules (especially linguistic rules). Hagendijk, like Giddens, fails to clearly locate rules and cultural conceptions in time and space. This is a critical problem because rules and resources are conceptualized in this theory to be the basic dimension of social structure (see item 4 below). The basic theory can be summarized as follows:

1. Social structures are simultaneously produced and changed by humans, and used as resources; structural properties are the medium and outcome of the practices they recursively organize.
2. Structure is a virtual order of rules and resources; the observable patterns of social interaction are conceptualized as systems.
3. Human agents are assumed to be knowledgeable.
4. The modalities of structuration are rules (normative and interpretive) and resources (political-authoritative, economic-allocative).

Science as an institutional order is theorized to be dependent for its reproduction on configurations of various types of rules and resources. Modern science, then, is supposed to have emerged out of the confluence of material, literary, and social technologies (following Shapin & Schaffer, 1985). Experimental science is viewed as a new "form of life" with material, cognitive, and moral aspects. The dynamics of boundary maintenance and collaboration are conceived in terms of discourse coalitions and discourse structuration. And national subfields of science are supposed to be created out of the intertwining of cognitive and social arrangements.

Hagendijk criticizes Merton for his normative bias but ends up defending a quasi-Mertonian view of science in which traditional dichotomies of cognitive-social, thoughts-institutions, and natural-social are resurrected. The theory work here demarcates science from politics and other nonscientific activities, and reflects explicit activities of boundary maintenance on behalf of science.[10]

Hagendijk's boundary work contrasts with Donald Fisher's (1990) effort to explain boundary work (see Gieryn, 1983). Fisher understands boundary work to involve processes through which knowledge units accrue legitimacy and cognitive authority. He is especially interested in the processes involved in demarcating science and nonscience. He links boundary work ideas in the sociology of science with two ideas taken from, respectively, the sociology of knowledge and the sociology of education: (a) the social construction of knowledge and (b) cultural reproduction. He concludes that the ruling class dominates production by deciding its means, contexts, and possibilities. All class cultures are then considered, in this theoretical framework, to be products of their place in the system of social stratification.

CONCLUSION

There are opposing forces at work in the theoretical science studies terrain. On the simplest level, one force tends to push the field in the direction of an increasingly sophisticated sociological analysis. This is behind the development of a sociological theory of mind and thinking Randall Collins and I are currently engaged in. An opposing force pushes the field away from sociological approaches toward more philosophical approaches. In some of these approaches, traditional philosophy is challenged by social epistemologies that could, from some points of view, be considered part of the first push I identified. But this part of the philosophical trend tends to be linked to a more strictly antisociological tradition, a positivistic backlash, and a protective orientation to the traditional ideology of science.

Social epistemology, for example, tends to sustain the distinction between the social and the cognitive that a more strictly and profoundly sociological approach tends to dissolve. It also tends to mollify the scientists and ideologues of science who are threatened by a sociology of science that can produce criticisms of scientific practice and revolutionary revisions of our conceptions of what science is (see Loughlin, 1990). I have in mind, of course, Fuller (1988), but the tendencies I identify here are stronger in Fuller's followers and imitators than in Fuller himself.

The resistance to a full-fledged social construction conjecture is, it seems to me, a reflection of resistance to a full-fledged sociological conception of self, mind, and knowledge as well as of truth, objectivity, and reality. The future of STS will not simply develop out of rational dialogues about the differences I have sketched or from professionally bounded arguments. Rather, it will work itself out of (just as it now reflects) the directions of cultural development that affect our views of who and what we are, what our values should be, and where we are headed.

NOTES

1. This defense was prompted by a rather lame criticism by G. Becker (1984) of Merton's claim concerning German pietism and science, which Becker tried unsuccessfully to extend to the general Merton thesis.

2. China is the crucial comparative case. The Karp-Restivo thesis depends on a comparative historical ecology that is supported by Braudel's (1977, pp. 33-35) remarks on China and the West as well as by other comparative historians cited in their paper.

3. The general conjecture that modern science is the mental framework of capitalism and the cognitive mode of industrialism is defended by Berman (1984, p. 37) and Geller (1964, p. 72).

4. For remarks on how to move in this direction, see Restivo (1990, pp. 79-98), and for an important move in this direction, see R. Collins (1988). Bernard Barber's independence as a general theorist of science within the Mertonian tradition is illustrated in B. Barber (1990).

5. Kuhn affirmed this in his acceptance speech when he was awarded the Bernal Prize by the Society for Social Studies of Science. The written version of this speech was weakened, but still makes the basic point. His strong criticisms of contemporary science studies should therefore be no surprise. Fleck's (1939/1979) theory, which influenced Kuhn, is a sociologically superior account that deserves greater attention and study. Pinch (1982) gives a rationale for a radical interpretation of Kuhn, and Barnes (1982) is the leading champion of Kuhn as a sociologist of knowledge.

6. Indeed, one could argue that much of what is innovative about Mertonian sociology of science owes a great deal to Marxist theory; note, for example, Merton's (1967, pp. 661-663) defense of Hessen against G. H. Clark's criticisms.

7. Two important contributions to the organizational theory of science that draw on R. Collins's work are Whitley (1984) and Fuchs (1992).

8. Traweek's (1988) work provides a "thick description" of high-energy physics culture and describes the "ground state" more or less implicit in the claims about the social construction of scientific facts and gendered science. See also the contributions of Susan Leigh Star (e.g., in press).

9. See H. Collins's (1990/1992) analysis of artificial intelligence and my review of this book (Restivo, 1992).

10. An even more extreme curiosity is the resurrection of a Parsonian model of an autonomous social system of science in the guise of a theory of self-organization imported from the physical and natural sciences (Krohn et al., 1990; and my review, Restivo, 1994). This theory even brings concepts of voluntarism and homeostasis back into the sociology of science.

PART III
SCIENTIFIC AND TECHNICAL CULTURES

It has long been understood that scientific knowledge is embedded in political economy and culture. But the exact interconnections between such social and cultural factors and the actual content of science remain opaque. Since the 1970s, S&TS scholars have used a powerful method—ethnography—to shed light on such connections. They examine science not as an abstract logic but as the activity of scientific communities.

In the first chapter in this section, Helen Watson-Verran and David Turnbull present an anthropology of knowledge. They examine wide-ranging cases: from the builders of Gothic cathedrals to past Amer-Indian cultures to still-existing aboriginal societies. Until recently, these and other indigenous knowledge-producing systems were characterized pejoratively as primitive, value laden, and local—nonscientific to say the very least. Watson-Verran and Turnbull find that indigenous knowledge systems have many of the characteristics of Western science; they claim that all knowledge systems, indigenous or Western, should be treated on an equal methodological footing.

Such ethnographic study and its attendant methods have clear implications for scholarship, and Watson-Verran and Turnbull present a challenge to the S&TS community. The strength of our scholarly work, they claim, is to show that what we accept as science and technology could be other than it is. But the great weakness of our approach is the failure to grasp the political nature of the enterprise and work toward change.

Karin Knorr Cetina's investigation of "laboratory studies" and Gary Downey and Juan Lucena's examination of "engineering studies" also employ an anthropological approach. However, rather than explicating little-known indigenous cultures of knowledge, these two chapters examine the day-to-day practices of mainstream scientists and technologists. It is an accomplishment of these two chapters that they impart to the routine practice of technique a certain anthropologically observed strangeness.

Knorr Cetina and Downey and Lucena summarize work that conceptualizes scientific and technical cultures as phenomena defined by local rules and local knowledge. In a departure from previous work in S&TS, these scholars address not only the institutional circumstances of scientific and technical work but also its technical content. Laboratories, or any places of work, contain (at least in part) and define (at least in part) scientific and technical culture. In that culture, so-called scientific facts are designed, constructed, packaged, and disseminated.

The laboratory, for Knorr Cetina, is far more than a place of work, it is a theoretical notion. For the entire natural and social order may be conceptualized—both cognitively and behaviorally—as a laboratory. Thus the ethnographic method may have scope and application far beyond the study of specific sites. The value of engineering studies, Downey and Lucena assert, is to elucidate not only the boundary between science and society but also the boundary between labor and capital. Engineering studies should also shed light on the theoretically suspicious notion of "applied research."

Local practice implies multiple voices. But not all voices have been equally heard, either in the history of science and technology or in our own history as S&TS scholars. For Judy Wajcman, such discrimination resides not only in institutional discrimination but in the gendered character of technology itself. Technology, for Wajcman, cannot be reduced to a set of neutral artifacts that merely happen to have been manipulated by men to women's detriment. Rather, she insists, as do all the contributors in this section, on treating technology as social relations.

From this perspective, a feminist analysis of reproductive technology must not only recognize but also transcend the view of technology as a patriarchal conspiracy. Instead, Wajcman sees technology as a set of social relationships dependent on many factors—for example, professional practices and capitalist enterprise—that form a matrix with gender.

As Mary Frank Fox points out, the previous STS handbook contained no chapter on gender, nor did its index list any entries for women and gender. This *Handbook*, on the other hand, has three chapters that consider gender as their primary topic, while numerous other chapters offer more abbreviated treatments of the subject.

Fox presents a quantitative analysis with two conclusions. Gender very much influences location, rank, and reward within the scientific community; and individual characteristics alone, ranging from ability to family status, do not account for these discriminatory patterns. Fox attributes these findings to a complex of factors related to the organizational features of science and social features of the workplace.

An ethnographic approach emphasizes local knowledge and multiple voices. It is therefore particularly suited for studying science and technology, subjects previously assumed to be universal. These chapters help us appreciate the multicultural and gendered nature of knowledge production.

6

□

Science and Other Indigenous Knowledge Systems

HELEN WATSON-VERRAN
DAVID TURNBULL

KNOWLEDGE SYSTEMS AS ASSEMBLAGES OF LOCAL KNOWLEDGE

Cross-cultural comparisons of knowledge and technology systems were a significant feature of STS studies during the 1960s and 1970s (Finnegan & Horton, 1973; Goody, 1977; Hollis & Lukes, 1982; Horton, 1967; Wilson, 1977)[1] but ceased to be an active site of STS work during the 1980s. This retreat from cross-cultural studies is currently being reversed as fresh insights are gained from the intersections of the social study of science with anthropology, postmodernism, feminism, postcolonialism, literary theory, geography, and environmentalism.[2] The characteristics of this renewed approach to the workings of systems of knowledge in disparate cultural contexts differ somewhat from those of past cross-cultural studies in science, technology, and society.

By and large, past cross-cultural work has taken Western "rationality" and "scientificity" as the bench mark criteria by which other culture's knowledges should be evaluated. So-called traditional knowledge systems of indigenous peoples have frequently been portrayed as closed, pragmatic, utilitarian, value laden, indexical, context dependent, and so on, implying that they cannot have the same authority and credibility as science because their localness restricts them to the social and cultural circumstances of their production. These were accounts of dichotomy where the great divide in knowledge systems coincided with the great divide between societies that are powerful and those that are not. Here was a satisfying explanation of the relation between knowledge and power.

This framework for comparative analysis can now be dissolved with an explicit focus on the local. Recent studies of science as social action[3] have identified local innovation as the implicit basis of scientific knowledge and have explored epistemological, ontological, and methodological consequences of this insight. What has generally remained unnoticed and unexplored in this new direction of science studies is that recognizing the localness of science subsumes many of the previously supposed limitations of other knowledge systems compared with Western science. Though knowledge systems may differ in their epistemologies, methodologies, logics, cognitive structures, or socioeconomic contexts, a characteristic that they all share is localness. Western contemporary technosciences, rather than being taken as definitional of knowledge, rationality, or objectivity, should be treated as varieties of knowledge systems.[4]

In this chapter it is argued that the ways of understanding the natural world that have been produced by different cultures and at different times should be compared as knowledge systems on an equal footing. We range widely across diverse examples of past and present knowledge systems, from the knowledge system within which the builders of Gothic cathedrals worked, to that of past Amer-Indian cultures (Inca and Anasazi), and the still existing Micronesian and Yolngu Aboriginal Australian knowledge systems. In doing so we explore the workings of knowledge systems in ways that can give us more useful understandings of power relations both within knowledge systems and between them.

Bruno Latour (1986) has pointed out that "rationality" is far too mysterious and thin a notion to be useful in accounting for differences between scientific and nonscientific knowledge systems. Instead he proposes many small and unexpected divides that he identifies as located in imaging craftsmanship and technologies of rhetoric. In science, in Latour's account, allies can be better aligned, and can more easily be shown as aligned, than in other systems. The difference between science and other knowledge systems is the result of differences in the effectiveness of technologies of surveillance. Latour has drawn attention to technoscience as a particular tension between the local and the global. Here we draw attention to other knowledge systems as alternative expressions of the necessary tension between the local and the global, involving different sorts of power practices.

Though scientific culture is now being more frequently recognized as deeply heterogeneous (see, e.g., Law, 1991c; Pickering, 1992b), there is, at present, no term in general usage that adequately captures the amalgam of places, bodies, voices, skills, practices, technical devices, theories, social strategies, and collective work that together constitute technoscientific knowledge/practices. Foucault's epistemes; Kuhn's paradigms; Callon, Law, and Latour's actor networks; Hacking's self-vindicating constellations; Fujimura and Star's standardized packages and boundary objects; and Knorr Cetina's reconfigurations—each embraces some of the range of possible components but none seems sufficiently all-encompassing (Bijker, Hughes, & Pinch, 1987; Callon, Law, & Rip, 1986; Foucault, 1970; Fujimura, 1992a; Kuhn, 1962/1970; Latour, 1987; Knorr Cetina, 1992a). Hence the proposed adoption of Deleuze and Guattari's (1987, p. 90) term *assemblage*, which in their usage is like an episteme with technologies added but that connotes the ad hoc contingency of a collage in its capacity to embrace a wide variety of incompatible components. It also has the virtue of connoting active and evolving practices rather than a passive and static structure. It implies a constructed robustness without a fully interpreted and agreed-upon theoretical framework while capturing the inherently spatial nature of the practices and their relations.

Assemblages constitute connections and contrive equivalences between locales in knowledge systems. In research fields and bodies of technoscientific knowledge/practice, otherwise disparate elements are rendered equivalent, general, and cohesive through processes that have been called "heterogeneous engineering" (see Law, 1987a). Assemblages are also power practices. Understanding them this way picks up on notions of power as strategic and involved with meaning making. Here the relations of power and knowledge are understood as invested in the material, social, and literary practices of discourse and representation, discipline and resistance.

Among the many social strategies that enable the possibility of "connecting up" are processes of standardization and collective work to produce agreements about what counts as an appropriate form of ordering, what counts as evidence, and so on. Technical devices that provide for connections and mobility are also essential. Such devices may be material or conceptual and may include maps, calendars, theories, books, lists, and recursive systems of names, but their common function is to enable otherwise incommensurable and isolated knowledges to move in space and time from the local site and moment of their production to other places and times. In exemplifying the tension between the local and the global, we look at a variety of knowledge systems. The next section briefly considers the knowledge systems of the Gothic cathedral builders, the Anasazi, the Inca, and the Micronesian Pacific navigators. All these are examples of systems that lack many of the elements often deemed essential to science—writing, mathematics, standardized measurement, laws, theory—yet are also systematic and innovative.[5] They differ from science and each other in the kinds of technical devices and social strategies through which local knowledge is mobilized.

The assemblages that we feature in the second section can be understood as "technologies" through which locales in knowledge systems are connected. In the third section we take two types of assemblages that many would consider to embody the universality of science, that is, theories and numbers; we consider work that has revealed these as assemblages of heterogeneous practices. Thus far we have presented assemblages as "entities" within various knowledge systems; here we look "inside" two such entities. Following this we consider work in the contemporary Australian context where contesting knowledge systems—an Aboriginal Australian knowledge system and the Western scientific knowledge system—are being worked together. Here there is mutual interrogation producing reinterpretations of how the systems might be understood with respect to each other. Though fundamentally different in their ontologies and epistemologies, the knowledge systems of Yolngu Aboriginal Australians and Western technoscience can be worked together to expand possibilities for choice by both Aboriginal and non-Aboriginal Australians. In concluding our chapter we briefly consider more general issues of power with respect to knowledge systems.

DISPARATE DEVICES AND STRATEGIES FOR MOVING AND ASSEMBLING LOCAL KNOWLEDGE

Gothic Cathedral Builders

The Gothic cathedrals, and in particular Chartres, have the appearance of the rationality, order, calculation, and uniformity that typify Western science. Our "forms of life" have so structured our understandings of the processes of knowing and making involved in the building of Chartres Cathedral that we take it as self-evidently necessary that such large, complex, innovative structures require an architect and plans. Simultaneously we feel constrained to attribute to its builders some mysterious and ineffable skill because they had no knowledge of structural mechanics.[6] A recent reanalysis of the building of Chartres Cathedral in the eleventh century shows that rather than being a uniform and coherent whole it is an "ad hoc mess" and was achieved without an architect, without plans, and without a standard measure (James, 1982). It was built in a discontinuous process by successive and different teams of masons using their own "local" geometries, techniques, and measures. The question is then: How was the work of all these people coordinated without the social technologies of planning, calculating, and designing that we take for granted? The answer lies mainly in the use of templates, which are patterns or molds, usually outlined on a thin piece of wood, that a stonemason uses to cut a stone to a particular shape.

The power of templates lies not only in the way in which they facilitate accurate mass production but also in the fact that simple geometrical rules of thumb will often suffice for the templates themselves to be accurately reproduced as often as required. Templates help to make possible the unified organization of large numbers of men with varied training and skill over considerable periods of time.

> On them were encapsulated every design decision that had to be passed down to the men doing the carving in shop and quarry. Through them the work of all the masons on the site was controlled and coordinated. With them dozens, and in some cases hundreds, of men were guided to a common purpose. They were the "primary instruments" of the trade. (James, 1989, p. 2)

In addition to the power to organize large numbers of workers, templates have the power to allow for great exactness of stonecutting and enable the building of a coherent structure, despite a discontinuous process and despite radical design and structural changes. The example of Chartres is especially important in enabling us to rethink the essential elements of a knowledge system. The work of groups of people with varied practices, skills, and understandings has to be rendered connectable and assemblable into a coherent whole whether the outcome is a cathedral, a body of theoretical knowledge, or an agricultural system. The case of Chartres shows that this can be achieved without structural theory, standard measures, plans, or architects; all that is required is a small piece of representational technology in conjunction with skills and constructive geometry. This example undermines some of the great myths about science, technology, and traditional knowledge. There is no great divide between the past and the present, between scientific and traditional knowledge, or between science and technology. Just as "Chartres was the ad hoc accumulation of the work of many men" (James, 1989, p. 2), so technoscience or any knowledge system can be ad hoc, ununified, atheoretical, lack a common measure, and still be effective—fundamentally because all knowledge systems are local and are the product of collective practice based on the earlier work of others.

The Anasazi

The Anasazi were a group of North American Indians who established themselves in what is now the Four Corners region (where Colorado, Utah, New Mexico, and Arizona meet) of the United States from around 200-700 A.D. They not only managed to survive in this most inhospitable region where the temperature ranges from 20°F below to 100°F above and where there is only 9 inches of rain often in destructive summer bursts, but they also created a complex society (Lekson, Windes, Stein, & Judge, 1988, p. 100). This society came to an abrupt end in about 1150 A.D. (possibly due to the drought between 1130 and 1180, though this is debatable). At its peak it consisted of 75 communities spread across 25,000 square miles of the San Juan Basin linked into a socioeconomic and ritual network centered on Chaco Canyon (Judge, 1984, pp. 1-12).[7] On the floor of Chaco Canyon were built massive stone buildings up to four stories high with hundreds of rooms including vast storage areas and huge round underground *kivas,* or temples. Chaco was connected to many of the outlying communities by over 400 kilometers of roads. In addition to the great buildings and the roads, the Anasazi built an enormous irrigation system with check dams, reservoirs, canals up to 50 feet wide, irrigation ditches, and leveled fields with banks (Frazier, 1986, pp. 95 ff.; Vivian, 1974).[8]

The key to supporting a population variously estimated at up to 10,000 in such a marginal environment was the development of an agricultural and storage system that enabled them to grow and redistribute a surplus. But by itself that would not have been enough. To successfully transform an almost totally arid environment, to coordinate the work of large numbers of people over a vast area, and to ensure the growth, storage, and redistribution of food, a large amount of knowledge and information had to be developed, sustained, and transmitted. This was achieved primarily with the calendar along with ritual, myth, poetry, and architecture.

The calendar was maintained by the sun priest's observation of the sun's seasonal passage past markers on the horizon and through the passage of light and shadow in buildings and structures like that on top of Fajada Butte. It was crucial that the solstice be accurately forecast because the timing of the planting calendar is of great moment in an environment with a short growing season and where the onset of frost must be anticipated.

McCluskey (1982) concludes of the contemporary Pueblo astronomy of the Hopi: "Considered as astronomy it shows all the concern with exact observation and the development of observational and theoretical framework that we would expect of modern astronomy" (p. 55; see also McCluskey, 1980).

This too can be said of the Anasazi even though the system was typically local. It strongly reflected its context of use in that it relied on specific horizon markers to record the sun's movements but it was nonetheless capable of movement to different places and times while simultaneously adapting to changing understandings and needs and providing for the growth of an extensive and complex society.[9] The potential for connectivity and equivalence is provided by the directionality that structures the calendar and their social life. All events, places, and people can be recognized, connected, and made equatable through the system of directions represented by the calendar. For its survival and transmission, this system is dependent on annual horizon observations and rituals organized by the sun priest—hence its limitations. The Anasazi knowledge system can move only as far as the priest can control.

The Inca

We turn now to the Inca, whose society and knowledge system has obvious parallels with those of the Anasazi but whose scale and power is very much greater. Indeed, their civilization is usually and quite justifiably referred to as the Incan Empire. At its height the Incan Empire extended over large areas of what is now Ecuador, Peru, and Chile. This organization of 5 million people in one state has been the subject of much speculation and admiration. It has been described as socialism, feudalism, despotism, a hydraulic society. Its coherence has been attributed to the hierarchy, the military, the tax system, laws, bureaucracy, land rights, political jurisdiction (Moore, 1985, pp. 1 ff.). However, we want to argue that, as in the case of the Anasazi, an essential element is the way in which local knowledge was moved and that, as for the Anasazi, the key device was the calendar. Further, the difference in scale and power between the two societies can be explained by the Incan augmentation of knowledge transmission through the use of the additional devices of stone alignments and knotted string, of *ceques* and *quipus*.

The Inca capital, Cuzco, was at the hub of the empire, which stretched over 2,000 kilometers from its most northern extremities to its most southern; but not only was the empire very far flung, it also incorporated a large number of preexisting cultures and covered very variable terrain from the highest parts of the Andes to the coastal plains. The key problem, as for the Anasazi, was how to coordinate a large population in an environment that was at best variable and at worst marginal, and furthermore how to administer it all from one center—Cuzco.

Spreading out from Cuzco were 41 radial lines marking significant rising and setting points on the horizon for the sun, moon, and stars (Zuidema, 1982a). These lines were the *ceques*, marked at intervals by stone cairns or shrines called *huacas*. These ceques not only integrated religious and astronomical knowledge but also provided the basis for the kind of precision calendar required by a state bureaucracy that had to record and correlate information about irrigation, agriculture, trade, warfare, and all the associated taxes, manpower, and resources—all of which operated in a intricate system of kinship, age, class, and social organization. The ceques were extended beyond the horizon to incorporate the whole empire and "formed a system of coordinates by which information of very different orders was organized, as is done in our maps" (Zuidema, 1982b, pp. 59-60), and in fact the Inca created very sophisticated three-dimensional maps of the landscape (de la Vega, 1961, p. 78). In addition to the ceques, the Incas developed a sophisticated system of tallying using knotted strings or *quipus.* On such knotted and looped strings it was possible to record a wide variety of information from instructions to details of taxes, labor obligations, and agricultural supplies, and the quipus could be carried by runners, or *chasquis,* over the extensive road network that ran the entire length of the country in two parallel systems.

Quipus have been extensively analyzed by the Aschers (1972, pp. 288-289; see also Ascher & Ascher, 1981), who conclude that

> To maintain a population that may have reached six million, knowledge of food production is indispensable. And in a land of steep mountains, knowing how to get enough food means discovering the altitudes where particular plants and animals flourish. We must postulate and indeed have evidence for, thousands of years of experimentation and the accumulation of information about plants, animals, and vertical landscapes as they relate to basic human requirements. The native Andeans dug irrigation canals, built bridges, and constructed community store houses. Clearly technical knowledge was needed to do these things, but knowing how to organize and direct large groups of people to do the work and keep the system going must also be postulated.

The orderly provision of knowledge capable of being used to organize and direct large groups was the role of the calendar, the quipu, and the ceques. The quipu and the ceques have a very strong set of similarities and redundancies of the kind that make for a very effective communication system. Zuidema (1977) points out that

> the ceque system has been compared to a giant quipu, laid out over the Cuzco valley and the surrounding hills that served in the local representation of the Incan cosmological system, in its spatial, hierarchical and temporal aspects. . . . Not only can the ceque system be compared metaphorically to quipu but every local group did in fact record its ceque system, that is, its political, religious and calendrical organization on a quipu. (p. 231)

Elsewhere, Zuidema (1982a) argues that

> as projected onto the landscape, the *ceque* system of Cuzco—with all the calendrical rituals carried out in relation to the huacas (places of worship) and ceques mentioned by it—was itself a table, like the *quipu* explaining it. The visibility of all the *ceques* from one centre meant that a person located in the Temple of the Sun had before him "an open book." The *ceques* organized space as a map and made reflection upon it as possible as if the person were seeing an actual map. (pp. 445-446)

The power of the Incan knowledge system lay in its capacity to provide connections for a diverse set of knowledges and to establish equivalences between disparate practices and contexts over a very large area. It was able to do this to a greater extent than the Anasazi's because the quipus and ceques were able to extend the range of their calendar beyond the horizon.

The Incan example also illustrates the failure of Jack Goody's dichotomy between oral and literate societies. Here we have a society that manifested an interest in abstract critical thought, empirical verification, lists, and tables, but without writing. Further, just as Shapin and Schaffer found that the European scientific revolution went hand in glove with the establishment of social order, so Zuidema (1982a) finds: "The Incan interest in exact and systematic knowledge springs not from a pragmatic interest in the measurement of volume or distance but from an interest in 'abstract and moral concepts such as "sin," "secret," "health," "obligation," and "order" ' " (p. 425).

The Pacific Navigators

The knowledge system of the Pacific navigators has much in common with that of the Anasazi and the Incas: It is embedded in an oral culture; it is structured on orientation; and while having large practical and astronomical components, it is an integrated body of natural knowledge. But it differs in some crucial aspects. It enabled the discovery and colonization of totally unknown territory, and its principal device for moving the knowledge is almost entirely abstract with no material manifestation.[10]

The Pacific navigators combined knowledge of sea currents, marine life, weather, winds, and star patterns to form a sophisticated and complex body of natural knowledge. This knowledge system combined with their highly developed technical skills in constructing large seagoing canoes enabled them to transport substantial numbers of people and goods over great distances in extremely hazardous conditions and to establish autonomous communities on distant islands—communities that were nonetheless able to return and maintain their cultural links.

One manifestation of the great divide is the claim that the finding of the islands in the Pacific by these early voyagers was accidental, as opposed to the "deliberate" discovery by the Europeans.[11] The Micronesians, in this account, had only a "traditional" knowledge system inadequate to the complex and difficult task of discovering the unknown; the scientific Europeans, by contrast, were able to plot a course and establish the position of unknown islands and were thus able to bring the knowledge back.

The ability to bring the knowledge back and enable two-way communication is the fundamental prerequisite for a knowledge system to transcend the "merely" local. There is now a good deal of evidence from archaeology, linguistics, anthropology, computer simulation of drifting, and experimental voyaging to show that the Pacific was colonized by one group of people with a complex and common culture. Such cultural integrity could not have been maintained if groups had drifted off, unable to return or communicate.

Thomas Gladwin (1970, p. 34) in his classic work on the Micronesian navigators has highlighted some very important characteristics of their knowledge system. First, their knowledge of the islands and star courses is like a map. In Bateson's (1980) evocative phrase, it is "the pattern that connects" (p. 4; see also Goodenough & Thomas, n.d., p. 15). Second, Gladwin (1970) makes the important observation that navigational knowledge is not an isolated system but is an intimate part of "a network of social, economic and often political ties" (p. 35). But it is not merely practical, "it adds a measure of meaning and value to every act, on land as well as at sea" (p. 35). Navigation is thus a major constituent of the "world of the Micronesians" and their distinctive way of knowing.

The three main practical skills of the Pacific navigator are the ability (a) to determine direction and maintain a course at sea, (b) to keep track of his position by dead reckoning, and (c) to have a system of expanding the island target to augment the chance of successful landfall. The major conceptual device used to determine direction and steer a course is the "star compass." But the star compass alone is not enough. It has to be integrated with the system of dead reckoning called *Etak.* A basic necessity for navigating is the ability to estimate how far you have travelled given the effects of current, drift, wind, and speed. The Micronesian solution is a mental mode of visual representation, of mapping the world in the mind.

On a given voyage between islands, an island to one side of the seaway is chosen as a reference point. These reference islands are part of the sailing directions learned by the apprentice navigator for each island passage. Given that the rising and setting points of the stars are fixed points on the horizon, it is easier for the navigator to mentally represent the actual line of travel of his canoe by breaking it up into conceptual segments. The navigator does this by conceiving his canoe to be stationary and the reference island as moving backward against the backdrop of the rising and setting points of the stars. As the reference island moves from one such point to another, it completes a segment of the voyage.

> *Etak* provides a framework "into which the navigator's knowledge of rate, time, geography and astronomy can be integrated to provide a conveniently expressed and comprehended statement of distance travelled." It is a tool "for bringing together raw information and converting it into the solution of an essential navigational question, "How far away is our destination?" (D. Lewis, 1975, p. 138; see also Hutchins, 1983)

The key point to recognize is that Micronesian navigation is more than a means of dead reckoning. It is a dynamic integrative conceptual framework. It enables the smooth meshing of the two conceptual devices, the star compass and *Etak,* so that the learned body of knowledge of star courses and sea-marks can instantaneously be summoned to the task of processing the observations of the moment. The total system forms a "logical construct or cognitive map" (Gladwin, 1970, p. 181).

The third, essentially strategic, element of the system is the technique of "expanding the target." Low islands can be easily missed so the target is expanded by looking for patterns of ocean swells, flights of birds, cloud formations, and reflections on the undersides of clouds. The islands are also in chains as a result of their formation at the edge of crustal plates, so the navigator can orient himself by intersecting the chain at any point.

Gladwin (1970) says that Puluwat navigation is "entirely a dead reckoning system" and "depends upon the features of sea and sky which are characteristic of the locality in which it is used" (p. 144). By "local," Gladwin means not only that the system depends on using knowledge and observations specific to the area but also that the techniques employed are specific to the individual island community. In the Marshall Islands, for example, they use wave interference patterns to maintain direction whereas the Puluwatans do not. In one reading this would seem to severely constrain the kind of knowledge deployed by the Micronesian navigators. However, while it is true that it uses dead reckoning, as we have already seen, it is not *merely* a dead reckoning system because at its core lies a dynamic cognitive map. It is this characteristic that enables it to move beyond the local. Much of our Western misunderstandings of how this can be may result from the embeddedness of the concept of plan. Just as the cathedrals could be built without one, so too can the navigators operate successfully in an ad hoc and planless way.

Lucy Suchman (1987) argues that

> in Micronesian navigation nowhere is there a preconceived plan in evidence. The basis for navigation seems to be instead, local interactions with the environment. The Micronesian example demonstrates how the nature of an activity can be missed unless one views purposeful action as an interaction between a representation and the particular contingent details of the environment. (p. 187)

Thus she concludes that

> the function of abstract representations is not to serve as specifications for the local interactions, but rather to orient or position us in a way that will allow us, through local interactions to exploit some contingencies of our environment and avoid others. (p. 188)

There are three major problems involved in the learning and use of a complex body of oral knowledge like that of Micronesian navigation. The first is the development of techniques to ensure that the vast body of detailed data is accurately retained and passed on over generations. The second is that the body of data must be instantly accessible to the user. It would be of no assistance to the navigator if he had to work through lists of items to get the desired bit. He must be able to instantly access any part of the system. The third is that the system must of necessity be local in nature but it must also be capable of moving beyond the local into the unknown. The first problem is resolved in part by a variety of strategies: the encoding of knowledge in songs and ritual, group learning and testing sessions, mnemonics, overlapping and redundant ways of connecting the knowledge, and constructing material models of the system, like the stick charts and stone arrangements (Farrall, 1981; Goodenough & Thomas, n.d.). The second problem is of course largely resolved through constant repetition and practice until the knowledge becomes completely tacit—an unreflective skill. But one of the most important components of this tacit knowledge or skill is the navigator's constant awareness of where he is on or, more precisely, in his cognitive map. It is that cognitive map that simultaneously provides a basis for solving the first two problems by providing the possibility of creating connections and equivalences and that also enables the knowledge to move.

The template of the Gothic cathedral builders, the calendar of the Anasazi, the calendar of the Incan empire in association with the working of ceques and quipus, and the complex cognitive "technology" of the Micronesian navigators of the Pacific—all are examples of the melding of quite heterogeneous and disparate practices to form stable assemblages that connect. Though quite disparate in their makeup, they can all be understood as giving impetus to the systemic aspects of knowledges. Thus they are "technologies" carrying the "power of the center," disciplining life at the local level in constituting a knowledge system. As quite different sorts of "technologies," they articulate systems in quite different ways. The micropower practices engaged in are qualitatively different; different degrees of "negotiation" between the local and the global are enabled through the "technologies."

THEORIES AND NUMBERS AS HETEROGENEOUS ASSEMBLAGES

So far we have described the working of various types of connecting assemblages as "entities" within disparate knowledge systems; now we look "inside" two such entities that can be understood as connecting locales of work in science. In this section we take two types of heterogeneous assemblages that many would consider to embody the universality of science—theories and numbers. We consider work that has revealed these as "technologies" enabling systematizing in science. Engaging Star's (1989a) treatment of theories in science as connecting assemblages that are both plastic and coherent, we take up an argument that one of us has previously made showing numbers as similarly heterogeneous (H. Watson, 1990).

Emphasizing the local in science necessitates a reevaluation of the role of theory. Typically philosophers and physicists have theory as providing the main dynamic and rationale of science as well as being the source of its universality. Karl Popper, for example, claims that all science is cosmology and Gerald Holton sees physics as a quest for the "Holy Grail," which is no less than the "mastery of the whole world of experience, by subsuming it under one unified theoretical structure."[12] It is this claim to be able to produce mimetic totalizing theory that Western culture has used simultaneously to promote and reinforce its own stability and to justify the dispossession of other peoples (Graham, 1991, p. 126). It constitutes part of the ideological justification of scientific objectivity—the "god-trick" as Haraway (1991a, p. 189; Nagel, 1986) calls it—the illusion that there can be a positionless vision of everything. The allegiance to mimesis has been severely undermined by analysts such as Rorty but theory has also been found wanting at the level of practice, where analytical and empirical studies have shown that it cannot and does not guide experimental research (Cartwright, 1983; Charlesworth, Farrall, Stokes, & Turnbull, 1989; Rorty, 1979). The conception of grand unified theories guiding research is also incompatible with what Leigh Star has pointed to as a key finding in the sociology of science: "Consensus is not necessary for cooperation nor for the successful conduct of work" (Star & Griesmer, 1989, p. 388).[13] In Star's (1988) view,

> Scientific theory building is deeply heterogeneous: different viewpoints are constantly being adduced and reconciled. . . . Each actor, site, or node of a scientific community has a viewpoint, a partial truth consisting of local beliefs, local practices, local constants, and resources, none of which are fully verifiable across all sites. The aggregation of all viewpoints is the source of the robustness of science. (p. 46)

> Any scientific theory can be described in two ways: the set of actions that meet those local contingencies . . . or the set of actions that preserves continuity of information in spite of local contingencies. These are the joint problems of plasticity and coherence, both of which are required for theories to be robust. Plasticity here means the ability of the theory to adapt to different local circumstances to meet the heterogeneity of the local requirements of the system. Coherence means the capacity of the theory to incorporate many local circumstances and still retain a recognizable identity. (p. 21)

Theories from this perspective have the characteristics of what Star (1989a) calls "boundary objects," that is, they are "objects which are both plastic enough to adapt to local needs and constraints of the several parties employing them, yet robust enough to maintain a common identity across sites" (p. 21). Thus theorizing is itself assemblage of heterogeneous local practices.

Star's treatment of theory as standardizing practice is worked out in her study of scientific work, which led to the development of the theory of physical localization of brain function. Star presents us with a picture of a theory growing from particular situations and "clotting" to become a form of standardized knowledge. As we see the theory "clot," we see values inherent in the work activities of the collective become encoded in the cohering yet heterogeneous form or assemblage that their work produces. The theory is as much prescriptive of practical action as it is an explanation of brain function; it demands belief and commitment to the values it encodes.

Theories as assemblages are the end result of many kinds of action, all involving work: approaches, strategies, technologies, and conventions. The component parts of a theory become increasingly inseparable as it develops; they become thicker or more clotted; events, observations, and assumptions come to be seen as connected. The most successful (i.e., the most robust) theories become so clotted, so multirooted, that they are in Latour's terms "black boxes," obligatory passage points in vastly different enterprises. In making this point at the beginning of *Science in Action*, Latour (1987) juxtaposes the theory of DNA structure in 1956 with the routine black box that people work through as the transparent technology that the theory had become by 1987.

While the term *black box* emphasizes the coherence of these forms, as a metaphor it does not adequately evoke the plasticity of these robust forms.

Black box implies design and fails to convey the notion that standardized forms of knowledge are the coagulated consequence of the work of a collective, where diverse interests expressed in a common situation are bound into the robust form enhancing (paradoxically) both its plasticity and its coherence. Failing to recognize the plasticity that goes along with the integrity and coherence of black boxes, Latour (1983), and likewise Rouse (1987), have standardized forms of knowledge swarming unimpeded out of the laboratory. As they see it, resistance is useless.

The robustness in theories with contradictory elements of flexibility and coherence, which we see through work like Star's, is important. It enables us to recognize the continuities between the relatively "freshly" assembled knowledge forms of our time, like the theory of DNA or the theory of physical localization of brain function, and other more pervasive and well-established standardized forms such as number and quantification, which are implicated in so much of Western life, particularly science.

Just as the theory of the physical localization of brain function has been revealed as a social product, so too have number and quantification (H. Watson, 1990). Quantification is a surprising weaving together of practices of ordinary talk and material practices. We can see this when we juxtapose "natural number" in two radically different language communities, for example, the Yoruba community of West Africa and the English-speaking community. Three quite different sets of practices are "clotted" together in both Yoruba and Western quantification.

One set of practices concerns the categories that language users adopt through engaging the particular method of predication in the language—a historical "accident." This set of practices has the speakers of a language constituting the universe with particular kinds of entities. The way we predicate and thus come to talk of "things" in English has us talking of entities individuated in space and enduring as such across time—entities in a spatiotemporal sense. The analogous set of practices in Yoruba has speakers talking of entities constituted on the basis of what in English we understand as qualitative properties, for example, the "waterness of water." A Yoruba language answer to the question, "What is it?—"*Kí nì yí?*" to which an English speaker replies, "It's water," could be "*Omi ni ó jẹ́* " (literally translated as "Watermatter [matter with the characteristics of waterness] here manifests its inner intrinsic and permanent nature"). We see that Yoruba *omi* is quite a different sort of category than English *water.*

These disparate categorizing practices that form part of ordinary language use are tied up with differences in what we would normally consider to be the practices of quantifying. In English language we understand qualities of spatiotemporal entities as constituting the basis of "unitizing" the material world prior to quantifying. If spatiotemporal entities have numerosity, then that is the quality we use, and we say we "count." With spatiotemporal entities that cannot be understood as having the property of numerosity, other qualities (length, mass, and so on) can be used to constitute temporary units understood as analogous to things, and we measure. For Yoruba speakers, with a world already categorized on what English speakers understand as a qualitative basis, the modes in which these sortal particulars manifest, with varying degrees of dividedness, constitute the basis of quantification. The sets of unitizing material practices incorporated into *kà* and *wọ̀n* might resemble the sets of unitizing practices built into counting and measuring, but because the categories taken to constitute the world differ, the sets of practices hold together in different ways.

Constituting a recursion of names is the third set of practices that contribute to the assemblage of "natural number." The role of fingers and toes in both the Western and the Yoruba assemblages is implicit. In both cases seriation in words is patterned on the scale of finger-toes, but there are significant differences. The contemporary numeral recursion that has developed with English has ten as its base—"ten" is the point in the series of words that marks the end of the basic set. As each ten is reached, the basic series is started again in a systematically modified form. The rule by which the series continues is addition of single units. Yoruba numerals are a multibase recursion. The most important base is twenty (*ogún*). Ten (*ẹ̀wà*) and five (*àrùún*) provide points at which the twenties are broken up. The rules for working the recursion make little use of addition; the processes of multiplication and subtraction are more important. We can explain the difference between English and Yoruba in the practices of numeral recursion by going back to the primary categories in the language. When the entities talked of are spatiotemporal objects, the linguistic code explicitly differentiates fingers from one another. When the primary entities talked of are sortal particulars, a linguistic code to report the position on the finger-toe scale must necessarily be more complex. With the primary categorical distinction of Yoruba, the fact that finger-toe matter ordinarily manifests in sets of twenty with inherent divisions into collections of ten and five is relevant—a sortal particular—and a person coincides with the manifestation of finger-toe matter in this way.

Thus a cross-cultural tension enables us to see number and quantification as the "clotted assemblage" of three quite heterogeneous sets of practices: linguistic practices of designating, material practices of unitizing matter, and practices of tallying units through linguistic analogy to fingers and toes. In the past, in communities speaking Indo-European languages and in those speaking West African languages, number and quantification have resulted from efforts of people to produce meaning. But this has been forgotten as we just go on using number as a standardized form of knowledge that has become so "clotted" as to now be considered part of our grammar.

We can extend the insights Star has given us into the development of theories as clotted assemblages that connect locales of work, to understand quantification as just another robust, clotted form of knowledge that originated in particular situations and enterprises. Par excellence, it displays both plasticity and coherence. Through number, other accomplishments are possible; it is "a technology." Understanding number as contrived in past work by people is likely to be a rather startling idea for some, yet it is a useful way to understand number as a social phenomenon. It is an understanding that has been crucial for the work we describe in the next section.

WORKING WHERE KNOWLEDGE SYSTEMS OVERLAP

This section describes work that is situated both within social science and within the Yolngu Aboriginal Australian community that holds lands in the northeastern section of the Northern Territory of Australia; it is work within the historically layered contestation between white Australia and Aboriginal Australia.[14] The Yolngu Aboriginal system of knowledge claiming reach over Yolngu lands understands itself as coherent and exhaustive; while seeing itself as a distinct entity, it does not assert incommensurability. Contemporary Aboriginal knowledge systems are less powerful than the contemporary scientific knowledge system; they survive in centers remote from the centers of scientific knowledge.

We can conceive of a knowledge production endeavor simultaneously located in dual contesting systems; boundaries between knowledge systems are vague and indefinable. Knowledge systems are polysemous so that where one system leaves off and another starts is a matter for strategic negotiation on the part of those involved in knowledge production enterprises. Locating across cultural traditions can render visible the strategies and technologies (i.e., the power practices) embodied in each of the systems. The research program described here involves practical mutual translation achieved through mutual interrogation. Part of the work is to display the standardized assemblages of heterogeneous practices of each knowledge system to the other, all the while resisting the production of new knowledge in both systems.

The work does not romanticize and/or appropriate the vision of the less powerful Yolngu knowledge system. The contemporary Aboriginal Australian systems of knowledge are no more innocent than contemporary scientific knowledge, yet they are a useful counterweight because, as subjugated knowledges, they are less likely to deny the critical and interpretive core of all knowledge.

> They are savvy to modes of denial through repression, forgetting and disappearing acts—ways of being nowhere while claiming to see comprehensively. The subjugated have a decent chance to be on to the god-trick and all its dazzling—and, therefore, blinding—illuminations. (Haraway, 1991a, p. 191)

The possibility of dual system knowledge production and the terms within which it might be accomplished are contested both within science-based Australian culture and within Yolngu culture, and in both argument is needed. But cogent argument is not the only aspect of the work. This is an endeavor of practical politics. What we are producing—practical criticism of past ways of understanding ourselves, and relations between the two peoples, and reinterpretation of the political and social processes of those relations—is of course subject to standards of theoretical coherence and empirical adequacy. But its overall adequacy is not determined solely by such criteria. The constructions that we are generating are "verified" also by participants engaging with the newly apparent sets of possibilities for social action.[15]

In contemporary Australian life, there are areas of continuing political contestation between Aboriginal and European traditions. In these places interaction between the knowledge production systems is still hot and controversial after 200 years of mutual involvement. Education of Aboriginal children is one such area; another concerns land ownership and usage. In the past, controversies in these areas have been closed by adjudication on the issues by non-Aboriginal authorities who have taken the view that there is only one legitimate knowledge system and that, insofar as claims are made from within other systems, these are taken as both illegitimate and inferior. Our research attempts to go beyond confronting adjudication. At present the research is focused on the generation of an education appropriate for Yolngu children. A particular emphasis within this is the mathematics curriculum.

The concerted use of three stabilized sets of practices in the Yolngu Aboriginal Australian community makes it possible for people and places to be joined in a formally related yet dynamic whole. These are analogous to the three sets of practices that we can understand as constituting quantification. All Australian Aboriginal peoples use a formalized recursive representation of kinship as the major integrative standardized form in much the same way that the formalized recursion of tallying—number—constitutes an integrative standardized form of knowledge in Western societies. The Yolngu Aboriginal Australians know their system as *gurruṯu,* which is an infinite recursion of a base set of names patterned on family relations enabling everything to be named and related and imposing an order on the entire world.

Gurrutu is a recursion of names that are understood as names of qualitatively different relations (or at least that is how we can characterize it in English), not as names of varying extents or degrees within a particular qualitative relation—hierarchy—as number is. This difference is associated with a profound difference in the primary categories of Yolngu language compared with English. As we have already noted, English has its speakers designating entities in the sense of spatiotemporal entities. In contrast, Yolngu language has speakers designating relations between connoted entities.

To understand this distinction better, imagine a photograph of some canoes drawn up on a beach. Asked to describe the photograph, an English speaker might say, "Canoes are lying on a beach." A Yolngu speaker might say, "*Rangi-ngura nyeka lipalipa.*" A close English translation of this statement would be something like "Beach-on staying canoe." Considering these further we can see that the English *canoes* countenances spatially separated units that can manifest in collections of one or more. In the Yolngu language statement, the types of elements in the scene are *rangi* (beach) and *lipalipa* (canoe) type elements. The suffix *-ngura* is one of many suffixes in Yolngu language that, when joined to another term, like *rangi,* names the relation between the elements in the scene. What is being talked about here is a relation—"beach-on" —between different types of elements. We can understand "beach-on-ness" or "beach-at-ness" as the subject of the sentence. The term *nyeka* implies "sitting at or staying at a place"; it tells us something about the *-ngura* (the "on-ness" or "at-ness").

Just as we saw the type of designating category having consequences for the type of recursion engaged in Yoruba quantification, we see that in the Aboriginal Australian Yolngu language talking of relations is associated with a recursive pattern of names of relations deriving from the material pattern of family relations. Thus we see how two of the disparate sets of practices hang together.

The third set of practices that helps constitute the working assemblage has to do with mapping the land and associating particular sections of *gurrutu* with particular places. This involves sets of material practices associated through idealized narratives of journeys made by idealized ancestors that relate particular places to contemporary Yolngu people. In much the same way sets of material practices are associated through "stories" of qualities inherent in spatiotemporal entities that enable the "application" of the number recursion to the material world. For Yolngu the travels and activities of the ancestors in creating the landscape constitute tracks or "songlines"—*djalkiri*—that traverse the whole country (see Watson, with the Yolngu, & Chambers, 1989, for a more detailed description of these two sets of practices). The use of *gurrutu* and *djalkiri* together accomplishes the same sorts of ends that the use of number and quantification accomplishes in the West.

Number-quantification juxtaposed with *gurrutu-djalkiri* focuses them both up as contrived systematizing "technologies" formed by and in turn shaping Western life and Yolngu Aboriginal life, respectively. Both normally remain invisible, their historicity stripped away; they have been naturalized to become part of the grammar of these forms of life. Judgments and choices, both individual and collective, are made through number-quantification and *gurrutu-djalkiri*, but as "technologies" they presuppose particular and qualitatively different distributions of power and open up possibilities for choice and judgment in different ways. And this can only be seen when their "naturalism" is stripped away through juxtaposition. For Aboriginal Australia, *gurrutu*—engaging the ties of kinship—is reestablished as a valid "technology" through which community and individual decisions can be rationally made and through which contemporary Yolngu community life is enhanced and extended. At the same time a "technology" of social order en- coding a set of values opposed to the rationality of numbers can be effectively revealed in the wider Australian community.

The possibility of reframing Yolngu concepts within Western knowledge and vice versa through the plasticity of number-quantification and *gurrutu-djalkiri* requires each side to assimilate something of the other. In this process Yolngu look for and emphasize metaphor in Western knowledge. Science looks for and emphasizes codification and develops a grid in which two systems can be seen in ratio. On the Yolngu side two metaphors have been developed as the framework to carry the practices of translation. They originate in the natural process of the Yolngu lands. On other hand, *Balanda*[16] researchers couch their framework in terms of metaphors of construction.

In science, "nature" and "society" are taken as quite different than each other and different than "knowledge"; scientific knowledge sees itself and all other knowledge systems as a representation of reality. What is taken as important in scientific knowledge is adjudication over true and good representations. This is in stark contrast to Yolngu knowledge, which is strongly antirepresentationalist and does not see nature-society-knowledge as constituted of distinct and different sorts of things. We might characterize Yolngu knowledge as idealist, as distinct from empiricist science, so that the forms of evidence considered relevant for Yolngu knowledge claims differ from those considered relevant in science. This goes along with recognition and reverence for the context of production of knowledge claims so that Yolngu knowledge celebrates itself as highly indexical (see Watson et al., 1989, p. 30).

The process of mutual interrogation and the negotiated making available of knowledge of one world in another is familiar practice for Yolngu. For their world has two mutually exclusive components: the *Dhuwa* and the *Yirritja*. These fundamental categories of Yolngu life are constituted by people and places, flora and fauna, words and songs, stories and metaphors, dances and graphic symbols. Everything, every person, every concept, every place that matters in the Yolngu world is either Yirritja or Dhuwa. Dhuwa *rom*[17] is made available to Yirritja clans for their use and vice versa; there are accepted ways of presenting Yirritja in the Dhuwa world and the Dhuwa in the Yirritja. Explicit acknowledgment of the process of mediation through use of metaphor is commonplace in the Yolngu world.

The metaphor through which the work proceeds on the Yirritja side of the Yolngu world, *ganma,* is the dialectic of the meeting and continual mutual engulfing of two rivers. The rivers have different sources and as they flow into each other their separate linear forces acquire the force of a vortex. This vortical flow gives deeper penetration into understanding and knowledge. In terms of the research project, the *ganma* metaphor is taken as the dialectic of a river flowing in from the sea (Western knowledge) and a river flowing from the land (Yolngu knowledge) continually engulfing and reengulfing each other as they flow into a common lagoon. In coming together the streams of water mix across the interface of the two currents and foam is created at the surface so that the process of *ganma* is marked by lines of foam indicating the interface of the two currents. In the terms of the metaphor, this text is part of the line of foam that marks the interface between the current of Yolngu life and the current of Western life.

On the Dhuwa side the research work proceeds through the *Milngurr* metaphor. This sees the dynamic interaction of knowledge traditions as the interaction of fresh water from the land bubbling up in fresh water springs to make water holes, and salt water moving to fill the holes under the influence of the tides. Salt water from the sea and fresh water from the land are eternally balancing and rebalancing each other. When the tide is high, the salt water rises to its full. When the tide goes out, fresh water begins to occupy the water hole. *Milngurr* is dual and balanced ebb and flow. In this way the Dhuwa and Yirritja sides of Yolngu life work together. And in this way Balanda and Yolngu traditions can work together.

Over the past few years negotiations have been conducted among Yolngu people (and are still continuing) about the use of these metaphors to underpin the enterprise of knowledge production involving both Yolngu and Western forms. For Yolngu this move is a highly contentious issue, because all metaphors are owned by particular clans and encode the interests of particular groups. In turning the metaphors to use in reframing the Western and Yolngu world in each other, we have elaborated the metaphors so that their life is not restricted to the Yolngu polity, yet particular Yolngu people still lay claim to them.

Those of us justifying the claims we are producing within the scientific knowledge production system are also using metaphors, although specific acknowledgment that we are dealing in metaphor within this discourse is not usual. Building and constructive/deconstructive metaphors have been used all through this chapter, which is itself within social science. Much of the deconstructive/constructive work involved in presenting evidence here lies in making analogies. Strict symmetry is essential; neither side is privileged in terms of producing true or good knowledge. We can give an account of the workings of both the scientific and the Yolngu knowledge production systems, showing that in each there are analogous processes of interrogation through which claims are generated, and analogous sets of stabilized standardizing practices through which claims can be mobilized.

Having learned how to see these analogies and understand things in new ways, we are answerable for what we do next. If we are to hope for transformations of systems of knowledge, for the construction of worlds less organized by axes of domination, we cannot present our claims to new knowledge as universal claims. Nor can we treat their mobilization in the dual knowledge production systems within which we work as unproblematic, using stabilized assemblages as though they were transparent technologies. In working through the dual sets of devices and strategies whereby claims are mobilized from Yirrkala, the site of our work, we must "focus up" the forms of association, the values, and the politics embodied in the products and the processes of our work.

We are generating an exemplar. We regard the principles and the processes of our work as generalizable, but the "facts" we produce, with assumptions, values, and ideology built in, should be treated with suspicion. Along with the explanations and practices we are producing that demand belief and that prescribe, we are attempting to make evident, and not transparent, the technologies we are using. The resistance inherent in our endeavor is shown and in turn invites informed resistance to our work.

We are engaged in the production of local knowledge but we are making its situatedness and its mobilization problematic so that the processes are recognizable. Others may consider and adopt our arrangements and understandings for their own purposes, but we are not attempting to enroll them as unwitting allies in our endeavor.

CONCLUSION

Throughout this chapter we have argued for the fundamental importance of the local while recognizing that, as far as knowledge is concerned, localness has paradoxical implications of systematicity. We cannot abandon the strength of standards, generalizations, theories, and other assemblages of practices with their capacity for making connections and at the same time providing for the possibility of systematic criticism. We need to recognize that "systemic discipline" and "local resistance" are two sides of the same coin; promoting systematicity is a local practice, and local resistance contains the impetus for systematization. If we do not recognize this joint dialectic of the local and the global, we will not be able to understand and hence establish conditions conducive to the possibility of directing the circulation and structure of power in knowledge systems, conditions for promoting redistributions.

Through recognizing the local-global tension of knowledge systems, we have considered the ways in which the movement of local knowledge is accomplished in different knowledge systems, and the consequent effects on the ways in which people and objects are constituted and linked together, that is, their effect on distributions of power. The challenging of the totalizing discourses of science by another knowledge system that we elaborated in the fourth section is what Foucault (1980, pp. 71 ff.) had in mind when he claimed that we are "witnessing an *insurrection of subjugated knowledges*." It corresponds to Clifford Geertz's (1973b) critique within anthropology that cultural meanings cannot be understood at the general level because they result from complex organizations of signs in a particular local context and that the way to reveal the structures of power attached to the global discourse is to set the local knowledge in contrast with it. Where knowledge systems abut and overlap are sites of cultural contradictions. These are local sites where collective resistance on the part of the marginalized is feasible. Such resistance is a challenge over distributions of power and can lead to increased freedoms; more and different choices over how we might live become possible.

The pervasive recognition, characterized as postcolonialism, that the West has structured the intellectual agenda and has hidden its own presuppositions from view through the construction of the "other" (see Clifford, 1988; Diamond, 1974; Nandy, 1988; Said, 1978) is nowhere more acute than in the assumption of "science" as a foil against which all other knowledge should be contrasted. Marcus and Fischer (1986) take this to be a general movement in the intellectual agenda; according to them we are at an "experimental moment" where totalizing styles of knowledge have been suspended "in favour of a close consideration of such issues as contextuality, the meaning of social life to those who enact it and the explanation of exceptions and indeterminants" (p. 8). In this emphasis on the local we are "postparadigm."

However, we should not be too easily seduced by the apparently liberatory effects of celebrating the local because it is all too easy to allow the local to become a "new kind of globalizing imperative" (Hayles, 1990, pp. 213-214). For all knowledge systems to have a voice and to allow for the possibility of intercultural comparison and critique, we have to be able to maintain the local and the global in dialectical opposition to one another (Said, 1990). This dilemma is the most profound difficulty facing liberal democracies now that they have lost the convenient foil of communism and the world has Balkanized into special interest groups, whether of genders, races, nationalities, minorities, or whatever. By moving into comparatist mode, there is a grave danger of the subsumption of the other into the hegemony of Western rationality, but, conversely, unbridled cultural relativism can only lead to the proliferation of ghettos and dogmatic nationalisms (see Adam & Tiffin, 1991, p. xi).

Analysis and critique of scientific knowledge, whether from the point of view of contesting knowledge systems, or any other, is part of science. In carrying out our endeavors, we are obliged to ask: What sort of politics do we want to characterize our knowledge systems? Part of the reason that it is important to identify the established assemblages of practices through which a knowledge system works is to be in a position to infer the forms of association and hence power relations they engender to make it possible to look for ways of remaking them.

The strength of social studies of science is its claim to show that what we accept as science and technology could be other than it is; its great weakness is the general failure to grasp the political nature of the enterprise and to work toward change. With some exceptions it has had a quietist tendency to adopt the neutral analyst's stance that it devotes so much time to criticizing in scientists. One way of capitalizing on the strength of social studies of science, and of avoiding the reflexive dilemma, is to devise ways in which alternative knowledge systems can be made to interrogate each other.

NOTES

1. For a discourse parallel to and more positivistic than the STS work, see the "ethnoscience project": Blaut (1979), Berlin and Kay (1969), Conklin (1964), Frake (1962), and Sturtevant (1964).

2. The renewed focus of interest in cross-cultural studies is indicated by the number of recent conferences featuring cross-cultural approaches: Comparative Scientific Traditions Conference, "Understanding the Natural World: Science Cross-Culturally Considered," held in Amherst, Massachusetts in 1991; the inclusion of two panels, "Ethnoscience" and "Non-Western Approaches to Science and Technology," in the 4S/EASST Conference, Gothenburg, August 1992; Science of the Pacific Peoples Conference, Fiji, June 1992; and Comparative Science and Culture Conference, Amherst, June 1992. For examples of the other intersections, see Krupat (1992), Haraway (1991a), Adam and Tiflin (1991), Said (1978), Clifford (1988), Hayles (1990), Marcus and Fischer (1986), Raven, Tijssen, and de Wolf (1992), Shiva (1989).

3. This has developed in rather different ways in various centers, as we would expect. In Britain the sociology of scientific knowledge (SSK) began the move; translation theory in France and symbolic interactionism in North America have different ways of posing similar puzzles.

4. The term *technosciences* is used to indicate the lack of a fundamental epistemological difference between science and technology as well as their strong interaction in the later part of

the twentieth century. The plural is used because there are no homogeneous entities "science" or "technology"; there are instead dynamically interacting sets of heterogeneous practices. On technoscience, see Latour, 1987. On heterogeneous practices, see Pickering, 1992b, for examples.

5. Similar systems of knowledge can be brought to light in a wide variety of cultures that have developed ways of organizing natural knowledge in conjunction with agriculture, irrigation, navigation, hunting, astronomy, and so on. For example, Inuit, Maya, Balinese, Indonesian megalith builders and many African cultures all have such contrived assemblages.

6. The following discussion of Gothic cathedrals is taken from Turnbull (1993).

7. According to the archaeologist Dwight Drager, the road system is linked with signal towers as part of a "vast communication network" (cited in Frazier, 1986, p. 125).

8. The most recent and comprehensive account of the "Chaco Phenomena" is Gabriel (1991; see also Crown & Judge, 1991).

9. There is some evidence for the North American Indian recording of accurate, complex, and detailed lunar and solar calendars in a mobile form as message sticks (see Marshack, 1989).

10. The following account is taken from Turnbull (1991).

11. "All the distant ocean islands in the world must have been discovered in the first place by accident, and not by deliberate navigation to those islands. Navigation implies that the existence and location of one's objective is known, and a course set for it. . . . Unless and until the objective has been discovered, navigation is not an issue. . . . [I]n the case of New Zealand, Hawaii and the other detached Polynesian islands, the prehistoric discoverers had no way of gaining the knowledge necessary for navigation back to their home islands. It will follow that the settlement of these detached islands was contemporaneous with their discovery" (Sharp, 1963, p. 33).

12. Allport (1991), a nuclear physicist, finds solace in Popper and Holton as he bemoans the appointment of Nancy Cartwright to Popper's old position at LSE.

13. Compare with the following: "Scientists can agree in their identification of a paradigm without agreeing on or even attempting to produce, a full interpretation or rationalisation of it. Lack of a standard interpretation or of an agreed reduction to rules will not prevent a paradigm from guiding research" (Kuhn, 1962/1970, p. 44).

14. This section is devoted to the work of Helen Watson-Verran with a group of Yolngu Aboriginal researchers. This research community, which calls itself an action group, was established in 1986. The most enduring products of the work of this group so far are the power to control the education of Yolngu children and also a fundamentally reformulated mathematics curriculum. The most accessible text produced by the group is Helen Watson with the Yolngu Community at Yirrkala and Chambers (1989).

15. We need also to produce demonstrable institutional change—to change the social conditions under which contemporary Australian life (Yolngu and non-Yolngu) is possible in terms of both individual experience and community development.

16. The term *Balanda* is a Yolngu term for non-Aboriginal Australian. It predates the British invasion of the continent and derives from the Macassan word *Hollander.* This word is borrowed from the Macassan traders with whom Yolngu had substantial dealings until the beginning of this century, when the White Australia Policy put a stop to such trading.

17. The Yolngu word *rom* is being used here to imply the extension of the category named by Dhuwa; other translations for *rom* are "the law" or "the logic and reasoning" of Dhuwa clans.

7

Laboratory Studies

The Cultural Approach to the Study of Science

KARIN KNORR CETINA

1. THE ORIGIN OF LABORATORY STUDIES

This chapter is about a perspective in recent science and technology studies (STS) that has come to be called "laboratory studies"—the study of science and technology through direct observation and discourse analysis at the root where knowledge is produced, in modern science typically the scientific laboratory. Laboratory studies became feasible in STS when, in the 1970s, the field became more possessive of its subject and more inclusionary—when analysts began to readdress[1] not only the surrounding institutional circumstances of scientific work but the "hard core" itself: its technical content and the production of knowledge. Touching the hard core, however, required a special methodological handle. It was not only necessary to give up the belief that science was the very paradigm of rationality—a belief put into question by philosophers/historians of science such as Kuhn (1962/ 1970) and Feyerabend (1975). One also needed to gain access to the technical content of science through channels other than those of accepted scientific "facts" and theories—for once knowledge has "set" (once it is accepted as true), it is as hard to unravel as concrete. The methodological handles deployed were the study of scientific controversies and the study of unfinished knowledge.

The study of controversies became the methodological focus of a sociology of scientific knowledge, which developed in the early 1970s and resulted in a thoroughgoing sociological contextualization of science (see Bloor, 1976);[2] it examined, for example, how internal scientific standards and experimental evidence fail to provide for scientists' beliefs (e.g., H. Collins, 1975, 1981a) and how the beliefs and knowledge claims of scientists are influenced by their social context (e.g., Barnes, 1977; MacKenzie, 1981; Pickering, 1984).[3] Unfinished knowledge—the knowledge that is yet in the process of being constituted—on the other hand, became the province of laboratory studies.[4] The real-time processes through which scientists, one of the most powerful and esoteric tribes in the modern world, arrive at the goods that continuously change and enhance our "scientific" and "technological" society are still hardly understood. In fact, these processes had not been systematically investigated by social analysts until the mid-1970s, when the first students of laboratories began their investigations.[5] Philosophers of science, who until then were the authority on matters of scientific procedure and content, showed a preference for "the context of justification" and treated the context of knowledge production, which they called "the context of discovery," with neglect and disdain.[6] Historians often defined issues of scientific content as questions in the history of ideas detached from local settings. To be sure, both groups also enriched their accounts with studies of crucial experiments, but experiments, as we shall see, are not laboratories. The whole process of knowledge production in the contemporary fact factory of the natural sciences, and the role of the fact factory itself, were, in the 1970s, untrodden territory in social studies of science.

Laboratory studies have turned this territory into a new field of exploration. The method used in laboratory studies, ethnography (participant observation) with discourse analysis components, has become something of a contemporary equivalent of the historical case study method that became popular in the wake of Kuhn. Ethnography furnished the optics for viewing the process of knowledge production as "constructive" rather than descriptive; in other words, for viewing it as constitutive of the reality knowledge was said to "represent." Constructionism is one of the major, perhaps the major, outcome of laboratory studies (see Sismondo, 1993); the origin of its emphatic use in STS lies in the attempt of students of laboratories to come to grips with their observational record of the "made" and accomplished character of technical effects. Another outgrowth of laboratory studies is the increased usage of ethnography in STS. Laboratory studies have stimulated studies of scientific work that are based on participant observation and now blend with the wider field of "ethnography of science and technology," which extends the ethnographic approach to the study of significant developments in whole fields and even to science policy (Cambrosio, Limoges, & Pronovost,

1990). Finally, there is one component of laboratory studies that needs to be mentioned specifically, and this is the notion of a laboratory itself. The laboratory is not just, I shall argue, a long-underexplored site of investigation newly "conquered by" students of science and technology. It is also a theoretical notion in an emergent theory of the types of productive locales for which laboratories stand in science. In sociology in general, localizing concepts are often associated with the small scale and the weak. Laboratory studies shed light on the power of locales in modern institutions and raise questions about the status of "the local" in modern society in general.

What is the theoretical power of the notion of a lab? How is it different from the older notion of experiment on which historians and philosophers of science placed a premium? What do laboratory studies mean when they claim that scientific facts are "constructed" in these settings? In this chapter, I shall seek an answer to these questions by first spelling out a notion of the laboratory that captures the laboratory's theoretical significance—I shall argue that this significance is linked to the reconfiguration of the natural and social order that in my opinion constitutes a laboratory (section 3). My second focus in this chapter will be to address the meaning of constructionism in laboratory studies (section 4) and to extract, from these studies, those elements and processes that demonstrate how facts are constructed (section 5). In this context, I will also summarize studies of scientific work and ethnographies of science and technology. In the last part of the chapter, I shall discuss some criticism of laboratory studies and needs for further studies of contemporary and historical science (section 6).

2. LABORATORIES ARE DISTINCT FROM EXPERIMENTS AND ORGANIZATIONS

How is the study of laboratories different from the study of experiments or from the sociology of organizations applied to facilities in science and technology? Both the study of experiments and the sociology of organizations have long been established and have constituted something of a tradition in science studies. Consider first experiments, which have until recently carried much of the epistemological burden in explaining the validity of scientific results and rational belief in science. They provided the frameworks within which "the scientific method" was deployed and bore fruit. They were the units in terms of which science proceeded empirically step by step, the rungs in a ladder of theory testing and empirical verification. Experiments were largely defined methodologically in earlier studies; notions like the testing of theories, experimental design, blind and double-blind procedure, control group, factor isolation, and replication are all linked to experiments. The advantages attributed to experiments include the fact that they disentangle variables and test each variable by itself, that they compare the results with those of a control group, that they avoid experimenter bias and subjective expectations, and that their results can be justified through replications that "anyone" can check or perform. With this methodological definition of experiments in place, the real-time processes of experimentation in different fields remained largely unexamined (Gooding, Pinch, & Schaffer, 1989).[7]

When the first laboratory studies turned to the notion of a laboratory, they opened up a new field of investigation not covered by the methodology of experimentation. For them the notion of a laboratory played a role that the notion of experiment, given its methodological entrenchment, could not fulfill; it shifted the focus away from methodology and toward the study of the *cultural* activity of science. In many ways, the notion of a scientific laboratory began to stand for what to the history and methodology of science has been the notion of experiment. The laboratory allowed social students of science to consider the technical activities of science within the wider context of equipment and symbolic practices within which they are embedded—without at the same time reverting to a perspective that ignores the technical content of scientific work. In other words, the study of laboratories has brought to the fore the full spectrum of activities involved in the production of knowledge. It showed that scientific objects are not only "technically" manufactured in laboratories but also inextricably *symbolically* and *politically construed*. For example, they are construed through literary techniques of persuasion that one finds embodied in scientific papers, through the political stratagems of scientists in forming alliances and mobilizing resources, or through the selections and decision translation that "build" scientific findings from within.[8] An implication of this has been the awareness that, in reaching its goals, research *intervenes,* to use Hacking's term,[9] not only in the natural world but also—and deeply—in the social world. Another implication is that the products of science themselves have come to be seen as cultural entities rather than as natural givens "discovered" by science. If the practices observed in laboratories were "cultural" in the sense that they could not be reduced to the application of methodological rules, the "facts" that were the consequence of these practices also had to be seen as shaped by culture.

If we now compare laboratory studies with earlier studies in the sociology of scientific and technical organizations, a similar change in perspective is apparent. Unlike traditional studies of scientific experiment, studies in the sociology of scientific and technical organizations had not concerned themselves with the technical content of the disciplines involved. Their topics were the classical questions of organizational structure and performance as a precondition for, and as possibly conducive to, scientific and technical achievements (see, e.g., Kornhauser, 1962; Pelz & Andrews, 1966). They were not how scientific facts are themselves produced in these settings or how the major mechanisms in the process of knowledge production can be described. Laboratory studies, to be sure, do not exclude organizational variables; for example, questions of resources, of communication within settings, or of linkages between organizational units reoccur in almost all laboratory studies. However, these questions play a role not with respect to a formally defined "organizational output" but as part of the cultural apparatus of knowledge production that becomes visible in laboratories. The shift in focus is significant in two respects: It suggests that isolating organizational structure variables and performance variables and studying them across organizations may not be sufficient to learn about constitutive aspects of organizational work, when this work involves the deep processing of complex information.[10] Second, it suggests that the power and productivity of organizations may reside not only with the general organizational structures they adopt but also with the specific differences they institute in organizational practice with respect to other organizations and with respect to their environment in general (see below).

3. THE LABORATORY AS A THEORETICAL NOTION: THE RECONFIGURATION OF OBJECTS AND SUBJECTS

The laboratory has served as the place in which the separate concerns of methodology and other areas such as organizational sociology could be seen as dissolved in cultural practices that were neither methodological nor social organizational but something else that needed to be conceptualized and that encompassed an abundance of activities and aspects that social studies of science had not previously concerned themselves with. The significance of the notion of a laboratory lies not only in the fact that it has opened up this field of investigation and offered a cultural framework for plowing this field. It lies also in the fact that the laboratory itself has become a theoretical notion in our understanding of science. According to this perspective, the laboratory is itself an important agent of scientific development. In relevant studies, the laboratory is the locus of mechanisms and processes that can be taken to account for the "success" of science. Characteristically, these mechanisms and processes are nonmethodological and mundane. They appear to have not much to do with a special scientific logic of procedure, with rationality, or with what is generally meant by "validation." The hallmark of these mechanisms and processes is that they imply, to use Merleau-Ponty's terminology, a *reconfiguration of the system of "self-others-things,"* of the *"phenomenal field"* in which experience is made in science.[11] As a consequence of these reconfigurations, the structure of symmetrical relationships that obtains between the social order and a natural order, between actors and environments, is changed. To be sure, it is changed only temporarily and within the walls of the laboratory. But it appears to be changed in ways that yield epistemic profit for science.

What do I mean by the reconfiguration of the system of "self-others-things" and how does this reconfiguration come about? The system of "self-others-things" for Merleau-Ponty is not the objective world independent of human actors or the inner world of subjective impressions but the world-experienced-by or the world-related-to agents. What laboratory studies suggest is that the laboratory is a means of changing the world-related-to-agents in ways that allow scientists to capitalize on their human constraints and sociocultural restrictions. The laboratory is an "enhanced" environment that "improves upon" the natural order as experienced in everyday life[12] in relation to the social order. How does this "improvement" come about? Laboratory studies suggest that it rests upon the *malleability* of natural objects. Laboratories use the phenomenon that objects are not fixed entities that have to be taken "as they are" or left to themselves. In fact, laboratories rarely work with objects as they occur in nature. Rather, they work with object images or with their visual, auditory, electrical, and so on traces, with their components, their extractions, their "purified" versions. Take the transition from agricultural science, a field science, to biotechnology described as a process of replacement by Busch, Lacy, Burkhardt, and Lacy (1991). Through the transition from whole plants grown in fields to cell cultures raised in the laboratory, the processes of interest become miniaturized and accelerated. Clearly, the growth of cells in a cell culture dish is faster than the growth of whole plants in the field. Moreover, these processes become independent of seasonal and weather conditions. As a consequence, natural order time scales are surrendered to social order time scales—they are sub- ject mainly to the limitations of work organization and technology. Astronomy provides another illustration: Through the switch to an imaging technology, to digitalization, and to computer networks (Lynch, 1991; Smith & Tatarewicz, 1985), astronomy has become a laboratory science though it is *not* an experimental science.

There are at least three features of natural objects that a laboratory science does not need to accommodate: First, it does not need to put up with the object *as it is*; it can substitute all of its less literal or partial versions, as illustrated above. Second, it does not need to accommodate the natural object *where it is*, anchored in a natural environment. Laboratory sciences bring objects "*home*" and manipulate them "on their own terms" in the laboratory.

Third, a laboratory science does not need to accommodate an event *when it happens*; it does not need to put up with natural cycles of occurrence but can try to make them happen frequently enough for continuous study. Laboratories allow for some kind of "homing in" of natural processes; the processes are "brought home" and made subject only to the local conditions of the social order. The power of the laboratory (but of course also its restrictions) resides precisely in its exclusion of nature as it is independent of laboratories and in its "enculturation" of natural objects. The laboratory subjects natural conditions to a "*social overhaul*" and derives epistemic effects from the new situation.

But laboratories not only "improve upon" the natural order; they also *"upgrade" the social order* in the laboratory, in a sense that has been neglected in the literature on laboratories. Traditionally, the social has been seen as external to the conduct of science—something to be brought into the picture only to explain incorrect scientific results (Bloor, 1976, p. 14). Laboratory studies and other approaches in the new sociology of science have eliminated this asymmetry—they have found it surprisingly easy to explain much of what goes on in knowledge production in terms of social factors, and stressed the inextricable linkages between the objects produced in science and the social world. Yet my point here is different; it refers to the fact that the reconfiguration model also extends to the social order. If we see laboratory processes as processes that "align" the natural order with the social order by creating reconfigured, "workable" objects in relation to agents of a given time and place, we also have to see how laboratories install "reconfigured" scientists who become "workable" (feasible) in relation to these objects. In the laboratory, it is not "the scientist" who is the counterpart of these objects. Rather, it is agents enhanced in various ways so as to "fit" a particular emerging order of self-other-things, a particular "ethnomethodology" of a phenomenal field. Not only objects but also scientists are malleable with respect to a spectrum of behavioral possibilities. Moreover, it is not at all clear that these scientists must remain stable individual entities that are separated out from other objects in the laboratory. Certain of their features may be coextensive with those of objects; they may be construed as "coupled" to objects and machines, or they may "disappear," as individual players, in epistemic collectives that match the objects in the laboratory. In the laboratory, scientists are, on the one hand, "methods" of going about inquiry; they are part of a field's research strategy and a technical device in the production of knowledge. But they are also, on the other hand, human materials *structured into* ongoing activities in conjunction with other materials with which they form new kinds of entities and agents.[13] Recently, Latour has asked that we consider not only how scientists construe nature but also how they "co-construe" society as part of their enterprise—for example, by inserting into it the products of their work (1988) and by defining the nature, ontology, and limits of social actors (1989). Laboratory studies show how the "constructors" themselves are reconfigured, not as a result of the political strategies of specific agents but as the outgrowth of specific forms of practice.

4. CONSTRUCTIONISM AND LABORATORY STUDIES

In the above, I have drawn upon bits and pieces from existing laboratory studies—the idea of the laboratory as an entity that brings epistemic dividends can also be found in Latour (e.g., 1987); the reconfiguration idea comes from Knorr Cetina (1992a, in press); the interest in symbolic construction is best represented in Traweek (1988). Nonetheless, I have so far neglected the diversity of existing laboratory studies, which do not share a single model of science. In fact, the five major book-length monographs on laboratories available in print display as many viewpoints and approaches to science studies as there are authors: Latour's and Woolgar's *Laboratory Life* of 1979 has evolved, in Latour's collaboration with Michel Callon, into a semiotically inspired actor-network approach (e.g., Callon, 1986b; Latour, 1988); Knorr Cetina presents, in *The Manufacture of Knowledge* (1981), a constructivist approach oriented toward the sociology of knowledge that is extended into a model of epistemic cultures (1991, in press); Michael Lynch's work *Art and Artifact in Laboratory Science* (1985a) can stand for the ethnomethodological orientation; and Traweek's monograph *Beamtimes and Lifetimes* (1988) represents the analysis of a symbolic anthropologist who enters the world of high-energy physics. Nonetheless, some threads run through more than one study; some orientations, like that of constructionism, have raised most comments from the outside[14] and have developed into fully fledged perspectives in their own right. Others, like the inspiration some laboratory studies drew from semiotics, rhetoric, and the metaphor of society as behavioral text, have led to specific literary models of how facts are constructed; still others, like the ethnomethodological concern with the fine-grained analysis of daily practices, have reinforced the interest in detailed description of scientific work. Consider first constructionism in laboratory studies.

4.1. What Is Constructionism?

Constructionism holds reality not to be given but constructed: It sees the whole as assembled, the uniform as heterogenous, the smooth and even surfaced as covering an internal structure. There are, for constructionism, no initial, undissimulatable "facts": neither the domination of workers by capitalists, nor scientific objectivity, nor reality itself. What generates this lack of deference for the solid entities in our systems of belief? Within laboratory studies, the insistence on direct observation and detailed description has consistently served as a device that calls forth and sustains the constructionist attitude. One of the first laboratory studies used as an epigraph a sentence from Dorothy L. Sayers: "My Lord, facts are like cows. If you look them in the face hard enough, they generally run away" (Knorr Cetina, 1981, p. 1). Detailed description deconstructs—not out of an interest in critique but because it cannot but observe the intricate labor that goes into the creation of a solid entity, the countless nonsolid ingredients from which it derives, the confusion and negotiation that often lie at its origin, and the continued necessity of stabilizing and congealing. Constructionist studies have revealed the ordinary working of things that are black-boxed as "objective" facts and "given" entities, and they have uncovered the mundane processes behind systems that appear monolithic, awe inspiring, inevitable. The deconstruction performed by constructionist studies is neither negative nor a turn toward "mere description"; rather, it is a turn away from the method of intuiting reasons for the so-called progress of science and toward the method of observing the real-time mechanisms at work in knowledge production. If these mechanisms are considered in sufficient detail, some form of constructionism ensues, whether one wishes for it or not. Constructionism was the answer laboratory studies gave to the microprocesses they observed in real-time episodes of scientific work.

Constructionist studies disassemble by multiplying—they multiply the players, the events, and the mechanisms associated with sustaining entities such as scientific facts. For example, they go from the fact of TRF (thyrotropin-releasing factor) to the agents implicated, the alliances mobilized, and the strategies involved in the making of this fact (Latour & Woolgar, 1979). Do they also disassemble the material world to which scientific "findings" refer? Yes, if we mean by this the real-world entities represented by scientific descriptions. No, if we mean the existence of a material reality, or the real-time intervention in and causal interaction with this world. Constructionism as exemplified in the first laboratory studies is neither nihilism nor skepticism, nor a doctrine that reduces objects to something like imputed and subjective meanings. Constructionist studies have recognized that the material world offers resistances; that facts are not made by pronouncing them to be facts but by being intricately constructed against the resistances of the natural (and social!) order. What constructionism departed from, however, is the idea that the laws and propositions of science provide literal descriptions of material reality, ***and hence can be accounted for in terms of this reality*** rather than in terms of the mechanisms and processes of construction.

Constructionism did not argue the absence of material reality from scientific activities; it just asked that "reality," or "nature," be considered as entities continually retranscribed from within scientific and other activities. The focus of interest, for constructionism, is the process of transcription.

4.2. Distinctive Features of Constructionism in Laboratory Studies

One root of constructionism surely lies in the idea that the world of our experience is structured in terms of human categories and concepts; for Kant, the basic categories of the human mind did the structuring; for Whorf and ethnoscientists, it was language and culture seen through language; and for recent microanalysts, it is often meanings infused in negotiations and in definitions of situations. A second root of constructionism can be seen in the idea that the world is created through human labor. Applied to human institutions, this notion is encapsulated in Marx's famous phrase that "men make their own history," even though, as Marx added, they do not make it free of constraints. Constructionism's recent history in sociology in general lies more with the phenomenological tradition; it is brought into focus in Berger and Luckmann's book *The Social Construction of Reality* (1967), which rekindled the interest of sociologists in constructionist ideas. Berger and Luckmann's main question—how it is that we experience social institutions as "natural" and unchangeable when they are, after all, created by society and consist of social action and social knowledge—reoccurs in STS, most succinctly perhaps in inquiries into the solid, monolithic, and awe-inspiring character of technological systems (MacKenzie, 1990a). But on the whole, the answers given and the thrust of constructionist arguments in STS are different.[15] Constructionist ideas were reinvented in studies of scientific laboratories rather than imported into them from phenomenology. There are several distinct uses and implications of these arguments that I take to be the following:

First, the construction metaphor takes on a much stronger flavor when it is applied to natural reality and to science. To say that kinship relations or gender relations are socially constructed and hence cannot be attributed to nature manifesting itself in these relations today hardly raises an eyebrow. But to make the same claim for scientific facts is still strongly contentious (Cole, 1991; Giere, 1988) and leads to endless arguments of the kind raised above about the constructionist understanding of the material world (e.g., Sismondo, 1993). When Pinch and Bijker (1984) asked that the same principles and methods that guide constructionist studies of science be applied to technology, little controversy ensued about whether technological artifacts could be seen as constructed. It seemed clear that they could; the term *construction,* Hamlin (1992) says in a recent discussion of technology studies, "would seem more at home in technology rather than in science" (p. 513). Thus constructionist ideas appear radical only with respect to natural reality, and with respect to our difficulty of granting two things simultaneously: the material world's independent existence and structure, and our locally successful[16] implementation of a world retranscribed from within science that may not be coextensive with the former. Constructionist arguments in studies of science do not settle such issues philosophically. It is rather that they carry them from one arena of discussion to another, from the arena of philosophical argument to that of empirical examination. A second characteristic of the respective studies surely lies in the specific project that results from this move, the project of an empirical epistemology of science and the social world. Such a project considers, for example, which relations to "nature" and material reality are implemented in different locations, what "material reality" means in the context of the respective work, how particular epistemic regimes implicate and manifest themselves in visual, literary, and other technologies, how technological complexes like detectors "mediate" nature and objectivity, and so on. For constructionist perspectives in the study of scientific work, unlike for constructionism in recent sociology, philosophical questions about the nature of knowing provide a resource. But the answer to the questions is sought elsewhere, in the study of the real-time processes of knowledge production.

The third characteristic of constructionism as it emerged from laboratory studies is the emphasis placed in the respective studies upon the phenomenon that knowledge is worked out, accomplished, and implemented through practical activities that transform material entities and potentially also features of the social world. This makes these studies continuous with the above conception of construction as the creation of the world through "labor" (Marx), and suggests a notion of practice that includes, but is not coextensive with, representations. One needs to emphasize that, in the face of recent, postmodernist epistemologies that see the world as constituted in terms of signifier chains and the deconstruction of science a part of a more general critique of representations (e.g., P. M. Rosenau, 1992, chap. 7).[17] Construction has not been specified in laboratory studies primarily in terms of linguistic, cognitive, or conceptual events. Finally, constructionist studies introduced, with the notion of a laboratory, a concept that suggests that, in modern society, at least some world constructions, those of natural science, occur localized in specific settings, in the "workshops" (Heidegger) and "topical contextures" (Lynch, 1991) described in these studies. Relocating the idea of construction into particular physical and epistemic spaces meant shifting the focus away from particular actors, the prominent individual scientists of the past and present who continue to populate historical and contemporary accounts of science.[18] In many areas of empirical natural science, the individual scientist is no longer the epistemic subject, and the role of individuals (or groups), and their status as human processors or workers fitted into a configuration of materials, needs to be reassessed from the vantage point of the settings in which they operate. The emphasis on the "bounded spaces" of world construction leads us back to what was said about laboratories earlier. But it also brings into focus a methodological principle that can guide laboratory studies but distinguishes itself from that of "following the actors" that Latour (1987) proposes for the actor-network approach in STS and that informs many historical and biographical studies of science. If construction is wrapped up in bounded locales, the ethnographer needs to "penetrate the spaces" and the stream of practices from which fact construction arises. In the next section, I want to address what the studies that have followed this principle have to say about how scientific results are constructed.

5. HOW ARE FACTS CONSTRUCTED?

5.1. Nothing Epistemically Special Is Happening

The thrust of the original laboratory studies was that they attempted to show how natural scientific facts are constructed. They considered the making of scientific knowledge open to social science analysis and proceeded to observe the creation of knowledge at the workbench and in notebooks, in scientific shop talk, in the writing of scientific papers. One immediate result of all laboratory studies was that nothing epistemologically special was happening in these instances. In Rorty's (1985) formulation, "no interesting epistemological difference" could be identified between the pursuit of knowledge and, for example, the pursuit of power. The emphasis here is on the epistemological; laboratory studies did not show that there were no interesting *sociological* differences between molecular biology bench work and a courtroom trial, or between the building of a detector and trading at the stock exchange. The experience that nothing epistemologically special was happening wiped out the doubts that had accompanied analysts' first steps into the lab; the making of knowledge was indeed amenable to empirical analysis, and more so than anyone had expected before.

5.2. Construction as Negotiation and Interactional Accomplishment

What was the second major result? Presumably that almost everything is negotiable in the making of scientific knowledge: what is a microglia cell and what is an artifact (Lynch, 1982, 1985a, chaps. 4, 8), who is a good scientist and what is an appropriate method (Latour & Woolgar, 1979, pp. 161 ff.), whether one measurement is sufficient or whether one needs to have several replications (Knorr Cetina, 1981, chap. 2.2), what one sees on an autoradiograph film and what one does not see (Amann & Knorr Cetina, 1989), what is the best environment for good physics (Traweek, 1988, chap. 5) and what counts as a proper experimental replication (H. Collins, 1985, chaps. 2-3). As the last reference shows, not only laboratory studies but empirical studies of scientific work in general have demonstrated the negotiability of the elements, the outcomes, and the procedures in knowledge production. The possibility of negotiation in consensus formation is rendered plausible by the observation, on the part of these studies, that scientific outcomes are empirically "underdetermined" by the evidence in the sense that they frequently do not have univocal outcomes;[19] in fact, experimental outcomes are often opaque, murky, ambiguous, and generally in need of interpretation and further experimentation. They are also made plausible by the discovery, in studies of scientific controversy and scientists' discourse, that scientific findings and scientific accounts are frequently contentious and meet with more than one interpretation (H. Collins, 1981a; Gilbert & Mulkay, 1984). Uncertainty affects not only scientific outcomes and their interpretation but also the process of investigation itself—as shown, among other things, in a recent wave of studies of scientific work undertaken by symbolic interactionists (e.g., Clarke, 1987, 1990a; Fujimura, 1987, 1988, 1992b; Gerson, 1983; Star, 1983, 1985, 1986, 1989a; Star & Griesemer, 1989). For example, Fujimura (1987, 1988, p. 263) analyzes how "doable" problems are constructed in a process marked by uncertainty and ambiguity that scientists cope with through "articulation work," for example, through negotiating tasks with funding agencies and others. Thus uncertainties and "interpretative flexibilities" open up the possibility for negotiation, and the resistance someone or something offers also creates and reinforces the openness of the process. Constructionist studies draw no line between these occasions. For them it is enough to show that the construction of facts includes recurrent elements of negotiation and that the straightforward process of putting "questions to nature" is interspersed with situations in which nature does not speak, or does not speak clearly and unambiguously enough to prevent contestation.

Who are the parties involved in these negotiations? Certainly they are scientists and groups of scientists, but also funding agencies, suppliers of equipment and materials,[20] clients, investors, Congress and ministries of science,[21] and so on. From the beginning of laboratory studies, it has been clear that external agents play a role in these negotiations—hence the notion that the "decision impregnatedness" of scientific results is anchored in "trans" epistemic arenas of research or "trans" scientific fields (Knorr Cetina, 1981, chap. 4, 1982, 1988a). Later studies (Fujimura, 1987, 1988, 1992b) and especially studies of scientific technologies have also forcefully argued that technical, social, economic, and political groups take part in the definition of scientific and technological developments, and thereby exemplify the flexibility in the way technologies are designed (e.g., Henderson, 1991a; Hughes, 1989b). Callon (1986b) and Latour (e.g., 1990), in their general argument for an analysis of the coproduction of society and nature, propose that we also include nonhuman actors as negotiating parties in our analysis. Nonhuman actors are, for example, the microbes, the scallops, or the acid rain investigated by science but also a door and an automatic door closer. Nonhuman agents include the world of things to which agency ("actant"ship[22]) is attributed on behalf of the constraints they issue upon human behavior (a door allows us to walk through only in a particular place), which is itself a consequence of the work, and power, that we delegate to them.[23] Collins and Yearley (1992a) criticized this extension of agency to things and the consideration of nonhuman entities as full parties to processes of negotiation—for example, it can be questioned in what sense our understanding of the making of knowledge is improved at all if we simply call any interaction with nonhuman agents a negotiation. But my point here is different; extensions such as the above show the degree to which the idea of construction as negotiation has penetrated all parts of the analysis of science and technology, even that of instrumental action (human action oriented toward things), which is redefined to mean negotiation with the entities to which it is directed.

What does negotiation mean when it refers to actors in the social world? It is unfortunate that most constructionist studies to date have failed to analyze the patterns and processes that turn an ordinary interaction into a "negotiation." We know that "what counts as a notable finding, a definitive anatomical entity, a thing's attributes, a procedure of measurement, an adequate display of data, and a plan of methodic action" may be asserted and modified in an interactionally sensitive manner (Lynch, 1985a, p. 264), but we know little about the rules and mechanisms that govern these modifications. An early attempt in this direction was made by the above author (1985a, chap. 7), who illustrates how scientists change their descriptions or accounts of scientific and technical objects in the face of expressions of disagreement by others. The "preference for agreement" Lynch discerns in the changes of descriptions accords well with ethnomethodological findings on conversations in general (e.g., Pomerantz, 1975), but their bearing is specific; conversational devices may be implicated in the production of scientific knowledge not just as passive vehicles of communication through which features inherent in natural objects are made explicit but as coproducers of these features. This is also the thrust of a series of articles Amann and Knorr Cetina (1988a, 1989; Knorr Cetina & Amann, 1990) published on the patterns and uses of specific conversational routines when scientists "think through talk" and when they perform analyses of technical images through shoptalk. Accordingly, conversational routines are part of an interactional machinery that produces emergent outcomes not identical with the contributions of individual participants and not reducible to "objective" features of objects. In other contexts, for example, in Traweek's study of high-energy physics laboratories, *negotiation* refers to practitioners' bids for equipment and access to beam time, to conflicts over which experiment is to be accepted and performed or to how to set up the best decision structures for such matters (Traweek, 1988, chap. 5).

Most meanings of the concept of negotiation bring into focus the *interactional* element in episodes of knowledge production: They show how the process and outcome of inquiry are sensitive to the process and outcome of social interaction. In that sense, "negotiation," more than other concepts, highlights the "*social*" character of the process of knowledge production. However, there are also other concepts with slightly different connotations that have been used to illuminate the idea of construction. One is the notion of *accomplishment* preferred by ethnomethodological studies of scientific work (e.g., Garfinkel, Lynch, & Livingston, 1981; Livingston, 1986; Lynch, 1985a). It includes the idea of shop work involving technical objects and instruments, and can be seen to encompass the kind of maneuverings practitioners engage in when dealing with anomalies and contradiction, for example, in the creation of a successful theory (Star, 1989a). Another is the notion of *"decision translation" and of the "decision impregnatedness"* of scientific results. This covers the results of negotiation but also includes selection processes that result from individual or structural and implicit selections (Knorr Cetina, 1981, 1982, 1988). How does a scientist decide to make a particular technical decision? By translating a choice into other choices. The point about these translations is that they often implicate nonepistemic[24] arguments and show how scientists continually crisscross the border between considerations that are in their view "scientific" and "nonscientific."

5.3. Construction as Literary Construction and Representational Craft

Processes of construction are often rhetorical processes; they involve representational techniques of persuasion, which have been investigated with respect to scientific arguments and papers. Early statements of the question of the literary rhetoric of science were provided by Gusfield (1976) and Mullins (1977). They are succeeded today by an abundance of analyses of scientists' written discourse covered in detail elsewhere in this volume, and by a broadening of the notion of "style" as a means of expression (see

Daston & Otte, 1991). Within laboratory studies, Latour and Woolgar (1979, chap. 6) described the larger process of stabilizing facts as an agonistic process in which "modalities" (modifiers of statements of fact that mark the degree of fact-likeness) are constantly added, dropped, inverted, or changed but in which the overall process is one of modality dropping. They also pointed out the significance of "inscription devices," "items of apparatus or particular configurations of such items which can transform a material substance into a figure or diagram" (Latour & Woolgar, 1979, p. 51). Following E. Eisenstein's (1979) arguments with respect to printed materials, one can identify the advantages of such inscriptions; inscriptions, especially printed inscriptions, can be more easily circulated, compared, and combined than the material objects of laboratory work. Inscriptions are, in Latour's terminology (1987, p. 227), "immutable and combinable mobiles,"[25] which, as Henderson (1992) claims with regard to visual representations, can also serve as a social glue between participants.

Other laboratory studies have pointed out a number of literary techniques of objectification that abstract away from interactional and other qualities of shop work in the laboratory. These include the use of the passive voice instead of the "I" or the "we" of the lab, the elimination of most if not all laboratory rationales for technical choices, strict sequencing techniques that reverse the sequence of events in the lab and make no reference to circular connections between stages of shop work, simplifications and extreme typifications of the experimental process that hide the idiosyncracies and the "know-how" of laboratory work, and the disembedding of the work with respect to its strategic components and motivational dynamics and its reembedding in a context of "grand" scientific and practical questions from which the work appears to flow (Knorr Cetina, 1981, chaps. 5, 6).

Questions of representational techniques were also addressed with respect to image design and processing (Amann & Knorr Cetina, 1988a; Henderson, 1991a; Hirschauer, 1991; Knorr Cetina & Amann, 1990) and can be exemplified by Lynch's (1988) work on visualization in the life sciences and Lynch and Edgerton's (1988) study of representational craft in contemporary astronomy. The study finds that an "ancient aesthetic" of perfecting nature through a crafting of resemblances is part of routine image processing work. Lynch (1991) sees technologies such as "opticism" and "digitalism" in image processing embodied in "topical contextures," by which he means the "spatiality of [a] situation bound to a technological complex." This notion takes us back to the idea of a laboratory as a space within which certain epistemic possibilities are bound up, by suggesting that techniques of perfecting nature associated, for example, with opticism may also be bound up and derive from certain local arrangements and contextual spaces.

Literary and rhetorical construction is construction through shifts of meanings and structure in texts similar to shifts of meaning in oral conversation. Inscription devices add to this by rendering the natural science laboratory into a workshop of text production. "Graphic" construction through the montaging of images elaborates further means of persuasion and shows how stylistic considerations inform the rendering of "realistic" images. However, the respective analyses to date have one drawback: They do not address the question of how scientific texts and technical images are decoded by an audience, and therefore cannot say how a given literary or graphic construction actually affects technical decisions. Because everyone within an audience of experts knows how statements, figures, and graphs are subject to techniques of representation, these entities are also easily and readily deconstructed, as discussions of scientific papers in laboratories demonstrate. While studies of scientific work occasionally make reference to such discussions, we are still lacking a systematic analysis of the deconstruction practitioners themselves routinely perform on images and texts. In other words, while we do have an elaborate picture of the scientist as author and writer, we lack a systematic analysis of the scientist as a reader.

5.5. Construction as Local Construction: The Reversals of Practice

There is another sense of construction that warrants special attention within the contexts of laboratory studies. This is that construction appears to be, always, local construction. As Rouse (1987) put it, paraphrasing Heidegger, "Science must be understood as a concerned dwelling in the midst of a work-world ready-to-hand, rather than a decontextualized cognition of isolated things" (p. 108). Reconfigurations of self-other things are local reconfigurations. The power of laboratories, as implied before, is the power of locales. But construction also means construction *with* local means and resources, with the equipment that stands around, the chemicals available, the technical skills and experience offered on the spot. There are numerous examples for this "opportunism" of inquiry, for its dependence and reliance upon local materials, the substitution of apparatus and chemicals for other apparatus and chemicals that are not in stock, the choice of one type of research animal rather than another depending upon the setting, the emergence of an "idea" suggested by local episodes or materials, and so on. Different analysts employ different terms to refer to the implied contingency; for example, Lynch, Livingston, and Garfinkel (1983, p. 212) report on the "embodied," "circumstantially contingent," and "unwitting" character of laboratory practices; Latour and Woolgar (1979, p. 239) refer to "circumstances" as that which stands around and becomes relevant in shop work; and Knorr (1977; Knorr Cetina,

1981, chap. 2) refers to the "indexicalities" manifest in "local idiosyncracies" and to emergent outcomes, variable rules and power, and the opportunism of research. Even philosophers who spend some time "watching" in a laboratory confirm this contingency (Giere, 1988, chap. 5). While there also are "standard" procedures that "work" successfully in many laboratories (see, e.g., the standardization of DNA technology described by Fujimura, 1988, 1992b), laboratory studies investigate how the successful working of a standard procedure is built out of painful processes of adaptation and learning that "fit" techniques to settings, and scientists to their methods. Furthermore, they examine how standard procedures appear to be standard only within specific contexts in which they are treated as a black box relative to other, "problematic" techniques, and how these black boxes nonetheless present persistent problems in the laboratory associated with variations in apparatus and materials (Amann, 1990; Jordan & Lynch, 1992).

Results on the *local* character of shop work fly in the face of received interpretations according to which claims and procedures in science are standardized and universal, and according to which the local environment is merely incidental to the generation of particular results. On the other hand, they are not really surprising in the face of the *reversals* of rules and general characteristics through the definition and dynamics of situations that we know from other settings[26] and through the reconfigurations laboratories breed and rely on. With respect to science, the finding is complicated by the fact that, within the old logic, if the particular specifications of a laboratory must be taken into account in interpreting its results, then it becomes possible to challenge these results as artifacts: Its properties are qualities of the setting rather than of the natural objects themselves (Rouse, 1987, p. 71). Because it would be difficult to argue that all the local choices implicated in fact construction are "nothing but" insignificant variations without impact on the properties of the results obtained, a new logic is needed. Constructionist studies suggest that results do not become weaker but more solid and interesting through local specifications.[27] Local configurations breed the specific advantages and opportunities that, when they are structured into a scientific object, may make it more successful in the wider context. In this sense laboratories are like environmental niches (Vinck, 1991).

5.6. Construction Machineries and Cultures of Fact Construction

Studies of the constructionist work of science have not until recently used a comparative approach; they ignored the potential *disunity* of science with respect to its epistemic strategies as well as the *cultural structure* of scientific methodology. On the other hand, the comparative approach has been used very fruitfully by Traweek (1988, 1992) to investigate cultural differences between nations (the United States and Japan) with respect to laboratory organization, approaches to detector design and building, leadership style, and models of good working conditions for the science she studied—high-energy physics. Thus the American "sports team" approach in which the leader is like a "coach" who has learned to locate highly skilled players and to design strategies for winning is contrasted with the Japanese "household" approach in which it is the responsibility of each member to keep the household and its resources intact and in which status is not determined by competition but by age (Traweek, 1988, p. 149, chap. 5). Other differences refer to the strong dependence of Japanese physicists on industry for building their detector, compared with the American physicists' preference for arranging (and later dismantling) their pieces of equipment on their own. Traweek (1988, chap. 3) also analyzes eminent physicists' stories about themselves, which define the virtues that are, in the different career stages of a physicist's life, associated with success.

Whereas Traweek's object of interest is less the making of physics than the making of (male) physics cultures, Knorr Cetina (1991, in press) recently used the same method in a first attempt to look not at the construction of facts but at the construction of the machineries deployed in fact construction. The machineries of fact construction include the skilful scientist, as investigated, for example, in studies of skills as social relational properties acquired and used in interaction (Amann, 1990; Hétu, 1989; compare Pinch & Collins, 1992). But they also include ontologies of organisms and machines that result from the reconfiguration of self-other-things implemented in different fields, the use of "liminal" and referent epistemologies in dealing with natural objects and their resistances, strategies of putting sociality to work through the erasure of the individual epistemic subject and the creation of social "superorganisms" in its place, or the use of equipment as "transitional" objects (Knorr Cetina, in press). A comparison between experimental high-energy physics and molecular biology shows that the meaning of the empirical changes as these machineries are implemented in different fields. In other words, it shows the *disunity* of the sciences with respect to construction mechanisms and the existence of *epistemic cultures*.

The containment of construction processes within epistemic (and national) cultures has consequences for the kind of results constructionist studies produced. Do we really find the same kind of negotiation in all experimental sciences? Are the basic categories we have developed (e.g., "contingencies," "translation," "decision impregnatedness," "actors," "networks") and the processes we have described equally salient in all fields? From a micro-empirical perspective, it is necessary to pay more attention to how these categories and processes are themselves dependent upon local practices. Models that generalize across all fields and disciplines are premature at best. The move toward considering construction machineries rather than single practices can point out a variety of models instantiated in different contexts; it multiplies the stories about how facts are constructed.

5.7. Construction "With" Laboratories and the Construction of Accepted Knowledge

Up to this point we have considered mainly construction processes within the laboratory. However, some of the concepts introduced to characterize inquiry in these settings such as negotiation and translation can also be deployed to relationships between entities in larger settings, for example, to relationships between laboratories or between entities that operate in different "social worlds" (e.g., Clarke, 1990a; Fujimura, 1992b; Star & Griesemer, 1989; Zeldenrust, 1985). Such larger settings are explored in Downey's (in press-a) ethnography of computer-aided design technologies, which, as Downey argues, reposition people and groups with vested interests in these technologies and renegotiate the institutional boundaries between universities, industry, and government. Cambrioso and Keating's (1991) ethnographic study of the establishment of the international classification of white blood cells displays a similar concern with nonlocal issues. With respect to such issues, construction often means the construction of an audience willing to believe in the knowledge produced by laboratories, willing to implement its results, and willing to link up with scientific "findings" by, in one way or another, reproducing them in further research and discussion. Latour (1983, 1988) and Callon (1980b, 1991; Callon & Law, 1982) have used the concept of translation to refer to this process. How do I persuade someone to accept my proposal, my method, my invention? By convincing them that it is in their interest to adopt my offer, by redefining ("translating") their interests such that they converge with mine, and thereby by "enrolling" a heterogeneous crowd of "actors" (in the sense of actants; Greimas & Courtés, 1979, p. 3) into a network of associations that stabilize the technical object (Latour, 1983). The notion of translation has been redefined over time to mean not just interest translation but the "definition" of an agent. "A translates B," according to Callon (1991, p. 143), "is to say that A defines B." With decision translations (see section 5.2 above), construction points at the selections, the local contingencies, and the "nonepistemic" considerations that are *structured into* a scientific object. With Callon and Latour's translations, the strong interpretation is that the establishment of a scientific or technical object depends not on the *inherent* usefulness or "truthfulness" of the result but on whether one succeeds in building a structure of associations between parties enrolled through mutual definitions that hold up.[28] However, the network includes the scientific object and associated technical components, and these objects must not give way if the whole system of associations is not to break down. Thus construction through network building by means of translation already presupposes and relies on "cooperating" technical objects such as microbes.

In an interesting twist to the argument, Latour (1983, 1988) in his work on Pasteur has claimed that the process of translation may also involve changing society through colonizing the world with extramural laboratories. In other words, recruiting an audience may mean reconstituting, in relevant places, the conditions that obtain in the laboratory and that sustain the "reproducibility" of scientific results. According to Latour, Pasteur changed the conditions at French farms in such a way that his vaccination procedure became reproducible in these settings. In addition, he also changed French society in a more general way—by adding to it the microbe he had discovered in the laboratory as a mediating agent in food processes and in the transmission of diseases and as a mediating agent in social relations in which processes of hygiene and infection are embedded. Thus the view upon the laboratory afforded in such studies is a view from the outside; the laboratory is itself seen as an agent of change, a device in the shaping and construction of society. It serves as a locus for the "definition" of society, for its dissolution into elements and contributing features. This view of construction as the social construction of society *with* the help of laboratories would seem to be most suitable for examining technology or scientific objects oriented toward practical applications. As Gooday (1992, p. 1) has argued using historical data, the extramural laboratory thesis has only limited explanatory power when it is applied to other disciplines practiced contemporaneously with Pasteur's work.

7. CRITICISMS OF LABORATORY STUDIES AND PROSPECTS FOR FURTHER RESEARCH

Laboratory studies and studies of scientific work have met with several criticisms. One issue consistently raised with respect to studies that make strong constructionist claims is the question of what status these studies accord to material objects. Many students of laboratory have claimed that scientific reality is "constituted by" the process of inquiry, that it is a "consequence rather than a cause" of scientific descriptions, or that scientific facts are "inseparable from the courses of inquiry which produce them" (Knorr Cetina, 1981, p. 3; Latour & Woolgar, 1979, pp. 180-183; Lynch, 1985a, pp. 1 ff.; Woolgar, 1988b). These claims are frequently objected to on the grounds that they suggest that material objects are snapped into existence through the accounts produced by science, which most readers find "wildly implausible" (e.g., Giere, 1988, pp. 56 ff.; Sismondo, 1993).

However, there is a more sympathetic reading of these claims according to which what does indeed come into existence, within, usually, a longer term process, when science "discovers" a microbe or a subatomic particle, is a specific entity distinguished from other entities (other microbes, other particles) and furnished with a name, a set of descriptors, and a set of techniques in terms of which it can be produced and handled. In other words, some part of a preexisting material world becomes specified and thereby real as something to be reckoned with, accounted for, and inserted in manifold ways into scientific and everyday life. This does not preclude the possibility that some physical correlate of this entity existed, unidentified, tangled up with other materials, before scientists turned their attention to this object. But what is *not* suggested is that science merely furnishes the conception for preexisting physical objects marked off in exactly the ways science later describes from other objects. Perhaps one could say that the "facts" produced by science provide for the encounterability of material objects within forms of life. But this encounterability of course remains problematic, sensitive to circumstances, and potentially dependent upon the reconfiguration of the rest of a form of life as the difficulties with knowledge transfer and the "diffusion of innovation" demonstrate. It should also be noticed that this encounterability, when it is projected into the past, requires work—for example, the work of historians who rewrite history.[29] Thus the question of how a physical entity comes into existence is indeed an interesting one raised by constructionist studies. But it is not a question that will be answered productively by an ontological sleight of hand, such as by assuming that specified material entities reside outside science and outside human experience.

There are other criticisms—most notably, perhaps, the criticism that laboratory studies, while well suited for the microstudy of scientists' day-to-day practices, are limited when it comes to the study of consensus formation (e.g., Pinch, 1986, p. 30).[30] This assessment is correct. Laboratory studies to date have mostly focused upon single laboratories or a few work situations. Many processes of consensus formation would seem to involve more than one laboratory, and potentially a whole scientific field. On the other hand, laboratory studies can shed light on the typical responses new knowledge claims encounter in scientific practice and on the circumstances under which such claims appear to be incorporated into further research. They can also shed light on how closure of debates over the interpretation of experimental results is reached within the laboratory, before scientific knowledge is published. If interpretative flexibility and the negotiation of experimental outcomes are an irremediable part of scientific work, as laboratory studies claim, the question of how negotiations end ought to be answerable by studies of this work. However, few students of laboratories have addressed these issues to date,[31] and those that have, such as Latour (1988) in his reformulation of the question as one of the "stabilization" of facts, have resorted to historical data.

Consider a third major criticism that is often voiced against laboratory studies. This is that these studies and studies of scientific work narrow their focus to the intramural lifeworlds encountered in laboratories but ignore the societal context in which laboratories operate as well as the political aspects of science (e.g., Chubin, 1992; Fuller, 1992). This critique, too, is warranted; but it would seem to be important mainly with respect to those aspects of the wider context and those relational properties of laboratories that recur in scientific work and influence the construction of knowledge. Laboratory studies profess to be interested in the making of knowledge, and one can hardly blame them if they do not, at the same time, profess an interest in matters such as science policy—except of course where science policy penetrates into laboratory culture or in some other way orients the outcome of knowledge production.

The above criticism nonetheless suggests areas of research into which laboratory studies need to be extended, and indeed are beginning to be extended —consider, for example, work that applies the laboratory studies' ethnographic approach to such politically sensitive issues as Canadian bio-safety regulations (Charvolin, Limoges, & Cambrosio, 1991) or to the travel of government "dossiers" along a path of institutional modifications within the framework of the Quebec government devising policy measures (Cambrioso et al., 1990, p. 206). Or consider Proctor's (in press) usage of the construction metaphor to study the social construction of ignorance with respect to the causes of cancer.

In addition, there is another area into which an extension of laboratory studies is sorely needed, and that is the area of the history of laboratories. Here, too, studies are beginning to emerge (e.g., Gooday, 1991, 1992; Hessenbruch, 1992; Kingsland, 1992; Shapin, 1988b), and historians have become aware of the need—and the methodological possibilities—for an investigation of the emergence, architecture, internal procedures, and different definitions of laboratories.[32] Historical studies of laboratories are joined by historical case studies of experimentation (e.g., Galison & Assmus, 1989; Gooding, 1989; MacKenzie, 1989; Pickering, 1989; see Gooding et al., 1989). These studies manifest a strong recognition of the need to investigate the real-time processes of experimental work and continue and reinforce a trend initiated when earlier controversy studies turned to examining experimental episodes (e.g., H. Collins, 1975; Collins & Pinch, 1982b; Pinch, 1986; Shapin & Schaffer, 1985).

Finally, a most interesting area of future extension would be one in which we move from science and technology studies to sociology in general and consider the possibility that laboratories exemplify features also present in organized settings such as the clinic, the factory, the garden, the government agency. Such questions would bring laboratory studies in contact with the "new institutionalism" in sociology (e.g., Douglas, 1986a; Powell & DiMaggio, 1991), a direction of research that unpacks institutional arrangements in modern society by adopting a cultural and symbolic approach. Work beginning in the area[33] suggests that processes of laboratorization can be found in a broad variety of settings and that it might be useful to distinguish between types of laboratories and patterns of laboratory formation. Modern society, after all, is a society of locales. What we learned from laboratory studies about the "situatedness" of knowledge (see also Haraway, 1991b) may be applicable to larger questions about the localization of experience in multiply embedded and varied sites. What different patterns emerge when we consider these sites in regard to their epistemic relations to an environment? How is the power of locales embodied in the scientific laboratory reflected in the power of other demarcated spaces? Localizing concepts have been of considerable importance to microsociological research in the last decades. Yet theoretical formulations of the relevance of locales are still sorely missing. The laboratory as studied within STS might help in focusing the many issues implied when we talk about situatedness and localization, and aid in producing the above theoretical formulation.

NOTES

1. I say "readdress" because there have been more inclusionary studies before, for example, Merton's (1938/1970) early work on science in seventeenth-century England.

2. During controversies, knowledge is deconstructed by practitioners themselves, and, as it comes apart, analysts can examine the functioning of the standards that are normally thought to hold it together and the contextual influences that inform the opponents and their work.

3. At the time, this influence was specified in terms of an interest model from which some sociologists of knowledge such as Pickering and Shapin have moved away in recent years. See, for example, Pickering and Stephanides (e.g., 1992) and Shapin and Schaffer (1985). For a critique of the notion of interests in science studies, see Woolgar (1981).

4. There is of course overlap between the two developments in that studies of controversies usually also concern knowledge claims that are "unsettled" and in this sense unfinished knowledge. The overlap is most apparent in studies that examine controversial experimental evidence and standards such as the aforementioned studies by Collins or the ones by Collins and Pinch (1982a, 1982b) and Pinch (1986).

5. I know of only one study from before 1977 that can qualify as a laboratory study. This is a study by the physicist and theologian Thill (1972) whose work was financed by a group of progressive Catholics not involved in academia. This work gives valuable technical insights in

the methods of physicists but does not address the issues central to social studies of science. The first "sociological" ethnographer of science to begin a study of scientific practice seems to have been Michael Lynch, who started his work in 1974 (see Lynch, 1985a, 1991, p. 51). The studies of Latour and Knorr Cetina were conducted between 1975 and 1977, like Lynch's study in California (Knorr, 1977; Knorr Cetina, 1981; Latour & Woolgar, 1979). Traweek also started her work in the late 1970s (see Traweek, 1988).

6. See Nickles (1988) for comments on these tendencies and for more recent changes in the philosophy of science.

7. This holds until the late 1970s and early 1980s, when the above-mentioned controversy studies turned to the investigation of the role of experimental evidence and of standards of experimentation; see Collins (1975), Travis (1981), Harvey (1981), and Collins and Pinch (1982b). The situation has changed in the wake of laboratory studies in the sense that there is now a new interest in all matters of relevance in real-time processes of experimentation (see, e.g., the studies collected in Gooding et al., 1989). See also the historical study by Shapin and Schaffer (1985) and section 6 of this chapter.

8. The laboratory studies that argued these points most forcefully are by Latour and Woolgar (1979), Knorr (1977), Knorr Cetina (1981), Zenzen and Restivo (1982), and Lynch (1985a). For an illustration of the political nature of science, see also Shapin (1979) and Wade (1981). For a more anthropological study of scientific laboratories, see Traweek (1988).

9. Hacking (1983) draws a distinction between experiments that "intervene" and scientific theories that "represent." This distinction, however, does not give adequate weight to the instrumental usage of theories in experimentation or to the fact that some experiments, as we shall see later, focus upon representation rather than intervention.

10. It may be enough when the content of the work is limited to digitalizable routine manipulations of the kind described by H. Collins (1990/1992).

11. Merleau-Ponty's (1945/1962) original notion in the French version of his book is "le system 'Moi-Autrui-les choses' " (p. 69). For the English translation and the exposition of this concept, see Merleau-Ponty (1945/1962, chap. 5 and p. 57). For a different implementation of the notion of a phenomenal field in recent ethnomethodological work, see the articles by Garfinkel, Bjelic, and Lynch in Watson and Seiler (1992).

12. This point needs to be distinguished from the ordinary language of scientists, which often preserves a distinction between the more or less "natural"—for example, between the "wild type" mouse that represents the "natural" mouse—and other types. However, the "wild type" mouse is as much a product of special breeding laboratories that supply mice to scientists as other laboratory animals; not only could it not survive in the wild, but the lab could not use real mice captured in the woods because of the "nonstandardized" biological variations and diseases these animals carry. Similarly, in vitro and in vivo are distinctions *of* the laboratory that refer to differences in degrees of intervention or to manipulations performed on plants and animals versus manipulations of test tube substances, and so on. In no case we encountered can "in vivo" be seen as an enclave of nature preserved in the lab. However, my point here is not that such enclaves cannot exist in laboratories or, phrased differently, that everything must be transformed at all times when it enters a lab. It is rather that the notion of a laboratory can be linked to the pursuit and dispute of central transformations and that the efficacy of the laboratory in modern science can be clarified if we consider these transformations. Note also that this notion does not furnish the laboratory with a strict boundary or require that the laboratory is a physical place.

13. For illustrations of such reconfigurations in experimental high-energy physics and molecular biology, see Knorr Cetina (in press, chaps. 3-5); for the production of reconfigurations in surgery, see Hirschauer (1991); for medicine in general, see Lachmund (1994).

14. Commentators who stand for many others are Giere (1988) and Cole (1991). For a general review, see Zuckerman (1988b).

15. Berger and Luckmann (1967) referred to processes of institutionalization, typification through language, and legitimation to account for the experience of social institutions as solid and "natural."

16. Whether the project of science is globally successful in the long run is at best contentious today.

17. Within science studies, deconstructionism informs perspectives such as "reflexivity" (Woolgar & Ashmore, 1988) and discourse analysis (Fuhrman & Oehler, 1986; Gilbert & Mulkay, 1984; Mulkay, Potter, & Yearley, 1983). See the respective chapter on discourse analysis in this volume.

18. Some students of laboratories, in particular Latour, retained a strong interest in actors, an interest that is codified in the "actor-network" approach (e.g., Latour, 1987).

19. This experience of empirical underdetermination is to be distinguished from logical underdetermination (e.g., Grünbaum, 1960).

20. This would seem to be more so in fields like experimental high-energy physics that depend heavily on external supplies of parts for the building of their detectors. The way a detector is built is influenced by what parts become available at what cost. The detector, on the other hand, determines together with other elements in the experiment (e.g., the energy and luminosity supplied by a collider) what "physics" will be seen as a result of the experiment.

21. Again, this is more relevant to areas of "big" science such as space research or experimental high-energy physics, where the equipment depends on the amount of financing negotiated with Congress and similar government agencies, and where these negotiations proceed without intermediaries such as funding agencies.

22. The term *actant* is borrowed from semiotics, where it includes nonhuman actors, only requiring that these "participate in a process" (see Greimas & Courtés, 1979, p. 3).

23. For example, the work people would have to perform if there were no doors, according to Latour, is the work of breaking a hole in a wall and then rebuilding the wall (Johnson [Latour], 1988).

24. I consider "epistemic" arguments those that, in the eyes of participants, promote the "truth" of the results.

25. Inscriptions, of course, are not really immutable. As with any statement or instruction, what matters is the meaning of these entities and how they are handled in practice. Because meanings change and are reconstructed between contexts, it is only the physical appearance of inscriptions that remains stable.

26. For a summary of these effects of local practices, see Knorr Cetina (1987).

27. In fact, from a Wittgensteinian perspective, it is hardly surprising that these specifications are needed if technical effects are to be made to work at all. I am referring here to Wittgenstein's treatment of rules and other instructions as receiving their meaning from a constant flow of practice and from the specifications embedded in this practice (1976).

28. The actual picture Callon (1991, p. 143) presents is more complicated because, to avoid idealism, definitions are seen to be inscribed in "intermediaries," which may be material objects.

29. Consider, for example, the various attempts to explain the cause of a historical person's death, which may reflect new potential causes of diseases "discovered" by science and then projected onto the past.

30. Models of consensus formation in science have been somewhat of a deferred agenda of many controversy studies—deferred because the first wave of these studies focused upon demonstrating how consensus in science cannot be explained, that is, by simply seeing it as an automatic consequence of the persuasiveness of experimental evidence or of the rationality of scientists (e.g., H. Collins, 1981a).

31. One paper that distinguished several types of consensus formation was published in German (Knorr Cetina & Amann, 1992).

32. As an example for this trend, consider the 1992 joint meetings of the BSHS, HSS, and CSHS societies in Toronto (July 26-28), which for the first time addressed the topic of laboratories from a historical perspective.

33. See also studies that begin to look at "laboratories" in fields like sexual research (Hirschauer, 1992) or astronomy (Gauthier, 1991), in which the notion of an intramural laboratory has to be replaced by an arrangement of participating localities. See also the applicability of the present notion of a laboratory to the development of the clinic in medical fields (Lachmund, 1994).

8

Engineering Studies

GARY LEE DOWNEY
JUAN C. LUCENA

PERHAPS the most significant contribution of research in engineering studies is that it provides case studies of life on the constructed social boundaries between science and society and between labor and capital. Engineering knowledge, for example, appears to be neither purely scientific nor only social but somehow a combination of the two. As the term *applied science* has long suggested, albeit misleadingly, knowledge-producing activities in engineering appear to occupy a double location both inside and outside of science. At the same time, engineering work appears somehow to be a combination of the activities of labor and the activities of capital. In positions that vary from country to country, engineers regularly find themselves grappling with ambiguities engendered by their double location as both objects and representatives of corporate power. The ambiguities in engineering knowledge and engineering work thus not only raise interesting conceptual problems about boundaries but also generate difficult power issues for engineers as persons.

Most of the differences within engineering studies derive from contrasting interpretations of these boundary locations. In this review, we take characterizing the different interpretations as our primary task, with the goal of bringing them into more productive dialogue. Our central argument is that, by throwing into question the conceptual status of the science/society and labor/capital boundaries while exploring the implications of those boundaries for engineers, engineering studies provide fertile ground for exploring and documenting diversities both within and between science and capitalism.

We bound our work in several ways. First, we focus on studies that interpret how engineers understand and do engineering. That is, rather than beginning with a definition of engineering knowledge or engineering practice and then looking for research that falls within that definition, we approach engineering as a cultural activity done by engineers and look for research that offers interpretations of engineers doing engineering. We thus position our field of vision around practicing engineers and limit our attention to work dealing with roughly the last 200 years. We do not inquire into the practices of Renaissance artisans or into post-Galilean, pre-industrial revolution negotiations between science and technology.

Second, despite declining to offer essentialist definitions of engineering, we distinguish engineering studies from technology studies, partly because this volume includes chapters on technology studies and partly to illustrate the conceptual point that demarcationist strategies are contingent acts of positioning. Just as we believe that extrapolating science forward into applied science does not capture sufficiently what engineers understand engineering to be about, neither does reasoning backward from technology into technological knowledge and design. Our contingent purpose in distinguishing engineering studies from technology studies is to enhance the comparative dialogue among studies of engineering.

Third, we confine our interests to work published since 1980, with a few exceptions. One of us has coauthored an essay that examines pre-1980 work in some detail (Downey, Donovan, & Elliott, 1989). As that essay argues, until roughly the last decade, the boundary location of engineering provided sufficient reasons for most scholarship in STS disciplines to ignore engineering, or at least to view it with yawns assertive enough to scare off untenured professors and most anyone else who did not revel in intellectual marginality and conceptual insignificance. Many scholars attempted to establish subdisciplinary traditions of inquiry into engineering, but these were generally unsuccessful. By repeatedly encountering conceptual problems that fell outside of disciplinary categories of analysis, such work typically found itself marginalized from disciplinary activities.

Fourth, we focus on empirical studies of engineering. In our judgment, further conceptual advancement in engineering studies depends critically upon continued, indeed accelerated, growth in the corpus of empirical work. So many surfaces have only been scratched.

The review is divided into four sections: engineering knowledge, engineering as technical work, gender studies, and studies by engineers. With these distinctions, we hope to help those of you whose interests are confined to certain areas, although take note that many studies cut across these boundaries. At the same time, these distinctions highlight what we think should be a central goal in the next generation of engineering studies: accounting for the interrelations among knowledge and power in engineering.

ENGINEERING KNOWLEDGE

Virtually all inquiries into engineering find some conceptual position for the content of engineering knowledge. This is no easy task, for the traditional model that devalues engineering knowledge as simply applied science (see Boorstin, 1978; Dorf, 1974; Lawless, 1977) is always present staring one in the face, challenging detractors to come up with equally clean alternatives. Much research explicitly focused on engineering knowledge has sought, in fact, to show that engineering does not simply involve the application of science to produce technology but has independent content. Researchers position themselves at various points around the question: How should one account for the distinction between engineering science and engineering design?

A 1955 committee of the American Society for Engineering Education classified the engineering sciences into six categories: (a) mechanics of solids, including statics, dynamics, and strength of materials; (b) fluid mechanics; (c) thermodynamics; (d) rate mechanisms, including heat, mass, and momentum transfer; (e) electrical theory, including fields, circuits, and electronics; and (f) nature and property of materials. This ASEE classification had a political dimension: It helped to legitimize educating engineers in each of these sciences beyond the "basic" sciences of chemistry, physics, and mathematics. In parallel fashion, most studies of the engineering sciences have distinguished them from the natural, physical, and biological sciences, claiming engineering science to be a distinct species of science. This "demarcationist" approach has served the important political goal of legitimizing direct scholarly attention to engineering.

The bulk of demarcationist work has been produced by historians of technology, for whom specifying in detail the character of technological knowledge has also helped to legitimize the history of technology as an independent activity alongside the history of science. The guiding spirit in this movement has been Edwin Layton. Ever since his 1971 article, "Mirror Image Twins: The Communities of Science and Technology in 19th-Century America," Layton has systematically defended what, following Barnes (1982a), he calls the "interactive model" of science and technology (Layton, 1971, 1974, 1976, 1984, 1987, 1988a, 1988b).

As a mirror-image twin of science, technology is "an autonomous, coequal community" whose relationship with science is "symbiotic, egalitarian, and interactive" (Layton, 1987, p. 598). Engineers are interesting in this model as the quintessential holders of technological knowledge, represented by the engineering sciences. According to Layton, the engineering sciences come in "two somewhat different types": (a) the "less idealized natural sciences," such as fluid mechanics, which transformed the empirical science of hydraulics by incorporating the study of viscosity, and (b) sciences that seek "to gain a scientific understanding of the behavior of man-made devices," such as portions of thermodynamics that advance idealized models of heat engines (Layton, 1984, p. 10; see also Böhme et al., 1978, p. 240).

While Layton has been the most persistent, the most thoroughgoing attempt at detailing the epistemological content of engineering science is found in the work of Walter Vincenti (1979, 1982, 1984, 1986, 1988, 1990). Viewing theoretical knowledge as a component of engineers' design knowledge, Vincenti (1990, pp. 207-224) adds to Layton's distinction on the one side purely mathematical methods and theories, ranging from analytical geometry to computer algorithms, and on the other side a cluster of device-related theoretical tools, including device-specific approximations (beam theory, modeling of transistors), phenomenological theories (modeling of turbulent flow), and quantitative assumptions (how rivets share loads).

Some studies have explored the genesis and early development of the engineering sciences in the nineteenth century. David Channell (1989) offers a comprehensive bibliography of core works and historical materials on applied mechanics, thermodynamics and heat transfer, and fluid mechanics. Channell (1982, 1984, 1986, 1988) also describes how Rankine consciously positioned engineering science between theory and practice during the mid-nineteenth century "in such a way that it would not threaten scientists, and . . . would avoid any competition with the offices of civil engineers, or the workshops of the mechanical engineers, or any interference with the usual practice of pupilage or apprenticeship" (Channell, 1982, p. 45). In his first call for investigations of technological testing, Edward Constant (1983) shows how the development and use of dynamometers to test water turbines contributed to the development of engineering science by relating scientific theory to waterpower practice. Eda Kranakis (1982) argues that grappling with conceptual anomalies in the injector, a nineteenth-century device for supplying water to steam engines, provided key contributions to the development of engineering thermodynamics. Ronald Kline (1987) explores how electrical engineers transformed Maxwell's electromagnetic theory in order to construct an engineering theory of an electromechanical device, the induction motor. Finally, Rosenberg and Vincenti (1978) outline the development of the method of parameter variation; Vincenti (1982), the development of control volume theory; and Layton (1988b), the development of the use of dimensionless parameters.

Other studies have explored interactions during the twentieth century between engineering theory and either science on the one side or design considerations on the other. Terry Reynolds (1986) describes the development of chemical engineering by exploring the efforts of production chemists to distinguish their knowledge from analytical chemistry. Bruce Seely (1984, 1988) examines the failed attempt by highway engineers to reconstruct highway research entirely in scientific terms during the period of proscience euphoria following World War I. By tracing the career of Irving Langmuir at General Electric Company, Leonard Reich (1983) details the boundary problems of industrial researchers, who find themselves to be neither scientists nor engineers yet both at the same time. In a reversal of the normal flow of theory from science to engineering, Joan Bromberg (1986) describes how, by applying circuit analysis to understand lasers, electrical engineers both provided conceptual help to physicists and generated boundary problems in their careers.

Several "internalist" histories have appeared that detail the evolution of engineering theories. For example, Stephen Timoshenko (1953) examines theories of strength of materials; Issac Todhunter (1960), theories of elasticity in relation to strength of materials; T. M. Charlton (1982), the theory of structures; and Donald S. L. Cardwell (1971), theories in thermodynamics. For a multitude of other works that conduct more limited forays into engineering science as a subset of technological knowledge, we also direct your attention to reviews by Staudenmaier (1985), G. Wise (1985), and A. Keller (1984), the symposium collection in Sladovich (1991), and to articles and books reviewed in the journal *Technology and Culture*. Finally, see Reynolds (1991) for an excellent comprehensive collection on the history of engineering practice and institutions.

Another source of demarcationist accounts is the philosophy of technology, which has sought to discern cognitive structuring in engineering theory at least since the 1966 debate among Mario Bunge, Joseph Agassi, and Henry Skolimowski over whether technology should be understood as applied science. In that debate, Bunge (1966) drew a sharp line between pure and applied science; Agassi (1966), between applied science and technology; and Skolimowski (1966), between science and social practice. As starting points for pursuing the study of engineering science further into the philosophy of technology, we suggest Michael Fores's (1988) critique of the concept of engineering science with responses by Layton (1988a) and Channel (1988), Joe Pitt's (n.d.) forthcoming theory of technology, the review in Downey et al. (1989), and annual issues of *Research in the Philosophy of Technology*. Finally, one philosopher of science, Ronald Laymon, has sought to examine the conceptual structure of engineering science. In an interesting account, Laymon (1989, pp. 353-355) examines engineers' strategies for using "idealization" to "manage [the] unavoidable complexity" that comes from applying scientific theories to design problems.

Research on engineering design has been more diverse, divisible into roughly three camps: demarcationist, constructivist, and actor-network interpretations. Each camp offers distinctive approaches to understanding the role of visual representations in engineering knowledge and practice.

A leader again among the demarcationists, Vincenti (1990, pp. 207-225) identifies five more categories of design knowledge beyond the theoretical tools described above. These include (a) fundamental design concepts defined in engineering terms, such as the central operational principles of static structures (bridges) or dynamics machines (aircraft); (b) criteria and specifications, such as those embodied in engineering standards; (c) quantitative data, including physical constants, properties of materials, and safety factors; (d) practical considerations, such as knowledge gained from accidents; and (e) design instrumentalities, including structured procedures and optimization strategies. In aesthetic approaches to demarcationism, David Billington (1979, 1983) argues that engineering structures have distinctive aesthetic characteristics, and Baynes and Pugh (1981) elaborate what they call the art of engineers. Finally, E. Ferguson (1992) critiques the twentieth-century replacement of visual knowledge in engineering with analytical knowledge.

Constructivist studies describe the design methodologies of engineers as shaped in various ways by social considerations. For example, Louis Kemp (1986) documents the aesthetic contributions of city planners to the design methodologies of highway engineers. Larry Owens (1986) explores Vannevar Bush's early work on the differential analyzer and how "[it] embodied an engineering culture belonging to the first decades of our century" (p. 95). L. L. Bucciarelli (1988) describes the engineering design process as an endless series of iterative loops as engineers respond to ever-changing problems and situations. Frederick Lighthall (1991) accounts for the failure of Morton Thiokol engineers to predict O-ring failure on the *Challenger* to insufficient training in statistical methods. Diana Forsythe (1993a, 1993b) examines the cultural construction of "knowledge engineering," the practice of transferring knowledge from experts to expert systems. Gary Downey (1992a, 1992b, 1992c, in press-a) shows how developers of CAD/CAM technologies (computer-aided design and computer-aided manufacturing) endow the technologies with agency by "transcribing" the activities of design engineers into computer code and then transforming those activities by inserting the technologies back into them as participants.

Finally, actor-network studies (see elsewhere in this volume) describe the activities of engineering design as situated practices that have both technical and nontechnical content and that combine to build networks of conceptual and political power. John Law's term *heterogeneous engineering* (Law, 1986b, 1987a, 1987b) encapsulates this dual conceptual and political point, for it claims simultaneously that all engineering is the product of heterogeneous factors and considerations and that all design activities in technology are forms of engineering. Michel Callon (1980a, 1980b, 1986a) initiated this line of argument with regard to engineers by describing how French engineers designing an electric vehicle simultaneously constructed the entire infrastructure within which the vehicle would work. Law and Callon (1988, p. 284; see also Law, 1988) also argue that engineers are "engineer-sociologists" in that they are "not just people who sit in drawing offices and design machines" but are "social activists" who design societies or social institutions to fit the machines. These latter claims draw inspiration, in part, from Langdon Winner's (1986b, pp. 19-39) arguments that technical manuals or designs for nuclear power stations imply conclusions about the proper structure of society, the nature of social roles, and how these roles should be distributed.

In other actor-network interpretations of engineers, Bruno Latour (1987, p. 104) reinterprets Lynwood Bryant's (1976) account of the development of the diesel engine by tracing Rudolf Diesel's early successes and later failures in building links to Carnot's thermodynamics, investors, and potential users. Susan Leigh Star (1990) accounts for how engineers use CAD (computer-aided design) programs to design computer chips by chronicling the tensions between CAD representations, "which are static and abstract," and engineering work, "which is real-time and concrete" (p. 128). Finally, Kathryn Henderson (1991a, 1991b) describes the visual communication practices of engineers by focusing on how engineering drawings and computer-aided design programs function as "conscription devices" that socially organize both workers and the structure of work.

One likely growth area that involves both engineering science and engineering design are new approaches to the knowledge content of engineering education. Noble's (1977) classic argument describes both the form and the content of engineering education unilinearly as "a major channel of corporate power" by providing the "immediate manpower needs of industry and the long-range requirements of continued corporate development" (pp. 47, 170). Carlson's (1988) study of academic entrepreneurship at MIT argues, however, that corporations "could not simply order entry-level engineers from engineering schools," and linkages between engineering education and industry were "marked by a clash of values and expectations" (p. 396). Similarly, Downey (in press-b) finds a "theoretical space between the notions of engineering control and corporate control" to show how engineering curricula empower engineers by shifting them to the boundary between humans and machines. In forthcoming work from a 3-year study of how engineering education constructs engineers as persons, Downey, Hegg, and Lucena (in press) show how the knowledge content of engineering problem solving generates identity conflicts according to gender, race, and class characteristics.

ENGINEERING AS TECHNICAL WORK

Over the past 10 years, as Chris Smith (1991) points out in a review essay, "there appears to have been a veritable explosion of interest in technical labour" (p. 452). Inquiries into engineering as a form of technical work generally focus on the social mechanisms and processes that shape and position engineers as workers. Excellent reviews by Peter Whalley (1991) and by Peter Meiksins and Chris Smith (1991) identify many of the changing issues and trajectories of research. A significant shift in emphasis has taken place from a focus on engineers as professionals to analyses of the class character and implications of engineers working in industrial organizations. Whereas the first, as Whalley (1991, p. 193) succinctly puts it, has been "largely Anglo-American and Weberian in influence," while the second has been "European and Marxist," the most significant recent trend has been a multiplication of class-based accounts. One of the most promising outcomes is a strong commitment to cross-national comparison.

From the 1950s through the mid-1970s, researchers typically took for granted the engineer's status as an autonomous, or free, professional akin to professionals in law, theology, and medicine and looked for answers to the functionalist question: What happens to professionals who work in organizations? When the answer that eventually emerged was that engineers were a different type of professional—that is, organizational professionals—the effect was to undercut the functionalist study of professions (see Downey et al., 1989, for a detailed account of work in the United States). At that time, however, as Whalley (1991) elaborates, a new sociology of professions developed with the goal of accounting for their autonomy and high class position. This perspective drew on the Weberian argument that one's position in the class structure depends upon the marketable skills that one possesses. Accordingly, it viewed professions as occupations whose possession of scarce and important knowledge enabled them to build power for themselves, effect closure, and achieve autonomy and high class status.

Developing alongside the sociology of the professions, Marxian accounts have treated engineers entirely in class-based terms. Prior to the 1970s, most Marxian studies described engineers as populating that part of the working class that was white collar, received high salaries, and was close to management. One strain of thought that developed among French sociologists, most notably Serge Mallet (1975), held that technical workers were part of a "new working class" that had potential for radical action. However, rapid growth in the number of such technical workers cut against the accompanying "proletarianization thesis" (Braverman, 1974), which held that white-collar workers were being "deskilled" and would be forced to merge with a growing proletariate (Whalley, 1991, p. 196). The key conceptual problem for Marxian analysis became how to account for a growing middle class of workers.

The Weberian and the Marxian accounts of engineers have thus converged on the problem of accounting for the ambiguous, middle-class character of engineers as technical workers. A key contributor to the rise of interest in the class characteristics of engineers was a large comparative analysis of engineers in three industrialized countries, carried out by three graduates of Columbia University under the supervision of Allan Silver: Peter Whalley (1984, 1986, 1987, 1991) in Great Britain, Robert Zussman (1984, 1985) in the United States, and Stephen Crawford (1989, 1991) in France. The original objective was to assess the implications of an international shift to a postindustrial economy built on knowledge-based industry. However, comparing the work and statuses of engineers in low-tech industrial companies with those in high-tech, knowledge-based companies, each study actually defuses the question of postindustrialism by finding little difference between the two. Rather, by documenting the varying class experiences of engineers in different national contexts, the case studies highlight contrasts in national traditions and focus attention on cross-national comparison as a method for formulating and evaluating alternative accounts of the class characteristics of engineers.

Understanding the contrasting national experiences of engineers can also help one to understand contrasting national patterns of research on engineering as technical work. The bulk of research on the role of professionals in organizations has been conducted in the United States, where the "professional" has perhaps been institutionalized most strongly as an occupational classification (Whalley, 1991, p. 202). That is, in the structural tension between workers and management, engineers clearly fall on the side of management. The Wagner and Taft-Hartley acts made them exempt from collective bargaining laws, and, armed with university training and credentials, they serve as technical specialists within corporations. Also, engineers in the United States have organized themselves into professional associations, although it is significant that these do not exclude corporate interests from their definitions of professional interests.

For example, in his republished *Revolt of the Engineers*, Edwin Layton (1986) traces how corporate-minded conservative elites in the early twentieth century succeed in overcoming attempts by profession-minded elites to establish engineering as an autonomous profession. This work established professional societies as a privileged site for tracing the political orientations of American engineers and the development of engineering as a profession (e.g., Downey et al., 1989, pp. 207-208; McMahon, 1984; Reynolds, 1983; Sinclair, 1986). Histories of engineering education in the United States also concentrate on the professional development of American engineers (Bezilla, 1981; Gordon, 1982; McMath, 1985; Ochs, 1992; Wildes & Lindgren, 1985). Trajectories of research in management studies, especially studies of "organizational culture," tend to focus on tensions between the individual professional orientations of engineers and the organizational orientations of their employees (Bailyn, 1980, 1985; Bailyn & Lynch, 1983; Kunda, 1992; Raelin, 1986). An extensive philosophical literature in engineering ethics seeks to enumerate ethical principles to guide engineers, given their status as professionals in organizations (e.g., Baum, 1980; Baum & Flores, 1980; Davis, 1991; Downey et al., 1989, pp. 202-203; Flores, 1989; Johnson, 1989; Layton, 1985; Martin & Schinzinger, 1989; Schaub & Pavlovic, 1983; Unger, 1982, 1989). Finally, the role of the military in shaping American engineering and the contemporary implications of post-cold war conversion strategies deserve radically expanded attention (e.g., Markusen & Yudken, 1992).

Yet the rise of interest in class is evident. Presenting the Marxian view that labor and capital stand in structural opposition, David Noble (1977, 1979, 1984) details histories of American engineers as domesticated servants of capital while Donald Stabile (1984) uses the experiences of mechanical and industrial engineers during the early twentieth century to describe engineers as situated in a contradictory position between labor and capital. In a Weberian account, Robert Zussman (1984, 1985) argues that the occupational identities of engineers are better understood by examining their trajectories through "careers." That is, the collective product of engineers' careers and their family lives in single-family dwellings and in neighborhoods alongside workers is a "working middle class," a notion that directly challenges the Marxian opposition.

Finally, in an interesting body of work on American engineers, Peter Meiksins has been a thoroughgoing proponent of a shift from profession to class. In survey research with James Watson, Meiksins argues that engineers are concerned less with professional autonomy than with the technical content of their work, concluding that researchers "need to shift the focus of research away from issues such as professional autonomy toward the nature of engineering work itself" (Meiksins & Watson, 1989; Watson & Meiksins, 1991, p. 165; see also Meiksins, 1982). Reconsidering the revolt of the engineers through a detailed case study, Meiksins (1986, 1988) explains the rise and fall of the American Association of Engineers during the early 1920s as the product of an alliance between elite progressives and rank-and-file engineers that dissolved for class reasons (see Sinclair, 1986, for another call to study rank-and-file engineers). Finally, in recent collaboration with Chris Smith (Meiksins & Smith, 1991, in press; Smith & Meiksins, 1992), Meiksins details the class characteristics of American engineering in comparative perspective.

In Great Britain, engineers have positioned themselves as higher status craft workers rather than as managers with autonomy. British companies have given little attention to professional credentials; the state has played a small role in engineering education; and engineers opt for unions rather than professional organizations. Success in engineering work has been based more on the acquisition of technical skills than on science-based education, producing a system that is now challenged by the expansion of high-technology industry. Yet, despite their comparatively low status (for a history of the status of engineers in Great Britain, see Buchanan, 1983), engineers have remained concerned in varying ways and at varying times with the possibilities and ambiguities of professional status.

Much of the research on British engineers has been, and continues to be, engaged with the problematic of professionalism. For example, historian R. A. Buchanan explores the history of the British engineering profession, with work on early civil engineers who sought the status of gentlemen (1983), institutional developments (1985a), the development of scientific engineering (1985b), engineers' roles in the colonial empire (1986), engineers and government (1988), and, most recently, a comprehensive history (1989). In the same spirit, W. J. Reader (1987) and Judy Slinn (1989) offer histories of professional institutions. Sociologists Ian Glover and Michael Kelly (1987) trace how engineering has developed as an "occupation" in the context of the "professional ideal." As summarized by Chris Smith (1991), this project is part of a collective effort by "British managerialists" to revitalize the British economy by repositioning engineers at the center of the manufacturing enterprise. Attempting to learn from abroad by identifying "the best practice" in other countries, this movement seeks to reform British management along German lines, including empowering engineers (Child et al., 1983; Hutton & Lawrence, 1981; K. McCormick, 1988).

A parallel body of work explores the trials and tribulations of British technical education in the context of employer scorn and state disinterest. For example, numerous works attribute Britain's economic decline, in part, to its overemphasis on craft training for engineers (Ahlstrom, 1982; Albu, 1980; Barnett, 1986; Locke, 1984; Wiener, 1985). Some justify the emphasis on craft training (e.g., Robertson, 1981). In a pair of recent empirical studies, Colin Divall (1990, 1991) argues that corporate firms significantly influenced both the development of engineering education as an elite entry into professional engineering and the changing curricular balance between engineering science and design (see also K. McCormick, 1989).

In the British case, the work on class runs in parallel and is constituted by a conflict between Weberian and Marxian approaches. On the Weberian side, Peter Whalley has produced a sustained body of work whose constructivist orientation links it to recent social studies of science and technology and has made him a strong advocate of cross-national comparison. Initially arguing more generally for the need to consider the labor market positions of engineers as key contributors to and components of their class positions (Whalley, 1984, 1986; Whalley & Crawford, 1984), Whalley (1986) maintains in an extended treatise that British engineers fall into a more general cross-national class of "trusted workers." Continuing that occupations are "socially constructed achievements" (Whalley, 1987, p. 3), he finds cross-national comparison a necessary strategy for identifying the practical distinctions that each tradition devises to define the boundaries around and between trusted workers, such as between jobs and careers, works and staff, and exempt and nonexempt. Whalley's constructivist approach also argues for extending the analysis of technical workers into the political domain, because every boundary drawn must be seen not as a "technical necessity" but as a "political achievement" (Whalley, 1991, p. 210).

On the Marxian side, Chris Smith (1991, p. 457) critiques market-based accounts of class as "condition-based" descriptions, which could easily be examining surface labels or attributes, that miss the "economic conflict between staff and management." In an extended account of British technical labor drawing on fieldwork at British Aerospace, Smith (1987) maintains that national differences in "political expressions" of class identities does not mean that class has no "global voice" (see also Meiksins & Smith, 1991, in press).

Both France and Germany have developed highly stratified systems of technical workers that, by closely linking divisions among educational institutions to divisions among employers, repeatedly raise the question of class. At the same time, the very diversities in hierarchical organization that these national traditions have produced expand considerations of class formation well past labor market activity and labor/capital opposition. A significant difference between the two is that, while French engineers have consistently placed highest value on theoretical work that derives analyses from first principles, German engineers have modified the French model by integrating and valuing practical training and knowledge alongside engineering science.

Much research on French engineering examines the evolution and organization of institutions of higher education. Terry Shinn (1980a, 1980b, 1980c, 1984; Shinn & Paul, 1981-1982) analyzes in detail the hierarchical structure of the French engineering community, emphasizing the scientific practices of the highest level institutions, especially the Ecole Polytechnique. John Hubbel Weiss (1982) explores the class origins of nineteenth-century students at a second-level institution (Kranakis, 1989, pp. 9-10), the Ecole Centrale des Arts et Manufacturers. Charles Day (1978, 1987) directs his attention below both levels in exploring the development of the ecoles d'arts et metiers. Finally, in a comparative study of engineering practice in France and the United States, Eda Kranakis (1989) shows how the scientific orientation of the top strata shaped interactions among all three.

The occupational analogue of this educational hierarchy is the legally sanctioned hierarchy of cadres, a social category that emerged during the 1930s to distinguish grades of technical managers. In contrast with the United States and Britain, graduates of French engineering institutions generally expect to take their place within management, even valuing their training in abstract math and science primarily for managerial purposes. Luc Boltanski (1987) applies Bourdieu's theoretical perspective in accounting for the development of this new category, showing how the meanings that various classes attribute to it have varied over time, such that giving it any particular definition becomes "in itself a political act" (p. 181). Cecil Smith (1990, p. 659) "demonstrates the continuity" that public engineering and planning have maintained from the eighteenth century to the twentieth century as the highest prestige occupation for elite engineers. Finally, Stephen Crawford's (1989, 1991) accounts of contemporary engineers in low-tech and high-tech companies argue that the cadre system plays a more important role in shaping the class experiences of engineers than does the "logic of industrialism" (Crawford, 1991, p. 190).

In parallel fashion, research on German engineers traces connections between the establishment and growth of a network of educational institutions and the structure of German industry. Karl-Heinz Manegold (1978) describes how a German academic elite consolidated itself in the mid-nineteenth century, although the form of technical science that developed was not understood as applied science, as was the case in France. Also, in an extended history of German engineers, Kees Gispen (1988, 1990) traces the class-based tensions between academics and industrial employers.

Starting points for studies of engineering in other countries include Earl Kinmoth (1986), M. Kumazawa and Jomoko Yamada (1989), and Kevin McCormick (1992) on Japan; Cornelis Disco (1990) on the Netherlands; Rolf Torstendahl (1982a, 1982b, 1985) and Boel Berner (1992) on Sweden; Kendall Bailes (1978) on the Soviet Union; Anna Guagnini (1988) and Anna Barozzi and Vittoria Toschi (1989) on Italy; Knut Sorensen and Anne-Jorunn Berg (1987) and Carol Heimer (1984) on Norway; M. D. Kennedy (1987) on Poland; and Rodney Millard (1988) on Canada.

GENDER STUDIES

In every country where engineering has established significant representation in the workforce, the proportion of women engineers has been exceedingly small. Authors discussed in this section argue uniformly that the content of engineering education and practice conveys and reinforces masculine values, yet such is rarely mentioned in other contexts. Arguably, the established traditions of engineering studies outlined above have tended to reproduce the gender content of both engineers and engineering. In parallel with the shift from profession to class in studies of technical workers, gender studies over the past decade have undergone a shift from purely functionalist to power-based perspectives. Further developments in gender studies are critically important to the future of engineering studies because these force awareness of and attention to forms of stratification and hierarchy within engineering that both extend beyond the dimensions of class and have clear knowledge content.

Existing data about women in the "engineering pipelines" of the world show that the proportion of women engineers entering the workforce over the past 5 to 10 years has either held steady at a low value or actually declined (E. Jamison, 1985; National Science Board, 1989; U.S. Bureau of the Census, 1988; Way & Jamison, 1986). In the United States, demographic data indicate that the traditional pool from which future engineers are recruited (i.e., 18-year-old white males) is shrinking. As a result, recruiters are increasingly seeking women and minorities for engineering to fill national needs, but, regardless of these efforts, women still remain underrepresented in engineering, where only 1 in 25 engineers is a woman (Baum, 1990). When compared with science fields, engineering has the lowest representation of women (e.g., 3.1% in engineering and 4.7% in physics and astronomy) (National Science Foundation, 1991a). Questioning current recruitment strategies, Stephen Brush (1991) says that

> we should consider the possibility that the young women who "leak out of the science and engineering pipeline" are behaving more intelligently than those who want to recruit them but refuse to provide adequate incentives. . . . The pipeline metaphor in itself is a clue to the problem: It suggests a factory-management attitude that treats people as raw material to be made into products, without regard for their own wishes or well-being. (p. 416)

Prior to 1985, virtually all research on women in engineering adopted what Judith McIlwee and J. Gregg Robinson (1992, pp. 13-18) call the "gender role perspective." Related to the labor market approach to class outlined above, such studies explained the nonparticipation of women in engineering as a product of their socialization as women. In some of the earliest studies, Alice Rossi (1965, p. 1201, 1972) describes women as preferring fields in which they can work "with people rather than things." Also, Carolyn Perrucci (1970; Hass & Perrucci, 1984), one of the first scholars to investigate the experiences of women in engineering, emphasizes the significance of family responsibilities in women's lives and the roles of socioeconomic background and education in shaping occupational choices. Finally, Mildred Dresselhaus (1984) argues that women faculty members in engineering institutions are necessarily political actors who have "extracurricular" responsibilities to encourage women students to form networks with women colleagues and to influence developments in national policies.

A correlate of this interest in the socialization of women engineers is that survey research has tended to focus on education rather than employment (e.g., Ott & Reese, 1975). Numerous studies show that women tend to enter engineering programs with higher grades and test scores than men (Gardner, 1976; Greenfield, Holloway, & Remus, 1982; Jagacinski & LeBold, 1981; Ott, 1978a, 1978b). Surveys also indicate that women are more likely than men to enter engineering if some family members, especially their fathers, were engineers (Auster, 1981; McIlwee & Robinson, 1992). To the extent that such work studies practicing women engineers, the focus has been on the significant salary gap between men and women (McAfee, 1974; Rossi, 1972; Vetter, 1981). See, however, Carolyn Jagacinski's (1987a, 1987b) recent work showing that women engineers are also less likely to be married and to have children than men.

Research on the history of women engineers has emphasized the heroic characteristics of such women. E. Rubenstein (1973) characterizes her early overview of women electrical engineers as "profiles in persistence." Martha Trescott (1979b, 1982, 1984) also emphasizes individual persistence in reporting the results of a survey of 500 practicing women engineers and tape-recorded interviews with nearly 50 older women engineers "who have made significant contributions to the theory, design, management, or education, or to the history of the Society for Women Engineers" (Trescott, 1984, p. 181). Examining in detail the life of Lillian M. Gilbreth, the mother in *Cheaper by the Dozen* and perhaps "the most well-known woman engineer in history," Trescott (1984, p. 192) also argues that women engineers tend to adopt a "holistic" approach to engineering problem solving.

The major impetus for integrating power-based perspectives into studies of gender in engineering has come in the work of Sally Hacker. Dorothy Smith and Susan Turner edited a collection of her papers on gender and technology, introducing each chapter with transcriptions from interviews that Smith did with Hacker prior to her death in 1988. This editorial approach systematically displays Hacker's central goal of constructing a "people's sociology" (Hacker, 1990, p. 2) that uses available theories and methods as tools to empower dominated or exploited people by critically examining how macrosocial processes affect them, including the organization and policies of government, corporations, and universities. Following a research method she calls "doing it the hard way," which means getting in and "being with people" to "know what it feels like," Hacker (1990, pp. 105-110, 157) draws on a year of fieldwork at MIT followed by engineering course work at Oregon State University to outline the "masculinist bases" of "engineering culture."

Hacker's (1990) central claim about engineering is that the "culture of engineering" in universities and in the workplace constitutes and reproduces "patriarchal" systems, that is, "sex-stratified systems in which men are dominant" (p. 50). For example, she explains and criticizes the emphasis that engineering education places on testing in mathematics as "embedded in a very masculine-shaped professional organization of knowledge and evaluation" (p. 109). She uses jokes told by engineering professors to argue that "persistent mind-body dualism" (p. 123) structures hierarchies in universities and the workplace to the detriment of women. Furthermore, she maintains that engineering training channels the passions of engineers, particularly erotic energies, into their experiences of technology (pp. 210-212; see also Hacker, 1989), joining this emotional experience with the emphasis on mathematics to produce engineering managers with a " 'blind spot' about social structure" (p. 127). Interpreting capitalist domination as a form of "patriarchy," Hacker (1990, pp. 177-179) adopts the deskilling hypothesis to argue that, like other workers, engineers are being disempowered by automation technologies, especially by computer-aided design and computer-aided manufacturing.

In a sensitive and systematic challenge to Hacker's analysis of patriarchy in engineering, Judith McIlwee and J. Gregg Robinson weave quantitative data from a sample of 407 working engineers with qualitative data from 82 in-depth interviews to assess the experiences of women in engineering from the precollege years to the workplace and beyond. Framing their arguments in a "conflict-structural" perspective, they link considerations of gender socialization to an understanding of universities, corporations, and families as structures of power relations. McIlwee and Robinson see Hacker's emphasis on abstract mathematics as but one variant of engineering culture that is found in the university that misses the cultural emphases engineers place on tinkering with technology, organizational power, and male presentational style in the workplace (McIlwee & Robinson, 1992, pp. 26-33). This leads to the interesting conclusion that women engineers tend to fare worse in organizations in which engineers are more powerful, such as computer and other high-tech firms, and better in large industrial bureaucracies, such as aerospace firms, within which affirmative action policies have become established practice. Along the way, McIlwee and Robinson offer many insights into how women "drift" into engineering careers, how the "interactional structure" of college presents barriers that women overcome with strong performances in math and theory, and how women encounter new sets of barriers in the workplace built around the valuation of tinkering and the appropriate presentation of self.

STUDIES BY ENGINEERS

By far the greatest volume of literature interpreting the experiences of engineers is produced, of course, by engineers. Such writings can be of use to other researchers in engineering studies in at least three ways. First, engineering committees and organizations regularly collect and publish massive amounts of quantitative data about engineers. Second, the writings of engineers typically defend engineers' points of view, thus providing a useful source of data about how engineers understand what they do. Third, numerous engineers have made the effort both to address nonengineering audiences and to get beyond self-justification to offer interpretive accounts that take their place alongside those of other analysts. Such accounts tend to focus on engineering ethics, the appropriate content and duration of education, and relations between engineering science and engineering practice.

Committee and organizational studies by engineers are generally organized along national lines. As a starting point, we recommend you identify by searching references at least one such body in the country you wish to study and use it to locate others. UNESCO publishes many useful international studies, such as reviews of continuing engineering education (Ovensen, 1980), engineering manpower (Van den Berghe, 1986), the environment in engineering education (Brancher, 1980), and engineering and endogenous technology (UNESCO, 1988).

In the United States, which we have reviewed in some detail, the rise of nationalist concern during the 1980s about "economic competitiveness" has elevated engineering to the status of a national problem. Numerous studies document American problems with productivity and point to developments in engineering education and engineering-oriented developments in process technologies as the pathway to solutions.

The National Academy of Engineering (NAE), in cooperation with the National Research Council (NRC), and the National Academy of Sciences (NAS) has taken the lead. A key starting point is a comprehensive NRC assessment, "Engineering Education and Practice in the United States," published in nine slim volumes on such topics as engineering technology education (NRC, 1985a), engineering graduate education and research (NRC, 1985b), engineering in society (NRC, 1985c), support organizations for the engineering community (NRC, 1985d), engineering employment characteristics (NRC, 1985e), continuing education of engineers (NRC, 1985f), engineering education and practice (NRC, 1985g), engineering undergraduate education (NRC, 1986a), and engineering infrastructure diagramming and modeling (NRC, 1986b). The NAS has published reports on the underrepresentation and career differentials of women and minorities in science and engineering (Dix, 1987a, 1987b) and engineering personnel needs for the 1990s (NAS, 1988). The NAE reports have dealt with education and employment of engineers (NAE, 1989) and engineering and competitiveness (NAE, 1983, 1985, 1986, 1987a, 1987b). The National Science Foundation also publishes a biennial report on the status of women and minorities in science and engineering (NSF, 1990).

Other useful sources include professional engineering societies, universities, and ad hoc organizations. For example, *Manpower Comments*, a monthly bulletin published by the Commission of Professionals in Science and Technology, provides information on supply and demand, salaries, representation of women and minorities, and education in science and engineering.

For writings by engineers that present and justify engineers' points of view, the best source of data is paging through the "comments" and "opinions" sections of professional journals and magazines as well as the semipopular technical publications, such as *Technology Review* and *Prism* (formerly *Engineering Education*).

Among those engineers who have presented interpretations of engineers in the context of other such interpretations, Samuel Florman has produced the most extended body of work. A civil engineer, co-owner of a construction firm in Manhattan, and holder of a master's degree in English literature from Columbia University, Florman has published several books (1968, 1976, 1981, 1987) and writes a monthly column for *Technology Review*. Although sometimes interpreted by critics as a literary apologist for engineers, Florman is diligent in formulating accounts on a wide variety of topics, especially liberal education for engineers, that directly engage and frequently challenge alternative accounts of engineers.

Henry Petroski (1985) makes an important contribution to demystifying engineering design by linking features of everyday life to the practices of engineers. By describing the multiple meanings that the term *failure* has for engineers, Petroski builds a framework for evaluating future changes in design practices, such as increasing reliance upon computer software. Peter Booker (1979) offers a detailed and insightful history of the genesis of engineering drawing concepts. His historical overview actually provides a thorough survey of contemporary concepts for it is a series of episodes whose sequence documents the sequential appearance of these concepts. Richard

Meehan (1981) presents an autobiographical account that systematically identifies links between nonengineering society and the content of engineers' knowledge, arguing, for example, that the "big questions" that ground and that threaten the engineers' authority "are the freshman questions, not the graduate school questions" (p. 43). Arthur Squires (1986) draws on his own experiences to document the inadequacies of "governmental management of technological change" in the United States and calls for managers to identify and promote "maestros of technology" from among the ranks of engineers and other technological "apprentices." Finally, Dan Pletta (1984) offers a call for an independent sense of "professionalism" in engineering.

For other work on engineering education, see Grayson's (1977) brief history of engineering education, Wakeland's (1990) proposal for internationalizing engineering education, and Abbott's (1990) account of the effect of American culture on engineering education. For work on popular attitudes toward engineers, see Vaughn (1990), A. M. Weinberg (1989-1990), Critchley (1988), and K. Strauss (1988).

KNOWLEDGE AND POWER IN ENGINEERING

Although research agendas in each of these four areas—engineering knowledge, engineering as technical work, gender studies, and studies by engineers—is reasonably well developed, the work in each area could benefit from enhanced critical dialogues. We believe that much could be gained if each researcher asked seriously: How might insights generated in other areas contribute to my accounts, and how might insights in my work contribute to accounts in other areas? In fact, judging from the array of studies we have reviewed, an implicit dialogue has already been emerging over at least one conceptual question that promises to position a great deal of work in the next generation of engineering studies: How do knowledge and power operate and interrelate in the activities of engineers? This conceptual question pushes each of the areas into new directions.

Much of the research on knowledge in engineering has been demarcationist in tenor and objective. We suggest that demarcationist approaches shift their orientation slightly, but significantly, from documenting the distinctiveness of engineering knowledge to investigating demarcationism as an engineering practice. The published work clearly establishes the empirical worthiness of exploring the contents of engineering knowledge; such is breaking entirely new ground. However, as a conceptual strategy, demarcationism has not held up well to critical examination elsewhere. Philosophers of science, for example, have found no characteristics that appear to be essential to knowledge in the natural and physical sciences. Accordingly, we believe that essentialist claims about the engineering sciences are likely to be of lesser lasting significance than the empirical value of documenting in detail the conceptual contents of engineering knowledge. By shifting strategies from defending logical relations within engineering knowledge to exploring the strategies through which engineers construct, maintain, and transform these relations, studies of engineering science could better account for the precise features of engineering theories and link with the constructivist and actor-network accounts of engineering design.

Such reorientation generates new research questions. For example, David Channell (1982) reports that Rankine purposely conceptualized engineering science in a way that did not look like applied science or challenge the apprenticeship system. This interesting insight could be elaborated in an account of how Rankine's strategies for demarcating engineering science incorporated his social strategies for legitimizing such activity in the scientific and engineering communities and established him as an authoritative figure. In other words, examining the conceptual strategies of designers as part of a broader set of social strategies also links accounts of engineering epistemology to biography, social history, and social studies. Some of the empirical questions that emerge include these: How do engineers develop, maintain, and assess boundaries among engineering disciplines (see Donovan, 1986)? How has the evolving legitimacy of engineering science been linked to the evolving legitimacy of engineering education? How do engineers in both universities and industry vary in valuing the distinction between engineering science and engineering design?

Almost all studies of engineering as technical work, as Peter Whalley (1991, p. 197) perceptively points out, view engineering knowledge as a defining feature of the professional or class status of engineers, but none of this work examines the knowledge content of engineering in a sustained way. If the class-based powers of engineers are linked to the knowledge contents of their engineering, then there is merit in shifting from taking for granted the knowledge-based authorities of engineers and exploring how knowledge strategies are related to class strategies. Empirical questions that emerge include these: How are status differences between design and manufacturing engineers linked to the contents of their knowledge? How are disciplinary differences related to the different class experiences of engineers, as indicated by the historical shift of intellectual hegemony from civil to mechanical to chemical to electrical engineering? What makes engineering work satisfying or not? What are the class implications of engineers doing satisfying or unsatisfying work? Following the call of James Watson and Peter Meiksins (1991), we support "shift[ing] the focus of research away from issues such as professional autonomy toward the nature of engineering work itself" (p. 165).

Perhaps the greatest contribution of research on gender in engineering studies is that it raises issues that clearly have both knowledge and power content, although the emphasis has been more on power than knowledge. While Sally Hacker points to the "masculinist bases" of testing in mathematics, for example, her work does not extend to consider the contents of the mathematical knowledge demanded of engineers, as is outlined in the research on engineering science and engineering design. Further questions that arise include these: What knowledge and power considerations shape the participation of women in industrial or mechanical engineering and their avoidance of electrical engineering? Despite the evident capabilities of women engineering students in mathematics, are those women who leave engineering expressing an unwillingness to accept professional identities built on "engineering problem solving"? What do the differing positions of men and women engineers in industry tell us about hierarchies of valued knowledge in the workplace?

Gender studies also force attention to two additional subjects of inquiry not engaged in the other areas: forms of hierarchy and stratification not based entirely in class, such as race, age, and ethnicity, and full accounting for the emotional dimensions of engineers' experiences (see also Sinclair, 1986). Research questions proliferate. Since 1983, more than 50% of the Ph.D.s awarded in engineering in the United States have been awarded to non-American-born students. The engineering faculties of universities in many countries have majority memberships of noncitizens. As multinational corporations gain in cross-national power, the ethnic makeup of their engineering staffs is changing significantly. What are the implications of increasing racial, ethnic, and cultural diversity in various arenas of engineering activity? Regarding the emotional status of engineers, how is it that engineers routinely feel powerless themselves but are viewed as highly empowered by outsiders? How is the "boring nerd" image of engineers constructed and maintained? Are engineers' pleasures in tinkering with mechanical technologies being replaced by computer hacking? What might be the gender implications of such a shift?

The engineers who represent the activities and practices of engineers to audiences of nonengineers could make a unique contribution to exploring questions of knowledge and power. Rather than presenting engineering communities as integrated wholes to explain engineering to outsiders, such work could provide insight into power relations within such communities. For example, how do new theories and methods, such as the rise of interest in numerical methods, gain acceptance among engineering researchers? What are the diverse considerations that shape the development of engineering curricula and the practices of instruction? What are the steps through which engineering students gain membership in engineering communities?

How do technical standards function in everyday engineering practices? What are the roles and implications of engineers participating in setting such standards?

Finally, we believe that the rise of cross-national comparative research on engineering introduces the potential for many new insights into the immense diversity of knowledge/power relations in engineering. Cross-national comparisons of engineers illustrate not only the commonality of capital-labor tensions in capitalist industrial systems (e.g., Smith, 1991) but also the contrasts among engineers' experiences within and between different systems. Given that we are all writing during a historical period in which the political economy of the world no longer appears to consist of a dualistic struggle between monolithic capitalist and communist systems, it seems likely that diversities among and within capitalist systems will become the object of greater scholarly and political attention. Both the nature of class conflicts and the very concept of class are nationally, even locally, variable. Research on the variable locations and experiences of engineers in both knowledge and power terms positions engineering studies at the center of such broader concerns. Overall, establishing enhanced dialogues within engineering studies about the interrelations of knowledge and power in engineering promises to take us a long way toward understanding the nature and organization of the diverse interrelations among science, technology, and capitalism.

9

Feminist Theories of Technology

JUDY WAJCMAN

OVER the last 15 years, an exciting new field of study has emerged, concerned to develop a feminist perspective on technology. The development of this perspective is more recent and consequently less theoretically developed than that which has been articulated in relation to science (see Keller, Chapter 4 in this volume). To date, however, most contributions to the debate on gender and technology have been of a somewhat specialist character, focused on a particular type of technology. Thus the area is characterized by many edited collections such as Martha Moore Trescott (1979a), Joan Rothschild (1983), Jan Zimmerman (1983), Wendy Faulkner and Erik Arnold (1985), Maureen McNeil (1987), Chris Kramarae (1988), and Gill Kirkup and Laurie Keller (1992), which do not necessarily share a theoretical approach.

While some feminists have been primarily concerned with women's limited access to scientific and technical institutions (see Fox, Chapter 10), others have begun to explore the gendered character of technology itself. This latter, more radical, approach has broadly taken two directions. There are those feminists who argue that Western technology itself embodies patriarchal values and that its project is the domination and control of women and nature (Corea et al., 1985; Griffin, 1978; Merchant, 1980; Mies, 1987). This approach finds political expression in the cultural feminism and eco-feminism of the 1980s, which calls for a new feminist technology based on women's values. Taking a different tack, there are a group of writers who adopt the methods of the social studies of technology, which is more historical and sociological in orientation (Cockburn, 1983, 1985; Cowan, 1976, 1979, 1983; Faulkner & Arnold, 1985; Hartmann, Kraut, & Tilly, 1986-1987; McGaw, 1982; McNeil, 1987). Much of their work has been concerned with the gender division of labor in both paid and unpaid work.

What I have attempted here, and developed more fully elsewhere (Wajcman, 1991), is to construct a framework that brings together these various perspectives. Instead of imposing an artificial uniformity, I will argue that different kinds of technology are shaped by specific constellations of interests, so that the male interests shaping reproductive technologies, for example, are different than those that form technologies in the workplace. In this chapter I have chosen to concentrate on the three most heavily researched areas—production, reproduction, and domestic technologies. In the final section I present a more general analysis of women's marginality, one that focuses on technology as a masculine culture.

THE TECHNOLOGY OF PRODUCTION

The study of technologies in the context of paid work has been and still is the main subject of research on technological change. Since the mid-1970s, feminist researchers and activists have addressed the effects of automation on women's employment. The introduction of computer-based technologies into offices is a prime site of this research, mainly because the majority of clerical and secretarial workers almost everywhere are women. Office automation forms the basis for many of the generalizations about women's work experience. This research examines the effects of technological change both on women's employment opportunities and on their experience of work (Crompton & Jones, 1984; Feldberg & Glenn, 1983; Hartmann et al., 1986, 1987; Webster, 1989; Wright, 1987).

Although word processors were initially seen as a threat to typists' skills and as leading to the fragmentation and standardization of work, a more complex picture is now emerging. With respect to skill levels required for given jobs, detailed empirical studies show that opposing tendencies of increased complexity and of greater simplification and routinization may coexist. Rather than the impact of automation being uniform across a range of office jobs, the effects of new technology operate within and reinforce preexisting differences in the patterns of work. It has been found, for example, that technological change tends to further advantage those who already have recognized skills and a degree of control over their work tasks (Baran, 1987).

Another strand of research has taken up the issue of divisions between men and women in the workplace and the implications of this for the sex typing of occupations. Of particular concern here is the remarkable persistence of the gender stereotyping of jobs, even when the nature of the work and the skills required to perform it have been radically transformed by technological change.

Much of this feminist research has been influenced by a theoretical perspective, mainly Marxist in orientation, that identified the connections between technologies of production and control over labor (Braverman, 1974; Noble, 1984). The basic argument of this literature is that, within capitalism, a major factor affecting the development and use of machinery is the antagonistic class relations of production. To control the labor force and maximize profitability, capitalism continuously applies new technology designed to fragment and deskill labor, so that labor becomes cheaper and subject to greater control (MacKenzie & Wajcman, 1985).

Although this theoretical approach has been sophisticated in its analysis of the capital-labor relation, feminists questioned the notion that control over the labor process operates independently of the gender of the workers who are being controlled. They pointed out that the relations of production are constructed as much out of gender divisions as class divisions (Beechey, 1988; Cockburn, 1983; Hartmann, 1976). Both employers as employers, and men as men, were shown to have an interest in creating and sustaining occupational sex segregation.

The Sex Typing of Technical Skills

Standard historical accounts of craft unionism have examined the role of technical skills in securing job control, that is, as a weapon in class conflict. Its role as a weapon in patriarchal struggles at work has been ignored. It is now well established in the feminist literature that, as exclusively male preserves, craft unions have played an active part in creating and sustaining women's subordinate position in the workforce. The identification of men with skill has been central to male dominance in the workplace.

Some authors have focused on the male domination of the skilled trades created with the introduction of machines during the industrial revolution (Cockburn, 1983, 1985; Faulkner & Arnold, 1985; McNeil, 1987). Male craft workers could not prevent employers from drawing women into the new spheres of production. Instead, they organized to retain certain rights over technology by actively resisting the entry of women into their trades. Women who became industrial laborers found themselves working in what were considered to be unskilled jobs for the lowest pay. Even when they did manage to enter technical/skilled industrial work, as in the two world wars, this was followed by a deliberate process of expulsion from that work once the immediate crisis had passed. Most women workers wanted to retain their jobs and were only removed through the combined efforts of management and unions (Milkman, 1987).

Thus male dominance of technology has largely been secured by the active exclusion of women from areas of technological work and it is fundamental to the way in which the gender division of labor is still being reproduced today. Let us now turn the focus around and look at how these gender divisions may themselves shape particular technological developments in the first place. If technology is designed by men, with job stereotypes in mind, then it is hardly surprising that sex segregation is being further incorporated into the workplace. Gender divisions in the workplace profoundly affect the direction and pace of technological innovation.

The Gendered Relations of Workplace Technology

One of the most important ways that gender divisions interact with technological change is through the price of labor, in that women's wage labor generally costs considerably less than men's. This may affect technological change in at least two ways. Because a new machine has to pay for itself in labor costs saved, technological change may be slower in industries where there is an abundant supply of women's cheap labor. For example, the clothing industry has remained technologically static since the nineteenth century with little change in the sewing process. While there are no doubt purely technical obstacles to the mechanization of clothing production, there will be less incentive to invest in automation if skilled and cheap labor power is available to do the job. Thus there is an important link between women's status as unskilled and low-paid workers, and the uneven pace of technological development.

There is a more direct sense, however, in which gender inequality leaves its imprint on technology. Employers may seek *forms* of technological change that enable them to replace expensive skilled male workers with low-paid, less unionized female workers. A good example of this comes from Cynthia Cockburn's (1983) account of an archetypal group of skilled workers being radically undermined by technological innovation. It is the story of the rise and fall of London's Fleet Street compositors, an exclusively male trade with strong craft traditions of control over the labor process. A detailed description of the technological evolution from the Linotype system to electronic photocomposition shows how the design of the new typesetting technology reflected gender issues.

The new computerized system was designed using the keyboard of a conventional typewriter rather than the compositor's traditional, and very different, keyboard. There was nothing inevitable about this. Electronic circuitry is in fact perfectly capable of producing a Linotype lay on the new-style board. So what politics lie behind the design and selection of this keyboard? In choosing to dispense with the Linotype layout, management were choosing a system that would undermine the skill and power basis of the compositors, and reduce them at a stroke to "mere" typists. This would render typists (mainly women) and compositors (men) equal competitors for the new machines; indeed, it would advantage the women typists. The keyboard on the new printing technology was designed with an eye to using the relatively cheap and abundant labor of female typists.

Although machine design is overwhelmingly a male province, it does not always coincide with the interests of men as a sex. As we have seen, some technologies are designed for use by women to break the craft control of men. Thus gender divisions are commonly exploited in the power struggles between capital and labor. In this way, the social relations that shape industrial technology include those of gender as well as class.

REPRODUCTIVE TECHNOLOGY

The area most vigorously contested at the moment by feminists, both politically and intellectually, is in the sphere of human biological reproduction. Much feminist scholarship has been devoted to uncovering women's struggle throughout history against the appropriation of medical knowledge and practice by men. Contemporary debates are fueled by the perception that the processes of pregnancy and childbirth are directed and controlled by ever-more sophisticated and intrusive technologies.

Reproductive Technology as Patriarchal Domination

Most vocal in their opposition to the development of the new reproductive technologies are a group of radical feminists who see these technologies as a form of patriarchal exploitation of women's bodies (Corea et al., 1985; Hanmer, 1985; Klein, 1985; Mies, 1987; Rowland, 1985). According to these writers, techniques such as in vitro fertilization, egg donation, sex predetermination, and embryo evaluation represent another attempt to control women's bodies. The technological potential for the complete separation of reproduction from sexuality is seen as a move to appropriate the reproductive capacities that have been, in the past, women's unique source of power.

Central to the radical feminist analysis is a concept of reproduction as a natural process, inherent in women alone, and a theory of technology as intrinsically patriarchal. In a similar vein to the work on reproductive technologies, eco-feminists have analyzed military technology and the ecological effects of other modern technologies as products of patriarchal culture that "speak violence at every level" (Caldecott & Leland, 1983; Griffin, 1978). These theories argue that women are more in tune with nature because of their biological capacity for motherhood. Conversely, men's inability to give birth has made them disrespectful of human and natural life, resulting in wars and ecological disasters.

Technology, like science, is seen as an instrument of male domination of women and nature. And, just as many feminists have argued for a science based on women's values, so too there has been a call for a technology based on women's values. From this perspective, a new feminist technology would be based on "a nonexploitative relationship between nature and ourselves" and would embrace feminine intuition and subjectivity.

This trend in feminism has been gathering force in recent years, and resonates with some feminist postmodernism that is largely concerned with an analysis of technology as a cultural phenomenon. It has been positive in taking the debate about gender and technology beyond the use/abuse model of technology and focusing on the political qualities of technology itself. It has also been a forceful assertion of women's interests, needs, and values as being different than men's as well as of the way women are not well served by current technologies.

However, there are clearly some fundamental problems with this idea of a technology based on women's values, including the representation of women as inherently nurturing and pacifist. The assertion of fixed, unified, and opposed female and male natures has been subjected to a variety of thorough critiques (Eisenstein, 1984; Segal, 1987). There is only space here to observe that the values being ascribed to women originate in the historical subordination of women. The belief in the unchanging nature of women, and their association with procreation, nurturance, warmth, and creativity, lies at the very heart of traditional and oppressive conceptions of womanhood. Rather than asserting some inner essence of womanhood as an ahistorical category, we need to recognize the ways in which both "masculinity" and "femininity" are socially constructed and are in fact constantly under reconstruction. The pursuit of a technology based on women's inherent values is therefore misguided.

The literature referred to above has surveyed the range of reproductive technologies and was important because it identified the sexual politics in which these technologies were embedded. Recently, there has been some attempt to make distinctions within the field of reproductive technologies by emphasizing the specific nature of each technology and the differing positions of women in relation to them. Edited collections such as those by Michelle

Stanworth (1987) and Maureen McNeil, Ian Varcoe, and Steven Yearley (1990) represent notable contributions to the social studies of technology. While maintaining that male interests have profoundly structured the form of reproductive technologies that have become available, this literature treats neither men nor the technologies as a homogeneous group. It also recognizes that women are not necessarily hostile to increased technical intervention. Indeed, many women, as patients, favor high-technology deliveries and want access to in vitro fertilization (Morgall, 1992).

While it is evident that all the stages in the career of a medical technology, from its inception and development, through to consolidation as part of routine practice, are a series of interlocking male activities, the male interests involved are specifically those of white middle-class professionals. The division of labor that produces and deploys the reproductive technologies is both sexual and professional: Women are the patients, while the obstetricians, gynecologists, molecular biologists, and embryologists are overwhelmingly men.

Technology and Professionalization

One of the key themes to emerge from these studies is that the "technological imperative" within reproductive medicine is intrinsic to the defense of doctors' claims of professionalism. The unequal power relations between medical practitioners and their female patients are based on a combination of factors, predominantly those of professional qualification and gender. The professional hierarchy means that doctors are regarded as experts who possess technical knowledge and skill that laypeople don't have. Technology is particularly attractive in the case of obstetricians because techniques such as the stethoscope and foetal monitoring enable male doctors to claim to know more about women's bodies than the women themselves (Oakley, 1987).

High-technology activities are not only the key to power at the level of doctor-patient relations but also to power within the profession. Status, money, and professional acclaim within the medical profession are distributed according to the technological sophistication of the specialty, and the new techniques of in vitro fertilization and embryo transfer are no exception. Before the introduction of these techniques, the investigation and treatment of infertility had long been afforded low status in the medical hierarchy. The new techniques of in vitro fertilization and embryo transfer provide gynecologists with an exciting, high-status area of research as well as a technically complex practice that only they can use (Pfeffer, 1987). Clearly, professional interests play a central role in determining the type and tempo of technological innovation in this area.

There are also wider economic forces at work. The commercial interests of the vast biotechnology industry are particularly influential. Much has been written about the "new medical-industrial complex" and the way in which resources are systematically channeled into profitable areas that often have no connection with satisfying human needs (Yoxen, 1986). However, there is as yet little detailed information about the financial interests of medical biotechnology corporations in the development of the new reproductive technologies.

The gender perspective presented in these studies has in some ways built on the earlier analysis of gender relations of production technologies. However, while the conflictual relations of the workplace provide the context for the analysis of technologies of production, reproductive technologies can only be understood in the wider context of the growing supremacy of technology in Western medicine. Although women are prime targets of medical experimentation, reproductive technology cannot be analyzed in terms of a patriarchal conspiracy. Instead, a complex web of interests has been woven here—those of professional and capitalist interests overlaid with gender.

DOMESTIC TECHNOLOGY

Just as women are the primary consumers of reproductive technologies, so are domestic technologies destined for use by women. In this case, considerable optimism has attached to the possibility that technology may provide the solution to women's oppression in the home. Since the 1970s, with the recognition of housework as work, feminist scholars have produced excellent material on the history of housework and domestic technology (Bose, Bereano, & Malloy, 1984; Cowan, 1983; Hayden, 1982; Ravetz, 1965; Strasser, 1982). Another body of writing on domestic technology has concentrated on the recent dramatic expansion of information and communication technologies in the home. It has been concerned with the cultural consequences for the family of their diffusion and consumption (Silverstone & Hirsch, 1992).

Such research has challenged the main orientation within the sociology of technology toward paid, productive labor in the public domain. Issues that have been central here are the relationship between domestic technologies and time spent on household labor, whether technology has affected the degree of gender specialization of housework, and gender bias in the use of new technologies. Dominating the debates is the knowledge that the amount of time women spend on household tasks has not decreased with "mechanization of the home."

"Labor-Saving" Appliances

In her study of household technology, Ruth Schwartz Cowan (1983) provides the following explanations for the failure of the "industrial revolution in the home" to ease or eliminate household tasks. Mechanization gave rise to a whole range of new tasks, which, although not as physically demanding, were as time consuming as the jobs they had replaced. The loss of servants in the early decades of this century meant that even middle-class housewives had to do all the housework themselves. Further, although domestic technology did raise the productivity in housework, it was accompanied by rising expectations of the housewife's role, which generated more domestic work for women. With a major change in the importance attached to child rearing and the mother's role, the home and housework acquired heightened emotional significance. The split between public and private spheres meant that the home was expected to provide a haven from the alienated, stressful technological order of the workplace as well as entertainment, emotional support, and sexual gratification. The burden of satisfying these needs fell on the housewife.

Much of the feminist literature has pointed to the contradictions inherent in attempts to mechanize the home and standardize domestic production. Such attempts have foundered on the nature of housework—privatized, decentralized, and labor intensive. The result is a completely "irrational" use of technology and labor within the home, because of the dominance of single-family residences and the private ownership of correspondingly small-scale amenities. "Several million American women cook supper each night in several million separate homes over several million stoves" (Cowan, 1979, p. 59). Domestic technology has thus been designed for use in single-family households by a lone and loving housewife. Far from liberating women from the home, it has further ensnared them within the social organization of gender.

Alternatives

There is a tendency to see the technologies we have as the only possible ones. This obscures the way particular social and economic interests have influenced their development. It is useful to ask how a particular technology or set of technologies might be redesigned with alternative priorities in mind. History provides us with examples of alternative technologies that have been developed but that have not flourished. In particular, studying paths not taken can illuminate the way in which ideologies of gender shape technology.

In the case of domestic technologies, this can be illustrated in a number of ways. For example, it is worth remembering that during the first few decades of this century a range of alternative approaches to housework were being considered and experimented with. These included the development of commercial services, the establishment of alternative communities and cooperatives, and the invention of different types of machinery (Hayden, 1982). Seeing that the exploitation of women's labor by men was embodied in the actual design of houses, a group of Victorian feminists believed that the only way to free women from domestic drudgery was to change the entire physical framework of houses and neighborhoods. The continued dominance of the single-family residence and the private ownership of household tools has obscured the significance of these alternative approaches.

Thus, when women have designed technological alternatives to time-consuming housework, little is heard of them. A contemporary example is Gabe's innovative self-cleaning house (Zimmerman, 1983). Although still premised on the single-family home, her design focuses on the need to relieve women of the burden of housework it generates. An artist and inventor from Oregon, Frances Gabe spent 27 years building and perfecting the self-cleaning house. In effect, a warm water mist does the basic cleaning and the floors (with rugs removed) serve as the drains. Every detail has been considered. "Clothes freshener cupboards" and "dish washer cupboards" that wash and dry relieve the tedium of stacking, hanging, folding, ironing, and putting away. The costs of her system are no more than average because it is not designed as a luxury item. Although ridiculed at the time, architects and builders now admit that Gabe's house is functional and attractive. One cannot help speculating that the development of an effective self-cleaning house has not been high on the agenda of male engineers.

Domestic Technology: A Commercial Afterthought

The fact is that much domestic technology has not anyway been specifically designed for household use but has its origins in very different spheres. Many domestic technologies were initially developed for commercial, industrial, and even military purposes and only later, as manufacturers sought to expand their markets, were they adapted for home use. For this reason new domestic appliances are not always appropriate to the household work that they are supposed to perform. Nor are they necessarily the implements that would have been developed if the housewife had been considered first or indeed if she had had control of the processes of innovation.

An industrial designer I interviewed put it thus: "Why invest heavily in the design of domestic technology when there is no measure of productivity for housework as there is for industrial work?" Given that women's labor in the home is unpaid, different economic considerations operate. When producing for the household market, manufacturers concentrate on cutting the costs of manufacturing techniques to enable them to sell reasonably cheap products. Much of the design effort is put into making appliances look attractive or impressively high tech in the showroom—for example, giving them an unnecessary array of buttons and flashing lights. Far from being designed to accomplish a specific task, some appliances are designed expressly for sale as moderately priced gifts from husband to wife and in fact are rarely used. In these ways the inequalities between women and men, and the subordination of the private to the public sphere, are reflected in the very design processes of domestic technology. Men design domestic technology with female users in mind and against the background of a particular ideology of the family.

THE INDETERMINACY OF TECHNOLOGY

Throughout this chapter I have been examining the way in which the gender division of our society has affected technological change. As I have argued that technology is imprinted with patriarchal designs, it may appear that the politics implicit in my account are profoundly pessimistic. A crucial point is that the relationship between technological and social change is fundamentally indeterminate. The designers and promoters of a technology cannot completely predict or control its final uses. There are always unintended consequences and unanticipated possibilities. For example, when, as a result of the organized movement of people with physical disabilities in the United States, buildings and pavements were redesigned to improve mobility, it was not envisaged that these reforms would help women manoeuvering prams around cities. It is important not to underestimate women's capacity to subvert the intended purposes of technology and turn it to their collective advantage.

A good illustration of how a technology can yield unintended consequences, and how women can disrupt the original purposes of a technology, is provided by the diffusion of the telephone. In a study of the American history of the telephone, Claude Fischer (1988) shows that there was a generation-long mismatch between how the consumers used the telephone and how the industry men thought it should be used. Although sociability (phoning relatives and friends) was and still is the main use of the residential telephone, the telephone industry resisted such uses until the 1920s, condemning this use of the technology for "trivial gossip." Until that time the telephone was sold as a practical business and household tool.

The people who developed, built, and marketed telephone systems were predominantly telegraph men. They therefore assumed that the telephone's function would be to replicate that of the parent technology, the telegraph. Telephone "visiting" was considered to be an abuse or trivialization of the service. The issue of sociability was also tied up with gender. It was women in particular who were attracted to the telephone to reduce their loneliness and isolation and to free their time from unnecessary travel. A 1930s survey found that, whereas men mainly wanted a telephone for business reasons, women ranked talking to kin and friends first (Fischer, 1988, p. 51).

A fuller feminist analysis of these processes is provided by Michele Martin's book (1991) on telephone development in Canada. She shows that women's access to the telephone came from their husband's phone connection between home and office and that they gained this access for functional purposes. Again, the author demonstrates that women consumers were quick to resist these limitations and to adapt and appropriate the telephone to their own needs.

Women's relationship to the telephone is still different than men's in that women use the telephone more because of their confinement at home with small children, because they have the responsibility for maintaining family and social relations, and possibly because of their fear of crime in the streets (Rakow, 1988, 1992). Although designed with working men in mind, the telephone has increased women's access to each other and the outside world. Thus, far from relating passively to male-designed technologies, this example shows that women can and do actively participate in defining the meaning and purpose of technologies.

Of course, the unintended consequences of a technology are by no means always positive for women. To take the same example, the diffusion of the telephone has facilitated the electronic intrusion of pornography and sexual harassment into the home. Not only are abusive and harassing telephone calls made largely by men to women, but new sexual services are being made available to men in this way. The point is that a technology can contain contradictory possibilities; its meaning will depend on the economic, cultural, and political context in which it is embedded.

TECHNOLOGY AS MASCULINE CULTURE

The ways in which technology is constructed as masculine, and masculinity is defined in terms of technical competence, have been alluded to frequently in this chapter. This is a good point at which to explore in more depth the interplay between the culture of technology and masculinity.

Hidden From History

To start with, women's contributions to technological innovation have by and large been left out of the history books, which generally still represent the prototype inventor as male. So, as in the history of science, an initial task of feminists has been to uncover and recover the women technologists who have been hidden from history. Some of this historical scholarship examines patent records to rediscover women's forgotten inventions (Stanley, 1992). In the current period, there has been considerable interest in the involvement of women such as Ada Lady Lovelace and Grace Hopper in the development of computer programming (Stein, 1985).

However, reassessing women's role in this way is limited by our understanding of what technology is. We tend to think about technology in terms of industrial machinery and cars, for example, ignoring other technologies that affect most aspects of everyday life. The very definition of technology, in other words, too readily defines technology in terms of male activities. It is important to recognize that different epochs and cultures had different names for what we now think of as technology. A greater emphasis on women's activities immediately suggests that women, and in particular black women, were among the first technologists. After all, women were the main gatherers, processors, and storers of plant food from earliest human times onward. It was therefore logical that they should be the ones to have invented the tools and methods involved in this work such as the digging stick, the carrying sling, the reaping knife and sickle, pestles and pounders. If it were not for the male orientation of most technological research, the significance of these inventions would be acknowledged.

Thus there is important work to be done not only in identifying women inventors but also in discovering the origins and paths of development of "women's sphere" technologies that seem often to have been considered beneath notice. By diminishing the significance of women's technologies, the cultural stereotype of technology as an activity appropriate for men is reproduced. We need to try and sever this link between what technology is and what men do. The enduring force of the identification between technology and manliness is not an inherent biological sex difference. It is instead the result of the historical and cultural construction of gender.

Masculinity and Machines

There are now a range of feminist analyses that focus on the symbolic dimension of technology and the way technology enters into our gender identity (Hacker, 1989; Kramarae, 1988; Turkle, 1984). Technology in this sense is more than a set of artifacts; it includes the physical and mental know-how to make use of those things. Appropriating these sorts of knowledge and practices is integral to the constitution of male gender identity. Men affirm their masculinity through technical competence and posit women, by contrast, as technologically ignorant and incompetent. That our present technical culture expresses and consolidates relations among men is an important factor in explaining the continuing exclusion of women. Indeed, as a result of these social practices, women may attach very different meanings and values to technology. To emphasize the ways in which the symbolic representation of technology is sharply gendered is not to deny that real differences do exist between women and men in relation to technology. Nor is it to imply that all men are technologically skilled or knowledgeable. Rather, it is the dominant cultural ideal of masculinity that has this intimate bond with technology.

In modern societies it is the education system, in conjunction with other social institutions, that plays a key role in the formation of gender identity. Schooling, in conjunction with youth cultures, the family, and mass media all transmit meanings and values that identify masculinity with machines and technical competence. There is now an extensive literature on sex stereotyping in general in schools, particularly on the processes by which girls and boys are channeled into different subjects in secondary and tertiary education, and the link between education and the extreme gender segregation of the labor market. This work has shown that discrimination against female students is compounded by exclusionary masculinist cultures within the scientific and technical classroom (Barton & Walker, 1983; Deem, 1980).

The durability of these cultures in the workplace has been the focus of another strand of feminist research. From school to workplaces, feminists have been frustrated by the limited success of equal opportunity policies and schemes to channel women into technical trades. This has prompted some writers to home in on men's workplace cultures so as to understand how it is that women experience them as alien territory (Cockburn, 1983; Hacker, 1981).

Engineering culture, with its fascination with computers and the most automated techniques, is archetypically masculine. Of all the major professions, engineering contains the smallest proportion of females and projects a heavily masculine image hostile to women. It is a particularly intriguing example of masculine culture because it cuts across the boundaries between physical and intellectual work and yet maintains strong elements of mind-body dualism.

Central to the social construction of the engineer is the polarity between science and sensuality, the hard and the soft, things and people. This draws on the wider system of symbols and metaphors in Western culture that identify women with nature and men with culture. These sexual stereotypes contain various elements such as that women are more emotional, less analytical, and weaker than men. Sally Hacker (1981) found that engineers attach most value to scientific abstraction and technical competence and least to feminine properties of nurturance, sensuality, and the body. The posing of such categories as "hard/soft" and "reason/emotion" as opposites is used to legitimate female exclusion from the world of engineering.

Engineering epitomizes another form of the masculinity of technology as well—that involving physical toughness and mechanical skills. All the features that are associated with manual labor and machinery—dirt, noise, danger—are suffused with masculine qualities. Machine-related skills and physical strength are fundamental measures of masculine status and self-esteem according to this model. The workplace culture of engineering illustrates a crucial point: that the ideology of masculinity is remarkably flexible. Masculinity is expressed both in terms of muscular physical strength and aggression and in terms of analytical power. "At one moment, in order to fortify their identification with physical engineering, men dismiss the intellectual world as 'soft.' At the next moment, however, they need to appropriate sedentary, intellectual engineering work for masculinity too" (Cockburn, 1985, p. 190). No matter how masculinity is defined according to this diverse and fluid ideology, it always constructs women as ill-suited to technological pursuits.

CONCLUSION: WHERE ARE WE NOW?

This chapter has looked at the connections between gender, technology, and society from the perspective of the social studies of technology. I have argued that a gendered approach to technology cannot be reduced to a view that treats technology as a set of neutral artifacts manipulated by men to women's detriment. Rather, this approach insists that technology is always the product of social relations. Although there are other equally powerful forces shaping technology, such as militarism, capitalist profitability, and racism, I have concentrated on gender. This means looking at how the production and use of technology are shaped by male power and interests.

Such an account of technology and gender relations, however, is still at a general level. There are few cases where feminists have really got inside the "black box" of technology to do detailed empirical research, as some of the recent sociological literature has attempted (Bijker, Hughes, & Pinch, 1987). Over the last few years, a new sociology of technology has emerged that is studying the invention, development, stabilization, and diffusion of specific artifacts. This literature attempts to show the effects of social relations on technology that range from fostering or inhibiting particular technologies, through influencing the choice between competing paths of technical development, to affecting the precise design characteristics of particular artifacts.

So far, however, this approach has paid little attention to the ways in which technological objects may be shaped by the operation of gender interests. Its blindness to gender issues is indicative of a general problem with the methodology adopted by the new sociology of technology. Using a conventional notion of technology, these writers study the social groups that actively seek to influence the form and direction of technological design. What they overlook is the fact that the absence of influence from certain groups may also be significant. For them, women's absence from observable conflict does not indicate that gender interests are being mobilized. For a social theory of gender, however, the almost complete exclusion of women from the technological community points to the need to take account of the underlying structure of gender relations.

A concept of power is by no means absent from sociological theories of technology, but it does not readily accommodate what feminist theory has come to understand by "male power." The process of technological development, and preferences for different technologies, are shaped by a set of social arrangements that reflect men's power in the wider society. Recent feminist work is providing new insights into the way that specific social interests, including men's interests, structure the knowledge and practice of particular kinds of technology. It is also enriching theoretical developments within feminism more generally. Empirical and theoretical work is now under way to show that gender relations are an integral constituent of the institutions and projects from which technologies emerge. It is my belief that the social studies of technology can only be strengthened by a feminist critique. Without it, we are not getting the full story.

10

Women and Scientific Careers

MARY FRANK FOX

WHEN the last handbook on science, technology, and society (Spiegel-Rösing & Price, 1977) appeared, it contained no chapter on women in science. In fact, the index of the volume contained no entries for either women or gender; the only kindred reference was to "sex roles of scientists," with a listing of merely three pages.

This reflects the scant attention given to issues of women in science in the years prior to publication of the volume. For the years 1960 to 1977 (1977 being the year in which the last handbook appeared), the subject index of the *Social Science Citation Index* (*SSCI*) lists only 16 articles under women or gender and science or women scientists.[1] In the mid-1970s certain landmark articles were published, notably Bayer and Astin's (1975) article on reward structures, Reskin's (1976) work on women in chemistry, and Zuckerman and Cole's early (1975) article on women in American science. And between 1978 and (August) 1991, 95 entries of articles appear under women or gender and science or women scientists in the subject index of *SSCI*.

Research on the topic had clearly multiplied and attention had increased. The appearance of articles on women scientists was concurrent with the surge during the last 15 years of scholarship on contemporary and historical roles,

AUTHOR'S NOTE: For his reading of and comments on this chapter, I thank Scott Long.

activities, and contributions of women in law, the economy, arts, literature, and politics, among other areas. This reflects the growth within academe of women's studies initiatives, fostering women as the subject of study and focusing upon gender as a fundamental factor in the development of knowledge and society. Yet research and attention to women scientists are not a simple response to increases in the proportion of women working in science. Women have long been "in science" though not necessarily "of science" in significant or influential roles (see Rossiter, 1982, and accounts in Zuckerman, Cole, & Bruer, 1991). Throughout this chapter we shall see that numbers may constitute presence, but not necessarily significant participation. In science as in other occupations, numbers of women may be present, yet limited or constrained in their occupational locations, positions, and rewards.

Although comparative data on employment of women in science over the century are not available (see the discussion in Vetter, 1981), data on doctoral degrees awarded are.[2] Doctoral degrees awarded do not correspond directly to professional participation, because, especially in the pre-World War II era, women had few employment options in science; essentially, the choices were a faculty position in a woman's college or a research associate position in a university lab, where the woman had little autonomy and little or no opportunity for advancement (Rossiter, 1982). While, in the pre-World War II era especially, it was not infrequent for women with doctoral degrees in science to face no employment prospects at all (Rossiter, 1982), recent data show that over 90% of women with doctoral degrees in science and engineering (S/E) are in the labor force at a given time (Vetter, 1981, p. 1315).

Despite the restrictions of data on doctoral degrees over time, they do indicate proportions of *professionally trained* women (compared to men) since the 1920s. Examining Table 10.1, we see that the proportions of doctoral degrees awarded to women do not represent a simple linear trend of increasing proportions of degrees over the past 70 years. The proportions of doctoral degrees awarded to women in the 1920s (12.3%) and in the 1930s (11%) are higher than the proportions in the 1940s (8.9%), the 1950s (6.7%), or even the 1960s (7.9%).[3] It was not until the 1970s that the proportions of doctoral degrees awarded to women equaled and surpassed those of the pre-1940 levels: In the 1970s women earned 14.9% of all doctoral degrees in science and engineering fields. Continuing gains were made in the 1980s, so that, for the period 1980-1988, women earned fully a quarter (25.8%) of all doctoral degrees in science.

Thus women have become larger proportions of the pool of doctoral-level scientists and engineers. But, as emphasized above, the issue at hand is not simply the presence or available pool of women but their relative status and rewards. In assessing the careers of women, the questions of this chapter are then these: (a) What are the employment locations and statuses—fields, sectors,

TABLE 10.1 Ph.D. Awards by Science and Engineering Field (S/E), Decade, and Sex: 1920-1988

	Decade						
Field	*1920-1929*	*1930-1939*	*1940-1949*	*1950-1959*	*1960-1969*	*1970-1979*	*1980-1988*
All S/E Fields	7,767	16,157	19,464	52,697	104,930	186,971	173,910
percentage women	(12.3)	(11.0)	(8.9)	(6.7)	(7.9)	(14.9)	(25.8)
Physical sciences	3,271	6,687	8,202	18,745	34,307	48,182	41,122
percentage women	(7.6)	(6.6)	(5.0)	(3.7)	(4.6)	(8.2)	(14.7)
Engineering	228	833	1,439	5,765	19,042	29,683	27,793
percentage women	(.9)	(.7)	(.5)	(.3)	(.4)	(1.4)	(5.5)
Life sciences	2,370	5,081	5,822	14,495	26,461	49,344	50,989
percentage women	(15.9)	(15.1)	(12.7)	(9.1)	(11.6)	(18.1)	(31.2)
Social sciences*	1,898	3,556	4,001	13,962	25,120	59,762	54,006
percentage women	(17.1)	(15.8)	(14.5)	(11.0)	(14.3)	(24.5)	(39.7)

SOURCE: Commission on Professionals in Science and Technology (1989, table 2-1).
*Includes psychology.

institutional types, ranks, salaries—of women compared to men? (b) What is the research performance—productivity—of women compared to men in science, and why does this matter? (c) What factors account for the relative status and performance, the career attainments, of women in science? Because space for this chapter is limited in a handbook designed to cover a broad range of topics, discussion of the data on and implications of these concerns will necessarily concentrate upon their most central aspects.

Taken together, the questions of this chapter address the issue of gender stratification in scientific careers. As we examine indicators of the profile of women compared to men, we shall see that, as in other occupations (Fox & Hesse-Biber, 1984; Reskin & Roos, 1990), women and men in science are differentiated in the fields they occupy, the places they work, the positions they hold, and the salaries they earn. Gender shapes location, rank, and rewards in science. Gender may also shape the meaning of science and of technology —the way in which questions are framed, data are interpreted, and knowledge and applications are created. Those considerations are the subject of chapters by Wajcman and by Keller.

In this chapter, the focus is upon doctoral-level personnel because that is the group for whom considered issues of research, research productivity, and their impact are most pertinent. Across fields, women are 15% of all scientists and engineers and are 16% of those at the doctoral level (Table 10.2). Between *fields*, we find that the employment of women is highly uneven (Table 10.2). The vast majority (82%) of women are in three fields—life sciences, psychology, and social sciences—in which 33%, 29%, and 20% of women are found, respectively. In contrast, just half of men are in these three areas. Men are almost twice as likely as women to be located in physical and mathematical science, and seven times more likely to be in engineering. It is for engineering that the disparity is greatest in the distribution of women and of men—2.5% of the women compared with 18.8% of the men are in this field. Because such small proportions of the women are in engineering, throughout this chapter references are made to women "in science." These do, in fact, refer to the vast preponderance (97.5%) of the doctoral-level S/E women who are in scientific, not engineering, fields.

The distribution of women by field suggests a nonuniformity of processes (self-selection of women into various fields and social restrictions on women's entry and participation that may vary by field) in science. Yet the factors governing variations in women's employment by field, particularly the higher proportion of women in life compared with physical sciences, are not well understood (Zuckerman, 1987, p. 128). Nor have issues and implications of subfield locations of women been addressed. In the study of women in science, we need to get beyond implicit assumptions of uniformity of women's status—across fields and within fields. Women's compared to

TABLE 10.2 Employed Scientists and Engineers (S/E), by Field, Sex, and Doctoral Status: 1987

Field and Sex	*Doctoral S/E*	*Distribution of Doctoral* *Men*	*Distribution of Doctoral* *Women*	*Total S/E*
Total, all fields	419,100	(100.0)	(100.00)	4,626,500
men	(84)			(85)
women	(16)			(15)
Physical scientists	68,600	(17.9)	(8.2)	288,400
men	(92)			(87)
women	(8)			(13)
Mathematical scientists	16,600	(4.3)	(2.4)	131,000
men	(90)			(74)
women	(10)			(26)
Computer specialists	18,600	(4.7)	(2.8)	562,600
men	(90)			(71)
women	(10)			(29)
Environmental scientists	17,800	(4.7)	(1.9)	111,300
men	(93)			(88)
women	(7)			(12)
Life scientists	107,400	(24.2)	(33.1)	411,800
men	(79)			(75)
women	(21)			(25)
Psychologists	56,400	(10.6)	(28.6)	253,500
men	(66)			(55)
women	(34)			(45)
Social scientists	65,900	(14.9)	(20.3)	427,800
men	(80)			(69)
women	(20)			(31)
Engineers	67,800	(18.8)	(2.5)	2,440,100
men	(97)			(96)
women	(3)			(4)

SOURCE: National Science Foundation (1990, tables 2, 4).
NOTE: Proportions may not add to 100.00% because of rounding.

men's subfield locations are particularly important because it is in subfields that research is conducted. Subfields can be characterized along scales or dimensions as new and emergent compared with old and established, as more compared with less well funded, as high growth compared with low growth, and as more theoretical (basic) compared with less theoretical (applied). These dimensions have consequences for the pace and impact of contributions. In new, emergent, well-funded, and high-growth areas, much remains to be studied; impact is more readily established; and resources (funding) are available for the work.[4]

In addition to patterns by field, women and men are clustered in somewhat different *employment sectors* (Table 10.3). Most notably, across fields, women

TABLE 10.3 Employed Doctoral Scientists and Engineers (S/E), by Field, Sex, and Sector: 1987

	All S/E Fields	*Percentage Total*	*Percentage Men*	*Percentage Women*	*Scientists**	*Percentage Total*	*Percentage Men*	*Percentage Women*
Total employed	419,118	(100)	(100)	(100)	351,350	(100)	(100)	(100)
Sector								
industry/ self-employed	131,699	(31.4)	(32.9)	(23.8)	94, 552	(26.9)	(27.8)	(23.0)
educational institutions	218,697	(52.2)	(51.4)	(56.3)	194,987	(55.5)	(55.2)	(56.9)
hospitals/clinics	12,158	(2.9)	(2.4)	(5.6)	12,134	(3.5)	(2.9)	(5.7)
nonprofit organizations	15,464	(3.7)	(3.4)	(5.3)	13,290	(3.8)	(3.4)	(5.4)
federal government and military	29,710	(7.1)	(7.4)	(5.4)	25,772	(7.3)	(7.8)	(5.3)
state and local government	9,223	(2.2)	(2.0)	(3.1)	8,697	(2.5)	(2.3)	(3.2)

SOURCE: Commission on Professionals in Science and Technology (1989, table 4-13).
*Does not include engineers.

are less likely than men to work in industry or to be self-employed. A quarter (24%) of the women compared with a third (33%) of the men in science and engineering work in this sector. Despite well-publicized initiatives (e.g., see Catalyst, 1992) within industry to recruit and retain women scientists and engineers, women are less likely than men to be found in industrial jobs. This may reflect the recency of these initiatives and of women's professional participation in industry (compared, for example, with the longer tradition of women's roles in higher education).

In addition, compared to men, women are more likely to be employed in hospitals, nonprofit organizations, the federal government, and state and local government. However, altogether these latter sectors employ only a min- ority of scientists/engineers of either sex—specifically, 19.4% of the women and 15.2% of the men.

It is the academic sector that employs the clear majority of doctoral-level scientists and engineers—51% of the men and 56% of the women. (Engineers, however, are much more likely than scientists to be employed outside of academia. Thus, if we confine the group to doctoral-level scientists, excluding engineers, the proportions of men and women in academia are even higher, 55% and 57%, respectively.) By field, the highest proportions of women working in academia (compared to other sectors) are in mathematical (81%), life (67%), and social (68%) sciences (data not shown). In computer science, environmental science, and psychology, women are more likely to be in nonacademic sectors. By field, the same patterns hold for men, likely reflecting job market conditions outside of academia that are stronger for computer and environmental sciences and for psychologists compared to mathematicians, life scientists, and (other) social scientists.

Because academia is the predominant employment sector for doctoral-level scientists/engineers, it is important to look at women's and men's institutional locations and ranks within this sector, in particular. The data on *institutional types* represent broad classifications of academic institutions (Table 10.4). However, they point to a pattern of lower proportions of women at higher level institutions. First, very small proportions of doctoral scientists and engineers (less than 2%) are in precollege-level educational institutions; but it is at this institutional level that the proportion of women (41%) compared to men is highest by far. Beyond this, the proportion of women (16%) compared to men is lower within universities than within 2-year colleges (23%) and 4-year colleges (20%). Of interest, the proportion of women (27%) compared to men is higher in medical schools than in other types of institutions in higher education. Data are not available on ranks of women compared to men within each institutional type. Thus we do not know whether women in academic appointments in medical schools are more apt to be in low ranks or even off-track (lecturer, instructor) positions.

TABLE 10.4 Doctoral Scientists and Engineers Employed in Academe, by Type of Institution and Sex: 1989

Type of Academic Institution	*Total*	*Percentage Men*	*Percentage Women*
Total academe	225,803	81	19
Two-year college	5,226	77	23
Medical school	31,711	73	27
Four-year college	31,693	80	20
Other university	153,154	84	16
Precollege	4,019	59	41

SOURCE: National Research Council (1991, table 6).

The lower proportion of women in universities compared to 2-year and 4-year colleges merits attention. It is at the university-level institutions that equipment is available, that graduate students are enrolled, that collaborative opportunities (with both faculty and students) are more likely to prevail, and, in consequence, that research is more likely to be conducted. The lower proportion of women in universities has implications for ways and means to research performance and for status as it is connected to productivity, as discussed later.

Because *academic ranks* are clearly specified and gradated, they are telling indicators of positions in science. Across fields, the higher the rank, the lower the proportion of women (Table 10.5). Overall, women are 28% of the assistant professors, 17% of the associate professors, and a mere 7% of the full professors. While the general pattern of lower proportions of women at higher ranks holds across fields, the proportional *levels* vary by field. Just as women are concentrated in three fields—life science, psychology, and social science—so correspondingly, in these compared with other fields, we find higher proportions of women at each rank. In life sciences, psychology, and social sciences, women are 32%, 48%, and 34%, respectively, of the assistant professors and 22%, 28%, and 18%, respectively, of the associate professors.

However, for every field except psychology, the proportion of women at the rank of full professor is very small. In half of the fields—physical, mathematical, and environmental sciences and engineering—women are 5% or fewer of the full professors. Only in psychology are women more than 10% of the full professors. Despite growing pools of women with doctorates in the 1970s (see Table 10.1) and the passage of 15-20 years during which those pools have matured in professional time, the proportion of women who are full professors has changed little over the past two decades. In 1973 women were 4% of the full professors in S/E fields, and in 1987 that proportion was still just 7% (also see A. Gibbons, 1992).

TABLE 10.5 Doctoral Scientists and Engineers in Four-Year Colleges and Universities, by Sex and Academic Rank: 1987

Field and Sex	*Total*	*Full Professor*	*Associate Professor*	*Assistant Professor*
Total, all fields	209,400	85,800	50,500	36,500
percentage women	(16.7)	(7.2)	(17.4)	(27.7)
Scientists, total	185,700	74,500	45,600	32,700
percentage women	(18.5)	(8.2)	(18.9)	(30.3)
Physical scientists	28,700	13,300	4,800	3,000
percentage women	(8.4)	(3.8)	(8.3)	(16.7)
Mathematical scientists	13,000	6,400	3,200	2,500
percentage women	(9.2)	(4.7)	(12.5)	(16.0)
Computer specialists	5,400	1,300	1,600	1,300
percentage women	(11.1)	(7.7)	(6.3)	(15.4)
Environmental scientists	7,400	2,800	1,500	1,300
percentage women	(8.1)	(3.6)	(6.7)	(15.4)
Life scientists	64,700	24,000	15,200	12,400
percentage women	(22.0)	(9.2)	(22.4)	(31.5)
Psychologists	22,000	8,500	5,700	4,000
percentage women	(30.5)	(14.1)	(28.1)	(47.5)
Social scientists	44,400	18,100	13,600	8,300
percentage women	(19.6)	(9.9)	(18.4)	(33.7)
Engineers	23,600	11,400	4,900	3,700
percentage women	(2.5)	(0.9)	(2.0)	(5.4)

SOURCE: National Science Foundation (1990, tables 20, 21).
NOTE: Total includes instructor, other, and no report. Proportions may not add to 100.00% because of rounding.

The data in Table 10.5 do not control for variables that affect academic rank, such as years since Ph.D. or productivity. However, studies that do take into account such variables point to academic rank as the area where disparity may be greatest in the statuses of women and men in science. Accordingly, Jonathan Cole (1979) found that, for scientists at each level of productivity in his classification, women were less likely than men to receive promotions. This held with controls for type of location (in better and in lesser departments). Likewise, among pairs of women and men matched on year in which Ph.D. was received, field of Ph.D., institution from which doctorate was awarded, and race, Ahern and Scott (1981) found large and pervasive differences in academic rank among natural, physical, and biological scientists. Ten to nineteen years past the doctorate, the men were 50% more likely than the women to have been promoted to full professor. Among younger matched pairs, women were less likely to have been promoted to associate professor, independent of marital status, the presence of children, or whether their work orientation was primarily research or teaching.

In addition, a study (Sonnert, 1990) of recipients of prestigious postdoctoral fellowships awarded between 1955 and 1986 reports that, controlling for years since doctorate, fields, and fellowship, the predicted academic rank of women is one third lower than that of men. However, in this study, the disadvantage for women in academic rank is outside of biological sciences, that is, in physical science, mathematics, and engineering. Among biologists in Sonnert's sample, women's progress through the academic ranks was similar to men's. This points again to the importance of disciplinary considerations.

These well-documented disparities in the rank and promotion of women and men contrast with the lesser disparity between the sexes in scientific awards, honors, and reputations. First, in a sample of 565 natural and social scientists, the correlation between sex (being male) and awards is low, although the association between being male and obtaining awards varies by field and is strongest in sociology (J. Cole, 1979, p. 59). Second, in the same sample, once research productivity is taken into account, sex has a very small effect upon reputation in science (J. Cole, 1979, p. 119). This suggests that it may be at the institutional—departmental, university—level that inequities are more likely to operate. Inequities may be more likely to operate in the organizational workplace, with its particular processes for promotion, than in the reward and reputational mechanisms of the wider scientific disciplines and community.

In science, scholarly recognition has been regarded as the equivalent of property (Cole & Cole, 1973) and recognition, rather than other rewards, the "coin of the realm" (Storer, 1973). Correspondingly, in academia in particular, salary has been thought to act "not as an inducement proffered to secure performance, but rather as a symbolic recognition of past, present, or promised performance" (Smelser & Content, 1980, p. 6). Still, in the analysis of stratification among gender groups, *salary inequality* is a telling indicator in science as in other occupations and institutions (Fox, 1981). For this reason and because, in science, salary has both particular symbolic as well as material value, the earnings of women and men merit scrutiny.

Among full-time doctoral-level scientists, women earn between 78% and 85% of that earned by men, depending upon the field (Table 10.6). These salary data (from the National Science Foundation) contain controls for field and education (doctoral level), but not for other salary-relevant characteristics of employees and their work setting. How much of the salary gap of women and men in S/E might be explained by taking into consideration scientists' individual characteristics (e.g., age and seniority) and employment characteristics (e.g., sector of employment)?

Few multivariate analyses of sex differences in salary focus upon scientists, specifically. Rather, the analyses have been based broadly upon academics (e.g., Barbezat, 1987; Fox, 1981) or doctorates (Ferber & Kordick, 1978)

TABLE 10.6 Median Annual Salaries of Doctoral Scientists and Engineers, by Field and Sex: 1987

	Annual Salary			
Field	*Total Employed (in dollars)*	*Men (in dollars)*	*Women (in dollars)*	*Women's Salary/ Men's Salary*
Total, all fields	49,600	50,700	40,200	.79
Scientists, total	47,800	49,200	40,000	.81
Physical scientists	51,400	52,400	41,900	.80
Mathematical scientists	46,600	47,500	39,600	.83
Computer specialists	54,400	55,400	43,400	.78
Environmental scientists	50,300	50,700	41,800	.82
Life scientists	45,700	48,000	39,600	.83
Psychologists	44,300	46,200	39,500	.85
Social scientists	45,300	47,000	39,600	.84
Engineers	58,100	58,500	48,200	.82

SOURCE: National Science Foundation (1990, table 26).
NOTE: Detail may not average to total because total includes "other" and "no report." Salaries are computed for individuals employed full time.

across fields. An exception is Haberfeld and Shenhav's (1990) study of (doctoral and nondoctoral) scientists and engineers, using longitudinal data from the Bureau of the Census. Controlling for numbers of human capital and job characteristics—including age, race, education, experience, citizenship, parenthood, type of employer, and field—they find that in 1972 women earned 12% less than men with comparable characteristics, and in 1982 women earned 14% less. These data would indicate that the salary gap of women and men in science reduces but does not close after standardizing for such individual and locational characteristics.

These analyses of Haberfeld and Shenhav do not include publication productivity as a factor in determining salary. Across fields (scientific and nonscientific), data do indicate that publication is related to salary—although the returns vary by level of publication, that is, are greater for first compared with subsequent articles and diminish with increased output (Tuckman, 1976; Tuckman & Hagemann, 1976). At the same time, these relationships are not uniform for women and men. The correlations between productivity and salary are much higher for men than for women, indicating that men are more likely to get economic returns for their publications (Bayer & Astin, 1975). Here again, level of productivity may be a factor. Data on academics in 22 disciplines (scientific and nonscientific) show that, for a single article, salary returns are twice as great for men as for women; for five articles, returns are still greater for men; at the highest levels of productivity, 15-25 articles, salary returns are more equitable (Tuckman, 1976). Consider, however, that the vast group of both men and women are at the lower levels of productivity, and thus, for the vast majority, the greater sex disparities apply.

In analyzing career attainments of women and men in science, assessments are incomplete without consideration of performance, particularly research performance. *Research productivity*, represented by publication productivity, is critical to assessment of persons and groups in science for two fundamental reasons. First, publication is the central social process of science, the way in which research is communicated, exchanged, and verified (Merton, 1973c; Mullins, 1973). Second, publication correlates highly with research impact and with scientific awards (Blume & Sinclair, 1973; Cole & Cole, 1973; Gaston, 1978). Thus, until we understand the productivity differences of women and men, we cannot adequately assess other sex differences in location, status, and rewards. The national data available across fields (from the NSF and other agencies, as they appear in the tables in this chapter) do not present comparisons of productivity or controls for productivity in relationships between sex and rank or location. Thus, in assessment of publication productivity, we must refer to studies based upon more limited samples—and in that way draw inferences about the connections between sex, productivity, and status in science, as in previous discussions of rank and of salary.

What do we know about the publication productivity of women compared with men? First, numbers of samples indicate that women in science publish less than men. Over a 12-year period, J. Cole (1979) found that, across fields of chemistry, biology, psychology, and sociology, the median number of papers published was 8 for men and 3 for women. Among a sample of women and men scientists in six fields who were matched for year of Ph.D. and doctoral department, Cole and Zuckerman (1984) report that women publish half as much as men—6.4 compared with 11.2 papers.

Between fields and disciplinary areas, levels of sex differences in publication are more variable. Among psychologists, men are significantly more productive than women, publishing 1.7 compared with .7 papers in a 3-year period (Helmreich, Spence, Beane, Lucker, & Matthews, 1980). For chemists, however, Reskin (1978a) reports slighter differences, which suggest "a true but small sex difference between the populations" (p. 1236) of the field. Likewise, in my national sample of social scientists in four fields (economics, sociology, political science, psychology), the sex difference in mean number of articles published in a 3-year period is modest (but statistically significant)—2.25 for women compared with 2.50 for men.

However, while men and women publish at different rates, the publication of both groups is strongly skewed. This means that, among both men and women, most work is published by a few while the majority publish little or nothing. In my sample of social scientists, nearly all full-time academics with

Ph.D.s, 14.5% of the men account for over half (52%) of all articles published in a 3-year period. Among women, 16% account for 52% of all articles published in this period. This skewed distribution of about 15% accounting for 50% of publications is a persistent pattern. It repeats across fields—among chemists (Reskin, 1977), in samples of men and women in natural and social sciences (J. Cole, 1979), and among matched samples of men and women in six scientific fields (Cole & Zuckerman, 1984). In fact, in the latter study, Cole and Zuckerman (1984) report that, in their matched sample, women's productivity is even more skewed than men's; that is, among women, the most prolific account for an even larger proportion of papers published by their gender group than is the case among men. Thus the pattern of highly variable and strongly skewed productivity, documented over 75 years ago in Lotka's (1926) analysis of articles published in physics journals, is characteristic of both men and women in science.

The first set of status indicators discussed (rank, location, rewards), together with research performance, constitute career attainments in science. The two are connected, although, as we have seen, the associations between status indicators and publication productivity are not equivalent for women and men. Still, the lower success of women in scientific careers refers to both status and performance and, accordingly, explanations of women's achievements refer to both. What factors then might account for women's lower attainments in science?

First, individual characteristics of ability, at least measured ability, do not provide an explanation. Although a high IQ may be a prerequisite for advanced training, once the degree is obtained, differences in measured ability do not predict subsequent levels of performance (Cole & Cole, 1973, p. 69). Further, although women's attainments in science are lower than men's, their measured ability (IQ) is actually higher. In chemistry, biology, psychology, and sociology, women Ph.D.s have slightly higher IQs than men Ph.D.s in these fields (J. Cole, 1979, p. 61); and for departments of varying ranks, there is less variability in average IQs for women than for men (J. Cole, 1979, p. 159). Data on IQs may not be the best indicators of ability or intelligence, but to the extent that they capture differences, they indicate that, if anything, women in science are a more selective group, intellectually, than men.

Additionally, for certain background characteristics that affect career attainments, women and men scientists are similar. Specifically, with some disciplinary variation, women and men are about as apt to have received their advanced degrees from top ranking universities (National Research Council, 1983). The most substantial gender difference is in mathematics, where 46% of the men compared with 37% of the women have received their Ph.D.s in departments rated as strong or distinguished. In physics, women are somewhat less likely to have graduated from a top ranking department, and in psychology and microbiology, women are more likely to have done so (National Research Council, 1983). For my national sample of social scientists, I find insignificant sex differences in the ratings of institutions from which women compared with men obtained their doctoral degrees; this is true for academics in each of the four major fields considered: economics, psychology, political science, and sociology. Thus, across scientific fields, the general pattern is one of similarity in doctoral origins of women and men.

Likewise, gender differences are small in certain indicators of financial support for graduate training, measured as percentages of men compared with women who had held research or teaching assistantships during graduate school (National Research Council, 1979, 1983). However, these data do not specify the quality or character of the assistantships. These factors may, in turn, indicate different opportunities of women and men to join research groups, to collaborate, and to gain access to the culture of science.

These areas of doctoral education may provide clues to the career outcomes of women. Studies of graduate students indicate that, compared with men, women see faculty members and advisers less frequently than do men (Holmstrom & Holmstrom, 1974; Kjerulff & Blood, 1973); that their interactions with faculty are less relaxed and egalitarian (Holmstrom & Holmstrom, 1974; Kjerulff & Blood, 1973); and that women tend to regard themselves as students rather than colleagues of faculty and say that they are taken less seriously than men (Berg & Ferber, 1983; Holmstrom & Holmstrom, 1974). This may signal the greater marginality of women graduate students—with consequences for their introduction to and apprenticeship in their disciplines.

In science, especially, models and modeling, apprenticeship, and teamwork have consequences for acquisition of taste, style, and confidence in research (Zuckerman, 1977b). Collaboration with a mentor is a particularly important factor, affecting predoctoral productivity, job placement, and later productivity (Long & McGinnis, 1985). In his study of biochemists, Long (1990) analyzes, in turn, the factors that affect collaboration with a mentor. He finds that, for women, having young children decreases odds of collaborating, while this effect does not exist for men. Having a female mentor increases the odds of women collaborating—but in biochemistry (as in other scientific fields), limited numbers of senior female mentors are available.

Aggregate data on rates of funding, assistantships, and financial aid fail to address the dynamics of graduate education as they may affect the career attainments of women. We need to know more about the patterns and consequences of student-faculty interaction, advising, training practices, and job placement as they may affect women students and their continuing prospects in science.[5]

Much has been made about the impact of marriage and children upon women's attainments in science. The mythology of science (Bruer, 1984) has it that good scientists are either men with wives or women without husbands and children. Yet much of the evidence contradicts this conventional wisdom.

First, there is the issue of presumed constraints upon the geographic mobility of married women. In their analyses of the residential patterns of both academics across disciplines and a sample of psychologists, Marwell, Rosenfeld, and Spilerman (1979) consider the relationship between geographic location, marriage, and women's academic careers. They find that women are located disproportionately in large, urban centers. This, they argue, reflects women's need to settle in areas with more opportunities for two-career households. Geographic constraints might, in turn, be expected to have consequences for women's career attainments and possibilities. Yet Marwell and his colleagues did not provide a direct link between geographic location and academic rank. And other data indicate that, while marriage negatively affects rank and salary of academic women, except in the case of salary for women in research universities, the effects of marriage are not significant (Ahern & Scott, 1981, tables 6.1, 6.5, 6.9, and 6.13).

Further, with respect to publication productivity, married women publish as much or more than unmarried women. This has been found across scientific and nonscientific fields (Astin & Davis, 1985); in physical, biological, and social sciences (Zuckerman & Cole, 1987); and in specific fields, as in psychology (Helmreich et al., 1980). Moreover, among various samples of scientists and scholars, we find that the presence of children has no effect on women's publication productivity (Cole & Zuckerman, 1987; Helmreich et al., 1980), a slightly negative, nonsignificant effect (Reskin, 1978a), or a positive effect (Astin & Davis, 1985; Fox & Faver, 1985).

Thus the perception may persist that marriage and motherhood govern women's status in science, but the data indicate otherwise. Single women and women without children do either no better or worse than women with husbands and children. At the same time, these data are based upon women who are active scientists. The women may be a selective group who have survived earlier trials. The effects of family status could take their toll along the way, in earlier stages, so that some proportion of women are deflected and not falling into databases of professional scientists (Long, 1987). This is unclear and remains to be determined.

Beyond marriage, motherhood, and other individual-level characteristics, we should look to the social and organizational context of science in explaining the career attainments of women in science—particularly their rank, salary, and research productivity.

Why and how is the environmental context important to assessing women's (and men's) career attainments? First, science is largely organizational work. It is performed within organizational policies; it involves the cooperation of persons and groups; it requires human and material resources. In its scope and complexity, scientific research relies upon facilities, funds, apparatus, and teamwork. Accordingly, performance can be tied to the environment of work—its signals, priorities, and resources (Fox, 1991). And rank and salary are determined in the workplace.

Further, scientists work in a larger, disciplinary (or interdisciplinary) community and environment that can bolster or hinder status and performance. More so than in nonscientific fields, the work of science "relates to, builds upon, and revises existing knowledge" (Garvey, 1979, p. 14). In science, one must continually shape, test, and update research. This takes place informally and interactively as well as formally, in conference presentations, for example.

In this way, science is more "social" than the arts or humanities, for example. Compared to the humanities, the sciences are more likely to be performed in teamwork than solo, to be carried out with costly equipment, to require funding, to be more interdependent enterprises. Yet more so than men, women are outside of the social networks of science in which ideas are exchanged and evaluated and in which human and material resources circulate (Fox, 1991). In the disciplinary communities of science, this is indicated in the lesser likelihood of women than men to have had connections as editors, officers of professional associations, and reviewers of grants (Cameron, 1978; Kashet, Robbins, Lieve, & Huang, 1974); to have appeared on programs of national meetings (Glenn et al., 1993); and to have been invited to lecture or consult outside of their institutions (Kashet et al., 1974). Because science is a social process, these factors (networks, memberships) may then have bearing on productivity and on the productivity differences between women and men.

In academic locations, in particular, we have seen that women are more likely than men to work in colleges rather than universities, with implications for allocation of time and availability of assistants and equipment. However, in explaining career attainments of women, it is not simply a matter of the distribution of women compared with men across types of institutions but also of organizational processes within institutions. The same type of setting (whether a major university or a 4-year college) may, in fact, offer different constraints and opportunities for one gender group compared with the other.

Organizational environments do not necessarily operate neutrally or uniformly (Fox, 1991). Resources and opportunities may operate differently for women and men. In my national survey of social scientists, I find, for example, that in both B.A.- and Ph.D.-granting departments, women report significantly less interaction with and recognition from faculty in their departments. In M.A. and Ph.D. departments, the women give significantly lower rankings to the resources available to them and, in Ph.D. departments, they report significantly higher undergraduate teaching loads. The factors of interaction, reported resources, and teaching loads correlate with productivity levels.

Nonuniformity of environment has been documented within the same setting as well as the same type of setting. Specifically, a study (Feldt, 1986) of scientists appointed as assistant professors at the University of Michigan reports that women and men received different treatment, even at this early career stage. The men got more start-up monies for their labs, better physical facilities, and better placement within existing projects with funds and equipment.

Such a factor as "placement in projects" is relevant for collaboration. Collaboration, in turn, is important because solo research is difficult to initiate and sustain in science. Further, collaborative work fares better in the publication process because it is more likely to result from funded research (Heffner, 1980); is more likely to be empirically based rather than theoretical work, which is said to be easier to review (Meadows, 1974; O'Connor, 1969); and, finally, because coauthored work may provide checks against error (Presser, 1980).

Is collaboration then a factor in explaining women's career attainments? The evidence and answer are mixed. Studies of sociologists have reported that women are both more likely (Mackie, 1977) and less likely (Chubin, 1974) to collaborate. In Cole and Zuckerman's (1984) analysis of publications among a matched sample of women and men who received their Ph.D.s in six different scientific fields, women were as likely as men to coauthor papers. Likewise, Long (1992) reports that, in his sample of biochemists, women and men have nearly identical levels of collaboration.

The issue may be more subtle, however, than simply rates of coauthorship. Women may have more difficulty finding and establishing collaborators and they may have fewer collaborators available to them. One study (Cameron, 1978) supports this notion, in reporting that men have significantly higher numbers of different collaborators. Further, Long's data (1992) show that in biochemistry women are much more likely than men to collaborate with a spouse. To the extent that this collaboration of women occurs with a spouse instead of multiple others (rather than with a spouse *and* multiple others), these data also may suggest a more limited range of collaborators among women compared with men. This remains to be assessed directly.

In the relationship between the social and organizational environment of science and the career attainments of women, another factor is the evaluative process in organizations—with consequences for rank and rewards. In one study, Fidell (1975) sent to 147 chairpersons of psychology departments 10 resumes with information on teaching, research, and service activities. Female and male names were randomly assigned to the resumes. Asked in a matter of hypothetical hiring to assign a rank to the persons represented in the resumes, most chairpersons recommended the rank of associate professor to male names and the rank of assistant professor to the same descriptions with female names.

How does one account for this? To begin with, J. Cole (1979, p. 75) has argued that "functionally irrelevant characteristics" such as sex are especially likely to be activated as a basis of evaluation and reward when there are few relevant criteria on which to judge. However, in Fidell's study and the real-life events it may represent, data on performance were available. In explaining such outcomes, consider that, in scientific (and scholarly) work, standards for performance are both "absolute" and "subjective" (Fox, 1991). Performance is judged against a standard of absolute excellence, which is, in turn, a subjective assessment. Experimental data (Deaux & Emswiller, 1974; Nieva & Gutek, 1980; Pheterson, Kiesler, & Goldberg, 1971; Rosen & Jerdee, 1974) indicate that, when standards are subjective and loosely defined, it is more likely that men will be perceived to be the superior candidates and that gender bias will operate. These evaluative processes may help explain the status of women in science.

In summary, this chapter points to these conclusions: First, in the assessment of women and scientific careers, we need to get beyond issues of numbers. Since the 1970s women have increased their proportions of doctoral degrees awarded in science and engineering and thus their proportions of professionally trained S/E personnel. Yet numbers do not ensure significant participation and status. We find that gender shapes location, ranks, and rewards in science. Women and men are differentiated in their fields and academic locations and ordered in their ranks and salary rewards. At the same time, the publication productivity of women is lower than that of men. Status indicators (location, rank, salary) and productivity are connected, although the associations are not equivalent for women and men. Thus, in explaining career attainments in science, one must consider both status and performance.

Second, among the factors explaining women's lower success in science, individual characteristics of ability as well as marriage and motherhood account modestly or little. Likewise, because women and men are similar in their doctoral origins and in their history of research and teaching assistantships, these aspects of background do not provide an explanation. However, because women and men may have different experiences in quality and character of graduate training, clues may lie in the nature and patterns of advising, collaboration, and apprenticeship in doctoral education. In like manner, I argue that the career attainments of women are conditioned by aspects of the social and organizational environment of science, social features of both the workplace and the larger disciplinary communities of science.

By implication, this means that improvement in women's attainments in scientific careers—their status and performance—will not depend merely upon the detection, cultivation, and enhancement of individual abilities, skills, and attitudes. Improvement is not a simple matter of correction of individual "deficits" and activation of individual performance. Rather, improvement in women's attainments means attention also to structurally based factors such as the allocation of resources, access to interaction and collaboration, and the operation of evaluative schemes, as discussed.

Finally, in understanding women's scientific careers, we should recognize that women scientists are not a homogeneous group. Although commonalties of women's status exist (and persist) in science, their status and performance vary, particularly by field. Women's career attainments in science may also differ in ways not yet disclosed by employment sector (particularly academic compared with nonacademic) and by types of institutional locations. The impact of social and organizational factors discussed—human and material resources, workplace practices and policies—may, accordingly, vary by field, sector, or institutional location in science. With further development of research then, the issue of "women in science" may become a matter of "women in sciences"—complex considerations of "where," in "what settings," and under "which conditions, practices, and policies" women do or do not attain significant participation and performance in scientific careers.

NOTES

1. Because interest here is in social studies of science, data were collected only from the *Social Science Citation Index.* Entries were counted under subjects of women and science, gender and science, and women scientists (articles on the subject can appear elsewhere, under salary or education, for example, but, for comparative purposes, counts were limited to these subjects). Because we are interested in research literature, specifically, on the subject, counts were limited to articles and excluded editorials, letters, book reviews, bibliographies, and notes less than three pages.
2. This chapter relies upon national data available through the National Science Foundation, the National Research Council, and the Commission on Professionals in Science and Technology. Thus the focus is upon American women in science. A cross-cultural treatment certainly would be valuable, but such an expanded focus is beyond the latitude, page limitation, and data available for this chapter.
3. Although such discussion is beyond the scope of this chapter, the historical factors associated with the proportions of women in science until 1940 can be found in Rossiter (1982).
4. Consideration of women's distribution by subfields is one of the subjects of the research of Etzkowitz and Fox: "Women in Science and Engineering: Improving Participation and Performance in Doctoral Programs," funded by the National Science Foundation.
5. These also are subjects of the current study cited in note 4.

PART IV
CONSTRUCTING TECHNOLOGY

One of the most notable developments in S&TS over the last decade has been called "the turn to technology." In the same way that sociologists of science earlier opened the "black box" of science, this new scholarship has opened the black box of technology. Rather than examining the impact of technology upon society, the arrow has been reversed; researchers now investigate how society gets into technology —in short, how technology is socially shaped or constructed.

As well as reviewing the burgeoning literature in technology studies, Wiebe Bijker offers a case study of the development of one of the most ambitious technological systems in the Netherlands—the complex of coastal defense systems that keeps the Netherlands as a country "afloat." Bijker's chapter is laced throughout with stories of disasters, near disasters, polders, dikes, sandbags, mattresses, engineers, scientists, politicians, and ordinary Dutch citizens. The metaphor of the "seamless web," first introduced by historian Thomas Hughes, in which technology is interwoven with society, science, politics, economics, and so on, is powerfully illustrated by Bijker's case study. Entities such as the Dutch coastal defense system are woven together from people and artifacts, from society and technology. Bijker calls such seamless webs "sociotechnical ensembles."

Bijker argues that the current enthusiasm for social constructivist and social shaping accounts of technology needs to be tempered somewhat by a return to the traditional question of technology studies: How do technologies affect society? The traditional question, he argues, can be given a new theoretical cast by examining how sociotechnical ensembles gain their obduracy and how they are taken up and modified within society.

Bijker notes that an adequate framework for analyzing the emergence, shaping, and development of sociotechnical ensembles is still missing. As in the rest of S&TS (see, for example, the chapters on theory and methods), healthy competition between different theoretical approaches seems to be the order of the day. Bijker concludes by pointing to the strategic place that technology studies occupy within the field as a whole. Because the political consequences of technology are so obviously manifest, impinging on the day-to-day lives of so many of us, it would be a pity if the activist political challenge originating in the science, technology, and society movement was not taken up by the new work in this area. Bijker warns that "the toll of academic respectability threatens to produce politico-cultural irrelevance."

The varied links between current S&TS and one of the most important modern technologies—computers—are examined in rather different ways by Paul Edwards and Harry Collins. Like Bijker, both Edwards and Collins use rich empirical examples to help make their argument. Edwards provides a review of three topics in the studies of computers: (a) military relations with computers in the post-World War II era, (b) the failure of computerized banking to lead to significant productivity gains, and (c) gender and computer use. Throughout, perhaps answering Bijker's plea not to neglect the impact of technology on society, Edwards is concerned not only to show the social forces shaping digital computers but also to examine some of the effects of their development.

In reviewing recent work on computers within S&TS, Collins takes a rather different tack from Edwards in examining a much more direct relationship between S&TS and computers. For Collins, computers, and particularly the search to build intelligent machines, provide an ideal test bed for ideas in S&TS. Collins does not advocate, as he has done in his work on science, that we investigate solely the social construction of computing; rather, he wants to explore the consequences of conclusions in S&TS about the nature of knowledge for fields such as artificial intelligence (AI) that try to model knowledge and knowledge acquisition processes.

Collins points out that such a research agenda draws attention to current epistemological and theoretical tensions within S&TS. The study of intelligent machines requires an unusually close connection between resource and topic. For example, knowledge is the topic of the sociology of scientific knowledge, but it is also AI's resource; from the point of view of those trying to make intelligent machines, the more they know about knowledge, the better; hence science studies expertise would seem to be particularly apposite. But, in producing recommendations for AI, the traditional agnosticism of the sociology of scientific knowledge—whereby researchers simply step back and study how the participants construct their own knowledge—seems in danger of being displaced. Such reflexive issues are never far from the surface of modern S&TS, but they seem particularly dramatic when intelligent machines become the object of study.

The last piece in this section, by Stephen Hilgartner, offers another case study. However, in this case it is drawn from the area of biotechnology—the Human Genome Initiative. As with computers, there is a rich S&TS literature developing for the genome project. In particular, the origin of the genome project in molecular biology draws attention to the difficulty of separating the study of science from that of technology.

Hilgartner, drawing on his own ethnographic laboratory studies, focuses in upon a specific aspect of the larger project—the attempt to provide mapping landmarks known as sequence-tagged sites (STSs). He shows that the activities of the genomic researchers in advocating and developing such sites can be described with the vocabulary of "heterogeneous engineering," "systems," and "actor networks" derived from the new technology studies. Perhaps the most noteworthy aspect of Hilgartner's case study is his conclusion that STSs served not only as a technical means to advance the genome project but also as a management tool, and ultimately as a way of exercising social control upon genome researchers.

The chapters in this section exemplify one of the most enduring strengths of S&TS—the close integration of theoretical analysis with empirical material. Throughout we see common themes such as the constructed nature of the boundary between science and technology; the mutual construction, hence also stabilization, of technology and society; the nature of the user of technology; the problem of reflexivity;

and so on. We also see links with other chapters such as those of Judy Wajcman on gender and technology; Karin Knorr Cetina on laboratory studies; Gary Downey on engineers; Ashmore, Myers, and Potter on discourse, rhetoric, and reflexivity; and Michel Callon on models for understanding S&TS.

11

□

Sociohistorical Technology Studies

WIEBE E. BIJKER

PROLOGUE

Friday, November 22, 1991, 0:30 a.m., in Zeeland, the Dutch coastal region south of Rotterdam, lights are still on in many of the houses and farms. The wind, force 11 on the Beaufort scale, howls over the countryside and drives huge waves against the dikes. Trees are uprooted and tiles fly from roofs. But it is not this that is keeping the Zeeland people from their sleep. At 1:30 a.m. it will be high tide. Not any high tide, but spring tide. The northwestern storm, which has raged for 2 days, has forced the North Sea into the narrow funnel of the English channel and high up against the Zeeland coast—some 3 to 4 meters higher than normal.

Fear of the sea is keeping the inhabitants awake. In February 1953, 39 years before, Zeeland was flooded during a similar combination of spring tide and storm surge. That night, 1,835 people were drowned, more than 750,000 inhabitants were affected, and 400,000 acres of land were flooded. Since then, Dutch engineers have closed the tidal inlets. They have raised the dikes and reinforced the dunes. Science-fiction types of high technology

---

AUTHOR'S NOTE: I am grateful to Eco W. Bijker, Karin Bijsterveld, Ben Dankbaar, Bruno Latour, Donald MacKenzie, Trevor Pinch, Pieter de Wilde, and Rein de Wilde for stimulating comments on previous drafts.

have been employed to defend this part of the Dutch coast. But who can guarantee that tonight will not be the 1 night in 4,000 when the sea sweeps over the dikes again?

INTRODUCTION

Technology and coastal engineering make it possible for some 10 million Dutch to live below sea level behind dikes. Without this technology, there would have been no Netherlands. Technology thus plays in this instance as in others a crucial role in constituting modern society. It seems pertinent to try to understand how technology can play such a role, how it influences society, how it develops, how it—in turn—may be controlled. This chapter will review studies that have aimed to do this. It will also say, by way of illustration, something about the dike-building technology that the Dutch have used to defend their coast against the sea.

In this chapter I will focus primarily on sociological studies of technology, thereby neglecting several other important bodies of literature. First, there is the history of technology. For a comprehensive review of, especially, the American work in this field, see Staudenmaier (1985). Troitzsch and Wohlauf (1980) provide an introduction to the German tradition. The second body of work, which is only briefly mentioned, is the economics of technical change. (But see also the chapter by Michel Callon in this volume.) The state of the art is represented by Dosi, Freeman, Nelson, Siverberg, and Soete (1988). Neither shall I discuss the emerging approaches in rhetorical and linguistic analyses of technology, nor shall I review the philosophical studies of technology. It is important to reestablish the links to philosophical studies (Winner, 1991). A central series of publications for this field is the various volumes of *Research in Philosophy and Technology* (edited by Paul T. Durbin). A good introduction is provided by Rapp (1981), and Mitcham (1980) gives a comprehensive history and bibliography. Anthropology and archeology are potentially relevant but they are not covered in this chapter.

I have not chosen a disciplinary or chronological review but a thematic ordering—technology's impact, modeling the development of technology, sociotechnical ensembles, controls, and intervention. In also presenting a picture of technology using Dutch coastal engineering, I shall be concerned to ask: How is life at the technological front? What are the activities of engineers and technicians? How is technology shaped and thereby how is society built?

It is a chutzpah to try neatly to define complex concepts such as "technology," "science," and "society" in a handbook like this. I shall therefore restrict myself to stressing the broadness of the word *technology* as I will use it in this chapter. *Technology* will have at least three different layers of meaning: physical *artifacts* (such as dikes), human *activities* (such as making the dikes), and *knowledge* (such as the know-how to build dikes and the fluid dynamics used to model them in the laboratory). Additionally, I will consider the word *technology* to apply not only to hardware technology (such as fascine mattresses) but also to "social" technology (such as the traditional dike management system used in the Netherlands).

This way of talking about technology cuts across the traditional distinctions between such words as *technique, technics*, and *technology*. (In French, German, and Dutch, such distinctions are used more coherently. In English, *technics* is equivalent to hardware; *technique* is associated with methods, skills, routines, and also with concrete instruments; and *technology* has two quite different meanings—first, the science of technics and techniques, and, second, the advanced science-based organizational system of technics and techniques [Rapp, 1981; Winner, 1977]). The first practical reason to provide these definitions is that, over the last two decades, most authors have indiscriminately used the word *technology* to cover all of these meanings and it would be rather pedantic to try to rephrase their work in the more differentiated, older terminology. The second substantial reason is that such distinctions seem to be rather spurious anyway. Where such distinctions seem to hold, they are the result of technologists' work rather than being based upon intrinsic properties of the technologies themselves. A dike is technics (hardware) constructed with a well-determined clay-sand-stone structure to withstand specific forces; it is also a technique (method) to keep out water; and, finally, it is a technology (organizational system) combining mattresses, sand, stones, sluices, measuring equipment, surveillance procedures, and management schemes. Which of these three dikes is the "right" one will depend on the questions asked. A way of defining technology that does not compel us to make a priori distinctions, but instead allows us to trace how actors make various distinctions, thus seems more fruitful. I shall return to this issue later.

Before continuing, I shall introduce some of the salient aspects of "coastal engineering and dike building technology" by briefly presenting the case of the 1953 Netherlands "Flood Disaster." This section is mainly based on published reports, eyewitness accounts, and interviews. See Veen's (1962, pp. 170-200) last chapter (under the pseudonym of Cassandra) for an eyewitness account in English by one of the leading Dutch engineers. Also in English is *The Battle of the Floods* (1953).

The 1953 Flood Disaster

A flood in the low countries of the Netherlands is different than floods in other parts of the world. When a storm flood breaks through ordinary seawalls, the salt water, after having done its often devastating damage, will return to the sea when the storm is over. This is not so in the Netherlands: Because the country is below sea level, the sea will not retreat at all. On the contrary, the tides will continue to run in and out through the gaps in the dikes, thereby further widening them every 6 hours. The technology and costs to reclaim such inundated lands are comparable to those for making completely new polders. (A *polder* is an area of land below sea level, created by building dikes around it and then pumping the water out. Some 40% of the Netherlands consists of polders, varying in size from only a few to many thousands of square miles.) The costs of reclamation of the most difficult parts that were flooded in 1953 were about five times as high as the ordinary selling price per acre. Viewed from a narrow economic standpoint, this is ludicrous.

In early Sunday morning, February 1, 1953, the radio warned: "Many dikes have been damaged. All soldiers on leave have to return immediately. There is a special danger at Ouderkerk. There is a wide breach at Ouderkerk." The dike on the river Rhine at Ouderkerk formed crucial protection for central Holland and its cities, such as Rotterdam, Delft, the Hague, Leiden, Gouda, and Amsterdam (see Figure 11.1). That central Holland was, nevertheless, not flooded was due to a combination of the quick reaction of a skipper and the existence of some of the small old dikes remaining from the network that had been woven through the Netherlands in earlier centuries. A small vessel navigating on the river near Ouderkerk in the early morning was steered into the breach, then 7 meters wide. She was grounded just in front of the breach on the dike's threshold; citizens and soldiers then filled the gap with sandbags and blocks of concrete waste. Before nightfall, the breach had been closed and central Holland saved. (This event emulates the mythical rescue of Holland by Hansje Brinker, an 8-year-old boy who rescued the city of Haarlem and surrounding polders by placing a finger in a hole in the dike when he saw the sea oozing through.) During that day, the old inner dikes held, which was considered by Van Veen (1962) as "sheer undeserved luck" (p. 177). "Undeserved" because engineers had warned the government a long time before that the dikes and sluices were in very poor condition. Van Veen himself, as chief engineer of the Rijkswaterstaat (see below), had been one of the principal spokesmen for upgrading the condition of the Dutch dikes. Almost a year later, on November 6, 1953, the last and largest of the 89 tidal breaches was closed—just before the winter storm season began again.

The whole gamut of technologies that had been developed during centuries of fighting the sea was employed to reclaim the lost land. First of all, *time* was crucial. Tidal currents quickly widen all gaps. The largest gap in the 1953 disaster was 100 meters wide and 15 meters deep on February 1, but a few months later it had grown to 200 by 20 meters. If the gaps had not been closed before the next winter season, this process of erosion might have been irreversible. Also on a scale of minutes, time is crucial. Currents rage at their fastest when gaps are at their smallest. So, the right moment to close off a final gap is in the few minutes of slack water. The age-old technique is to build the dam by working continuously, then close the last gap in the minutes of slack water with a dam, when the water is only a few inches above sea level. (During "slack water," the direction of the tidal current changes from ebb to flood or vice versa.) Later the height can be increased and the dam finished.

For centuries the key *material* used to strengthen and repair dikes was sandbags. On the night of February 1, 1953, sandbags were made available from emergency depots and played a crucial role. Only since the closure of the Zuiderzee in the 1920s has boulder clay come to be used to build dikes that are too large for sandbags. The allied invasion of Normandy was made possible by caissons that have increasingly been used to close a tidal gap suddenly and to build the basic structure of the dike. A caisson is a large floating structure, often made of concrete, that is towed into the right position and then sunk onto the sea bed to form the skeleton of a breakwater or dike. Since the 1920s huge supplies of "unpacked" sand, dredged or sucked from the sea bottom, has been used to strengthen the framework of a dike. More recently (in the 1960s) huge blocks of concrete have been employed—dumped by special ships, from cable trains, or even from helicopters.

In 1953, as in the centuries before, *human power* was the energy source that did most of the work. This was the only power source that was distributed widely enough throughout the Dutch coastal area to act adequately at short notice. The sacks were transported by human chains and dumped in the right positions. For the final closures of the 1953 breaches, motor dredges, tugs, ships, and cranes were used, but on that 1953 February night it was human power that saved the day.

An *armored foundation* is necessary to build a dike in a gap with tidal currents—the natural seabed of sand will not do as it will be scored away, resulting in the undermining of the dike. For centuries, fascine mattresses have been used for this foundation. Such a mattress consists of a net structure about 20 centimeters thick, some 100 meters long, and 20 meters wide. A series of such mattresses is lowered onto the seabed to provide a foundation for the dike to be built upon. Until the 1970s, when synthetic mattresses were used for the first time for large tidal closures, dikes in the Netherlands (and on all sandy coasts throughout the world) were built on hand-woven mattresses made from branches of willow (or similar) trees. After the mattresses have been prefabricated on land, they are towed out to sea, where they are sunk by the careful dumping of quarry stone on them. This used to be done by hand to guarantee a gradual and controlled lowering into the right position.

Scientific research started to play a role in the 1920s: The physicist Lorentz was asked to make mathematical predictions about the tidal effects caused by a closure of the Zuiderzee. Hydraulic research received a boost during World War II when it was called upon to investigate the conditions under which harbors could be built for the invasion of Normandy. Empirical research with scale models had already started in the 1930s but was intensified immediately after the war. Such scale models played a crucial role in the closure of the last gap in the 1953 disaster. The closure with caissons was carried out some hundred times beforehand in the scaled-down reality of the laboratory. The engineers did not have much "real life" experience with these caissons. Researchers using force-measuring instruments held the cables and acted as tugboats holding the last caisson in place to plug the gap. To be finished before the tide returned, the tugboats had to start while the flood was still rather strong. On the day of closure, at scale 1:1, the young engineers who had practiced in the laboratory were standing on deck behind the old experienced tugboat captains. When one of the cables snapped and control of the caisson was about to be lost, they intervened because they had seen that snapping rope a dozen times and had worked out a scenario to save the caisson. With a series of unusual commands, they took advantage of the queer characteristics of the currents they had discovered in the lab. The last caisson was eased down into the final gap during the crucial few minutes of slack water, and the breach was closed. (Research using modeling is no guarantee for success. For one thing, it depends on whether you have modeled all relevant aspects. Though the Zierikzee closure first seemed a success, a few days later the caissons started to shift. Because the engineers had not wanted to lose the time necessary to lay a fascine mattress foundation, the ground was too slippery and the caissons were pushed out of the gap.)

TECHNOLOGY'S IMPACT

The first theme of technology studies to be discussed is the impact of technology. Technology's effects on society and nature are so ubiquitous that they frequently become invisible. In public discussions about the closure of the Zeeland tidal inlets in the 1970s, for example, it was often argued that "technology is destroying nature." In the Netherlands, however, as in many other densely populated areas of the world, "nature" is nothing more than an environment created by previous technologies. If it were not for a variety of coastal engineering technologies, Dutch society and nature simply would not exist. God created the world, but the Dutch made Holland!

The effect of technology on society is one of the oldest themes in technology studies. A classic case is the alleged effect of the stirrup in establishing feudal society (White, 1962). The stirrup, according to White, effectively welded horse and rider together into an unprecedented fighting force. The maintenance of such military technology, however, required large investments in men, horses, and equipment—investments that could only be made after a fundamental reorganization of society. (See, however, Hilton & Sawyer, 1952, for a critical review of this claim.) Another example of such an approach is Giedion's (1948) plea for "anonymous history," by which he meant an empirical inquiry "into the tools that have molded our present-day living": locks, bread-baking technology, furniture, household technology, bathing equipment.

Marx's analysis of technology as affecting the forces of production and thus determining society in general presents a similar claim. (See Marx's *Das Kapital*, 1867.) MacKenzie (1984) has argued for the continuing relevance of Marx's analyses of the machine; the article also provides a comprehensive bibliography. MacKenzie, though, shows that Marx cannot be classified as a simple technological determinist (see below). (See also Bimber, 1990.) In addition to Marx, only a few other economists—all outside neoclassical economics—have considered technology's impact on society. Within neoclassical economics, technology is considered to be important but it is taken for granted that technological innovations are always unproblematically available, "off the shelf." Neoclassical economic models try to capture the influence of specific labor, capital, and technical changes on economic growth. Technology itself, however, stays outside of the analysis—an exogenous variable. This work consequently does not contribute to a *specific* understanding of the impact of technology on society. The factor technical change is "a rag-bag for all social, managerial, structural, educational, political, psychological and technological changes other than the purely quantitative increase in the volume of labor and capital" (Freeman, 1977, p. 244). Next to Marx, Schumpeter (1934/1980, 1939, 1942/1975) has been the most influential economist to have analyzed the role of technology. (See Hagedoorn, 1989, for an introduction to early economic technology studies and their relevance for modern work.) For Schumpeter, radically new innovations could create completely new industries and thus lead to new economic and societal directions. In his earlier work, he assumed these innovations to be available off the shelf—technology was still an exogenous variable. In his later work, technology became an endogenous variable when he analyzed the role of the entrepreneur in creating innovations. Schumpeter's position has been labeled "technology push" and the opposed view as "demand pull"—increases in inventive activity respond to increases in demand.

(Schmookler, 1966, is the *locus classicus* for this hypothesis. Many empirical studies were carried out to settle this issue of demand pull *versus* technology push. See Coombs, Savioti, & Walsh, 1987, and Rosenberg, 1982, for reviews and critique.) This situation changed around 1980 when economists no longer accepted this dichotomy and started to find ways of analyzing technology as an endogenous variable. Examples are Rosenberg (1976, 1982), Freeman (1977, 1982), Nelson and Winter (1977), Dosi (1984), Coombs et al. (1987), Clark and Juma (1987), Hagedoorn (1989), and the work reported in Dosi et al. (1988). Blume (1992) addresses similar issues from a sociological starting point.

Another fairly general analysis of technology's impact is given by Mumford (1964) in his categorization of technologies as authoritarian versus democratic. Much of Mumford's work can be understood, retrospectively, as underpinning his claim that these two types of technology have recurrently existed side by side since neolithic times—"the first system-centered, immensely powerful, but inherently unstable, the other man-centered, relatively weak, but resourceful and durable" (Mumford, 1964, p. 6). He warns against the hold that distributed technologies exert over all members of society: "Our system has achieved a hold over the whole community that threatens to wipe out every other vestige of democracy." According to Mumford, it does so in a much more subtle way than previous forms of overtly brutal authoritarian technological systems. One important focal point for Mumford has been the city as a technology (see, e.g., Mumford, 1934, 1938, 1961). Mumford (1967, 1970) later extended this analysis of the city into an account of the "megamachine"—an elaborate network of human and nonhuman components used to order the world and to exert power. (For a recent introduction to Mumford's work, see Hughes & Hughes, 1990.) Ellul (1954/1964) has followed similar lines of argument.

In Mumford's analysis, technological development does have a general political, cultural thrust. In other studies, investigating the impact of technology in more specific terms, the same claim is summarized in the adage "artifacts having politics"; not only does technology have an impact on society, its effects are also strategically directed. Winner (1986b, pp. 19-39) coined this phrase in his 1980 article "Do Artifacts Have Politics?" One of his cases was formed by the bridges over the Long Island parkways. These were deliberately designed by Robert Moses so as to discourage the presence of buses on the parkways leading to the beaches:

> Automobile-owning whites of "upper" and "comfortable middle" classes, as he called them, would be free to use the parkways for recreation and commuting. Blacks and other poor people who normally used public transit were kept off the roads because the 12-foot-tall buses could not pass underneath the bridges. One

> consequence was to limit access of racial minorities and low-income groups to Jones Beach, Moses' widely acclaimed public park. (Winner, 1986b, p. 23)

Whose politics do artifacts have? One specific answer is offered by studies of the labor process. Marx's analysis, in which the capitalist owns the means of production and products while the worker owns her or his labor power, is extended especially by Braverman (1974) in recent "labor process theory." (See also Blackburn, Coombs, & Green, 1985; Knights & Willmott, 1988; Noble, 1984; Thompson, 1983; Wood, 1982; Zimbalist, 1979.) The capitalist (and his chief instrument, management) are analyzed as seeking control over the workers' labor power. Following F. W. Taylor's principles of scientific management, the capitalist tries this by deskilling the worker. Technology proves especially useful in this process, according to labor process theory. It enables capitalists to acquire a monopoly over skills and thus to control labor. Much of early German sociology of technology is focused on similar questions. The volume edited by Jokisch (1982) provides a comprehensive overview, which also addresses, in addition to the impact theme, questions concerning the shaping of technology.

How do artifacts acquire their politics? Is it bestowed upon them by their users or is it "baked into them" during their construction? Have the fascine mattresses leaped into existence because of a brilliant inventor, or have they emerged by a slow, gradual accumulation of small improvements? Different kinds of answers have led to considerable bodies of literature to which I shall now turn.

MODELING THE DEVELOPMENT OF TECHNOLOGICAL ARTIFACTS

In this section I want to discuss three classes of models for the development of technological artifacts—materialistic, cognitivist, and social models. The first class focuses on the *technology itself* as its explanatory basis, thus stressing the relative autonomy of technological development; the second class consists of evolutionary and rationalistic models, taking technological *knowledge* as the key characteristic for modeling technology; and the third class finds its starting point in the *social practices* related to technology and focuses on the social shaping of technology. The three classes of models are to some extent complementary—they stress different but important aspects of the development of technology. In other respects, however, they are contradictory, and I will argue that the models presented below give successively more adequate descriptions of the development of artifacts. They do so because they provide successively better instruments to analyze the development of technology *in its relations to society*. For, if the ultimate goal of sociohistorical technology studies is to understand the interplay between technology and society, those models that explicitly conceptualize (parts of) the relationship between the social and the technical are to be preferred.

Another way of comparing theories of technology is to ask whether they are technologically determinist. Bimber (1990) distinguishes different versions of technological determinism—the "norm-based account" (associated with Jürgen Habermas and Jacques Ellul), the "unintended consequences account" (used by, for example, Langdon Winner), and the "logical sequence account" (most clearly advocated by G. A. Cohen, Richard Miller, and Robert Heilbroner). It is only the latter, Bimber argues, that is both technological and deterministic. I will use the term in this sense. *Technological determinism,* then, comprises two ideas—technological development is autonomous, and societal development is determined by technology. Following Edgerton (1993), it seems wise for analytical purposes to reserve the term *technological determinism* only for the second idea, a theory *of society*. This thesis of technology determining society should then be kept separate from the idea of an autonomous technology, a theory *of technology*. In practice though, the two views often are held in combination. *Technological determinism* has become a term of abuse among scholars. Even those who take the view that technological change is the prime mover of socioeconomic change vehemently reject the label. This is probably partly caused by the important ideological role the technological determinist view plays in public and political discussions about technology. There it results in a displacement of causation from human agency to machines, which is detrimental to attempts to create instruments for more democratic control of technology and society. The following models provide successively better tools for avoiding such an autonomous view of technology.

Materialistic Models

The development of fascine mattresses, especially in this century, seems to provide an ideal example of a technology that "logically" was improved, and thus requires no further explanation for its development. By small steps the production yard where the mattresses were prefabricated was improved; new materials such as polypropylene were used as a substitute for willow branches; mechanization was introduced to fabricate various elements of the mattresses; special ships were constructed for sinking the mattresses by mechanically dumping stones in a controlled way. The apex of this development was the design of a special mattress for use in the Oosterschelde in the 1970s. Here, a synthetic mattress was manufactured in a specially constructed plant and rolled onto a huge drum that could be picked up by a special ship. The ship was manoeuvred into the right position and the mattress was gradually rolled off and lowered into the sea—as if dwarfs were playing with a gigantic paper toilet roll. Developments in new materials and mechanization provided the impetus for this seemingly "automatic" development.

Most materialistic models that stress an "autonomous logic" of technological development are not very elaborate and often can be found implicitly in much of the work discussed in the previous section on technology's impact. These models are often also technologically determinist. Other models in this materialistic class stress the contingent character of technological development rather than a unilinear logic. Generally these models are evolutionary. An evolutionary model of technology would, first, highlight the variations generated by engineers and others, and then focus on the process by which some of these variations are selected. The fascine mattresses, for example, passed the selection process because in the wet Netherlands there was an abundant supply of willows, and thus for centuries this technology hardly changed. Many innovations in making these mattresses found a niche when after 1953 the demand increased dramatically—a mechanical "spinner" to make a cable of willow branches, an artificial yard to do the weaving without being bothered by the tide, and synthetic materials to substitute for wood. Several of the variations that were invented in the 1950s would have been possible, from a narrow technical point of view, decades earlier.

Various authors have formulated evolutionary models to get away from the image of the heroic inventor. Ogburn (1945) and colleagues (Gilfillan, 1935a, 1935b; Ogburn, with Gilfillan, 1933; Ogburn & Nimkoff, 1955) and Usher (1954) are among the earliest of these. They all stress the accumulation over time of small variations that finally yield novel artifacts. Recently, Basalla (1988) has presented an evolutionary theory of technological development that also focuses primarily on the development of technical artifacts. The economic model developed by Nelson and Winter (1977) also falls within this rubric, though the unit of analysis in their model is the firm rather than the artifact. They identify "technological regimes" that capture engineers' beliefs about what is feasible and thus give technological development an explicit place. Their model has some overtones of technological autonomy—they see technology as developing along a "natural trajectory."

Cognitivist Models

One step away from an autonomous view of technological development is to stress the role of technological knowledge. Such a more cognitivist view of the fascine mattresses development avoids the assumption of a logical (or evolutionary) chain of successive artifacts. Rather than a fairly automatic process induced by mechanization and material innovations, or a more contingent process governed by blind variation and selective retention, the cognitivist view stresses the role of problem solving. Mechanization thus was a solution to the increased demand for mattresses for dike foundations; new materials were a solution to the scarcity of willow trees; and preloaded mattresses with fixed ballast were a solution to the problem of sinking mattresses on steep slopes from which the loose ballast stones would have rolled down.

Cognitivist models of technology come in many shapes and forms. They take their starting point in the observation that the most important aspect of technology is technological knowledge (see Laudan, 1984, for a comprehensive review and several contributions to this approach). Some of these models are primarily rationalistic (see, e.g., Vos, 1991). Most, however, try to capture the contingent character of technological knowledge in evolutionary models. (See Elster, 1983, for a discussion of several varieties of models, including the evolutionary one.) Constant (1980) and Vincenti (1990) both find their starting point in D. Campbell's (1974) work, which is part of a large tradition in trial-and-error and variation-and-selection models for the growth of scientific knowledge (see Radnitzky & Bartley, 1987, for a selection, including Campbell, 1974). Constant (1980), in an analogy to Kuhn (1962/1970), distinguishes between normal and radical technological activity. He then analyzes the development of normal technology by stressing the role of the relevant technological community. Constant identifies the possibility of coevolution of two closely linked but separate technologies that exert powerful mutually selective pressure. He describes the turbojet revolution as a case of coevolution: "New airframes partially defined the evolutionary environment of aircraft piston engines, while creation of the turbojet ultimately would compel a revolution in airframe design and construction" (Constant, 1980, p. 14). By focusing on the design process, Vincenti (1990) studies the "epistemology of engineering." He identifies six categories of technological design knowledge and subsequently describes several knowledge-generating activities. Finally, Vincenti proposes an evolutionary model to describe the growth of this technological knowledge as a nested hierarchy of blind-variation and selective-retention processes in which the knowledge produced at one level is used in the process at the next outer level.

What is the relationship between science and technology? All technology studies problematize the relationship between science and technology. The old image of technology as being merely applied science was already dismissed in the first STS handbook (Spiegel-Rösing & Price, 1977)—it is no adequate description of the entanglement of the two (Layton, 1977). Pinch and Bijker (1984) argue for an empirical approach, not assuming a priori any inherent distinction. Vincenti (1990) draws some distinctions between technological knowledge and scientific knowledge but grants autonomy to the realm of technological knowledge—it is no longer to be considered an appendage to science.

Social Shaping Models

The idea that technology is socially shaped, rather than an autonomously developing force in society or a primarily cognitive development, is not entirely new (Mumford, 1934, 1967, for example, argues that sociocultural conditions precede the development of specific technologies), but its present momentum and precise formulation are quite recent. This view is most clearly advocated by MacKenzie and Wajcman (1985). It takes issue with the autonomy view head on. Social shaping models stress that technology is not following its own momentum or a sort of rational, goal-directed, problem-solving path but is instead shaped by social factors.

One early strand of this theme again is inspired by Marx and Braverman. Noble (1984) argues that, in the automation of machine tools, the choice of the numerical-control type over the record-playback type can be understood as the management's social shaping of this technology: The first would allow more control over the workforce than the latter, where highly qualified workers still play a crucial role.

At a more general level, state intervention is a form of social shaping of technology. Military technology is one example where such shaping plays an important role with its direct consequences for electronic, nuclear energy, and transportation technologies. At a similar general level, domestic technology and house building are socially shaped by the gender relations in a society. (See the chapters by Smit and by Wajcman in this volume. MacKenzie, 1990a, uses his analysis of guided missile technology as a platform to contribute to an understanding of sociotechnical development more generally.)

Focusing on the communities of practitioners, the sociological versions of some of the work discussed under the previous heading also contributes to an understanding of the social shaping of technology. In Constant's (1980) analysis, for example, the "community structure of technological practice" plays a crucial role. The primarily economic Nelson/Winter model was first extended by Dosi (1984), who introduced the concept of "technological paradigm." It was then sociologized by Belt and Rip (1987), who added that the variation and selection processes are not independent but linked by a "nexus"—examples of such nexuses are the patent system and the test departments of large R&D laboratories.

Pinch and Bijker (1984) have argued that technology studies could benefit from recent social studies of science. In a general methodological argument concerning the explanatory tools to be applied to the analysis of technology, they first argue, with Staudenmaier (1985), that, to avoid a Whiggish account of technological development, more attention should be paid to failed technologies rather than exclusively doing the success stories. Then they suggest, with Bloor (1976), that, in analyzing the failure and success of artifacts, the same conceptual framework should be applied. This "principle of symmetry" can be formulated as follows: Do not, in explaining the success or failure of an artifact, refer to the working or nonworking of that artifact as explanation. The working of an artifact is not an intrinsic property from which its development stems but is a constructed property and the outcome of its development.

SOCIOTECHNICAL ENSEMBLES

Students of technology have not been satisfied with the partiality of the *impact* and shaping themes. One of the problems of the social shaping thesis is, for example, that there seems to be too little room for the obvious effects of technology on society. Much recent work has been devoted to integrating the impact and shaping themes within one analysis of what I will call "sociotechnical ensembles" (see Law & Bijker, 1992, for a review of recent work along this line). Before briefly reviewing this work, I will again turn to a coastal engineering case—one that particularly seems to call for such an integrated analysis.

The Deltaplan

The 1953 disaster in the Netherlands hastened political discussions over the deterioration of the dikes. Did the solution lie in repairing and heightening the dikes or in pursuing more radical options? (Before World War II, Van Veen had actually proposed several plans that included closures of the tidal basins in the Zeeland delta. None of these plans had been discussed very seriously.) On February 18, 1953, a government committee was set up. Its most important report, published in February 1954, recommended a radical solution. Work started unofficially in August 1955, and in 1957 the Dutch Parliament finally followed the committee's advice and opted for the radical solution. A law was passed that specified "the Deltaplan"—all tidal inlets in the Zeeland delta, where the Rivers Rhine and Maas flow into the North Sea, were to be closed except the Westerschelde, connecting Antwerp to the sea, and the Waterweg, connecting Rotterdam to the sea (see Figure 11.1).

This radical choice entailed a partial abandonment of the preexisting decentralized system of dike and water management. The large "high-tech" closures of the tidal inlets would come under national jurisdiction rather than under the command of the local water boards. One of the conclusions after the 1953 disaster formulated by the central authority, the Rijkswaterstaat, was that this decentralized control system was partly to blame for the bad quality of some of the dikes. (The Rijkswaterstaat is the national governmental body for water and dike management. Because of its huge size and related political power, it was sometimes called "a state within the state.") Another possible conclusion, though at that time formulated by only a few, was that the decentralized dike management system had provided the crucial skills, knowledge, and technology to react adequately, immediately, and on the spot after dike breaks. In the Deltaplan, coastal protection previously provided by some 1,000 kilometers of dikes under decentralized control was to be taken over by less than 30 kilometers of dikes under centralized control. By the late 1950s, the faith in the Rijkswaterstaat was such that this centralization met with no resistance.

The work proceeded well. It was explicitly set up as a learning process—the "Delta-school." The first project was a flood barrier in the Rhine just downstream of the 1953 breach at Ouderkerk to protect the central region of Holland. (A flood barrier is a structure that normally is open to let the water pass but can be closed in case of a dangerously high tide.) Other projects followed in order of increasing magnitude and complexity. In addition to the four main closures, a series of secondary inland dams and sluices was needed to control the tidal streams during the main closure works and to manage water levels afterward. The four main closures were to be three dikes and one discharge sluice for the water of the Rivers Rhine and Maas. Safety criteria had overriding priority in decisions on the details of the Deltaplan. The buzzword in both political and technical discussions was *Delta level*—the water level during a storm surge that had a chance of occurrence in 1:10,000 years. Dikes were designed with this level in mind. In design practice, this so-called basic level with a transgression chance of 1:10,000 did not have the unambiguous hard character it had in the public debate. Various differential factors, among which hydrological coastal variables and an "economic reduction factor," resulted in different design levels along the Dutch coast. The resulting design level for the Zeeland coast was, for example, 1:4,000 years. Flood barriers or sluices to allow the tide to continue were ruled out because these would be too expensive. The Deltaplan would thus create several huge new freshwater lakes without any tidal movement. Though detailed insight into the ecological effects was neither available nor deemed very important, it was explicitly recognized that oyster farms and saltwater fishing would have to be abandoned. This price seemed reasonable enough in the light of the required safety and there was no discussion of it, especially because the freshwater would be beneficial to agriculture in the region.

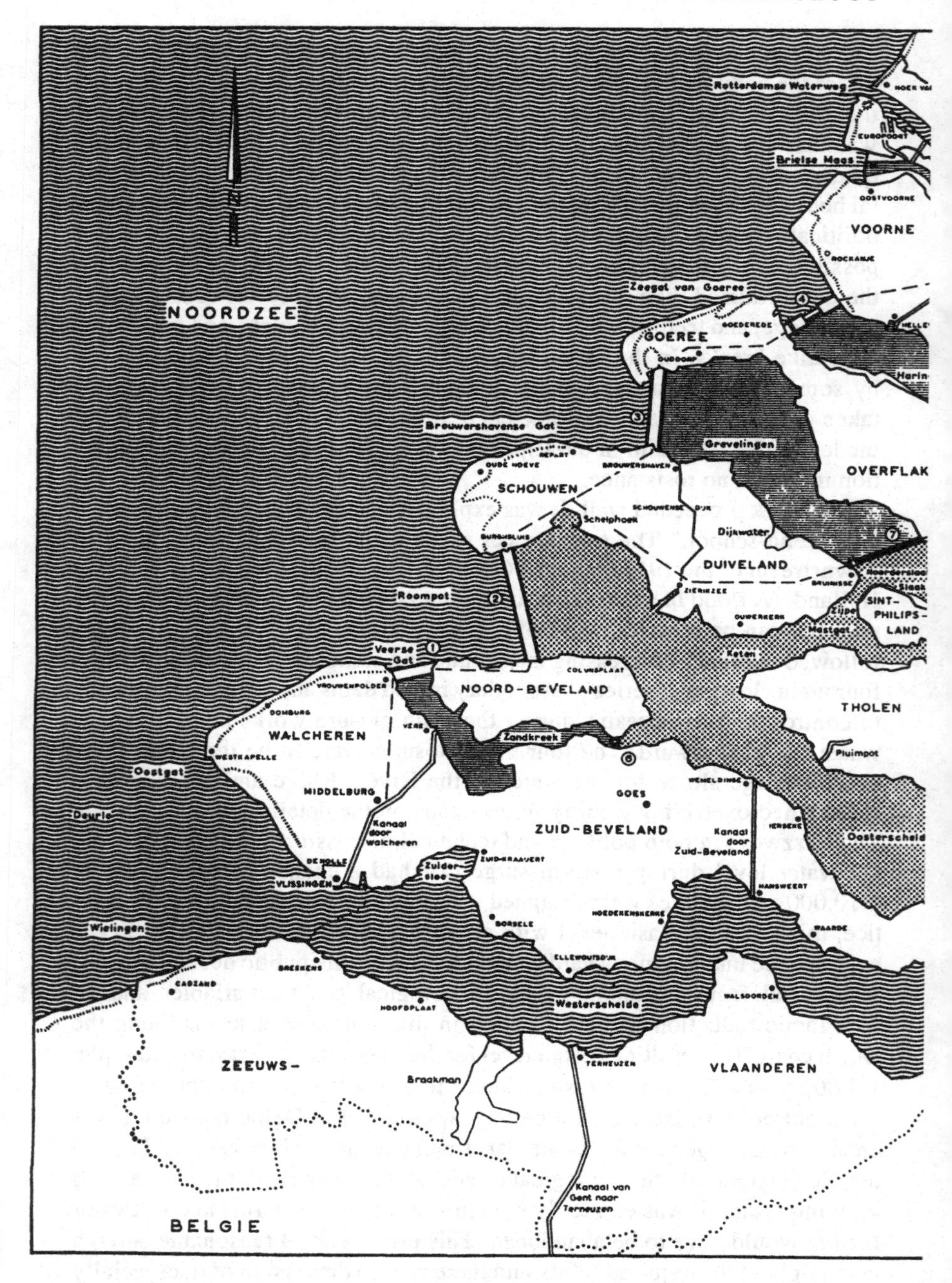

Figure 11.1.The Deltaplan as Adopted in the Delta Law of May 1958

NOTE: The northern boundary of the map is formed by the Hook of Holland and Rotterdam; the southern, by the Belgian border (Antwerp is at the end of the most southern water way, De Schelde).

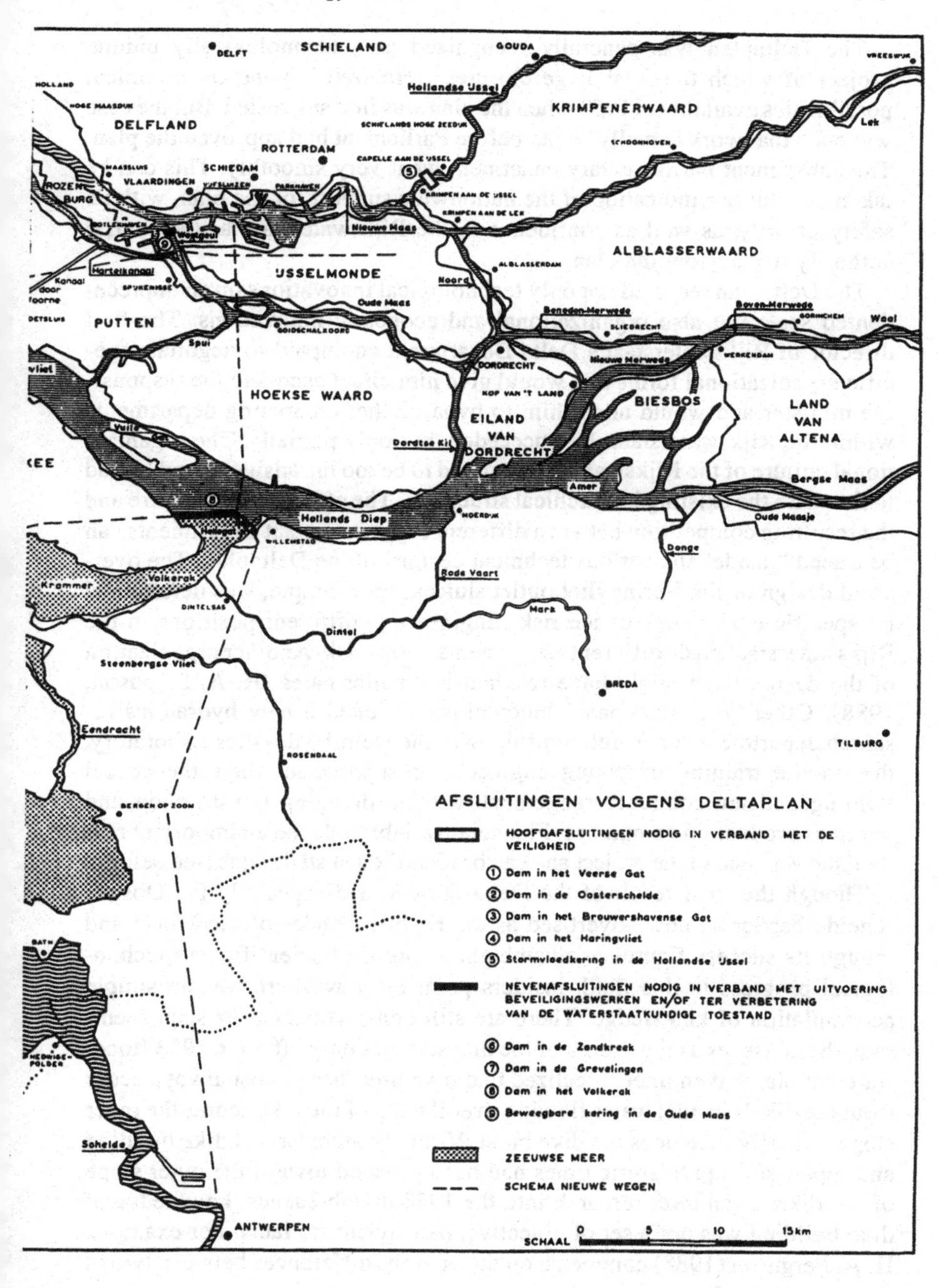
SCHIELAND
DELFT
GOUDA
Hollandse IJssel
KRIMPENERWAARD
Lek
DELFLAND
ROTTERDAM
SCHIEDAM
VLAARDINGEN
MAASSLUIS
ROZENBURG
Nieuwe Maas
ALBLASSERWAARD
IJSSELMONDE
Noord
Oude Maas
PUTTEN
Spui
Beneden-Merwede
Boven-Merwede
Waal
SLIEDRECHT
GORINCHEM
DORDRECHT
Nieuwe Merwede
WERKENDAM
HOEKSE WAARD
KOP VAN 'T LAND
BIESBOS
LAND VAN ALTENA
EILAND VAN DORDRECHT
Dordtse Kil
Bergse Maas
Amer
Hollands Diep
Oude Maasje
MOERDIJK
WILLEMSTAD
Donge
Krammer
Volkerak
Roode Vaart
DINTELSAS
Dintel
Mark
Steenbergse Vliet
BREDA
Eendracht
TILBURG
ROSENDAAL
BERGEN OP ZOOM
BATH
Schelde
ANTWERPEN
AFSLUITINGEN VOLGENS DELTAPLAN
HOOFDAFSLUITINGEN NODIG IN VERBAND MET DE VEILIGHEID
① Dam in het Veerse Gat
② Dam in de Oosterschelde
③ Dam in het Brouwershavense Gat
④ Dam in het Haringvliet
⑤ Stormvloedkering in de Hollandse IJssel
NEVENAFSLUITINGEN NODIG IN VERBAND MET UITVOERING BEVEILIGINGSWERKEN EN/OF TER VERBETERING VAN DE WATERSTAATKUNDIGE TOESTAND
⑥ Dam in de Zandkreek
⑦ Dam in de Grevelingen
⑧ Dam in het Volkerak
⑨ Beweegbare kering in de Oude Maas
ZEEUWSE MEER
SCHEMA NIEUWE WEGEN
SCHAAL 0 5 10 15km

The Deltaplan was generally recognized as a technologically unique project of which the final large closures were well beyond the technical possibilities available in 1953 when the plan was first suggested. But the case was such that work actually began before Parliament had approved the plan. The subsequent parliamentary enactment went very smoothly. This can be taken as a further indication of the nationwide support for the plan, with its safety priority, as well as confidence in the Rijkswaterstaat as the central authority to carry out the plan.

The Deltaplan required not only technological innovations on an unprecedented scale but also organizational and economic innovations. The first director of Rijkswaterstaat's Delta Department attempted to negotiate specific organizational forms that would give him direct access to the responsible minister and would allow him to bypass other competing departments within the Rijkswaterstaat. He succeeded, but only partially. The organizational culture of the Rijkswaterstaat proved to be too intransigent, and he had to live with the existing hierarchical structures. The effects of this failure and the resulting competition between different Rijkswaterstaat departments can be traced "inside" the various technical designs of the Deltaplan. The oversized design of the Haringvliet outlet sluices, for example, was determined by specific evaluations of ice risk. Engineers at different positions in the Rijkswaterstaat made different assessments of this risk. Another organization of the design team might have resulted in smaller gates (H. A. Ferguson, 1988). Other "organizational" innovations included a new hydraulics research department, a new relationship with the Delft Hydraulics Laboratory, the on-site training of young engineers to supplement their theoretical training, and new contract arrangements with the dredging, construction, and mattress production companies. Financial variables played an important role until the very end of the project and can be identified in all designs (see below).

Though the final result of the Delta project, and especially the Oosterschelde barrier, is now advertised as the Eighth Wonder of the World and though its success figures as unambiguous proof of scientific and technological progress in this field, insiders point out how there was no simple accumulation of knowledge. There are still controversies over such seemingly basic issues as the design of the dikes. It was only after the 1953 flood, for example, that engineers realized that dike breaches almost always occur from the dike's inside; water flowing over the top of the dike scores the inner slope and only then does the dike burst. Virtually all efforts at dike building and repair since prehistoric times had been directed toward the outer slope of the dike. Even after research into the 1953 dike breaches, knowledge of dike building was not a set of objective, hard scientific facts. For example, H. A. Ferguson (1988) comments on the obvious differences between two of the main Deltaplan closure dikes:

> One may think that these [differences] have been caused by important technical considerations. The reason however is, that the designers could not agree about the benefit of an outside shoulder; in the one case the opponents of the shoulder won and the Veere Dam did not get a shoulder; some time later the proponents won, and the Brouwers Dam got a broad shoulder. We still do not know what is best. (p. 60)

Research using models played a crucial role in carrying out the Deltaplan. Since the 1920s mathematical models had been used, and since the 1940s these were complemented by physical models in which physical dimensions were scaled down by factors of 100 and 400; time was scaled up by a factor of 40; sand was scaled down by using finely ground Bakelite; and water stayed water at 1:1. (Vertical down scaling, for example, 100:1, cannot be as large as the horizontal down scaling, for example, 400:1, because the water's behavior changes fundamentally when flowing in more shallow streams. This is one example of the complicated scaling laws that are involved in all technological modeling. These scaling laws mean that results from the model cannot be translated into scale 1:1 in any unambiguous or "objective" way, just as the results of scientific experiments cannot be taken to provide an unambiguous answer about Nature's state.) The most complicated model of the Oosterschelde used a combination of salt and freshwater. For detailed studies of dikes and structures, wind and wave flumes were used. The organization of this model research was as difficult and crucial as interpreting the scaling laws. Managing the relations between Rijkswaterstaat, the Delft Hydraulics Laboratory, and the private construction firms was thus as much part of the Deltaplan closures as was the weaving of mattresses or the design of the discharge sluices. Most research was carried out by the Delft laboratory, but the translation into design and construction was the Rijkswaterstaat's responsibility. At some point, the private dredging firms also carried out some scale model studies in Delft. Because, for reasons of competition, they wanted to keep these results secret from other firms and the Rijkswaterstaat, the hall in which the models were located was as carefully guarded as any military research facility.

By the end of the 1960s the Deltaplan had been extremely successful and proceeded to its final phase—the closure of the largest tidal basin, the Oosterschelde. The pylons for the cable way that would dump the large concrete cubes for the core of the dike were put in position in 1971. However, the almost nationwide support that the Deltaplan had received in the 1950s now started to wear thin. The special quality of the tidal ecology of the Oosterschelde became valued more than before; the unwelcome prospect loomed of the closed estuaries becoming huge sinks for the polluted waters of the Rhine and Maas; growing attention to the "mountains" of butter and wheat in the European Community lessened the importance of stimulating agriculture with new freshwater supplies—food production no longer seemed to be the primary problem it had been immediately after World War II. Also other societal changes in the 1970s affected the Delta works—the Rijkswaterstaat's authority (as in so many other political institutions in the Netherlands) was disputed. During the general elections in 1972, the Oosterschelde closure became a political issue. The alternative of leaving the Oosterschelde open and increasing the height of all its dikes was proposed. As a result of this debate, the Netherlands were almost torn apart. In Parliament the almost unanimous support for the project from 15 years before was now reduced to only 50%, and some of the coastal engineers were transformed from respected builders of a safer nation into ecology wreckers, even to their own families.

In 1974 the government decided after resolving a deep internal controversy to close the Oosterschelde with a flood barrier. This would delay for some 7-8 years the finishing of the Deltaplan, and thus the reaching of the "Delta-level" safety criterion of a 1:4,000 years flooding chance. To many, especially in Zeeland, this was an unacceptable risk given that the present dikes had a flooding chance of 1:100 years. In Parliament a motion to continue the complete closure was rejected by 75 to 67 votes. Those who were in favor of closure called this "a purely political decision."

The pylons for the cable way were dismantled and brand-new mattresses were dredged away. Intensive research on the tidal streams and sand bottom started. To increase the safety of Zeeland during the delay period, the dikes were heightened to a flooding chance of 1:500. It soon became clear that the project was of unprecedented complexity. The Oosterschelde bottom was found to be of an unusually fluid sort of sand and the tidal velocities and volumes encountered were larger than in any previous project. One of the consequences was that the flood barrier had to be constructed on the spot, in open water. The only other big mechanical construction for the project—the discharge sluices—had been built in a working polder in the inlet; after completion of the sluices, the polder was flooded, its dikes removed, and the remaining gap between sluices and shore closed with a dike. Such a solution was now impossible because the resulting decrease of tidal volume would irreversibly affect the ecology of the Oosterschelde. Several radically different designs were discussed and tried in model research.

The two main problems turned out to be the foundation of the dam (a shift of only a couple of inches would jam the sliding doors of the dam, reducing the flood barrier to a useless ruin) and control of the budget (Parliament had stipulated maximum costs; if this condition could not be met, the closure had to be made with an ordinary dam). The foundation problem was solved partially by building a new ship that used four 40-meter-long vibrating needles to condense the seabed, and partially by using prefabricated mattresses that were sunk by a specially designed ship with a horizontal and vertical accuracy of a few inches. The budget problem was solved by a combination of measures: a design adaptation in the latest stage (one sliding door less in the barrier), violation of the no-interference-with-ecology principle during construction (engineers were allowed "to flipper": to close the barrier several times to facilitate works), and creative accounting.

The Oosterschelde flood barrier dam was opened by the queen of the Netherlands on October 4, 1986. This virtually finished the Deltaplan. It was completed some 10 years behind schedule with a budget overrun on the Oosterschelde barrier of 30%. At the same time, this technological system is widely acclaimed to be the crown of Dutch coastal engineering. It is, however, still mistrusted by a few experts because of the unprecedented critical nature of its foundation. In the Deltaplan, as perhaps best illustrated by the Oosterschelde barrier, values of safety and environmental protection are built into the very technologies. But the organization of Rijkswaterstaat, the relationships between dredging firms and laboratories, and the financial politics of the Dutch Parliament are also embedded in the technology—just as much as are mattresses, concrete pillars, steel doors, and control computers. Dutch society shaped the technology of the Deltaplan as much as the Deltaplan shaped Dutch society for the future.

The approaches in technology studies that are to be discussed in this section do not, however, stop here—with this conclusion that the impact of technology on society and the shaping of technology by society are complementary phenomena. There is an additional argument—that even the very distinction between social and technical cannot be made a priori. Is the closure of the Oosterschelde by a flood barrier a technological solution? It certainly is seen as such by most visitors to the dam, engineers and others alike. Or is it a political solution? It certainly was seen as such by important spokespersons of both contesting parties in the late 1970s. Is the Deltaplan a technology to shorten the coastline and thus protect Zeeland against the sea? Or is it a social system that centralizes dike management and thus safeguards Zeeland? When zooming in and analyzing the sociotechnical ensemble of the Deltaplan and the Oosterschelde barrier (or, for that matter, fascine mattresses), it becomes clear that these are not constructed from elements that are a priori and intrinsically social, technical, economic, or cultural. These various elements form a seamless web.

Sociotechnical Ensembles: Three Approaches

In the 1980s research got under way that analyzed technology and society as an intimately interconnected, heterogeneous ensemble of technical, social, political, and economic elements. There are several approaches that view technology in this way. Some themes are common to all the approaches. (Some recent German work defies a categorization in these terms, because it explicitly draws on all of them, and more. See, for example, Hennen, 1992; Joerges, 1988; Rammert, 1990; Weingart, 1989.) Though, for example, not accepting a technological determinist or an autonomous technology view, they all try to account for the apparently obdurate character of many technologies. MacKenzie and Spinardi (1988) differentiate the concept of technological determinism and thus mold it into a more usable concept. Misa (1988) demonstrates how the occurrence of technological deterministic views within technology studies cannot be simply dismissed as an inappropriate usage of the old standard image of technology. He demonstrates how technology studies on a micro level of analysis tend to support the contingent nondeterministic character of technology, while macrostudies tend to produce deterministic images.

The *systems approach* analyzes technology as heterogeneous systems that in the course of their development acquire a technological momentum that seems to drive them in a specific direction with a certain autonomy. (See Hughes, 1983, for the fullest presentation of this approach, which is developed in a comparative analysis of the electricity distribution systems in New York, London, and Berlin.) Hughes (1987b) explicitly makes the argument against a priori distinctions between the social, the technical, the scientific, and so on. Shrum (1985b) also studies large technical systems and networks, but from a more sociological perspective, combining qualitative and quantitative methods. The Deltaplan can be considered such a system. It consists of a heterogeneous set of elements such as dikes, flood barriers, channels, lakes, water levels, salt- and freshwater regimes, navigation routes, and rules for closing and opening the sluices and barriers. The social shaping theme is further elaborated by Hughes's identification of national styles in the development of technological systems. In the case of coastal engineering, this would amount to the existence of clear differences between, for example, English and Dutch flood barrier design styles. The concept of "technological momentum" nicely captures the seemingly autonomous nature of technological systems, while at the same time showing that it is not an intrinsic property but is slowly built up during the system's development. Though the approach was developed by analyzing technological systems that are in a physical sense large, such as electricity distribution systems, railway systems, or telephone networks, the approach can be applied to physically small technologies such as missile guidance systems (MacKenzie, 1990a) and space shuttles (Pinch, 1989). The thrust of the research program, however, is still in studying large technical systems (La Porte, 1989; Mayntz & Hughes, 1988; Weingart 1989). Hughes's specific framework relates to (private) business systems—such as Edison and Siemens as entrepreneurial system builders.

Also, one of the ways in which these systems build up momentum is through economies of scale. Neither aspect seems directly applicable in the Deltaplan case.

The *actor-network approach*, associated with Callon, Latour, and Law, describes sociotechnical ensembles as heterogeneous networks of human and nonhuman actors. Clear examples of this approach are Callon (1980, 1981, 1986b), Law (1987a), and Latour (1984, 1987, 1992). For more references, see Law (1986c), Bijker, Hughes, and Pinch (1987), Law (1991c), and Bijker and Law (1992). The development of these networks is analyzed as a concatenation of translations—efforts by actors in the network to move other actors to different positions, thereby translating the meaning of these actors as well. The concept of translation is the crux of the actor-network approach (see the chapter by Callon in this volume). It is used to analyze how an ordering of society is brought about by reshuffling and transforming machines, institutions, and actors. The power of actors (such as a captain of industry, or the Rijkswaterstaat, or a Mafia leader) does not consist of something inherently special in those individuals or institutions but originates from the networks they can control. A characteristic of the actor-network approach, and one by which it can be distinguished from the numerous other network vocabularies in STS studies, is its ontological basis. By not accepting a fundamental distinction between human and nonhuman actors, which is central to Western sociology and indeed to most post-Kantian thinking, the actor-network approach is based on a premodern footing. (See Collins & Yearley, 1992a, 1992b, and Callon & Latour, 1992a, for a debate on this foundational question.) A "principle of generalized symmetry" is adhered to: Analyze the human and the nonhuman world with the same conceptual framework; in other words, the explanation of the development of sociotechnical ensembles involves neither technical nor social reductionism. The relevance of this approach to the Deltaplan case will be clear. The case could be described (though I did not attempt this) in a vocabulary that builds a network from such diverse actants as engineers, the Rijkswaterstaat, environmentalists, dikes and barriers, oysters, tidal streams, and storm surges. (The actor-network approach draws on semiotics. By using the term *actant*, the traditional social science connotations of the word *actor* are avoided.) An explanation of the development of the Oosterschelde barrier would thus avoid drawing one-sidedly on either "social factors," such as politics, finances, or fear, or on "technical factors," such as increased engineering knowledge, better mattresses, or faster computers.

In the *social construction of technology approach*, Pinch and Bijker (1984) take "relevant social groups" as their starting point. (This "SCOT" approach originally appeared under the banner of "social shaping of technology." After some hesitation, I have decided to present it here, in the section about sociotechnical ensembles, because of its ongoing theoretical development and the focus on the concept of "technological frame"—see below—as a hinge between the social shaping of technology and the technical shaping of society.) Artifacts are, so to speak, described through the eyes of the members of relevant social groups. The interactions within and among relevant social groups constitute the different artifacts, some of which may be hidden within the same "thing." In that case, the "interpretative flexibility" of that "thing" is revealed by tracing the different meanings attributed to it by the various different relevant social groups. This demonstration of interpretative flexibility is a crucial step in arguing for the feasibility of any sociology of technology—it shows that neither an artifact's identity nor its technical "working" or "nonworking" is an intrinsic property of the artifact but is subject to social variables. The next step then is to describe how artifacts are indeed socially constructed, thus tracing the increasing (or sometimes decreasing) degrees of stability of that artifact. Pinch and Bijker (1984) use the high-wheeled bicycle of the 1870s as an example. The first step in their descriptive model is a *sociological deconstruction* of the artifact (by the analyst). In this case, this results in showing that "hidden within" the highwheeled Ordinary were at least two vastly different artifacts: the Unsafe Machine for women and older men, and the Macho Machine for "young men of means and nerve." The second step is to trace the *social construction* of these artifacts (by the relevant social groups of actors). Several "closure mechanisms" can be identified that bring the interpretative flexibility to an end and start the process of stabilization of the artifact within the relevant social groups. The third step is to generalize beyond one case study to form *a theory of sociotechnical ensembles.* The concept of "technological frame" is then proposed to explain the development of heterogeneous sociotechnical ensembles, thus avoiding social reductionism (Bijker, 1987, 1992). A technological frame structures the interactions between the actors of a relevant social group. A key characteristic of the "technological frame" is that it is applicable to all relevant social groups—technicians and others alike. It is built up when interaction "around" a technology starts and continues. The early Deltaplan technology can, for instance, be said to have shaped the coastal engineers' technological frame, strongly influencing the choice for an ambitious "high-tech" solution for the Oosterschelde rather than merely heightening all dikes around the Oosterschelde basin. Existing practice does guide future practice, though without logical determination. The concept of "technological frame" forms a hinge in the analysis of sociotechnical ensembles; it is the way in which technology influences interaction and thus shapes specific cultures, but it also explains how a new technology is constructed by a combination of enabling and constraining interactions within relevant social groups in a specific way.

CONTROL AND INTERVENTION

How do the approaches discussed in the previous sections deal with control and intervention? In most impact studies, technology is conceived of as a separate entity that follows a linear path. Technology is like a train running on a track that is fixed though not known in detail; one cannot hope to change the train's direction, only to check its speed and improve the safety of the crossings. Orthodox "technology assessment" as exemplified by the early work of the American Office of Technology Assessment seeks to predict technological development and its impact on society and thus hopes to avoid some of its negative effects by "early warning." (See Leyten & Smits, 1987, for a comparative review of different concepts of technology assessment.) Cost-benefit analyses are another example of control and intervention instruments that are based on this image of technology. Such an approach to technology implies a control dilemma (Collingridge, 1980)—either it is too early to foresee the implications of a new technology or it is too late to intervene because the technology has become so entrenched in society and culture that it cannot be changed anymore.

Linked to the social shaping and evolutionary views of technology are models of technology assessment developed in Sweden and the Netherlands. Leyten and Smits (1987) present a comparative study of the different notions of technology assessment in these places. (For more information about Sweden, see SFS, 1982; for more about the Dutch "constructive technology assessment" approach, see Daey Ouwens, Hoogstraten, Jelsma, Prakke, & Rip, 1987, and Schot, 1991). Schwarz and Thompson (1990) examine "constructive technology assessment" on the basis of their cultural theory, which is an extension of the work of anthropologist Mary Douglas (e.g., 1970, 1982). In these models the possibility of continuously shaping and reshaping a technology during all its stages of development is recognized. A framework has been developed to encourage a positive interplay between the formal institutional technology assessments and the more informal technological evaluations carried out by other relevant social groups, such as consumers. In giving a role to these groups, it is recognized that technological development involves more than just engineers and politicians.

What form can control and intervention take in approaches that use sociotechnical ensembles as their unit of analysis? It is immediately clear that Collingridge's control dilemma now disappears. The seamless web character of technology and its pervasive socially constructed character indicate a multitude of opportunities to influence the development of technology (and society and so on) at all stages. But more studies of how this can be brought about are needed. Studies of ethical arguments made by proponents and opponents of a specific technology may be a valuable way of coming to grips with technology, as might be studies of technical norms and regulations. Ethnographic studies of accounting practices could document the process of shaping a technology by the management of an organization (MacKenzie, 1990b). So, for example, was it possible to "handle" the overrun on the Oosterschelde barrier budget by shifting some of the costs to other secondary projects, by redistributing the costs to other ministries, and by arguing that the costs were actually 20% lower than planned but that "the political decision to leave the Oosterschelde open" had caused the increase. Applied on a more macro-economic scale, the reclamation of flooded polders may not now seem as ludicrous as I suggested in the introduction. There is more to socioeconomic analysis than narrowly defined costs.

CONCLUSION

In this chapter I have tried to review the body of technology studies by presenting the various themes that are addressed by scholars in the STS community. The implicit thrust of the argument has been that "underneath" these themes, and relatively independent of their specificities, a general pattern can be recognized in which the study of technology and society has been developing. This pattern can, very schematically, be characterized as a sort of slow pendulum movement—a dampened oscillation.

Before the 1940s the social sciences did not pay much attention to the study of the detailed development of technical artifacts and society. (William Ogburn and Lewis Mumford were important exceptions; see Westrum, 1991, chap. 3, for a discussion of the "old" sociology of technology.) Except as an abstract concept, technology was almost nonexistent in the social sciences. The pendulum started to swing, and especially historians, some economists, and, later, philosophers and sociologists discovered technology. The swing went too far, however, and technology was viewed as an autonomous factor to which society had to bow. Technology was all important. With the rise of social shaping models, the pendulum swung back from this technological determinist conception. But again the swing went a little too far. The impact theme almost disappeared from view and technology seemed merely a social construct that could not appear in an obdurate, transformations-resisting, and society shaping form. Recently the pendulum started to swing back again to redress this imbalance. Technology recaptured some of its obduracy without completely losing its socially shaped character. The swings are smaller now. Perhaps we should say that the pendulum is not moving anymore in a flat plane but moves in Foucaultian circles.

Technology studies should not stop at the conclusion that sociotechnical ensembles have emerged from the seamless web of technology and society.

Research is called for in at least two directions. First, do we need a new theoretical framework to analyze the emergence, shaping, and development of ensembles? Simple technical or social reductionism will not do, but what will? (Woolgar, 1991b, addresses this question and argues for reflexively analyzing technology as a text. The focus, then, would not be on an analysis of different representations of technology but on the representational activity itself—by actors as well as by the analyst. See also Pinch, 1993, and the chapters in this book by Callon and by Restivo.) Second, we need to return to some of the political questions that informed early STS studies. Technology studies as a field seems to have lost some of its direct relevance for the plethora of problems our societies at present face. The toll of academic respectability threatens to produce politico-cultural irrelevance.

EPILOGUE

Happily, in November 1991 the dikes did indeed resist the storm surge; in February 2053 there may be such a storm surge again. Small chances may combine into big effects. Will the Dutch dikes and flood barriers stand up to this? Will the global sea level rise as a result of the greenhouse effect and cause dangers that were not foreseen when the Delta level was fixed? Will the gradual lowering of Dutch land levels due to the exploitation of fossil gas supplies add more to this than currently expected? Will there be lights on during that stormy February night in 2053? How much will the Zeeland inhabitants then trust the Deltaplan technology?

The Deltaplan project was closely and critically followed by various relevant social groups—this technology certainly was not developed in isolation. The general public probably knows more about these coastal defense technologies than about any other modern technological system. It seems to make perfect sense, for example, to have a parliamentary discussion about the operating criteria of the Oosterschelde barrier. Public interest and insight into the sociotechnical system are such that the recently published evaluation of the first 5 years of operation will be discussed intensively in public hearings before leading to a decision about operational criteria of the barrier: At what water levels should the doors be closed? What tide regime should be realized during these closures? The old parties take position: Environmentalists propose to increase the level so that the barrier will be closed less frequently; regional political authorities will probably opt for a lower norm for greater safety.

Can the type of technology studies discussed in this chapter contribute to the safety of the Netherlands in 2053? Certainly not in the form of presenting straightforward instruments of policymaking or technology assessment.

Technology and society are entangled in much too complex an ensemble to hope for context-independent instruments or recipes. If technology studies can be expected to contribute, it will more likely be in the spirit of the ideals of the Enlightenment—to provide insight into fundamental processes underpinning the development of societies and technologies. An analysis of how literature and art mirror technological and societal developments and thereby contribute to their mutual shaping could also be a fruitful entrance for such enlightening studies (e.g., Hughes, 1989a; Marx, 1964; Williams, 1990). There are no privileged actors anymore—neither engineers, nor managers, nor policymakers. All contribute, knowingly or unknowingly, to the shaping of sociotechnology. Of course, we can choose to develop organizational and regulatory frameworks that make more explicit the involvement those specific relevant social groups have (see also Jasanoff, 1990a, who investigates the regulatory process from a similar perspective). It is encouraging that the public can participate in the decision process to set the operation procedures of the Oosterschelde flood barrier. A decade ago most people probably would have agreed that democratic control and "high-tech" systems are a *contradictio in terminis*. It was argued (e.g., Winner, 1980) that nuclear energy would only be possible in a centralized "police state." Democratic energy distribution technology then was synonymous with decentralized, small-scale energy systems. Mathews (1989a, 1989b) addresses problems of democratic control in a social-democratic system by taking recent technology studies into account. We have to continue to study sociotechnology in all its heterogeneity. We can only hope to contribute to the democratic control of technology by continuing to pry open the black box of technology and to monitor the evolution of sociotechnical systems.

12

□

From "Impact" to Social Process

Computers in Society and Culture

PAUL N. EDWARDS

A couple of years ago I received from a publisher, unsolicited, a copy of a new textbook on computers and social issues. It was a sleek large-format paperback with a beautifully designed computer graphic on the cover. In imposing black type, the title read: *Computers and Society—IMPACT!*

The sensationalism of this title, with its billiard-ball imagery, nicely encapsulates what is probably the most common view of the relationship between information technology and the social world. Computers are arguably among the half dozen most important post-World War II technologies, an impressive list that might include television, jet aircraft, satellites, missiles, atomic weapons, and genetic engineering. The proliferation of cheap, powerful information processing and computerized control systems has unquestionably altered—and in some cases deeply transformed—the nature of warfare, communication, science, offices, factories, government, and certain cultural forms. This point hardly requires substantiation; reportage on the "information revolution" has become a virtual cottage industry.[1]

But the exact nature of these "impacts" of computing, as well as the details of how computers are supposed to produce them, remain in dispute. The utopian/dystopian character of much of the analysis in this area is aggravated by its generally ahistorical character. The basis for claims of "impacts" lies more often in broad economic or cultural analysis than in the detailed exploration of local effects characteristic of some of the best science studies literature (Dertouzos et al., 1991; Garson, 1988; Roszak, 1986; Weizenbaum, 1976).

This chapter explores some of the significant social effects of digital computers and some of the social forces shaping their development. Because even a cursory overview of such an immense arena is beyond available space limits, the chapter focuses on three cases: military relations with computing in the post-World War II era, the "productivity puzzle" of computerization in banking, and the relationship of gender identity to computer use. The chapter has two goals. First, it offers the uninitiated a point of entry into some of the vast literature on computers and society. Second, and more important, it provides a historical and social analysis that treats computers not merely as causes but also as effects of social trends.

In this I take as a given that technological change is, as Merritt Roe Smith (1985) has put it, a *social process*: Technologies can and do have "social impacts," but they are simultaneously social *products* that embody power relationships and social goals and structures. Social impacts and social production of artifacts in practice occur in a tightly knit cycle. The three cases presented here show how any full-blooded analysis must reflect the complexity of this interaction.

COMPUTERS AND THE MILITARY AFTER WORLD WAR II

The U.S. armed forces have been the single most important source of support for advanced computer research from World War II to the present time. How did this support affect the technology itself? How did the new technology affect military doctrines and institutional structures? The historical analysis presented here demonstrates how military needs and priorities guided computer development, especially in its first two decades, and shows how computers, in turn, shaped the military.

Historians now generally recognize John Atanasoff of Bell Laboratories as the inventor, in 1940, of the first electronic digital computer. But while Atanasoff and others created this and other prototypes just before the United States entered World War II, their significance went for the most part unrecognized. This was largely because *analog* computers, such as the differential analyzer of Vannevar Bush, were already a well-developed technology.[2] Bush built a series of these machines, which were highly though not perfectly accurate for solving complex differential equations, culminating in one built in 1942 at MIT that was fully programmable using punched paper tape (Goldstine, 1972, pp. 92-102 passim). New analog computers, such as those used in antiaircraft weapons, were among the decisive technical achievements of the war (Fagen, 1978). But the feverish technical developments of World War II weaponry generated demand for huge numbers of computations to solve ballistics and coding problems—and, because of their urgency, for unprecedented rates of speed. It was to this end that programmable, electronic digital computers, capable of dramatically faster calculation, were developed.

The first of these were created by U.S. and British military forces. The Electronic Numerical Integrator and Calculator (ENIAC) was constructed at the Moore School of Engineering in Philadelphia between 1943 and 1946 by the U.S. Army Ordnance Department. Its purpose was to automate the tedious calculation of ballistics tables, on which antiaircraft weapons and artillery then depended for accuracy. During the war these calculations were performed by a mostly female corps of young mathematicians, known as "computers," using hand calculators. When the ENIAC project began, some of these women became its first programmers—hence the sobriquet "computer" for the new machine. The ENIAC was not completed until after the war's end, when it was immediately put to work on physics equations connected with thermonuclear weapons for the Los Alamos laboratories. (It failed to solve some of them, producing demands for more powerful machines.) Among the many influential members of the ENIAC development team were John von Neumann, who developed the serial control architecture that now bears his name, and J. Presper Eckert and John Mauchly, who proposed and directed the project and were responsible for most of the ENIAC's key design features. Eckert and Mauchly started their own company—UNIVAC, the first commercial computer producer—in 1946 using knowledge gained from working on the ENIAC and its successor, the EDVAC.[3]

Credit for the first operational electronic digital machine, however, belongs to the British "Colossus," constructed at Bletchley Park with the participation of Alan Turing, the mathematician who had invented the theory of digital computation in 1936 (Turing, 1936). The first Colossus was completed in 1943 and used throughout the rest of the war to break the Enigma and Fish ciphers used by the German high command. The machine's great speed and accuracy, compared with existing hand calculation techniques and automated analog computation, enabled it to crack the cipher quickly enough for intercepted messages to be useful to the Allies. The Colossus thus played a major—perhaps even a decisive—role in preventing Britain's defeat and assuring a subsequent Allied victory (Hodges, 1983).

World War II era computer development, then, may be characterized as *need-driven research.* Ideas for automating calculation came from scientists and engineers. They were adopted by the military because of specific, preexisting needs for calculation. World War II era computers produced only limited impacts on the military, because they were used simply to speed up existing processes. But these military projects did produce local concentrations of researchers working on electronic digital techniques, and these groups persisted after the war, providing the social and organizational nucleus for future research. At this point, computers were clearly more a social product than a driver of social change.

Computer development in the 1945-1955 period occurred very rapidly, with projects such as the National Bureau of Standards' SEAC, von Neumann's Institute for Advanced Studies (IAS) machine and its several copies, and Eckert and Mauchly's BINAC (built as a guidance computer for Northrop's Snark missile). Almost every new machine incorporated new innovations. The UNIVAC team struggled to create and introduce a production computer (it finally succeeded in 1951 and subsequently sold 46 UNIVAC Is), but most machines were one-of-a-kind, experimental prototypes. Then as now, technical advancement occurred with astonishing speed. Indeed, statistical measures of computer development, such as the rate of doubling of random access memory capacity and the halving of cost per computation, became and remain virtual tropes of progress and technological "revolution" (see Figures 12.1 and 12.2, below).

Perhaps bedazzled by this muscular technical progress, most historiography of computing has focused on three things: (a) the technical characteristics of devices, (b) the biographies of individuals responsible for important innovations, and (c) the intellectual history of computing as a problem of mathematics and engineering (Mahoney, 1988). Until recently (Flamm, 1987, 1988; Noble, 1984; Winograd, 1991), few historians had much to say about the social relations involved in computer R&D—in particular, the meaning of military sponsorship.

Did military needs influence computer technology after World War II, when the wartime research laboratories were dissolved or returned to civilian control? The new U.S. status as a superpower, the central role of science and technology in the war effort, the massive wartime federal funding and the associated advancement of communal aims for science, and other factors all contributed to the emergence of a powerful scientists' lobby for continued federal sponsorship, on the one hand, and a wholly new sense within the armed forces of the importance of science and technology—and the potential contribution of "civilian" scientists and engineers—on the other. The incipient cold war was the final element that allowed military organizations, especially the Office of Naval Research (ONR), to become the default federal sponsors of science and technology R&D in the 1940s and 1950s (Dickson, 1984; Edwards, 1989, in press; Forman, 1987; B. L. R. Smith, 1991). Still, most computer R&D projects took place not in military facilities but in industrial or university laboratories. This was consistent with the general pattern of postwar federal sponsorship of science and technology (B. L. R. Smith, 1991). Because so many areas of science and technology benefited from the ONR's relatively nondirective funding, many historians have neglected military influences because of the idea that "everyone was feeding from the same trough."

But military sponsors did not need to undertake detailed direction of research projects to achieve their goals, which were in any case of a very general character in relation to new technologies such as the computer. They could rely, instead, on the mere requirement of a plausible military justification for research projects (Winograd, 1991). The civilian scientists' and engineers' own imaginations, combined with their wartime experience of military research problems, generated new military ideas in large numbers. These frequently proved far more ambitious and farsighted than those of the military's own leaders, wrapped up in a military traditionalism rendered problematic by new technologies of war (Gray, 1989, 1991).

At least in the computer field, a process of *mutual orientation* occurred, in which engineers constructed visions of military uses of computers to justify grant applications, while military agencies directed the attention of engineers to specific practical problems computers might resolve.

The most sophisticated leaders, both military and civilian, had an explicit understanding of this mode of *enrollment* of civilian scientists, engineers, and other intellectuals (Callon, 1987; Latour, 1987). Vannevar Bush, for example, in his famous report on postwar science policy, *Science: The Endless Frontier,* cited the secretaries of war and the Navy:

> This war emphasizes three facts of supreme importance to national security: (1) Powerful new tactics of defense and offense are developed around new weapons created by scientific and engineering research . . . (3) war is increasingly total war, in which the armed services must be supplemented by active participation of every element of civilian population. To insure continued preparedness along farsighted technical lines, the research scientists of the country must be called upon to continue in peacetime some substantial portion of those types of contribution to national security which they have made so effectively during the stress of the present war. (Bush, 1945, p. 12)

Another indirect channel for military influences on technology was the marketplace itself. The sheer size of the increasingly high-technology armed forces ensured corporate investment in military-related R&D projects. The development of the transistor—privately financed by Bell Laboratories, but with military markets its major rationale—is the best-known example. But there are others of equal importance. The DoD sponsored the development of integrated circuits in the 1950s and purchased the *entire* first-year output of the integrated circuit manufacturing industry, worth $4 million, mostly for use in Minuteman nuclear missile guidance systems. Two major programming languages, COBOL (in the 1960s) and Ada (in the 1980s), were products of standard-setting efforts initiated by the military to assure software compatibility among different projects. Military sponsorship of and specifications for very-high-speed integrated circuit (VHSIC) fabrication in the 1980s led to initial American leadership in the field—followed by failures due to poor cost performance of equipment designed for the military's "high-spec," small-lot production needs (Brueckner & Borrus, n.d.; Flamm, 1987, 1988; Jacky, n.d.; Rosenberg, 1986; Winograd, 1991).

Military influences on computer technology were thus widespread but were frequently the product of *indirect* forms of intervention that go unnoticed in traditional historical analysis.

Project Whirlwind and the Sage Air Defense System

Probably the single most important computer project of the decade 1946-1956 was MIT's Whirlwind, which, under the direction of engineer Jay Forrester, actually began in 1944 as an *analog* computer for use in a flight simulator, funded by the Navy. News about the ENIAC and EDVAC digital computer projects led Forrester to abandon the analog approach in early 1946. But the original application goal of a flight simulator remained. Flight simulators of the day were servo-operated, mechanical imitations of airplane cockpits that simulated an airplane's altitudinal changes in response to its controls, giving novice pilots a chance to practice without the expense or the risk of actual flight. In theory, flight simulators were, and remain, what is known as a "dual-use" technology, equally useful for training military and civilian pilots. But the urgency of the World War II air war made them in practice, in 1940-1945, a military technology. This practical goal distinguished Whirlwind from almost all other digital computer projects of this era, because it required a computer that (a) could be used as a control mechanism and (b) could perform this function in real time.

It is important to emphasize that at this historical juncture these were not obvious goals for a digital computer.

- Analog computers and control mechanisms (servomechanisms) were well developed, with sophisticated theoretical underpinnings. (Indeed, Forrester began his work at MIT as a graduate student in Gordon Brown's Servomechanisms Laboratory.)
- Analog controllers did not require the then-complex additional step of converting sensor readings into numerical form and control instructions into waveforms or other analog signals (Valley, 1985).

- Mechanical or electromechanical devices were inherently slower than electronic ones, but there was no inherent reason that electronic computers or controllers should be digital, because many electronic components have analog properties. Numerous electronic analog computers were built during and after the war.
- Most other projects saw electronic digital computers as essentially giant calculators, primarily useful for scientific computation. Their size, their expense, and this vision of their function led many to believe that once perfected only a few—perhaps only a couple—of digital computers would ever be needed. Even Forrester at one time apparently thought that the entire country would eventually be served by a single gigantic computer (G. W. Brown, March 15, 1973).

The technology of digital computation had not yet achieved what Pinch and Bijker call "closure," or that state of technical development and social acceptance in which large constituencies generally agree on its purpose, meaning, and physical form (Pinch & Bijker, 1987). The shape of computers, as tools, was still extremely malleable, and their capacities remained to be envisioned, proven, and established in practice.

By 1948, with its interest in a supersophisticated and by now extremely expensive flight simulator rapidly declining, the ONR began to demand immediate, useful results in return for continued funding. This dissatisfaction was largely due to Whirlwind's truly colossal expense. Where the cost range of computers like the UNIVAC lay typically between $300,000 and $600,000 current dollars, the Whirlwind group was planning to spend a minimum of *$4 million.* "MIT's funding requests for Whirlwind for fiscal 1949, almost $1.5 million, amounted to roughly 80% of the 1949 ONR budget for mathematics research, and about 10% of the entire ONR budget for contract research" (Flamm, 1988, p. 54). The actual budget for that year was $1.2 million—still an amazing level of investment, by any standard, in a single project.

Whirlwind's "estimated completion costs . . . were about 27% of the total . . . cost of the entire DoD computer program" (Redmond & Smith, 1980, p. 154). By March 1950 the ONR had cut the Whirlwind budget for the following fiscal year to only $250,000. Compared with the $5.8 million *annual* budget Forrester had at one point suggested as a comfortable figure for an MIT computer research program including military and other control applications, this sum was virtually microscopic.

Forrester therefore began to cast about for a new institutional sponsor—and for a new military justification. He was in a special position to do this for a number of reasons. First, Forrester's laboratory entertained a steady stream of visitors from both industry and military centers, each with questions and ideas about how a machine like the Whirlwind might be used to automate his operations. Forrester's notebooks indicate that between 1946 and 1948 these visitors raised dozens of possibilities, including military logistics planning, air-traffic control, damage control, life insurance, missile testing and guidance, and early warning systems (Forrester, 1946-1948). Second, Forrester "shared the apprehensions of Navy Special Devices Center (SDC) personnel regarding confidential projections of a Russian atomic strike capability by 1953" and believed his work could make a personal contribution (Redmond & Smith, 1980, p. 150).

Finally, Forrester and his group had been deeply concerned with the issue of military applications all along. In early 1946, when Forrester first reported to the Navy on his plan to switch to digital techniques, he had included several pages on military possibilities.

> In tactical use it would replace the analog computer then used in "offensive and defensive fire control" systems, and furthermore, it would make possible a "coordinated Combat Information Center," possessing "automatic defensive" capabilities, an essential factor in "rocket and guided missile warfare." (Redmond & Smith, 1980, p. 42, citing Forrester)

In October 1947 Forrester, SDC leader Perry Crawford, and Whirlwind coleader Robert Everett had published two technical reports on how a digital computer might be used in antisubmarine warfare and in coordinating a naval task force of submarines, ships, and aircraft. That year, in frequent meetings at its Sands Point headquarters, Crawford and other SDC personnel had encouraged Forrester and Everett "to see more ambitious prospects of the sort that had stimulated the forward-looking systems-control views represented by their L-1 and L-2 reports" (Redmond & Smith, 1980, p. 120).

The following year, as continuation of ONR support became increasingly uncertain, MIT president Karl Compton requested from Whirlwind a report on the future of digital computers in the military. The group produced a

> sweeping vision of military applications of computers to command and control tasks, including air traffic control, fire and combat control, and missile guidance, as well as to scientific calculations and logistics. The estimated cost of this program was put at $2 billion [current dollars], over 15 years. The . . . flight simulator [project] was replaced by the broader concept of a computerized real-time command and control system. (Flamm, 1988, pp. 54-55)

Indeed, the report discussed most of the areas where computers have eventually been applied to military problems (Forrester, Boyd, Everett, & Fahnestock, 1948).

Finally, working with the so-called Valley Committee (headed by another MIT professor, George E. Valley), Forrester constructed a grand strategic concept of national perimeter air defense controlled by central digital computers (Jacobs, 1983; Redmond & Smith, 1980). These would monitor distant-early-warning polar radars and, in the event of a Soviet bomber attack, automatically assign interceptors to each incoming plane, direct their flight paths, and coordinate the defensive response.

Military research budgets took a steep upward turn as a result of the Soviet explosion of an atomic bomb in 1949 and the outbreak of war on the Korean peninsula in 1950 (Forman, 1987). By that time, because of its control of nuclear weapons, the Air Force had emerged as the military focus of the cold war, the most forward-looking and technologically oriented of the armed services. In 1950 the Air Force took over Whirlwind's support from the ONR. Under Air Force sponsorship, the Valley Committee plan rapidly evolved into the SAGE (Semi-Automated Ground Environment) air defense project.

However, the Air Force's primary commitments were to *offensive* strategic forces. Commanders at the highest levels believed that an effective defense against a full-scale Soviet nuclear attack—even without missiles—was a virtual impossibility. They preferred to rely on a policy of "prompt use" of nuclear weapons, a euphemism for preemptive strike (Herken, 1983). Under this strategy, air defense would naturally be unnecessary. Forrester's group was ridiculed as "the Maginot Line boys from MIT." General Hoyt Vandenberg called the project "wishful thinking" and noted that

> the hope has appeared in some quarters that the vastness of the atmosphere can in a miraculous way be sealed off with an automatic defense based upon the wizardry of electronics. . . . I have often wished that all preparations for war could be safely confined to the making of a shield which could somehow ward off all blows and leave an enemy exhausted. But in all the long history of warfare this has never been possible. (General Hoyt S. Vandenberg, cited in Schaffel, 1989, p. 15)

The Air Force especially feared that emphasis on air defense would reduce budgets for the nuclear-offensive Strategic Air Corps (SAC). But it was essentially forced by political pressures to produce something that looked like an active air defense so as to assuage public fears of nuclear attack. These fears, combined with the "can-do" technological mind-set of the MIT engineers, generated the momentum necessary for the SAGE project. Eisenhower ended up supporting both the SAC and the continental air defense program under his high-technology New Look defense strategy.

Valley's group quickly became convinced of the effectiveness of Forrester's digital techniques. But the digital approach involved a major restructuring of Air Force command systems, because it was centralized and automated rather than decentralized and pilot oriented. A competing project at the University of Michigan, based on analog technology, would have retained the basic command structures but speeded up the calculation process with analog computers. The Air Force continued to fund the Michigan project until 1953. Even then, the Air Force only canceled the project when MIT threatened to quit if it did not commit to the digital approach.

The first SAGE sector became operational in 1958. Its control center consisted of a windowless four-story building with 6-foot-thick blast-resistant concrete walls. The Whirlwind machine became the prototype for its contents, the FSQ-7 production computer, manufactured by IBM. "Composed of seventy cabinets filled with 58,000 vacuum tubes, the FSQ-7 weighed three hundred tons and occupied 20,000 square feet of floor space, with another 20,000 square feet devoted to display consoles and telephone equipment." By 1961 all 23 sectors were working. The total cost of the project in the 1950s was somewhere between $4 and $12 billion. Parts of the system operated—using the original vacuum-tube computers—until the mid-1980s (Jacobs, 1983).

Whirlwind and SAGE were responsible for many, many major technical advances. The list includes the invention of magnetic core storage, video displays, light guns, graphic display techniques, the first algebraic computer language, and multiprocessing. Many of these advances bear the direct imprint of the military goals of the SAGE project and the political environment of the postwar era—another example of the social shaping of technology. I will mention just three examples.

First, as Paul Bracken (1984) has pointed out, the cold war, nuclear-era requirement that military systems remain on alert 24 hours a day for years on end represented a completely unprecedented challenge not only to human organizations but to equipment. The Whirlwind computer was specifically designed for the extreme reliability required under these conditions. It was the first duplexed computer (i.e., it was actually two computers running in tandem, one of which could take over from the other on the fly in case of failure). For the same reason, the machine had a fault-tolerant architecture and pioneered methods of locating component failures. Whirlwind research also focused heavily, and successfully, on increasing vacuum-tube lifespan, a major cause of breakdowns in early computers. Downtime for FSQ-7 machines was measured in minutes per year—other computers of that era were frequently down for numbers of weeks.

Second, SAGE was the first large control system to use a digital computer. It translated radar data into fighter-interception coordinates and flight paths, relayed to pilots by radio. Real-time operation was a demand imposed by the control function of the SAGE system. This required, first, much faster operating speeds than any other machine of that period, not only for the central processing units but for input and output devices as well. Second, it required the development of methods of interconverting sensor and control signals from analog to digital form. For example, radar signals were converted to digital impulses for transmission over telephone lines.

Finally, this long-distance digital communication was used both for transmission of data from radars and for coordination of the SAGE centers. SAGE was thus the first computer network, a requirement of the centralized command structure. But this centralization was itself a product of SAGE. It was both a technological impact, because, without its high-speed communication and coordination, central control on such a scale would not have been possible, and a social product, because SAGE was envisioned by "system builders," in Thomas Hughes's phrase, who constructed technologies to fit a visionary ideal (Hughes, 1987a).

How do the Whirlwind and SAGE projects exemplify social process in the history of computers? Three important points may be made.

First, considered as a politico-military venture, the value of the SAGE project—like its 1980s counterpart, the "Star Wars" strategic defense system—was almost entirely imaginary and ideological. Its *military* potential was minimal, but it helped create a sense of active defense that assuaged some of the helpless passivity of nuclear fear. Civilian political leaders, the incipient corps of military technocrats, and engineers with an almost instinctive belief in technological solutions for politico-military problems—all riding on the technological successes of World War II—thus allied against the Air Force around an essentially ideological program of technological defense. Real-time control computers were a product of these social forces.

Second, in discussions of military contracts, it is common to dismiss "grantsmanship," or the deliberate tailoring of grant proposals to the particular aims of funding agencies, as insignificant to research outcomes. Supposedly, grant proposals that justify basic research in terms of applications are simply a vehicle to obtain funds that both recipients and agencies know will really be used for something else.

In the case of Whirlwind, at least, a much more complex relationship between funding justifications and technology obtained. Their studies of possible military applications and their contacts with military agencies expanded the Whirlwind group's sense of possibilities and unsolved technical problems. At the same time, they served to educate the funding agency about as yet undreamt of possibilities for centralized command and control. While the ONR was not ultimately convinced, the thinking and the documents produced in this exchange kept funding going for several years and later proved of enormous value in convincing another military agency, the Air Force, to offer support. The source of funding, the political climate, and their personal experiences directed the attention of Forrester's group toward military applications, while the group's research eventually directed the military toward new concepts of command and control.

We could call this a process of *mutual orientation,* in which each partner oriented the other toward a new arena of concerns and solutions. Negotiations over funding, at least in this case, became simultaneously negotiations of the eventual technical characteristics of computers and of military command structures and strategic goals.

Through this process, within the space of a very few years, the Air Force traditionalists who had opposed the computerized air defense system either became, or were replaced by, the most vigorous proponents of high-technology, computerized warfare anywhere in the American armed services.

Finally, SAGE set a pattern, repeated incessantly in subsequent years, of computerized command and control of nuclear defenses. Over two dozen large-scale, computerized, centralized command-control networks were built by the Air Force between the late 1950s and the middle 1960s—the so-called Big L systems, including the Strategic Air Command Control System and the Ballistic Missile Early Warning System (Bracken, 1984). In 1962 the World-Wide Military Command and Control System, a global network of communication channels including (eventually) military satellites theoretically enabling central, real-time command of American forces worldwide, became operational.[4] The distant early warning systems used by SAGE were ultimately connected with central computer facilities at the headquarters of the North American Air Defense Command in Colorado for ICBM detection and response.[5] President Reagan's Strategic Defense Initiative was thus only the latest in a long series of computer-controlled, centrally commanded schemes for total defense (Edwards, 1987, 1989, in press; Franklin, 1988). In this sense, SAGE technology had major impacts on military doctrine and organizational structure. SAGE technology was also used by IBM to build the Semi-Automatic Business-Research Environment (SABRE)—a direct reference to SAGE—the first computerized, centralized airline reservation system.

COMPUTERS AND WORK: BANKING AND THE "PRODUCTIVITY PUZZLE"

Computers have had equally massive effects on the nature, quality, and structure of work, where they are said to be largely responsible for the emergence of "postindustrial society" and for an "information revolution." Here, too, we find that an ideology of technological determinism is commonplace, reflected in managers' frequent belief that both productivity gains and social transformation will be automatic results of computerization. This section attempts to balance this view against the idea of a "web of computing," in which computers are only one of a variety of social and technical factors affecting organizational efficiency and culture (Kling & Scacchi, 1982).

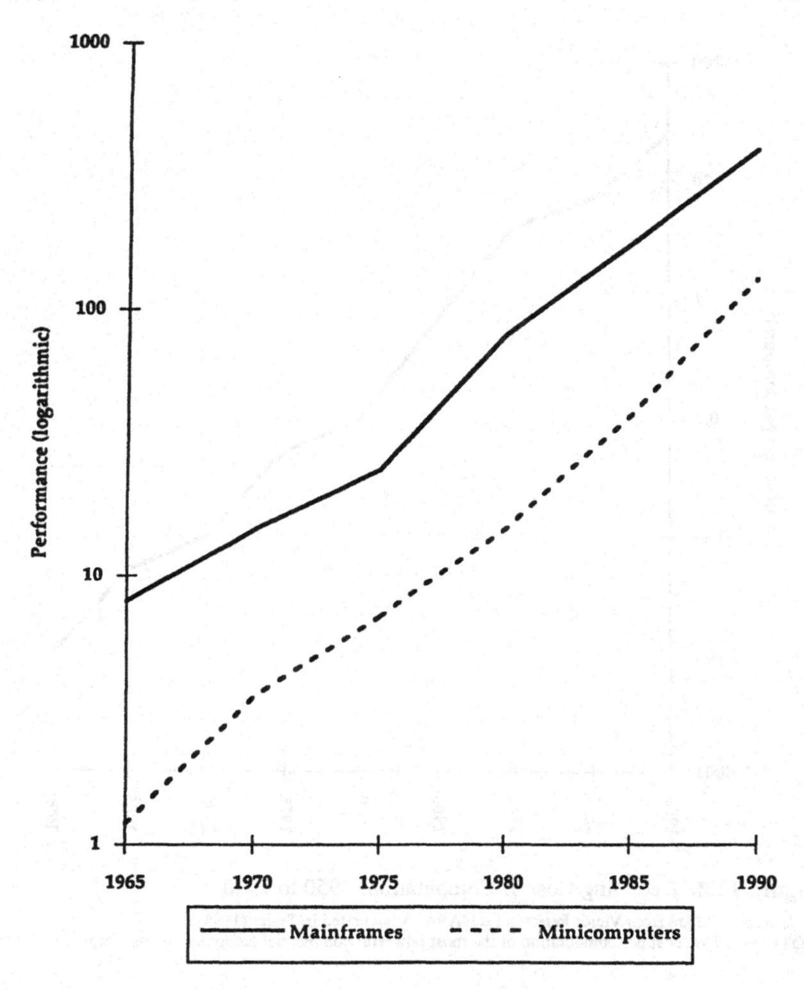

Figure 12.1. Computer Performance Growth: 1965 to 1990
SOURCE: Data are from Jack Worlton (Los Alamos National Laboratories).

Figures 12.1 and 12.2 show the dramatic trends in expanding computing power and decreasing cost per computation (note the logarithmic scales of both charts). Computers began to be widely used in nonmilitary industry and business in the late 1950s. At that point they were still so expensive that only large corporations could afford them. A decade later they were enough smaller and cheaper to be practical for middle-sized firms, and by the end of the 1970s almost any business with significant data processing needs either

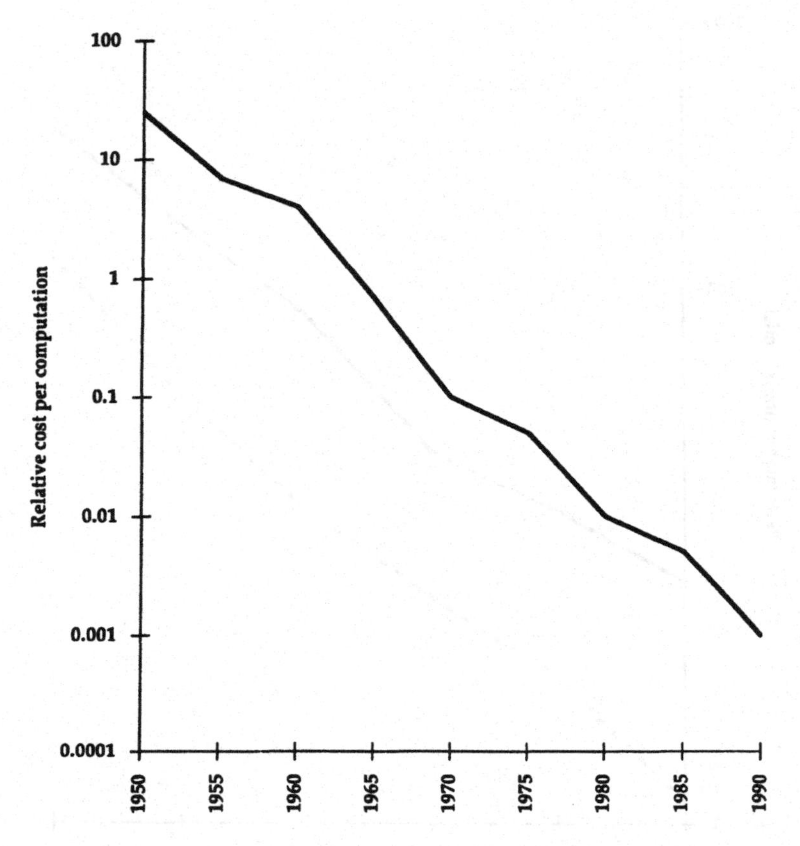

Figure 12.2. Declining Cost of Computation: 1950 to 1990

SOURCE: Data are from Victor Petersen (at NASA, Ames) cited in Tesler (1991).
NOTE: Based on cost per computation of the most powerful commercial computers of each era.

owned or leased a computer. In the past decade, of course, the introduction of personal computers (PCs), workstations, and powerful networking devices has put computers on the desks of a huge proportion of the American workforce, especially in offices. Somewhere in the range of 50 million personal computers are installed in American homes, offices, and schools as well as millions of other kinds of computers and terminals (Dertouzos et al., 1991, p. 63).

Along with computing and communication technologies have come dramatic increases in the size of the service and information sectors of the economy, as shown in Figure 12.3. It is often assumed, commonsensically, that the reason for the rush to computerize must lie in the benefits of computers

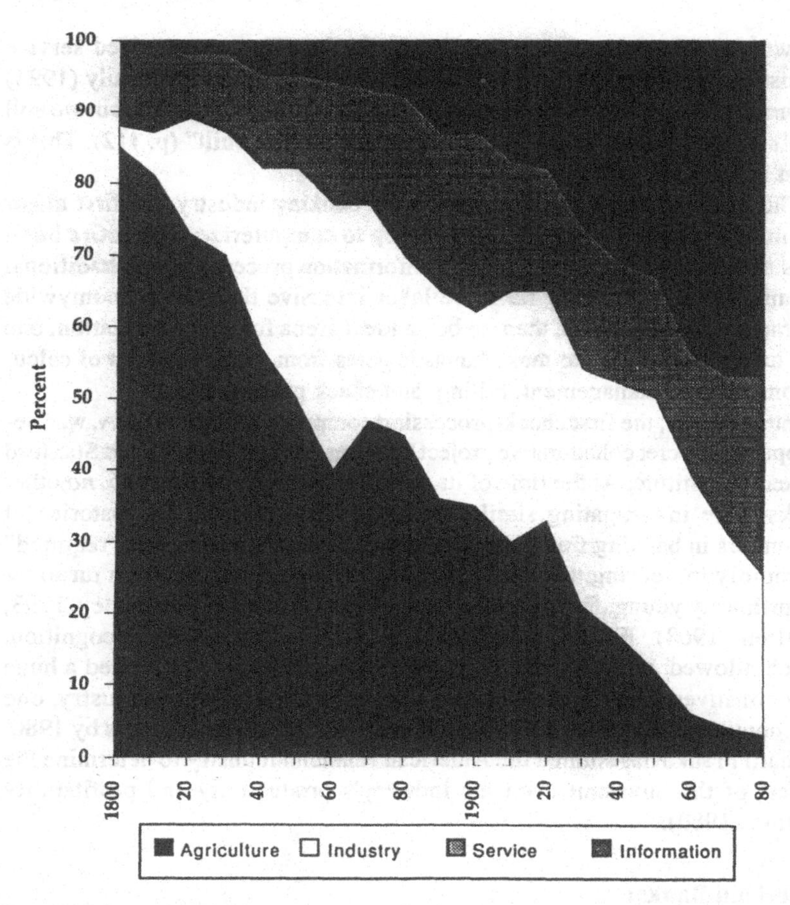

Figure 12.3. U.S. Civilian Labor Force by Sector: 1800 to 1980

SOURCE: Reprinted by permission of the publishers of *The Control Revolution: Technological and Economic Origins of the Information Society* by James Beniger, Cambridge, Mass.: Harvard University Press, Copyright © 1986 by the President and Fellows of Harvard College.

to productivity (defined as the ratio of output to hours worked), and indeed automation is frequently urged as the key to productivity growth (Cohen & Zysman, 1987). But despite an enormous scale of investment, the expected benefits have materialized in a way that must be characterized as at best spotty and fragile. Since the end of the 1960s American productivity growth has been weak, and with the rise of Japan in the late 1970s this became, and remains, a major policy concern (Baily, 1991; Cohen & Zysman, 1987).

The role of computers in this problem is a strange one. The computer manufacturing sector has been the *greatest* single contributor to productivity growth in American commerce. But in the heavily computerized service industries, productivity growth has been very poor. As Martin Baily (1991) observes, "Apparently we are getting better at making computers, but we still don't really know what to do with them once they're built" (p. 112). This is what is known as the "productivity puzzle."

The example I will consider here is the banking industry, the first major nonmilitary sector of the world economy to computerize. The entire business of banking is in effect a form of information processing, and traditional techniques made banking far more labor intensive than the economywide average. It would appear, then, to be an ideal arena for computerization, one that could be expected to make fantastic gains from the automation of calculation, account management, billing, and check processing.

Interestingly, the first check-processing computer system, ERMA, was developed in a secret collaborative project between Bank of America and Stanford Research Institute. At the time of its public announcement in 1955, *no* other banks were investigating similar computerized systems. Yet histories of computers in banking frequently claim that computerization was "required" by rapidly increasing transaction volumes, labor costs, and high turnover of (primarily young, female) tellers and clerks (Fischer & McKenney, 1993; O'Brien, 1968). ERMA introduced magnetic ink character recognition, which allowed partially automatic processing of checks. It initiated a huge wave of investment in computer equipment by the banking industry, one that continued such that 97% of commercial banks used computers by 1980. Richard Franke has studied the American financial industry to determine the effects of this investment on the industry's productivity and profitability (Franke, 1989).

American Banks: Heavy Investment, Slow Growth

Franke found that, between 1948 and 1983, American banks' output rose fourfold, though the strongest period of output growth was 1948-1958, before computers were introduced (see Figure 12.4). Labor input (that is, hours worked) also rose steadily, though more slowly, to three times its 1948 level. After 1958 labor input rose slightly more quickly, rather than less. And capital input—as might be expected—rose to 14 times its 1948 level, jumping from a 2.7% per year rate of growth to a 9.1% rate after 1958, the bulk of the jump attributable to computing and its indirect effects, such as the increased convenience of branch banking.

Yet this immense investment had virtually no effect on labor productivity. Figure 12.5 shows that productivity rose more quickly before 1958 than afterward, peaked in 1975, and declined slightly thereafter. This meant, of

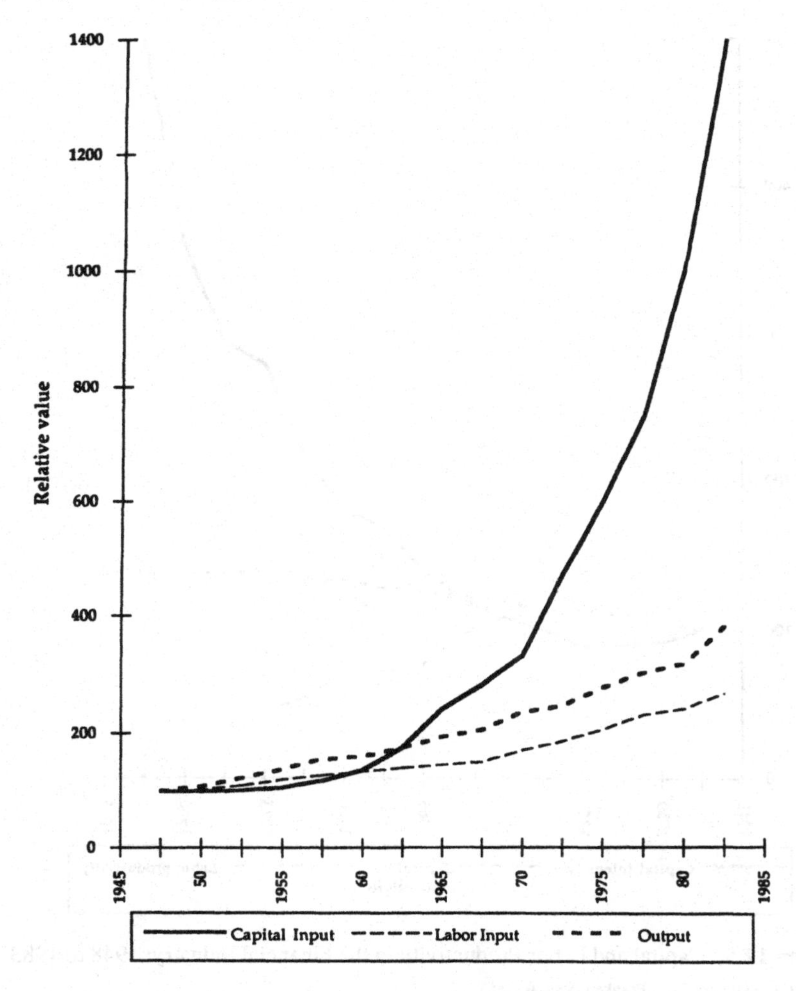

Figure 12.4. Financial Industry Inputs and Output: 1948 to 1983
SOURCE: Data are from Franke (1989, p. 284).
NOTE: In real terms: 1948 = 100.

course, that, while the capital intensity (the ratio of labor to capital inputs) of the industry quintupled, its capital productivity (the ratio of output to capital) declined to a mere one fifth of its 1948 level. Data from the 1980s show productivity growing again—but only at the unimpressive rate of 2% per year.

This investment did, of course, take place during a period of very rapid technological change, when banks found themselves frequently replacing

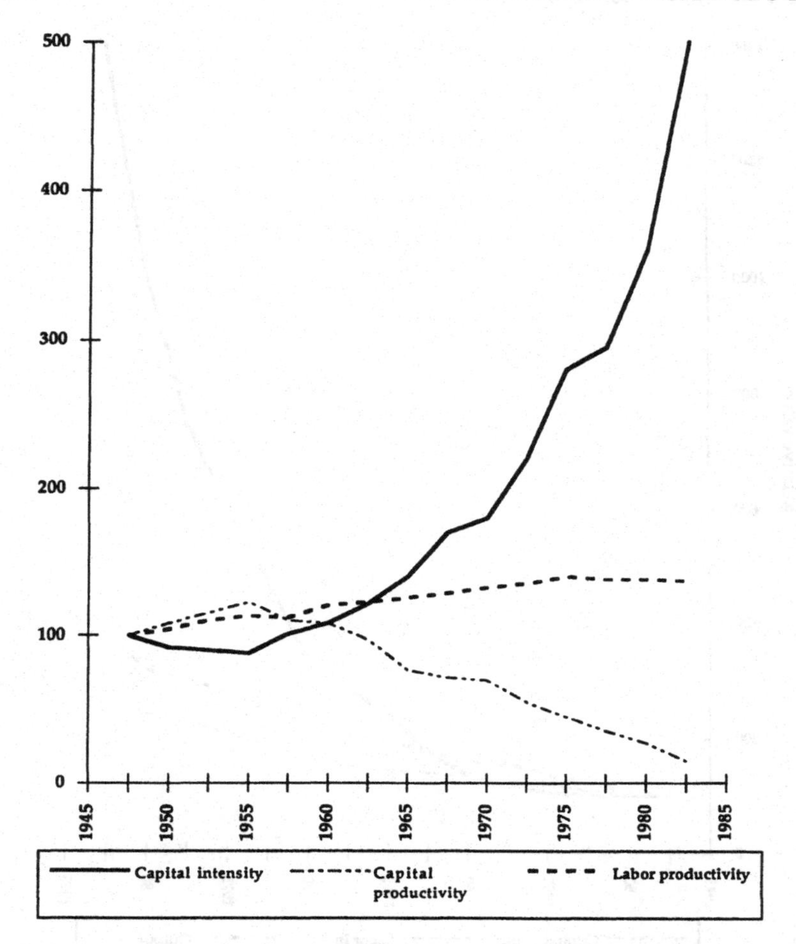

Figure 12.5. Capital and Labor Productivity in the Financial Industry: 1948 to 1983

SOURCE: Data are from Franke (1989, p. 285).
NOTE: In real terms: 1948 = 100.

obsolete equipment that had been new a few years before. However, Franke used statistical regressions to allow for these effects and still found that productivity did not begin to improve until the fourth generation of computer technology, and even then not much.

What explains the paradox of massive automation with nearly nil results? One possibility is Franke's own view. He concludes that "fundamental changes in the distribution and organization of work, *due to the new technology,* result initially in diseconomies. Only with time can enterprises adjust to become productive" (Franke, 1989, p. 288, italics added). This macroeconomic explanation relies on the familiar "impact" model of the relations between technology and society: Computers, colliding with the banking industry, split it apart like a fissioning atom that is only now beginning to restabilize into a new coherence. "Diseconomies" were the result.

But a look at the micro level—at what has actually happened in individual banks—shows that a diametrically opposite explanation may, at least in some cases, be more appropriate.

"Global Bank Brazil"

Shoshana Zuboff, who carried out detailed longitudinal studies of computerization in several factory and office settings between 1982 and 1986, examined the development of a database environment at the Brazilian branch of a major U.S. bank (Zuboff, 1988). She calls the institution "Global Bank Brazil."

At Global Bank Brazil, a group of farsighted young managers had determined to leapfrog other banks by developing and installing an information system. The new computers would allow them not only to automate existing procedures but to develop and sell a wide range of new information-based products. For example, they envisioned integrated real estate sales in which the bank would provide a "package" of information about properties, loans, and insurance, or "smart" loan brokerage based on continually updated knowledge of clients' cash positions. The bank's computers would link one company's need for cash with another's excess, and bankers would mediate the deal.

These same managers also subscribed to an "impact" view of the database environment. They believed that, once the system was installed, bankers would automatically become more involved in analysis and decision making based on information the system provided. Instead of spending their time on the phone or golfing with clients, maintaining personal relationships, and getting an intangible "gut feeling" for their situations, bankers would work with hard data. The key to their new jobs would be effective exploitation of information.

In the words of one manager: "Service, excellence, and innovation are only buzzwords right now. As we push the technology, people will realize that they have a really valuable tool on their hands. Then they'll be forced to use it. Then we can change the way they think and do their work."

Another said,

> We're on a learning curve now, trying to understand the technology. But at some point we'll have a revolution. The technology will prove that the current organization

> is inadequate. Some people will accommodate to the new environment, and some won't. In every revolution a lot of people are killed. And some people will be dead at the end of this one, too. (Zuboff, 1988, p. 214)

But instead of causing a "revolution," the database environment became mired in an institutional backwater, automating some routine clerking tasks and having very little effect on the way the bank did business.

The reason had to do with the fact that senior managers, from the beginning, had been resistant to the database project. To avoid the senior managers' interference and the watering down of their own "revolutionary" goals, the database developers had decided on an implementation strategy that would sneak the technology in through the bank's figurative back door. Instead of installing it first in the bank's marketing department or some other high-visibility area, they chose to introduce it into the central liabilities section, the oldest, least automated, and one of the most deeply internal of the bank's operations—a "back office."

Central liabilities maintained records of customers' credit balances and client histories. Here the database served simply to automate an existing task. Clerks were trained to enter data on the new system but were not told how it functioned. The algorithms it used were deemed too difficult for clerks to understand, and even the meaning of the term *database environment* was never explained to the group. A cursory training period totaling 8 days left no one in the department in a position to understand the "revolutionary" potential on which the designers relied.

In consequence, the database became understood by those not privy to the designers' goals as a control function, not a product development function. Worse, it was associated with the dreariest of the bank's tasks. The project, still going on when Zuboff's study ended in 1984, had stalled far short of its original visionary goals.

The project managers had chosen this arena because they believed the technology would force a reorganization of the bank's functional divisions and power structures. But the implementation strategy they chose produced a particular *social* role for the new system. They isolated themselves from the bank's senior management and effectively concealed the nature of their project even from its first users. Relying on an impact model of social change, the database developers avoided raising organizational issues directly—and guided their project into an organizational black hole. They did not understand that the database "environment" was not self-contained but only one element of a larger sociotechnical system that Kling and Scacchi (1982) have called the "web of computing."

Let us now look at an opposite case, where system developers understood very well the social purpose of what they were doing but failed to take into account some of the social impacts of the technology. This case is the computerization of British banking in the 1970s and early 1980s.

British Banks: Computers as Strategies for Organizational Change

In British banking the traditional mode of training prior to computerization was based on a master-apprentice model, according to Steve Smith (1989). Employment began at age 15 or 16, and one then rose level by level through a pyramidal hierarchy. Ultimately, with luck and aptitude, any employee could hope to become manager of a branch bank or even a general manager at corporate headquarters. Branch banks under the old system were full-service banks under a decentralized corporate system. Branch managers, by virtue of their apprenticeships, were capable (at least in theory) of performing any operation at any level of the branch's hierarchy. Senior managers were thus generalists whose decision-making skills and authority were held to result from a broad and deep personal experience.

Along with this career structure went an ethos of employee flexibility. Clerks had a relatively wide range of skills, allowing them to shift from task to task during the banking day, which might require posting of transactions and billing in the morning, when few customers were coming in, and cashiering toward the end of the day, when customers came in to cash paychecks and withdraw funds.

Computerization, in this case, was introduced largely *in order* to restructure work. Smith quotes the managing director of Olivetti:

> Information technology is basically a technology of coordination and control of the labor force, the white-collar workers, which Taylorian [i.e., F. W. Taylor's scientific management] methods does not cover . . . [E]lectronic data processing (EDP) seems to be one of the most important tools with which company management institutes policies directly concerning the work process conditioned by complex economic and social factors. In this sense EDP is in fact an organizational technology, and like the organization of labor, has a dual function as a productive force and a control tool for capital. (Franco de Benedetti, 1979, cited in Smith, 1989, p. 383)

British bankers installed computers as part of a general plan to move away from the craft-apprenticeship model toward a rationalized industrial-production model. Computers facilitated, for example, progressive specialization of tasks and automation of a great deal of work once done by hand.

Along with this specialization went a deliberate restructuring of career paths. Today not one but several tiers of entry are recognized, and more horizontal and vertical segmentation of functions has occurred, resulting in a more differentiated structure where not all paths begin at the bottom or lead to the top, and more specialized jobs mean greater expertise but less flexibility.

Some banks also used computers to centralize operations into a hub-and-satellite configuration called "branch network reorganization." Satellite branches, in the new scheme, offer limited services, mostly to individuals. Some satellites have no managers. The central office houses the data processing services as well as specialized services for corporate clients and investors. This centralization reinforces the segmentation of banking work and creates a class of specialist managers.

But the outcome of this computer-based restructuring of bank organization was mixed. While productivity in such repetitive tasks as data entry rose, numbers of clerical staff did not decline and frequently rose. A new gender division of labor also emerged, with more women working in the low-ceiling role of clerks and men clustering in what was known as the "accelerated career program." Smith (1989) cites the frequent "complaint that staff who joined to be bank employees find themselves 'dedicated' to repetitive 'factory work': 'This isn't banking, it's factory work' " (p. 385).

The repetitive nature of more segmented work, together with the corresponding reduction in sense of collectivity and community, caused declines in morale in some (though not all) banks. This finding has been replicated in other studies of computerization in office work where managers' goals have been similar (Attewell, 1987; Garson, 1988; Zuboff, 1988). The decreased flexibility of less skilled workers led to inefficiencies because of the variable work structure of the banking day. Finally, tensions arose between old-school generalist managers and the younger specialists over the very nature of banking. The younger group tended to treat branch operations as mechanical or industrial processes. Generalists felt this to be an insult to a formerly dignified career and also believed that the younger group lacked an intuitive understanding of bank operations, relying too heavily on analysis. The overall result, as in the American case, was a surprisingly low growth in productivity.

The British case shows computers used to facilitate the creation of a social product—in this case automation of traditional craft work and a centralization of the formerly decentralized branch system. These ends were embodied in the design of the computing systems they incorporated, especially in the centralization of data processing (reducing branch banks to input-output devices) and the segmentation of work, with data entry tasks separated from other, more complex banking assignments. As the reorganization and the investment in computing equipment proceeded, they had impacts upon the social space of the organization—many of which were neither foreseen nor desired by the designers. "Rationalizing" an existing process introduced new irrationalities, partly because it treated the organization as a machine without taking into account such social factors as job satisfaction and gender, and partly because computer systems rigidified a less flexible work structure (Kling & Iacono, 1984; Kling & Scacchi, 1982).

From these two examples we can see that there may be sociocultural as well as technological-economic reasons for the productivity puzzle. When computers were introduced at Global Bank Brazil in the hope that they would "impact" the organization, the result was partial failure due to inertia—a failure to treat the social context directly. When computers were introduced as part of a direct treatment of the organizational context, based on an automation model, they had unforeseen impacts on the culture of work that led to inefficiencies and social dislocations.

GENDER AND COMPUTERS

Computer work is stratified in an almost linear way along an axis defined by gender. Women are overwhelmingly dominant in the lowest skill, lowest status, and lowest paid areas, such as microchip manufacture and computer assembly (especially in "offshore," or foreign, factories) and data entry, where women account for up to 95% of the workforce. While statistical evidence in this area is problematic, a general trend is unmistakable: Numbers of women begin to decline as skill levels rise, with somewhere on the order of 65% of American computer operators, 30%-40% of programmers, and 25%-30% of systems analysts being female. Gender imbalances in European countries are more dramatic (Frenkel, 1990; Gerver, 1985).

A similar pattern exists in education, in a way that closely parallels gender differentiation in mathematics. Girls and boys display roughly equal interest and skill in the primary grades, but starting around age 11 or 12 girls gradually begin to stop enrolling in computer courses. By high school age boys outnumber girls in such courses roughly 2 to 1. During the 1980s roughly this same ratio of men to women persisted through undergraduate college, with about 35% of bachelor's degrees in computer science awarded to women. But there is some evidence that this ratio has declined substantially, perhaps to as little as 20%, in the last 2 or 3 years, without a corresponding drop in other technical majors.[6]

By the Ph.D. level the situation is much more dramatic; the percentage of computer science Ph.D.s awarded to women has remained steady at 10%-12% since 1978. The situation in engineering is worse, with women receiving only 8% of Ph.D.s, though the numbers there have been rising. For comparison, note that the percentage in the physical sciences and mathematics is now about 17% and rising.

The imbalance is most severe at the level of faculty employment. Only 6.5% of tenure-track faculty in computer science departments are female (7% in computer science and 3% in electrical engineering). One third of Ph.D.-granting departments have no women faculty at all.

Sexism in Educational Settings

One possible version of this story relies for an explanation on bias and systematic oppression. High school-age boys have frequently been observed to harass girls and demean their skills, sometimes deliberately to keep enrollments in computer classes low. Illustrations in computer science text-books typically show a 10:1 ratio of men to women, and computer advertising is strongly male oriented. Women students at all levels have reported oppression in many forms, ranging from overt statements by senior professors that women do not belong in graduate school to more subtle and probably unconscious mistreatment, such as having their own ideas ignored or patronized in the classroom while similar ideas of their male colleagues receive praise. The following quotations from students and research staff illustrate the sometimes very direct nature of this sexism.

> While I was teaching a recitation section, a male graduate student burst in and asked for my telephone number. Men often interrupt me during technical discussions to ask personal questions or make inappropriate remarks about nonprofessional matters.
>
> I was told by a secretary planning a summer, technical meeting at an estate owned by MIT that the host of the meeting would prefer that female attendees wear two-piece bathing suits for swimming.
>
> I was told by a male faculty member that women do not make good engineers because of early childhood experience . . . little boys build things, little girls play with dolls, boys develop a strong competitive instinct, while girls nurture. (Anonymous interviewees, cited in Frenkel 1990, pp. 36-37)

Such factors as the lack of female role models and the so-called impostor phenomenon, in which minorities feel themselves not to be "real" members of the dominant group, distrusting their own skills and avoiding public display so as not to be caught out "impersonating" a "real" computer scientist, are among the other ways gender stratification perpetuates itself (Leveson, 1989; Pearl et al., 1990; S. Weinberg, 1990).

These are real and important mechanisms in creating gender imbalance. At the same time, there is evidence to suggest that in the computer industry, far from a systematic exclusion, many companies have made active efforts to recruit more women, and that, compared with other, older industries, computing has been a more favorable environment for women. In academia, the very scarcity of women Ph.D.s makes finding qualified candidates difficult. So while more subtle bias persists, *direct* discrimination against women is probably somewhat less of a factor in computing than in other careers (Leveson, 1989).

Cultural Construction and Gendered Tools

But another approach to the issue of gender differences is to ask the question of whether or not computers, *as tools,* are gender neutral. I will argue that they are not: In fact, computers are culturally constructed in such a way as to stamp them with a gender and make them resistant to the efforts of women to "make friends" with them (Edwards, 1990, in press; Perry & Greber, 1990; Sanders & Stone, 1986).

Scientists tend to think of computers abstractly as Turing machines, universal machines capable of doing anything from controlling a spaceship to balancing a checkbook. But people always encounter technology in a particular context and develop their understanding from there. If they first meet computers in a course, they are likely to be introduced to them in a theoretical mode that emphasizes their abstract properties and their electronic functioning. If they meet computers in an office, they may understand them as word processors or spreadsheet calculators. In every context they will be surrounded by a sort of envelope of other people's talk, writing, attitudes, images, and feelings about them. The formal content of a course or a training session or a conversation with another user is only part of what is communicated.

Many investigators have suggested that computer avoidance in girls is connected with differences between what can be loosely termed the *cultures* of men and women. (Of course, there is great variability within the generalizations I am about to describe.) Men learn to value independence—the ability to do things on their own, without help. They are most comfortable in a social hierarchy in which their position is relatively clear. They are trained early on for roles as competitors and combatants, and they value victory and power. Abstract reasoning is, for men, an important value, partly because of its connection with power. Carol Gilligan's well-known study of men's and women's morality, *In a Different Voice,* revealed that men tend to see the highest form of morality as one based on a reasoned adherence to an overarching moral law that treats all actors as equals (Gilligan, 1982).

Women, by contrast, tend to prefer interdependence. Reliance on others is valued because it continually maintains a social fabric or network, seen as more important than individual self-sufficiency. Instead of hierarchy, women's culture practices social "leveling," in which an underlying goal of conversations or games is to keep everyone at the same level of status. Similarly, competition and winning are less important than keeping a game or conversation going (Tannen, 1990). Practical skills rather than abstract reasoning tend to be primary values, and this goes along with a morality that perceives particular relationships as superseding abstract rules—people are treated differently depending on their needs and relationships to other actors, rather than similarly based on their moral equivalence (Longino, 1990).

In her studies of children learning to program in the LOGO language at a private school, Sherry Turkle observed two basic approaches to computer programming. Students she calls "hard masters" employed a planned, structured, technical style, while "soft masters" relied on a more amorphous system of gradual evolution, interactive play, and intuitive leap. In her words,

> Hard mastery is the imposition of will over the machine through the implementation of a plan. A program is the instrument for premeditated control. Getting the program to work is more like getting "to say one's piece" than allowing ideas to emerge in the give-and-take of conversation. . . . [T]he goal is always getting the program to realize the plan.
>
> Soft mastery is more interactive . . . the mastery of the artist: try this, wait for a response, try something else, let the overall shape emerge from an interaction with the medium. It is more like a conversation than a monologue. (Turkle, 1984, pp. 104-105)

Note the similarity of these two modes with the two cultures I have described. In fact, Turkle found that the majority of hard masters were boys, and the majority of soft masters were girls. But both styles produced some consummate programmers.

Both Turkle's hard and soft mastery and my descriptions of men's and women's cultures are, of course, caricatures of immensely flexible and complicated processes rather than hard-and-fast rules. A culture is not a program but a subtle set of nudges in particular directions that not everyone receives to the same degree or responds to in the same way. Many men are more at home in what I have described as "women's" culture, and vice versa. Some learn to be equally at home in both modes. And it is important that excellent programs can be written by people of both sexes using both methods, something Turkle saw in men and women of all ages (Turkle, 1984; Turkle & Papert, 1990).

Nevertheless, these two dichotomies are suggestive. Consider, for example, the fact that many if not most video games emphasize violence, often with a military metaphor. The first video game was "Space War," written by MIT hackers during the early 1960s (Levy, 1984). (But the first commercial game was the benign "Pong," and one of today's most popular games is the equally unmilitaristic Tetris.) Still, the bulk of the games that led the video arcade craze of the early 1980s were combative in nature, and it was partly as a belated response to the potential market among adolescent girls that less violent alternatives such as Frogger and Pac-Man were introduced.

Hacker culture, to give another example, is strongly male oriented. Hackers frequently work in independent isolation. Many say their fascination with hacking is related to the sense of control and power, an elation in their ability to make the machine do anything (Hafner & Markoff, 1991; Weizenbaum, 1976). While the so-called hacker ethic described by Steven Levy theoretically values programming skill above all else including physical appearance and gender, in practice hackers frequently avoid women and exclude them from their social circles (Levy, 1984; Turkle, 1984). Turkle's ethnographic study of MIT hackers revealed a powerful competitive side in such phenomena as "sport death," the practice of staying at one's terminal until one drops, achieving fame through a kind of monumental physical self-denial. In the 1960s and 1970s, and to some extent still today, hackers played an important unofficial role in the development of system software and computer games. So their conceptions of the nature of computing were, in a sense, embodied in machines.

Another source of gender differentiation may be the nature of computer instruction in schools and colleges. Computer science, with its marginal disciplinary position between mathematics, cognitive psychology, and engineering, has to a certain extent relied for institutional survival on laying a claim to mathematical-scientific purity, and one place this claim is expressed (and students are weeded for correct skills and orientations) is introductory computer science courses. Traditional programming courses, partly for this reason, are taught in a highly theoretical mode that emphasizes abstract properties of logic, computation, and electronics rather than practical uses. Girls report disinterest and frustration in classes with this orientation and get better grades in courses with a more practical bent.

In a major 1989 debate in the pages of the computer science journal, *Communications of the ACM,* University of Texas at Austin computer scientist Edsger Dijkstra proposed that introductory computer science be taught in an even more formal model, emphasizing its fundamentally mathematical core (Dijkstra, 1989). Rather than use real computers, students in Dijkstra's program would have to write programs in unimplemented languages and prove their validity logically (instead of debugging them by trial-and-error methods). Many of his colleagues objected to this excessively formalistic view—but it unquestionably reflects one important strand of thought about computer learning. To the extent that this teaching strategy holds sway, it tends to inhibit women's entry into the field (Frenkel, 1990).

These last three examples—video games, hacking, and computer instruction—all show the process of cultural construction in action. Interaction with combative video games constructs the computer as a site of conflict and competition, a game where winning is a matter of metaphorical life and death. Hacking uses the computer as a medium for a social process of self-construction in which young men compete with each other and with the machine and achieve independence and power. The computer, as the site of this self-construction, receives a gender association. Computer instruction that emphasizes abstract rationality is more appealing for boys and facilitates the association of computers with men. Thus computers have frequently been culturally constructed as male-gendered objects.

Here, then, is another social process through which computer technology and the social production of knowledge and values interact. The tendency is to think of these cultural factors, because they are so flexible and variable, as separate from and independent of design. But people encounter them in their experience of computing as necessary presences that structure the computer they perceive. Social "context" and design interpenetrate; no element is purely essential and no others purely accidental (Winograd, 1987).

CONCLUSION

These three brief case studies bring into relief the interaction of technology with politics, society, and culture as computers increasingly permeate industrialized societies. Computers rarely "cause" social change in the direct sense implied by the "impact" model, but they often create pressures and possibilities to which social systems respond. Computers affect society through an interactive process of social construction.

Who will get computers? What new kinds of access to information will they allow? Who will benefit, and whose activities will be subject to more detailed scrutiny? How will these actors react to such changes? These questions are especially important precisely because the computer is not only inserted into an organization or a culture but frequently embodies particular images of how the organization or culture functions and what the roles of its members should be. Once introduced, a computer system, by embodying these images, can help institutionalize and rigidify them. What is needed is an awareness of the "web of computing" (Kling & Scacchi, 1982), that is, of the ways in which a new computer system will be inserted into an existing network of social relationships. Neither a "social impacts" nor a "social products" approach will produce an adequate picture of this interaction; only an image of technological change as a social process is likely to be robust enough to capture the flavor of how computers work in society.

NOTES

1. The introduction to Beniger (1986) provides an interesting synoptic view of this literature, and the introduction to Dunlop and Kling (1991) gives a very helpful critical analysis. As Dunlop and Kling argue, notions of utopian and "revolutionary" effects—or their converse, the Orwellian idea of computerized Stalinism—have been substantially, even hysterically oversold. This is especially true in the areas of office automation and computing in government, where their effects on productivity and panoptic power have been considerably less than many imagine.

2. Analog computation represents variables using continuous physical quantities such as electrical resistance, motor speed, or voltage, which are physically combined to yield a result. Everyday examples of analog devices are volume controls (variable resistors) and ordinary clocks (motor speed). The once ubiquitous slide rule is a commonplace example of an analog computer; mathematical operations are performed on numerical quantities (represented as positions along the length of the rule's scales) by sliding the rule's moving middle section back and forth. The rule's length is a continuous quantity. Digital computation represents variables as discrete quantities such as whole numbers, switch positions, or magnetic polarity. Everyday examples of digital devices are light switches (on or off) and digital clocks (which unlike ordinary clocks show hours and minutes as discrete, unit quantities).

3. For fuller accounts of wartime and postwar developments, see especially Edwards (1987, 1989), Goldstine (1972), Redmond and Smith (1980), and Rees (1982). My account in the rest of this section relies heavily on Flamm (1987, 1988), who gives the best informed history of military involvement, though his perspective is generally technical and economic. For a book-length social and cultural analysis, see Edwards (in press).

4. But see Jacky (n.d.) for a description of the system's shaky record of reliability.

5. Borning (1987) presents a lengthy history of NORAD computer failures, some serious enough to lead to escalations in the alert status of nuclear forces. Problems of complexity and reliability in these systems became a social trope for nuclear fear, as reflected in films and novels from *Dr. Strangelove, Fail Safe,* and *Colossus: The Forbin Project* in the 1960s to *War Games* and *The Terminator* in the 1980s, all of which involved some variation on the theme of computer-initiated nuclear holocaust (Edwards, in press).

6. Most of the statistical information in this section is drawn from the National Science Foundation (1990), Frenkel (1990), Gerver (1985), and Pearl et al. (1990).

13

Science Studies and Machine Intelligence

H. M. COLLINS

MACHINE intelligence is a relatively new topic in science studies, therefore a simple survey would not necessarily represent the promise of this fast-growing field. My review, then, is as much a forward-looking setting out of the ground as it is a survey. I will try to develop a classification of the relationship between science studies and machine intelligence, using illustrations along the way, while worrying less about complete coverage of every publication. I apologize in advance to all those absentees who ought to have been mentioned.

There is another problem—defining the rubric. Construed one way, machine intelligence is a narrow topic, coextensive with artificial intelligence (AI). Construed another way, pocket calculators are intelligent machines as are slide rules and logarithm tables. Written works in general and even mechanical diggers, door closers, and doorstops share aspects of our collective life in some sense and most of what is puzzling about complicated intelligent machines is also puzzling about at least some of these simple things.[1] Then there is the problem of defining science studies. For example, is the offshoot of ethnomethodology—conversational analysis—which is important in the study of human computer interaction, to be included? I have not

AUTHOR'S NOTE: I am particularly grateful to Barrie Lipscombe for help in preparing this chapter.

agonized about these problems too much; I have simply wandered around the edges of the disciplines as seemed right at the time. No doubt different choices would have been made by others.[2]

I will argue that science studies' relationship to machine intelligence recapitulates many of the concerns of the field as a whole. There are rationalist and relativist approaches; there are approaches that see the real world as something apart from human forms of life and those that see only one sphere of actants with no internal boundaries; there are those concerned to model the individual scientist and those concerned with the collectivity; there are those concerned to improve machine intelligence and those concerned only to describe it; there are those who work almost within the science of artificial intelligence and those who work without it, and those who take up different positions at different times.

In the work on artificial intelligence, science studies has been surprisingly successful in invading the established institutions of science. The daily newspaper tells us:

> For its first 30 years, computing was driven by technological developments. Computer science was first seen as a branch of mathematics, and later as an engineering discipline. The results were systems that showed technical brilliance, but their use was limited to that small number of people with the desire, the aptitude and training to use them.
>
> Meanwhile, Parc was busy recruiting not mathematicians and computer scientists but linguists, philosophers, cognitive psychologists and social scientists. . . . That is why Parc's ideas . . . are starting to dominate the computer world. . . . Mathematicians and engineers are having to share their pre-eminence with all those "humanities and social science types". . . . The successful computer scientist of tomorrow will be both mathematician and sociologist. (Devlin, 1991, p. 31)

Among the social scientist "types" are Lucy Suchman and her team at the famous Xerox PARC in Palo Alto, and a host of ethnomethodologists at "EuroPARC" in Cambridge, England. At least some of these consider themselves members of the science studies community, attend the relevant conferences, and would find themselves solidly connected to any sociometric map of the subject. Other contributors include Nigel Gilbert and Steve Woolgar, who were early members of the community of sociologists of scientific knowledge; they obtain major funding from AI or information technology-oriented sources. Leigh Star, an American sociologist of science, has worked with computer scientists at MIT and spent a large part of her academic career in a computer science department. Bruno Latour's flexible vocabulary concerning networks of actants has been "enrolled" by many forward-looking information scientists. Finally, one of my own papers on the topic of expert systems was awarded the prize for technical merit at a meeting of the British Computer Society (Collins, Green, & Draper, 1985). This invasion of a prestige, near-market, relatively big-money science by ethnomethodologists, discourse analysts, reflexivists, relativists, and so forth can only remind us that academic life still has the potential to astonish—especially when it comes to applications. It has turned out that the first 10 seconds of telephone conversations *are* as interesting as the formation of galaxies, are probably harder to analyze, and anyone who can exploit their analysis will be rich beyond the dreams of emperors. The beggars, then, have come to the feast (the most striking example being the ex-Manchester Polytechnic radical "ethno," Bob Anderson, now directing EuroPARC). Let us hope they can pay their share of the bill.[3]

Looking at the relationship from the other end, the attempt to develop artificially intelligent machines can be said to represent a kind of natural laboratory for testing science studies' ideas. If isolated mechanisms can accomplish that which sociology of scientific knowledge (SSK) claims to be the preserve of the human collectivities, then its practitioners will have to think again. There is a lot to be learned from our new laboratory.

COMPUTER MODELING OF SCIENTIFIC DISCOVERY

Since the 1980s claims have been made that computer programs can reproduce scientific discovery (for a standard reference, see Langley, Simon, Bradshaw, & Zytkow, 1987). To many sociologists of scientific knowledge, discovery programs appear to be such a shallow representation of scientific action as to be of little interest. In 1989, however, Slezak (1989) suggested that, because discovery programs discovered without being social, the sociology of scientific knowledge had been disproved. The odd thing about this was that it was not a hypothetical claim—"if there were isolated programs that could make discoveries, they *would* disprove SSK"—but a claim about existing programs—"they *have* disproved SSK."

The resulting debate showed that the builders of discovery programs claimed less for them than Slezak claimed on their behalf. Especially noteworthy was the contribution by Herbert Simon (1991), a pioneer of the field, who said that discovery programs were just as contextbound as individual scientists.[4] The apparent disappearance of context in discovery programs turns on an implicit "digitization" of the notion of experiment; experiments are taken to provide unambiguous data that can then be fitted to a theory. The refutation of this model is, of course, the very stuff of sociology of scientific knowledge; the interpretation of experiment has long been shown to be a collective activity (Collins, 1989). As computer programs are constructed that try to model more of the scientific process, the topic will, no doubt, resurface.

COMPUTER MODELING OF INDIVIDUAL SCIENTISTS' WAYS OF THINKING

Given that existing computers cannot model the whole process of scientific discovery, including the reception of discoveries by different scientific communities, there remains the question of whether they can model the thought processes of individual scientists.[5] This is a much more difficult problem.

It is clear enough that even ordinary computers can do some things that we regularly speak of as "discovering." For example, an old-fashioned factor analysis program might pull "factors" out of a mass of data that could only be found as a result of an enormous amount of calculation by a human being. A fortiori, the same applies to more intelligent programs such as rule-inducing expert systems.[6] These are given a set of problems and solutions and induce a rule of action that seems to underlie them. Often these rules will be new to the operator and it has been said that interesting discoveries have been made this way. Even more might be expected of neural net computers, which "learn" to make generalizations from repeated examples. It is the surprise value of the "discoveries" made by these machines that makes them interesting.

We may note that an intriguing feature of the debate about the effectiveness of discovering machines is that there is one class of machines (factor analysis, automated induction, and so on) where the warrant for success is the novelty of their results, and another class (the Simon type) where the warrant is their ability to rediscover established results—that is, their lack of novelty. When these two types of programs are compared with the model of discovery most typically found in the sociology of scientific knowledge, the "friction" is found in different places.[7] Simon-type machines, as we have noted, are suspect at the input stage; their "discoveries" may be a little too immanent in the data that is fed to them—it has been cleaned by reference to the desired output. On the other hand, where outputs are novel we need to pay close attention to the output stage. If it turns out that the machines produce a large number of different outputs and the operator picks just those that seem to be plausible, then the machine is not doing much of what is involved in discovering except the calculating. It is scientists who have to establish that a relationship that they have found in some data should count as a discovery; they have to change the order of scientific social life.

It is very useful to have a tool that can quickly generate a large number of potential "discovery candidates" that might otherwise have been overlooked, but this is not the whole of discovery. I would suggest that even novelty-producing computers are not sufficiently embedded in sociai life to do more than mimic a small part of the scientific discovery work, and it may be more accurate to describe this class of computer as a "discovery assisting tool" rather than a discoverer.[8] To establish *this* claim, one would need to look at the whole range of outputs prior to the weeding out process.

On the other hand, it seems that computers can be of assistance to those trying to describe or map individual scientists' discovering behavior. For example, Gorman and Carlson are using computer techniques such as hypermedia to model the way scientists work on multiple simultaneous projects while Gooding is using computers to simulate discovery processes (see, e.g., Gorman & Carlson, 1990).[9]

Gooding is using the computer to represent Michael Faraday's manipulation of ideas, images, bits of apparatus, and so forth. The source of information on these manipulations is Faraday's detailed laboratory notebooks studied in conjunction with repetition of some of Faraday's experiments. The way this differs from other "discovery" programs is that it follows Faraday's procedures and their development rather than claiming to reproduce them from scratch or reconstructing them according to a "logic of inference." Nevertheless, Gooding hopes to discover a heuristic of experiment that fits at least Faraday and his contemporaries' style of experimentation. Gooding claims that he will be able to "model the socially mediated nature of the inputs" to his computer program but this seems hard to credit because it would imply that the program itself had been socialized. There is a world of difference between retrospective descriptions of social behavior as rule following and real-time embedded social action. Gooding might find the difference more salient if his program were intended to make real-time discoveries rather than reproduce existing ones. This theme will be touched on several times in this review.

Individual and Social Models of Discovery

It may be that it is too simple to separate the social side and the individual side of discovery. Woolgar (1989) makes the point clearly. He observes that some argue as though there were a choice between social and individual models of scientific discovery, whereas for the true social analyst of science, the actions of each individual are already shot through with the social. For example, in their every utterance, scientists draw on their, inherently social, natural language. The same applies to the way each problem is framed and each action is undertaken.

Given this way of looking at things, it is not a matter of more or less social but all social or not social at all. In these terms, the ideas of Gorman (1989) or of Thagard (1989), that a thorough understanding of science requires cooperation between an individual cognitive perspective and a social, sociology of knowledge perspective, seem misplaced. Curiously, what follows is that the existence of *any* discovery program, even one that mimics an individual scientist, is fatal for any social theory of science.

I have suggested something similar, pointing out that a sharp critic of the sociology of scientific knowledge would argue that even the existence of, say, pocket calculators—asocial individuals that appear to do what only socialized arithmeticians ought to be able to do—disproves social models of science (Collins, 1989). My response to such a critic would be that the premise is wrong; that is, much of the time calculators do not do the same sort of arithmetic as humans—the two types of arithmetic only coincide when the humans are doing a special kind of action that I refer to as "behavior-specific action." According to this theory, to which I will return later, computer models of individual scientists could contribute to our understanding of only one component of scientific action—the part that is carried out in a behavior-specific manner. Discovery programs, factor analysis programs, and the rest mimic behavior-specific action rather than the whole of scientific discovery.

There are, then, three possible models of the contribution of studies of individual scientists to the scientific process. *Possibility one takes it that all action is essentially social.* This possibility has two branches: *Branch one* claims that computer models of individuals are simply irrelevant to the question of what science is; *branch two* claims that those programs that appear to model individuals are actually already impregnated with society—in the way the programmers handle the inputs and the outputs or the way inputs and outputs encode retrospective accounts of social rules—whether the programmers know it or not.[10]

Possibility two in this scheme comes out of the theory of behavior-specific action; it *suggests that there can be computer models of special, highly circumscribed parts of the scientific process.* This possibility would allow computers and computer modeling a small role in these special areas—the same role they fill when they are used as tools to aid scientists.

Possibility three—the essentially cognitivist model—*takes it that there are "social factors" affecting the progress of science but that a large proportion of the essentials are rational and individualistic.*[11] In this case much of science can be modeled by computer programs but, because both individual and social perspectives are components in the process of science, both must be encompassed if science is to be completely understood.

Attractive as is this last view, I believe that authors who adhere to it, such as Gorman and Thagard, have a little more work do in explaining what they mean by a "complete" explanation. There is a rhetorical force to the notion that if there is more than one component to a process then all components are vital in an explanation of that process. But this would dictate that their own explanations ought also to encompass, say, the biology of the brain. After all, without the brain's blood flow, oxygen use, energy transport, and all that, there would be no science. Gorman and Thagard seem content to black-box brain biology (and brain chemistry, brain physics, and so on), so it is not entirely clear why they should insist that those who want to hold even a "social factors" model of science *have* to include cognitive factors. Here we see, sharply defined, the shadow of the larger debate between sociological and psychological explanations of action in general.[12]

SCIENCE STUDIES' CONTRIBUTION TO UNDERSTANDING AND IMPROVING INTELLIGENT MACHINES

Probably the most decisive intervention of a sociologist into the world of computer science so far has been made by Lucy Suchman with her book *Plans and Situated Actions* (1987). This book was predominantly a critique of those psychologists' models of human actions that took them to be governed by plans. Suchman showed that plans could only be superimposed on actions retrospectively while in real time actors responded to the details of the situation. Real-life situations were too unpredictable to be encompassed by preformed plans. Suchman illustrated her point by showing the failures of people to interact with a very simple "computer"—an interactive photocopier. The assumptions upon which the photocopier's instructions were based did not mesh with the actions of new users; the fact that she videotaped the frustrated struggles of some famous computer scientists attempting to use the copier lent Suchman's study a certain *frisson*.

One of the main planks of Suchman's argument—the difference between prospective and retrospective descriptions of action—is, of course, also to be found in sociology of scientific knowledge. Suchman, it is true, developed her ideas independently, drawing heavily on anthropology and expressing them with different metaphors. Nevertheless, she is properly included here because of the sympathy between what she says concerning plans in general and what sociologists of scientific knowledge say concerning rationality in the process of science.

Suchman and her coworkers at Xerox PARC do not seem to be primarily interested in improving the performance of machines in their interactions with human beings except insofar as their work disabuses computer scientists of certain simplistic models. Nevertheless, various computer scientists have tried to use her perspective in a more direct way. It does seem that as soon as a social scientist says something along the lines, "Your computer cannot model human action because human action is situated," someone will immediately try to build a "situated inference engine" or some such. Often, these attempts miss the whole point of the critique but (being self-consciously unevaluative about what counts as progress in science), we have to say that this work does move computer science on.

Leigh Star (1989b, see also 1989a, 1991a) discovered in her collaborative work with computer scientists that sociological metaphors were adopted wholesale. Adopting a metaphor allows the difficult parts of the problem to be ignored while the terminology makes it seem as though social life *is* being encapsulated in programs. What happened to Star seems to be the sociological equivalent of what Drew McDermott (1981) called "wishful mnemonics." McDermott complains about computer scientists' tendency to use psychological terms such as *learn* and *understand* as though they could be transferred without loss from human to computer—solving the problem by fiat, as it were; a more sophisticated generation of AI researchers are now doing the same for terms such as *community, network,* and *situated.*[13]

Nigel Gilbert describes some work intended to model the growth of human societies using networked programs to model collections of individuals (Gilbert & Wooffitt, in press). This work seems to be wide open to the employment of wishful mnemonics. It seems unlikely that the computer-instantiated "individuals" in these programs bear much resemblance to members of society; it is more likely that they are rational abstractions of the sort used by economists and rationalist philosophers of science.[14]

Suchman and her team have studied the way that humans work interactively both on a small scale (using a "whiteboard") and on a large scale (in an airport). A fine interactive study of ship navigation that may have a bearing on human interactions with machines has also been completed by Ed Hutchins at San Diego (1989). It may be that new devices that aid interactive work will emerge from these studies, though the route may not be direct.

Something of the same ambiguity surrounds the relationship between conversational analysis (CA) and computer design. Again, we may count ourselves entitled to include at least some CA under our rubric because of the input to science studies of discourse analysis and because of the developing sociometric links between the communities.[15] We may also say that the study of conversation has to do with artificial intelligence in that the "Turing Test"—a test of a computer's ability to converse in written language—is widely accepted as a test of computer "thinking" (Turing, 1950).

Conversational analysis promises to provide some generalizable rules of conversation but again it is not clear whether these are the type of rules that can be programmed into computers.[16] It may be that CA shows the hopelessness of modeling human interaction rather than providing practical answers to computer scientists' problems.

Probably the most self-conscious *application* of CA/discourse ideas to computers has come from Nigel Gilbert and his group at Surrey University. Gilbert, a former sociologist of science, has been one of the chief links in Britain between computer scientists and sociologists of science.[17] In a recent paper, Gilbert and Wooffitt recommend the application of conversational analysis techniques to the improvement of human computer interaction and Gilbert and Luff have edited a book on the same topic (Button, 1990; Luff, Gilbert, & Frohlich, 1990).

The interest of Gilbert in natural language interfaces seems to have more to do with making machines usable than modeling intelligence. These are quite different projects. The 1960s program ELIZA, or Winograd's SHRDLU, could be mistaken for conversational partners but they have little to tell us about language use between humans (Weizenbaum, 1976; Winograd & Flores, 1987). The difference between making a passable program and making a program that uses language in a way indistinguishable from humans under severe examination circumstances is enormous in terms of the question of artificial intelligence though there may be only a small difference at first sight. Thus, at first sight, a computer spell-checker does the same job as a human but closer examination reveals that, not only do they work in quite different ways, there are significant differences in what they take to be error.[18]

Conversational analysis and ethnomethodology have also been applied by the group associated with Xerox EuroPARC in Cambridge. They include Bob Anderson, Christian Heath, Graham Button, and Wes Sharrock. The work they have done on air-traffic control (here a group at Lancaster were also involved), the London Underground, and studies of various types of human-machine interaction is easily recognizable for their continuity with what they did before moving on to computers. As already intimated, it is not yet clear whether the payoff will be in terms of improved communication with machines—as Gilbert believes—or in disabusing computer designers of their ambitions. In either case, the study of human-computer interaction could not sensibly continue without input from those who have studied the details of human conversation most closely.

RELATIVIST/ACTANT-NETWORK ANALYSES OF THE BOUNDARY BETWEEN PEOPLE AND MACHINES

We can go no further without defining some fundamental concepts belonging to science studies—"relativism," "constructivism," and "symmetry." *Relativism* is the prescription to treat the objects of the natural world as though our beliefs about them are not caused by their existence. The idea is to discourage analyses of scientific history in which the existence of some phenomenon, such as the inverse square law of gravitation, is taken as sufficient explanation for its widespread acceptance. To explain the establishment of a view, it is important to proceed as though the truth of the view was not a causal factor. This can best be accomplished by a self-conscious relativism—treating what seems to exist as being relative to the social group in which it is taken to exist. The social group *constructs* its beliefs about the natural world in much the same way as it constructs its beliefs about art, poetry, or politics—hence the term *constructivism.*

David Bloor first expressed the principle of "symmetry." He said that for the purposes of historical and sociological analysis one must treat what comes to be seen as true in science in the same way (symmetrically) as what comes to be seen as false; never must analysts allow themselves to explain what comes to be believed *now* by reference to what is discovered to be true *later.* What counts as true is the *outcome* of social processes, not the cause of that outcome.[19]

In everything said so far, I have assumed that it is sensible to distinguish between humans and machines. It follows quite naturally, however, from relativist/constructivist approaches to science studies that the distinction is itself constructed. Woolgar was the first to make this point explicitly in his provocatively titled paper, "Why Not a Sociology of Machines?" (1985). Woolgar's four-part typology of the sociology of AI comprises the sociology of AI researchers; the social construction of intelligent machines (in which machines are seen as the social products of humans); a sociology of machines, in which the interactions of machines alone would be the subject of study (this is a straw man); and his favored sociology, which takes as its topic the maintenance of the distinction between humans and machines. The latter project requires, of course, a symmetrical approach to machines and humans and therefore a deep commitment to treating the humans and machines as members of a single community, any apparent ontological differences resting on divisions *constructed* by the joint community.

This kind of approach is generalized in the actant-network program of Michel Callon (1986b) and Bruno Latour (1987). Callon and Latour do not have intelligent machines in special focus but "things" in general; they have extended the principle of symmetry to all dichotomies including the social and the natural. Both for them and for Woolgar, we must not even say that the social and the natural are *socially* constructed because that would be to use an element of the dichotomy that is to be explained—the difference between the social and the nonsocial—as the starting point of the explanation.[20] Rather than seeing human actors and nonhumans competing for power within our theories, Callon and Latour try to treat both equally—as *actants.*

I have argued elsewhere that this "radical symmetry," in spite of its philosophical innovativeness, leads to more conservative analyses of science than old-fashioned symmetry.[21] Radical symmetry and "actant-network theory" can have nothing to say to engineers that might lead to an improvement in the power of machines, nothing to say to programmers or to psychologists about the potential of machines as mimics of human intelligence, and nothing to say to those who want to use AI as a laboratory for testing out the ideas of SSK. To be more correct, what they have to say is that anything might happen given the appropriate "alignment of forces" in the network of undifferentiated humans and things—the network of actants.[22]

"CLASSIC" SOCIOLOGY OF SCIENTIFIC KNOWLEDGE-STYLE *ANALYSES* OF THE WORK AND WORLD OF COMPUTER SCIENTISTS

"Traditional" sociology of scientific knowledge can be applied to various aspects of the work of computer scientists just as it can be applied to various other areas of science. Sometimes the area examined might be related to artificial intelligence and sometimes not; the general approach will not differ.

Among those essaying "ethnological," "laboratory study"-type approaches to computer science are Diana Forsythe and, stepping from one compartment to another, both Steve Woolgar and myself.[23] Forsythe (1987) has applied her anthropological skills to expert systems laboratories, first at Stanford and latterly at Pittsburgh. Forsythe has watched the leading knowledge engineers at their work; she has seen them discovering just those problems that one would expect them to discover from an SSK perspective and, in collaboration with knowledge engineers, she has worked out some of their implications (Forsythe & Buchanan, 1988, 1991).

Woolgar has spent some time working in a computer firm participating in the development of a new personal computer. This is not a study of intelligent machines per se, but doubtless the results will be applicable in areas of AI too. A number of Woolgar's associates are engaged in related studies, for example, studies of the research culture of software houses and of information systems development (that is, the way information is organized in a company). On a different tack, Woolgar and Russell have written a paper discussing computer viruses in terms of moral panics (Woolgar & Russell, in press).

As for my own work in this vein, I describe cooperative work with a knowledge engineer and our joint development of an expert system for semiconductor crystal growers. The difficulties involved in articulating the rules of a skill for incorporation into a program are described as is the relationship between the finished program and the technician's skills (Collins, 1990/1992, chaps. 10-13, see also in press-a).

Sherry Turkle has investigated the way computers are treated by children—the process of anthropomorphization, as one might say—in her book *The Second Self* (1984), which also provides an excellent history of AI debates.[24] More recently, Turkle has argued that the opacity of the rules of neural nets lends them a fascinating mystery and accounts for the hopes that have been invested in them.

A number of authors have used SSK or general science studies perspectives in critiques of AI. For example, Phillip Leith, who is now based in the Law Department of Queen's University Belfast, fought a heroic campaign against those who wanted to use the programming language PROLOG as a substitute for legal decision making (see, e.g., Leith, 1987).[25]

At Edinburgh University, Eloina Pelaez (in press) has written a graduate dissertation on the software crisis while Mikel Olazaran's history of neural nets is being considered for publication. Donald MacKenzie, who supervised both theses, has himself written an account of the development of arithmetic algorithms, showing that what counts as correct computer arithmetic is relative to its scientific milieux. The argument turns on appropriate ways of handling rounding errors (MacKenzie, n.d.; see also, Collins, 1990/1992, chap. 5). MacKenzie (1991) has also published an analysis of the notion of "provability" of computer programs, focused on the British "Viper" computer. Using science studies-related analytic skills, MacKenzie has shown that the claim that this computer's program had been fully "proved" could not be supported in terms of the claimants' own criteria.

THE APPLICATION OF THE FINDINGS OF THE SOCIOLOGY OF SCIENTIFIC KNOWLEDGE TO THE QUESTION OF THE MECHANIZATION OF KNOWLEDGEABILITY

There is an ambivalence when science studies is applied to computing and to AI. Because academics in science studies are computer users in the way that they are not users of, say, gravity wave detectors, the problem of symmetry makes itself felt in a very direct way. For example, computer viruses, Woolgar and Russell's topic, are a potential practical problem as well as an object of study (irrespective of whether computer viruses exist in large numbers). The standard sociology of science tactic—turning resources into topics—encounters a resistance; some of the topics are our resources in a very obvious way. At the same time, our topic—knowledge—is AI's resource; from the point of view of those trying to make intelligent machines, the more they know about knowledge the better, and science studies, if it knows about anything, knows about knowledge.

Given this confused relationship, a number of authors coming from SSK and similar fields have found themselves taking a more evaluative position regarding artificial intelligence than would otherwise be expected. The Wittgensteinian foundations of SSK make it a natural bedfellow of other philosophical critiques of AI such as that of Hubert Dreyfus (1979/1972) or Winograd and Flores (1986). Notable sympathizers with the general view that programs cannot substitute for human practice because "rules cannot contain the rules for their own application" include Bloomfield (see Leith, 1987), Leith, Suchman, and Collins. Taking this view, however, means taking an essentialist view of the difference between people and machines—a departure from SSK's symmetry and relativism.

The dilemma is very clear in my own work. I claim to be applying the accumulated understanding of knowledge generated by SSK to artificial intelligence. One part of this enterprise is to explain the nature, limits, and design principles of expert systems, while another is to establish the overall limits of intelligent machines. I argue that, because knowledge is socially located, no social isolate, such as a computer, can exhibit the full range of human abilities; indeed, given the neo-Wittgensteinian analysis of knowledge, the difficult question is to explain how computers can exhibit any knowledge at all. My answer is that there is a special form of knowledge that can be computerized. The relationship between computerizable knowledge and knowledge in general is the same as between computerizable physical activities and physical activities in general. The common thread is the class of action to which I have already referred—behavior-specific action—in which the intention is always to instantiate actions with the same behaviors.[26]

This kind of approach is diametrically opposed to other modern approaches in science studies such as actant networks and radical symmetrism because it rests on two types of human action. In my approach, humans are given primacy while the fundamental difference between human and machine is assumed, investigated, and further established. Actant networks, as already explained, treat humans and machines as indistinguishable as far as their ability to contribute to the network is concerned.

The actant-network approach avoids the necessity for epistemological compartmentalization—or "meta-alternation" as we have called it elsewhere (Collins & Yearley, 1992a, 1992b)—but at the cost of critical impotence. It does seem as though there is a stark choice to be made. The necessity of this choice is clearly visible where the author tries to do more than develop a philosophical system. Thus Woolgar (in press), who endorses the general drift of the actant-network approach in the first part of his paper "Configuring the User," is forced to change tack in the second half when he studies computers in the field.[27] The link between the two halves—in my view somewhat unsatisfactory—consists of the claim that, though the meaning of a text (machine) is in the hands of the reader (user), this does not mean that any reading is possible. A text will be organized in such a way as to "make a reading possible." Likewise, the machine (as text) is organized in such a way that its purpose is available as a reading to the user.[28]

CONCLUDING COMMENT: THE DEBATE ABOUT AI REFLECTS DEBATES WITHIN SCIENCE STUDIES AS A WHOLE

I have tried to show the sometimes tortuous relationship between science studies as a whole and one of its new areas of study. The fascinating thing about this new area is that it represents an experimental test bed for some of our own ideas about knowledge. I doubt that we sociologists of scientific knowledge would have been able to claim the resources to develop our own experimental science in this way, but now that it has been done by the computer scientists and cognitive scientists, it would be foolish not to take advantage of it.

As well as being an experimental test bed for our ideas, the new area of study throws up some problems for the sociology of scientific knowledge that are less pressing in other substantive areas. These problems are salient because of the unusually intimate relationship between topic and resource in studies of intelligent machines. I have suggested that there is a choice between developing a coherent position that has little leverage on the world and accepting a compartmentalized epistemological universe. Choosing the latter, we will hold open a compartment in which we reflect on our scientific practice but will not close off the new, collectivist "knowledge science" growing out of SSK without which cognitive science will remain a shadow.[29]

The resource-topic dilemma is always there beneath the surface of sociology of science and that is why I believe that what we learn as we study intelligent machines in our knowledge science compartment will eventually inform the whole subject. In studying intelligent machines, we are exposing ourselves to the greatest risk of being torn apart by the opposing forces of essentialism and relativism, and that is no bad thing.[30]

NOTES

1. For a discussion of pocket calculators and written works, see Collins (1990/1992). For door openers, see Latour (i.e., J. Johnson, 1988). For other mundane artifacts, see Woolgar (1985).

2. Of course, there is a sense in which all boundaries are constructed (see Gieryn, this volume), but one simply cannot worry about this all the time.

3. For reasons that will become clear, I, personally, do not think we are going to solve the big problems. But at least we can reveal what they are.

4. The challenge of AI for SSK and phenomenological approaches in general had been noted some years earlier: "Observed by only a few commentators, the pigeons of philosophical skepticism and phenomenology are coming quietly home to roost in the nest of artificial intelligence. The next decade or so, which will witness the first large scale attempts to develop artificially intelligent machines, are going to be of monumental interest to philosophers and philosophically minded sociologists—optimists and pessimists alike" (Collins, 1992, p. 20).

5. For an "attributional theory" of discovery, see G. Brannigan (1981), who says that what counts as a discovery is a matter of retrospective attribution by the receiving community. See also Collins (1985) for a related set of ideas.

6. Paul Edwards has made a similar point about the "scientific contribution" of hypertext in his paper presented to the Society for Social Studies of Science (1991). For a favorable view of automated induction, see Michie and Johnston (1985). For an interesting technical critique, see Bloomfield (1986).

7. The typical sociology of scientific knowledge analysis of discovery may be found in G. Brannigan (1981) and Collins (1985). In these theories, scientists offer many potential discoveries but what comes to count is a matter of negotiation within the scientific community; it is the community that changes scientific order, not the individual. It is difficult to see how either kind of discovery program can be made compatible with this model.

8. For a discussion of the difference between tools and autonomous machines, see Lipscombe (1989, 1990).

9. For an account of the theory behind Gooding's work, see Gooding (1990). For an account of the computer-based work, see Gooding and Addis (1990).

10. Note that, under "possibility one," understanding science by studying individual human scientists—who are already social through and through—is quite a different matter than trying to understand science by supposedly modeling individual scientists with computers. Hence there is a warrant for, say, the ethnomethodologists' claim that the detailed study of individual passages of scientific work can shed light on the social nature of science. (See Lynch, Livingstone, & Garfinkel, 1983.) Computer models of scientists, on the other hand, reproduce at best an abstract model of a discoverer, so cannot furnish the same insights as the study of individual humans.

11. I borrow the term *social factors* from Steve Woolgar (1989a).

12. Bear in mind that we are talking only of "possibility three." "Possibility one" would not recognize the social/individual dichotomy.

13. Actant-network theory and the postmodernist critique encourage this tendency.

14. Indeed, it turns out that they are the sort of individual that one encounters in "game theory." (See Gilbert, 1991.)

15. As represented in the United Kingdom's long-running "discourse analysis workshop," currently organized by Malcolm Ashmore from Loughborough University.

16. For a paper that argues that CA rules are more like Wittgensteinian social rules—that is, only retrospectively applicable—see Button (1990).

17. Though it is not clear to what extent Gilbert is an intellectual migrant to computer science rather than an academic bridge between communities.

18. Is "weird processor" incorrectly spelled? See Collins (1990/1992, chaps. 13, 14, in press-b). Gilbert (personal communication) claims to be little interested in the philosophical/sociological problems of artificial intelligence and, given that much can be done with a slightly flawed mimicking of human abilities, one can see why the deeper problems need not affect his practical concerns. I suspect, however, that his work on speech recognition would soon encounter

the deep problems if it were to step outside a narrow domain of application (Gilbert & Wooffitt, in press).

19. Bloor (1976): The same sentiment can be found in the words of McHugh, quoted in my 1975 paper, "There are no adequate grounds for establishing truth except the grounds that are employed to grant or concede it." See also Collins (1985).

In the "Afterword" to the new edition of Bloor's *Knowledge and Social Imagery* (1991), he concedes that natural causes, along with our natural propensities to induce in certain ways, may make the formation of one social institution that reflects nature more likely than others that do not. This seems somewhat asymmetrical. I prefer a less ambiguous version of Bloor's symmetry thesis.

20. The point is made crystal clear in the preface to the second edition of Latour and Woolgar's *Laboratory Life* (1986), where they discuss the change in their subtitle from "The Social Construction of Scientific Facts" to "The Construction of Scientific Facts." The first edition of their book, they now believe, was haunted by an asymmetry that they had at last exorcised.

Shapin and Schaffer, though they explain the emergence of the separate categories of the social and the scientific, provide a social explanation. This is too asymmetrical to be seen as a part of the Latourian program in spite of attempts to co-opt them. (See Shapin & Schaffer, 1985.)

21. Collins and Yearley (1992a, 1992b): This latter article also explains why I use the term *actant networks* rather than the more common *actor networks* to describe the Latour/Callon model.

22. We should remember that having nothing to offer regarding the truth of the matter is precisely the trademark of old-fashioned symmetrical analyses of science. Where the subject under study is something like, say, the paranormal, symmetrical analyses are irritating to those who want to know "what is really going on." Just being irritated is not a good enough reason in itself to dismiss radical symmetry, but it does mean that if one chooses to try to say something that will help or explain the problems of AI one must choose a different philosophical starting point. This is not a matter of antagonism to the symmetrical approach, it is simply doing something belonging to a different compartment of epistemological life.

23. For the idea of epistemological compartments, see Collins and Yearley (1992a, 1992b).

24. For a discussion of the role of talk about computers in the anthropomorphization process, see Bloomfield (1989).

25. In the same volume edited by Bloomfield can be found an article on the history of the development of AI (Fleck, 1987), a paper by Bloomfield (1987), another by Schopman (1987), and a paper by philosopher S. G. Shanker (1987).

26. For related work, see Hartland (in press). Hartland is completing a doctoral thesis on this topic at the University of Bath. See also Collins, de Vries, and Bijker (in press).

27. For an argument similar to Woolgar's in its focus on the constructed boundary between machine and human, see Peter Dear (1991).

28. Of interest, this is a position almost identical to that adopted by David Bloor in the "Afterword" to the second edition of his *Knowledge and Social Imagery* (1991). I believe these formulations create more problems than they solve and beg more questions than they answer. My own view, ill-developed at the moment, is that the concept of behavior-specific action offers a way of explaining how a text can be arranged to make one reading more available than another. This, I hope, will be the contribution of this idea to the sociology of scientific knowledge.

29. The term is taken from Collins (1990/1992).

30. For a courageous assault on the same dilemma coming from a completely different direction, see Fuss (1989).

14

The Human Genome Project

STEPHEN HILGARTNER

THE leaders of the Human Genome Initiative (HGI) argue that in the next century genetic map and sequence data will utterly transform biology, biotechnology, and medicine. They see map and sequence data as providing *the* major database for biology in the next century. They envision, as Walter Gilbert (1991) puts it, a dramatic "paradigm shift": With all the genes identified and available in computerized data banks, biologists will cease to attack problems in a "solely experimental" manner; increasingly, the field will become a "theoretical" science. The computer terminal will grow as important as the lab bench, as sophisticated algorithms are used to search through masses of sequence data, identifying patterns, comparing structures, and tracing the evolution of biological mechanisms across species. New scientific questions will not concern single genes but will explore how groups of genes (and their protein products) function together as complex assemblages. Knowledge of diverse organisms—many with their genomic structure fully characterized—will lead to an exponential increase in the power of biotechnology to manipulate nonhuman life.

AUTHOR'S NOTE: This chapter is based on research supported by the National Center for Human Genome Research (NIH Grant No. HG00417) and the National Science Foundation (NSF Grant No. DIR-9012342).

The leaders of the genome project also foresee fundamental changes in medical practice, changes that will reach far beyond an ability to detect and eventually to treat things now classified as genetic disorders. Indeed, some predict that the most important impacts will stem from improved understanding of normal human biology. Physicians will use computerized analysis of each patient's personal genetic profile to tailor medical care and lifestyle recommendations to fit the individual. They see the goals of the HGI—to improve mapping and sequencing technology and to map and ultimately to sequence the genomes of the human and of several "model organisms" (e.g., yeast, the mouse, and the worm *C. elegans*)—as providing the "infrastructure" for supporting these transformations.

Obviously, not all observers and not all biologists share this optimistic view of the future of the genome project. Some critics are concerned about resource allocation in science, and they question whether the data produced by sequencing entire genomes will in fact be useful or be worth its cost (Davis et al., 1990; Leder, 1990; Lewontin, 1992b; Tauber & Sarkar, 1992). There is concern about the HGI's impact on the training of young scientists, on drudgery in the lab, on the distribution of scientific rewards, on "small labs" versus "large centers," on international and interlaboratory collaboration, on the bureaucratization of research, and on the ethos of science. Others are worried about broader societal impacts, and the genome project has become a lightning rod for concerns about the applications of biotechnology or about the prospect of a new eugenics.[1] Even among those who support the basic goals of the HGI, there are differences of opinion about the particular directions that the project should take. Competing technologies and models of how the project should be organized are in play. It is thus unclear, as the genome project continues to evolve, what "the Human Genome Initiative" will end up becoming.

This chapter provides an early and preliminary view of the efforts of the genomics community, especially in the United States, to construct this emerging sociotechnical system.[2] To accomplish this succinctly, it focuses on a particular technological artifact, known as an STS, or sequence-tagged site. The concept of an STS was presented in an influential position paper, published in 1989. STSs began as a generic definition but quickly grew deeply embedded in the practices of the genomics community. In the United States, they became a key component of official policy. Examining STSs and exploring the multiple ways of "reading" this artifact provides a window on the world of genomics and illuminates some of the science, technology, and society issues that permeate the genome project.

MAPPING AND SEQUENCING THE HUMAN GENOME

In the final years of the 1980s, the idea of mapping and sequencing the human genome evolved from a visionary proposal into an international scientific effort. Proponents of mapping and sequencing the genome convinced most biologists that the project is feasible and won substantial financial commitments from the U.S. Congress. Genome programs were established at the National Institutes of Health and the Department of Energy (Barnhart, 1989; Cook-Degan, 1991; Watson & Jordan, 1989). The European Community (and a number of its member states) and Japan set up genome programs (of widely varying sizes and configurations), and an international scientific society, the Human Genome Organization (HUGO), was founded to help coordinate the effort worldwide. The new specialty of genomics emerged, as reflected in a growing research effort, increases in funding, rapid technological change, attendance at scientific meetings, and the birth of new journals (McKusick & Ruddle, 1987).

In the United States, the leadership of the effort set the goal of having the complete sequence in hand by the year 2005 at a cost of $3 billion (roughly $200 million per year). The leadership maintains that it is too early to develop detailed plans for accomplishing this goal because it is unclear what technologies will prove most useful. The estimate of $3 billion assumed that technology would improve roughly 100-fold, predicating the whole initiative on technical advances. At present, the genomics community is working to develop mapping and sequencing technology. A period of experimentation with various technological approaches is expected to lead to efficient techniques and systems for sequencing on a massive scale.

By any measure, the genome project is an ambitious undertaking. Using the linguistic metaphors that prevail in the field, genomics researchers sometimes describe the human genome as "a string of text written in a four letter alphabet." The letters—A, C, G, and T—correspond to the four nucleic acid "bases" of which DNA molecules are composed. The human genome is roughly 3 billion base pairs long, making it an enormous "text" equivalent in length to 1 million printed pages (National Research Council, 1988). To sequence a genome means to identify all of the "letters" in the text in order. Mapping a genome means to locate recognizable "landmarks" on the genome that can serve as points of reference. Physical maps display the spacial relations among these landmarks. As the field of mapping evolved, people began extending and experimenting with several different technologies, and hence there are several kinds of physical maps—including restriction maps, contig maps, in situ hybridization maps, and radiation hybrid maps.[3] These types of physical maps use different landmarks, and they produce different kinds of information. In contrast to physical maps, which are based on spacial relations, genetic linkage maps are based on patterns of genetic inheritance.[4]

The task of assembling a technological system for mapping and sequencing large genomes can best be conceptualized as a problem of constructing systems or networks, a concept that links several recent approaches to the study of technology (Callon, 1980b, 1986b; Hughes, 1983, 1986, 1987a; Latour, 1987; see also Pinch & Bijker, 1984). The task is, in Law's terminology (1987a), a "heterogeneous engineering" problem of immense proportions. This problem entails building a network—of researchers, techniques, organizations, laboratories, databases, biological materials, funding sources, political support, and so on—that can operate at high "throughput" and produce map and sequence data with few errors.[5]

This effort is complicated by significant uncertainties. With current technology, sequencing the human genome would be prohibitively expensive, and mapping remains a laborious and time-consuming process. Techniques for mapping and sequencing are numerous and rapidly evolving, and the brisk pace of change makes it impossible to predict where important technological opportunities will arise.[6] The genomics community therefore must develop its research program and organizational structures in this environment of uncertainty; its leaders sometimes frame the problem as attempting to maintain flexibility and experiment with diverse approaches, without dissipating resources. As the HGI evolves, the community may face controversial decisions about which approaches merit more work and which should be abandoned.

Coupled with efforts to develop techniques are a host of organizational challenges. The genome project differs from other biological research, most of which is conducted by small groups, in that the HGI will require many researchers to contribute to a coordinated effort to reach a clearly defined, long-term goal. This type of long-term "infrastructure" project is new for biology, so biologists lack an established set of off-the-shelf organizational mechanisms or cultural resources to draw on. The coordination problem is all the more complex because research is not being conducted at a single site but is taking place at labs distributed throughout the world.[7]

Holding this emerging sociotechnical network together and garnering resources for expansion requires much work that has a political dimension. The leadership of the project must obtain a reliable and steady flow of funds and must demonstrate that it is making progress to the Congress and the Office of Management and Budget (OMB). It must deal with critics, who argue that the HGI is not cost effective and that it will adversely affect other biological research. More broadly, the possibility of sustained political conflict over the societal impact of genetic technologies looms on the horizon. Many genomics researchers are concerned about what they classify as the "ethical, legal, and social" issues that genetic technologies entail, and some also worry that controversies could provoke opposition to the HGI.[8]

As is typical of the development of a technological system, the task of constructing the genome project involves an inextricable mix of problems with technical, economic, organizational, and political dimensions. Problems and obstacles, solutions and opportunities, are likely to arise from numerous directions in a relatively unpredictable manner, and no doubt they will interact in complicated ways. The emergence of sequence-tagged sites, an artifact that, some argued, "changed the whole outlook" for genome mapping, is a case in point.[9]

SEQUENCE-TAGGED SITES

The brief history of sequence-tagged sites (STSs) provides a look at the heterogeneous engineering efforts of the leadership of the genome project. The concept of an STS was first fully articulated at a retreat of a scientific advisory group that was working, at the request of Congress, to develop a joint plan for the first 5 years of the NIH and DOE's human genome programs. A few weeks later, a two-page position paper outlining the STS concept appeared in *Science* under the names of four prominent members of the genomics community (Olson, Hood, Cantor, & Botstein, 1989). In this paper, known informally as "the STS proposal," they defined the concept of an STS and argued that STSs offered a way to solve several critical problems. STSs won a salient place in the NIH-DOE 5-year plan, and they were quickly incorporated into laboratory practice. Mapping strategies were reconfigured around them, and laboratories were adjusted to speed the development of STSs. Researchers soon were giving progress reports at genomics meetings that measured headway in terms of the number of STSs mapped. In a matter of months, STSs became an important entity in the genome project.

What is an STS? From the perspective of science studies, STSs are rather complex entities that play a heterogeneous collection of roles in the world of genomics. According to the authors of the STS proposal, an STS is a new kind of mapping landmark. Maps of any kind are constructed by defining the relations among a set of recognizable landmarks. The authors of the STS proposal argued that this new type of landmark had fundamental advantages over landmarks of other kinds. To their proponents, STSs solved one of the critical problems that plagued genome mapping efforts: "the inability to compare the results of one mapping method directly with those of another and to combine maps constructed by two different techniques into a single map" (NIH-DOE, 1990, p. 13). Restriction maps could not be readily merged with contig maps; genetic maps could not be related to physical maps;

sequence data—which some genomics researchers viewed as a physical map at the highest possible resolution—could not be neatly integrated with map data. But STSs provided a way to take data from "any of a variety of physical mapping techniques" and report it "in a 'common language' " (NIH-DOE, 1990, p. 13).

The idea was "to 'translate' all types of mapping landmarks into the common language of STSs" (Olson et al., 1989, p. 1434). To accomplish this, each mapping landmark would be examined to find a sequence-tagged site, or STS, which is basically a short DNA sequence that only occurs once in the genome. This short, unique sequence would identify that landmark and distinguish it from all others. The site on the genome where the landmark appeared would thus be "tagged" by the STS.[10] The STS proposal, as its authors put it, thus "redefines the end product" of physical mapping (Olson et al., 1989). The ultimate goal would be to create an "STS map" that would show the order and spacing of sequence-tagged sites. This new kind of physical map would tie together all other kinds of maps and would become "the centerpiece" of the mapping effort (Olson et al., 1989, p. 1435).

Why, according to their proponents, could STSs link together diverse forms of data into a single, integrated map? Because STSs provided a way to represent mapping landmarks wholly as "inscriptions" (Latour, 1987; Latour & Woolgar, 1979) rather than relying on biological materials. Prior to the development of STSs, all mapping landmarks were based on clones. In a map of a human chromosome, for example, each landmark would consist of a recognizable piece of human DNA that had been cloned into bacteria or yeast. These clones would have to be carefully archived, for researchers who wished to use such a map require access to them. The only way to determine whether a particular landmark existed in a DNA sample was to perform a test using the clone containing the landmark.

STSs, however, do not depend on clones but are based on the polymerase chain reaction (PCR)—a technique for making copies of specifically targeted DNA segments without cloning. Because they rely on PCR, STSs can be expressed in terms of sequence information. For example, the sequence AGTTCGGGAGTAAAATCTTG/GCTCTATAGGAGGCCCTGAG describes an STS on human chromosome 17. With this information and some standard PCR equipment (available in off-the-shelf kits), one can test a DNA sample for this STS.[11] As the STS proposal put it:

> The overwhelming advantage of STSs over mapping landmarks defined in other ways is that the means of testing for the presence of a particular STS can be completely described as information in a database. No access to the biological materials that led to the definition or mapping of an STS is required by a scientist wishing to assay a DNA sample for its presence. (Olson et al., 1989, p. 1434)

But the authors of the STS proposal saw STSs not only as a way to manipulate data, they also intended STSs to alter the social world of genome mapping by simplifying the problem of coordinating its activities. In the HGI the problem of coordination—an important one for any technological activity—is complicated by the fact that research is being conducted at labs dispersed throughout the world. Moreover, because techniques are rapidly evolving, the problem also entails allowing for coordination across time; data and materials (e.g., clones) produced in the early 1990s may ultimately be used by future researchers equipped with different tools. STSs were intended to address these coordination problems. The "common language" of STSs was expected not only to tie together diverse types of data but also to simultaneously tie together *laboratories* that were geographically, and even temporally, dispersed.

The STS proposal suggested that the new landmarks would facilitate the smooth evolution of an STS map "with inputs from many laboratories and methodologies," a process that would allow "the dichotomy between 'big' and 'small' laboratories" to "disappear" (Olson et al., 1989, p. 1435). The STS proposal also argued that "international discussions should be initiated to maximize the likelihood that the new common language would cross national boundaries" (Olson et al., 1989). STSs thus were seen, in Star and Griesemer's (1989) terminology, as a "boundary object" that could tie together several diverse categories of labs—"small" ones and "big" ones, genetic mapping and physical mapping labs, restriction mapping labs and contig mapping labs, today's labs and the labs of the future, labs in the United States and labs elsewhere. They would permit a diversity of scientific strategies—including those based on the techniques of the future—to proceed together toward a common goal.

THE STS OF STSs

The leadership of the American genome effort believed STSs could address a variety of technical, organizational, and political problems. The scientific advisory committees charged with shaping the U.S. genome effort quickly threw their weight behind the STS concept. The NIH-DOE 5-year plan set targets that, in effect, required mapping groups to express their maps in terms of STSs. The plan thus altered the definition of "good," "fundable" work, and the new definition, as one might expect, was soon reflected in the texts of grant proposals and the structures of laboratories. STSs were rapidly woven into the sociotechnical fabric of genome mapping, where they were intended to help control many diverse entities: STSs were a means of linking data, but they also linked laboratories. They tied together pieces of DNA, but they also tied together mapping techniques. And, as we will see below, they could be used both to perform tests on DNA samples and to "test" the performance of research groups. STSs were not only a biotechnology, they were also a social technology, designed to act on the genomics community as well as on DNA. The new landmarks were intended simultaneously to enhance the mobility, stability, and the combinability of data (Latour, 1987, p. 228); to tighten control over a dispersed network of laboratories; and to integrate the huge and growing number of entities connected to the world of genomics into a more coherent whole.

At the time of this writing, STSs are less than 3 years old, and their role in the world of genomics is still in flux. Some members of the genomics community, especially in Europe, where official policy does not stress the new landmarks, argue that the Americans have oversold STSs. Whether STSs will live up to the expectations of their proponents remains an open question. But what lessons can analysts from the field of science, technology, and society take away from the (incomplete) story of sequence-tagged sites? What can we say about the STS of STSs?

Clearly, we can read these artifacts in many ways. A "technical" reading might emphasize how PCR, "a method that has only come into widespread use during the past two years" (Olson et al., 1989, p. 1434), solves the problems of linking data and archiving clones. A "managerial" reading would stress how STSs provide a tool for improving communication, controlling quality, measuring progress. A variety of "political" readings are also possible, as we will see below.[12] STSs can also be read as a "case" that speaks to three critical issues in science studies: managing science, access to scientific data, and the evolution of technology.

Managing Science

STSs can be read as the technological embodiment of a particular philosophy of how to manage the HGI. The question of how science should be managed, and even whether it should be centrally "managed" at all, is a perennial issue in science policy and has been an area of debate regarding the genome project. Many genomics researchers in the United States argue that, unlike most biology, the HGI requires active central management. They argued that the normal practices of the NIH—which one influential genomics researcher characterized as waiting for Federal Express trucks full of grant proposals to "pull up at the loading dock" on the day of the application deadline—should be adjusted for the HGI.

As a management tool, STSs were not only seen as a way of enhancing coordination by providing a way for a lab, say, in Baltimore to communicate and cooperate with one, say, in Boston. STSs were also intended to provide a means for the NIH and DOE to hold laboratories accountable to Washington. The new landmarks not only facilitated communication among the nodes of the research network, they also tightened the link between the center and the periphery. On one dimension of accountability, STSs offered a way of measuring the accuracy of data. Because STSs made it possible to compare maps produced by different mapping methods, they simultaneously made it possible to discover discrepancies and identify the locations and sources of errors.

Along a second dimension of accountability, STSs provided a metric for measuring a laboratory's progress, regardless of the mapping method used. One physical mapper explained:

> You could theoretically down the road sit there in '95 and say, "Well, okay, these two groups have gotten comparable amounts of money, and group A has got 5,000 of these things [STSs] and group B got 500; what's going on here? Has group B worked one-tenth as hard and squandered the money?"

Similarly, the "completeness" of a map in a given region of the genome could be measured in terms of the density with which it was covered by STSs. STSs thus provided the NIH and DOE with a tool for evaluating and comparing not only data but also research groups. STSs provided a way, as one mapper put it, "to hold people's feet to the fire."

The issue of accountability extended beyond individual labs; STSs could also contribute to an effort to hold the genome project accountable as an enterprise.[13] The need for accountability and for mechanisms to ensure that genome funds would not be dissipated has been a recurring theme in Washington.[14] Congress's request that the DOE and NIH develop a 5-year plan for the U.S. human genome project can be read as an instruction to set program goals that would not only focus the project but could also be used as a yardstick of progress. When the 5-year plan was completed, its goals for genetic and physical mapping were explicitly stated in terms of STSs. One 5-year goal, for example, read: "Assemble STS maps of all human chromosomes with the goal of having markers spaced at approximately 100,000 base-pair intervals" (NIH-DOE, 1990, p. 14). These specific goals provided Congress and the OMB with tools for measuring the HGI's progress (Cook-Degan, 1991, p. 164).

STSs thus were embedded in a complex politics of accountability. They can be seen as devices for appealing to the Congress and the OMB, who control the genome project purse strings, by enhancing accountability. Pressing the argument in a slightly different direction, one might read STSs as a means of appeasing various critics of the genome project, such as those concerned about the cost of the project (work will be efficient, money well spent), or those concerned about the role of "small laboratories" in the genome project (the big/small dichotomy will "disappear").

STSs can also be read as a comment on the emerging "technological style" (Hughes, 1987a) of the American genome program. This type of reading is an example of what Woolgar (1991b, pp. 37-38) classifies as "technology as text 2." From this perspective, one might see STSs—and, more important, those things that the leadership sees as the *advantages* of STSs (e.g., a metric for measuring performance, a means for merging many kinds of maps into a single, dominant one)—as reflecting a managerial style that favors centralized coordination.

In general, the proponents of STSs see this managerial style as appropriate to genome mapping. But some of the Europeans who are most critical of the American emphasis on STSs argue that STSs are symptoms of an "obsession" with central management on the part of the Americans. They argue that this obsession led the American advisory committees to fail to consider several significant disadvantages of STSs (e.g., STSs are costly to generate and use; an STS map does not provide the means to easily perform certain types of important experiments, in cases where clone-based maps would). The appeal of STSs to the Americans, in this view, stems from a preoccupation with control and a profound "lack of trust" of the genomics community. Some even criticize this approach to science—with what they call its "top down" approach and "production quota" mentality—as reminiscent of the failed command economies of eastern Europe.

Access to Scientific Data

STSs can also be read as a means of making it harder for a research group to monopolize access to scientific data. The politics of access to scientific data is a critical issue for science studies,[15] and it is important to understand how scientists (and others) construct and erase boundaries that shape who gets access to what data, when, and under what terms (Hilgartner & Brandt-Rauf, 1990). In highly competitive fields with commercial potential, such as human genetics, data access issues are perhaps especially significant. Thus, in the genome project, questions about access to map and sequencing data—and to biological materials, such as clones—are prominent concerns. As the NRC (1988) report on the genome project put it:

> The human genome project will differ from traditional biological research in its greater requirement for sharing research materials among laboratories. . . . Free access to the collected clones will be necessary for other laboratories engaged in mapping efforts and will help to prevent a needless duplication of effort. (p. 76)

In the competitive fields of molecular biology and human genetics, tensions surrounding access to clones have been common. Access to biological materials often conveys a decisive competitive edge, and interlaboratory negotiations over access often play a significant role in setting the terms of a nascent collaboration. Clones, and other scarce or unique resources, can be conceptualized as bargaining chips used in cutting deals. (The terms of collaborations vary considerably. The following is a possible schematic example: I will provide these clones; your postdoc will do these experiments; we will all review the data; if everything goes as expected, she will get first authorship on the paper, I will get second authorship, and you will get last authorship.)

Researchers' practices in releasing clones have varied greatly. Some journals require as a condition of publication that any clones used be provided (post-publication) to any scientist who requests them, but waiting until publication introduces a significant delay. In human genetics, researchers often complain that clones they have been promised never arrive in the mail. Also, in some labs, there have been bitter disputes about the disposition of clones following the departure of personnel. However, because STSs do not depend on clones, they were expected to ease such problems. Once the inscriptions describing an STS have been published, there is no mechanism for restricting access to the STS. Insisting that people report data as STSs (e.g., at the time of a grant review or in published articles) thus tightens data-sharing requirements and moves a step closer to a world of "free access" to map data.[16]

For the leadership, STSs were a tool for altering data access practices within the genomics community. The new landmarks therefore provide a strong example of how social control in science (e.g., over data sharing) can be constructed: not simply through reliance on an ethos of shared norms but through technical and procedural *means* that build social control—directly and materially—into the very fabric of scientific production.

Evolution of Technology

Finally, it is instructive to read STSs as a comment on the evolution of technology. STSs became the centerpiece of a dramatic change in the long-term vision of the sociotechnical system for genome mapping. Because all mapping landmarks except STSs rely on clones, early visions of the genome project pictured the construction of a huge clone repository, responsible for archiving hundreds of thousands of clones. This clone repository would need an extensive quality control program (because clones can be contaminated or can spontaneously change form) and would entail running a major mail order operation to manage requests for materials. Estimates of the cost of operating such a repository reached as high as $250 million over 30 years.[17]

Because STSs are expressed as inscriptions, the American leadership saw them as a way to eliminate the giant clone repository from the genome project blueprint. Gone were the banks of freezers, the army of technicians and robots, the mail order house. These would never have to be designed, assembled, and disciplined, and no one would need to be persuaded to pick up the $250 million tab. An STS database—which could be accessed electronically from anywhere in the world—was inserted in the clone repository's place.

In the shorter run, publication of STSs (in journal articles or in public databases) promised to ease the immediate problems that some of the large physical mapping labs expected to encounter handling requests for clones. Preparing even single clones for shipping can grow time-consuming; in an extreme case, one lab recently found that it had a technician working half time filling orders for a single, very popular clone. Physical mappers thus anticipated being "inundated" with requests for materials, causing a backlog that would delay research and provoke conflict between the users and the producers of map data.[18] STSs were expected to greatly diminish these problems.[19]

The elimination of huge clone repositories from the long-term vision of the HGI sheds light on the evolution of technology. As STSs were integrated into the world of genomics, their mode of embodiment was altered. Initially they were only a generic definition (e.g., "a short piece of DNA that has been shown to be unique"; NIH-DOE, 1990, p. 13), the stuff of discussions at meetings, proposals, plans. When the STS proposal was first published, for example, sequence-tagged sites remained a concept, expressed in written and spoken texts;[20] no one had actually constructed an STS map. STSs won a central place in the official NIH-DOE 5-year plan even though, as the plan stated, "few, if any, mapping projects have started to use the STS system" (NIH-DOE, 1990, p. 14). Not long thereafter, many mapping projects were preparing to generate STSs, and soon STSs were no longer merely a definition or even simply a generic type of landmark. They also existed as specific landmarks, such as AGTTCGGGAGTAAAATCTTG/GCTCTA-TAGGAGGCCCTGAG, which, according to the Second International Workshop on Human Chromosome 17, can be found on the chromosome's q arm.

This variety of modes of embodiment represents a key dimension of the heterogeneity of sociotechnical networks. Technological systems exist in a state of flux as a mixture of blueprint and hardware, plan and practice, the nearly on-line and the almost obsolete. As they evolve, entities that can be found only in a semiotic world (such as the huge clone repository that now, apparently, will never be constructed) may get displaced by objects (such as the computerized STS database) that at first are also only expectations, plans, or proposals. Much of the change in sociotechnical networks takes place in portions of the network that have not yet been built in material form or embodied in routine practice. To understand technological evolution, then, one cannot limit attention to "successful" artifacts and artifacts that "fail" *after* people begin trying to construct them. One must also consider "artifacts," like the clone repository, that occupy the attention of technologists but get eliminated or significantly modified while they exist on paper.

As STSs evolved—from a definition to become something embodied in official policy and material practices—their roles in the world of genomics both grew in number and achieved specificity. A similar process of translating generic visions into particular implementations is under way, on a larger and grander scale, with the human genome project more generally. In this manner, the ultimate form of the project, and its eventual role in reconfiguring biology, medicine, and society, is now being fashioned as "the" Human Genome Initiative continues to take shape on the ground.

NOTES

1. For an overview of some of these concerns, see Nelkin and Tancredi (1994), Holtzman (1989), and Duster (1990).

2. The pace of change in the genomics community is very rapid. In 1988 the author began a long-term, prospective study of the genomics community, using field research methods. Data for this paper are from participant observation in genomics laboratories and interviews. Unattributed quotations are from these sources. Research subjects were promised anonymity.

3. Restriction maps show sites on the genome where specific enzymes will cut the DNA. Contig maps are built out of clones that partially overlap one another, thus "covering" contiguous portions of the genome with a set of clones. In situ hybridization uses microscopy to see where specific sequences (flourescently or radioactively tagged) appear on intact chromosomes. Radiation hybrids are constructed by breaking chromosomes with ionizing radiation; the pattern of breakage allows one to ascertain the order of markers and estimate the distance between them.

4. Genetic linkage maps measure the frequency with which landmarks are inherited together; landmarks that are often inherited together are said to be "linked." All of these types of maps can be constructed at varying levels of resolution. As in traditional cartography, low resolution maps have relatively few landmarks per unit of mapped territory, whereas high resolution maps have more landmarks.

5. In the genomics community, the term *throughput* is used to describe the rate of production of data per unit of time, cost, delay, and so on. Increasing throughput is a central concern.

6. Significant developments include cloning very large DNA fragments into artificial chromosomes in yeast, the widespread availability of automated sequencing machines, multiplex DNA sequencing, the development of automated systems for identifying overlapping clones, and high resolution in situ hybridization. Perhaps most important is the integration of the polymerase chain reaction (PCR) into many new strategies (including STSs). In addition, novel sequencing schemes (e.g., using scanning tunneling microscopy) are being explored.

7. It is likely that a few large "production centers" devoted to "cranking out" huge volumes of map and sequence data will someday be set up, perhaps quite soon. But at present even the

larger genome centers are not massive operations. Efforts to increase "throughput" are currently not centered on amassing in single sites armies of scientists and technicians but entail more subtle reconfigurations of sociotechnical networks.

8. The NIH and DOE have both established research programs on the "ethical, legal, and social implications" of the HGI. It is worth noting that, genomics researchers rarely use the word *political* to refer to these issues, although it seems likely that many of them will be bitterly contested in courts, legislatures, and government agencies.

9. The quote is from Norton Zinder, who chaired the NIH Program Advisory Committee on the human genome (quoted in Roberts, 1989).

10. In addition to "translating" previously mapped landmarks, STSs can be generated de novo and used in mapping directly.

11. One also needs to know the size of the PCR product (in this case, 630 bp) and the conditions for running the PCR reaction.

12. Of course, these readings are intertwined with one another into a fabric of mutually reinforcing and conflicting meanings.

13. The genome project remains quite vulnerable to changes in its budget at the hands of the Congress or the OMB. The genomics community thus has only partially achieved what Cozzens (1990, pp. 167-168) defines as autonomy: "the insulation of decisions about research from control by nonscientists."

14. See, for example, Office of Technology Assessment (1988b) and NRC (1988).

15. See Nelkin (1984b) and Weil and Snapper (1989).

16. STSs did not solve, for once and for all, the problem of data access. On the contrary, data access will be an area of ongoing tension and negotiation as the genome project proceeds.

17. On costs, see the discussion in NRC (1988, pp. 81-82). Some critics of the American emphasis on STSs argue that (a) the new landmarks will not really eliminate the need for clone repositories and (b) the costs of operating such a facility would be much less than $250 million.

18. In interviews, some of them used parody to dramatize their plight. "Great, Andy, can I have all 3,000 of your arrayed clones. Oh, and I want all your YACs [yeast artificial chromosomes], and blah, blah." Even more modest requests were expected, because of sheer numbers, to require labs to "totally turn over all of our resources right now to sending out DNA." Robotics could ease the burden some, but "you still have to have pretty highly trained techs to babysit these things."

19. In Fujimura's (1987, 1988) terms, procedures for manipulating clones are part of standardized "packages" (e.g., of "cookbook" techniques and theory), shared among molecular biologists, making clones readily "transportable" among labs and facilitating scientific work. See also Cambrosio and Keating (1988). The case of STSs underlines how the extent to which standard packages make an entity, such as a clone, transportable is (a) a matter of degree, (b) relative to alternatives (STSs are more transportable than clones), and (c) dependent on a particular sociotechnical context (clones may be sufficiently transportable for many purposes but not transportable "enough" for physical mapping). In Hughes's (1983) terminology, the solution to one "critical problem" can become the next "reverse salient."

20. These texts, of course, were conceptually linked to material items, such as PCR machines, automated DNA sequencers, and so on.

PART V
COMMUNICATING SCIENCE AND TECHNOLOGY

The processes by which science and technology are communicated have provided the basis for some of the most innovative analyses by practitioners of science and technology studies. Scholars have challenged, and in some cases turned on its head, our "commonsense" understandings about how scientific and technical knowledge is constructed and transmitted. Particularly useful insights have been provided by those S&TS scholars who have treated science just as any other literary form through an analysis of its language, style, and rhetorical structure.

Recent works on the discourse and rhetoric of science have explored the power of unconventional or "new" literary forms to break down our expectations about distinctions between form and content and to provide opportunities for reflexivity. Malcolm Ashmore, Greg Myers, and Jonathan Potter exemplify this approach in a literature review that takes the form of a "diary" of a fictional female graduate student. Her diary recounts an intellectual odyssey during a week-long escape from her research on rodent behavior in which she explores the library's holdings on a variety of topics including discourse analysis, rhetorical studies, and reflexivity.

While clearly a departure from the familiar review essay, the unorthodox form provides Ashmore, Myers, and Potter with some unique opportunities. The fictional "author" of the diary can serve as a means of identification for those coming to this literature as outsiders. Furthermore, this device provides a legitimate frame for setting out a description of the processes that resulted in the review. At the same time, readers are directly exposed to the multiple ironies of a self-referential approach in which the authors have been transmuted with respect to status, gender, and number.

In his chapter on "science and the media," Bruce Lewenstein points to the inadequacies of the traditional view that the role of the science journalist is to popularize science. The notion that the principal obligation of the science journalist is to simplify and disseminate to the public knowledge that scientists have previously communicated among themselves has lost adherents in recent years. Instead, S&TS scholars have pointed to the absence of a clearly defined boundary between science and popularization. Scientists, however, frequently claim to be able to make this distinction because the ability to label someone as a popularizer can serve as a useful political resource in public disputes.

Lewenstein also reviews the literatures on science writers, scientists in the media, and the presentation of scientific images in such media as films, television, and science fiction. In examining survey research on public knowledge of science, he observes that the emerging concern with science literacy in the United States has resulted in a "growth industry" based on prescriptions to solve lack of knowledge of mathematics, physics, geography, and other fields. In Lewenstein's view, however, such prescriptions are generally flawed because they pay little attention to either the processes of science or the complex manner in which science is communicated.

Brian Wynne, in his chapter on "public understanding of science," asks us to recognize the extent to which public discourse on this issue has been shaped by the ideological programs of science itself. By placing the issue in historical and cultural context, he demonstrates the degree to which research and policy on public understanding of science have assumed that scientific and technical knowledge, practices, and institutions were unproblematic. Recent constructivist research on public understanding of science has, however, problematized science as well as publics. By examining how people experience science in their lives and by investigating how scientific constructions embody certain models of social relationships, S&TS scholars have expanded the range of options available for constructive public engagement with science.

Authors in this part do not suggest that we kill the messenger. Rather, they assert that the message cannot be read without the messenger, that a different messenger produces a different message. Even more, syntax, grammar, and word are not merely the form, they ineluctably affect the content of the communication.

15

Discourse, Rhetoric, Reflexivity

Seven Days in the Library

MALCOLM ASHMORE
GREG MYERS
JONATHAN POTTER

. . . I'll start it with my Plan for the Week just to let you know right away what's going on.

Monday: I will find the Canonical Footnote and discover Discourse Analysis.
Tuesday: I shall look at Rhetorical Studies.
Wednesday: The day for Historical Studies and Science Education.
Thursday: When I intend to deal with Other Texts: Visual, Mathematical, Material.
Friday: The occasion for looking at Gender Studies.
Saturday: Will find me immersed in Studies of Social Science.
Sunday: A rest, surely? But no, I will be getting into Reviews, Reflexive Work, and Decisions.

But before we can start I must explain why I have written this "review" in the form of a "diary." (Why the accredited authors of this piece are three men, while I am not, must wait until the end to be fully explained.) I have four very specific reasons for foregrounding "form."

First, studies of discourse and rhetoric have broken down easy distinctions between form and content as well as showing the historical contingency and rhetorical orientation of the literary genres used in technoscience. The form of science writing has thus been made problematic. It is not just a matter of how *it* is put; the *it* is mixed up with the *putting*. This can, of course, simply be *said*, if not said simply. How much better, then, to *show* the mutual constitution of form and content in a form that, through its unconventional character, makes it visible.

Second, reviews, as I conclude on Sunday, are a particularly interesting and underanalyzed genre. As far as I can see, the only way to write a review, while at the same time attempting an analytic treatment of "the review," is to use a form that is "self-commenting." One prominent strategy I use is to point up certain noticeable absences in normal review discourse by foregrounding their presence in my text. For example, in this diary I situate myself in space and time, in networks of relationships, as a person with specific interests and a particular history.

Moreover, the task of reviewing, as we all know but seldom tell, is riddled with accident, serendipity, and loss. The rules of selection, the criteria for inclusion and exclusion, are only made explicit, if at all, post hoc; pre hoc, they are ad hoc. The appearance of coherence and order that is the achievement of the text is built on extremely shifting sands. My review, through its metaphor—if you insist on being Borgesian about it—of the working Library, with all its practicalities, frustrations, and miracles, attempts to display not just an achieved order but the processes contributing to that improbable result. I believe that work stressing the *achievement* of order is an important corrective to "final versions" that deny their own production conditions; I am with ethnomethodology *and* SSK on this.

Third, a point about my first person role as narrator in/of the text. This device is a method of reader identification. The progress of the narrator during her Seven Days in the Library mirrors the progress of the ideal reader of the text (and of the *Handbook* as a whole). Starting the week (text, book) as an interested but ignorant outsider, she ends it as a newly sophisticated paraparticipant (even to the point of being able to write this inappropriately placed explanation). Thus the writer and reader merge in a single textual figure representing the point of the project.

Last, an argument from tradition: In research in scientific discourse and rhetoric, one of the current orthodoxies is the use of "new" literary forms (Myers, 1992). People have written dialogues, plays, encyclopedias, lectures, fragments, and parodies; and now we have a diary. (And now I've found I'm not the first—I came across Rényi, 1984, on a table in the math section.)

And one more thing—gender: Gender has been a notorious "blind spot" in studies of technoscience, as we will see on Friday. Here it is, up front. For me, the form/content being inseparable, I *need* to be gendered, which means, doesn't it, *this* gender, to make the review work appropriately. I have just finished Jeanette Winterston's *Written on the Body* (1992) and am acutely sensitive to the strange literary effects that can be produced by characters without gender; in most science writing and writing on science, it is the presence of characters *with* gender that makes for strangeness. Prepare, then, to be "stranged"—and may you enjoy the experience.

The Epigraph

And now for the epigraph, which I hope you will find as appropriate as is appropriate (Katriel & Sanders, 1989). A *vademecum,* by the way, is a handbook. A classicist friend objects to my conflating *handbook* with *vademecum.* Apparently the *OED* defines *vademecum* as "a book or manual suitable for carrying about with one for ready reference." That sounds like a handbook to me, though perhaps not one this heavy.

> The vademecum is not simply the result of either a compilation or a collection of various journal contributions. The former is impossible because such papers often contradict each other. The latter does not yield a closed system, which is the goal of vademecum science. A vademecum is built up from individual contributions through selection and orderly arrangement like a mosaic from many colored stones. The plan according to which selection and arrangement are made will then provide the guidelines for further research. It governs the decision on what counts as a basic concept, what methods should be accepted, which research directions appear most promising, which scientists should be selected for prominent positions and which consigned to oblivion. Such a plan originates through esoteric communication of thought—during discussion among the experts, through mutual agreement and mutual misunderstanding, through mutual concessions and mutual incitement to obstinacy. (Fleck, 1935, p. 119)

MONDAY: THE CANONICAL FOOTNOTE AND DISCOURSE ANALYSIS

I never want to see another gerbil sand bathing again. I think I may give up on my dissertation; what really interests me now is not the behavior of rodents but the behavior of the psychologists and zoologists studying them. Specifically, what does this dissertation contribute to knowledge? What does a study like this do for me or anyone else? Why does it have to be written in that strange style and form?

Jamie said I should be thinking of switching to philosophy—there's lots of stuff on epistemology, she said. Well, I tried looking at that last week and only lasted a couple of days. I just couldn't see the point of most of it. Of interest though, from a "rhetorical perspective" (I suppose), a lot of the more recent philosophy of science (such as Hacking, 1983, and Galison, 1987) seemed to be engaged in heavy "intertextual negotiation" (does this kind of language make *any* sense?) with relativistic work in science studies. And to judge from the concessions made in such work, the philosophers are in retreat. Though traditional claims for realism and method are still made, they appear increasingly weak and residual. I enjoyed some of Kuhn (1962/1970)—very suggestive and obviously highly influential in science studies—and Feyerabend (1975) was fun though frequently infuriating. But really, I don't want to know what some philosopher thinks, I want to know what people have found out about scientific arguments and writing and talk. So I am now thinking I might switch to science studies, and I've promised myself, this time, a whole week in the library to see if I've got a new research topic.

This morning was awful. After queuing to get on a terminal, the first searches I tried were on scientific TALK and WRITING. Very little joy there—just a load of "how-to" books addressed to novices. So I browsed the stacks for a while to sophisticate myself and then tried again with RHETORIC and DISCOURSE, and hit the jackpot! For a while there I had gotten quite excited. There was a profusion of works with titles like *The Rhetoric of ____ or Discourse and ____* about almost every science and academic discipline. The same topic seemed also to be addressed in terms of texts and literary production as well as, yes, talk, writing—and even wrighting, whatever that is. There were titles that were conjugated from a set of esoteric and oh-so-trendy terms like *epistemology, representation, reflexivity, postmodernity.* There were journals called *Philosophy and Rhetoric* and *Discourse and Society.* It was just too much. Of this wealth of writing (about writing), what was Important and Central and what unnecessary and peripheral? Anyway, I got myself a random selection of these texts and began to read. It didn't take me long to realize that I was no longer enjoying myself. In fact, a feeling of depression set in. And a tell-tale headache began (why are libraries so *airless*?). Much of what I was plowing through was so . . . *pointless*, somehow. Full of airy discussion of the possibility and likely significance of a rhetorical and/or discourse analytical study of science yet devoid of any actual *analysis.* "If things don't improve," I thought, "it's back to the gerbils."

I decided to keep my sanity by looking for some *empirical* studies of scientific discourse and some review articles to help me find them. So I made myself a list:

1. Studies that actually analyze some text (instead of talking about "discourse" or "rhetoric" in general)

2. Studies that make generalizations extending beyond the text analyzed (instead of just giving a clever reading of *The Origin of Species*)
3. Studies that have something to say about science (instead of assuming an understanding of science at the outset)

Those guidelines, I thought, should clear out a lot of the undergrowth so that I could see the fauna.

I felt a lot better at lunchtime, sitting out on the square in the sun eating my sandwiches and skimming through a sheaf of photocopies of science studies reviews and introductory chapters. Though many of these were far from complimentary about what was usually referred to as "discourse analysis," they were helpful in one very important respect. They all appeared to carry a version of what I have come to call "The Canonical Footnote," which represents mainstream science studies' standard gesture toward *all* studies of "scientific language." Looking back, I am unsure now if the note as I write it here ever existed in precisely this form—though as I (now) understand it, such "intertextual uncertainty" can only enhance its canonical status (I think!). (Of course, I [have] also made a list of the full bibliographical details of these texts.)

[26]See, for instance, Medawar (1964); Gusfield (1976); Woolgar (1976, 1980); Latour and Woolgar (1979); Knorr Cetina (1981); Yearley (1981); Law and Williams (1982); Mulkay, Potter, and Yearley (1983); Gilbert & Mulkay (1984); Latour (1985/1986/1990, 1987); Lynch (1985a); Mulkay (1985); Shapin and Schaffer (1985); Potter and Wetherell (1987); Bazerman (1988); Ashmore (1989); Myers (1990a).[1]

Anyway, the CF was definitely the best find of the day so far and proved to be very fertile. I spent the rest of the afternoon collecting all the available CF texts. Unfortunately, Potter and Wetherell was on loan, Ashmore was on order, and they didn't have Myers, but I managed to find most of the others. Leafing through them I noticed that many had pictures (scientific diagrams and photographs mostly, though Latour's book, 1987, had a much greater variety). Most of them seemed to analyze conversations transcribed in comic detail. Several (Latour and Woolgar, Knorr Cetina, Law and Williams, and Lynch) were apparently *laboratory studies*, a term I learned from the science studies reviews. Of interest, the life sciences were by far the most popular site for study and only Shapin and Schaffer and part of Bazerman dared to study physics. Incidentally, there also seems to be a flourishing industry in analyses of medical discourse, to judge from Soyland (1991).

The impression that most of these texts deal only with the "softer" end of science was confirmed by a glance at Gusfield's piece. Though called "The Literary Rhetoric of *Science*," it was about some form of social policy research on drunk drivers! Still, maybe I shouldn't be too snobbish—gerbil watching is hardly atomic physics. Gusfield's paper was most peculiar—it is set out almost as a play with three Acts, a Prologue, and an Epilogue. However, it was not half so weird as Mulkay's book, which not only includes a "real play" with characters and everything but also has a whole series of other "alternative textual forms," which I really failed to see the point of. I'm clearly not up to this sort of thing yet and I'm far from sure I want to be. Another rather odd text is Latour's, which is full of funny diagrams and jokes.

Finally, I settled in a quiet corner of the library bar with a coffee and Nigel Gilbert and Michael Mulkay's *Opening Pandora's Box* (1984) for a more detailed read. Their main idea, as I understand it, is that scientists' discourse is structured by the use of two basic "accounting repertoires"—the "empiricist" and the "contingent." In a chapter called "Accounting for Error," they analyze various extracts from their interviews with biochemists involved in a controversy and find that the scientists' use of each repertoire is overwhelmingly asymmetric with respect to their assessments of each other's correctness. In particular, Gilbert and Mulkay claim the empiricist repertoire, in which the facts speak for themselves, is used to account for correctness (such as each scientist's *own* position) while the contingent repertoire, in which social and personal factors play a part, is used in accounts of (others') error.

> Each speaker who formulates his own position in empiricist terms, when accounting for error, sets up the following interpretative problem: "If the natural world speaks so clearly through the respondent in question, how is it that some other scientists come to present that world inaccurately?" . . . [T]he introduction of the contingent repertoire resolves the speaker's dilemma by showing that the speech of those in error . . . is easily understood in view of "what we all know about" the typical limitations of scientists as fallible human beings. . . . . Because contingent factors are mentioned only in the case of false belief, because they are directly contrasted with the purely experimental basis of the speaker's views and because their power to generate and maintain false belief is taken as self-evident, the contingency of scientists' actions and beliefs is made to appear anomalous and as a necessary source of, as well as an explanation of, theoretical error. (Gilbert & Mulkay, 1984, pp. 69-70)

This analysis made some sense to me: I had always been rather puzzled about how my colleagues could so easily explain away competitive results while holding fast to the "empiricist version" of their own. What puzzled me now was why Mulkay should have gone on to write plays—but that would have to wait for another day. The light was fading, the library was closing, and discourse analysis had had its allotted time. Tomorrow was rhetoric. Tonight was . . . well, none of your business, really.

TUESDAY: RHETORICAL STUDIES

Perhaps I should get up later. I have to report that the morning was awful—again! The phrase *rhetoric of science* is so suggestive and I was really looking forward to finding out what it might mean. So I started with a recent review article by Randy Harris (1991) with that title and then hunted up some of his references. Harris spends a lot of time categorizing the various approaches, rather like rhetorical treatises do, but he's funnier than a treatise. The big problem, though, is that he had the kind of bibliography that almost sent me running back to the gerbils. Most of his references were by people in the Communication or Composition programs of big North American universities. There were lots of grand theoretical statements of the "Rhetoric as Epistemic" sort (R. L. Scott, 1976; Weimer, 1977), which I skipped. Then there were studies for which rhetoric means various features of style and studies for which rhetoric means arguments. There was almost a whole shelf of studies of scientific style (C. L. Barber, 1962; Gopnik, 1972; Huddleston, 1971). But they didn't seem to be interested in what scientists do; they just tabulated the verb types or sentences as if they'd been handed some literary work. So I looked around for something that had more to do with science.

There are lots and lots of people finding that scientists use metaphor or other figures of speech (D. McCloskey, 1985) or that there is some other way that the framework of classical rhetoric can be applied to them. The rhetorical studies seem to begin by asserting that there is something striking about the very idea of a rhetoric of science, and assume their readers will think of science as unrhetorical and unliterary (J. A. Campbell, 1987; Gross, 1990; Halloran, 1986; Myers, 1985). Most of these choose canonical figures like Newton, Darwin, or Einstein, rather than the kinds of psychologists and zoologists I know and love. I found the *Quarterly Journal of Speech*, which seems to have rhetoric of science articles frequently, and turned to a paper by John Lyne and Henry Howe (1986) on Stephen Jay Gould and Niles Eldredge. But Lyne and Howe hardly refer to the texts at all, offering their own rather loaded paraphrase instead. And what's worse, in their account rhetoric is contrasted with scientific truth—the good guys don't need rhetoric. Hmm, well . . .

Later, and as something of an antidote, I went back to the science studies section on Level 4 and found an issue of *Science, Technology, & Human Values* with a section on rhetoric (Bazerman, 1989; Fahnestock, 1989; C. Miller, 1989; Waddell, 1989; Woolgar, 1989b). These articles focus on specific, current controversies. Jeanne Fahnestock, who looks at a controversy over when people first crossed the Bering Straits to the New World, has a rather more appealing definition of the rhetorical approach that emphasizes forms of argument and textual detail:

> An analysis of texts from a rhetorical perspective asks what tactics and topics of argumentation are used, how the arguments are arranged sequentially as a series of effects, and how they are actually expressed, their precise wording, their qualifications (or lack thereof), their indirection, their use of figures (tropes and schemas). The rhetorician is primarily interested in explaining textual features as an arguer's creative response to the constraints of a particular situation. (Fahnestock, 1989, p. 27)

This seemed a tall order! I looked for an example and found she talks about something called *copia*, "the technique of enumeration or listing, creating a series that suggests a large number of things, too many for the writer or speaker to specify" (Fahnestock, 1989, p. 37). This is the sort of commentary Fahnestock gives on a passage from one of the articles in the controversy that uses a long list of sites as part of its argument.

> The first three items in the series are fuller, specifying the artifact as well as the location. From then on only sites are named. The impression created is one of increasing quickness as though the arguer were speedily recalling instances from a vast store. The speed creates the impression that the author could name many more sites if either he or the audience had the time. That is precisely the impression *copia* is supposed to create. (Fahnestock, 1989, p. 38)

At first I thought this was just an exercise in dusting off a classical term and giving some modern examples (and as Woolgar, 1989b, might argue, Fahnestock's enumeration and listing of examples of *copia* must themselves be examples of *copia*). But Fahnestock goes further, comparing specialist publications and popularizations, and relating the choices of devices to the dynamics of the particular controversy.

But in the end I was left wondering what the status of this system was supposed to be. What does it mean that there are examples of *copia* in the work of physical anthropologists (or economists, or whatever)? Surely not that these scientists must have been taught classical rhetoric as Shakespeare or Milton were. Was it that the classical notions of persuasive approaches define universals, or perhaps define a particular cultural line of what is persuasive? Or perhaps this kind of work merely shows that, given a long tradition of lists, terms, and broadly defined features, a trained analyst can find them anywhere. For Fahnestock, "rhetoric" provides the terms, and thus a way of organizing the complex data of comparisons. But as I drifted out of the main entrance and headed for the bus stop, I was left thinking, "So what?" However, tomorrow is another day, as they say.

WEDNESDAY: HISTORICAL STUDIES AND SCIENCE EDUCATION

Last night something else started to bother me about these rhetoricians; they don't try to deal with the ways texts are treated by the scientists in context. All they have is this ahistorical, technical system for analyzing them. They didn't tell me much about science that I didn't already know. So this morning I headed straight for the history of science books—except that they were not with the science studies collection on Level 4. It turned out they were in the general science and engineering section on Level 1. So I'd traversed the entire library before I started. After getting my breath back, I tried looking through the last few years of indexes in *Isis* and found a large amount of history of science consists of very detailed accounts of how a text was read in its time and of its influence in later times (Bazerman, 1991; Cantor, 1987; Dear, 1985, 1987; Golinski, 1987; Le Grand, 1986; Schaffer, 1986, 1989; Yeo, 1986; R. Young, 1986). I was lucky to find a new collection (Dear, 1991) that led me back to some cases from the seventeenth, eighteenth, and nineteenth centuries. Despite the fact that (I admit it) I hadn't heard of a lot of the famous scientists discussed, and despite the rather oppressive weight of the footnotes, I enjoyed it. After the first shock, I found the articles dealt with more general issues than their titles seemed to indicate: ideas of genre (Thomas Broman), visual representation (Lissa Roberts), and narrative (Peter Dear and Frederick Holmes). I was surprised these current issues went so far back, and there was a satisfyingly solid feel to the cases.

I found a sort of a review in Markus (1987), but it said it was about *hermeneutics*. Lots of these articles aren't about *rhetoric* or *discourse*; they say they are about things like *documents, method, context*, or *tradition*. But they do have detailed readings of texts. These historians allow none of the grand generalizations of the rhetoricians about ahistorical categories, about likely responses, about ideas that were in the cultural soup at the time. Everything is pinned down to some letter or marginal annotation, a manuscript somewhere, in a way that is alarming to people working far from a major library. (I don't want to be insulting; this place is fine, no really, but . . .) Lots of these history of science articles point to the need to go beyond published sources to notebooks, letters, drafts (Bazerman, 1988; Herbert, 1991; Holmes, 1987; Myers, 1990a, chap. 3; Rudwick, 1985). Others blithely infer authors' intentions from the texts (Block, 1985; Sapp, 1986) without using evidence for earlier stages. The method in most of these studies is to focus on change: from notebook to publication, or from draft to draft, or from publication to publication. For instance, Frederick Holmes (1987) looks in detail at Lavoisier's notebooks. Living scientists only occasionally get this attention (Myers, 1985). Perhaps historians' self-identities would be compromised were they to deal too often with the living.

One of those terrifyingly detailed articles had a tempting subtitle: "Saving Newton's Text: Documents, Readers, and the Ways of the World" (Palter, 1987). A quick skim showed Palter was interested in the dating and interpretation of a text by Newton I'd never heard of. He is a true historian of science in his concern with the documentary details, like whether Newton had Galileo in his library, or had access to it, or knew Italian (fn. p. 392)—quite a change after the grand generalizations of the literary-rhetorical types. But he is struck by the problem of how historians and philosophers disagree in their interpretations. That leads him to some reflections on how interpretation is possible at all, with no access to the intended meaning, to the readings of the time, or to the worldview, except through texts (what he calls "pantextuality"). (I saw there was the same sort of division of the problem in Marcus, 1987: "The inscribed author," "The intended reader," and "The work in the context of its tradition.") Palter points out the circularity of dating the Newton manuscript as early because of its immaturity, and then finding support for its immaturity in its early date. He ends up by explaining the differing interpretations of historians and philosophers of science, in terms of their distinctive disciplinary orientations and styles. The task in his view is to place this text in relation to other texts, other conceptions of space and time, even later ones. What I don't get is how he then decides the other interpretations are wrong, which is an approach that seems to ally him with philosophers, after all.

About midafternoon, I decided to give myself a break. I had been wondering whether there had been any relevant work done on science education. I suppose my interest in this area was based on an analogy with infancy: If you can help explain adult capacities through studying infant development, you should be able to cast some light on scientific practice through an understanding of scientists' socialization in science education. On the basis of my brief search, however, there seemed to be very little. It seems education, as that Delamont (1987) piece in *Social Studies of Science* put it, is indeed a notable "blind spot" in current work in science studies. Most of the stuff I found seemed either not to take much notice of the discourse (Collins & Shapin, 1983; Millar, 1989) or to be more concerned with education generally than science in particular (Mehan, 1979). There was a new book for science teachers that focused attention on language (Sutton, 1992). Not exactly trailblazing, but good examples. A book by Valerie Walkerdine (1988) looked at mathematics teaching and made some nice points about gender issues. Apparently if you use the "three bears story" to teach size relations to a mixed class of kids, you come unstuck because of hidden assumptions about gender and size. An ethnography by Derek Edwards and Neil Mercer (1987) focused on the way teachers deal with the contradictory requirements of "child-centered pedagogy" and of experiments having "right" outcomes. You have to subtly lead the kids to the right answer but always give the impression that they found it themselves, which is also the conclusion of a "nondiscourse" study by Atkinson and Delamont (1977). The authors don't make the point, but I wonder at the problems that this generates for any of these kids going on to do real science when there will be no one subtly pushing them in the right direction. I know that my (ex?) Ph.D. work is worryingly hard to square with the stories I learned about science at school. And the worst thing is how the aura of certainty surrounding those old stories is perpetuated in the textbooks I still have to read. While some theorists of science recognize this disparity, most (Brush, 1974; Kuhn, 1962/1970) rather insultingly consider that it is necessary for our morale to be told comforting stories.

In the end, libraries are so oppressive. Full of words, words, words. Zillions of them! I need a hot bath and a cup of tea.

THURSDAY: OTHER TEXTS—VISUAL, MATHEMATICAL, MATERIAL

It was as much as I could do to drag myself in this morning. "If there's one good thing about gerbils," I thought, "it's that they don't write and don't talk." Word sickness, you could call it. A bad case of. So today I'm avoiding them—as much as possible, that is. I have settled myself near the art history stacks; the atmosphere suits my mood.

And really, this isn't as silly or self-defeating as it sounds. After all, most of what I was doing with the gerbils involved persuasion through numbers or graphs. There were some chapters or comments on illustrations by authors in the Canonical Footnote (Gilbert & Mulkay, 1984, chap. 7; Latour, 1985/1986/1990; Lynch, 1985a). But there were lots of other studies, scattered all over the library, from history, visual arts, and cultural studies (Fox & Lawrence, 1988; Ivins, 1973; Jacobi, 1985; Rudwick, 1976; Silverstone, 1985) and some collections (Fyfe & Law, 1988; Latour & de Noblet, 1985; Lynch & Woolgar, 1990). These read quite differently than the studies of written texts, even those by the same authors. Academics seem to know what to do with writing—find metaphors in it, or find an argument, or look at the pronouns. There seems to be no general agreement about how to approach visual images; for example, some use art history (Latour, 1985/1986/1990) and some use semiotics (Bastide, 1985/1990; Myers, 1990b). Or maybe the problem is that people don't see any need to *read* visual texts; they are just there.

I found another set of stuff that had been shelved under semiotics; puzzlingly, most of this work also went under the name of "actor-network theory" or the "sociology of translation" (Callon, Law, & Rip, 1986; Latour, 1987; Law, 1986c, 1991c). I got quite excited. It seemed to promise a much more *complete* framework of analysis than the straight rhetorical analyses I had been looking at. It claims to deal with power, politics, agency, knowledge, the association of people and things, and no doubt everything else there is. However, when I showed the stuff to Karla (who has nearly finished her doctorate on Baudrillard, the State, and modern architecture), she said it wasn't what she thought of as semiotics, and then lost me with a lot of talk about paradigmatic oppositions, free-floating signifiers, and the relation between poststructuralism and postmodernism. Rather reluctantly, I concluded that actor-network theory should not be in this review.

And, to tell the truth, I was getting distracted from the books by the realization—was this trivial or significant?—that peoples' clothes here on Level 3 (Art and Humanities) were so much more interesting than on Level 1 (Science and Engineering). As I walked up the stairs to Level 4 (Social Sciences), I began to wonder if I looked the part. How *did* science studies researchers dress? And, just as important, would any of them find this an interesting question?

I finally settled down at my favorite table in the science studies section with a piece by Michael Lynch (1990), because I liked the choice of pictures, lots of related examples. He talks about how images like photographs get made into images like graphs, talking about *selection* and *mathematicization*. First, he looks at pairs of photographs and diagrams, which were like those I had in my own textbooks, analyzing what happens in the transformation. Then he looks at how quantitative data are translated from visual data; I particularly appreciated the example from a field study of lizards. He says these processes are not just a matter of reducing complexity; they *add* visual features as well. He says, "Specimen materials are "shaped" in terms of the geometric parameters of the graph, so that mathematical analysis and natural phenomena do not so much *correspond* as do they *merge* indistinguishably on the ground of the literary representation" (Lynch, 1990, p. 181). Now what does that mean? First, I think he's saying we see the gerbils in terms of the graphs we are finally going to make of them. Sand bathing becomes a countable behavior, not something they do "in real time." But also, when we make an image, it is hard then to separate that image from the object it represents, maybe like the way I think of hormone levels.

Now that Lynch has made me aware of mathematicization, I see there are other, mainly historical, studies that deal with this issue (Dear, 1987; Kuhn, 1977b; Restivo, 1990; Shapin, 1988c). Most of these, though, seemed interested in mathematicization in the abstract. In contrast, a paper by Potter, Wetherell, and Chitty (1991) tried to show how different sorts of mathematicization could be put to different rhetorical purposes. For example, they draw a rather nice analogy between the way a table of cancer death statistics is constructed and the way a market trader constructs an item as a bargain. I particularly liked the idea that numbers, which are often treated as hard, solid things in contrast to airy rhetoric, can themselves be viewed as *the* most effective form of rhetoric.

Also, I am starting to note a theme that crops up repeatedly in many of the studies I have been looking at. Researchers use variations between stretches of discourse as a guide to what the discourse is used to do. I can also see that very soon I am going to actually do some of this sort of analysis so I can really understand how it works. I get the impression that there are some important craft skills at work here that do not come over very clearly in the published papers.

If images and numbers can be read as "text," why can't everything else? No reason at all, it seems. Latour (1987) writes about laboratory apparatus as an "inscription device"; Woolgar (1985, 1991b) advocates a "sociology of machines"; and someone called Jim Johnson (1988) has studied the semiotics of a door-closer! I also saw some titles of articles and chapters that dealt with space, especially with the space of laboratories. Latour and Woolgar (1979) start with a tour of a biochemistry lab, and Traweek (1988) has a chapter that describes all the buildings of a high-energy physics lab, including the offices and cafeteria. There was a piece on the design of British university science buildings (Forgan, 1989) and, on the current periodicals shelves, which I looked at before I left, there was a whole issue of *Science in Context* (Ophir, Shapin, & Schaffer, 1991), with yet another piece by the ubiquitous Michael Lynch (1991)! But I'm getting confused. If authors and readers are texts, and gerbils are texts, and labs are texts, what isn't a text? A meal and a bottle of wine, perhaps? No, *don't* tell me about Mary Douglas (1975).

Oh yes, I picked up a splendid-looking book called *Incorporations* (Crary & Kwinter, 1992) from the new acquisitions on the way out. It is full of photos printed on different-color paper and contains lots of short pieces on machines, architecture, and science from such as Félix Guattari, John O'Neill, and the novelist J. G. Ballard. Jamie dismissed it as a postmodernist coffee-table book. It made me uncomfortable, though. Too many distinctions blurred, perhaps, just as I was starting to get them clear. And although it is not *about* the rhetoric of science, it succeeds in commenting obliquely on such matters, through, I am beginning to think, the way it foregrounds form even, perhaps especially, in its self-conscious stylishness.

FRIDAY: GENDER STUDIES

This morning was—pretty good, really, if rather puzzling. Though I've been slogging away for 5 days and coming across studies of scientific discourse from lots of different disciplines, I've seen remarkably little in women's studies. So today, that's what I've been concentrating on. Right off I found a good review in *Theory and Society* (Jansen, 1990) of books by people like Sandra Harding and Evelyn Fox Keller, and it said that feminist critics of science had made use of the same sociologists of scientific knowledge (mainly Bloor and Barnes of the Edinburgh/Strong Programme tendency) as these discourse people. On the other hand, Delamont (1987) argues that sociology of scientific knowledge ignored feminism, which on my reading is quite correct. So what is the relationship? Or is this an irrelevance for discourse studies?

Though there are a number of books on science and gender (Keller, 1985), on women's careers (or lack of them; Kelly, 1987), on famous women scientists (Keller, 1982), on the biases of various disciplines, especially biology (Brighton Women and Science Group, 1980), on the way sciences construct women (MacCormack & Strathern, 1980), and on the masculine orientation of scientific epistemology (Harding, 1986; Harding & Hintikka, 1983), what I don't find are detailed readings of texts. Is this because they are often looking at a broader level? On the other hand, the people combing texts for signs of social construction don't seem to be much interested in gender. They raise epistemological questions but aren't interested in alternative epistemologies. What's the problem? Surely this area should not be impervious to feminism. So I spent the rest of the morning skimming tables of contents in collections on scientific discourse and looking through the feminist studies for a long quotation.

It turns out that feminist studies do raise questions about texts but don't usually foreground the method of discourse analysis. For instance, Jordanova's *Sexual Visions* (1989) has close readings of Michelet and of wax dummies. She points out the dangers of superimposing contrasts of male and female on contrasts of culture and nature, with a sharp conclusion. There's another collection of detailed studies of particular issues by Schiebinger, *The Mind Has No Sex?* (1989), that has a nice chapter on allegorical representations of science as a woman, especially in the frontispieces of seventeenth- and eighteenth-century books; it neatly complements Jordanova's chapter on the image of unveiling and dissections. She too sees a complex interaction of various contrasts revolving around masculinity and femininity, for instance, in a brief comment on scientific and literary style in the eighteenth century. Traweek's essay "Border Crossings" (1992) raises the issue of texts by giving her own text a narrative form that dramatizes what she calls her marginality, her difference; there is no analysis of texts but there is an implied critique of existing scholarly forms. Star's (1991b) essay "On Being Allergic to Onions" tries to reorient the actor-network studies I read about yesterday, to see these networks from the margins rather than from the perspective of a masterful strategist. Haraway's huge and fascinating *Primate Visions* (1989) deals with some alternative forms of texts explored by primatologists like Zihlman and Hrdy. It's interesting that all these authors deal as much with nonverbal as with verbal texts. Maybe this is just the way people work in cultural studies in the late 1980s. But maybe pictures play off against texts in a way that is useful to them. Maybe the reason I haven't found more close readings of scientific writing in feminist studies is that there is something about the assumptions underlying this kind of "literary" work that is inconsistent with recent feminist approaches; if close reading is the attempted provision of a One Best Reading, it is not going to appeal to those who wish to legitimate a diversity of readings.

A collection edited by Jacobus, Keller, and Shuttleworth (1990) has lots of articles with some textual analysis, though again none on the methods used. In it, Emily Martin (1990) has a piece about kinds of language used to describe menstruation. She analyzes metaphors in textbooks, and as one might expect she finds a male view in and as the scientific view. But this does not lead her to a juxtaposition of false male science with true female experience. She reports a study in which she asked various women two questions: "What is your own understanding of menstruation?" and "How would you explain menstruation to a young girl who didn't know about it?" She contrasts one kind of language women use to talk about menstruation—"internal organs, structures, and functions," with another she calls "the phenomenology of menstruation"—what it feels and looks like to the woman experiencing it. The working-class women interviewed "share an absolute reluctance to give the medical view of menstruation." It's not just that there's a scientific mode and a popular mode but that there are significant and analyzable ways of slipping into or out of the scientific mode of talking about one's own body.

With a tip from Susan, who was also spending time in the library, I finally found a direct link between feminism and discourse analysis in the sociology of scientific knowledge. It was in what I thought was a surprising place, Celia Kitzinger's *The Social Construction of Lesbianism* (1988). She draws on Gilbert and Mulkay (1984) in a critique of the rhetoric of earlier psychological work on gays. Now, I thought, that should be like shooting fish in a barrel, ironicizing the scientism of all that stuff on homosexuality as an illness. But Kitzinger does something more complicated and much more interesting; she also analyzes the line of research on gayness as "lifestyle" that I was naively expecting her to favor and thus to exempt from scrutiny. She even includes quotations from her own work as she shows how the two sides invoke scientific authority. In her next chapter she confronts the issue of political commitment and concludes that it is still possible (as her own work suggests) to work for a clear political aim while still insisting on analyzing *all* relevant discourses, however they may be evaluated.

The idea of competing discourses or registers or styles seems consistent with the largely separate line of work in sociology of scientific knowledge, such as Gilbert and Mulkay's (1984) "repertoires" (so why are they so often separate?). I'd come across similar sorts of juxtapositions in reading earlier in the week, like Walker (1988) on therapy talk and Brodkey (1987) on attempts to collaborate on a paper in feminist literary criticism. There was a recent piece by Mulkay (1989) that tried to link his own work to feminist critiques of science, largely through the valorization of multiplicity. So maybe what I am seeing is similar methods—bringing out the multiplicity of discourses—put to different ends—the critique of scientific authority and of patriarchal knowledge. Maybe that's my problem. They use similar methods sometimes, but feminist studies of science and social studies of scientific knowledge seem better defined by their separate missions than their common methods. Question: How Bad a Thing can this be, if diversity is what they both value?

On the bus home I read Donna Haraway's "Manifesto for Cyborgs" (1985), which Susan had lent me; I must remember to make a copy of it tomorrow. Now, this could not be fitted into any one of my day's plans; nor did it fit easily with what I am coming to understand as the usual forms of theorizing or analysis in studies of technoscience. It mixed political polemic, feminism, sociology of science, and cultural studies—and quite uniquely. It could have been a mess—it *should* have been a mess—but it really wasn't. I was rather inspired by it. And again, the text was not so much *about* rhetoric as it was a very conscious, some may say "arch," display of a different rhetoric—indeed, a rhetoric of difference.

SATURDAY: STUDIES OF SOCIAL SCIENCE

I was late in this morning because I had to get to the shops as I'd run out of . . . Sorry. Start again. On Monday, which feels a long time ago, I was appallingly dismissive of the Canonical Footnoters' tendency to analyze only the softer end of science. It has struck me forcibly since that this attitude is ridiculous—and not only because I am now trying to become a much softer scientist than I used to be. The real point, of course, is that the division of the sciences into "hard" and "soft" is itself a historically analyzable phenomenon and a rather obviously gendered rhetorical device. So to make amends, I am going to focus on social science analyses today. My favorite table had been taken so I carried my new heap of books to a quiet corner of the top floor that had a soothing view of the still frosty countryside just off campus.

One thing I saw immediately is that it is much more difficult to tell who this work on social science is for. The natural/biological science stuff is typically published in journals like *Isis* or *Social Studies of Science*, and you don't get the feeling that particle physicists or plant biologists are meant to read it. However, the social science work is often published in places where the readers are going to be the people the studies are about and often it does not address the theoretical concerns of science and technology studies at all. Take psychology, for example. There is some work dealing with themes like repertoires and categorizations, which reminded me of sociology of science discourse analysis (McKinlay & Potter, 1987; Potter, 1984, 1988; Potter & Mulkay, 1985). Yet there are also plenty of papers—like Harré (1981, with a neat section here called "The Rhetoric of Social Psychological Theory Considered as Talk"), Lubek (1976), Billig (1991), and many in the volume edited by Parker and Shotter (1990)—which seem to be mostly concerned with effecting change in the discipline itself.

A lot of the most interesting stuff seemed to work on both levels, talking to some group of social scientists as well as to sociologists and rhetoricians of scientific knowledge. Bazerman's (1988, chap. 9) piece on the American Psychological Association style manual tried to show how the assumptions of behaviorism were embodied in the prescribed textual form of the discipline. I liked Steve Woolgar and Dorothy Pawluch's (1985) argument about the way researchers into social problems rig the outcome of their discussions by a process of "ontological gerrymandering."

> The successful social problems explanation depends on making problematic the truth status of certain states of affairs selected for analysis and explanation, while backgrounding or minimizing the possibility that the same problems apply to assumptions upon which the analysis depends. By means of ontological gerrymandering, proponents of definitional explanation place a boundary between assumptions which are to be understood as (ostensibly) problematic and those which are not. This "boundary work" creates and sustains the differential susceptibility of phenomena to ontological uncertainty. (Woolgar & Pawluch, 1985, p. 216)

I could see that this kind of analysis could be more generally applied to a lot of areas of scientific work.

I found a rhetoric of economics (D. McCloskey, 1985), a poetic for anthropology (Clifford & Marcus, 1986), and one of each for sociology (R. H. Brown, 1977; Edmondson, 1984). D. McCloskey's work is interesting as it appears to be part of a large and booming project known as "rhetoric of inquiry" (Nelson & Megill, 1986), which seems to have covered every known discipline in the humanities/human sciences area and now boasts at least three large and diverse collections (Nelson, Megill, & McCloskey, 1987; Simons, 1989, 1990). It is anthropologists, though, who seem to have undertaken the most comprehensive and radical look at the way their own discourse works, both through rereadings of the classics (Marcus & Cushman, 1982; Stocking, 1983) and rewritings of ethnographies (Tyler, 1987; S. Webster, 1982).

However, I was particularly taken by a book by Paul Atkinson (1990). This looks at sociological ethnography as a textual form using particular conventions and devices to produce a "reality effect." It draws on some of the stuff I encountered earlier in the week like literary theory, ethnomethodological ideas, and sociology of science. For example, Atkinson closely compares the introductory paragraphs of a short story by Hemingway (actually, he gets this from Fowler, 1977) and a well-known ethnography. He shows how both use very similar devices to get the reader into the story and suggests that these devices "serve to warrant the subsequent sociological discourse by establishing its *vraisemblance* [naturalness or genuineness]. It furnishes the 'guarantee' of an eyewitness report, couched in terms of the dispassionate observer, using the conventional style of the realist writer of fiction, or documentary reporter" (Atkinson, 1990, p. 70).

I wondered if some of the descriptions in my diary were able to generate this effect. Probably not! Atkinson did not seem to have any particular story to tell about the general role of this reality construction in ethnography, except that it makes it convincing and we should pay more attention to it. In contrast, a piece by Peter Stringer (in Potter, Stringer, & Wetherell, 1984) tried to show how in a social science setting the rhetoric of the researcher can become mixed up with the rhetoric of the people who are being researched. Rather neatly, it shows the way President Kennedy's key advisers' excuses and justifications for the Bay of Pigs fiasco become incorporated in the "scientific" account of this group's "group processes" given by a social psychologist. I noted that Stringer was insistent that his own text should not be treated as immune from such problems.

The next thing of his I read made me see exactly how insistent he was. This was a paper called "You Decide What Your Title Is to Be and [Read] Write to That Title" (1985). Tucked away in a book on a rather obscure psychological theory, it is an . . . I am rather at a loss for the word here . . . an *exploration* of various textual forms in social science: the review, the title, the textbook, informal talk, exam papers. For example, it shows how the version of a particular theory in textbooks is a product of the particular organizational scheme used, and that textbooks "demonstrate despite themselves the inevitability of the re-writing that reading evokes" (Stringer, 1985, p. 221). Through a set of textual "fragments," it questions standard ways of understanding the relation between authors and readers, and between truths and fictions. Its final challenge to the reader is to be more active in your own readings—hence the title. I now recognize this as a "reflexive piece," a "new literary form." Ashmore, Mulkay, and Pinch's (1989) book on health economics is another good social science example. On Monday this all seemed to be idle navel-gazing and too-clever-by-half tricks—as most critics insist (Baber, 1992; Doran, 1989)—but after nearly a week in the library I am starting to see the point. I have also realized that I have visited every level and probably every set of stacks in this library.

SUNDAY: REVIEWS, REFLEXIVE WORK, AND DECISIONS

Last night I had a long talk with Jamie—I had hardly seen her since I started all this. She said that the way I'd been going at it all week—first one thing and then another, and another—was crazy. And that I was looking worn out. She said that if I *insisted* on coming back here tomorrow (today), I should spend the day thinking more clearly about what I'm going to *do* with all this stuff. She's right, of course. At the moment, my project looks set to be a discourse analytical cum rhetorical and historical study of natural and social scientific texts of all possible kinds that is greatly concerned with gender and slightly concerned with science education! This clearly won't work. I think I'm getting depressed again.

In an effort to cheer myself up, I spent the morning reading about reflexivity. I had been thinking about this on and off ever since Monday when I came across that funny book by Mulkay (1985). And looking at all those social science analyses yesterday—and Haraway's cyborgs, Traweek's border crossings, and Star's onions on Friday—brought it to a head. All week, the most interesting things I have read are those that foreground their authors' own involvement in the text. And it seems that the closer one's *topic* is to one's *method*, the more important it becomes to devise some way of coming to terms with the implications of that similarity. In this argument, writing about writing *has* to be a self-consciously circular process and its practitioners must learn to live with the (rhetorical) consequences—such as my own initial negative reaction. The earliest argument on these lines I have found in science studies is in a piece by Latour (1981), though he certainly doesn't advocate writing plays. (Which is odd, come to think of it, given his earlier, 1980, experimental piece written as "a sociologist's nightmare.")

One of the most entertaining pieces of reflexive writing I read was by Trevor Pinch and Trevor Pinch (1988), which masquerades as a critique of reflexivity and "new literary forms" (or "unconventional texts," as the Pinches put it) and yet that does so in the form of a playful and intensely self-referential dialogue. After a discussion of two versions of reflexivity as formulated by Woolgar (1983, 1988c), this is how the two versions of the author (fail to) present Woolgar's third version of reflexivity (admittedly, things do get a bit complicated):

> *By the way, I take it that we do not have to mention here the third form . . . the "benign introspection" version? I know that neither you nor I would ever take that sort of psychological talk seriously.*
>
> Quite—no need to mention that here.
>
> *Good. Back to self-reference.* (Pinch & Pinch, 1988, p. 183)

What this kind of work does, it appears to me, is to put in question some of the most basic and taken-for-granted desiderata of scholarly or academic or scientific writing—such as the distinction between the serious and the nonserious, the important and the trivial. Perhaps, after all, this kind of thing did have something to offer. Perhaps, I too (no less) could learn to play. (But would I be allowed to? Would an outsider be accepted? After all, this kind of stuff did have a rather closed and cliquey feel to it. Where does the *community* end up in all this "hermeneutic hyperconsciousness"—Beer & Martins, 1990a, p. 172?)

After a quick lunch at the pub around the corner, I realized I would have to start at the beginning again so I took a new look at the collection of reviews I had amassed during the week. I began to notice that these texts were highly argumentative even (especially?) when they presented themselves as "innocent" and above the battle. For instance, a general survey of linguistically turned history of science by J. V. Golinski (1990a) rejects the work of Gilbert and Mulkay (1984) by contrasting it with that of another Canonical Footnote member (Shapin & Schaffer, 1985), which is made to stand for the acceptable face of rhetorical analysis: "[Shapin and Schaffer] have used techniques of the rhetorical analysis of scientific discourse, but have not shirked the historian's duty to place that discourse in its historical context" (Golinski, 1990a, p. 120). I love that "but" (which is elaborated by Shapin, 1984, in his own critique of discourse analysis, that is, Gilbert and Mulkay). There's a similar sort of move evident in political critiques of the sociology of scientific knowledge. For instance, Aronowitz (1988b) reviews the field fairly sympathetically but then withdraws, saying that the primacy of discourse is itself historically conditioned. If the use of this kind of move serves to ally Aronowitz with Golinski, their alliance does not last for long—and they fall out, significantly, on the issue of what their common discipline of history is all about. For Golinski, "history" is what historians do; for Aronowitz, it is what the working class does.

One source—or better, resource—for review rhetoric appears, then, to be disciplinary membership. Though as befits the concept of "resources," accounts and attributions of disciplinarity appear to be highly "occasioned" (I think I'm getting the hang of this). An interesting example of this feature is a pair of reviews (D. McCloskey, 1987; Potter & Wetherell, in press) that come, independently as far as I can tell, to a similar conclusion about a major piece of work in the sociology of scientific knowledge: Collins's (1985) study of the gravity wave controversy.

> Collins . . . writes on rhetoric without knowing it. . . . This is not sociology of knowledge but rhetoric of knowledge. (D. McCloskey, 1987, p. 14 [draft])[2]

> Rhetoric is, arguably, one of [Collins's] major concepts for understanding social life. However, he does not theorise the concept and explore its senses. (Potter & Wetherell, in press: chap. 1, p. 23 [draft]; see note 2)

Both of these examples attempt a reconstruction of Collins's work in alternative disciplinary terms more in line with the reviewers' own concerns. And in both cases, Collins is chided for not already having done this work himself.

In all three examples, we can see that the rhetorical resource of disciplinary membership is used to do what is known as "boundary work" (Gieryn, 1983). But whereas Golinski, as spokesperson for history of science, attempts to exclude inappropriate work, McCloskey and Potter and Wetherell, on behalf of rhetoric of science, use an inclusionary strategy. It is tempting, though doubtless premature, to conclude from these differences in reviewing strategies the existence of concomitant differences in the status of the disciplines concerned—and so on. Of course, this is only a sketch of an outline of a single aspect of what might be involved in a fully fledged analysis of "review rhetoric." I haven't touched at all on other aspects of the construction of authority, let alone the ways that reviewers justify their selections and judgments. Nevertheless, this *could* be just what I've been looking for: an interesting and worthwhile dissertation topic. Perhaps something like "The Rhetorical Structure of Review Discourse."

I'm going to take a break now while I think over what I've done. After all, what do I really know about writing reviews? Surely, if I'm serious about taking on board the reflexive argument, the first thing I would have to do is to write one myself. Now. Hang on. What was it that Jamie mentioned about this new *Handbook of Science and Technology Studies*? Something about a bidding process for all the chapters. But if that's true, then why shouldn't I use my week's diary . . .? No. They'd never accept it. I've got no standing, no credibility, no . . . GOT IT! I'll attribute it to someone—or better, several people—who have. Not anybody whose work I've reviewed, of course. That would be unethical. Yes. Of course. The three absentee (and male!) authors from the Canonical Footnote: in alphabetical order—Ashmore, Myers, and Potter. Perfect! And I'll write it as a new literary form—as a diary of a fictional graduate student's 7 days in the library. Jamie will hate it. "Overcute and pointless," she'll say. I must do some careful work on that. Everything must be justified. So . . .[3]

NOTES

1. See, for instance, Gilbert and Mulkay (1984, p. 194, n. 27); Lynch (1985a, p. 17, n. 2); Schuster and Yeo (1986, p. xxxii, n. 5); Bazerman (1988, p. 156, n. 3); Golinski (1990b, p. 497, n. 11); Ashmore, Myers, and Potter (1995, n. 1).

2. I now find, on going back to the library, that this quote does not appear in the published version of McCloskey's review. Let me explain. My new colleagues—Malcolm, Greg, and Jonathan—were kind enough to let me see the draft of McCloskey's piece (and of Potter and Wetherell's forthcoming book). A pity, though, that they neglected to tell me of the important changes in the published text; perhaps they didn't know themselves? In any case, the evident "illibrarality" of these two texts provides an occasion for two interesting textual manoeuvres: the acknowledgment of debts, as above, and the confession; here, of having resorted, in the interests of realism, to the occasional fiction.

3. We now go back to the beginning—as weeks do, of course.

16

□

Science and the Media

BRUCE V. LEWENSTEIN

"SCIENCE and the media" can mean many things. Most commonly in the scholarly literature, it has meant science *journalism*—that is, nonfiction portrayals of science in newspapers, magazines, books, and television news and documentary shows. (Radio is hardly ever mentioned.) Only rarely has the science and media literature moved beyond these formats to include other media in which nonfiction information about science can be presented, such as museums and plays; nor has scholarship on science and the media connected well with research on the portrayals of science in literature or dramatic film and television.

The distinctions between fiction and nonfiction, or between print and broadcast, may have meaning for scientists, journalists, writers—and academic researchers. But the goal of research on science and the media is often to understand what the general public knows, thinks, and feels about science.

AUTHOR'S NOTE: The ideas in this chapter were deeply influenced by discussions with, writings and (in some cases) student papers of, and comments on earlier drafts by Steven Allison, Christopher Dornan, Sharon Dunwoody, Stephen Hilgartner, Todd Paddock, Kerry Rodgers, and Wade Roush. Many of the citations were tracked down with the help of bibliographies prepared by Sharon Dunwoody and her colleagues. Secretarial and research assistance has been provided by Michele Finkelstein, Bonni Kowalke, and Lisa Buzzard. Financial support has come from Hatch grant NYC-131403.

Because it is not clear how the public distinguishes among the various media and formats, the value of retaining traditional distinctions is also unclear.

Thus my goal in this chapter is twofold. First, I want to summarize the existing literature both in science journalism and in related fields. Second, I want to suggest some attributes of a model for thinking about science and the media —indeed, for thinking about all science communication—that allows these disparate fields to be usefully integrated.

TRADITIONAL SCIENCE AND THE MEDIA LITERATURE

Comments by Working Journalists

As the new profession of science journalism arose in the United States in the years between the world wars, science writers began to publish articles arguing for the importance of their own field, stressing the validity of their own techniques, and inveighing against the obstacles placed before them by scientists. Dornan (1988) argues that an ideology of science writing was created during this period and provides references to relevant texts; two useful examples were produced by Krieghbaum (1963, 1967). Although science journalism developed at about the same time in England, fewer works of creation myth have been published there; the most well known is autobiography of J. G. Crowther (1970), who called himself the first science writer in England. The literature is even sparser in other countries.

Reflecting the realities of journalistic practice, the creation myths generally collapse science, medicine, and technology into a single category, called "science." These works call for an increase in the "quantity and quality" of science writing for the general public. They take scientific knowledge to be an unproblematic compilation of the findings of scientists and define the goal of science writing as disseminating and "interpreting" scientific knowledge for nonscientists. By *interpretation,* practitioners have traditionally meant finding the contemporary practical applications that make basic scientific research relevant to people without deep interest in the intellectual exploration that is taken (by these authors) to be at the core of scientific research.

To the science and technology studies researcher, the best articles provide useful information and reliable evidence; the worst articles unhesitatingly repeat the cants of the day and are themselves evidence, at least of particular cultures. The problem lies in distinguishing between the two. Those who believe in the objective nature of historical reality will often have little trouble distinguishing between "reliable" evidence and mere opinion. But those more inclined to sociologically sophisticated positions will insist that practitioner analyses should be held apart, treated as evidence of contemporary interests, not as loci for historical facts.

Science Writers

Because science journalists are among the most visible actors in the interplay between science and the media, many studies have looked at their characteristics and work habits. A series of studies begun in the 1930s has shown that science writers have gradually become better educated in general, and specifically more committed to having scientific training as part of the background they bring to science journalism (e.g., Fraley, 1963; L. Z. Johnson, 1957; Krieghbaum, 1940; Ryan & Dunwoody, 1975). This change toward a professional body with stronger ties to the scientific community has intensified the potential for conflict between the values of journalists and the values of scientists as well as creating a cadre of individuals who are trying to merge the sets of values. Dunwoody (1980) described this phenomenon as the operation of an "inner circle" of influential science journalists; other descriptions of the meshing of values appear in B. J. Cole (1975), Nelkin (1987), and Lewenstein (1992c). Useful collections that include both statements by working science journalists and analyses of how scientists and journalists work together include Alberger and Carter (1981) and Friedman, Dunwoody, and Rogers (1986).

These works have shown that science writers, despite claims to journalistic independence, are often under both personal and institutional pressures to conform to scientific values. Thus their reporting tends to reflect the concerns of the scientific community rather than those of the "public" that they often claim to represent. (Dornan, 1990, has argued that this same bias affects *studies* of science journalism as well.) The result has been the production of stories about science that do not challenge the positivist ideology held by most scientists, and thus do not challenge the role or status of science in society. In particular, this line of research has suggested that popular science produced by professional science journalists is likely to reinforce the vision of science as a coherent body of knowledge about an underlying natural reality produced by carefully controlled methods; this image of science is fundamentally at odds with contemporary ideas about the social construction of scientific knowledge.

Surveys of Content

Many systematic studies of science content in the mass media confirm that the public presentation of science rarely is shaped by "objective" scientific issues but instead presents images of science shaped by—and sometimes intended to shape—particular cultural contexts. The ideological goals of media presentations are often clearest in historical cases. Ziporyn's (1988) examination of press coverage of diphtheria, typhoid, and syphilis in American magazines from 1880 to 1920, for example, showed that social values were more important than technical discoveries in changing the character of coverage. Similar conclusions emerge from various surveys of general science coverage in magazines, such as Sheets-Pyenson's (1985) history of low-culture magazines in London and Paris in the mid-1800s and LaFollette's (1990a) detailed content analysis of 11 major American magazines from 1910 to 1955. Caudill (1989a) also demonstrated the ideological drive behind much popular science in his analysis of Darwinism in the media.

Surveys of more contemporary periods, although usually only descriptive in character, also suggest that cultural contexts shape the coverage (e.g., Weiss & Singer's, 1988, thorough examination of recent social science coverage in American newspapers and newsmagazines, and Hansen's, 1990, study of media coverage of the environment). In addition, a number of studies have questioned the focus on the mainstream "quality" press, showing that the tabloid press in the United States provides useful information on science, despite its reputation for sensationalism (Evans, Krippendorf, Yoon, Posluszny, & Thomas, 1990; Hinkle & Elliott, 1989). Virtually all of these works look only at printed media; no systematic look at television science journalism has been published.

Until the recent recognition that cultural context was important, many studies had focused on more "objective" measures of media content, such as accuracy and readability (see, e.g., Bader, 1990; Bostian, 1983; Dunwoody, 1982; Hayes, 1992). These studies had reinforced the idealized vision of "simplification" that had come with the commitment to a "diffusion" model of scientific popularization. But with growing attention to science journalism from "critical studies" researchers in communication studies (e.g., Hornig, 1990; Thomas, 1990) and in science and technology studies (e.g., Collins, 1987, 1988), the idealized vision of science popularization is unlikely to remain viable.

Controversies

In addition to general studies of media coverage of science, much research over the last 25 years has looked at coverage of particular issues in science. Many of these case studies have involved scientific controversies, and so this literature overlaps with the controversy literature (Mazur, 1981; Nelkin, 1992). In general assessments, B. J. Cole (1975) showed an increase in coverage of controversy from 1951 to 1971, while Goodell (1980, 1986) and

Nelkin (1987) both suggested that this tendency may have abated in the late 1970s.

The number of detailed studies of media coverage of controversy is too large to list in this chapter. Some of the areas that have been investigated include technological accidents such as Three Mile Island, Bhopal, and Chernobyl (Durant, 1992; Friedman, 1989; Wilkins, 1987); health issues such as smoking and AIDS (Kinsella, 1989; Klaidman, 1991); catastrophes and natural disasters (Walters, Wilkins, & Walters, 1989); and environmental issues (Hansen, 1993), including risk (Wilkins & Patterson, 1991).

The research on media and controversies has tended to focus either on particular communication problems (providing information to communities at risk from a particular public health problem) or on general communication theories (such as agenda setting). As with most research based on the simple diffusion model of science popularization, these studies generally conclude that not enough information was published, and that what was published was not provided in sufficient quantity or detail to have been useful. More usefully, they frequently identify the social groups involved in a controversy and show how structural relationships among these groups affect coverage. B. J. Cole (1975), for example, identified the importance for a newspaper of the social structure within a community affected by a controversy; Goodell (1980, 1986) and Nelkin (1987) stressed the importance of scientific culture and the ability of scientists to impose their values on journalists; and Wilkins and Patterson (1991) emphasize the importance of shifting political relationships in shaping the coverage of technological risks.

Taken together, these studies clearly imply that media coverage of controversies cannot be "improved" by better "dissemination" of scientific or technological information. Rather, media coverage is shaped by structural relationships within communities (including political relationships) as well as by the media's need to present "stories" that have "conflict" embedded in them. As Hansen (1993) has pointed out, one goal for future researchers must be to integrate science and media studies into broader theoretical contexts.

Scientists in the Press

One final area of traditional science journalism literature involves the enduring myth that scientists do not like to cooperate with the media (Goldstein, 1986). Virtually every systematic attempt to look at this question has shown that scientists are extremely willing to work with the press (Boltanski & Moldidier, 1970; Dibella, Ferri, & Padderud, 1991; Dunwoody & Scott, 1982). Though it may serve the traditional norms of science to declare that science should be communicated only among scientific peers, both scientific institutions and individual scientists have long recognized the value of communicating with the public (Burnham, 1987; Lewenstein, 1992b, 1992c). Goodell's (1977) study of "visible" scientists, though it reports comments that support the original myth, shows that major scientists have often found a role that allows them to speak out about science. A confounding issue is that the accessibility of some scientists can lead to their being used as sources in situations outside their own area of expertise—an issue raised by Goodell and explored in detail by Shepherd (1979, 1981) in his study of coverage of marijuana.

Complaints about scientists in the media usually come when scientists communicate with the public before their work has been fully assimilated by their scientific peers. Much of the conflict over the announcement of cold fusion, for example, was about the premature public announcement (Lewenstein, 1992a). This conflict is caused by the rhetorical commitment of many scientists to a standard definition of scientific communication as something separate from dissemination of information to the public (Garvey, 1979; Ziman, 1968). The frequency of the conflict, however, suggests that the traditional model does not accurately portray the process of science communication. Hilgartner (1990) has argued that the traditional model endures, however, because it gives scientists a rhetorical tool useful for discrediting unwanted interpretations of their work.

Models of Science Communication

The traditional model of science communication includes the idea that "popularization" is a "diffusion" process, in which scientific or technical information is "disseminated" to broad, uninformed publics. In the science and media literature, the model is most evident in several useful collections of articles on science journalism (Alberger & Carter, 1981; Friedman et al., 1986), although these volumes do not present any explicit statement about an underlying model of communication. However, in recent years, several strands of research have come to question the model.

One thread came from research that questioned all linear models of communication. Working in that tradition, Grunig (1980) proposed a "situational" theory of science communication, in which issues of accuracy, simplification, topic, risk information, audience knowledge, and so on could be seen as interacting, depending in part on the level of active information-seeking interest on the part of audience members. While Grunig's model did not question the authority of information developed by scientists, it did recognize the complexity of the contexts in which science communication takes place.

Another thread came from work in the sociology of scientific knowledge. Breaking the distinction between communication among scientists and popularization, Shinn and Whitley (1985) edited a collection of essays on "expository science." Drawing on case studies across two centuries on both sides of the Atlantic, these essays argued that all writing by scientists and about science could be understood rhetorically as attempts by the scientific community to marshall resources for its own purposes. Thus James Watson's exuberant and contentious autobiography, *The Double Helix* (1968), was seen as an argument intended to recruit students and other researchers to a new, more competitive, molecular-based style of biological research (Yoxen, 1985). The popularization of political economy in the nineteenth century was shown to be shaped by the desire to control working-class behavior through the dissemination of only selected ideas from political economy (Goldstrum, 1985). In this model, a common set of rhetorical tools can be used to describe all communication about science, no matter what source or what audience is involved.

The most complete elaboration of this model for dealing with the continuum of science communication is Hilgartner's (1990) analysis of the controversy over the link between diet and cancer. He showed how estimates about the degree to which diet contributes to cancer risk were recast and restated in various contexts; Hilgartner demonstrated that it is theoretically impossible to find a boundary between "science" and "popularization." Hilgartner argued that scientists do claim to distinguish between "science" and "popularization" because those claims can be used as a political resource; by retaining the ability to label some texts as "popularizations," scientists are able to discredit work with which they disagree. Fundamental to Hilgartner's analysis is the concept of a continuum of science communication occurring in various contexts.

Another thread to be woven into a new model is the interactive approach that has been developing in risk communication and health communication literatures for some years. This approach reminds us that communication is (at least) a two-way process, depending as much on the interests and concerns of the audience as on those of the scientists or others in positions of social authority. Drawing on this tradition, Logan (1991) argued that the dissemination model is flawed because it depends on a naive interpretation of social learning theory, in which single social variables affect whether members of the public acquire new information. A more realistic model, according to Logan, is one that recognizes the multiplicity of variables that affect the acquisition of knowledge. He proposed therefore a new model for understanding science and the media, which he called "secularization." Logan stressed the dialogue that takes place among the public, the media, and scientific sources such as the health care delivery system (in the case of news about public health issues), pointing out that dialogue involves give-and-take, not a one-way flow from the scientists to the public. For public communication of science to be successful, Logan argued, it must reject scientific authority and acknowledge the value of opinions, beliefs, and values held by the audience. In that sense it is secular; it rejects the almost-religious primacy of science. Wynne (1989b), in his analysis of Cumbrian sheep farming and Chernobyl, demonstrated how a model similar to Logan's works in practice.

Though the idea that science's role in the media can best be understood in terms of a continuum of science communication is clearly gaining ground, it has not yet fully integrated the disparate literatures that address popularization of science. Nonetheless, a model that recognizes both the interactive nature of communication and the multiple contexts in which science communication takes place offers the best hope for making sense out of issues in science and the media.

FIELDS RELATED TO SCIENCE AND THE MEDIA

A new model of science communication that stresses its interactive, contextual nature may be useful for integrating into the field much of the literature that is clearly relevant but that doesn't fit well into the traditional model of science as fixed, objective knowledge being disseminated to either "scientists" or "the public." A new model will be useful for establishing links between science and media literature, science and politics literature, traditional studies in sociology of science, and newer studies in the sociology of scientific knowledge. In the sections that follow, I will describe some of the literature that seems most relevant to science and media, along with some indication of how that literature might be integrated into a general field of science communication using a model like the one sketched above.

Instrumental Science Writing

One of the ancillary literatures most directly related to science and media studies is the "instrumental" literature—works directed to journalists, scientists, and others who must learn how to accomplish their goals of public communication of science and technology. A tremendous literature exists in such fields as health communication (e.g., Pettegrew & Logan, 1987) and agricultural extension (U.S. Department of Agriculture, 1979). Only a single textbook in general science journalism exists (Burkett, 1986), although many journalism textbooks now include a brief section on covering science, and several texts directed toward scientists have appeared (Gastel, 1983; Shortland & Gregory, 1991). In addition, many handbooks, workshop proceedings, and collections have been produced as journalists struggle with problems of explaining technological risk, dealing with complex numbers, and similar issues (e.g., Cohn, 1989; Environmental Health Center, 1990). At least two of these books deal explicitly with non-Western contexts (Amor, Icamina, & Laing, 1987; Friedman & Friedman, 1988). Although these works are often thoughtful and full of valuable practical information, they are almost always based on the simple diffusion model of science communication. The practical knowledge embodied in these books needs to be incorporated into any new model of science communication; at the same time, a more complex model may help suggest why some of the exhortations in these texts are so difficult to abide by.

Some researchers have combined studies of how members of a particular discipline deal with the media with suggestions about how to proceed. The most notable discipline in this regard is psychology (Goldstein, 1986; McCall & Stocking, 1982). This literature is interesting because it notes the essential need for communicating the findings of psychological research to the public—not just to generate support for psychology but because the research itself can be directly useful to members of the public. Thus the psychologists have begun to recognize that communicating with the public is not simply a dissemination of research results but can be construed as an integral component of the research process. A similar point has been made about seismology (Fiske, 1984; Gaddy & Tanjong, 1986; Meltsner, 1979). These findings clearly support the concept of a continuum of science communication, which is part of a more complex model that stresses contextual issues.

Commentaries on Public Understanding of Science

A second literature that is closely related to studies of science and the media consists primarily of commentary. Over the past 100 years or more, prominent scientists have produced a series of essays and commentaries on public understanding of science (Burnham, 1987; LaFollette, 1990a; McGucken, 1984; Werskey, 1978), and the tradition continues to this day (Bodmer et al., 1985; British Association, 1976; Conant, 1947; Goodfield, 1981; Leavis, 1963; Snow, 1959; Weaver, 1955). One strain of research linked to these commentaries, but based more on secondary analysis than primary declamation, deals with the influence of science on social culture (Holton, 1965, 1992; Holton & Blanpied, 1976; Shils, 1974). Other contributions to this genre have been made by conferences that were originally designed, in part, to provide an opportunity for scientists to express concerns about public understanding of science (Doorman, 1989; Evered & O'Connor, 1987). These works often use the media as evidence, without deeply exploring the ways in which media presentations of science are developed. Nonetheless, because both the commentaries and the analyses are often produced by scientists who are presenting their work in complex rhetorical contexts, these literatures again suggest the value of a continuum model of science communication.

What the Public Knows

Surveys of knowledge. Beginning in the 1950s, several studies in the United States looked at how specific groups—such as politicians or high school students—viewed science (Mead & Metraux, 1957). The most comprehensive surveys of the audience for science news were conducted late in the decade, in two studies bracketing the launch of Sputnik (Survey Research Center, 1958, 1959). They showed, in general, that members of the public wanted more science news. At the same time, the surveys showed the need to distinguish between various audiences; for example, women were more likely to read about medical topics, while men were more likely to read about space.

For nearly 30 years, no study attempted to replicate the 1950s' research, although secondary analysis and more focused surveys did allow pieces of the original studies to be confirmed (Pion & Lipsy, 1981; Wade & Schramm, 1969). In a 1981 review, Siwolop argued that a variety of surveys had all shown that science was in high demand, at least when it went by the label of "health and science" or "environment" or "energy" or "space exploration." Finally, in 1988 a new national survey was conducted based on the original survey instrument (National Science Board, 1991). Its results essentially confirmed the stability of the original findings.

The 1988 survey was conducted as part of an ongoing series of studies on public knowledge and attitudes about science conducted for the National Science Board's *Science Indicators* volumes. Those studies have been the basis of the widely reported interpretation that only 20% of the American public is "attentive" to science, and only 5% of the American public is "scientifically literate" (J. D. Miller, 1983a, 1987). In the late 1980s, the "science indicators" surveys were replicated in many countries (Durant, Evans, & Thomas, 1989, 1992; Einsiedel, 1992; National Science Board, 1989, 1991). The results reveal broad patterns of similarity in all the industrialized, developed countries. Topics that concern individuals in their daily lives are always better understood than more remote subjects; medicine is always better understood than basic physics. Although indices like the "science literacy" measure, which require respondents to answer correctly 7 of 10 questions about specific scientific facts, do allow researchers to divide the survey respondents into dichotomous categories, the raw data tend to show a more continuous distribution of knowledge.

Much of the literature on what the public knows about science has been interpreted to indicate that there is a single audience that enjoys science, while another, much larger audience poses a "problem" for those committed to improving public understanding, knowledge, or attitudes toward science (Dornan, 1990). As Grunig (1980) pointed out, this fits in well with general communication theories about the "knowledge gap" or the "communication effects gap," which suggest that, the more information an individual has, the more likely it is that an individual will gather more information.

However, both greater attention to the details of the studies listed above, as well as several narrower studies, suggest that it is a mistake to conceive of the audience in such monolithic terms. Grunig's (1980) "situational" model for science communication showed how the audience for technical information will change with different contexts. Using empirical evidence, many studies have shown that, when a scientific and technological issue has direct impact on a community, members will quickly and accurately acquire significant amounts of technically sophisticated information (see, e.g., Donohue, Tichenor, & Olien, 1973; Fitchen, Fessenden Raden, & Heath, 1987; Scherer & Yarbrough, 1988).

Another difficulty with the surveys of knowledge is that the literature is primarily American. With the exception of the "science indicators" series mentioned above, only a few studies have attempted systematic looks at other countries (see, e.g., Dulong & Ackermann, 1971).

Science literacy. Tied to concerns about the level of public knowledge of science has been an emerging concern with "science literacy." This literature usually decries the lack of public knowledge about particular scientific issues. It then sets out a series of definitions or categories delineating what the public should know. These categories are often supported by surveys showing the lack of knowledge of mathematics, physics, geography, history, or other topics. An early attempt to define science literacy was Shen (1975), who divided the topic into "practical" science literacy, "civic" science literacy, and "cultural" science literacy. Most subsequent efforts (such as Graubard, 1983; Laetsch, 1987; Thomas & Durant, 1987) have merely restated or extended Shin's categories. In the late 1980s, with the legitimacy given to "literacy" concerns in general, more prescriptive measures appeared (Hazen & Trefil, 1991).

While these surveys and definitions provide useful raw data and foils for argument, they are frequently skewed by the ideological biases of the authors. Moreover, these works are often deeply flawed by a simplistic commitment to scientific "facts," with little attention either to the processes of science or to the complex nature of science communication (H. H. Bauer, 1992; Levy-Leblond, 1992; Trachtman, 1981). Unlike most of the literature discussed in this chapter, they focus attention on the formal educational system rather than on the informal media exemplified by newspapers and magazines. Thus this literature represents a field that is important for understanding the audiences for science in the media while at the same time being as much primary evidence as secondary analysis.

What the public reads and sees. One area rarely investigated carefully is the content of public reading material. A few authors have made some attempt at such a survey (Kidd, 1988; Macdonald-Ross, 1987), and several commentaries (mostly essays accompanying collections of book reviews) have explored the area, but the field is still wide open. Many, many books have been published that attempt to popularize some aspect of science; as yet, we know little about either the images conveyed by these books or their reception by various audiences.

Almost equally missing is a systematic look at science on television. One set of studies, by Gerbner, Gross, Morgan, and Signorelli (1981), has included science in its surveys of what people see on television. But these studies have been conducted in the context of a particular communication theory that is still in debate, and the raw data for others to use has not been fully presented.

Taken together, all of the studies on what the public knows suggest that the question can only be intelligently answered when it is qualified: Which audience in what context looking for what information? Such a statement of the problem reinforces the importance of more fully articulating an interactive, contextual model of science communication.

Images

A relatively new area of research involves the images of science and technology conveyed by metaphors and other writing strategies. Although these studies have used a variety of methods, from highly quantitative content analysis through impressionistic reporting, they agree that images tend to appear in several clusters: "devil," "out of this world," "wizard/genius," and "war."

Two early studies were a survey of high school students by Mead and Metraux (1957) and a review of comic books and other "pop culture" items by Basalla (1976). Both identified the stereotype of white-coated, male scientists with extraordinary powers, often divorced from social responsibility. These tentative analyses have been expanded by LaFollette (1990a) and Nelkin (1987). In a content analysis of 11 major magazines in the first half of the twentieth century, LaFollette identified several major themes, especially a "myth of scientific differentness" that stressed the "extraordinary intelligence, persistence, foresight, [and] modesty" of scientists. "Piercing eyesight" reflected the superb capabilities of insight attributed to scientists. LaFollette found the stereotypes of scientists divided into "magician or wizard," "expert," and "creator/destroyer." Nelkin (1987) used a more qualitative analysis when looking at newspaper and newsmagazine coverage of science and technology in the last generation; she emphasized the number of "pioneering," "frontier," and military images to be found in writings about science. More detailed studies of particular sciences and technologies include Boyer (1985), Marvin (1987), Knight (1986), S. Douglas (1987), Weart (1988), and Ziporyn (1988).

A somewhat different approach to examining images of science has come from the sociology of scientific knowledge, where the concern has been to show how scientific certainty depends on contingent, rather than objective, factors. Collins (1987, 1988) has shown that science on television is usually interpreted as displays of certainty, even when the displays themselves are of uncertain outcomes with multiple possible interpretations. Silverstone (1985) explored the constraints and contingencies operating on a television documentary program, showing how uncertain outcomes end up as displays of certainty. Other works in this tradition include Murrell (1987), Jacobi (1985), and Jacobi and Schiele (1989).

Few of these studies of images have attempted to distinguish between science in the mass media and science in more specialized media or contexts. Again, a contextual model of science communication may help sort out the implications of these studies by stressing the changes in images in different contexts.

Science Outside the News

Scientific images are presented to the public in many contexts other than newspapers and magazines, but these contexts have not been well studied. A few works have appeared on museums (Bud, 1988; Hein, 1990; Macdonald & Silverstone, 1992; Samson & Schiele, 1989; Serrell, 1990). In addition, a literature has begun to develop on science in movies (Dubeck, Mosher, & Boss, 1988; Lambourne, Shallis, & Shortland, 1990; Shortland, 1988, 1989) with special attention being paid to the image of psychiatrists (Sharf, 1986; Shortland, 1987; Schneider, 1987) and to "mad" scientists in science fiction films (Tudor, 1989). Gerbner and his colleagues (1981) have looked at images of scientists on television, and Turow (1989) has looked at doctors on television. A few works have looked at science fiction from the perspective of science and technology studies (Friedman, 1987). As yet, the link between these studies and models of science communication based on science journalism are unclear; it seems likely, however, that a contextual model that stresses the links between different outlets, rather than sharply delineating "science," "news," and "science fiction" (as traditional models do), will be useful.

A related area clearly relevant to studies of science and the media is research into the audiences for information about science. Indeed, some recent academic commentaries have suggested that research into *production* of science popularization should be abandoned in favor of research into audiences (Lewenstein, 1992d).

Historical Contexts

One of the most serious failings of contemporary analysts and commentators on science and the media is the failure to understand the broad range of activities and contexts in which the public gains access to scientific knowledge. Unfortunately, the scholarly literature that sheds light on this problem is varied and scattered. Broad historical surveys include Burnham's (1987) history of popular science in America and Thackray, Sturchio, Bud, and Carroll's (1985) discussion of chemistry in American mass culture. Throughout his books on physics and eugenics, Kevles (1978, 1985) discusses public presentations of these sciences.

A few patterns do recur. First, much of the literature deals with ways in which particular national or cultural contexts shape public communication of science and technology. Thus Rydell (1984, 1985) looks at how American culture shaped the presentation of science at world's fairs, while Pauly (1979) examines *National Geographic* magazine for the same purpose. Caudill, in his examination of Darwinism in the press (1987, 1989a, 1989b), is concerned less with how science was presented than with the implications of that presentation for cultural discourse. Cotkin (1984), in an examination of socialist uses of science at the turn of the twentieth century, shows how ideals of scientism meshed with the need to find immutable laws of history. Rhees (1987), Tobey (1971), and Kuznick (1987) are all concerned with how science was presented in the years between the world wars as well as the relationship with broader aspects of both American and scientific culture; Lewenstein (1992c) looks at similar concerns in the post-World War II period. Each author finds that scientists used public communication to build resources that would enhance the cultural and political authority of science. In a recent European historical context, Fayard (1988) examines the development of a new, more critical mode of popularization (what he calls "public science communication") as a result of changes in the broad social structure of France associated with the 1960s. Many other works look at fairly limited episodes or instances of public communication of science, providing descriptive information or displaying the fine detail of how particular contexts shaped presentations.

Second, some of the literature (especially those works inspired by contemporary studies in the sociology of scientific knowledge) focus attention on the ways in which public displays of science influence debates and arguments within science itself. These works contribute substantially to the idea that science communication can be considered a continuum (or even a web of interconnected discourses), for they suggest the multiple interactions between various presentations of science and other human activities. Cooter's book on phrenology (1984), for example, shows how popularizations could be used to create support for a science that was increasingly suspect among some scientific quarters. Shapin (1984) shows the importance of public displays of science for mobilizing belief in particular experimental results. Many of the essays in Shinn and Whitley (1985) also demonstrate the power of public presentations of science for reshaping a range of debates within and about the scientific community, as do works by Shapin (1974), Shapin and Barnes (1977), Sheets-Pyenson (1981), Schaffer (1983), and Secord (1985).

Direct Influence of Media on Science

Despite the implications of recent historical and sociological research, very few studies have looked at instances in which science in the media has directly influenced the activities of scientists. One series of independent studies has shown the importance of newspapers for communication among scientists, especially physicians (O'Keefe, 1970; Phillips, Kanter, Bednarczyk, & Tastad, 1991; Shaw & Van Nevel, 1967). A provocative article by Clemens (1986) claims that the media helped direct the debate on asteroids and the extinction of dinosaurs (see also Raup, 1986). Kwa (1987) explores the role of media presentations of nature in the relationship between the scientific discipline of ecology and science policy decisions, and Ashmore, Mulkay, and Pinch (1989) look at the role of media presentations on the development of health economics. These works all argue that media presentations are not merely part of the social context in which science exists but instead are the direct causes of some aspects of scientific work. Such conclusions obviously support a model of science communication more complex and interactive than the traditional, unidirectional model.

Literary and Rhetorical Analyses of Science

Two rapidly growing areas of study look in different ways at science, technology, and literature. The "science and literature" field, which is well served by a comprehensive bibliography compiled by Johnson, Schatzberg, and Waite (1987), looks at the interactions of scientific ideas with literary texts. Typical works, such as Tichi (1987), explore the ways in which authors of fictional works have incorporated or responded to scientific ideas as guiding metaphors for social or intellectual activity. While works in this tradition are not often concerned with direct evidence of how literature affects its readers, they operate on the assumption that such effects are present.

The newer field of science and rhetoric is more directly concerned with how rhetorical effects are produced by scientific texts. The impetus for these works comes from new ideas in rhetorical studies generally. Although the focus of much of the work has been on how primary scientific literature achieves its rhetorical goals, the methods of these researchers are clearly applicable to texts of popularization and some authors have started that application process (Bazerman, 1988; Gross, 1990; Myers, 1990a; Prelli, 1989). In addition, some of the insights from the literature on science and rhetoric have been applied to instrumental studies of scientific writing, including those that bridge the gap between writing *in* science and writing *about* science (R. L. Anderson, 1970; Rowan, 1988, 1992).

CONCLUSIONS

How are we to understand the disparate literatures that deal with science and the media? In the past few years, a new and fruitful area of research has emerged that treats science in the media as merely a component in the overall *communication* of science. Heavily influenced by developments in literary theory, deconstruction, the sociology of scientific knowledge, and other elements of postmodern academic interest, these studies have shown how scientific rhetoric and forms of discourse serve to establish the authority of science in complex social contexts.

At the same time, a vibrant and somewhat disordered and diverse literature has developed in the study of science journalism and the popularization of science. Some of these studies are clearly influenced by the same sorts of intellectual questions as the literature on rhetoric in science. But much more of the literature comes from another tradition, that of concern with the practical aspects of popularizing science.

To bring these literatures together, we must ask: For what purpose are we trying to bring them together? To understand the structure of science? To understand relations between science and society? To improve the public understanding of science? To improve the technical flow of information? To improve science literacy?

Simply to ask many of these questions is to take a position on the structure of science. The best way to resolve the problem is to treat science and the media as a subset of issues in a more general model that describes science communication as an interactive, multidirectional activity occurring in many contexts. The literature dealing with science and the media makes clear the difficulty of dividing up what "scientists" do into neat little packages of "science" and "nonscience." The goal for future researchers must be to integrate studies of science and the media into our broader understanding of the contexts of science communication—and the role of science communication in the relationships between science and the societies in which it is embedded.

APPENDIX: PREVIOUS BIBLIOGRAPHIES AND REVIEWS

No previous bibliographies have explored the topic of science and the media from the perspective of science and technology studies, although entries to relevant topics can be found in the bibliographies published in *Isis, Technology & Culture*, and elsewhere. A recent essay by Shapin (1990) begins to suggest what a bibliographic review might look like from the perspective of science and technology studies but admits to finding the field still thin and unplowed. A new journal in the field, *Public Understanding of Science*, launched in 1992, hopes to eventually publish regular bibliographic compilations.

However, several authors in communication studies have made attempts to collate the literature. The most current bibliography was prepared by Dunwoody and Long (1991); with more than 200 annotated entries, it updates bibliographies previously published by Dunwoody with several coauthors during the 1980s and is expected to be updated approximately annually. Focusing on traditional journalism issues (audiences, accuracy, readability, and the interaction of sources and journalists), these bibliographies also take excursions into coverage of social sciences and the teaching of science journalism. Other detailed bibliographic lists in this tradition include Bowes, Stamm, Jackson, and Moore (1978), which looks largely at technical issues such as readability and accuracy, and Guillierie and Schoenfeld (1979), which is directed specifically at environmental communication.

Several bibliographic essays in communication have explored topics related to science and the media. Cronholm and Sandell (1981) examine much of the literature identified in the Dunwoody bibliographies, providing general summaries. Grunig (1980) uses his survey to propose a model of science communication that stresses the interaction among the various actors involved in science communication as well as the importance of recognizing the situational context of science communication behavior. Like so many other reports, however, Grunig acknowledges that much of what he proposes must remain speculative in the absence of more coherent and substantive research studies.

17

Public Understanding of Science

BRIAN WYNNE

PUBLIC understanding of science (PUS) is a wide and ill-defined area involving several different disciplinary perspectives. Although no coherent paradigm has gained sovereignty over this area, a dominant political paradigm exists that shapes a particular framing of the "the PUS problem." Vague though powerful concerns about "public understanding of science" have been woven into ideological programs of various kinds ever since science entered public discourse. A common thread has been anxiety among social elites about maintaining social control via public assimilation of "the natural order" as revealed by science. On a wider front, this anxiety is reflected in repeated laments about the so-called failure of traditional Third World societies to absorb Western "scientific" programs and their founding culture of rationality.

Despite the long career of these discourses, it is only since the mid-1980s that the PUS issue took on the trappings of institutionalization. Apart from large-scale public attitude surveys, which began in the 1950s, systematic research on PUS also dates only from the 1980s. This chapter reviews the main research perspectives that have been developed and offers comparative observations about their different commitments and implications. A further but necessarily attenuated aim is to outline the relationships between PUS research and several other areas with which it has potentially fruitful connections.

Claims about "the public" or "society" have long been embedded within scientific discourses, and there is a similar history of recurrent concern about public acceptance of "scientific" authority (Layton, 1973). Over the last decade or so, research in history of science has systematically exposed how the tacit rhetorical constructions of the social order help constitute scientific knowledge (Golinski, 1992; Shapin & Schaffer, 1985) and how this knowledge helps shape the social order, in processes of mutual construction of science and society. The modern PUS issue should be seen in this historical context.

Until recently, the dominant agenda of PUS research (and practice) was shaped by problematizing publics, and their cognitive processes and capabilities, thereby implying scientific knowledges, practices, and institutions to be unproblematic. Indeed, PUS research that problematized science was misrepresented as raising only the limited problem of scientists' lack of interest or skill in public communication. By contrast, critical research approaches informed by the sociology of scientific knowledge (SSK)—what I call here the constructivist perspective—have attempted to investigate how people experience and define "science" in social life, and how particular scientific constructions incorporate tacit, closed models of social relationships that are or should be open to negotiation. In other words, scientific knowledge is seen as encoding taken-for-granted norms, commitments, and assumptions that, when deployed in public, inevitably take on a social-prescriptive role. Thus dominant internal criteria of "valid knowledge" or "good science" may be legitimately open to question when science is used in public arenas.

In thus problematizing science as well as publics, the constructivist strand of PUS research raises the issue of reflexivity, but less introspectively than is sometimes the case. Moreover, it offers a richer repertoire of ideas about the different potential forms of constructive public engagement with science. Before surveying the main research paradigms for PUS, I review the different ways in which the issue has been defined and bounded.

Agendas, Boundaries, and Confusions

In many dominant formulations (e.g., Royal Society, 1985), PUS is automatically equated with public appreciation and support of science, and with the public's "correct" understanding and use of "technical" knowledge and advice. Thus, when publics resist or ignore a program advanced in the name of science, the cause is assumed to be their misunderstanding of the science. The PUS research agenda is thus confined to measuring, explaining, and finding remedies for apparent shortfalls of "correct understanding and use" as if this were free of framing commitments that have social implications.

The field of risk perception research, for example, was defined by the assumption that the public opposed technologies like nuclear power because they misunderstood the "real" risks as known to science; this view has resisted many substantive critiques (Otway, 1992; Wynne, 1992e).

Other confusions of the meaning of PUS flow from frequent neglect of the distinctions between public appreciation of, interest in, and understanding of science. There are, for instance, important differences of meaning between public "understanding" of science-in-general and of science-in-particular (Michael, 1992). Understanding may mean the ability to use technical knowledge effectively, but inability to use such knowledge does not necessarily mean lack of understanding. Understanding science may also mean understanding its methods rather than its specific content (Collins & Pinch, 1993; Collins & Shapin, 1989; Wynne & Millar, 1988), and it may mean understanding its institutional characteristics, its forms of patronage and control, and its social implications (Wynne, 1980, 1991a). As is shown later, there are important interactions among these three different dimensions.

A most important, unrecognized factor is the role of different tacit models of social agency underlying encounters between science and public groups. For example, even a technically literate person may reject or ignore scientific information as useless in the absence of the necessary social opportunity, power, or resources to use it; yet scientists may assume, in a different model of social agency, that such public "neglect" reflects technical ignorance or naïveté. The implict structures of social agency will also affect people's interest in available knowledge, hence their "literacy" in the first place. Thus an indigenous social parameter—the tacitly perceived usefulness or relevance of scientific knowledge in the layperson's own social context—directly shapes public uptake of science, and hence the public's observed "understanding" of science. The separation of the cognitive and social dimensions that inform dominant approaches to PUS must accordingly be seen as an artifact.

The scope and meaning of PUS also varies importantly across political cultures. In the more paternalistic European political cultures, chronic anxiety about institutional authority is projected onto the public, whose apathy, ambivalence, or hostility is attributed to misunderstanding of "rational" principles. As Layton, Davy, and Jenkins (1986) describe the quintessential U.K. situation: "Public involvement in science, in so far as it occurred, meant socialization into the perceptions of scientists and their ways of interpreting the natural world" (p. 34). Here, they could have omitted the qualifying term *natural*. In the more populist U.S. political culture, vigorous and active public advocacy in all sorts of scientific and technical issues mitigates this tendency, even though, as Lewenstein (1992c) has noted, the paternalist rhetorical definition has also been influential there.

Comparative analyses of science in public policy (Brickmann, Jasanoff, & Ilgen, 1985; Gillespie, Eva, & Johnson, 1979; Jasanoff, 1986) also show the same distinctions. The less deferential U.S. culture encourages more diverse, active attempts by various lay groups to affect the course of technical decision making; indeed, this greater participatory access is enshrined in U.S. legal principles and institutional mechanisms. There has been no comparative research that attempts to explore whether this generally greater sense of political agency in the United States influences the public uptake and understanding of science or scientists' views about public understanding. Beneath the cross-cultural differences, however, lie fundamentally similar issues of the legitimation of science not only as instrumental knowledge but as a corresponding universalist culture.

The identifiable research approaches to PUS fall into three main classes:

- large-scale quantitative surveys of selected samples of "the public," which have been used to elicit attitudes toward science as well as to measure levels of public scientific literacy or understanding of science
- cognitive psychology, or the reconstruction of the "mental models" that laypeople appear to have of the processes that are the object of scientific knowledge
- qualitative field research observing public contextualization of scientific expertise, exploring how people in different social contexts experience and construct its meaning

The "social representations" paradigm (Moscovici, 1984) falls somewhere between the latter two. These traditions of research, as I outline below, also parallel and overlap with the fields of risk perception research and "ethnoscience" research from cognitive anthropology.

In this chapter I adopt the view, informed by SSK, that a proper approach to PUS has to problematize what is meant not only by "science" but also by "understanding"; in other words, scientific meaning cannot be taken for granted as if deterministically provided by nature or some other privileged authority. This also automatically problematizes the "public" in well beyond the obvious sense that there are countless "publics" of science. Recent research on public responses to scientific expertise suggests that "understanding" is a function, inter alia, of social identification with scientific institutions, and these processes of identification or alienation are multiplex, often fractured, and chronically open to redefinition (Michael, 1992; Wynne, 1992a).

This perspective highlights the culture of control and standardization that characterizes scientific knowledge and that engenders ambivalent responses from those who encounter it in public. By constructing the public as ignorant, when that public may in its own idiom be expressing legitimate concern or dissent, scientific institutions inadvertently encourage yet more public ambivalence or alienation. Thus the whole rationalist temper of modern society may undermine itself by the nonreflexivity of science about its own constructions of "the public" and the institutional factors that give rise to these constructions (Ezrahi, 1991; Wynne, 1987). In the concluding section I attempt to place the PUS issue in a wider context of social and cultural change as this has been analyzed in sociological debates about late modernity and its transformations (Beck, 1992; Lash & Friedman, 1992; Lyotard, 1984; Ross, 1991).

SURVEY RESEARCH: ATTITUDES, LITERACY, AND UNDERSTANDING

Using large-scale surveys for research on public understanding of science is associated with the longer standing quantitative sampling of public attitudes to science, which goes back to at least 1972, in the "science indicators" program of the U.S. National Science Foundation (NSF). This built upon an earlier initiative by the U.S. National Science Writers' Association (NSWA) in 1957 (Withey, 1959). Jon Miller, author of the NSF surveys, argues that "it is likely that few individuals actually possess attitudes to science" (NSF, 1991b, p. 175) because these require attentiveness to science and that is a minority pursuit. Miller's surveys thus attempt to gather data on levels of attentiveness to science as well as understanding of science and attitudes toward it.

Attempts to measure and compare levels of public understanding of science, over time, between countries, or between subpopulations obviously require a stable system of measurement. In its most precise form, this has been developed into regular measurement and comparison of national levels of "scientific literacy" (Miller, 1983b, 1991). Since the late 1980s, the U.S. NSF "science indicators" measures of public attitudes and understanding of science have also included a comparison with parallel coordinated Japanese and European surveys. These methods, their research protocols, and the outputs have become the established international framework for measuring "public understanding of science."

In Miller's work, "attentiveness" to science is defined by an index that combines self-reported interest in scientific issues, self-reported level of knowledge, and regular use of different information sources. On this basis, Miller (1991) was able to conclude, for example, that "about 8% of US adults are attentive to new scientific discoveries," compared, for example, with about 20% "attentiveness" to environmental pollution. What respondents themselves meant by saying they were "interested in" science was not addressed.

Attempts to measure public understanding of science have asked closed questions about substantive issues such as atomic structure or differences between viruses and bacteria and then scored responses against "correct" answers. These have also been coordinated in an international comparative study since 1988. Some of the results have occasioned surprise, even shock, concerning the apparently low levels of understanding they exposed. Thus the 1991 U.S. *Science Indicators* report (NSF, 1991b) shows that only 6% of the population could give a "scientifically correct" answer to a question about the causes of acid rain. Even among the "attentive public," 76% could not give a more specific answer than "pollution." Indeed, the so-called attentive public scored no better in this survey than the general public in its understanding of scientific aspects of environmental issues. It was concluded that attentiveness and level of understanding as defined and measured by the surveys were not well correlated. These results have usually been simplistically interpreted, for example, in national league tables of scientific literacy, sometimes encouraged by the authors of such studies themselves. For example, Durant, Miller, and Tchernia (1991), in referring to their own work, claim that "studies carried out by Miller in the US and in the UK in 1988 show that over 90% of the population is to be considered scientifically illiterate" (p. 1).

These interpretations, however, are far from straightforward. To take acid rain, for example, the survey results ignore the fact that scientists themselves are in considerable uncertainty as to the precise causes of acid rain; thus the "correct" answer against which to measure public responses is not simple. For other questions, such as whether the earth goes round the sun, or the sun round the earth, the "obvious" correct answer is the former. But both in fact are true; the "obvious" true answer is a simplification. On yet other questions, such as whether hot air rises, or whether light travels faster than sound, answers are less ambiguous. Even so, to construct a bald measure of "scientific literacy" out of context raises serious questions about what such measures actually mean and about the inadvertently prescriptive and political role of such research.

The U.S. NSWA 1957 survey (Withey, 1959) first attempted to measure public understanding of scientific method; respondents were asked to describe, in their own words, what it means to study something scientifically. This open question has been retained in subsequent surveys (Miller, 1991; NSF, 1991b). Yet responses have been evaluated for "correctness" against a highly Popperian model of science. As M. Bauer (1992) put it,

> [This method of analyzing the open question] arbitrarily takes a Popperian view of science as the baseline . . . [it] does not measure people's understanding of science, but the diffusion of a certain notion of science among the public. . . . The-

ory, deduction and falsification of data by experimental test are the answers that rate high on this coding frame. If we do not want to know to what extent the public is Popperian, we have to take a more open approach.

The same unquestioning approach was built into the NSF "science indicators" program. Miller (1991) used it as part of his multicomponent index of scientific literacy. Other surveys, however, have indicated that, when asked more concrete and closed questions, for example, about statistical tests or clinical trials, people scored much more highly in questions about scientific process. Durant, Evans, and Thomas (1989, 1992) expressed doubt about the reliability of the open question when they found that only 14% of a sample of 2,000 respondents referred to experiment in reply to the open question; yet 56% did so in reply to a concrete closed question about investigating a medical issue.

In further pursuit of the process dimension, Durant and colleagues split a survey sample (for EuroBarometer) randomly between a question about the role of experiment in testing new drugs and one about the role of experiment in testing the materials for a machine. In each case the format was identical, asking the respondent to choose among "methods" involving (a) reference to patient or machinist opinion, (b) reference to existing theory, (c) experimental test, or (d) don't know. The distribution of responses was quite different for the two cases. In the medical case, experiment was the majority answer, whereas in the metallurgical case, reference to existing theoretical knowledge took over. The authors interpreted this to mean that medical research practices are much more familiar to the public and that clinical trials as experimental method act as a general exemplar for the public understanding of scientific method. F. Price (personal communication, 1992) has questioned this on the grounds that it fails to distinguish public familiarity with clinical medicine from familiarity with medical research.

This survey went some way toward convergence with the insights of qualitative research. Durant and colleagues concluded that it is probably misleading to look for context-independent notions of scientific method, a conclusion already resoundingly endorsed by the qualitative PUS studies (see below).

Yet even this study highlights unacknowledged problems in survey approaches. There are other possible interpretations of the different patterns of response than Durant et al.'s conclusion that medical trials are more familiar to the public than methods in other areas of science. For example, the novelty of the item in question, in one case a new drug, in the other case a machine, might affect judgments about whether existing knowledge is worth drawing upon. So too might different perceptions of the error—costs in each case.

Also, Durant et al. postulate that metallurgical theory may be regarded by the public as more "advanced" than "medical theory," but then they reject this interpretation as implausible. Yet metallurgical knowledge may be seen as more established and mundane than pharmaceutical knowledge of new drugs, which appears to be at the cutting edge of uncertainty in biomedical research. This may be good reason for people to choose "experiment" for drugs, and "existing knowledge" for the metallurgical case. These variant public understandings are essentially invisible to the survey method.

The point here is not that these alternative interpretations are necessarily superior to those of Durant et al. Rather, it is that the survey approach is intrinsically unable to control for such variables as what people mean by reference to "existing knowledge," "theory," or "experiment." Even an identical research format cannot show whether and how variable respondent perceptions and judgments of the kind indicated above affect the data gathered. These are just the kind of informal contextual factors for which qualitative methods are necessary.

Similar basic difficulties are encountered in survey attempts to ask people to define what is meant by science or to rank the scientific status of various disciplines, including astrology (Durant et al., 1989). In aggregating a field like "medicine," which encompasses meanings as diverse as biomedical research, prenatal ultrasound monitoring, or a general practitioner's diagnosis, it becomes impossible to control for what the respondent means by, for example, "medicine," "psychology," or "scientific."

Miller (1983a, 1991) tested perceptions of the scientific status of astrology as part of his measurement of people's scientific literacy. If they classified astrology as scientific, they were marked as scientifically illiterate. Yet Bauer and Durant (1992) found no particular correlation between a person's expressed belief in astrology and other accepted measures of their understanding of science.

Several survey approaches have attempted to examine public understanding of scientific institutions, as a dimension beyond scientific contents and processes. Miller's (1991) attempts have largely focused on the relationships between science and technology and the effects of scientific research. M. Bauer (1992) has tried to examine understandings of the effects of science. Again, these surveys attempt to measure public understanding by evaluation against prescribed baseline "correct" models, which are themselves tendentious. Neither of the above models does even approximate justice to the complexity of the institutional dimensions of science, such as the perceived accessibility, control or ownership, and wider accountability of science in diverse domains.

The U.S. NSF initiatives, and the U.S.-U.K. and European parallel coordinated surveys, have stimulated a rash of attempts to survey public attitudes to science in many countries. Some of these are extensive; for example, one study in India involved 16,000 respondents (Raza, Dutt, Singh, & Wahid, 1991); another in China involved over 4,000 (Zhongliang & Jiansheng, 1991). The latter was designed following the Miller-Durant, U.S.-U.K. collaborative surveys. Other Indian studies have focused more upon the institutional and cultural dimensions of science (Raina, Raza, Dutt, & Singh, 1992; Vasantha, 1992).

Quantitative surveys of public understanding of science or public attitudes in relation to specific technological sectors like energy, the chemicals industry, or biotechnology are widespread (e.g., Cantley, 1988; Couchman & Fink-Jensen, 1990; Marlier, 1992; Martin & Tait, 1992; Office of Technology Assessment, 1987; Otway & Gow, 1990; RSGB, 1988). Most of these surveys combine questions on attitudes—for example, about the risks and needs for control or the justifiability of further research—with questions about understanding. It is usually uncritically assumed that "better public information" will lead to greater "understanding" and that this means greater acceptance; but the competing kinds of information or "understanding" that might be in play are rarely discussed. Again, many of these surveys do not control for what the respondent may mean by various core concepts, for example, by "further research" in eliciting whether she believes that further research in genetic engineering is worthwhile.

The EuroBarometer survey (Marlier, 1992), the most extensive and elaborate quantitative study of public "opinions" on biotechnology, found very high percentages (circa 30%) of "don't knows" even though the questions were not, on the whole, testing knowledge. It also found that the most trusted sources of information in this field were consumer and environmental organizations; only in Denmark were government agencies trusted nearly as much as these NGOs.

Martin and Tait (1992) found complex relationships between the "understanding" variable and attitudes. Whereas most surveys assume knowledge or understanding to correlate positively with attitude, their study hinted at the many contextual reasons for gaining understanding or for maintaining ignorance (see below). Those surveys that differentiated between areas of biotechnology also found that people were able to treat the different risks, benefits, and other implications separately, and not according to some undifferentiated composite notion of "biotechnology." The same point has been noted about public understanding of "energy" (e.g., Hedges, 1991; Kempton, 1987; Kempton & Montgomery, 1982).

In summary, large-scale surveys of public attitudes toward and understandings of science inevitably build in certain normative assumptions about the public, about what is meant by science and scientific knowledge, and about understanding. They may often therefore reinforce the syndrome noted earlier, in which only the public, and not science or scientific culture and institutions, are problematized in the PUS issue. Such surveys take the respondent out of social context and are intrinsically unable to examine or control analytically for the potentially variable, socially rooted meanings that key terms have for social actors. The survey method by its nature decontextualizes knowledge and understanding and imposes the assumption that their meaning exists independently of human subjects interacting socially. Evidence of internal coherence among survey data is not itself evidence of wider validity—only of internal consistency. Too often the latter is mistaken for the former.

MENTAL MODELS AND COGNITIVE APPROACHES

A significant volume of work in public understanding of technical knowledge has used the concept of mental models. These can be defined as simplified models of the world that organize new information into recognizable patterns; help generate inferences, causal connections, and predictions; and solve problems. They are thought to structure systematically the cognitions of laypeople about most areas of experience, including nature and technology, but major questions have been raised about their stability and their independence of context.

Mental models have found analytical use in a variety of disciplines ranging from cognitive psychology, artificial intelligence, and social psychology to sociology and anthropology. In the work of cognitive sociologists such as Cicourel (1974), mental models are seen as negotiated constructs that serve to reify and thus stabilize particular social relations in "natural" categories—such as "juvenile delinquent" or "epileptic." Other approaches also treat mental models as a medium of social interaction, thereby stressing their fluid, contingent, and shared character—their structure is a function of the underlying social relationships. In many cases, of course, such social relations are themselves sufficiently stabilized to render the corresponding natural categories and concepts more "hard" and stable.

Mental models have a similar explanatory role in both cognitive and social psychology. In these disciplines, unlike in sociology and anthropology, the socially negotiated origins and roles of mental models are not examined—

once identified, their existence is taken for granted. They operate here more as independent explanatory variables in the reconstruction of why people think the way they do. Thus mental models are conceived as resources that allow people to (try to) predict and control the domains in which they are engaged. In social psychology, mental constructs such as prototypes, schemas, analogies, and metaphors represent social and cultural contents such as racial stereotypes and relations, social class characteristics, cultural typologies, or "natural" gender relations. One version of such "social" models that has gained particular influence is the "social representations" theory of Moscovici (1984).

Cognitive psychology applies the same analysis to individual constructs of a nonsocial kind, for example, about physical and technological phenomena (Collins & Gentner, 1987; Gentner & Stevens, 1983). The basic assumption here is that, for reasoning about new situations, people use familiar analogies that appear salient. These are often physical models because that is most often the domain of stable and controlled experience, especially in technological society (Holland & Quinn, 1987).

Collins and Gentner (1987) describe the mental models approach thus: "A major way in which people reason about unfamiliar domains is through analogical mappings. They use analogies to map the set of transition rules from a known domain (the base) to a new domain (the target), thereby constructing a mental model which can generate inferences in the target domain" (p. 247).

Many such examples of lay mental models have now been analyzed, among them navigation (Hutchins, 1983), electricity (Gentner & Gentner, 1983), home thermostats (Kempton, 1987), evaporation (Collins & Gentner, 1987), and global warming (Kempton, 1991). Pointing to the assumptions underlying much of this work, M. McCloskey (1983) termed lay models of motion "naive theory," though Kempton (1987, p. 223), for example, explicitly adopts a less prejudicial view, analyzing where and how "folk theory earns its keep in everyday use."

Kempton's study of lay models of home heating identified, by inference from interview data, two different implicit theories in play. One was a feedback model, in which the thermostat was thought to measure temperature at a relevant place and to turn the furnace on and off to maintain the set temperature. The other was a valve model, in which the thermostat was believed to control the amount of heat flowing, so that a higher setting allowed a higher flow, like the familiar gas pedal on a car.

Kempton notes that according to expert knowledge the feedback model is correct and the valve model is wrong. To obtain more effective energy management in homes, it might be assumed, the "valve theorists" should be educated into the correct model. However, Kempton (1987) observes:

> Both folk theories simplify and distort as compared to a full physical description, each causes its own types of operational errors and inefficiencies, and each has certain advantages. . . . Heating engineers are fairly comfortable with the folk theory [of feedback]—they consider it simplified but essentially correct . . . however, their evaluation of correctness may be based on irrelevant criteria. (p. 224)

In other words, the predictive validity of the two competing theories or mental models in real homes depends upon their supplementation with further knowledge. It is not at all simple to predict their relative validity in use, even though one is clearly more "correct" according to science. For example, the feedback model would predict that turning up the thermostat is unnecessary in colder weather. The valve model would not. In practice, the feedback model leads to the wrong conclusion because it does not take account of infiltration and distribution asymmetries in marginal rooms, whereas the valve model leads to the right conclusion without requiring these supplementary models.

How should these competing understandings be evaluated? The valve theory, though "technically" incorrect, gives at least as many correct predictions and practical responses as the feedback theory. To give guaranteed correct responses, the "correct" model would have to be supplemented by several extra models and theories that would so complicate existence for the typical householder as to become unmanageable. So if the criterion of validity is to be "does it work?" then it depends upon what is agreed as the context for "working": a typical householder's life, involving many competing issues and demands, or a life assumed to be akin to that of a specialist, with only that problem to attend to or "optimize." A much larger issue is exposed here.

A more general way of stating the same issue is that theoretical knowledge, handed down from science as the "correct" knowledge against which to measure public understanding, invariably has embedded within it tacit assumptions about the *conditions* under which the theoretical models are valid and useful. Whether these conditions actually prevail in ordinary-user situations is an open question, but one that is usually overlooked. Thus the extra knowledges (and practices), often specific to the situation, that are necessary to make the theoretically correct knowledge work in practice are typically unrecognized and undervalued. This point is developed in the next section. In picking up these points from the mental models perspective, Kempton's functionalist approach connects with the wider anthropological debate over the cultural universalism or localism of scientific knowledge, in that beliefs about nature—indigenous "mental models"—that are incompatible with science can nevertheless be defined as valid *in their context* because they may help maintain important social, technical, and ecological practices (Douglas, 1966; Horton & Finnegan, 1973).

Collins and Gentner (1987) provide another perspective on mental models and their relationship to PUS research when they note that most lay reasoning necessarily involves combinations of several models, because situations are typically more complex than a single model can handle. The models may be combined in ad hoc fashion to reason about complex processes; these combinations may then contain inconsistencies that have to be managed in some way. Collins and Gentner implied that inconsistent and ad hoc uses of models were a problem with lay reasoning or "public understanding of science" itself, but their observation is a starting point for identifying three more general issues about the mental models approach and its inherent limitations.

First, the founding notion of "domain" is problematic. "Home energy management" is taken as a coherent domain for which lay mental models should exist; but, as Kempton (1987) found, this is too simple and unrefined to correspond with the practical concepts of energy that householders use in context. "Energy" is multidimensional and comprises many separate, cross-cutting activities, relationships, and agendas, such as cooking and diet, family lifestyle, "do-it-yourself" work, and overall money management. "Insulation" likewise is differentially contextualized into more detailed, practically meaningful terms (Hedges, 1991). It is therefore not surprising to find multiple models in use in "the same" domain as defined by the researcher, because to the lay respondent they may not be in the same domain but in several different if intersecting domains. What might look like "inconsistency" (and implied public incapacity) may thus reflect a more complex social existence than the mental models approach recognizes.

A related issue concerns the assumed stability of mental models and their social roles. The very methods of mental models research implicitly assume stability, because they are based upon eliciting such models from one-off structured interviews. Thus there is no follow-up opportunity to test whether or under what conditions the inferred models might or might not continue from one situation to others in the life of the respondent. In addition to Collins and Gentner's observations about ad hoc and "inconsistent" use, DiSessa (1985) and Kempton (1987, 1991) found evidence for questioning the cardinal assumption that mental models, as elicited, are essentially cognitive structures independent of social relations.

Whether mental models are dependent on, or independent of, the social relations from which they are abstracted turns upon methodological and interpretive issues. It remains a question, for instance, whether mental models approaches have inadvertently focused empirical attention on situations and respondents who happen to have relatively stable social relationships cohering around a "domain," and have unwittingly assumed these to be the general case when they may be atypical. Thus a question hangs over this approach that does not falsify it so much as place it in a different interpretive perspective. The question is whether "mental models" are more socially derived and situational than primary entities. The field of cognitive anthropology, or "ethnoscience" research, has addressed this directly.

Ethnoscience: Anthropology of Everyday Cognition

Lave (1988) and others (see also Lave, Murtaugh, & de la Rocha, 1984; Rogoff, 1990) addressed the "consistency" or stability of lay mental models from one social situation to another. They wished to test the cardinal assumption of cognitive psychology that the experimental situation, as framed by one-off structured interviews, is an adequate model of the complexity, variety, and indeterminacy of "real-life" reasoning situations.

Accordingly, Lave and colleagues studied people's mathematics in a variety of contexts, ranging from grocery shopping, simulation experiments of these same calculations, dietary calculation in the kitchen, formal arithmetic tests, and home money management. They found that people do adapt their mathematical reasoning idioms from one practical context to another to take account of relevant situational conditions. For example, experts and researchers may assume—wrongly—that "the problem" in the supermarket *is to* get the grocery list ticked off as fast and as cheaply as possible, when the shopper has several other objectives to juggle, ones that may even vary from one shopping trip to another. The same multidimensional variability is true for most real-world problems to which people apply their reasoning. Formal testing and research methods ignore these extra dimensions, decontextualizing the knowledge by imposing an artificial unidimensional problem definition (such as least-cost grocery harvesting). This allows the identification of one problem situation as identical to another, hence consolidating the problematic assumption that the requisite "true" knowledge is universal.

Lave also introduces the important dimension of indeterminacy. In typical complex contexts, the combination of intersecting problems demanding attention means that a final sovereign problem definition is never achieved until a solution is reached. In implicitly assuming given problems and ignoring contextual complications, mental models approaches, like survey research, reproduce some of the science's own self-defeating cultural prescriptions. They attempt, cognitively or materially, to reorganize the diversity and open-endedness of problems and settings into a uniform, quasi-laboratory version that can be subjected to standardized, universal, and precise analysis and solution. Ordinary lay knowledge typically eschews this epistemology of control and universalism and the prescriptive commitments that attend it. Ethnoscience work thus recognizes the "rationality of ambivalence," which has become a central element of feminist critiques of science, as discussed in the next section.

CONSTRUCTIVIST SOCIAL AND ANTHROPOLOGICAL RESEARCH

This area of research shares a commitment to avoiding a priori assumptions about what "proper" science is. Through ethnography, participant observation, and in-depth interviews, it attempts to examine the influence of social contexts and social relations upon people's renegotiation of the "science" handed down from formal institutions as if already validated and closed. This general approach immediately opens to question the very notion of what counts as a scientific-technical issue or as scientific-technical knowledge. A common thread in all this research is the encounter of different cultures: on the one hand, scientific culture, which tends to reduce issues to those of control and prediction within the terms of the scientific field in play, and on the other hand, social worlds that reflect fundamentally different models of agency and also recognize many more crosscutting and open-ended agendas and interests beyond those embodied in scientific discourse.

Despite these underlying convergences, research that can be called constructivist has developed in many, largely separate areas and includes

- medical sociology and sociology of public health (Bakz, 1991; Blaxter, 1983; Helman, 1978)
- community responses to technological accidents, emergencies, and other forms of expert intervention (P. Brown, 1987; Edelstein, 1988; Levine, 1982; McKechnie, in press; Paine, 1992; Wynne, 1992a)
- women's studies, especially in relation to reproductive technologies (E. Martin, 1989; F. Price, in press; Strathern & Franklin, 1993)
- environmental controversies, campaigning, and regulations (Dietz, Stern, & Rycroft, 1989; F. Fischer, 1990; Jasanoff, 1986; Yearley, 1992a)
- studies of lay uses of expertise, such as decision makers' use of toxic waste chemistry (Layton, Davy, & Jenkins, 1986; Layton, Jenkins, MacGill, & Davy, 1993), lay residents' relationships with "building science" (Shove, 1992), or lay constructs of deviant knowledge such as UFOlogy (Westrum, 1977, 1978)
- anthropology of "Third World" encounters with scientific culture (Arce & Long, 1987; Bailey, 1968)

Although much of this work takes science as its focal point, it follows from the constructivist perspective that "science" as such may never appear explicitly. Irwin and colleagues have discussed how science "goes underground" in social life (Irwin, Dale, & Smith, in press; also Michael, 1992; Wynne, 1991a), in the sense that its interpretive commitments are tacitly renegotiated and become encoded into "natural" forms of social and cultural practice.

Knowledge in Social Context

Research in this genre shows emphatically that people always encounter "science" imbued with social interests of some kind, regardless of the motives of specific disseminators. Thus it inevitably has implications for existing social relations, values, and identities.

E. Martin (1989) has argued that working-class women reject dominant medical views of menstruation because the whole medical-scientific construction of menstruation represents it, in social terms, as "failed reproduction." This implicit social construction reflects middle-class concerns for working careers and inadvertently denigrates working-class social relations and values. An alternative cultural construction—of menstruation as the very process that creates "the lifestuff that makes us women"—is alive and well among working-class women, who more positively value noncareer social relations, including motherhood.

Martin shows that the working-class women had been exposed to ample biological knowledge but rejected the scientific knowledge because it was associated with the pejorative "failed reproduction" construct. Rejecting the idea that these differences of favored explanatory idiom were a measure of differences of understanding (working-class women were well able to offer articulate explanations, including "book" knowledge where relevant), Martin offered this as an example of healthy and legitimate resistance to dominant ideologies carried in scientific idioms. Testing women's understanding of the biology of menstruation could not be separated from testing their assimilation of the dominant cultural model within which the biological constructs had been integrated—in this case, of women as "failed reproducers."

This analysis of public scientific knowledge, embodying particular social relational norms as if they were natural, is of quite general significance in the field of public understanding of scientific knowledge. It illustrates a fundamental impossibility of defining the boundary between nature and culture or between the domains of sovereignty of natural knowledge and social-cultural knowledge. Many constructivist analyses converge on this key point, whether they see scientific knowledge as a vehicle of central domination of marginal cultural identities (McKechnie, in press), as a scientistic reduction of kinship relations and cultural patterns to laboratory manipulability (Strathern & Franklin, 1993), as a risk of being ensnared in nuclear propaganda (Michael, 1992; Wynne, 1992a), as an alien culture of assumed control and standardization (Wynne, 1992b), or as a tacit imposition of unacceptable social relations with one's own child (Davy, Layton, & Jenkins, 1993). Yet science appears unable to recognize these social dimensions of its own public forms or the fact that public readiness to "understand" science is fundamentally affected by whether the public feels able to identify with science's unstated prior framing.

These observations about the socially laden meanings embedded in any scientific communication and the unreflexive nature of scientific responses to those public responses have been borne out in the work of Wynne and others in the field of public risk perceptions (Freudenburg, 1992; Royal Society, 1992; Wynne, 1980, 1987, 1992e). The scientific framing of "natural" discourse about "objective" risk magnitudes is treated by public groups in terms of social questions about the basis of trust in the controlling institutions. The scientific discourses conceal founding assumptions about those institutional dimensions, although there is usually accessible historical experience of relevant "institutional demeanor" that can be used as evidence to test the assumptions framing the science. Hence scientists patronizingly describe public reactions as "subjective" irrationality even though science may legitimately be rejected on grounds different than technical ignorance.

Trust

Sociologists have long recognized the pervasiveness of trust relations in society (Cicourel, 1974; Gambetta, 1988). Work in the areas of public responses to science and technology, and public risk perceptions, asserts that the basic framework of public responses rests upon the experience and perception of the relevant institutions or social actors, not upon the understanding of technical information framed in ways that implicitly take trust for granted (Renn, 1992; Wynne, 1980, 1982, 1992b). Jupp (1989) found in her study of public reception of risk and emergency information around chemical sites that people differentiated their relative trust in social sources of information from their ranking of which source they would first turn to for information. The company was very low in terms of public trust but ranked first among sources that would be looked to for information. Respondents replied that they trusted the company "so long as it is well-policed"; their priority consultation of the company was more to register the demand that it take public information seriously than out of any naive expectation that the information provided would be adequate. People seemed to have plenty of informal "scientific savvy" (Prewitt, 1982), attuned to their own social capacities and needs. From this vantage point, public "acceptance" or "understanding" of science depends upon implicit trust in and identification with the institutions controlling and deploying it. The vast majority of processes of social assimilation of science are so routine and taken for granted as to pass unnoticed as, in effect, exercises of social trust (though see below regarding reification of trust). Nevertheless, the trust dimension has been shown to be critical, and often neglected, whether its practical manifestations are toward apparent public acceptance or rejection of science.

Relevance

Laypeople may ignore scientific knowledge because they regard it as irrelevant, even though scientists assume it ought to be central to them. P. Hughes (1992) found emergency information around nuclear sites to be routinely ignored, even ridiculed, by ordinary people because to them it was naive and unrealistic about real conditions that would prevail after a major accident (see also Wynne, 1989). Lambert and Rose (in press) likewise found that lay sufferers of hypercholesterolemia dismissed medical advice about dietary fats because it did not sufficiently differentiate among different kinds of fats. These authors also noted that survey methods of measuring public understanding of science inevitably reduce "understanding" to simple indices that cannot do proper justice to the complexity of what is being "understood" in real-world contexts. Prewitt (1982) observed that scientists' own unreflective social assumptions about what is relevant to lay people are built into scientific knowledge for public communication and, furthermore, into the design of survey instruments to test public understanding. For these reasons, Layton and colleagues (1986) questioned whether such surveys test public scientific literacy or whether, instead, they measure the degree of the public's social conformity to a stereotype held by scientists of a "scientifically literate public."

Models of Agency

Laypeople often have no social freedom or power to use available scientific knowledge in the way they are assumed to have by experts. In other cases they may deliberately eschew the freedom even if they have it, or they may be uncertain whether or not they have it and therefore be unwilling to risk the consequences of testing their power. In other words, tacit models of social agency underlie assumptions about what people can or should understand about science. If available knowledge is useless, or even (socially) dangerous, there is no point in taking on the often considerable costs involved in assimilating it. Furthermore, these contours of social agency become so much a part of people's very identity that they shape the boundaries of recognizable natural knowledge. As Michael (1992) observed, people find the relevance or usability of scientific knowledge chronically problematic, partly because their social agency is chronically uncertain. They may thus shift between different registers of understanding, negotiating agency in the process. This diffidence may then be manifested, misleadingly, as simple ignorance or resistance.

Dickens (1992) examined the effect of fatalism upon people's appetite for scientific knowledge and their judgment of the worth of that knowledge. After the hurricane that hit the south of England in October 1987, he found many people who rejected "scientific" weather forecasting on the moral grounds that it is socially dangerous and unacceptable to allow this scientistic culture of control and management to expand yet further. Dickens argued that people reflected their tacit sense of agency in the moral stance that too much control of nature was bad medicine and risky. Thus "ignorance" reflected an active, legitimate moral stance, countering what was seen as an irresponsible culture of scientism around forecasting.

In many public encounters with science, this basic cultural-epistemological conflict is obliquely played out between, on the one hand, scientists' assumptions of certainty, control, and management and, on the other hand, popular assumptions of intrinsic indeterminacy, need for adaptation, and the dangers of control. Like Wynne's work on public definitions of technology and risk (1980, 1987), Dickens's work also saw people's fatalism toward *nature* as a tacit projection of a sense of the unfathomable complexity and impenetrability of the *social forces* that surround them. These are not merely passive reproductions of social experience in natural form—they are actively espoused and articulated.

Another research tradition that implicitly includes models of agency in its purview is Moscovici's (1984) theory of "social representations," which tacitly shape lay concepts of specific domains. However, this perspective appears to exclude the crucial idea that science also embodies "social representations." This exclusion inevitably impoverishes its conception of the basis of public responses toward science and expertise.

Social Construction of Ignorance

Work by Wynne (1992b) and Michael (1992) has further explored the roots of public ignorance of science. Using discourse analysis, Michael examined how laypeople in interviews reflected various possible relationships between themselves and "science" as well as various meanings of "science." They showed a rich and active reflection on their own social position (and agency) in relation to what they took to be science. This underlying relational process shaped the scope of their interest in the cognitive contents of science and their sense of trust in or identification with it. As Michael observed, this reflexive social positioning in relation to science took different forms depending on circumstances. In some cases it resulted in mistrust and deliberate avoidance of radiation science because it was thought to be imbued with the interests of "selling" nuclear power. Thus ignorance was actively constructed and maintained even though some respondents were scientifically qualified. Ignorance was not a cognitive vacuum, or a deficit by default of knowledge, but an active construct, and one with cognitive content, about the social dimensions of science. It was part and parcel of the dynamic construction of social identity.

In other instances, ignorance was constructed more positively. For example, Wynne, McKechnie, and Michael (1990) found that volunteers in a program for monitoring household radon levels constructed a "*positive*" ignorance around a collaborative division of labor with the scientists, having offered their homes as laboratories in the public interest. This vicarious participation exempted them from having to understand the technical details. Again, the scientific ignorance was not a mere vacuum but a construct built on a particular tacit model of social relations and identity.

In the case of radiation workers at the Sellafield nuclear reprocessing plant, the same authors found a similarly unsuspected, positive social construction of ignorance. The researchers expected that these workers would have an especially powerful self-interest in understanding the science of radiation risks, but they found to their surprise that this was not the case. Indeed, the workers vigorously justified their ignorance, on several grounds. First, there was plain economy—if they started to follow the disputed scientific arguments, they would never finish. Second, to follow the science would only mean they had to confront endemic uncertainties, which would not only be unsettling but might even be dangerous. Finally, they emphasized that there were already several bodies of experts in their own company and in regulatory bodies whose job it was to know the science and to fold it into design and working procedures. Thus to show an active interest in understanding the scientific knowledge of radiation risks would be a direct threat to existing social arrangements, signaling mistrust of those actors and arrangements that were meant to protect them.

If ignorance of science is actively constructed in tacit accordance with the contours of existing relationships, divisions of labor, dependency, and trust, then it would be perverse to treat "lack of understanding" of the science as a sign of intellectual or social deficit. Technical ignorance thus becomes a function of social *intelligence*, indeed of an *understanding* of science in the sense of its institutional dimensions. Of course, the Sellafield workers were not naive enough to imagine that they could trust existing social arrangements without any critical attention; in fact, relations between management and workers at the company were by no means totally harmonious. However, the point is that their attention was devoted not to the science of radiation but to the evidence for or against trust in those other actors on whom they knew they were inevitably dependent.

Trust and Dependency: Identity and Ambivalence

Critical examination needs to be given to the basis of trust as a dimension crucially affecting public uptake of science. As noted above, Jupp's work suggested that people may hold conditional notions of trust. Wynne (1992a) has argued that apparent trust may be more realistically viewed in "as if" terms. If people feel they are dependent on particular institutions for their safety or other valued conditions, they may feel the need to act as if they trust them, even while skeptically monitoring their behavior for evidence to support or undermine the "trust hypothesis." Reifying trust into an objective parameter that can supposedly be measured (and manipulated), like understanding, is in this view fundamentally mistaken.

Wynne's analysis of Cumbrian farmers' responses to post-Chernobyl radiation science further supports this point. The farmers had good reasons to believe that the radioactive contamination blighting their farms had come from the nearby Sellafield site, undetected or unadmitted by the authorities since long before Chernobyl. Against the assertions of scientists that they had unambiguous "fingerprints" that proved it was from Chernobyl, not Sellafield, the farmers could advance several cogent pieces of evidence and logic supporting their skepticism. This was reinforced by regular conversations with farmers from beyond the affected area, who placed the blame firmly on the local source. Yet ultimately the local farmers showed ambivalence about what they believed, expressing the fear that, if they said what they suspected, it could fracture their communities and even kinship patterns, because many of the hill-farming families and communities were dependent on Sellafield for work. The farmers, it seems, were trying to reconcile contradictory social identities corresponding with opposing beliefs about the origins of the contamination that served as part of a wider set of interpretations of the behavior and integrity of controlling organizations. These interpretive stances in turn integrated with different social networks, interdependencies, and affirmations of identity. Treating these multiple elements of identity as matters of "choice" may not adequately reflect the extent of human commitment and risk involved in such processes.

Reflexive Understanding

Research in the "constructivist" idiom suspends any automatic privileging of science's own presumptions of its universal remit. In taking seriously people's indigenous problems, relationships, and "givens," it allows a richer perspective from which to see the place of scientific interventions and constructions. It also allows recognition of the complex skills and different legitimate moral commitments that people exercise in "managing" conflicting demands in contexts where they cannot assume control of all or even most of the relevant variables. As such, it potentially offers the way to more productive negotiation and mutual accommodation between scientific and lay cultures.

It becomes clear that local knowledges—both lay and scientific—may be couched within different epistemic commitments—for example, about the legitimate scope of control, standardization, uncertainty, or assumptions about agency. Thus levels of resolution or standardization in structures of knowledge, primary parameters in models, expressed levels of certainty, or criteria of adequate evidence or inference may need to be reproblematized and reopened for negotiation with new "extended peer groups" in the arena of application. This would involve renegotiation of what counts as "good science" in different contexts of use, and of the boundaries of science. What is often taken—by scientists but also by social science researchers—to be public misunderstanding of science instead comprises the markers of people's tacit negotiation of their own social relationships with "science." To avoid provoking alienation, mistrust, and lack of uptake, scientific cultures and institutions need to be more open and self-reflexive about their own framing assumptions and commitments.

Local Studies and "Representative" Surveys: A Reconsideration

The interpretive-descriptive claims of constructivist work intertwine with a normative commitment, which holds that such pluralism in public science would be possible, even beneficial. The problem with this approach is not so much the implied normative commitment—all approaches involve this in some way, albeit with varying degree of explicitness. It is more the mundane question of how far it is possible to draw general conclusions from what are often microsociological studies of localized histories and interactions.

It cannot be assumed, however, that detailed, "thick-description," qualitative studies from constructivist perspectives can only offer interstitial elaboration within the framework of objective accounts of public knowledge as measured in large-scale quantitative surveys. The role of qualitative studies is not to identify more refined and "sensitive" questions for surveys to test on an objective level. Rather, the relational constructions of "understanding," "science," "knowledge," and "trust" that the qualitative studies identify and explore are simply not accessible to large-scale survey methods, which cannot avoid imposing exogenous, standardized, and determinate definitions of the crucial terms on the social actors from whom they extract responses. There are some potentially fruitful correspondences and complementarities between the survey and constructivist approaches (Jasanoff, 1993), but the relationship between them mirrors that between local cultures and "universal" science. The constructivist approach identifies attempts by local actors to negotiate more reflexive institutions of science, and hence to renegotiate and dissolve the very idea of a "center" with a dominating universalist discourse.

CONSTRUCTIVIST PUS RESEARCH AND SSK

It is valuable here to explore the relations between constructivist PUS research and two mainstream perspectives in SSK. The two perspectives are the actor-network theory (ANT) of Latour and Callon (Callon, 1986b; Latour, 1987, 1991b) and the constructivism of Collins (1988). Both inspired the constructivist perspectives in the PUS field, but there is potential for further development.

The ANT approach is well exemplified in Callon's (1986b) study of the enrollment by marine biologists of (inter alia) the scallop fishermen of St. Brieuc. As Callon and Latour show, the enrollment process, which can here be seen as "public assimilation of science," involves not merely the understanding and uptake of knowledge but a renegotiation and restructuring of identities and interests. This is entirely consistent with constructivist PUS research.

However, ANT has tended to overinterpret the extent to which ostensibly fully enrolled actors align their identity with the enrolling framework. In so doing, ANT ignores the engagement of the enrolled public (here, the fishermen) in multiple other crosscutting networks that involve different identities. In other words, ANT overlooks and conceals the *ambivalence* that an actor may tacitly hold toward a network with which she apparently completely identifies.

Thus Callon refers to the sudden cataclysmic withdrawal of the scallop fishermen's support for the marine biologists' program, into which they had previously been enrolled, as a "betrayal." This suggests a sudden reversal from complete monovalent identification to complete monovalent alienation. Yet, what appears to be a complete switch of loyalty may, when the substructures of ambivalence are in focus, be recognized as the result of a relatively minor shift in a fine balance of conflicting elements of identity, reflected in competing social networks. Star (1991b) has offered similar observations from an attempt to bring consideration of power into ANT.

Singleton and Michael (1993) also analyzed the positive role of ambivalence on the part of actors in the sociotechnical network of the U.K. cervical smear test (CST) program. In essence, ambivalence allows the network to be maintained (see also Singleton, 1992). She also relates this point to feminist perspectives (e.g., Harding, 1986) that criticize, as a patriarchal culture of control, science's dismissal of ambivalence as "imprecision."

Similar problems are also evident in Collins's (1988) work on public "experiments" as rhetorics of public witnessing to "demonstrated fact," such as the claimed integrity of nuclear fuel flasks against "worst case" accidents. His key point is that these demonstrations are socially constructed, in the sense that they can never be complete replications of real-world operating conditions in all their contingency. Noting in passing that understanding scientific process is at least as important as understanding contents (Collins & Shapin, 1989; Shapin, 1992b; Wynne, 1987; Wynne & Millar, 1988), Collins argues that these factual demonstrations need to be deconstructed in public to ensure that the public understands how they are socially shaped.

While this is a valuable argument, it also falls short in one important respect. Collins assumes that without a technical deconstruction the public is doomed to a false certainty about the result. Yet PUS research shows that people usually have several dimensions by which they can relate to and judge such claims, for example, by the institutional demeanor, interests, and historical experience of the authors (Wynne, 1982). It is misleading to assume that, without such technical deconstruction, people are bound to be dupes of the tacit way the science was constructed. A richer understanding of how authority is negotiated for science is offered by closer convergence of mainstream SSK and constructivist PUS research.

CONCLUSIONS

The combined effect of anthropological approaches to lay cognitive processes ("ethnoscience") and the diverse sociological studies of science in public contexts, which I labeled "constructivist," is to demonstrate resoundingly that problematizing science is a central part of any serious attempt to define the overall research and public policy issues of public understanding of science. This is not to imply that there are no problems to be researched and resolved concerning either publics or media; on the contrary, there are plenty. However, the dominant definition of PUS problematizes *only* these latter, and thus in effect helps to propagate the existing institutionalized cultures and boundaries of science as natural and given, as a universal standard of judgment and "rationality."

Research in the SSK vein poses an important challenge to the dominant issue framing in PUS. In the associated field of public risk perceptions, some headway can be discerned, for example, in the "risk perception" chapter of the London Royal Society's 1992 report on risk. Yet the depth of resistance to constructivist ideas is marked by the same Royal Society's COPUS (n.d.) observation that the public is positive about science but, as many "outside the scientific community" have observed, has been alienated from it by most scientists' reluctance to communicate with the public.

This oblique COPUS reference to PUS research radically misrepresents it as showing the problems of public "misunderstanding" to lie in scientists' overcautious approach to popularization. Thus it also fundamentally misrepresents "the public" as only ambivalent toward science because they do not hear enough from scientists. The more challenging sociological finding that PUS problems are as much to do with institutional and epistemic characteristics of dominant forms of science is thus systematically deleted.

PUS research has found ample evidence of the reflexivity of laypeople in problematizing and informally negotiating their own relationship to "science." It has also identified the silent alienation created by the unreflexive ways in which scientists construct the public in their interactions with them. The unreflexive responses of the scientific establishment appear to reflect a deep institutional insecurity about actually encountering lay publics on their terms and negotiating valid knowledge with them. There are strong resonances with an equivalent situation in the field of risk, where legitimate public ambivalence and resistance to expert presumptions about the framing of risk issues was first interpreted as simple ignorance, then "misunderstanding," and latterly as a naive wish for an impossible "zero-risk" environment. These constructions of the public ignore sociological evidence that shows that people are by no means naive about the existence or complete eradicability of risks, and that points rather to scientists' unacknowledged insecurity about recognizing the conditionality of their own knowledge and the prescriptive commitments it embodies (Irwin & Wynne, in press; Turner & Wynne, 1992).

The lack of reflexivity of science is also indicated in its resistance toward reopening "closed" bodies of knowledge to explore whether embedded levels of standardization, resolution, certainty, inference rules, or other commitments might need to be renegotiated. Funtowicz and Ravetz (1990) have aptly called this extended peer review. It would imply the reopening of cognitive commitments that have already achieved closure in the privileged contexts of validation within science. There are structural similarities here with the issues identified in sociology of technology (Law, 1991c) and risk (Wynne, 1992e) about constructs of the user or other downstream actors embedded in design or risk-analytic knowledge. Exposing the *social* closure of these apparently deterministically closed commitments, and the reorganization of the social world that goes on in order to validate the forms of "technical" closure, has been a crucial general contribution of SSK, and it is here that the fields of PUS, participation, and SSK could most fruitfully converge in the future.

Very little of the voluminous work down the years on participation in science and technology has addressed these dimensions, remaining instead at the cruder conceptual level of competing interests and rights, where scientific knowledge remains substantially unproblematized except in notions of deliberate political manipulation (Schwarz & Thompson, 1990). One exception is Krimsky (1984). Other related work is from Helman (1986), Freudenburg (1992), F. Fischer (1990), Merrifield (1989), and Ross (1991). A more substantial body of parallel work comes from Third World research involving "peasant" encounters with Western scientific culture (Arce & Long, 1987; Bailey, 1968; Geertz, 1983; Mamdani, 1972).

History of science, by contrast, has long provided a rich seam of research insights and historical studies supporting the general SSK orientation. Hilgartner's (1990) study of postwar popularization in the United States identifies the tacit anxieties of scientific elites about social order and public affirmation. Similar observations have been made about the reemergence of a strong concern about public understanding of science in the 1980s (Wynne, 1992e). Layton (1973) and Barnes and Shapin (1979) noted similar implicit agendas, respectively, in movements of "science for the people" in the nineteenth century and the eighteenth-century English Mechanics' Institutes.

Haraway's (1989, 1992) historical studies of primatology, and other "externalist" histories of science (e.g., Barnes & Shapin, 1979; Berman, 1974; Shapin & Schaffer, 1985; R. Young, 1986), expose recurrent concerns to have the public assimilate a proper understanding of science as part of the establishment of ideological versions of the natural (social) order. Much of this work, however, implies by default that such elite discourses are automatically effective. Desmond's (1987) study of the forms of resistance of working-class people to dominant discourses of "natural knowledge" adds important depth to this perspective because it emphasizes that public recipients of science are not mere dupes of dominant discourses. As Desmond (1987) notes:

> Pinpointing the scientific trade that [social elites] were plying in working class markets is different from revealing the sorts of intellectual commodities that the artisans themselves were prepared to buy—or make; for we might picture the artisan-craftsmen not as passive recipients of bourgeois wisdom but as active makers of their own intellectual worlds, their own really useful knowledge.

As Desmond recognizes, public defense of autonomous realms of meaning negotiation against expropriation by science (including social science) may or may not be explicit. It may produce an alienation or ambivalence—hence rejection of dominant forms of knowledge—that is not seen in overt behavior.

A central thread of this review has been the importance of recognizing the multivalent nature of "public understanding of science." Scientific institutions are, despite what some assume (e.g., Giddens, 1990), very weak on the self-reflexivity that would allow them to recognize their unwitting role in their own crises of public credibility and support. Beck (1986, 1992) has referred to this as the "self-refutation" of scientific modernity, and Wynne (1987) has called it rationalism's "self-delegitimation." In defining the issues of public risk perceptions and public understanding of science, the taken-for-granted framing of the problems inadvertently patronizes and denigrates lay publics and tacitly covers up people's concerns about the institutional forms and culture of science and its implicit construction of them as human beings. Thus a self-defeating dynamic is built into both research and practice that does not acknowledge and integrate the problematization of science itself. The wider implications of this lack of self-reflexivity are worth addressing.

Beck (1986, 1992) has argued that modern society is fundamentally transformed by the pervasive, inescapable nature of risks that have been generated by science and technology. Others (e.g., Lyotard, 1984) see modernity crumbling due to its own scientistic hubris, which no longer enjoys cultural purchase and legitimacy. Whereas some celebrate what they see as a radical postmodern transformation and liberation from the yoke of scientistic universalism, Beck identifies a more complex path of reflexive modernity. By this he means a negotiated diversification of the grand univocal principles of rationality and natural order that have dominated modernity, into more modest forms that reflexively recognize their own conditional foundations and intrinsic limits.

The problem for the conventional forms of defense of a legitimate role for science in an increasingly decentered "postmodern" context is that they are inherently self-defeating. Control over public definitions of what is to count as science is still anxiously clung to (as reflected in the dominant PUS problem definition) as if there is only one natural version; yet scientists routinely negotiate definitions of "good science" among themselves, including that for public consumption (Jasanoff, 1990). The innumerable attempts by ordinary publics in effect to negotiate what counts as legitimate public knowledge are frequently defined by those anxious elites as "antiscience." Holton's (1992) historical review is a case in point, treating any questioning of the modernist project, with its epistemological principle of instrumental control, as a fundamental threat to order and reason per se. Yet as Cozzens (1992) has put it, many so-called antiscience groups are proknowledge—

and, she could have noted, "proscience" assertions of the parochial kind are in important respects antiknowledge.

Why has the concern about PUS seen another of its periodic flowerings in the last decade? I would suggest that it reflects another intensification of the weight placed upon public credulity and legitimation by major bursts of commercialization in leading areas of scientific research, notably biotechnology and biomedicine. Science has thus placed immense demands on what is always an ambivalent and incipiently alienated relationship with the public. Trying to plug the gap with more of the same basic discourse only fuels the sense of social risk and furthers alienation and mistrust. This is a somewhat more social-cultural model of the concept of "risk society" than Beck's. My formulation suggests that there is a tacit cultural politics of legitimation of science, and of related institutions, being conducted under the language of "public understanding of science." This might ultimately be more fruitfully engaged via more explicit questioning, and renegotiation, of the dominant institutional forms of science and the monovalent epistemology of instrumental control.

PART VI
SCIENCE, TECHNOLOGY, AND CONTROVERSY

In our introduction to this volume, we employed an analogy between the map maker's task and our own. This analogy has particular utility for understanding the chapters in this section, for most controversies in science and technology are disputes over the authority of individuals and groups to conduct and interpret these activities—in other words, disputes over boundaries.

Using this metaphor, Thomas Gieryn begins his chapter with several questions. Where is the border between science and nonscience? Which claims or practices are scientific? Who is a scientist? What is science? Gieryn is clear about his own agenda; he seeks to make these borders messy and contentious. States of war, or temporary truces at best, rule Gieryn's map.

Gieryn uses four case studies to illustrate the messiness of borders that scientists draw: The argument between Hobbes and Boyle illustrates the creation of monopolistic cultural authority; D'Alembert and Diderot's *Encyclopedia* shows an effort to legitimate the expansion of scientific boundaries; the Cyril Burt case is used to examine the expulsion of a deviant scientist; and a study of science advisers in the United States illustrates an effort of scientists to protect their autonomy from political outsiders. Each of Gieryn's cases shows that boundaries are not fixed by logic or time; rather, they are drawn by local actors, and they are temporally episodic.

Dorothy Nelkin, who pioneered the study of controversies, sees these as ideal research sites to study various problems in postindustrial society: tensions between individual autonomy and community needs; ambivalence between science and other social institutions (media, regulatory systems, and courts); and, perhaps most important, the appropriate role of government.

Nelkin, in short, examines the social consequence of the boundary between science and society. What does scientific expertise permit? What control do various publics have over the scientific enterprise? What is (or ought be) the role of government in defining the boundaries of science? In an examination of the animal rights controversy, Nelkin finds that contemporary controversies typically are not disputes between individuals but are more like complex social movements, driven not only by vested economic interests but by a strong moral imperative.

Steven Yearley examines the subset of scientific controversies that have developed around the environmental movement. In these disputes, ambiguous knowledge claims, the authority of science, and social movements, political power, and vested interests have met in yet unresolved conflict.

For Yearley, S&TS scholars and activists have much to learn from one another through a study of environmental science controversies. S&TS provides a framework for explaining the difficulty in using knowledge claims to adjudicate such disputes. Through the S&TS perspective, we understand why the Green movement has found scientific knowledge to be at once an appealing and an untrustworthy ally. S&TS scholars should note the systematic critique of science and technology —by proxy, the entirety of postindustrial society—as constructed by contemporary environmentalists.

In the postindustrial world, argue Henry Etzkowitz and Andrew Webster, intellectual property is the engine of development, displacing in part traditional elements such as monetary capital, natural resources, and land. Thus not only do disputes over the capitalization and commercialism of knowledge arise, but such controversy assumes central importance in the political economy.

These trends, Etzkowitz and Webster argue, will broaden the context of academic-industry relations to include regional, national, and international government. The boundary between science policy and industrial policy will be obliterated as the two heretofore distinct institutional efforts merge. As a consequence, science and technology policy will move from the periphery toward the center.

Brian Martin and Evelleen Richard's chapter—the final contribution in this part—supplies a coherent framework for the preceding contributions. Their chapter reflects the methodological and reflexive cast of this *Handbook*. As well as studying actors in controversy, they study how ST&S scholars study controversy, identifying four approaches—positivist, group politics, constructivist, and social structural.

Having presented this classification schema (and no doubt having aroused the curiosity of ST&S scholars to see where they are placed), Martin and Richards conclude with a plea to cross, indeed obliterate, boundaries and create an integrated approach. Using the controversies over fluoridation and vitamin C as examples, the authors call for further studies that will emphasize the relationships between cognitive judgments, professional interests, political power, and financial concerns.

Just a few decades ago, scholars typically characterized scientific and technical controversy pejoratively. Disputes, in this view, arose from the activities of deviant scientists—ones who cheated, sought political power, profited inappropriately, or were just odd. Today we recognize—as these chapters attest—that controversy in science and technology is not only typical and common but a normal (perhaps even desirable) product of the encounter between science and society.

18

Boundaries of Science

THOMAS F. GIERYN

THE working title of this *Handbook* presumed three neatly bounded territories: science, technology, and society. This chapter makes those territories and especially their borders into objects for sociological interpretation and seeks to recover their messiness, contentiousness, and practical significance in everyday life. Its focus is on the "boundary problem" in science and technology studies: Where does science leave off, and society—or technology—begin? Where is the border between science and non-science? Which claims or practices are scientific? Who is a scientist? What *is* science?

The chapter begins with two perspectives on the boundary problem, essentialism and constructivism. Essentialists argue for the possibility and analytic desirability of identifying unique, necessary, and invariant qualities that set science apart from other cultural practices and products, and that explain its singular achievements (valid and reliable claims about the external world). Constructivists argue that no demarcation principles work universally and that the separation of science from other knowledge-producing activities is instead a contextually contingent and interests-driven pragmatic accomplishment drawing selectively on inconsistent and ambiguous attributes. Research in the sociology of science has raised doubts about the ability of any proposed "demarcation criteria" to distinguish science from non-science. Attention has thus shifted from criticisms of essentialism to examinations of when, how, and to what ends the boundaries of science are drawn and defended in natural settings often distant from laboratories and professional journals—a process known as "boundary-work."[1] Essentialists *do* boundary-work; constructivists *watch* it get done by people in society—as scientists, would-be scientists, science critics, journalists, bureaucrats, lawyers, and other interested parties accomplish the demarcation of science from non-science.

The second part discusses four theoretical precursors and extensions of the elementary constructivist idea that the boundaries of science are social conventions. Each draws upon scholarly literatures and theoretical traditions initially developed outside STS: sociological studies of professions, symbolic interactionism and its recent interest in social worlds, anthropologically driven studies of cultural history focused on the significance of classification practices, and feminist theories of knowledge.

The chapter concludes with empirical research illustrating constructivist studies of the boundary problem. I examine four episodes of boundary-work, different in the *goals* sought by those who claimed their place *inside* science. The dispute between Hobbes and Boyle (Shapin & Schaffer, 1985) illustrates pursuit of a monopoly over cultural authority through exclusion of those offering discrepant and competitive maps of the place of science in the intellectual landscape. D'Alembert and Diderot's *Encyclopedia* (Darnton, 1984) is an example of how insider-scientists construct other knowledge-producing systems as foils to legitimate the expansion of science. Third, the affair surrounding Cyril Burt's alleged misconduct (Gieryn & Figert, 1986) points to a familiar kind of boundary-work: expulsion of deviant scientists, as a means of social control and of preserving professional reputation and public confidence. Finally, Jasanoff's (1990a) study of science advisers involved in government policy making presents the contested border between science and politics as an occasion for scientists to protect their autonomy and authority from usurpation or control by outsiders—government bureaucrats, elected officials, and lobbyists.

ESSENTIALISM AND ITS CONSTRUCTIVIST CRITIQUE

The analytic problem of demarcating science from non-science has attracted the attention of three major figures in science studies—Karl Popper in philosophy, Robert Merton in sociology, Thomas Kuhn in history—and each opted for an essentialist solution. No doubt a desire for definitional tidiness led them to the demarcation problem. Each was an interpreter of *science,* and how could one know what to look at—and what to look away from—without a workable definition of the phenomenon at hand? Demarcation was perhaps vital as well for efforts to explain what these three took as a singular achievement of science: an improving validity and reliability in its models of the world. Criteria of demarcation became, in effect, explanations for the superiority of science (among knowledge-producing practices) in producing truthful claims about the external world. There was, as well, an ideological dimension in these classic demarcation tries. Merton's wish to dismiss Nazi science as perverse, or Popper's wish to dismiss psychoanalysis and Marxism as pseudoscience, created the need for a standard to identify the real McCoy.

Falsifiability

Popper gave the demarcation problem its name, and his philosophical solution—falsifiability—remains the most familiar one. He arrived at demarcation while dealing with *logical* problems discerned in philosophers' prior attempts to "justify" theories or generalizations. For Popper, demarcation is an epistemological matter, to be resolved by finding something in the *Methodology* of science—in its "epistemic invariants" (Laudan, 1983, p. 28)—that accounts for the superiority of science in providing reliable and valid knowledge about the world.

When Popper started out in the 1920s, a once-popular philosophy held that science seeks justification for its theories through the accumulation of confirmatory empirical evidence. Scientists try to verify theories with corroborating facts, and, if they succeed, science progresses toward truth. Popper recognized that this verificationist strategy ran aground on logical problems of induction. The accumulation of corroborating evidence could tell scientists what *was*. But reliable predictions of the future (something that general theories in science are expected to offer; Popper, 1972, pp. 349 ff.) are uncertain: The next observation could always in principle yield a refutation. Popper's watershed response not only offered an escape from the inductivist illogic of verification but a criterion for demarcating scientific from nonscientific statements.

In place of verification, Popper prefers falsification; in place of certain truth, he offers conjecture. Science advances toward truth (though never arriving at certainty) by a combination of bold conjecture and severe criticism. Scientists work from problems to theoretical generalizations and basic statements that are in principle *falsifiable*; potentially, some empirical observation could logically contradict or refute them (Popper, 1959, pp. 40 ff.). The bolder such conjectures, the more rapidly science moves ahead: Theories that are more easily falsified (i.e., a larger number and wider variety of observations are potentially able to refute the claim) contribute more to the advancement of objective knowledge than more timid claims. Bold conjecture alone does not make science: Scientists must also subject such conjectures to severe critical scrutiny as they try their hardest to refute the theory. No theory is immune from such criticism. Science thus is not a confirmation game (looking for evidence to corroborate a generalization) but a refutation game (looking for evidence to shoot it down). The result is not certain truth but ever bolder conjectures that have (for the time being) survived critical refutation, and thus assume better (but still fallible) approximations to reality and enlarged credibility.[2] Scientific knowledge grows—progresses—more through the elimination of error (actual falsification) than through the cumulative repetition of corroborative evidence.

Popper made falsifiability do double duty. First, in contrast to verification and its problems with induction, conjecture-and-refutation put science on a steadier logical footing and appeased some nervous philosophers. Second, falsifiability provided Popper with a wedge to drive between science and non-science. Any of three conditions would suffice to dump a claim, a practice, a belief, or their adherents into non-science: if the claim were not potentially falsifiable (i.e., no conceivable empirical observation would logically contradict the assertion), if there was no sincere and severe effort to refute the claim, or if a claim is not rejected when refuting empirical evidence is presented. Non-science includes metaphysics, ideology, and pseudoscience, and Popper insists that these activities may have meaning or practical significance, but they simply are not *scientific* (which is why Popper can put such worthy but non-empirical pursuits as mathematics and logic outside science; Popper, 1963, pp. 253 ff.). This same demarcation criterion enabled Popper (1972) to locate the historical turn from prescience to science in ancient Greece of the fifth or sixth century B.C., where a "critical attitude" (falsification and the rest) replaced a "dogmatic hanging on of the doctrine in which the whole interest lies in the preservation of the authentic tradition" (p. 347).

Popper (1963) traces his interest in demarcation to the intellectual swirl of 1919 Vienna, involving Marx's theory of history, Freudian psychoanalysis, and Adlerian "individual psychology" (pp. 34 ff.). He asked himself: What is *wrong* with these theories, and what made them so different than those in physics? The theories were not "wrong" because they lacked verification; indeed, it was precisely the abundance of confirming observations that made Popper suspicious: "Once your eyes were thus opened you saw confirming instances everywhere" (Popper, 1972, p. 35). Popper's eventual conclusion that Marxism and psychoanalysis lacked falsifiability—and thus were outside empirical science—was reached via astrology. Astrologers did indeed accumulate much confirming evidence, but they put their theories and prophecies in such a vague language that any evidence could be made to fit. That soothsaying is then pinned on Marxists, who reinterpret not just the theory but the evidence as well in efforts to salvage their general theory of history: "Marxism has established itself as a dogmatism which is elastic enough . . . to evade any further attack" (Popper, 1963, p. 334). Psychoanalysis shares this inadequacy; both Freud and Adler cast their original theories in a form that could not be falsified by any observation of human behavior. Popper sought a methodology for science that would at once rescue its theories from the logical problems of induction and permit the demarcation of empirical science from impostors and from other non-empirical but still meaningful belief systems such as metaphysics. Falsifiability, conjecture, and refutation fit the bill.

Constructivist criticism of Popper's try at essentialist demarcation centers on the *reproducibility* of falsifying empirical evidence. Falsifiability is a logical condition but falsification is a practical accomplishment of observation and experiment, and on this Popper and his constructivist critics agree. They disagree on whether falsification is a straightforward unambiguous lineup of evidence against theory, or whether the process is shrouded in interpretative ambiguities that get resolved only through complex social negotiations (that go beyond logic and methodology). Popper is no naive empiricist, and the idea of "theory-laden facts" is as much a part of his legacy as falsification. Science cannot be built from immediate and unstructured sense perceptions; rather, observations are guided by problems-at-hand and by theories that act as nets for sifting through the infinite details of reality. And though observations must be sufficiently independent of theories if they are to perform their occasionally necessary role of falsification, the question remains: *When* does observation justify the abandonment of a theory? Popper worried about the possibility that worthy theories would be too quickly dumped in the face of any old falsification report: "We shall take [a theory] as falsified only if we discover a *reproducible* effect which refutes [it]. In other words, we only accept the falsification if a low-level empirical hypothesis which describes such an effect is proposed and *corroborated*" (Popper, 1959, p. 86, italics added). Nothing hard about that, for Popper: "Any empirical scientific statement can be presented (by describing experimental arrangements, etc.) in such a way that anyone who has learned the relevant technique can test it" (p. 99).

No one has done more than Harry Collins (1985) to expose the problematic nature of what Popper takes as open and shut—namely, *when* is an empirical claim successfully replicated, reproduced, or corroborated? Collins argues that there is no unambiguous and impersonal algorithm for reproducing an experimental procedure. Scientists routinely face the problem of deciding when a replication is competent and authentic.[3] "Experimenters regress" is a paradox for those like Popper who want to "use replication as a test of the truth [falsification] of scientific knowledge claims" (p. 2) because negotiation of the competence of a replication attempt is, at once, negotiation of the reality of phenomena at hand. In his study of gravity wave experiments, Collins reports that scientists' judgments about the competence of a replication experiment hinged on whether the results of that experiment were consistent with their theoretical assumptions going in. This research challenges an inherent part of Popper's demarcation criteria, for how can refutation or falsification occur if scientists sometimes exploit available rhetoric (human error, machine failure, extraneous circumstances, technical infelicity) to neutralize potentially falsifying observations by attributing them to incompetent or unauthentic replications? Thus Collins sets the stage for post-essentialist efforts to ascertain *how* some replications are deemed authentic and authoritative, a cultural and rhetorical game that is often tantamount to labeling them "scientific." Popper's demarcation criteria become a matter for scientists and others to negotiate.

Social Norms of Science

Merton's (1973b, pp. 267-278) four social norms of science require only brief rehearsal here, for they have energized debate in sociology of science for a half century. His argument is as essentialist as Popper's, with the institutionalized ethos of science replacing falsifiability as a criterion for demarcating science from non-science. For Merton (1973b), "the institutional goal of science is the extension of certified knowledge," that is, "empirically confirmed and logically consistent statements of regularities (which are, in effect, predictions)" (p. 270). The theoretical problem is to identify a social and cultural structure for science that aids pursuit of this goal. Part of that structure is methodological, encompassing technical norms of empirical evidence and logical consistency. Another part is "moral" or social, consisting of four affectively toned norms "held to be binding on the man of science" (p. 269). These norms take the form of prescriptions and proscriptions, they are communicated and internalized during the socialization of scientists, and they are reinforced by sanctions levied against transgressors and by rewards heaped on successful conformists. Empirical evidence for their existence is found in moral indignations expressed by scientists in reaction to violations, along with their positive behavioral and rhetorical endorsement of the norms.

Scientists of course will violate this moral code on occasion, but here is what they are institutionally expected to live up to as an ideal. *Communism* asks scientists to share their findings, and the institution promises "returns" only on "property" that is given away. *Universalism* enjoins scientists to evaluate knowledge claims using "preestablished impersonal criteria" (say, prevailing theoretical or methodological assumptions), so that the allocation of rewards and resources should not be affected by the contributor's race, gender, nationality, social class, or other functionally irrelevant statuses. The norm of *disinterestedness* does not demand altruistic motivations of scientists, but channels their presumably diverse motivations away from merely self-interested behavior that would conflict with the institutional goal of science (extending certified knowledge). *Organized skepticism* proscribes dogmatic acceptance of claims and instead urges suspension of judgment until sufficient evidence and argument are available.

I have not yet found the words *demarcation* or *boundary* in Merton's classic paper (originally written in 1942), but the ideas behind them are expressed in its rationale. Merton (1973b) creates an image of wolves at the door of science: "local contagions of anti-intellectualism" and "a frontal assault on the autonomy of science" (pp. 267, 268). Few readers in 1942 could think of anything but Aryan Science serving Naziism (see Hollinger, 1983). Such "incipient and actual attacks upon the integrity of science have led scientists to recognize their dependence on particular types of social structure" (Merton, 1973b, p. 267). The social structure on which science *depends* is one where the four social norms are institutionalized. Without reading too much into the word *dependence,* this casts the norms as a kind of institutional sine qua non, as essential for extending certified knowledge as, say, falsifiability. In effect, the four social norms of science save the autonomy of science from external political or cultural interferences by arguing that such intrusions compromise the necessary moral conditions, which in turn make possible the extension of certified knowledge. The very words *Aryan science* contradict universalism, for example.

If the norms are read as demarcation criteria, then knowledge-producing activities not ensconced in that institutionalized moral frame must be non-scientific. And so they are, when Merton uses the norms to distinguish scientific claims from mere ideology, such as racist assertions from Hitler's Germany. "The presumably scientific pronouncement of totalitarian spokesmen on race or economy or history are for the uninstructed laity of the same order as newspaper reports of an expanding universe or wave mechanics. . . . The borrowed authority of science bestows prestige on the *unscientific* doctrine" (Merton, 1973b, p. 277). What makes those claims unscientific? Not their substantive content so much as the anything-but-disinterested ambitions of their promoters: The "authority [of science] can be and is appropriated for *interested* purposes" (p. 277). When politics get inside the door of science, the autonomy and regulative force of the norms is breached, and the resulting claims are unscientific. The implication, of course, is that when the same claim is served up in the court of real science—with scientific ethos intact—then error will be exposed: "The criteria of validity of claims to scientific knowledge are not matters of national taste or culture. Sooner or later, competing claims to validity are settled by universalistic [and disinterested] criteria" (p. 271, n. 6).

Of the many criticisms of Merton's theory of the normative structure of science, one in particular is akin to Collins's criticism of Popper: It takes the supposed essential qualities of science—those that distinguish it from non-science—and makes them into matters for people in society to construct, interpret, negotiate, and deploy. Cicourel (1974, pp. 11-41) argues that rules or norms in general are not things for definition by sociological analysts but are available for definition by actors in everyday life. He suggests that a tacit layer of interpretative or basic norms guide actors as they try to decide which surface rules (i.e., "structural" norms) are relevant and appropriate for the situation at hand. In this scheme, Merton's social norms of science are surface rules that do not translate into behavior patterns in an immediate and direct way. Rather, the process is mediated by interpretative norms through which actors collectively decide what universalism would mean in a given setting or even whether universalism is pertinent at all for the evaluation of a scientific claim or action.

Mulkay extends Cicourel's general argument into a criticism of the norms of science in particular. Using Woolgar's study (1976) of the discovery of pulsars, Mulkay suggests that scientists' decisions to make findings public—as they are implored to do by the norm of communism—are caught up in momentary and situational contingencies such as "any result which might be of interest to the press must be kept as secret as possible" or "researchers must be particularly careful not to release information in such a way that the first achievement of a graduate student is jeopardized" (Mulkay, 1980, p. 121). These mitigating circumstances are more than "counternorms" (Mitroff, 1974a) balancing out the injunctions of the institutionalized norms identified by Merton. Rather, interpretative work (grounded in identifiable tacit assumptions or interests) is required before the meaning of a surface norm such as communism is decided, and before reaching a judgment on whether it is pertinent as a guide for one's own behavior or as a standard for evaluating the behavior of others. Mulkay does not argue that scientific norms do not exist; something like communism may well be used rhetorically in certain situations as scientists make decisions and justify their practices to others. The norms could become stable elements of some scientists' boundary-work, useful at the time for separating their science from others' non-science.

Paradigmatic Consensus

It may be difficult to argue that Kuhn sits comfortably as an essentialist-demarcationist alongside Popper and Merton. The distance between Kuhn and Popper is substantial, as measured by Kuhn's (1977a, p. 272) argument that critical refutation and incessant falsification attempts may simply not be present in "normal science," and by Popper's (1970, pp. 51-58) dismissal of

Kuhn as a historical relativist and (worse) one who would allow the sociology and psychology of science to settle issues (like demarcation) that are more properly settled by logic and methodology. And some readers see Kuhn's emphasis on the moral force of *cognitive* norms in opposition to Merton's emphasis on *social* norms (see Pinch, 1982), although the principals involved see their work as complementary (Kuhn, 1977a, p. xxi; Merton, 1977, p. 107). Moreover, Kuhn (1970) has expressed doubts about the possibility of demarcation criteria: "We must not, I think, seek a sharp or decisive" "demarcation criterion" (p. 6). Still, looking back on Kuhn from a perspective shaped by 10 years of constructivist empirical studies of boundary-work, a case can be made that he set that line of inquiry in motion more as foil than as pioneer.

Kuhn (1962/1970) sets up the argument in *The Structure of Scientific Revolutions* with two puzzles, each announcing that the demarcation issue will run as leitmotif throughout the book. While hanging around social and behavioral scientists, Kuhn noticed that their arguments differed fundamentally from those he observed among physicists. Social scientists could not agree on first principles; they fought incessantly to distinguish significant from trivial problems for study, acceptable from unacceptable solutions, legitimate from illegitimate frameworks. Physicists at any particular time seemed to agree on first principles and reached consensus on answers to those domain questions. So, Kuhn wondered, what did the hard sciences have that the soft ones lacked? The second puzzle arose from obvious historical changes in the content of certified scientific claims—what was held as fact or a good explanation *then* is *now* preposterous, simplistic, crude, or just wrong. Can those old, now-rejected beliefs about nature (and associated investigative practices) be called scientific? Kuhn is no Whig: Rather than dumping rejected beliefs and practices into the non-scientific dustbin of myth, error, superstition, and ignorance, he asks instead how reasonable people *doing science* at earlier times could accept as valid beliefs and practices that look so obviously wrong to us today (Kuhn, 1962/1970, pp. 2-3, 1977a, p. xii). Kuhn needs criteria for science independent of the content of provisionally valid beliefs and legitimate practices, because something identifiable as science persists through often revolutionary changes in its content.

Both puzzles are solved with Kuhn's "discovery" of paradigms, and the degree and kind of consensus among practicing scientists that paradigms engender. In place of a progressive, linear, and cumulative movement toward the present and best comprehension of nature, science for Kuhn moves through fits and starts along distinctive historically specific trajectories. Periods of normal science are punctuated by revolutions and these, with time, are closed off by a new normality—which compares with earlier normal science not as a progressive improvement toward more accurate or encompassing models of nature but as an incommensurable way of thinking about nature and how it should be understood. Kuhn's model of the history of science hinges on the multivalent concept of paradigm (Masterman, 1970), which, in a sense, consists of background assumptions about the way the natural world works—coupled with methodological and theoretical exemplars or models that translate those deep assumptions into working rules to guide the selection of problems and acceptable procedures for their solution. Research in normal science is puzzle solving where the perimeter frame, the cut-out pieces, and the spaces to be filled in are specified by a paradigm. On occasion, some anomalous pieces cannot be made to fit, and when this happens at a time when another puzzle paradigm becomes available, science undergoes a temporary period of revolutionary alternation between frames of meaning. Eventually, a new paradigm attracts most practitioners and wins, not by convincing scientists with its superior logic or empirical evidence but through a non-rational gestalt-switch conversion grounded more in the psychology of perception and the sociology of commitment than in methodology.

The periodically successful achievement of paradigmatic consensus within a research community separates mature science from immature science, social science, Baconian science, art, technological craft work, astrology, and other realms of non-science. Referring to investigators of electricity (Cavendish, Coulomb, Volta) in the late eighteenth century, Kuhn (1962/1970) writes: "They had . . . achieved a paradigm that proved able to guide the whole group's research. . . . it is hard to find another criterion that so clearly proclaims a field, a science" (p. 22). And of Newtonian physics, he writes: "Paradigms prove to be constitutive of science" because they "provide scientists not only with a map but also with some of the directions essential for map-making" (Kuhn, 1962/1970, p. 109). Paradigmatic consensus is both unique to science and essential for its successes at filling in the puzzles (though not sufficient; Kuhn, 1977a, pp. 60-65, 231, n. 3). We arrive at a third essentialist demarcation principle: "Work under the paradigm can be conducted in no other way, and to desert the paradigm is to cease practicing the science it defines" (Kuhn, 1962/1970, p. 34, cf. pp. 17, 76, 60). "That commitment and the apparent consensus it produces are prerequisites for normal science, i.e., for the genesis and continuation of a particular research tradition" (p. 11).

In *The Structure,* Kuhn spends considerable time discussing knowledges and practices that are not part of mature and normal science, to show that the presence or absence of paradigmatic consensus can be used to distinguish real science from something else. He constructs the intriguing category of "sort of" science (Kuhn, 1962/1970, p. 11), which, at the level of individual practice, mimics mature science but in its collective accomplishments fails to achieve the coherent progress that characterizes fully developed science (filling in the puzzle). Speaking of pre-Newtonian optics, Kuhn argues that "though the field's practitioners were scientists, the net result of their activity was something less than science," because they could "take no common body of belief for granted, [and] each writer . . . felt forced to build his field anew from its foundations" (p. 13, cf. pp. 101, 163). The lack of consensus is generally apparent in the "interschool debates" of "immature sciences," and in the "prehistory of science" (p. 21). Moreover, when a paradigm does come along to offer a more rigid definition of a field, those who choose not to conform "often simply stay in the departments of philosophy from which so many of the special sciences have been spawned" (p. 19). And philosophy is not science because "there are always competing schools, each of which constantly questions the very foundations of the others." Once a paradigm is in place, for a researcher to abandon its worldview without hopping to an alterative puzzle is tantamount to leaving science: "To reject one paradigm without simultaneously substituting another is to reject science itself" (p. 79). Kuhn's demarcation of science from social science suggests *how* (in part) paradigmatic consensus makes real science mature. The paradigm provides an agenda of interesting research problems for scientists to attack and insulates their work from possibly competing agendas coming into science from the wider society. With no paradigm in place, social scientists lack this insulation, which, by extension, feeds their incessant and ideological arguments over first principles and prevents their coherent progress (Kuhn, 1962/1970, pp. 21, 37). Finally, Kuhn agrees with Popper that astrologers were non-scientific, but not because they failed to put their claims in a falsifiable form; rather, "they had no puzzles to solve and therefore no science to practice" (Kuhn, 1977a, p. 276).

Having worked hard to set up paradigmatic consensus as an essentialist demarcation principle, it is puzzling then to see Kuhn dismiss the practical problem of *defining* science. He asks rhetorically: "Can very much depend upon a *definition* of science? Can a definition tell a man whether he is a scientist or not? If so, why do not natural scientists or artists worry about the definition of the term?" (Kuhn, 1962/1970, p. 160). Kuhn implies that scientists rarely worry about definitions of themselves. But as the upcoming review of constructivist studies of boundary-work will make abundantly clear, scientists do on occasion worry about the definition of science, because they use those definitions to tell others that they are not scientific, in episodes where much depends on how a definition is played out. Kuhn is an essentialist not only because he offers paradigmatic consensus as a demarcation principle but because he dismisses as unimportant, merely "semantic," those questions that animate constructivist studies of boundary-work.

What Kuhn chose not to consider is that the degree of consensus in science itself might be a matter of interpretation, negotiation, and settlement—by scientists and sometimes other involved parties. When does a research community reach the level of paradigmatic consensus that moves it automatically from immature to mature science? And who answers that question, participants or their analysts? Gilbert and Mulkay's study of the reception of chemiosmotic theory in biochemistry treats consensus as part of scientists' interpretative work as they construct and give meaning to the history of their field. Consensus cannot be treated "as a typical collective phenomenon, that is, as a potentially measurable aggregate attribute of social groupings which exists separately from the interpretative activities of individual participants" (Gilbert & Mulkay, 1984, p. 139). Instead, consensus is a contextually contingent product of scientists' variable interpretative procedures, which means that, for Kuhn to conclude analytically that consensus *exists* in a research community at a designated time, he must ignore potentially wide discrepancies in scientists' own sense of the degree and kind of consensus they supposedly share.

In particular, Gilbert and Mulkay suggest that scientists must themselves solve three interpretative problems as they consider what consensus means and whether their field has it. First, they must decide the limits of membership of their research community, for inclusion or exclusion of certain individuals could easily affect their conclusions about the extent of consensus. Second, they must reach judgment on the changing beliefs of other scientists in regard to chemiosmosis: Who accepts it, and when did their conversion to the new framework occur? Third, scientists must decide the cognitive content of the chemiosmotic theory: If there is consensus, just what does the community agree on? From interviews, Gilbert and Mulkay find considerable variation among scientists (and even within a scientist in different situations) in how these interpretative problems are worked through. The boundaries of the research community were discrepantly drawn, beliefs were attributed to other scientists in varying ways, and chemiosmotic theory came to mean quite different things. Gilbert and Mulkay's (1984) findings raise doubts about whether Kuhn's paradigmatic consensus can essentially distinguish science from non-science if "a given field at a particular point in time cannot be said to exhibit a specifiable degree of consensus" (p. 140). The stage is set once again for empirical investigations of how consensus, among many other possible attributions, is used by people in society to construct a border between science and something else.

From Demarcation to Boundary-Work

Constructivist studies of scientific knowledge and practice[4] raise doubts about the ability of criteria proposed by Popper, Merton, and Kuhn to demarcate science from non-science—but in their wake comes a paradox. If there is nothing inherently, universally, and necessarily distinctive about the methodology, institution, history, or even consequences[5] of science, then why and how is science today routinely assigned a measure of "cognitive authority" rarely enjoyed by other cultural practices offering different accounts of reality? Paul Starr (1982) defines *cultural authority* as "the probability that particular definitions of reality . . . will prevail as valid and true" (p. 13). Few would doubt, in modern Western societies, that science has considerable cognitive authority: "Science is next to being *the* source of cognitive authority: anyone who would be widely believed and trusted as an interpreter of nature needs a license from the scientific community" (Barnes & Edge, 1982, p. 2). *On what grounds* is this authority warranted, if not for some epistemological or social quality essential to science and not found outside it? The challenge is to explain the cognitive authority of modern science without attributing to it essential qualities found by sociologists to be anything but essential.

The boundary problem in science studies is, in effect, an attempt to get around this paradox. The object of sociological study is no longer practices of scientists at the lab bench or their accounts of nature in professional journals; no explanation for the cultural authority of science could be found there without succumbing to the essentialism of Popper, Merton, or Kuhn. Instead, attention shifts to representations of scientific practice and knowledge in situations where answers to the question, "What is science?" move from tacit assumption to explicit articulation. The task of demarcating science and non-science is reassigned from analysts to people in society, and sociological study focuses on episodes of "boundary-work": "the attribution of selected characteristics to the institution of science (i.e., to its practitioners, methods, stock of knowledge, values and work organization) for purposes of constructing a social boundary that distinguishes some intellectual activity as non-science" (Gieryn, 1983, p. 782).

Boundary-work occurs as people contend for, legitimate, or challenge the cognitive authority of science—and the credibility, prestige, power, and material resources that attend such a privileged position. Pragmatic demarcations of science from non-science are driven by a social interest in claiming, expanding, protecting, monopolizing, usurping, denying, or restricting the cognitive authority of science. But what *is* "science"? Nothing but a *space*, one that acquires its authority precisely from and through episodic negotiations of its flexible and contextually contingent borders and territories. Science is a kind of spatial "marker" for cognitive authority, empty *until* its insides get filled and its borders drawn amidst context-bound negotiations over who and what is "scientific."

Put another way, the authority of science is reproduced as agonistic parties fill in the initially empty space with variously selected and attributed characteristics, creating a cultural map that, if accepted as legitimate, advances their interests. In these cartographic contests over *distributions* of scientific authority among diverse people, practices, and knowledge claims, the link between authority and the space marked "science" is made ever more secure. Whatever ends up as inside science or out is a local and episodic accomplishment, a consequence of rhetorical games of inclusion and exclusion in which agonistic parties do their best to justify their cultural map for audiences whose support, power, or influence they seek to enroll. Crucially, the "essential features" of science are provisional and contextual *results* of successful boundary-work, not determinants of who wins. Why are some maps more persuasive than others? There are no *general* determinants of success at cultural cartography, but clearly it helps if your depiction of the edges of science makes the interests of powers-that-be congruent with your own plans.

"Unique" features of science, qualities that distinguish it from other knowledge-producing activities, are to be found not *in* scientific practices and texts but in their representations. Boundary-work stands in the same relationship to what goes on in laboratories and professional journals as a topographic map to the landscape it depicts; both *select* for inclusion on a cultural or geographic map those features of reality most useful for achieving pragmatic ends (legitimating authority to knowledge claims or hiking through wilderness). Geographic cartographers often make new maps without constant reinspection of reality outdoors by "copying" and selectively editing extant maps to suit changing needs or wants, and the same is so for cultural cartographers. Those contesting the borders of science select from and creatively reconstruct past episodes of boundary-work, often using old maps to legitimate the validity of their own. Importantly, neither actual scientific practice and discourse in labs or journals nor earlier maps showing the place of science in the culturescape *determine* how the boundaries of science will get drawn next time the matter comes up for explicit debate. Rather, opposing maps are better understood in terms of immediate (and dynamic) interests and goals of their cartographers and the uses to which they are put (i.e., convincing people of one's cognitive authority or denying it to somebody else). In this sense, then, the space for science is empty because, at the outset of boundary-work, nothing of its borders and territories is given or fixed by past practices and reconstructions in a deterministic way.

But that idea could easily be exaggerated into a silly conclusion that every episode of boundary-work occurs de novo, and that there are no patterns at all from one episode to the next. Scientific practices and antecedent representations of it form a *repertoire* of characteristics available for selective attribution on later occasions. That repertoire is presumably not limitless; it might be extremely challenging these days to persuade others that eye of newt and toe of frog make witches purveyors of "good laboratory practice." Interpretative flexibility in the boundaries of science need not imply infinitely pliable;[6] some maps of science are easier than others to defend as bona fide representations, in part because some cartographers are more easily able to point to specific concrete practices or to earlier mappings as rhetorical "evidence." Indeed, some maps achieve a provisional and contingent obduracy that may preempt boundary-work. Borders and territories of cultural spaces sometimes remain implicit, matters of personal belief or of such apparent tacit intersubjective agreement that people working together need not explicate "what everybody knows" about the meaning of science.

Such stability in the borders of science could itself be overestimated, as if the issue were settled centuries ago once and for all. *Boundary-work abounds* simply because people have many reasons to open up the black box of an "established" cartographic representation of science—to seize another's cognitive authority, restrict it, protect it, expand it, or enforce it. A survey of historical instances of boundary-work would turn up a science with no consistent shape, no necessarily enduring features. One would find massive diversity in the characteristics attributed to science and used to demarcate it from something else (e.g., theoretical, empirical, certain, uncertain, useful, useless, finite, endless, quantitative, qualitative, precise, imprecise, inductive, deductive) just as one would find in "something else" a massive diversity of non-sciences (pseudoscience, amateur, bad science, fraud, Marxism, popularizations, politics, technology, management, religion, philosophy, art, mechanics, craft, social science, and so on). It is precisely the emptiness of science—a space waiting for edging and filling—that best accounts for its historically ascendent cultural authority.

The turn from essentialist studies of demarcation to constructivist studies of boundary-work is well under way[7], as the next two sections will report.

THEORETICAL NOURISHMENT

Contingent constructions of boundaries among cultural domains—like science|non-science—have been signaled as consequential by disparate lines of inquiry in the social sciences and humanities. None of the four theoretical frameworks discussed in this section were developed within the sociology of science, but they intersect with it especially well when generic problems of cultural categories, classifications, and boundaries are raised. Especially mysterious is the neglect of recent, much improved, studies of the professions.

Sociology of Professions

Sociological study of the professions once was marked by its own brand of essentialism. From Carr-Saunders and Wilson's (1933) pathbreaking work to Parsons's (1939/1954) functionalist explorations, the hunt was on for distinctive characteristics that set the professions (paradigmatically, medicine and law) apart from other lines of work and occupations. This analysis of professions consisted first of their definition: An occupation is a profession *if* its practices are grounded upon a body of abstract theoretical knowledge, applied skillfully to case-specific solutions of paramount problems of human existence (death, disease, injustice, sin), all done in interactive contexts of altruistic service where professional practitioners become fiduciary trustees of clients' interests and welfare (see Merton & Gieryn, 1978). When analysis proceeded along functionalist lines, the goal was to explain the occurrence of these distinctive attributes in terms of their consequences for effective provision of expert services to clients. For example, institutionalized norms of altruism (or service) created an atmosphere where clients could trust professionals in spite of an asymmetry of expertise that otherwise would make clients vulnerable to exploitation. The parallel to Popper, Merton, and Kuhn—with their definitive lists of distinctive attributes of science functionally necessary for progressive advance toward increasingly valid and reliable knowledge claims—needs little amplification.[8]

A sea change in sociological thinking about professions occurred when the by-products of professional work—control over tasks and knowledge, esteem, authority, material wealth, power, and autonomy from outside evaluation—moved to the analytic foreground. How have practitioners of some kinds of work managed to secure relatively large amounts of these resources in the historical course of their becoming professions? How have other kinds of workers lost those resources, or failed to attain them in the first place? Various theoretical strategies have developed to answer these questions.[9] The "professionalization" model focuses on the organizational structure of occupations, emphasizing key moments in the natural evolution of some work toward the status of profession. Licensing legislation, national associations, professional schools, codes of ethics, and specialized journals each increase the professions' control over individual practitioners (e.g., Wilensky, 1964). The "monopoly" model sees professionalization as the successful pursuit of dominance, authority, and power by a corporate group of workers seeking upward collective and individual mobility. Organizational features such as licensing legislation or codes of ethics enable the monopolization of a market for professional services, which in turn enlarges the prestige, wealth, and power of practitioners (Larson, 1977) and enlarges as well the security of their domination in professional-client relationships (Freidson, 1970). Monopolization is achieved not only through internal organization and control of practitioners and tasks, but also through the linkage of these skills and services to values and interests of powers-that-be: an elite class, corporate capital, or the state.

The question of boundaries runs throughout this literature on professions but moves to center stage in Andrew Abbott's *The System of Professions* (1988). Abbott rejects the idea that the development of a profession is independent of the practices and claims of *other* professions or differently organized occupations. Instead, the history of one profession is best understood through its contests with other professions for jurisdictional control over tasks—the professions together constitute a system that is the proper unit of analysis for sociological theorizing. Abbott focuses on *inter*professional competitions created when more than one occupational group lays claim to the legitimate provision of three tasks—diagnosis, inferential interpretation, and treatment of problems. The book offers a structural model (tempered by sensitivity to historical contingencies and actors' initiative) to explain the causes of jurisdictional contests among professions, the mechanisms through which one or another side gains advantage, and the variety of settlement patterns that restore the system to equilibrium. The boundary problem is ubiquitous: How do tasks and competences map onto the ecology or geography of professions and their abutting occupations?

Implications of Abbott's model for understanding the constructed boundaries of science are several and significant. First, he identifies three *arenas* in which jurisdictional contests are fought, each with different mixes of combatants, patterns of competition, and settlement: the legal world of legislatures and courts, the public world of media representations and popular opinion, the actual work site of professional practice. The precision with which interoccupational borders are drawn (as tasks are assigned to some individuals and not others) is clearest in legal settings and murkiest in the workplace. The law (through licensing, for example) sets up neat homogeneous boxes of professional jurisdiction that get blurred in the pressures of providing heterogeneous services to heterogeneous clients. Public understanding of professional jurisdictions seems to change more slowly than their definition in law or on the shop floor, requiring professionals occasionally to account for discrepancies in popular accounts broadcast through the media. It would no doubt be profitable to compare representations of science as they arise in, say, the mass media (when a new scientific discovery—or infelicity—is announced), in the courts (when control over the science curriculum is adjudicated), and in the laboratory (as tasks are variously assigned to principal investigators, technicians, students, and clerical staff).

Second, Abbott describes the range of factors that bring to a head jurisdictional disputes among workers and, in doing so, successfully avoids the unwarranted implication that border wars are merely instrumentalist games of greedy professions wanting to extend control to new or somebody else's tasks.[10] Beyond individual or corporate intentions in making a jurisdictional claim, identifiable social structural and cultural circumstances affect the frequency and distribution of interprofessional contests. These include technological changes, changes in organizational settings (such as bureaucratization), co-optation of professions by external powers, shifting values that legitimate (or not) jurisdictional claims, and, finally, the growth of modern universities (as sites for credentialing and for the creation of new knowledge). Episodes of boundary-work are not random occurrences, nor do they result simply from desires of incumbent scientists (or their professional rivals) to expand their jurisdiction. How do changes in the rest of society—technological, cultural, organizational—create situations where boundary-work involving science becomes more or less likely?

Third, Abbott describes the processes through which jurisdictional competitions among professionals are settled. Evidently, there are no surefire predictors of success in professional boundary disputes. Abbott restricts himself to identifying diverse factors related to success or failure in something less than a deterministic way. For example, in most circumstances, the profession with a better-organized national association will compete more effectively than one without, simply because of the importance of "getting the word out" about a jurisdictional challenge or opportunity—both to persuade publics and to mobilize practitioners. Abbott also attaches importance to the form and content of abstract knowledge on which professional expertise rests. Too much formalization or codification of expert knowledge makes tasks vulnerable to routinization, and they may be easily usurped by less expensive paraprofessionals (or delegated to subordinates). At the other extreme, too little formalization makes its jurisdiction vulnerable to technological or other exogenous changes. Railroad dispatchers grounded their practice on knowledge specific to the running of trains, and so were not able to move their jurisdiction to possibly related tasks when railroading went into decline (they disappeared). A more abstract knowledge base might have allowed that profession to redefine its expertise and transform itself into "operations research" with jurisdiction over any transportation system that might come down the pike.

Fourth, Abbott identifies rhetorical strategies that professionals use to argue jurisdictional claims: *Reduction* reasserts professional control over a task by showing it to be the same as other tasks more obviously within the professions' domain; *metaphor* extends professional control to new tasks by demonstrating how they are "like" other tasks within the professions' jurisdiction; *agency* appeals to previously demonstrated utility of professional practice; *gradient* argues that control of severe instances of a problem translates into control over more mild problems (e.g., psychiatrists extending their jurisdiction from psychosis to neurosis). Are these same rhetorical patterns used by speakers-for-science (or by their opponents) when they seek to discriminate their work and its products from something else?

Finally, Abbott offers a useful typology for describing settlement patterns in jurisdictional competitions. Each of the seven resolutions differently assigns tasks to professions. Two outcomes stand at opposite extremes: A profession might be left with *full jurisdiction* over a task (complete and exclusive control of its exercise) or it might be left with *no jurisdiction* at all (either because the tasks have been assumed by other professions or because there was no one who wanted or needed that task done). Four other settlement patterns involve divided jurisdiction. *Subordination* involves the delegation of some tasks to non- or other professionals under the control of a dominant profession (physicians' delegation of bedside care to nurses; scientists' delegation of "routine" experimental procedures to "invisible" technicians—Shapin, 1989). *Division of labor* here refers to splitting tasks between two interdependent professions with more or less equal structural power and resources (generally, architects and engineers are both needed for different phases of getting a building up). *Intellectual* control splits the production and control of the abstract knowledge base from its applications, and divides these tasks among different professions or assigns them to separate parts of one. *Advisory* control means that a profession may participate in another's jurisdiction, but only to advise or interpret for clients decisions made by the profession in charge—ministers or priests on hospital wards is the paradigmatic case. The final settlement pattern—*clientele differentiation*—is distinctive in that once-competing professions in effect continue to do the "same" tasks but do them for different clients or markets (perhaps stratified by social class) and often under different labels. Does this typology of settlements make sense as a description of jurisdictional boundary disputes involving scientists? Are some resolutions in the history of science more common than others? Why, and with what effects?

In applying Abbott's model to the case at hand, the question is clearly not whether science is or is not a profession. That would move back in an essentialist direction in which some analytic definition had the force to make an occupation (like science) into a profession. Far better to ask the same questions of science that Abbott (1988) asks of law, medicine, and librarianship. Why do the professions (and science) "dominate our world"? "Why should there be occupational groups controlling the acquisition and application of various kinds of knowledge? Where and why did groups like medicine and law [or science] get their power?" (p. 1). Do his concepts and models—with their central focus on jurisdictional competition among professions—help to answer those questions when asked about science?

Social Worlds

The transition from Abbott to the investigation of "social worlds" is easy to find, for each can be traced to Everett Hughes and, more broadly, to symbolic interactionism and pragmatist philosophy. Hughes pushed sociologists to examine *how work gets done,* focusing attention on the work site as a place "where diverse people meet" (see Hughes, 1971). The meaningful construction or representation of "the work itself" (i.e., its location in cultural space) is here assumed to be *part of* that work (whether of doctors and lawyers, or artists and scientists). The implication is that boundary-work should be found at the work sites of science as well as sites where those activities are more visibly reconstructed, such as courts, mass media, public speeches, and legislative bodies (Clarke & Gerson, 1990).

Those who examine science as a social world make three contributions to the boundary problem in STS. First, by revealing the diverse and often unexpected set of people needed to "get work done," these symbolic interactionists challenge conventional categories of art or science, and thus sensitize researchers to the problem of how and why boundaries of those categories came to be "conventional." Second, through detailed ethnographic and historical studies of scientific or artistic practices, these sociologists show *how* boundary-work and other representational activity constitutes the work itself. Third, starting from the observation that science is the intersection of multiple social worlds, they turn away from Abbott's interest in the jurisdictional *competition* engendered by such contact and instead examine how the *cooperative* pursuit of tasks is accomplished in spite of boundaries that could prevent separated social worlds from achieving ends collectively.

A "social world" is a group with shared commitments to the pursuit of a common task, who develop ideologies to define their work and who accumulate diverse resources needed to get the job done. Empirical embodiment of a "social world" is a function of researchers' interests and problems-at-hand; science itself may be a social world, made up of many social worlds, or part of a more encompassing social world. Gerson (1983) identifies three kinds of social worlds: Production worlds make something (science manufactures facts); communal worlds pursue community and shared values; social movements seek change in the society beyond the borders of their world. Three properties are common to all social worlds: segmentation (potentially endless division into subworlds), intersection (where social worlds meet), and legitimation (definition and enforcement of standards and boundaries of a social world). Because membership in social worlds is fluid and non-obvious, those who study such collectivities appreciate the significance of boundary problems: "Very important activities within all social worlds are establishing and maintaining boundaries between worlds and gaining social legitimation for the world itself. These processes involve the social construction of the particular world and a variety of claims-making activities" (Clarke, 1990a, p. 20; cf. Strauss, 1978, 1982).

Who does art—or science? The production of works of art necessarily involves people not customarily thought of as artists; the bored man in the photograph on the cover of H. Becker's *Art Worlds* (1982)—hauling down a portrait from the museum wall for storage—is as much a part of the art world as the painter, curator, critic, or patron. Susan Leigh Star and James Griesmer (1989) show an equally eclectic mix involved in the production of knowledge at the Museum of Vertebrate Zoology at Berkeley: professional biologists, trappers, a wealthy patron, university administrators, amateur naturalists, and environmentalists. The crucial analytic move is to begin inquiry *not* with the conventional categories of art or science but with tasks or activities that generate promiscuous lists of people with commitments to their achievement. If museum custodians are found in art worlds, and animal trappers in science worlds, the sociological question becomes: Why are such individuals not *conventionally* defined as doing art or science? Through what boundary-work are such people defined as peripheral, marginal, or outside art or science —why, how, when?[11]

Doing art involves the negotiation of who and what is art, as part of the productive activity of this social world. The same goes for science. In her study of nineteenth-century British neurophysiologists, Star describes their local uncertainties and strategies for managing them. For example, neurophysiologists faced technical uncertainties (experiments that did not work consistently) as well as political uncertainties (unreliable funding sources), many of which were managed effectively (standardization of protocols reduced experimental failures, while restricting experiments to research materials at hand was a useful response to sporadic funding). *Management* of these uncertainties is a vital part of doing scientific work, says Star, but it is not defined as such by scientists themselves. See the boundary-work described by Star (1985): "Scientists in the field often characterize the management of local uncertainty as 'not really science' (and with pejoratives including 'mere administrative work,' 'bean-counting,' 'mere logistic,' or even 'sociology')" (p. 393; see Star, 1989a). Scientists are more likely to define real science as moving claims outward from local uncertainties toward global certainties—the production of Truth. Importantly, boundary-work to locate "strategies for managing local uncertainties" *outside* science contributes to this goal of claiming global certainties; as the management of uncertainties is jettisoned in the course of writing up Real Science for publication in professional journals, contingencies disappear and claims no longer appear as produced in a local setting but as found in a universal Nature (a process described by Latour and Woolgar, 1986, as "splitting and inverting"; p. 176). Exclusion of "management of local uncertainty" *from* science is as "central to research organization" (Star, 1985, p. 393) as the management strategies themselves.

Cambrosio and Keating (1988) offer a second example where boundary-work becomes an integral part of making science. Techniques for producing monoclonal antibodies in contemporary biotechnology are variously defined as art, science, and magic; the location of this production process in one or another cultural category (that's the boundary-work) has important consequences for how the work is evaluated and controlled. For example, characterization of hybridoma technology as "scientific" putatively frees it from hands-on, artisanal, and place-specific skills so that it can circulate easily from lab to lab. Moreover, to call the latest manufacturing process for antibodies "scientific" enhances its superiority over processes burdened with the idiosyncracies and uncertainties of art or magic. On the other hand, to describe the technology as art allows scientists who make hybridomas to escape the routinizing control of industrial practice—as art, such techniques require skill and technical virtuosity found in university labs but rarely in bioindustrial assembly lines. Obviously, the issue is not whether the manufacture of monoclonal antibodies is really art or science but the consequences of such boundary-work for how the process is defined, carried out, and controlled.

Abbott's model made competition for jurisdictional control a necessary by-product of the meeting of two professional domains. Enthusiasts of social worlds analysis see such contacts as opportunities for cooperative work toward collective ends. Indeed, for Clarke (1990b), Fujimura (1987), and Star and Griesmer, science is the intersection of bounded social worlds: How do people from distinctive social worlds—with commitments to activities and interpretations different than those across the border—come together to get something done? Star and Griesmer in effect ask how Berkeley's Museum of Vertebrate Zoology was accomplished, and how it became a vehicle for doing science. For the museum to become a going concern, commitments (of time, skill, vital resources) were needed from people inhabiting diverse social worlds each with its own activities and interests: professional biologists, amateur naturalists, trappers, university administrators, the general public, philanthropists, conservationists, and taxidermists all became part of this particular science world. But how? The useful concept of "boundary object" begins an answer, defined by Star and Griesmer (1989) as objects "which inhabit several intersecting social worlds and satisfy the informational requirements of each of them." They are objects "plastic enough to adapt to local needs and the constraints of several parties employing them, yet robust enough to maintain a common identity across sites" (p. 393). Boundary objects may be ideas, things, people, or processes; the requirement is that they be able to span boundaries separating social worlds, so that those on either side can "get behind" the boundary object and work together toward some goal. For example, standardized forms for recording information about the ecological circumstances in which a zoological specimen was found served as a boundary object linking professional scientists to amateur naturalists, conservationists, and even trappers. The form was set up by biologists at the museum to ensure that the necessary ecological information was attached to each specimen so that it would be scientifically usable, but the form was kept simple and short enough so that non-professional biologists could easily fill it out. The boundary-work here is double-edged in interesting ways. The creation of the museum was in effect a means for *professional* zoologists to claim greater scientific authority for their work by distancing it from activities of amateurs and conservationists; but amateurs were a vital element of the museum's scientific success, as providers of specimens and information that incumbent scientists were too few to gather by themselves.[12] Standardized procedures for handling specimens, and forms for recording information, become boundary objects in that they sustained a boundary between professional and amateur, while allowing the boundary to be bridged through mutually cooperative work in service of the museum's goals. The State of California itself becomes another boundary object, linking up the interests of local conservationists in preserving the wildlife of the state, the University of California administration with its mandate to serve its residents, and professional zoologists seeking worldwide prestige by collecting research materials immediately at hand for them but no one else. In sum, Star and Griesmer point to a different sociological dimension of the boundary problem in science, one worth further exploration: What are the consequences for doing science when boundaries are laid down in such a way that those people with vital parts to play find themselves in separated social worlds? How are the potential *barriers* of constructed boundaries among social worlds overcome in cooperative pursuits of science?

History of Cultural Classifications

Sociological studies of professions and of social worlds open up new perspectives on the boundary problem, but they share one big limitation. Negotiations of boundaries need not necessarily be tied so instrumentally to pursuit of occupational interests and resources, as the literature on professions often has it. Nor must such negotiations be restricted to concrete actors working together on a project, as portrayed by the social worlds' tradition. A different set of questions is raised by seeing science as a *cultural space,* that is, as part of enduring cartographic classifications of cultural territories that people use to make sense out of the world about them. Science becomes a bounded space on everyday pragmatic cultural maps—with other territories such as politics, religion, superstition, markets, family—that serve as relatively sedentary interpretative frameworks and guides to practical action by people located throughout society.

Cartographic metaphors are everywhere in the anthropology of Clifford Geertz (1973a, p. 21): "Culturescape" forces one to think of beliefs, actions, and institutions in the manner of mountains, rivers, buildings, and other features of terrain. Maps are needed to get around these lands, physical and cultural, and what is a map without boundaries? "It is in country unfamiliar emotionally or topographically that one needs poems and road maps" (p. 218). Sometimes the cultural map is a poem, but it can also be an ideology: "Whatever else ideologies may be—projections of unacknowledged fears, disguises for ulterior motives, phatic expressions of group solidarity—they are, most distinctively, maps of problematic social reality and matrices for the creation of collective conscience" (Geertz, 1973a, p. 220). To see science as a bounded space in the culturescape affords not just an explanation of the rise of a professional power but a better understanding of historically changing "webs of significance" (p. 5). "The whole point of a semiotic approach to culture is . . . to aid us in gaining access to the conceptual world in which our subjects live so that we can, in some extended sense of the term, converse with them" (p. 24). What does science mean for those who *use* it (not just produce it)?

Mary Douglas suggests that the classification of natural or cultural things is vital for whatever might be called "structural" in social life. Mapping out domains of culture (like science or art) is an important intellectual (Douglas, 1986a, p. 45) part of their "institutionalization": "The labels [classifications] stabilize the flux of social life and even create to some extent the realities to which they apply. . . . It is much more of a dynamic process by which new names are uttered and forthwith new creatures corresponding to them emerge" (p. 100). Once labeled, the "entrenching of an institution" (p. 45) involves an increasing fixity of the cultural classifications, so that eventually "all the classifications that we have for thinking with are provided ready-made, along with our social life" (p. 100). But these classifications—these institutions, structures, maps—retain a measure of malleability in all cases, so that people can challenge and perhaps change even the most hardened of them:

> This is how the names get changed and how the people and things are rejigged to fit the new categories. First the people are tempted out of their niches by the possibilities of exercising or evading control. Then they make new kinds of institutions, and the institutions make new labels, and the new label makes new kinds of people. (p. 108)

Douglas's sequence of steps is not how I would line things up; institutions themselves must be labeled as such before they in turn can get on with their own subsequent labeling of things and people.

For Geertz and Douglas, cultural classifications are significant for the problem of meaning: What does science *mean* for people of different times and places? It took the French to move the boundary problem from meaning to power. Foucault (1980) admits to "spatial obsessions" (p. 69), which enable him to see connections between power/knowledge: "Once knowledge can be analyzed in terms of region, domain, implantation, displacement, transposition, one is able to capture the process by which knowledge functions as a form of power and disseminates the effects of power" (p. 69). Fundamental to this spatialization of knowledge are processes of boundary drawing, which Foucault talks about in terms of inclusion/exclusion: At any historical moment, what can or cannot be said, who is entitled to say what, what will count as truth? Answers to these questions must be sought through interpretation of the changing borders and territories of what is taken as science. "What was striking in the epistemological mutations and transformations of the seventeenth century is to see how the spatialization of knowledge was one of the factors in the constitution of knowledge as a science" (Foucault, 1984, p. 254). That process is never far from the exercise of power:

> The spatializing description of discursive realities gives on to the analysis of related effects of power . . . [T]he formation of discourses and the genealogy of knowledge need to be analyzed, not in terms of types of consciousness, modes of perception and forms of ideology, but in terms of tactics and strategies of power . . . deployed through . . . demarcations, control of territories and organizations of domains. (Foucault, 1980, pp. 70-71, 77)[13]

Bourdieu's idea of a scientific field—as the locus of competitive struggle—takes up Foucault's interest in spatializations of knowledge and links it to power but veers back toward an interest-based reductionism:

> What is at stake is in fact the power to impose the definition of science (i.e., the delimitation of the field of problems, methods and theories that may be regarded as scientific). . . . The definition of what is at stake in the scientific struggle is thus one of the issues at stake in the scientific struggle. (Bourdieu, 1975b, pp. 23-24)

A better understanding of classifications in social space has also been pursued by the "new" cultural history. Roger Chartier (1987, 1988), Natalie Zemon Davis (1975), and Robert Darnton (1984) have restored the problem of meaning to centrality in a discipline that—under Marxist influences—had buried it under the presumed primacy of economics or politics (Hunt, 1989, pp. 4 ff). "The deciphering of meaning rather than the inference of causal laws of explanation, is taken to be the central task of cultural history" (Hunt, 1989, p. 12), a job requiring explicit attention to "the classifications, divisions and groupings that serve as the basis for our apprehension of the social world as fundamental categories of perception and evaluation of reality" (Chartier, 1988, pp. 4-5). Recovery of cultural categories—*where* is science?—is not set apart from study of politics and economics; cultural classifications constitute those other realms: "Economic and social relations are not prior to or determining of cultural ones; they are themselves fields of cultural practice and cultural production" (Hunt, 1989, p. 7). Chartier (1988) defines the new cultural history in a way that puts Marxist notions of base/superstructure to rest: "Culture is not over and above economic and social relations . . . all practices are 'cultural' as well as social or economic, since they translate into action the many ways in which humans give meaning to their world" (p. 11). Finally, the new cultural history identifies the back-and-forth relationship between institutionalized categories and those momentary, contingent, discursive, and pragmatic enactments of categories that make up boundary-work. "We [must] accept the schemata found in each group or milieu (which generate classifications and perceptions) as true social institutions incorporating . . . mental categories and collective represen*tations*." But these institutions are neither universal nor beyond reconstruction, as Chartier (1988) continues: "Behind the misleading permanence of [categories of intellectual objects], we must recognize not objects but objectifications that construct an original configuration each time" (pp. 6, 46). Intellectual fields or spaces are not constituted once and for all but continuously reconstituted in discursive practices through which they sometimes achieve institutional stability and obduracy.

Georges Canguilhem and Michel Serres bring this free-range discussion of cultural space back to the case of science. Canguilhem (1988) sets up his historical project in a language by now familiar:

> The history of science is the history of a certain cultural form [space?] called "science." One must then specify precisely what criteria make it possible to decide whether or not, at any given time, a particular practice or discipline merits the name science. (p. 27)

Shifting the task of defining science from analyst to actor apparently has deep roots that extend well before the emergence of constructivism in science studies (Canguilhem was in his seventies when he wrote this passage). Cartographic imagery is found here as well, as Canguilhem (1988) warns historians against "reducing the history of science to a featureless landscape, a map without relief [or boundaries]" (p. 39). Canguilhem links his work to that of Serres (1982), who also sees himself doing a "cartography of knowledge" (p. xx). Both see the

> history of science [as] the victim of a classification [i.e., "science"] that it simply accepts, whereas the real problem is to discover why that classification exists, that is, to undertake [now quoting Serres] a "critical history of classifications." (Canguilhem, 1988, p. 18)

In Serres's own work, the history of science merges easily into histories of myth and literature, as he opens up the common borders of these cultural spaces to show their arbitrariness and historicity.

What do these anthropologies and histories of cultural classification add to our boundaries-of-science soup? Four ingredients, mainly: First, cartographic metaphors offer a robust language for thinking about relations among cultural phenomena such as science. Other useful metaphors have been developed for interpreting science, technology, and society—notably "networks" (T. P. Hughes, 1983; Latour, 1987). What features of science are exposed by seeing it as a place on a map? What is gained by looking for borders and spaces? Does it help to note that the boundaries of native categories like science vary in three ways—some are ambiguous, others clear; some are permeable (people or beliefs frequently crossing), others tight; some are malleable, others obdurate? Does it help to think of cultural spaces as distanced, conterminous, overlapping, or nested? What more can be learned about science by transporting additional concepts from the cartographers' lexicon—contours, landmarks, scale, orientation, coordinates, points of interest, and legend?

Second, this perspective does not privilege "what-has-become-science" as the object of inquiry in sociology of science, but instead compels analysis of how such a distinctive space was slowly carved out from a mixed bag of cultural practices and products. As a cultural space, science cannot be understand apart from the ground against which—with centuries of boundary-work —it now appears in relief. Historical *changes* in cultural classification of science—at both epochal and episodic levels—are the key.

Third, this orientation shifts attention from what once was fashionable in sociology of science (study of the social institution and scientists' careers) *and* from what has become more fashionable since (study of claims making and related practices). The focus is instead on the shifting borders and territories of science, on maps that people use to assign meaning, evaluate circumstances, plan strategies—as they read the newspaper, argue politics, decide who to trust or what to believe. Studies of boundary-work and science ought to examine not just busy cartographers who have immediate stakes in drawing and promoting a certain cultural map, but also consumers who use those maps selectively to get where they want to go.

Fourth, cultural historians offer help with a befuddling theoretical problem. Does science get constructed de novo each time someone is forced to decide which claim to believe? Of course not: Some maps of cultural spaces achieve a degree of stability and portability that allows them to be enacted in actual situations where people need help finding their way through domains of belief and knowledge. These on-hand maps never determine the content or outcome of particular instances of boundary-work, but provide a repertoire of familiar characteristics available for selective attribution to science or to some contested claims (claimants, investigative practices, and so on) but not others. To complete the circle, these maps get their familiarity—their obduracy—precisely from repeated or instantiated (Giddens, 1979, p. 64) attributions of the same qualities. Identifying this interactive and "structuring" relationship between an enduring cultural space and its episodic reconstruction makes it easier to explain why some science-wannabes are so easily and routinely kept outside *without* resorting to any essentialist, transcendent definition of science grounded in what it "really" is. The analytic danger is to reify the cultural space of science into something so stable, so "structural," or "institutionalized," that the significance of episodic reproductions in boundary-work is lost altogether. The antidote is careful empirical attention to processes through which cultural categories achieve or lose their obduracy—their more or less taken-for-granted reality. Transitional cultural maps are an accomplishment in need of sociological explanation: Why are some characterizations of the borders-and-territories of science more portable through space and time than others?

Feminist Studies of Science

Feminist scholarship brings a fresh take on the study of boundaries, one that combines a historical interest in how culturally rooted definitions of science have affected women, with a reflexive and pragmatic interest in rethinking the received boundaries of scientific methodology in light of feminist intellectual and political critique. Feminists who ask "the science question" (Harding, 1986) examine two sets of boundaries—gender and knowledge—and find compelling evidence for their intimate coevolution; centuries of double-boundary-work have moved whatever counts as science toward the masculine, and whatever counts as feminine away from science. Two questions in particular link feminist studies of science to the boundary problem in STS. First, why are women underrepresented in science, especially so at the highest levels of honor and power? Second, are the boundaries conventionally drawn around "legitimate" scientific practice themselves "androcentric" or "masculinist," so that their redrawing now becomes a task of top significance for feminist projects?

What has been the place of women in science? A small one, by most accounts, and typically near the margins. Or is that just an appearance? Margaret Rossiter's survey of women scientists in America through 1940 finds enough women doing science to fill a 400-page book but not so many to avert the conclusion that their place has been "historically subordinate" and (thus) invisible to most historians who tell what science was (Rossiter, 1982, p. xv). The task of understanding women's underrepresentation in science is two-fold: first, to suspend present-day, conventional boundaries of "real" science so as to recover the manifestly excluded or just overlooked contributions of women; second, to examine changing cultural boundaries drawn around both female and science that discouraged women from pursuing careers in science, made their journey difficult when they started down that road, and made the trip end far from the prestige, power, and resources achieved disproportionately by men (see Noble, 1992). Simply put, increasingly stabilized cultural boundaries of science excluded or marginalized women scientists, first, by pushing their inquiries and contributions to the edge (or beyond) and, second, by providing a normative context that legitimated stereotypical assumptions that any woman doing real science was not a real woman.

A greater number of women seem to be contributing to science, and a greater variety of investigative work seems to be scientific, once the boundaries of science become contingent pragmatic constructions used by people in society to interpret and change their worlds. Londa Schiebinger (1989) offers the well-known example of midwifery—and less familiar ones of medical cookery, home economics, and debates in Parisian salons—to illustrate how some investigative, discovering, and practically useful activities dominated or controlled by women get defined as prescience, superstition, applied arts, or the humanities while boundaries of real science get laid down in their professionalization. In tracing out the emergence of a feminine style of discourse and inquiry, Schiebinger wonders why the sociability found in French salons came to be defined as a poetic style distinctively feminine—even though many men, and important men of science, participated in these gender-integrated places of scholarly and scientific discussion. The developing boundaries of science—made increasingly obdurate through their institutionalization in credentialing and training systems, and through the shift of scientific work from private settings (like the home, where women have always been) to public settings at universities and national academies (where, for a long time, women were not)—chopped up traditions of inquiry, discovery, and practical applications along gender lines. Some traditions where women participated or even prevailed became non-science.

The same maps that moved these science-like contributions to the margins also had prior effects on individual choices and institutional arrangements that thinned the flow of women into science, and steered it toward contributions that came to be seen as ancillary. Co-evolving demarcations of male/female and science/other-knowledges overlaid masculine and scientific spaces in a way that discouraged many women from taking even the first steps toward careers in science, and structured the institutions of science (training centers, research organizations, universities) to make subsequent steps difficult (see Rossiter, 1982; Schiebinger, 1989). That boundary-work drew mythological maps (that were real in their consequences) that "cast objectivity, reason and mind as male, and subjectivity, feeling and nature as female." Such spatial gendering of values, continues Keller (1985), has led to a prevailing situation where "women have been the guarantors and protectors of the personal, the emotional, the particular, whereas science—the province par excellence of the impersonal, the rational and the general—has been the preserve of men" (p. 7). There is nothing necessary or historically universal about the coupling of science images with the masculine. The restriction of women to "soft, delicate, emotional, non-competitive, and nurturing" activities (nothing here sounding like real science) took full force in the nineteenth century, just when the feminization of the labor force—and of science along with it—was starting to gain a momentum that (by century's end) had dissipated (Rossiter, 1982, p. xv). Women were now constructed to be inappropriate for science, and (not by accident) science was trotted out to lend its growing authority to that myth. Biological studies of sex differences (especially in the late nineteenth and early twentieth centuries) discovered in the female *body*—brain, skull, pelvis—a machine less effective for the intellectual demands of science than for mothering (Schiebinger, 1989, p. 215). Harding (1991) concludes that "women know very well that knowledge from the natural sciences has been used in the interest of our domination and not our liberation" (p. 8).

The myth becomes a kind of self-fulfilling prophecy (Merton, 1948/1968) if one considers the typical career of women who chose to enter science. As Rossiter shows (1982, p. 314), the stereotypical cultural incongruence "female scientist" is a backdrop against which two forms of marginalization took place. Some women scientists were (are) hierarchically subordinated by assignments to tedious and anonymous data-mongering tasks thought to be ancillary to the real work of creating certain knowledge. Others were (are) segregated into generally lukewarm-to-cold scientific fields for which their feminine talents were thought to be especially appropriate: "home economics, botany, and child psychology."

To raise questions about how culturally contingent processes of boundary-work help to explain the modest and marginal place of women throughout the *history* of science has led some feminist philosophers and sociologists to ask an even more challenging question. Has the mapping of gender onto knowledge—male onto science—bounded scientific knowledge and practice into a space too small, or poorly located, for the emancipatory projects that complement the analytic projects of feminism? Does feminist inquiry require a different methodology for science, a kind of knowledge or way of knowing different than "science" but no less valid, useful, or credible (see Figert, 1992)? The question has taken on urgency in the wake of characterizations of "feminism" as itself involved in something other than science. Geertz shows himself to be not only a sensitive observer of others' boundary-work but an adept practitioner of the rhetorical form. He has recently written about feminist science and scholarship: "The worry is, of course, that the autonomy of science, its freedom, vigor, authority, and effectiveness will be undermined by the subjection of it to a moral and political program—the social empowerment—external to its purposes" (Geertz, 1990, p. 19). See the map Geertz just drew? Freedom, vigor, authority, and effectiveness become points of interest in the territory "science," placed across an inescapable border from another territory—"feminism"—with its own landmarks: moral and political programs for the social empowerment of women. Must this be the topographical relationship between science and feminism? Why? Is value-free science an androcentric plot?

The range of opinions about the need for (and promises of) distinctively feminist methodologies for science (and for its social analysis) is too wide to fit into this chapter. At the center of the fray is a boundary between science and politics (values, ethics) as volatile as it is ambiguous. Much of the credibility and cultural authority of science has been won through scientists' attempted exclusion of values from their space, a bit of boundary-work never sufficiently convincing to prevent incessant challenges. Feminism is one more mapping of the knowledgescape that relocates things scientific and not: On which side of the border go objectivity, values, disinterest? Two tactics have emerged from feminist studies, in effect, as a response to Geertz. First, the border between science and values is erased, on grounds that historical and ethnographic examinations of scientific *practice* render the separation chimerical. Second, a new cultural territory is staked out that overlays science and politics, which is said to produce knowledge every bit as credible and useful as that coming from science *sans* values or interests (see Longino, 1989). Haraway (1991) draws "a *biopolitical* map of the chief systems of 'difference' in a postmodern world" with "odd boundary creatures" such as simians, cyborgs, and women. Where is science located, and what are its contours? The "union of the political and physiological" is the focus of a chapter Haraway (1991a, p. 7) writes on evolution and domination. No indel-ible borders remain between spaces for science and for politics, as situated, positioned, partial, multiple, passionately detached, critical, and objective +subjective *knowledges* transcend both territories.

In sum, feminism advances the boundary problem in STS by exposing the gendered configurations of science (and the scientific configurations of women) and by showing the practical utility of such boundary-work for excluding or marginalizing women's place in the scientific enterprise. Feminism is also a robust specimen of boundary-work in practice, a project seeking emancipation in part through reconfigurations of science and politics, culture and nature, object and subject, male and female.

EMPIRICAL STUDIES OF BOUNDARY-WORK: FOUR SPECIMENS

Rather than survey empirical studies of the boundary problem in STS, I have chosen exemplary works that display four types of boundary-work: monopolization, expansion, expulsion, and protection. Together, they suggest how precipitous Barnes (1982) was to declare—after theoretically setting up the boundary problem in 1974—that "from a sociological point of view there is little more to be said . . . about the boundary of science in general. The boundary is a convention" (p. 93). Much has been learned about science since, from continuing scrutiny of its margins.

Monopolization: A Cartographic Contest for Cultural Authority

As Shapin and Schaffer tell it (1985), the debate between Robert Boyle and Thomas Hobbes in the 1660s was "about" an air pump in the same way that *Moby Dick* is "about" a whale. At stake was the delineation of authentic and authoritative knowledge: How was it made? Who could make it? What was it for? The dispute was more than one between Boyle's experimentalism and Hobbes's rationalism, for at issue as well was the constitution of the social order itself in Restoration England. The debate is a classic specimen of a kind of boundary-work involving science, where contending parties carve up the intellectual landscape in discrepant ways, each attaching authority and authenticity to claims and practices of the space in which they also locate themselves, while denying it to those placed outside. What makes *Leviathan and the Air-Pump* such a useful guide to the controversy is its sustained attention to places and spaces, borders and territories, insiders and outsiders—as the authors note, "the cartographic metaphor is a good one" (Shapin & Schaffer, 1985, p. 333).

Hobbes versus Boyle was a battle of the maps—cultural maps on which authoritative and authentic knowledge could be assigned or denied to knowledge producers by *where* they were located (and how). Crucially, there was no single map, as if Boyle and Hobbes glared at each other across a common frontier that moved to and fro depending upon who was "winning." Each sought to occupy and control a space for authentic and authoritative knowledge, but it does not follow that Boyle's "outside" (excluded and delegitimated kinds of knowledge making) mirrored Hobbes's "inside" (and vice versa). The two maps depicted *alternative* cultural universes, with important places and landmarks given different labels and with distinctive grounds for locating a border here or there. The maps guided presumed users to where they could find authentic and credible knowledge and told them why they could not find it outside that space.

Boyle's map put authenticity and authority on the side of his experimental physiology, carefully bounded from such unworthies as metaphysics, politics, and religion (Shapin & Schaffer, 1985, p. 153). "Science" (the inside space) sought matters of fact—fallible, provisional, corrigible—whose authenticity was decided by nature, as the community of competent interveners and observers collectively witnessed its goings-on. Assent to an experimental fact was, at once, a collective accomplishment of those committed to a certain discursive style *and* the result of procedures through which collective judgments were objectified into questions that "nature decides." Experimental discourse was limited to matters of fact; carefully contained dissent over theoretical explanations of the observed was tolerated but such interpretations and hypotheses were treated as undecidable. Facts gained authenticity through their collectively being seen, via a multiplicity of witnesses extended through three means: opening up the house of experiment—the nascent laboratory—for public view, so that like-minded experimentalists could see for themselves; replicating the experiment by building air pumps throughout Europe; and allowing for "virtual witnessing" of elaborate textual and faithfully detailed graphic representations of the experimental apparatuses and procedures. This community of experimenters saw themselves both as humble craftsmen, modest in the range of what they said they knew, and as priests poring over the book of nature. Both the technical and the priestly were arguments legitimating the cultural authority of Boyle and his fellow experimenters.

Boyle's "outside" was a raucous mix of illegitimate forms of knowledge making, held together only by their tendency to exacerbate dissent rather than contain it. Knowledge grounded in private and personal experience—what came from the alchemist's closet or from the passionate dogmatism of secretists and religious enthusiasts—was excluded. Because proper experimental discourse was nescient about causes (say, of the spring of air), all conjecture and speculation were located out there as well. If the matter was not decidable on experimental grounds, it was put outside—and such was the case for politics and human affairs in general, and in particular those philosophies proffering certain truth (described on this map as "tyrannical dogmatism"). Hobbes was a behemoth landmark in that region of imperious, egotistical dogmatists—the systematists, the authorities who would not submit their claims to the discriminating trial of nature-in-experiment.

Hobbes's "inside" is not labeled tyranny and dogma, of course, but neither does his "outside" contain anything that would command assent to a claim (as Boyle thought collectively witnessed experiment could do). On the philosophy or "science" side of Hobbes's border was certain knowledge, rational deductions capable of securing irrevocable, universal, and obligatory assent. Geometry was the model for knowledge making that carried authenticity and authority: One must accept that any line drawn through the center point of a circle divides it equally. Philosophy sought not facts but explanations in causal form, where cause is traced back to the motions of contiguous bodies. The compelling force of such logical deductions does not reside in their objectification in nature but precisely in their conventional, artifactual qualities: People *made* knowledge certain by deducing inescapable explanations as they proceeded rationally from agreed-upon definitions. Once properly deduced, there was little room for continued dissent. These principles for the manufacture of certain knowledge were as applicable to politics as to mechanics—Hobbes had no boundary separating studies of nature from studies of human affairs. As Leviathan commands assent in matters political, so philosophy—causal, rational, deductive, certain—commanded assent in matters epistemological.

Hobbes's "outside" includes some territories also "outside" on Boyle's map. Neither made room for religion on the inside, but for different reasons. Theologians did not participate in the experimental form of life. Their knowledge was certain—not fallabilities grounded in nature but timeless truths grounded in proper readings of sacred texts—and that sounded sufficiently dogmatic to warrant exclusion by Boyle. Hobbes excluded theologians—along with all other exclusive professional groups including Boyle's circle of experimenters—because they *impeded* the pursuit of certain truth and the order (social and intellectual) that only a philosophy commanding universal assent could secure. The church, the professions, those who practice "physics" and "natural history" (where Hobbes put Boyle, with derision) were sources of divided authority; each offered competing grounds for assent and so together could never reach the settlement commanded by authentic philosophy. What Boyle did was simply not philosophy for Hobbes; its refusal to decide causes made it incomplete, the tentativeness of its claims (always provisional and corrigible) attested to its variability and weakness, the inductive move from experimental observation to causal explanation was obviously fallacious, and its reliance upon experiment anchored so-called knowledge on the shifting passions of personal experiences and professional interests (guarding one's exclusive turf). What anyone saw was inherently and endlessly contentious. Worse, the obvious technical ingenuity of experimenters did not count as philosophical wisdom, and so Boyle went on Hobbes's outside along with gardeners, apothecaries, workmen, quacks, machine minders, and other banausic pursuits possessing little moral authority.

Several generic features of boundary-work are illustrated by this episode. The maps are those of Boyle and of Hobbes, as read by Shapin and Schaffer (1985, p. 342). They are not interpreted in terms of how well or poorly they correspond to putatively universal or essential qualities of science or to the "science" we might map today. The Hobbes-Boyle debate is a contest for what kind of knowledge making would be accepted—at that moment—as authentic and authoritative. Shapin and Schaffer (1985) seek to recover cultural circumstances especially difficult for the modern person to grasp: It seems to us as if authoritative knowledge about nature—science—has always been tightly coupled with experiment, yet there was nothing inevitable or self-evident about that link in the seventeenth century (p. 13). By going back to when the bounds of authentic natural knowledge were less formed, Shapin and Schaffer discover particular historical conditions in which that connection between experiment and authoritative knowledge first took hold.

Importantly, the use of experiment to divide legitimate from illegitimate knowledge is an accomplishment of *representations* of Boyle's practices; it does not flow directly from those practices in some unmediated way. It was less important what Boyle "actually" did with his pump than how those activities and that machine were represented, described, located on a map drawn precisely to legitimate them. The authors state their purpose to explore "the historical circumstances . . . in which experimentally produced matters of fact were *made into* the foundations of what counted as proper knowledge" (Shapin & Schaffer, 1985, p. 3, italics added; cf. pp. 52, 91). The actual practices of "science"—whether Boyle's experiments or Hobbes's philosophy—lay open for multiple interpretations and offer a bank of qualities from which cultural cartographers make selective withdrawals as they construct maps that give meaning (in this case, authority, credibility, authenticity) to those practices.[14] Actual scientific practices underdetermine the maps of it; the remainder of variance must be explained by circumstances of the boundary-work itself: What was at stake? Who needed to be convinced? What arguments did the adversary present?

The Boyle-Hobbes spat also shows cultural maps *in action*, as something more than coffeehouse chatter. Boundaries were not only drawn but policed (Shapin & Schaffer, 1985, p. 135)—a task Boyle made easier by translating his inside space into a physical place in the built environment (the nascent lab). The game for Boyle was not only to legitimate experiment but to deny authenticity and authority to the kind of knowledge Hobbes made. Success would be likely if Boyle could move everyone—rivals, audiences, bystanders—onto his playing field, with borders and territories that he drew and labeled (pp. 173-174). He did that in a crafty move, in effect arguing that only those who were *in* the experimenter's community—who went into Boyle's laboratory or built one for himself and performed competent experiments—could challenge claimed facts. But, catch-22, the price of admission to the lab (and to the Royal Society, as Hobbes found out) was a commitment to Boyle's program. "External critics" (p. 222) were instantly delegitimated! But Boyle's policing tactic opened up a new attack for Hobbes: Wasn't his exclusion from experimental space and the laboratory place evidence enough for the private nature of Boyle's sect? Hobbes challenged the claim that experimental facts achieved their authenticity from public witnessing by pointing to Boyle's insistence that, because it was his air pump, everyone had to play by his rules or go home (he was the "master" of the lab; p. 39). The significance of this for studies of boundary-work is clear; maps alone do not win authority for those inside a space, but they provide rules for making real-world decisions—denying admittance or membership, for example—that selectively allocate cultural authority in terms more palpable than cartographic.

Shapin and Schaffer eschew the question: "Who won and why?" Boyle comes out a leg up on Hobbes: Although boundary-workers in the four centuries since the first air pump have used Hobbes's right reason to distinguish their science from others' unreason, Boyle's experiment has perhaps more often been employed as the sine qua non of science. If Boyle's experimental space has achieved legitimacy, it is not because its contours—against those of Hobbes—more closely corresponded to "real" science; nor should we assume that Boyle's cartographic arguments would necessarily work on any other occasion when the boundaries of science are contended (there are no universal determinants of success). Boyle "won" because his space was better able to hold the diverse interests of powers-that-be in Restoration England—indeed, the laboratory became an "idealized reflection of the restoration settlement" (Shapin & Schaffer, 1985, p. 341). There, competent practitioners could publicly gather to manufacture facts in cool technical and professional calm, and agree to their validity in nature without lapsing into endlessly contentious debate over hypothetical causes. Experimental space was what Restoration England could become: a "peaceable society between the extremes of tyranny and radical individualism" (hadn't England had enough of both during the Commonwealth?) (p. 341). Hobbes was portrayed as the dogmatist whose certain knowledge would undo the fragile Restoration polity. Boyle rode to victory on the coattails of the Restoration as a political achievement, even as he and his experimental space contributed to that settlement. But his "victory"—as in any boundary-work—is a "local success," for there is no "unbroken continuum between Boyle's interventions [and his cultural cartography] and twentieth-century science" (p. 341). In no way did the borders and territories of science get settled for all time by Boyle and Hobbes.

Expansion: Enlightenment Encroachments

A second type of boundary-work occurs when insiders seek to push out the frontiers of their cultural authority into spaces already claimed by others. Such was the task of the *philosophes* in the eighteenth century, who sought to extend their mixture of rationalism and empiricism into a domain of questions and problems owned by religion and embodied in the institution and dogma of the church. A manifesto for the Enlightenment project was D'Alembert's *Preliminary Discourse,* written as a warm-up to Diderot's *Encyclopedia* and wonderfully interpreted by the historian of French culture Robert Darnton (1984). That text and its accompanying map are boundary-work nonpareil, spatial representations of a cultural landscape that provide grounds for a belief that "philosophy"—D'Alembert's inside authoritative space—could swallow up whatever counts as genuine knowledge, leaving only poetry and memory outside.

D'Alembert called his *Preliminary Discourse* a *mappemonde,* a map of the world of knowledge "to show the principal countries, their position and their mutual dependence, the road that leads directly from one to the other" (in Darnton, 1984, p. 195). The map has three continents, one much larger than the other two and with many more nested countries and provinces. In this "Detailed System of Human Knowledge," all understanding is divided into three spaces depending upon its source: memory and imagination (the two little continents) and reason (the big one). *Imagination* is the origin of poetry, both civil and sacred, and includes narrative, dramatic, and parabolic (allegorical) forms. *Memory* is the source of history, divided into civil (memoirs, antiquities, literary history), sacred (history of prophecies), ecclesiastical (conspicuously empty of further subdivision), and natural (uniformities and deviations of nature—"monstrous vegetables"—along with arts, trades, and manufactures). *Reason* (in the center, of course) is the fount of philosophy, which holds everything else, and divides into the science of being (metaphysics), the science of God (theology, religion, and, "whence through abuse, superstition"), the science of man (logic and ethics), and the science of nature (mathematics and physics—each subdivided into many tiny regions that have since grown to be "sciences" on their own).

Only reason yields knowledge; imagination and memory must produce something else. D'Alembert put up a boundary between the unknowable (poetry and history) and the known (what science tells us). He shrinks the territory given to poetry and history: Ecclesiastical history has nothing identifiable within it, in sharp contrast to the highly differentiated—and visually more capacious—space for mathematics, divided first into pure and applied and then broken down into optics, ballistics, hydraulics, and dozens more. Most significant, however, are the kinds of understanding moved under the umbrella of philosophy; natural and revealed theology were now knowable in the same sense as botany and zoology, as were questions of good and evil. Matters of ethics and morals, questions about the spiritual, were located within the compass of reason, which put them under the control and the authority of those men who epitomized the rational spirit—D'Alembert, Diderot, and their kindred *philosophes.* The church had no claim on such issues as these and was in effect forced to choose between speaking unknowables or speaking the rational and empiricist tongue of the Enlightenment. "Pigeon-holing is therefore an exercise in power" says Darnton (1984, p. 192), and this mapping of knowledge identifies winners and losers: Philosophy is now in charge of problems once assumed to be ecclesiastical, as rationality and empiricism are extended to matters once known only through tradition or revelation.

It is one thing to claim cartographically an expansion of territory under the cultural authority of science; it is quite another to legitimate those bulging frontiers so that people in society accept them. D'Alembert shows himself the master rhetorician, moving through three steps to show *why* reason, philosophy, and science are justified in staking their new claim—and each step is a response to potential difficulties raised by the one before it. He begins by rewriting human history in a classically progressive idealist way —life is better now than before thanks to philosophers' accomplishments. Bacon, Descartes, and Locke, along with scientific allies Newton, Galileo, and Harvey, are not just Great Men but Heroes, carrying along the "progressive march of reason" right up to its culmination in the *Encyclopedia* itself. The *philosophes* and their precursors were "cast in the grand role" and became the "moving force in history" (Darnton, 1984, pp. 199, 206-209). The tune is familiar enough and played still as scientists engage in border wars—"better living through chemistry." But this legitimation of the expanded authority of science through appeals to its salutary effects on the human condition carries a risk: Isn't the progressive, idealist reconstruction of history nothing but a self-interested polemic designed to justify what is really just a sectarian grab for power by the *philosophes* from the churchmen? Because "no map could *fix* topography of knowledge" (Darnton, 1984, p. 195), D'Alembert's *mappemonde* was vulnerable to delegitimizing arguments that it was a not-so-transparent effort to render authoritative precisely that which the *philosophes* uniquely offered (rationality tempered with empiricism), and thus intended to put them on top of an expanded reign. How could D'Alembert lessen the appearance of self-interest?

Scientists then as now are skilled at objectifying the referent and making it real, and that is what D'Alembert needed to do with his map of knowledge.

The second step in his legitimation of the expanded frontiers of philosophy is another familiar feature of scientists' boundary-work: drawing independent authority for one's own map by linking it to cartographic efforts of earlier generations of boundary-workers (objectification by attributing authorship *elsewhere*). If no map can fix the topography of knowledge, some achieve greater stability and obduracy from their repeated redrawings—at later times, in other places. D'Alembert's map is not merely of his own making; it is sufficiently close to Bacon's outline in *The Advancement of Learning* for some critics to accuse him of plagiarism. No matter: It is less the content of Bacon's map that D'Alembert exploited than his authority as a cartographer distant in time and place who represented the world of knowledge in similar ways.

Or *not* so similar—and that is as well an endemic feature of boundary-workers' reproduction of maps from the past. D'Alembert drew selectively from Bacon's map as he drafted his own and, in certain places, undermined it entirely. Where Bacon had made a large space for ecclesiastical history, D'Alembert's was tiny; where Bacon was at pains to separate divine learning from human learning, D'Alembert "submitted God to reason"; where Bacon showed natural history (i.e., science) as then deficient and embryonic, D'Alembert shows it in full flower. The discrepancies between Bacon's and D'Alembert's maps raised yet another problem for the latter's efforts to legitimate his *mappemonde.* His selective, even creative, redrawing of Bacon's outline is evidence itself of the arbitrariness and frailty of all exercises in cultural cartography. Darnton (1984) writes that "what one philosopher had joined another could undo" (p. 195), and D'Alembert needed once again to find a way to prevent the undoing of his own map.

His third step is legitimation through a kind of self-referential authority, yet another familiar move in boundary-work. *Who* has the authority to draw (or undo) maps of the world of understanding? D'Alembert in effect argues that the authority of his philosophy extends not just to morals, ethics, revelation, theology, logic, oratory, rhetoric, jurisprudence, mechanics, astronomy, geology, medicine, and falconry—but to cultural cartography as well. "Setting up categories and policing them is therefore a serious business," and "the boundary keepers turned out to be the philosophes" themselves (Darnton, 1984, pp. 193, 209). Drawing boundaries around the knowable is itself a knowable task, best accomplished through the same reason and evidence that marks all philosophical inquiry. D'Alembert's rearrangement of our "mental furniture" (Darnton, 1984, p. 193) worked reasonably well, for in retrospect one sees in them the seeds of the academic disciplines—some scientific, others left out—that came along in the next century. Enlightenment maps of the knowable—with contingent modifications, of course—then took on an even more obdurate and stable form as curricula that spread everywhere when modern universities were built.

Expulsion: Banishing Burt

A common kind of boundary-work involves insiders' efforts to expel not-real members from their midst. The labels attached by insider scientists to those booted out vary: deviant, pseudoscientist, amateur, fake. Those excluded typically give off the appearance of being "real" scientists, and may believe themselves to be so. But insiders define them as poseurs illegitimately exploiting the authority that belongs only to bona fide occupants of the cultural space for science. Such processes of social control no doubt foster a homogeneity of belief and practice within science by threatening insiders with banishment for perceived departures from the norm (Zuckerman, 1977a). Sanctioning deviants is also an opportunity for corrective public relations campaigns, restoring among powerful constituencies elsewhere in society a belief that science on its own is capable of weeding out impostors (so hands off) and restoring confidence that science is really only what genuine insiders say it is (nothing dirty going on).

All this comes together in the posthumous trial of Sir Cyril Burt, where continuing negotiations over the propriety of his scientific conduct ramify into self-conscious anxieties among psychologists about whether their discipline itself belongs inside the space for science. Gieryn and Figert (1986) examine, as a specimen of boundary-work, *psychologists' responses* to Oliver Gillie's visible 1976 accusation in *The Times* (London) that Burt had engaged in fraud. A short list of Burt's supposed misconduct includes the following: He concocted statistics that had been presented as results of experiments evidently never carried out; he invented fictional authors as allies to bolster support for his claims; he made inappropriate use of research conducted by students; he exaggerated his stature in the history of psychology by claiming priority to another's discovery. Gillie's accusations forced psychologists to reconstruct the boundaries of science: On what grounds should Burt remain inside or be expelled? At the time Gieryn and Figert concluded their analysis,[15] Burt was widely recognized by insiders to have been guilty of fraud and was banished from real science. His first biographer L. S. Hearnshaw wrote the sentence:

> The verdict must be, therefore, that . . . beyond reasonable doubt, Burt was guilty of deception. He falsified the early history of factor analysis . . .; he produced spurious data on MZ twins; and he fabricated figures on declining levels of scholastic achievement. . . . Neither by temperament nor by training was he a scientist . . . . His work often had the appearance of science, but not always the substance. (in Gieryn & Figert, 1986, p. 80)

So much for the man whose science was once so respected that he became the first from his discipline to be knighted.

Hearnshaw's summary judgment is of less interest to sociologists of boundary-work than the heated negotiations among psychologists that came before it. As is probably the case in most allegations of pseudoscience, amateur science, or deviant science, decisions about inside-outside forced debate over where that border shall be drawn—debate that becomes contentious because of diverse interests attached to the map that eventually wins out as (provisionally) accurate. Visions of Burt's misbehavior were refracted through the controversy between hereditarians and environmentalists over the determinants of intelligence. His data on twins lent support for genetic factors, and predictably those psychologists aligned with hereditarian theories argued that charges against Burt were trumped up. Several suggested that suspicions about Burt voiced by Leon Kamin (an environmentalist) even before the Gillie article were little more than a political attack on all research suggesting hereditary determinants of intelligence. But Kamin and others on the environmental side argued that the politics were all Burt's, that his views about intelligence were ideology masquerading fraudulently as science. In the mid-1970s, Burt's standing inside or outside genuine science depended upon who drew the map, a hereditarian or an environmentalist.

When it became clear to almost everybody—even hereditarians—that something fishy was going on, those who sought to defend Burt and his work redefined his conduct as on the margins of legitimately scientific practice—but not over the edge. There was much speculation about Burt's motivations: Was there intent to deceive (necessary, it seems, to sustain the charge of fraud) or was he simply a sloppy and negligent methodologist (reprehensible and even embarrassing perhaps, but hardly grounds for expulsion from science). One supporter even suggested that the oddities found in Burt's statistics would turn up with considerable frequency in examination of a random sample of similar articles in educational psychology. Though methodological sloppiness is hardly a flattering image for science, in the noisy context of the Burt controversy this lesser charge functioned for a time as a plea bargain (misdemeanors do not warrant death sentences).

When matters started to look even worse for Burt and guilt seemed the imminent conclusion, psychologists set aside their theoretical differences to salvage the cultural authority of psychology that would come from securing its place inside science. The rhetoric changed its focus from "What did Burt do and did he mean to do it?" to "What are the implications of the Burt affair for the science of psychology?" If Burt's guilt or innocence were simply a matter of the *politics* of intelligence, or if his alleged sloppiness is typical of psychological research, the professional risk for the discipline is clear: Is it really a science at all? Securing psychology's authoritative place within real science goes through three steps. First, responsibility for cultural cartography—where would Burt be located?—is claimed for psychologists themselves. An outsider, like medical reporter Oliver Gillie, could not possibly know enough sophisticated psychology to judge Burt's conduct as proper or not, and he is condemned by psychologists for trying. Hearnshaw the insider is their preferred border guard, because a real science like psychology can patrol itself. Moreover, the authority of science itself is reproduced when its procedures alone are cast as the unique means to determine just what Burt was all about. One psychologist suggested that the whole affair be "thrashed out in the leading scientific journals" that "operate a proper refereeing system" (in Gieryn & Figert, 1986, p. 75). Such peer review would get to the scientific truth of the matter.

Second, psychologists located the cause for Burt's misbehavior in personal and idiosyncratic troubles (outside the borders) rather than in structural problems inside the science of psychology. Burt's life after World War II was a long string of crises, we are told, from failed marriages to (accidentally) destroyed research records, from debilitating Meniere's disease to humiliation at being dismissed from the editor's post at a leading journal. The cognitive authority of scientific psychology is hardly touched by the misconduct of a "sick and tortured" man whose behavior was "not the act of a rational man" (in Gieryn & Figert, 1986, p. 79).

Finally, psychologists employed the rhetorical strategy that "the truth will out" (Gilbert & Mulkay, 1984) by distinguishing Burt's behavior and even his intentions (put outside science) from the possible validity of his claims and findings (inside). Hereditarians and environmentalists converged on the assumption that only further research on intelligence in twins would prove whether Burt's work has a lasting place in the science of psychology. Just as scientific claims are customarily detached from their authors (and linked instead to nature), Burt's data are detached from the seamy conditions of their creation: Science (as this episode of boundary-work concludes) is not in the making but in the made. The putatively self-correcting procedures of science guarantee that the truth will out; the rest is gossip (and not science). When people in society are persuaded of this belief, the cultural authority of science is not weakened by a potentially embarrassing case like Burt's, but strengthened.

Protection: Keeping Politics Near but Out

A final type of boundary-work by scientists involves the erection of walls to protect the resources and privileges of those inside. Successful boundary-work of this kind is measured by the *prevention* of the control of science by outside powers—or, put the other way, protection of the autonomous control *of* science *by* scientist-insiders. Threats are all around. When the Animal Liberation Front challenges the right to conduct certain experiments using laboratory animals, scientists work hard to distinguish their research from the use of animals in testing by the cosmetics industry, thereby appealing to higher values to justify their practices (new medicines or therapies save lives, but a new mascara will only make your eyes less itchy). In cases of technological failure such as the *Challenger* explosion, science is practically demarcated from "management decisions" or from "manufacturing" to shift blame for the disaster away from scientists and onto these other culprits (Gieryn & Figert, 1990). In these instances, boundary-work is employed in the struggle for *control* of science among scientists and outside powers, a struggle with high stakes for scientists; they stand to lose autonomy in setting research agendas or deciding among methodological strategies, and they risk loss of prestige, credibility, or even funding if blamed for unwanted technological developments.

Nowhere is this struggle for the control of science more apparent than in the endless negotiations of the boundary between science and politics, as recently examined in Jasanoff's (1990a) study of science advisers in the American federal government. The cartographic challenge for scientists is to draw science near enough to politics (ideally, as adjacent cultural territories) without risking spillover of one space into the other or creating ambiguity about where the line between them should fall. That challenge is heightened (as in other kinds of boundary-work) by cartographic efforts of *others* seeking advantage from possibly different cultural maps—elected officials, government bureaucrats, journalists, interest groups, and many other folks who stand ready to draw science|politics in ways congruent with their particular interests and programs.

For scientists, the mapping task is to get science close to politics, but not too close. Why? A key to the legitimation of scientists' cultural authority is the perceived pertinence of science for political decision making: As government officials turn to scientists for expert advice before promulgating regulations or statutes, they are simultaneously measuring and reproducing the authority of science over claims about reality (Mukerji, 1989). Too great a distance between science and politics threatens a critically important route for scientists' legitimation via their perceived political utility—and in particular their claim on government funding for their research. The relationship is symbiotic, of course; just as scientists draw legitimation from the use of science in government deliberations, so government officials (and others) are better able to legitimate their policy decisions by attaching to them the cultural authority of scientific expertise. The territories of science and politics converge not as a matter of some structural necessity or faceless rationality but because insiders to both have good reason to keep the other near at hand.

But not *too* close, and certainly not overlapping or interpenetrating. Only good fences keep politics and science good neighbors. Politicians, government bureaucrats, interest groups, and involved citizens keep science at arms' length to preserve their discretion and thus their power. If policy can be fully determined by facts under control of scientists, what place is left for political choice—whether democratic, bureaucratic, or judicial? The dilemma for policy makers (in the broadest sense) is clear: Bring science near enough so that political choices are legitimated by their perceived grounding in authoritative and objective understandings of the facts as only science provides, but not so close that choices and futures become exclusively "technical" and beyond the grasp and thus control of non-scientists. Scientists also need to keep the fence on their "politics" frontier well mended. After all, what makes scientific knowledge useful for politics is not just its content but its putative objectivity or neutrality.[16] Science can legitimate policy only if scientists are *not* treated as just another interest group and their technical input is *not* defined as just another opinion. Spillover in the other direction—from politics into science—is just as dangerous for scientists' autonomy: When politicians themselves make facts, the professional monopoly of scientists over this task is threatened. An even more likely threat is the capture of science by policy-making powers—a loss of scientists' control over their research agendas and, in the limiting case, over what is represented as "scientific" knowledge.

Jasanoff's book moves from these general observations to specifics of boundary-work in scientific advisory proceedings. She neatly summarizes the above tensions:

> When an area of intellectual activity is tagged with the label "science," people who are not scientists are *de facto* barred from having any say about its substance; correspondingly, to label something "not science" [e.g., mere politics] is to denude it of cognitive authority. (Jasanoff, 1990a, p. 14)

Although those on both sides have reason to keep the two cultural territories close but not too close, Jasanoff finds over again that the "social construction" of the science|politics border is a crucial strategy through which distinctive interests of diverse players are advanced or thwarted. Scientists may or may not have a stake in any particular policy outcome, but they do have a professional interest in protecting their claim to authority over fact making, which thus shapes how they variously distinguish scientific matters from political (and how they respond to others' maps as well). Politicians, bureaucrats, and others involved in policy making may worry little about the cultural authority of scientists except insofar as it legitimates preferred policies and programs without preempting their discretionary choices about which ones to enact—and so they too draw maps showing variously located borders between science and politics.

Four examples will illustrate occasions for science|politics boundary-work —and the diverse interests this rhetorical game can serve—as scientists are brought near (if not fully within) the policy-making machinery. First, industry lobbyists or citizens' interests groups use boundary-work to discredit or delegitimate an unwelcome policy initiative by challenging the credibility of the science on which that policy is based—all the while preserving (and, obviously, reproducing) the cultural authority of science as an abstract space. The trick is to create a different cultural space for ersatz scientific practices and findings—variously called "regulatory science" or, bluntly, "bad science" —that do not measure up to standards established across the border in "real" science (pure, academic, basic). Such a move leaves science untouched as the cultural space capable of producing expert, credible, and authoritative knowledge pertinent for policy making and, moreover, points out the road to policy initiatives more salutary for the interests of the critic: more *better* science.

In one case considered by Jasanoff, boundary-workers challenging the science behind a policy initiative turned out to be other scientists themselves. The Environmental Protection Agency (EPA) was asked to assess public health risks at several sites used by Hooker Chemicals for dumping toxic waste, including the infamous Love Canal neighborhood near Niagara Falls. The EPA commissioned a scientific study from a private consulting firm, whose results were leaked to the press and seemed to indicate unusually high rates of chromosomal abnormalities among neighborhood residents. The findings were immediately challenged by panels of scientific experts, who argued that proper research protocols were not followed (e.g., some critical controls were absent) and concluded that the consultant's report was not worthy of "science." Jasanoff (1990a) writes that "in a politicized environment such as the U.S. regulatory process, the deconstruction of scientific facts into conflicting, socially constrained interpretations seems more likely to be the norm than the exception" (p. 37). It is not clear whether these deconstructions of facts from inside-real-science critics were shaped mainly by their interest-based opposition to the policy implications of initial findings, or (possibly in addition) by their concern as professional border guards for the EPA's unauthorized use of the authoritative adjective "science" to legitimate their policy initiative by grounding it in research that falls short of insiders' standards of good science. Importantly, the sociological issue is not some essentialist set of standards for good science (forever up for negotiation) but struggles between scientists and a government agency over who has the power to draw boundaries between good science and bad, and thus control the allocation of cultural authority attached to that space.

Jasanoff (1990a) sets up the second example of boundary-work when she writes: "Participants in the regulatory process often try to gain control of key issues by changing their characterization from science to policy or from policy to science" (p. 14). Here, specific regulatory tasks or responsibilities are moved around the map as different players—agency bureaucrats, politicians, the courts, scientific advisors, scientists working as consultants to industry or interest groups, the outside scientific community—vie for authority to control a decision and steer it toward their own interests. When industry representatives depicted the promulgation of risk-assessment guidelines as a scientific matter, they were simultaneously defining the borders of science to include that task and announcing who alone could do the job.

> Industrial groups were convinced that these technocratic organizations [such as the National Academy of Sciences] would reach conclusions that were scientifically more conservative, hence more sympathetic to business interests, than those advocated by administrative agencies or the courts. (Jasanoff, 1990a, p. 59)

On other occasions, responsibility, and thus control, is moved away from scientists back to agency administrators. For a time, the EPA administrators argued that the development of a "criteria document" for assessing ozone risks was a *separate* matter from "standard setting," and then used the argument to limit involvement of their Scientific Advisory Board to the former task while retaining discretionary control over the latter. Standard setting for ozone risks was mapped onto the politics side, so that EPA administrators could proceed without approval or review from scientists (Jasanoff, 1990a, pp. 107-113). In still other instances, border disputes over the location of designated responsibilities result in a loss of control of the task by *all* of the players involved. That happened in a controversy over the risks of Alar, a growth regulator used on apples. As the EPA and Uniroyal argued over maps of good|bad science, *60 Minutes* and Meryl Streep (testifying before Congress) rendered their arguments largely irrelevant (p. 149).

The third example, in contrast to the others, suggests that porous and ambiguous fences *sometimes* make better neighbors than impermeably crisp ones. Looking at the regulatory activities of the Food and Drug Administration (FDA), Jasanoff suggests that blurring boundaries between science and politics can be effective for attaching the authority of science to policy initiatives while retaining political, administrative, or judicial control over their direction. "The line of demarcation between FDA's decision-making authority and that of its scientific advisers [was left] ill defined." For example, in considering possible approval for certain antiarrhythmic drugs (for treating angina pectoris), the FDA's scientific advisory network did not limit its work to a review of available scientific literature; it also "prepared the equivalent of a preliminary position statement": Isn't that "politics"? Although the agency exploited the cognitive authority of this scientific position statement as it justified its eventual policy, it did not cede ultimate control to its scientific advisers. The FDA employed "creative boundary-work to expand the extent of its authority and gracefully deviate from committee recommendations." Jasanoff (1990a) concludes: "The ambiguity of the boundary between science and politics is strategically useful to FDA, permitting the agency to harness the authority of science in support of its own policy preferences" (p. 178).

But the fourth example is a reminder that hyperexis can occur in boundary-work as in most other aspects of social life; if blurring the boundary between science and politics is a good thing, there can—at times—be too much of it. Attention shifts now to scientists working for government agencies such as the EPA or the FDA, and their efforts to *distance* their science from the agency's politics. Boundary-work by these scientists is motivated less by an abstract concern for the cultural authority of science than by a quotidian pragmatic concern for producing a useful, valued product for an employer. Their "job description" is to provide an informed and objective context of expert knowledge for policy deliberations—and that requires the manifest *appearance* of a clean break between research and politics (even if edges are really rather sloppy). As a means to bolster its scientific credibility, the EPA created the Scientific Advisory Board (SAB) in 1974 to review studies carried out by the agency's in-house Office of Research and Development. A challenge for the SAB was to "maintain its scientific authority on the one hand" and on the other to avoid capture by the "partisan political interests" inherent in the EPA's mission (Jasanoff, 1990a, p. 95). The very name of the board attested to its concern for research rather than policy, an image that was reinforced by the timing of SAB's reviews; they typically occurred early in the deliberations, safely upstream from eventual policy initiatives. Of course, the socially constructed distance between science and policy is helpful for maintaining the appearance of objectivity among science advisers, but even that can go too far. Returning full circle to a point made at the start of this section, Jasanoff (1990a) notes that "the very success of the Board's rhetorical strategy of distancing itself from policy creates a risk that its advice will be seen as irrelevant to policy" (pp. 97-98)—far enough to be objective and authoritative, close enough to be useful.

CONCLUSION

There is, of course, no single "boundary problem in STS." To appreciate the socially constructed, contingent, local, and episodic character of cultural categories such as science opens up new perspectives on a vast terrain of issues. Examination of how and why people do boundary-work—how they define "science" by attributing characteristics that spatially segregate it from other territories in the culturescape—could be the first step toward a cultural interpretation of historically changing allocations of power, authority, control, credibility, expertise, prestige, and material resources among groups and occupations. Boundary-work here becomes an important feature of professionalizing projects of scientists, a rhetorical form well suited to the seizure, monopolization, and protection of those goodies. From a very different direction, exposing the contingent, flexible, pragmatic, and (to a degree) arbitrary borders-and-territories staked out for science is the start of a critical evaluation of the consequences of such cultural classifications—not just for intellectual but real-world lives. Undermining the givenness, naturalness, necessity, universalism, and essentialism in cartographic representations of science/non-science opens the possibility of their reconfiguration.

Whether studies of boundary-work are designed to interpret the world or to change it, available methodological options are many and varied. One could home in on the rhetorical tropes used by boundary builders to distance science from non-science—with careful analytic attention to the particular and possibly unique discursive formulations in a local episode. Or one could move from the local and episodic to examine cultural categories in their obdurate, stable, and enduring forms, in order to understand cultural and social change over the long haul. If science need not be defined from scratch each time the question comes up, and if the question does not even come up because people tacitly agree that *this* is science (and *that* is not), then some analytic attention must be given to cultural spaces that transcend their momentary instantiation, to extant maps that get unfolded (not drawn fresh) in an episode of boundary-work.

This chapter carries the following prescription for STS. Get constructivism "out of the lab" to release its interpretative potency on claims and representations where the referent is not nature but culture. If science studies has now convinced everybody that scientific facts are only contingently credible and claims about nature are only as good as their local performance, the task remains to demonstrate the similarly constructed character of the cultural categories that people in society use to interpret and evaluate those facts and claims.

Getting constructivism out of the lab assumes that the actual practices of scientists in laboratories—and their "professional" inscriptions of nature—are surrounded by an interpretative flexibility that allows for multiple and variable accountings. Getting constructivism out of the lab moves science (the practices, claims, and instruments in need of mapping) closer to places where matters of power, control, and authority are settled. Constructivism is not just a stick for beating up old-fashioned epistemologies; it is as well a theoretical challenge to just-as-old sociologies—Marxism, functionalism, rational choice theory—seeking explanations for things at the top of that discipline's agenda: uneven distributions of authority, power, control, and material resources. Political economy is cultural, interests rest on shifting meanings, pursuits of power or wealth are carried out through interpretative categories—cultural maps—that people use to arrange their worlds.

NOTES

1. Steve Woolgar offered the concept of boundary-work in response to some half-baked ideas I discussed with him at a conference in 1981, and I first used it in Gieryn (1983). Since then, I have found too many others who have used synonymous tags for the same idea, and one goal of this chapter is to bring these common but fragmented efforts together. Perhaps the earliest mention in science studies of the boundary problem as characterized here is Barnes (1974): "We should not seek to define science ourselves; we must seek to *discover* it as a segment of culture already defined by actors themselves. . . . It may be of real sociological interest to know how actors conceive the boundary between science and the rest of culture, since they may treat inside and outside very differently" (p. 100). By 1982, when Steven Shapin (1982) reviewed sociologically informed work in the history of science, he was able to group 20 studies under a heading "Interests and the Boundaries of the Scientific Community" (pp. 169-175). Shapin (1992a) continues the review up to the minute.

2. There is irony in my lumping together Popper with those whose solution to the demarcation problem is *essentialist,* for he railed against essentialism: "I do not think that we can ever describe, by our universal laws, an ultimate essence of the world" (Popper, 1972, p. 196, see 1963, pp. 103 ff.). He was steadfast in his rejection of the essentialist idea that "if we can explain the behavior of a thing in terms of its essence—of its essential properties—then no further questions can be raised, and none need be raised" (1972, p. 194; see pp. 309-310). Evidently, Popper's anti-essentialism did not extend from "scientific explanations of the *world*" (i.e., of nature) to "philosophical-cum-sociological explanations of *science,*" for it is precisely my argument that falsifiability (bold conjecture, severe critique, and so on) constitutes the essence of science for Popper and, moreover, explains for him why science is superior to any alternative in advancing objective knowledge toward an ever closer approximation to reality. When talking *about* science, Popper's (1972) essentialism is inescapable: "For once we have been told . . . that the most satisfactory explanation will be the one that is most severely testable and most severely tested, we know all that we need to know *as methodologists*" (p. 203).

3. Amidst his study (with Trevor Pinch) of parapsychology, Collins (1982b) explicitly denies the ability of falsifiability (tied as it is to replicability) to serve as a criterion demarcating science from non-science: "To a philosopher of science who follows Karl Popper's views the events must seem to reveal a classic case of unfalsifiability akin to the case of astrology. Proponents cannot agree on a procedure that will render falsifiable the existence of the phenomenon in which they are interested. Reasons or excuses are always provided to explain away failures of the subjects to perform in any set of circumstances. Yet, the scientists involved would not want to say that they were not doing science. It is difficult to see what they are doing if it is not science" (p. 134; see Collins, 1975).

4. Surely more ethnographic studies of laboratory life and textual analyses of scientific discourse are needed to better understand the ambiguities of replication, the interpretation and deployment of social norms, and the construction and reconstruction of scientific consensus. For additional discussion of the contrast between essentialist and constructivist views of demarcation, see Woolgar (1988b, pp. 11-21).

5. "Consequences" is smuggled in here to preempt an obvious (but wrong) explanation for the cultural authority of science: Scientific interpretations are accorded credibility because they *work* when translated into technologies with demonstrated effectiveness in getting things done (airplanes fly thanks to aerodynamics and the rest of physics; polio rates are reduced by biomedical science). The error should be apparent to those familiar with the literature on the social construction of technology: the workability of a machine is itself a matter for negotiation, as is the attribution of technological success *to* science (as opposed to engineering, manufacturing, karma, or some other cultural realm). See Bijker, Pinch, and Hughes (1987) and Bijker and Law (1992).

6. This is a contentious issue, of course. The relationship between scientific practices or claims and subsequent reconstructions parallels exactly the relationship between nature and scientific knowledge as described in classically constructivist works (e.g., Pinch, 1986). Relativistic indeterminacy between referent and reference does not mean that "anything goes." Rather, it becomes an empirical question—of the first order in significance—to understand how *some* scientific claims assume the kind of obduracy that leads people to see them as real (i.e., given in nature). To shift now from construction of knowledge to construction of science, it is also an empirical question—of equal significance—to understand how some representations of scientific practice attain the same obduracy and perceived reality. Studies of this question are, sadly, in their infancy.

7. But a backdoor essentialism appears even among stalwart constructivists. Collins and Yearley's (1992a) call for "social realism" verges dangerously toward *definitive* conclusions about the boundaries between science and other social institutions: "Close description of human activity makes science look like any other kind of practical work. Detailed description dissolves epistemological mystery and wonder. This makes science one with our other cultural endeavors" (pp. 308-309). And elsewhere, Collins (1992a) writes: "The sociology of scientific knowledge has only one thing to say about the institution of science: it is much like other social institutions" (p. 190). For sociological analysts to erase the cultural boundaries of science *is as essentialist* as Popper, Merton, and Kuhn's laying them down. It is not for analysts to decide (for example) that science *is* politics by other means, in some definitive sense. On occasion, people in society (scientists, for example) do build a Berlin Wall between science and politics that demarcates authority and human action every bit as effectively as the real one once did. Studies of boundary-work begin with constructivist findings that essentialist demarcation principles are poor descriptions of SSK-observed scientific knowledge and practice, but this line of inquiry would be stillborn if sociologists were to stop at Collins and Yearley's conclusion that science is much like other social institutions. That judgment rests with historical actors, who will make science into something "much like other social institutions" only if that cartographic representation serves their interests. Collins and Yearley (1992b) are on more fertile ground when they write: "None of this excludes an examination of the initial work of demarcating science and non-science" (p. 378). (Presumably, this "work" is done by someone other than sociologists, historians, or philosophers.)

8. For an example of this essentialism applied to the study of science as a profession, see Shils (1968b).

9. This typology comes from Abbott (1988), whose reliable synthesis of sociological work on the professions obviates the need for a longer review.

10. A powerful case against the notion that "professionalization" is merely the result of collective and individual self-interest is made by Kuklick (1980), who also draws attention to the historical significance of "charter boundaries" around what came to be the social sciences.

11. The same idea is found in Latour (1987), working from a French tradition of "actant-network" theory that has filiation with social worlds analysis: "Those who are really doing science are not all at the bench; on the contrary, there are people at the bench because many more are doing science elsewhere" (p. 162). For two studies of how the press plays vital roles in the production of natural facts in the science world, see Gieryn and Figert (1990) on Richard Feynman's part in the investigation of the *Challenger* explosion and Gieryn (1992) on Pons and Fleischmann's claim to cold fusion.

The point is clearly made by Becker (1982) for art worlds: "I am not concerned with drawing a line separating an art world from other parts of society. Instead, we look for groups of people who cooperate to produce things that they, at least, call art; having found them, we look for other people who are also necessary to that production, gradually building up as complete a picture as we can of the entire cooperating network that radiates out from the work in question. . . . One important facet of a sociological analysis of any social world is to see when, where, and how participants draw the lines that distinguish what . . . is and isn't art, . . . who is and isn't an artist" (pp. 35-36).

12. The boundary between professional scientists and amateurs has received considerable attention (Lankford, 1981; Westrum, 1977, 1978).

13. D. Fisher (1990) draws out the implications of Foucault for matters of boundary-work and power, bringing in recent work by sociologists of education (e.g., Bernstein) on the classification of knowledge, all in the service of a history of the Social Science Research Council.

14. The same point is made elsewhere in *Leviathan*: "The category of mechanical philosophy was an interpretative accomplishment; it was not something which resided as an essence in the texts"—or in their experimental practices (Shapin & Schaffer, 1985, p. 205).

15. Since 1986, not one but two biographies of Burt have been published (Fletcher, 1991; Joynson, 1989), each challenging Hearnshaw's guilty verdict and seeking some vindication of Burt's scientific image. Their appearance is evidence itself for the potential endlessness of debates surrounding the boundaries between good science and bad. Whatever Burt might have done or not done, his scientific practices are forever available for reconstruction as proper or not—and those reconstructions are best interpreted sociologically in terms of the contexts in which they appear than judged for accuracy against some elusive yardstick of what the man really did.

Gieryn and Figert base their analysis primarily on exchanges among psychologists in the *Bulletin of the British Psychological Society*.

16. For a case study of how scientists seek to retain objectivity for their claims while still appearing useful for obviously partisan deliberations of acid rain regulation, see Zehr (1990).

19

Science Controversies

The Dynamics of Public Disputes in the United States

DOROTHY NELKIN

IN 1976 animal rights activists picketed the American Museum of Natural History in New York City to block experiments that seemed to impose unnecessary cruelty and pain. A decade later protesters were trashing laboratories, stealing animals, and demanding on moral principles the end of all animal research. The controversy over animal research is but one of the many disputes over scientific developments and technological applications that have proliferated over the past 20 years. And like the animal rights protests, they have escalated their tactics and increased their demands. Antiabortionists succeeded in banning federal funds for research using the human fetus from 1981 to 1994. Gay rights activists have challenged the procedures and guidelines surrounding the use of HIV tests. Protesters with moral reservations have joined farmers with economic concerns to obstruct the applications of biotechnology. Religious groups have opposed the teaching of evolutionary biology in public schools. And just as environmentalists confront corporate policies that threaten global resources, so corporate interests have disputed scientific commissions that threaten to impose regulations.[1]

AUTHOR'S NOTE: Research and writing for this chapter were supported by a grant from the Alfred P. Sloan Foundation.

These and other controversies over science and technology are struggles over meaning and morality, over the distribution of resources, and over the locus of power and control. Over the past few decades, science has increasingly become an arena to battle out deeply contested values in American society. We prize efficiency yet value political participation. We insist on individual autonomy yet expect social order. We value scientific knowledge yet fear the influence of scientific ways of thinking on widely accepted beliefs. Controversies over science and technology reveal tensions between individual autonomy and community needs. They reflect the ambivalent relationship between science and other social institutions such as the media, the regulatory system, and the courts. They highlight disagreement about the appropriate role of government, concerns about the increased role of technical expertise, and discomfort with the instrumental values so fundamental to the scientific endeavor (Ezrahi, 1990).

Controversies offer a perspective on the politics of science and a means to explore public attitudes. As disputes have proliferated so too have studies that document and analyze them. In this chapter, I draw from this growing literature to analyze the dynamics of science controversies as they have come to express political tensions as well as moral reservations about the value of certain scientific practices.

SOURCES OF PUBLIC AMBIVALENCE

Controversies over science and technology have often focused on the question of political control over the development and applications of science. But in the last decade protests against science have assumed an increasingly moralistic spin. Many recent disputes are framed in terms of moral absolutes. Fetal research is "wrong" and should be abandoned regardless of the clinical benefits. Animal experimentation, likewise deemed immoral, should be banned regardless of its contribution to medical knowledge. Many of the critics of science—creationists, antiabortionists, ecologists, animal rightists—are uneasy with instrumental activities that turn nature, fetuses, women, or animals into resources or tools. Their moral concerns have radicalized many of the protests that began as political challenges in the 1970s.

The development of science and technology remained largely unquestioned during the period of rapid economic growth that followed World War II. But by the 1970s belief in progress was tempered by growing awareness of risks. Technological improvements were threatening neighborhoods and causing environmental problems; drugs to stimulate the growth of beef cattle were causing cancer; efficient industrial processes were threatening worker health. Even efforts to control technology seemed to pose inequities as new standards and regulations pit economic expectations against the quality of life.[2]

Ironically, surveys indicate little change in the level of public support of technology over the past 20 years (Miller, 1990). When questioned, most people perceive science and technology as instrumental in achieving important goals and believe that the benefits of technology outweigh the risks. However, in the early 1970s, concerns about environmental problems began to generate political efforts to obstruct specific projects and to increase public participation in technology policy decisions. A decade later, even scientific research lost its exemption from political scrutiny. Antiabortionists blocked the availability of federal funds for fetal research; a rapidly expanding animal rights movement brought regulation and constraints on biomedical research practices; and challenges from whistle-blowers brought congressional investigation and oversight to the research community, threatening its long-standing autonomy from political regulation and public control.

Many scholars have addressed the significance of these trends. In March 1978 an issue of *Daedalus*, titled "The Limits of Scientific Inquiry," examined the proposition that some kinds of research should not be done at all. The same year, a conference, "Social Assessment of Science," examined international efforts to impose regulations on research.[3] Throughout the 1980s there was talk about the "crisis" in science as research seemed faced with attacks by forces from both the left and the right. Some regarded the activities of protest groups as a form of nineteenth-century Luddism—a wholesale rejection of science and technological change. One observer, Zbigniew Brzezinski (1970), called such opposition "the death rattle of the historically obsolete." Another, Theodore Roszak (1968), believed that protest is a positive and necessary force in a society that "has surrendered responsibility for making morally demanding decisions, for generating ideals, for controlling public authority, for safeguarding the society against its despoilers" (p. 22).

Today's controversies reflect a long history of ambivalent public attitudes toward science in American society (Mazur, 1981). The acceptance of the authority of scientific judgment has long coexisted with mistrust and fear, revealed, for example, in the early opposition to innovations such as vaccination or to research methods such as vivisection. The romantic view of the scientist as "a modern magician," a "miracle man who can do incredible things," has been paralleled by the negative images of mad scientists: the Dr. Frankensteins and Dr. Strangeloves that pervade popular culture (Roszak, 1974, p. 31).

In part, public ambivalence has been a response to the obscurity and complexity of science that appears to threaten the power of the citizen. The growing importance of expertise in policy decisions seems to limit the democratic process (Goggin, 1986).[4] Activists demand greater involvement in decisions about science and technology, seeking participation in review boards and decision-making groups. However, only about 5% of American adults are both attentive to science policy issues and sufficiently literate scientifically to understand and assess the arguments underlying the disputes (Miller, 1990). Thus disputes often have less to do with specific technical details than with broad political issues: They represent the growing polarization between those who see scientific and technological developments as essential to social progress and those who see these developments as driven by political or economic interests (Richards, 1988); between those with programmatic agendas seeking to implement specific goals and those with moral lenses concerned about accountability, responsibility, and rights. Some controversies (e.g., over the superconducting supercollider) remain mainly at a policy level where issues are debated by experts, ethicists, and policy elites. Others (e.g., over the use of animals in research) are public protests engaging social movements and citizens groups. Sometimes concerns have less to do with the implications of science and technology than with the power relationships associated with them. Protests may be less against specific technological decisions than against the declining capacity of citizens to shape policies that affect their interests; less against science than against the use of scientific rhetoric to mask political or moral choices (Fischer, 1990).

Questions of power, responsibility, and accountability continue to drive disputes. But controversies have also changed over time. In the 1970s and early 1980s, they represented the so-called crisis of authority that prevailed in the political life of that time (Salomon, 1977a). And they indicated the willingness of local groups to mobilize against decisions that affected particular interests. By the end of the 1980s, protesters increasingly framed their attacks on science in the moral language of rights.

By 1990 Yaron Ezrahi, in *The Descent of Icarus*, suggested that the attacks against science represent a major conceptual change in the role of science in society: "In the closing decades of the 20th century, the intellectual and technical advances of science coincide with its visible decline as a force in the rhetoric of liberal democratic politics" (Ezrahi, 1990, p. 13).

A TYPOLOGY OF DISPUTES

Studies of controversies suggest their origin in a range of political, economic, and ethical concerns (Engelhardt & Caplan, 1987; Graham, 1979; National Academy of Sciences, Institute of Medicine, 1991). First, the most intense and intractable disputes concern the social, moral, or religious implications of a scientific theory or research practice. The controversy over the teaching of evolution in public schools has persisted at the level of local school districts, even after a U.S. Supreme Court decision seemed to bring closure to the issue (Nelkin, 1984a). The practice of animal experimentation has spawned a belligerent animal rights movement morally opposed to the use of animals as tools (Jasper & Nelkin, 1992). The 1970s' opposition to fetal research was amplified with the development of new medical uses for fetal tissue (Maynard-Moody, 1992). And the creation of transgenic animals through techniques of biotechnology provoked the opposition of groups convinced that tampering with "natural" forms of life is morally suspect (Krimsky, 1991). These and other disputes reflect the preoccupation with morality in American society. Even as biomedical research brings about dramatic improvements in medical care, critics question and, indeed, try to stop certain areas of science that threaten their moral convictions. As new therapeutic possibilities emerge —such as the ability to intervene in the reproductive process or to use fetal tissue for transplantation and research—they too become the focus of moral disputes. Some defend these practices for their therapeutic benefits; other see only the potential abuse. For critics, the use of women as surrogate mothers, or the use of animals or fetuses in research, are morally questionable activities threatening concepts of personhood and violating fundamental beliefs. These critics are not simply questioning specific research practices; they are challenging the basic values underlying research.

A second type of controversy reveals the tensions between environmental values and political or economic priorities. Many disputes arise when the interests of citizens are threatened by decisions to site noxious facilities—power plants, toxic waste disposal dumps—in their neighborhoods (Brown & Mikkelski, 1990). Such ubiquitous conflicts engage communities in prolonged political actions expressing what has been called the NIMBY (not in my back yard) syndrome (Freudenburg, 1984). They raise questions about equity in the distribution of risks, the role of the citizen in technological decisions, and the access of local communities to expertise. Similar tensions over environmental values are expressed in the growing concerns about the global implications of technological decisions. The threat of ozone depletion (Brown & Lyon, 1992) or oil spills from supertanker disasters (Clarke, 1992) pose problems that can hardly be considered in a local political context. Yet local political structures and economic interests often support controversial policy choices that reflect short-term economic or political priorities. Such issues have generated a "new environmentalism" that has been refocusing controversies on the global dimensions of the environmental problems caused by technological change (McGrew, 1990).

A third type of controversy focuses on the health hazards associated with industrial and commercial practices, and the resulting clashes between economic interests and those people concerned about risk (Rosner & Markowitz, 1991). We are deluged with warnings about "invisible" hazards (PCBs, freon, radiation, the carcinogens in food additives—the list is long and ever growing). Uncertainties about the extent and the nature of risk have aggravated public fear. Gaps in technical information inevitably leave considerable leeway for conflicting interpretation. While new technologies have increased our capacity to detect potential risks, the public is confused by disputes among scientists. What food should we eat? How dangerous is the workplace? How should the possibility of risk be weighted against potential benefits? Risk disputes focus on balancing competing priorities in decisions about regulation and the setting of safety standards, and on ways to best protect the public and those working in hazardous occupations (Nelkin, 1985).

A fourth type of controversy over technological applications reflects the tension between individual expectations and social or community goals. Characteristically, such controversies, reflecting the usual disputes over government regulation, are framed in terms of "rights," and they frequently revolve on the introduction of science and technology. If a water supply is fluoridated, universal vaccination required, or a course of study mandated in the public school curriculum, everyone must comply with the decision and share its consequences. If the use of a pharmaceutical product like AZT is constrained, those who want it are denied. Government bans on alternative cancer therapies may infringe on a patient's right to choose her own medication (Markle & Petersen, 1980). Gun control legislation threatens the individual's freedom of choice. Governments impose constraints on individual behavior to protect the community, but constraints on individual freedom may also be interpreted as protection of professional turf or as unnecessary government paternalism.

Scientific developments are sometimes perceived as threatening to individual rights. Advances in the neurosciences, for example, may be used to impose social controls over individual behavior (Nelkin & Tancredi, 1994; Valenstein, 1980). Theories suggesting the biological basis of human behavior evoke fears that genetic determinism will be used to justify state control over reproductive rights (Hubbard & Wald, 1993). Creationists see the teaching of evolution as a threat to their right to maintain the religious faith of their children. AIDS patients see requirements for HIV testing and partner notification in the case of a positive diagnosis as a threat to their right to privacy. And scientists themselves view external controls over research as an infringement of their right to free inquiry. Many of these disputes play on tensions in American society over the appropriate role of government and regulation and the extent to which community values or, in some cases, public health requirements may intrude on the rights of individuals.

There are other types of disputes. Megascience projects such as the superconducting supercollider, the Human Genome Project, and the space program have generated conflicts over questions of equity in the distribution of resources within science (Dickson, 1984). The growing commercial interest in biotechnology products and the expansion of industry-university collaboration in this field have become a source of disputes over patenting and property; those advancing technological innovation in a competitive market conflict with those who believe the public interest would be better served by more open communication of new ideas (Krimsky, 1991). And incidents of scientific misconduct, from fraud to misappropriation of research funds, are generating disputes over the accountability of science and the ability of scientists to regulate themselves.

Controversies over science and technology represent in part a loss of public trust, a declining faith in the ability of representative institutions to serve the public interest. Critics are asking about research priorities: Is science for the public or simply for the advancement of scientific careers? Are technological developments benefiting society or simply fulfilling narrow economic goals? The significance of controversies lies partly in their expression of political concerns; but they are also moral statements about the role of science. These two aspects of disputes—their political and moral dimensions—call for further analysis.

Controversy as a Political Challenge

The political challenge presented by science disputes varies, depending on the issue and the affected community. Some people become involved in protest because of their immediate and pragmatic interests. Living near a noxious facility or working in a chemical plant, they are directly affected by health risks or social disruption. But some issues have no natural constituency, attracting people who have no direct or pragmatic interest. In the debate over ozone depletion, for example, the affected interests are, for the most part, future generations. Animal rights protests attract those who are morally committed to the cause of animal protection. While some critics of biotechnology are concerned about specific economic or environmental impacts, others worry about the moral implications of "tampering" with life. Some risk disputes are motivated as much by ideological agendas as by fear of harm (Douglas & Wildawsky, 1984; Downey, 1986). The nuclear debate as it developed in the 1960s and 1970s, for example, had ideological overtones that had less to do with the technology than with its political context (Jasper, 1990). In such cases, political challenges come from people with a moral or social mission.

Most activists in science-based policy disputes are middle-class and educated people with sufficient economic security and political skill to participate in a social movement (McCarthy & Zald, 1973). Their involvement is not necessarily tied to traditional political alignments. The environmentalists who oppose technological projects and many animal rights advocates come from social movements associated with liberal values. But the prolife groups who oppose fetal research are politically conservative, as are the fundamentalists who seek to block the teaching of scientific theories that offend their faith. Focused on particular problems, science controversies attract people whose concerns rest more on the nature of the issue than on their prior political orientation as liberal or conservative, left or right.

Linking these diverse groups is their demand for greater accountability and increased public control. Technological controversies, as sociologist Alain Touraine has described them, represent a reaction against technocracy in the search for a more human-centered world (Touraine, 1980). This is the central political challenge of technological disputes.

Controversy as a Moral Crusade

Cutting across nearly all of these controversies are ubiquitous claims of "rights." In the individualistic culture of America, nearly every political demand becomes cast in the moral rhetoric of rights—a rhetoric with deep roots in American history (Jonsen, 1991). The tendency to formulate problems in terms of distinct, overarching moral principles apart from their social context was nurtured by the religious tradition of Calvinism, and moralist thinking later permeated secular thought through the tradition of Puritanism (Miller, 1962).

Today, this tendency is reflected in the revival of bioethics as an influential profession. And it emerges in the discourse of social movements with their insistence on moral absolutes and claims for "rights." Animal advocates call for animal rights, antiabortionists make claims for fetal rights, scientists claim the right to conduct their research without unwarranted intervention, creationists claim their right to choose the theories taught to their children, and environmentalists advocate the rights of future generations.

Some claims to rights are based on obligations; rights may be a practical condition necessary to fulfill certain tasks. Thus government agencies claim the right to constrain individual freedom in order to carry out their mandated responsibilities. Other claims to rights are based on utilitarian arguments; certain rights are valued because they maximize the public interest. Scientists, for example, argue that the acquisition of knowledge is so important for the long-term interests of society that freedom of inquiry must override other considerations. Others, like animal rights advocates or creationists, base their claims to rights on basic moral or religious premises; and still others base their claims on the libertarian assumption that individual autonomy is an ultimate value in itself. But, whether justified in terms of natural rights, obligations, or traditions, rights claims become a moral imperative. Based on beliefs or deeply held intuitions that are considered nonnegotiable, they leave little room for compromise or accommodation.

Rights claims inevitably exacerbate conflict, for they are, as philosopher H. L. A. Hart (1955) observes, "moral justifications for limiting the freedom of another." The claims on behalf of the rights of animals limit the freedom of inquiry that scientists believe their due. The rights of future generations constrain the actions of today's consumers. And rights to individual privacy conflict with the government's need to regulate for social ends.

In some controversies, claims to rights are little more than ad hoc responses in competitive situations, confusing moral categories with strategic goals. Indeed, the rhetoric of rights may be simply a way to elevate instrumental behavior to the level of a moral imperative so as to limit negotiation. Thus rights claims may be the central issue in a dispute, or simply a tactic, a way to gain public support in a controversial political context.

TACTICAL CONSIDERATIONS

The ideological complexity of technological controversies is often matched by strategic complexity as moral arguments are combined with extensive use of technical expertise. In some cases scientists have initiated controversies by raising questions about potential risks in areas obscured from public knowledge. Scientists were the first to warn the public about the possible risks of recombinant DNA research. They were the first to call public attention to the problem of ozone depletion. And they have been active on all sides of the diet-cancer disputes. But technical expertise is a crucial political resource in all policy conflicts; for access to knowledge and the resulting ability to question the data used to legitimize decisions are an essential basis of power and influence (Benveniste, 1972).

The authority of scientific expertise has rested on assumptions about scientific neutrality (Proctor, 1991). The interpretations and predictions of scientists are judged to be rational and immune from political manipulation because they are based on data gathered through objective procedures. Thus scientists are enlisted by all sides of disputes. Just as industrial advocates use technical expertise to support their projects, so too do the protest groups who challenge them. Environmentalists hire their own experts, who expose potential risks. Among the animal rights advocates are scientists who debunk the need to use animals in research. Even the creationists present themselves as scientists and claim that creationism is a valid scientific theory that should be taught in the schools.

Though political values or moral issues may motivate disputes, the actual debates often focus on technical questions. Quality of life issues are discussed in terms of the physical requirements for a disputed facility or the accuracy of risk calculations rather than the needs or concerns of a community. Concerns about the morality of fetal research are reduced to debates about the precise point at which life begins. This displacement of issues can be tactically effective, for in all disputes broad areas of uncertainty are open to conflicting scientific interpretation. When decisions must be made in a context of limited knowledge, and there is seldom conclusive evidence to dictate definitive resolution, power may hinge on the ability to manipulate knowledge and to challenge the evidence presented to support particular policies. But as technical expertise becomes a resource, exploited by all parties to justify competing moral and political claims, it becomes difficult to distinguish scientific facts from political values. Debates among scientists reveal the value premises that shape the data considered important, the alternatives weighed, and the issues regarded as appropriate (Hilgartner, 1992).

Ironically, the willingness of scientists to lend their expertise to various factions in widely publicized disputes has undermined assumptions about the objectivity of science, and these are precisely the assumptions that have given scientists their power as the neutral arbiter of truth. Disputes among experts have thus brought growing skepticism about the policy role of scientists and awareness of the political dimensions of decisions commonly defined as technical. The very fact that experts disagree—more than the substance of their disputes—has forced disputes into the public arena, firing protest and encouraging demands for a greater public role in technical decisions.

Beyond seeking technical resources, those engaged in controversies over science and technology must organize their activities to broaden their political base. Many protest organizations—such as animal rights and ecology groups—rely on the backing of a direct mail constituency to provide political and financial support for their causes. To attract this support, they must generate dramatic and highly publicized events. Moving beyond routine political activities such as lobbying or intervention in public hearings, they engage in litigation, laboratory break-ins, street demonstrations, and other civil disobedience actions.

Litigation has been an important strategy, not only to block technologies but to mobilize constituents. In the 1970s the role of the courts in environmental decisions expanded through the extension of the legal doctrine of standing—private citizens without alleged personal or economic grievances could bring suit as advocates of the public interest.[5] The courts have since been used by citizens not only in environmental cases but also in challenging research practices, as in the litigation over fetal research and animal rights.

Such cases attract the media, bringing public visibility to the issues and amplifying the disputes.

Capturing public attention and political interest requires attracting the media (Mazur, 1981; Nelkin, 1994). Thus science protest movements engage colorful or visible writers and activists, such as Jeremy Rifkin, Peter Singer, Ralph Nader, and Paul Brodeur, and the support of film personalities and politicians. Visual imagery also brings attention to the cause. The gruesome photographs projected by the animal rightists attract public sympathy. Pictures of nearly full-term fetuses that look like infants fan public concerns about fetal research and tissue transplants. The television images of oil-coated birds inspire public outrage about supertanker oil spills. Rhetorical imagery is also strategically important. To scientists engaged in fetal research, the fetus is a "tissue"; to opponents, it is a "baby." Such verbal and visual images have helped turn abstract concerns about science and technology into moral missions.

THE RESOLUTION OF CONFLICT

How one perceives science and technology reflects special interests and personal values. The social and moral implications of a particular practice may assume far greater importance than any details of scientific verification. Perceptions therefore differ dramatically:

- Will advances in biotechnology bring significant medical or agricultural benefits, or are they simply serving commercial interests?
- Is genetic screening in the workplace a way to protect vulnerable workers or an excuse to avoid cleaning up the environment?
- Are genetic engineering experiments creating humane and beneficial therapies or tampering with nature?
- Are nutritional guidelines and regulations of dietary claims a necessary and scientifically justified consumer protection or a form of government paternalism?
- Is animal experimentation essential to medical progress or an unnecessary moral affront to the rights of living beings?

The means to resolve disputes will depend on the nature of such perceptions. If disputes reflect competing interests as in many siting controversies, negotiations and compensation measures may reduce conflict and lead to resolution. But where moral principles are at stake, efforts to negotiate and compromise may fail to sway those who are committed to a cause.[6]

In some cases, dramatic events such as a major oil spill or the accident at Chernobyl have changed the terms of debate about certain technologies. In the late 1980s the growing concerns about global environmental issues projected the problem of ozone depletion onto the national agenda, and the tragedy of AIDS has forced the Food and Drug Administration to reconsider policies of drug approval and regulation. If the underlying stakes in disputes are economic or political, the discovery of new evidence may change the character of disputes. The arguments over the relationship between diet and cancer, and over the environmental affects of chlorofluorocarbons, have shifted over time. But in moral disputes, there is little evidence that technical arguments affect the position of protagonists, for conflicting visions preclude closure. Thus disputes over animal rights or fetal research continue despite changes in research practices intended to accommodate public concerns.

Resolution of conflicts necessarily reflects the relative political power of competing interests. In some cases industrial interests prevail: Chemical firms are clearly important in framing the principles that shape the use of chlorofluorocarbons and the applications of biotechnology. But through persistence, protest groups have wielded considerable influence; antiabortionists and animal rightists have had a striking affect on research practices ranging from reforms to outright bans on certain types of research. And critics, calling attention to occupational carcinogens and the discriminatory implications of genetic testing in the workplace, have influenced legislation. Some controversies have resulted in government withdrawal of funding from projects (e.g., fetal research in the 1980s). And some scientists, hoping to avoid conflict, have voluntarily moved away from certain areas of research (e.g., on the XYY chromosome).

Ultimately, the implementation of science policy depends on public acceptance—or, at the least, on public indifference. Efforts to foster greater acceptance of science and technology have proliferated in the United States. Legislation provides public access to information through public hearing procedures and extended opportunities for intervention in rule making and adjudicatory procedures. Agencies have organized experiments in negotiation and mediation (Susskind & Weinstein, 1980). Citizens are often included in the advisory committees and institutional review boards overseeing research. Peer review groups, consensus panels, and special commissions are appointed to build public trust (Jasanoff, 1990a).

At the same time, controversies have sometimes led to the suppression of information that might arouse public concerns about potential risks. Secrecy can be a way to divert criticism, reduce the intrusion of burdensome regulations, prevent panic, and avoid costly delays. After the Chernobyl accident, federal agencies issued gag orders to energy agency officials and the several thousand scientists at national laboratories. They feared that disclosure of information to the press would result in hasty and inappropriate public responses to the controversial American nuclear power program (Nelkin,

1989, 1994). Chemical companies seek to restrict information about accidents until there is certainty about risk. In the context of controversy, public communication of information has become an increasingly sensitive issue (Jerome, 1986; Stevenson, 1980).

Based on competing social and political values, few conflicts are really resolved. Even as specific debates seem to disappear, the same issues recur in new contexts. Environmentalists' concerns about the instrumental use of nature were taken up by feminists and animal protectionists. The opposition to experimenting on vulnerable research subjects in the fetal research dispute extended to protests against research on the human embryo and on helpless animals. Risk issues are contagious; a problem in one location creates a public awareness that turns local issues into national disputes.

As conflicts persist, they continue to raise questions of control. What is the relevant expertise? Is responsibility for decisions to rest with those with technical know-how or with those who bear the impact of technological choices? But controversies are increasingly expressing moral judgments as well as economic interests, and they are becoming crusades. The social movements organized to challenge science and technology are driven by a moral rhetoric of good and evil, of right and wrong. They are attracting constituents who fear the misuse of science by major social institutions, who see the need to reassess the social values, priorities, and political relationships underlying scientific and technological progress, and who see themselves as preserving the moral values lost in the course of technological change. Thus controversies matter and must be taken seriously as an indication of public attitudes toward science.

NOTES

1. Case studies of these and other controversies can be found in Nelkin (1992).
2. For a review of the literature on science and technology policy emphasizing increasing concern with the problems of technology in the 1970s, see Nelkin (1977a).
3. This conference, organized by the International Council on Science Policy Studies, was held in Bielefeld, Germany, in May 1978.
4. For a literature review, see Nelkin (1987).
5. For a review and bibliography on citizen litigation, see Dimento (1977).
6. Engelhardt and Caplan (1987), *Scientific Controversies*, includes many cases demonstrating the difficulties of resolving disputes.

20

□

The Environmental Challenge to Science Studies

STEVEN YEARLEY

SCIENCE STUDIES AND THE RISE OF ENVIRONMENTAL ISSUES

Scientific understanding of the environment and public debates over "green" issues have not figured prominently in the science studies literature to date. This is not to say that there have been no analyses of ecological science or of environmental controversies from the perspective of science studies; on the contrary, it is possible to identify numerous, high-quality—though generally unconnected—contributions (e.g., Collins, 1988; Cramer & van den Daele, 1985; Nelkin, 1977c; Worster, 1985; Wynne, 1982). Rather, it is to suggest that "science and the environmental movement" or "the greening of science" have not become as well established focal points of research interest in science studies as, for example, laboratory ethnographies or replication studies.

AUTHOR'S NOTE: I should like to express my gratitude to Sheila Jasanoff for her clear and insightful editorial guidance. Thanks are also due to Dale Jamieson and to participants in seminars in the Science Dynamics Department, University of Amsterdam, the CRICT Centre, Brunel University, and the Sociology Department of Trinity College Dublin; they all suffered early versions of the chapter with good humor.

I have two principal aims in writing this chapter. One is to review and organize the small but growing science studies literature on the environment. My second aim is to make the case that the environment should be recognized as a key site for social studies of science. There are four important intellectual reasons for making this argument. First, scientific expertise is increasingly at the forefront of environmental policy formulation and of contests over policies. Whether in relation to global warming, ozone depletion, pesticide residues, or the release of genetically engineered organisms, scientists have been sought out, or in some cases have succeeded in putting themselves forward, as sources of authoritative advice. This advising work has not, however, gone smoothly. Frequently, experts have fallen out over their recommendations or their analytic assumptions, and their expertise has come to be viewed as inadequate; in some cases the public has rejected experts' interpretations. There may be strong cultural influences on the conceptions of nature that underlie scientists' discourse about the environment (Rayner, 1991, pp. 86-89). Political organization and interests can also shape the content and forms of scientific assessment (Jasanoff, 1986). In general terms, such phenomena are familiar to social studiers of science. Science studies can offer its experience in interpreting such issues and at the same time learn from the specific features of environmental disputes.

The second reason that social studies of science should demonstrate a special concern for environmental issues is that many environmentalists have employed science in assembling a systematic critique of modern industrial society. Thus environmental activists increasingly insist that the "hard facts" of science now support their contention that the planet is being poisoned by our industrial society (Porritt, 1989, p. 353). The independent scientist and originator of the Gaia hypothesis, James Lovelock, is wont to claim that his electron capture detector (which offers to detect extremely minute quantities of pollutants) provided the background information for the beginning of the environmental movement.[1] The popular German sociologist Ulrich Beck has nominated environmental protest as a prime example of what he terms "reflexive modernization"—the destructive application of modernist principles to themselves. For Beck, the crisis known as "postmodernity" arises because modernist analyses can be used to undermine themselves. Thus the systematic (that is, scientific in the broadest sense) analysis of science itself—for example, in the philosophy of scientific method—has identified limitations in science's cognitive authority. Disputes between experts and counterexperts over the scientific assessment of risk and safety exemplify this problem:

> Not only does the industrial utilization of scientific results create problems; science also provides the means—the categories and the cognitive equipment—required to recognize and present the problems *as* problems at all, or just not to do so.

> Finally, science also provides the prerequisites for "overcoming" the threats for which it is responsible itself. (Beck, 1992, p. 163)

In the last two centuries, science has increasingly been funded by the state, and the "scientific establishment" has often come to be criticized for pursuing the state's aims. The use of science as ecological critique provides analysts of science with an important and potentially enlightening counterexample.

The third reason for science studies to take a special interest in the environment is closely related to ecological science's possible critical role. For scientific knowledge claims and expertise not only influence the environmental debate through methodological capacity (the electron capture detector) and technical expertise, science also offers important cognitive models for environmentalists. Lovelock's notion of Gaia, for example, offers a possible ecocentric metaphysic for interpreting the planet as well as a scientific legitimation for that metaphysic's plausibility. I do not wish to imply that this relationship between (purported) scientific reality and worldview is unique to the environmental sciences. Clearly it is not; Darwinism and natural selection too have been taken as imparting metaphysical lessons. But these relationships are not frequent in the history of modern science, and they merit close analysis when they arise.

The final reason for taking environmentalism to the heart of science studies is that, because of national government initiatives, international programs, and the actions of private funding bodies, scientific study of the environment is climbing rapidly in scope and significance. University courses are expanding, along with student numbers. Other associated activities, such as scientific consultancies and even expert-led wildlife holidays, have also benefited from this growth (Yearley, 1992b). Because more of this kind of science is being done, it seems only reasonable that a rising share of science watching should be devoted to it.

I organize my review of environmentalism and science studies around influences working in two complementary directions: First, we observe that the green movement and the politics of environmentalism are to a peculiar degree tied to scientific authority and, second, we consider the impact that the greening of politics and public opinion has had on the scientific community. While these influences are not independent of each other, it is helpful to separate them for purposes of presentation.

SOCIAL PROFILES OF ENVIRONMENTALISM

The environmental movement and environmental policies have developed differently in different countries (see, e.g., Jamison et al., 1990; Jasanoff,

1990b; Vogel, 1986). The United States assumed an early lead both in environmental research and in institutional innovations. Admittedly, the United States is regularly criticized as the country that makes the heaviest demands on the earth's resources; its trade and defense policies have also encouraged environmental despoliation in other parts of the globe (Leonard, 1988). At the same time, possibly in response to the very intensity of pollution and resource use, the treatment of environmental issues within the United States has at times been distinctly more "progressive" than elsewhere in the industrialized world. Established in 1970, the U.S. Environmental Protection Agency (EPA) was a landmark in the institutionalization of environmental concerns. Its regulatory activism in its first decade quickly earned it the opposition of many businesses, which saw it as the enemy of commerce and wealth creation—a criticism that persists to this day.

By contrast, the ecological concerns of the late 1960s and early 1970s did not generally result in such significant institutional changes in Europe's largest economies, most notably in Britain, Germany, and France (see Brickman, Jasanoff, & Ilgen, 1985). The economic adversity of the late 1970s further reduced the political attractiveness of environmental issues. In Britain in particular, environmental pressure groups were more or less alone in criticizing state policies and pressing for reform (Lowe & Flynn, 1989).

Smaller continental European nations followed more closely the U.S. route, with official agencies playing a leading role in environmental reform; in the absence of the conservative and free-market policies adopted under presidents Reagan and Bush, countries such as Denmark and the Netherlands, by the late 1980s, took the lead in many areas of environmental reform and in the pervasive integration of green thinking into public policy (Jamison et al., 1990). Indeed, so advanced was the greening of the leading Dutch political parties that there was no electoral room for a distinctive Green party.

Finally, there has been a significant growth in international pressure groups (notably Greenpeace International, Friends of the Earth [FoE] International, and the World Wide Fund for Nature), bringing the styles of environmental campaigning that originated in the United States and the United Kingdom to international forums. In recognition of these differences, I treat in separate sections the characteristic positions of pressure groups and of regulatory agencies, focusing largely on the United States and the United Kingdom as contrasting "ideal types" of environmental regulatory practice.

SCIENCE AND GREEN CAMPAIGN GROUPS

The environmental movement has much in common with other critical social movements, both progressive and reactionary (such as antiabortion groups or moral rearmament campaigns); it has administrative and organizational needs, it must raise funds, it must hold on to its supporters, and it must choose attainable yet plausible campaign goals. But there is one sense in which the green movement stands apart; more than other movements, it depends on science to support its claims about the problems that confront society (see Yearley, 1992c). Unlike nationalist movements, anticensorship movements, or movements urging support for particular disadvantaged groups, environmentalists root their arguments in science. Thus greens argue that if we continue to release chlorofluorocarbons (CFCs) into the atmosphere ozone depletion will worsen, causing an increased incidence of cancers and disruption of the food chain. Similarly, they assert that unless we reduce the output of carbon dioxide into the atmosphere the earth's temperature will rise, provoking natural disasters and the extinction of plant species. Such claims, environmentalists often assert, are not just statements about values; they are matters of fact (for an example, see Porritt, 1989, p. 353). The evidential basis for their movement is thus stronger, they imply, than if it were exclusively moral. This scientific orientation lends them certain strengths but a reliance on science for authority fails to deliver the decisiveness and moral certainty that activists desire. The view of science that has emerged from the science studies tradition allows us to make sense of the environmentalists' predicament.

The scientific orientation of some environmental organizations was present from their origins. Thus, in the British case, the Royal Society for Nature Conservation—Britain's leading general nature conservation charity—continues from a nineteenth-century precursor dedicated to the promotion of nature reserves. The reserves in question were intended not for recreation, nor even for the benefit of nature per se, but as sanctuaries for scientifically important wildlife. Equally, the British Trust for Ornithology (founded in 1933) institutionalized a naturalist's interest in wildlife. Similarly, in Sweden environmentalism first emerged as a "defensive ecology, aimed not at challenging industrialization but rather at protecting parts of nature from encroachment, *preserving them for scientific research* and aesthetic appreciation" (A. Jamison, 1988, p. 81, emphasis added), while in the United States the Society of Mammalogists provided an early institutional focus for scientists opposed to the pest-control programs of the Bureau of the Biological Survey (Worster, 1985, p. 275). This naturalist's perspective persisted into Rachel Carson's historic critique of pollutants in *Silent Spring* (1965), and it still pervades British legislation, where the most frequently used conservation designation is the Site of Special *Scientific* Interest.

In addition to these relatively "establishment" organizations, the younger, more radical groups formed during the upsurge in environmental interest at the end of the 1960s—such as Greenpeace and Friends of the Earth—also perceived the importance of science as a resource. In the United States, partly because of the early prominence given to scientific expertise in the public actions of the EPA, groups such as the Environmental Defense Fund and the Natural Resources Defense Council recruited scientific talent early on, often coupling it with legal expertise. In Britain such development was slower. Still, Greenpeace appointed an academic scientist as its director of science in London in 1989 (*Times Higher Education Supplement*, April 7, 1989, p. 8) and its authorized historians proclaim that it is equipped with the "most sophisticated mobile laboratory in Europe" (Brown & May, 1989, p. 150). Both Greenpeace and FoE (U.K. Cabinet Office, 1990a) now supply scientific references in support of their campaign material and press releases.

There are two reasons for this convergence on the use of scientific expertise. First, some environmental campaign issues themselves arise through our scientific culture. Without science we would know nothing of the ozone layer or have any way of assessing whether or not it was thinning. (Of course, this invites the ironic observation that without our technological civilization we would not have the CFCs that are responsible for much of the thinning either.) Second, in the modern, industrialized world there are no practical cognitive alternatives. The growth in environmental consciousness has promoted a great deal of philosophical and religious writing, some of it aimed at gaining a deeper knowledge of nature's needs and purposes through spiritual or metaphysical enquiry. But for green groups to offer remedies based on these kinds of knowledge claims would be an ineffectual strategy. For one thing, the bodies with which environmentalists must deal respond better to technical and apparently objective claims than to arguments couched in spiritual or moral discourse. Second, even when people are primarily concerned about the interests of whales or even oak forests and wish to count them as the moral equals of humans, one still has to find someone with expert knowledge to determine what it is that whales or trees want (Stone, 1988). Natural historians end up serving as the mouthpiece for endangered mammals or for cherished birds whose breeding grounds are disappearing under developers' plans.

This use of scientific knowledge and the resulting links with sectors of the scientific establishment has lent social authority to the movement. But there are also drawbacks. For one thing, a perception that movement organizations are becoming closer to establishment science can itself be a handicap because many of the more radical supporters of the green movement are ideologically opposed to technological society and its scientific helpmates and are alienated from things that scientists have done (such as research on nuclear power). In this sense a too-close affinity to science can be a hindrance.

Furthermore, as work in science studies has shown, the very characteristics that give scientific knowledge its strength also make it vulnerable. The avowed empirical basis of science means that the knowledge environmentalists desire may sometimes elude them. Cramer (1987, p. 55) gives a lighthearted but illustrative example of ignorance about ducks on a Dutch polder. On this occasion, environmental management plans depended on reliable knowledge about where ducks that whiled away their days on the nearby coast spent their nights. Naturally enough, however, the darkness inhibited observation and the production of a believable number. In the similar case of rare ferrets studied by Clark and Westrum (1987), routine methods of estimating populations were confounded by the personal habits of the ferrets and their prey. Other knowledge claims may be difficult to substantiate because techniques have not been developed or are disputed—as with knowledge about transfer of gases across the sea-air boundary (needed for modeling the circulation of carbon dioxide)—or because the relevant data have not been collected—as with information on past water temperature around the globe.

Scientists acknowledge that science is revocable and provisional (Yearley, 1989). This is a great cognitive strength because it allows innovation and facilitates the exposure of error, but it prevents scientists from having an authoritative view on every question in the way that a charismatic leader might. It also means that in crucial cases large and seemingly inexplicable reversals in opinion may occur, further undermining the credibility of science. Official knowledge about the mechanisms by which radioactivity may promote cancer in people living near or working in nuclear power plants has undergone dramatic changes in the last 30 years (see Arnold, 1992, p. 150). On another front, recent meteorological work suggests that surface warming in the northern hemisphere may be associated primarily with variations in the sun spot cycle (*Science*, November 1, 1991, p. 652). If accepted, this interpretation could provoke a huge reevaluation of views on global warming. Thus something that scientists "know" with great certainty at one time may come to be denied with equal confidence at a later stage.

The customary ways in which scientific knowledge is produced also render it less than ideally suitable for environmentalists' uses, because most ecological problems are multidisciplinary while most research is not. European disputes over acid rain in the 1970s and 1980s revealed this problem with some clarity (Irwin, 1993). There was little room for doubt about the acid output of fossil-fuel-burning power stations but the exact connection between these emissions and the acidification of a particular lake or forest was far harder to detail. Few scientists possessed, and institutional arrangements in universities did little to promote, the combination of atmospheric chemistry, meteorology, soil science, and ecology needed to address these issues. Green groups have generally been reluctant to fund original research to counteract this problem because they see it as a demand on financial resources for which there is a more pressing need. Of course, even if they did undertake research, their knowledge claims would not escape the generic problems already discussed.

That scientific knowledge often does not lend itself to immediate application is a source of frustration for environmental campaigners. It affords comfort to their opponents who can use the uncertainty of science to argue for delays in introducing reforms. To take the acid rain example again, in Britain the government and the (then) state-owned electricity-generating company joined together to argue that, because there was no compelling evidence that *British* acid rain was causing damage in continental Europe, they had a moral obligation to the taxpayer not to introduce costly abatement measures prematurely (Rose, 1990, pp. 125-129). Exactly similar problems divided the United States and Canada on this issue, with Canadians pressing for controls while "US scientists argued that more research was needed before strict emission controls could be imposed" (J. McCormick, 1989, p. 186).

Finally, although central to the green movement, science is insufficient to meet all its cognitive demands. The thinning of the ozone layer may lead observers to suppose that scientific enquiry will both help to identify the problem and to supply technical options for overcoming it. Science seems to be all that is needed. But, as Wenzel suggests in his study of the controversy over seal hunting in North America, environmentalists often have to supplement their scientific arguments with appeals to other grounds for environmental protection. In this case, green groups initially invoked scientific arguments to show that the taking of seal skins had a detrimental effect on populations. But when later research indicated that culling, at a certain level, presented no threat to populations at all, the argument was rejected and new grounds had to be found (Wenzel, 1991, pp. 46-48).

Scientific reasoning may seem to provide an apparently incontestable basis for handling morally complex arguments. In the case of hunting, it is hard to draw agreed *moral* distinctions between unacceptable and acceptable forms of killing. By contrast, an argument that hunting is driving a species to extinction seems irrefutable. But such scientific arguments are actually insufficient to support all aspects of the greens' case. Scientific expertise cannot tell us whether whale hunting is legitimate or whether elephant herds should be managed for ivory production. Thus, in any contested arena, scientific claims are likely to lose their credibility not only because of the contingent character of scientific knowledge but also because environmental controversies have moral and political components that cannot be resolved by scientific inquiry.

SCIENCE IN THE WORK OF ENVIRONMENTAL REGULATORY AGENCIES

The opportunities that environmental groups have to advance their scientific arguments vary from country to country, depending on differences in their political organization and institutional structures (Brickman et al., 1985). The U.S. situation is unique in the number and variety of forums where science can be debated. Not only were regulatory bodies such as the EPA established relatively early, but they took steps that were seen as radical at the time, including pressing for the adoption of scrubbers in coal-fired power-station chimneys and the fitting of catalytic converters to cars. Extensive proposed reforms stimulated rearguard action on the part of industry, which argued that suspect claims about environmental damage were being allowed to justify the introduction of commercially damaging regulation. This theme won political support at the highest levels during the Republican administrations of Presidents Reagan and Bush.

Aside from political manoeuvering, both industry and environmental groups pursued their regulatory interests through the courts, marshalling counterexpertise to combat the judgments and technical assessments adopted by the EPA and other bodies (Brickman et al., 1985; Jasanoff, 1986; Vogel, 1986). The availability of citizen suits and other judicial remedies meant that pressure groups in particular found themselves in a very different context than that prevailing in Europe. The obvious role for them to adopt was as a prod to the EPA, and sometimes as an explicit counterweight to industry interests, marshalling expert arguments either to press for further tightening of EPA policies or to support the EPA in the face of industry's opposition.

Given the resources that industry could devote to challenging environmental regulations and the high stakes involved in these challenges—for example, a ruling that formaldehyde is carcinogenic to humans would have affected a billion-dollar industry in the early 1980s (Jasanoff, 1990a, p. 195)—it is no surprise that disputes over scientific evidence were fought tenaciously and with great inventiveness. As challenges were channeled through the courts, technical disputes over health, safety, and environmental hazards were opened to judicial—and hence public—scrutiny.

From the viewpoint of science studies, these arguments ran a familiar course. Studies of controversy, both within academic science and in the public arena, have alerted analysts to ways in which scientific knowledge claims can face deconstruction. For example, if toxicity tests are repeated, leading to new and different results, we enter the domain of the experimenters' regress (Collins, 1985, p. 2). The "correct" results are, by definition, the ones produced by the better test, but there are no independent means of determining which test is better unless the "correct" outcome is known in advance. This problem is bad enough in "pure" science, where the reasons for distrusting others' results are disciplinary or occasionally personal. In disputes over environmental safety, huge commercial and political motivations may also be involved, creating further incentives for discrediting the opposing side's claims to scientific knowledge.

U.S. environmental disputes were further complicated because agencies were supposed to maintain a separation between the technical and political (or policy) aspects of their decisions. This led to repeated attempts to operationalize the line between science and policy during the long course of the EPA's attempts to regulate toxic chemicals. The agency suffered a number of setbacks and embarrassments, which commonly led to calls for the EPA to become more scientific. A typical response was to argue that the EPA's errors were attributable to the temporary confusion of science and policy; consequently, the approved remedy was to take various administrative steps to segregate the activities.

However, as Jasanoff has convincingly argued, such segregation can only be achieved through a form of boundary work (Gieryn, 1983; Jasanoff, 1990a, p. 14). The search for a boundary gives the impression that the problem is being addressed, but there is no "true" boundary to find apart from the constructions of the participants. In making this argument, she points first to contradictions in attempts to keep science and policy apart. In a well-known National Research Council report, for example, alongside calls for a clear distinction "between the scientific basis and the policy basis for the agency's conclusions," one finds admissions that policy considerations affect the way scientific evidence is drawn (Jasanoff, 1992a, p. 17). The advocates of the distinction cannot even specify it for themselves. Second, she employs the argument from science studies that all knowledge (whether theoretical or embedded in practices) rests on assumptions that cannot be tested independently of that knowledge. Thus evidence of risk to humans comes—in part at least—from animal toxicity studies; yet the question of whether evidence about deceased rats should be taken as probative at all is both a policy matter and an issue of methodology.

Attempts have been made to shore up the boundaries around "proper" science by creating new theoretical categories. Weinberg (1972) proposed "transscience" as the correct designation for the liminal enterprise that engages scientific expertise but is not fully scientific; he urged that such activity should not be construed as science in the accepted sense. But in many cases there appear to be no grounds for allocating disputes to a special, transscientific category (Jasanoff, 1987). The only thing that the issues have in common is the very thing that the category is supposed to explain—the fact that they face public controversy. We know from social studies of science that all manner of scientific issues have been hotly disputed, even ones of an impeccably concrete and scientific nature. Debates over the age of the earth, a simple mathematical quantity and one apparently wholly accessible to science, raged for decades (Burchfield, 1975). Had billions of dollars rested on the outcome, and had the procedures for assessing planetary age been exposed to judicial review, no doubt the conflict would have lasted far longer. In this context, it is noteworthy that concern over climate change—currently a contested mix of science, policy, and ethics (see the report from the Indian Centre for Science and Environment, dealing with the allocation of responsibility for climate change: Agarwal & Narain, 1992)—was until recently a matter for basic science.

There are of course some peculiar features of the scientific issues that the EPA and other environmental agencies often have to determine. They deal with quantities that are hard to measure, physical phenomena that are highly interactive, and diseases that occur over the course of a lifetime and for which there may be many plausible causes. The science involved in such determinations lends itself to controversy (see Collingridge & Reeve, 1986). But the fundamental insight that science studies brings to the analysis of environmentalism and environmental policy concerns the disputability of scientific knowledge per se, not the special disputability of the science of cancer, climate change, or pesticide toxicity.

To summarize this section, one general conclusion that social studies of science can draw from the experience of regulatory agencies concerns boundary making in science. In the past this has typically been a subject for philosophical consideration although sociologists have not been silent (Wallis, 1985); in this case, however, the experience of EPA scientists and of science advisers has provided a veritable ethnomethodology of the identification of the bounds of science. Agencies have had to determine when they are doing science (which is permitted) and when they are doing policy (which is not). Equally, in reviewing evidence about any environmental hazard, the regulatory authorities must decide which kinds of scientific evidence to take into account (on debates over the validity of "pharmacokinetic" modeling, for example, see Jasanoff, 1990a, pp. 203-205). The study of environmental disputes highlights in this way not only negotiations over the validity of scientific findings but also the social construction, indeed the "structuration," of the boundaries of science itself.

PUBLIC DISPUTES OVER ENVIRONMENTAL KNOWLEDGE

Environmental science is at its most fickle when official scientists, and even political interest groups, lose control over the terms of a controversy that they were seeking to control through science. Two brief case synopses will illustrate what I mean.

In the first, the story of the plant growth regulator Alar, a protracted dispute arose between the EPA and industry scientists over the validity of research suggesting that the substance caused cancer. In 1989 the EPA proposed that the manufacturers should discontinue the chemical's use because new evidence appeared to bear out earlier concerns. However, the debate was taken out of government's hands when the Natural Resources Defense Council, a leading U.S. public interest group, succeeded in attracting massive publicity for its own assessment that Alar was a leading cause of cancer risk to American schoolchildren (Jasanoff, 1990a, p. 148). Although the EPA, the manufacturers, and segments of the food industry tried to argue that the new results were invalid, the public remained unconvinced and the industry was forced to withdraw the substance from the market.

The second case centers on sheep farmers in Cumbria, in northwest England, whose land was contaminated by radioactive fallout from Chernobyl. Scientists from government agencies assured the farmers that the polluting cesium would quickly pass from the food chain, but, years after these predictions, official scientists still measured high levels of contamination and the farmers were still unable to sell their stock. In this case the apparent failure of the official scientists' version was put down to the difference between cesium mobility in the clay soils of upland northern England and the lighter soils of the test pasture where the official measures were derived. In Wynne's study of the controversy, the farmers felt they were more knowledgeable about their soils and particularly about the context in which contamination had to be handled; for example, animals could not be grazed on the lower pastures, which might correspond most closely to official calculations, because the use of this land would exhaust the winter feed. The inaccuracy of the scientists' predictions was seen as proof of the fallibility of all scientific expertise and generated severe skepticism about the scientists' subsequent recommendations. Wynne (1991a) concludes that "people do not use, assimilate, or experience science separate from other elements of knowledge, judgment, or advice. Rarely, if at all, does a practical situation not need supplementary knowledge in order to make scientific understanding valid and useful in that context" (p. 114).

Environmental disputes such as these help highlight the extent to which social actors other than scientists are involved in negotiating the depiction of natural reality. In the case of Alar the "logic" of media presentation supplanted scientists' control; in the case of the Cumbrian sheep farmers, the farmers became active participants in negotiating the reality of contamination.

Difficulties allied to those highlighted in Wynne's study have recently risen to prominence in U.S. literature on "environmental racism." Commentators have claimed that hazardous plants and dangerous wastes have been sited in areas predominantly occupied by ethnically disadvantaged groups (Bullard, 1990). In part, the argument here is straightforwardly one about discrimination. But the debate has become complicated by suggestions that supposedly universal "safety standards" or "safety procedures" (standards legitimated by appeals to science) may actually be insensitive to local conditions. In such circumstances, it is argued, appeals to universal standards may actually legitimate unjust outcomes, just as the assumptions about consistent cesium mobility disadvantaged Cumbrian farmers. However, once critics mobilize "deconstructive" arguments against universal standards, they run the risk of finding themselves in the same uneasy situations as the environmental NGOs described earlier; they are skeptical about the universal pretensions of scientific knowledge but lack alternative bases for cognitive authority (Yearley, 1992c, pp. 529-530).

If the study of regulatory agencies allowed us to see how negotiable are the boundaries around what counts as science, the cases of public controversy described in this section reveal just how wide and how variable the network of reality constructors can be.

A GREENING OF SCIENTIFIC LIFESTYLES?

In this part of the chapter I review the influence of environmentalism on the scientific community. One obvious effect is the growth of new employment opportunities for scientists through working or consulting for government agencies, industry, and environmental organizations. But there are grounds for supposing there may be a more diffuse influence on the practice and culture of science. In this and the following sections I explore four distinct possibilities: first, that there may be an impact on scientific conduct; second, that a special relationship may develop between certain scientific theories or paradigms and environmentalism; third, that environmental concerns may affect research policy; and, finally, that environmental debates may bring questions about the objectivity of science into particularly sharp focus.

Turning to the first possibility, has the large role for scientists in the environmental movement stimulated a disproportionate green awareness in the scientific profession? Because the green message is to an extent a question of lifestyle politics—it demands that glass objects be taken to the "bottle bank," that gas-guzzling cars be avoided, and so on—we might reasonably anticipate evidence of such thinking in the everyday work of scientists. But universities and research institutes (where scientific thinking is presumably concentrated) have been no more "progressive" in using recycled paper, for example, than large-scale commercial organizations. Scientists in Third World countries have long been accustomed to reusing apparatus, to inventive recycling, and to careful husbanding of resources; by comparison, Western universities remain profligate in the consumption of materials and equipment. As individuals, scientists may commit their time to voluntary work for environmental groups, but universities, faced with strong financial demands, and academics' trades unions have rarely fashioned environmental sensitivities into a major part of their central mission.

The area of scientific practice most directly affected by green concerns is probably the treatment of laboratory animals (Yoxen, 1988, pp. 39-47). While animal rights are not at the center of the leading green groups' agenda, this issue represents the area of most acute contact between the scientific community and natural world-based politics. University and industry researchers have repeatedly been targeted as the perpetrators of harm to animals. The response of most scientific groups has been to oppose the more extreme forms of protest and to reaffirm the moral and practical value of animal experiments. There are numerous sociological reasons for this response. For example, for the majority of experiments it will be easier to gain professional and commercial recognition by using accepted methods (animal testing) than by insisting on alternatives (Sharpe, 1988, p. 115). Animal testing has, moreover, been strongly institutionalized within scientific practice. As Rupke (1987) puts it, in the last century, vivisection became

> of great methodological and symbolic significance to experimental physiology. . . . Experimental physiology was used as the horse which drew the cart of the biomedical sciences close up behind the exact sciences of chemistry and physics, and it was animal experimentation that gave physiology the strength to do so. (p. 6)

With regard to animal rights, therefore, universities are caught between a desire to protect the interests of their members and the liberal and progressive dictates of environmentalism.

A GREEN PARADIGM: ECOLOGY AND ITS CLIENTS

If the scientific community has not embraced the green movement at the most individualistic level, there still remains a deeper level at which cognitive developments in science may be linked to the advance of environmentalism. Since the "scientific revolution," scientists have been careful to balance two competing rhetorics of science (see Bimber & Guston, this volume). On the one hand, they have stressed its utility but have not wished it to be regarded as so much a practical art that it should fall under the direct control of its paymasters. On the other hand, they have wanted independence, but not so much as to be perceived as engaged in a merely academic exercise. The science of ecology faced this dilemma in especially acute form. Support for ecology, after its institutionalization at the beginning of the twentieth century, was frequently related to practical management issues (McIntosh, 1976; E. Worthington, 1983). But ecological thinkers typically harbored complex political and philosophical commitments that informed their work (Bramwell, 1989, pp. 64-91). As McIntosh (1985) puts it: "The conflict between the image of science as objective and value-free and that of ecology as intrinsically value-laden and a guide to ethics for humans, animals and even trees is difficult to reconcile" (p. 308).

By the end of the 1960s, the British and American ecological societies, founded in 1913 and 1915, respectively (Worster, 1985, pp. 205-206), were confronting the consequences of this dilemma. Public concern about environmental degradation furnished a context into which it was tempting for ecologists to introduce their expertise. Ecology could be offered as *the* science of the environment. McIntosh shows how senior ecologists held back from the commitment to intervene at the level of public policy, but Cramer, Eyerman, and Jamison (1987), arguing chiefly from research on scientists in northern continental Europe, offer a different picture, finding that not all scientists sought to insert the insulation of professionalism between their research and public concerns:

> Environmental researchers [are] a group of scientists who, at least [at] first sight, appear sympathetic to a closer relation between science and the solving of social problems. Banding together with environmental activists, they have called for the development of a new kind of science which can contribute to the solving of environmental problems. (pp. 89-90)

For Cramer and her colleagues, the interconnectedness of species, documented and insisted on by the science of ecology, lent legitimacy to the environmentalists' ideology. Moreover, this connection was made with the clear connivance of scientists.

Such a close relationship to ideological client groups is unusual for a modern, institutionally accepted, basic natural science. While Bunders (1987) has argued that ecological and biological scientists have become especially adept at cooperating with outsiders in devising their research work ("piggy-backing" for funding purposes and so on), ideological clientship is out of the ordinary. Yet ideological clients for ecology abound. As Dobson suggests (1990, p. 39): "Any book on Green politics" will advert to ecology. Ecology is routinely used to supply "ammunition against hierarchy and discreteness." It implies that all species are equally constituents of one large, systemic web, and it thus appears to endorse both egalitarianism and holistic thinking. Dobson describes ecology as "the science most obviously connected with the Green movement."

The other side of the coin is that exponents of the "new" ecological science had a vested interest in presenting their interpretation of ecology as uniquely associated with benign environmental intervention (see Cramer & Hagendijk, 1985, pp. 492-493). Previous research traditions had tended to focus on descriptive and taxonomic work, whereas the "new" ecology was mathematical and economistic in its models (see Worster, 1985, pp. 294-315). It claimed relevance because of its supposedly unique ability to examine environmental problems in their entirety. The dynamics of this development lend themselves to description in terms of Latour's (1983) metaphor of coalition building. Proponents of the new ecology were happy to publicize the problem to whose solution they believed they held the sole key.

According to Kwa (1989), the new ecology was importantly promoted through the International Biological Program (IBP). Originally conceived as a study of the biological basis of productivity and human welfare, the program could potentially have embraced many components of biological science, including genetics, molecular biology, and human population studies. Yet by the time the program secured funding from Congress, it had been narrowed down to ecology and, more specifically, ecosystems ecology. The peculiar success of ecosystems ecology derived in Kwa's view from the central position it allocated to cybernetic analogies (Kwa, 1989, p. 24). The portrayal of nature as a machine that could be understood and managed set the field apart from other biological interpretations of the natural world and gave it much greater appeal to U.S. political leaders.

The success of the supporters of mathematical ecology in mobilizing resources meant that they had promoted high expectations among politicians and the public. Yet, as Cramer et al. (1987) record, in Europe there was a decline in the closeness of the relationship between ecological science and the environmental movement. In part, this was due to a growth in the availability of fundamental information to activists in the movement. They knew enough to concentrate on holding governments to their stated pollution reduction targets without needing to draw on fresh scientific information.

In the late 1970s and into the next decade, there was also a reduction in governmental support for ecological work, as research became more focused on projects with a likely commercial return. With economic stagnation, research budgets were directed to fostering a small set of technologies (in the biological area, notably biotechnology). In the Dutch case, scientists trained in the new ecology acquired jobs within the enlarged state research and environmental protection system. They were thus introduced into an ethos that left them much less free to engage in campaigning activities. The professionalization of the discipline also led away from the immediate concerns of activists. In the Swedish case, Söderqvist (1986) reports that "ecologists who leant towards experimental studies were not in the forefront in identifying environmental problems" (p. 272). Finally, Cramer et al. (1987) cite the importance of a perceived failing of "cybernetic ecosystems ecology itself" (p. 105). This European experience was matched in the United States, where in the early 1970s, "after a brief burst of public activity, many of the [ecological] scientists went back to the lab, 'pure scientists' once again, their traditional attitude towards politics an 'alien element, essentially destructive of scientific endeavour,' reinforced" (Nelkin, 1977c, p. 81). After the establishment of the EPA, their "idealism was overwhelmed by practical problems" (1977c, p. 81); back in their labs, "many of them found little spillover from practical ecology to their fundamental research" (1977c, p. 84).

A significant departure from this pattern of declining closeness is recorded by Haas (1989) in his analysis of the Mediterranean Action Plan, an international effort to analyze, regulate, and then reduce the pollution of the Mediterranean. Haas's analysis suggests that an "epistemic community" of ecologists and ecologically trained public officials persuaded successive governments in the area to act on sewage treatment, river-borne pollution, and the like. The growing agreement among experts was, in his view, decisive: "Consensual knowledge proved compelling to the uninitiated" (Haas, 1989, p. 397). Governments increasingly fell in line, ceding power over their policy toward the Mediterranean to environment ministries, themselves dominated by members of this epistemic community. While acknowledging that governments' actions were influenced by international pressure and by domestic public opinion, Haas (1989) maintains that unified expert testimony was crucial: "In the face of uncertainty, a publicly recognized group with *an unchallenged claim to understanding the technical nature of the regime's substantive issue area*" was able to see its interpretation win out (p. 401, italics added). It is curious, given the experience of the EPA described earlier, that these ecologists' views went "unchallenged." Haas does acknowledge that countries persisted in banning different substances and in setting varying limits, so perhaps the force of argument was less overwhelming than the quoted claim asserts. Still, Haas's analysis suggests that ecologists are influential in certain international forums and that, perhaps under very specific conditions, carefully developed expert consensus may escape deconstruction.

GAIA: A SCIENCE OF THE PLANET AND A NEW GREEN PARADIGM

In the 1980s another science offered to assume the foundational role once played by ecology: geophysiology, or the new Gaian science of life. As first outlined in 1972 by its originator James Lovelock, the Gaia hypothesis proposes in essence that life on earth is in some way coordinated so as to maintain conditions suitable for life. The most striking achievement of the Gaian system, in Lovelock's view, has been to maintain the earth in a habitable condition while the sun's heat has increased. Planetary life and its products have allowed the earth to stay reasonably stable despite huge changes in the extraplanetary environment. Lovelock concludes that life can only be thoroughly understood if analyzed at a planetary level. For him, the earth is effectively a superorganism; it is this superorganism that he calls Gaia.

Despite Lovelock's considerable scientific credentials, his work on Gaia initially received a dismissive response from fellow scientists (Yearley, 1992a, pp. 145-146). His thinking was criticized as circular. The claim that the earth is actively maintained in a state fit for life was problematic because it was unclear how "fitness" can be assessed. There were also difficulties in identifying specific cases of Gaian feedback. As one scientific objector put it, "Bugs produce gas, gas has climatic and environmental effects, but does this result have a direct effect on the bugs?" (Joseph, 1991, p. 79). Opponents also focused on the apparently anthropomorphic aspects of the theory. Lovelock seemed to imply that the earth had a purpose, but there was no mechanism for relating this purpose to the billions of living organisms on the planet.

The Gaian hypothesis has been taken up in two directions that underscore its interpretive flexibility. On one hand, it was "normalized" as science. Some natural scientists, most famously Margulis and Schneider, put forward a tempered interpretation, whose main effect was to ascribe far more importance to biological factors than was previously common among geologists, oceanographers, or meteorologists (Yearley, 1992a, p. 145). On the other hand, the idea exerted a strong appeal over elements of the green movement, in particular those who wished to displace humans from the center of the moral universe. Gaia goes beyond ecology in arguing not only that other organisms are important but that there is something more important than any one species—our planetary superorganism.

As Dobson points out (1990, pp. 42-45), however, the green movement's embrace may be ill-considered, because the Gaian hypothesis does not necessarily regard the human environmental predicament as urgent at a planetary level. If people make their world uninhabitable by excessive global warming, the human race may decline disastrously. But this is no proof that the net impact on Gaia will be negative.

The two divergent developments of the Gaia hypothesis have been held together by a growing and sophisticated "New Age" literature exemplified by Joseph's work (1991), cited above. Standing outside the professional scientific literature, these books are nevertheless demandingly technical. Just as with the publications of parapsychologists, their moral and practical force is intimately associated with claims to cognitive authority. Again as with parapsychology, there is a large enough audience to sustain them without their authors having to put enormous effort into constructing their work as "science." They do not *need* the legitimacy that would be conferred by recognition in the traditional academies of science.

ENVIRONMENTAL INFLUENCES ON SCIENCE POLICY

Not only the green movement but also government and commerce have sought to enroll the scientific profession. Furthermore, such interest in enrolling the sciences has spread well beyond ecology and geophysiology; other sciences have come to be seen as at least equally important to analyzing our environmental concerns.

Promotion of research proved to be an attractive option for governments as they acknowledged the seriousness of environmental problems. When former U.K. Prime Minister Mrs. Thatcher turned her government's attention to the environment, among her first concrete commitments was financial support for a center for climate change prediction. Critics argued that promoting research was a way of being seen to do something without taking radical steps (see Yearley, 1992a, p. 20). Nonetheless, increases in environmental research expenditure found few opponents.

The category of environmental R&D is a new one for many governments, and many nations are only just beginning to enquire into their expenditure patterns. In 1990 the U.K. Cabinet Office published a classification of environmental research for the first time (U.K. Cabinet Office, 1990b, p. 217) and trends are accordingly hard to spot. Clearly there is a temptation for agencies to redescribe their existing research commitments as environmentally related, with agricultural productivity research turning into countryside management and genetic engineering transforming itself into the study of clean disposal technology. Moreover, Brian Wynne's (1992f) research on environmental R&D indicated that such R&D is dominated by a physics-based "sound science model." Hierarchies within the scientific profession appear to have been reproduced in scientific approaches to the environment, leading to conservative standards of proof and inflexibility in addressing complex environmental problems.

In the European context the institutions of the European Union (EU) are also a spur to green research (an introductory view is presented in MacKenzie & Milne, 1989). Environmental issues can be presented as inherently supranational. Moreover, environmental research can be offered as advancing the common good, so it offers an attractive moral high ground to EC representatives. So far, for example, the EU Framework program has included initiatives on nonnuclear energy alternatives, marine science, and climatology. The politics of European scientific collaboration have also led to numerous nongreen initiatives (such as work on fusion; Jackson, 1991); however, environmental groups, effective lobbyists at many EU levels, have neglected to press for radical reforms of R&D policy.[2]

Because environmentalism is a politics of lifestyle, we might expect radical research initiatives to appear in a bottom-up way from the scientific community as well as top down from funding agencies and their political controllers. A. Jamison (1988) argues, for example, that "the ecology movement has put participation and democratization on the science policy agenda" (p. 84). But there are few examples to which one can point. Dutch science shops made environmental expertise available to community groups and trades unions in the 1970s and a Science for Citizens Program in the United States in the late 1970s was first diluted by Congress and then evaporated under the Republican administration (Dickson, 1988, pp. 229-231). The dearth of such organizations can probably be ascribed to three factors: the success of environmental groups in developing their in-house technical skills and research capabilities; the increasing practice of consultation by bodies such as the EPA, aided in the United States by the Freedom of Information Act (Jasanoff, 1990a, p. 48); and the absence of incentives for scientists to engage in public interest work. Cramer et al. (1987) argue that there has been a tendency to move away from "the democratization of science towards the demand for practical research, carried out . . . by scientists in the movement's name" (p. 110). Scientists in government or private industry are generally unable to participate in public interest work, while university scientists are deterred from doing so through pressures to improve their productivity in traditional ways.

ENVIRONMENTALISM AND THE INDEPENDENCE OF SCIENCE

Despite the absence of concerted bottom-up movements, academic and institutionally independent scientists still play a central role in analyzing environmental issues. In a BBC World Service radio interview, Mark Simmonds, a scientist working with Greenpeace, voiced his concern about threats to the independence of academic researchers.[3] His argument was that, as universities become more dependent on building external earnings, a situation already well advanced in the United States, there will be pressures on scientists not to offend their institutions' sponsors. And because it is the big companies who can afford large-scale sponsorship, there is a danger that their vested interests will be opposed to those of the environmental movement.

Of course, the suspicion of bias can work in the opposite direction as well. The fact that the Greenpeace International Science Unit is hosted by the University of London could be used to support analogous reasoning, but with implications critical of the green movement. Indeed, Malcolm Grimston, energy issues adviser for the U.K. Atomic Energy Authority, made exactly this argument to the British Association for the Advancement of Science at its 1991 meeting. He accused activist organizations of promoting a "Green science" in which "a scientific theory is judged by its success in *promoting* the Green view of society." Green scientists, he suggested, would give undue credit to findings supportive of their general worldview but would seek to devalue adverse findings. For industry to claim the scientific high ground is, of course, not new; Hays (1987, p. 361) records a similar attack by Dow Chemicals on the EPA soon after its formation. Industries, activists, and agencies all work hard to construct the boundaries of science in ways that show their preferred views in the most favorable light. Against this backdrop, the identity of the "disinterested expert" is extremely difficult to construct and maintain. This view from the social studies of science casts doubt on the possibility that environmental disputes at the national or international level can be resolved apolitically through expert bodies such as epistemic communities.

CONCLUSION: SCIENCE STUDIES AND THE ENVIRONMENT

I have implicitly advanced a twofold argument about the relationship between science studies and environmentalism. First, science studies provides a framework for explaining the ambiguous role of science in illuminating problems or resolving conflicts about the environment. Thus the analytic apparatus of science studies explains the difficulties that the EPA has experienced in determining the toxicity of chemical compounds. The science studies view of knowledge and expertise allows us to appreciate the reasons the green movement has often found scientific knowledge an untrustworthy ally. The sociology of the scientific community permits us to understand the complex, multilayered, and unstable relationship between ecological science and environmentalism. And studies in the sociology of marginal science shed light on key aspects of the sociocognitive development of the Gaia hypothesis.

The second strand of my argument is that the study of environmentalism is equally beneficial to science studies. This is true not only because the science of the environment occupies an increasing percentage of scientists' attention but because features of environmentalism throw issues about science and scientific knowledge into particularly sharp relief. Thus, returning to the case of the EPA, the problems of applying science to stochastic and speculative but politically urgent matters disclose important lessons about science and policymaking, about scientific and legal canons of reasoning, and about the conduct of public controversy. Wynne's example of lay expertise about environmental matters also highlights questions about the identity and limits of science itself. These cases show how tenuous and contested are the claims of science to serve as the sole legitimate depictor of reality. The history of ecology reveals with what complexity expertise is dispersed over environmental groups, public officials, and the scientific community. The international character of many environmental problems, particularly when seen in the light of differences between countries' environmental protection apparatuses, encourages a fully comparative sociology of science.

Finally, the wide public concern about environmental degradation and the perceived urgency of the threats to human and ecological health and welfare compel science studies to be less at ease with methodological relativism than in examining the social construction of eighteenth-century optics or of gravity waves. The breadth of environmental controversies—dealing not just with disputes over natural science but with social scientific and metascientific conflicts as well—invite (some might say tempt) scholars in the social studies of science to embroil themselves as experts on expertise itself or as authorities on scientific authority. The typical response in science studies to date has been to remain as detached as possible. An alternative, though, is for science studies to seek a moral identity without letting go of its skeptical agenda. This challenge has faced science studies before; perhaps the challenge of environmentalism will stimulate analysts of science to address it head on.

NOTES

1. This is from an interview on BBC Radio 4, August 21, 1992.

2. This is from a personal communication with Sonia Mazey and Jeremy Richardson, based on their research on EU lobbying by nongovernmental organizations.

3. This is based on a BBC World Service program in the "Topical Tapes" series, broadcast to various parts of the world on various dates in September 1991.

21

Science as Intellectual Property

HENRY ETZKOWITZ
ANDREW WEBSTER

THE Mertonian norms of science depicted scientists as unwilling to involve themselves directly in transforming research results into objects of monetary value (Merton, 1942, 1973b). Academic scientists who marketed their research were defined as deviant. Since at least 1980, however, a significant number of academic scientists have been making contributions to the literature into marketable products, broadening their interests from a single-minded concern with publication and peer recognition. Moreover, these scientists are looked upon as role models by peers who are contemplating business opportunities. Thus the conduct of academic scientists in relationship to the economic value of their research is undergoing a process of a redefinition and normative change (Etzkowitz, 1983, 1989). On the other side of the equation, the growing importance of basic scientific research for economic development has increased the importance of the university to the economy (OECD, 1984, 1990).

Attention to the economic value of academic research has meant that research results have been defined as property (Nelkin, 1984b) and that property in knowledge is contested not only for its symbolic but for its monetary value (Samuelson, 1987). Science and property, formerly independent and even opposed concepts referring to distinctively different kinds of activities and social spheres, have been made contingent upon each other through the concept of "intellectual property rights" (Ganz Brown & Rushing, 1990). Formerly, academic scientists were content to capture the reputational rewards and leave the financial rewards of their research to industry; this division of institutional labor is breaking down, hastened by financial pressures as professors and universities view their research enterprises as akin to businesses that must generate revenues to survive. Scientists, newly interested in the pecuniary significance of their research, retain their interest in its theoretical significance. Indeed, they often find that the two are compatible because the same discovery can serve both purposes. Although such congruence is increasing, it certainly existed in the past (Etzkowitz, 1983).

In recent decades new areas of technology such as microelectronics, computers, and biotechnology have largely developed outside of the older industrial corporations, often based on university research conducted with government support for achieving military or health goals. Spinoff activities from academic research have also called into question the unilinear model of basic and applied research in which theoretical and practical questions and activities occur in sequence and in separate institutional spheres. This has led to the formulation of a spiral model of innovation in which theoretical and practical questions and activities are often interrelated, cross over the boundaries, or occur at the interstices of heretofore relatively distinct institutional spheres (Webster & Etzkowitz, 1991). Indeed, it can be argued that the world economy has embarked upon a new stage of economic growth with knowledge and therefore intellectual property as the engine of industrial development, displacing traditional elements such as monetary capital, natural resources, and land as the driving force. As the capitalization of knowledge becomes the basis for economic growth, science policy and industrial policy merge into one. In this chapter, we shall discuss the new relationships between science and property in academia, government, and industry, both nationally and internationally.

THE CAPITALIZATION OF KNOWLEDGE

The secularization of knowledge production from its magical and religious origins, earlier viewed as a defining characteristic of Western civilization (Weber, 1920), has expanded into the contemporary global phenomenon of international science (Sorlin et al., 1993). In a concomitant development, the function of knowledge in society has evolved from its specialized role in the cultural sphere as, for example, in Buryat shamanism (Eliade, 1972) into a central feature of economic life; the social role of creators of knowledge has undergone a similar transformation (Gouldner, 1979). The quantification of data (Dijksterhuis, 1961), the organization of knowledge into professionalized disciplines (Ben-David, 1968), and the exponential increase in scientific publications (D. J. S. Price, 1963) has, at the same time, made knowledge production an economic activity in its own right with measurable "inputs" and "outputs" (Machlup, 1962). The scientific movement of the seventeenth century (Merton, 1933) has diversified from a limited number of topical areas into the highly differentiated research enterprise that we are familiar with (Fusfeld, 1986) while its practitioners have grown from a small coterie of persons into a measurable proportion of the population (Bell, 1974).

Technology has grown up as a parallel world, intersecting with science at crucial points. The origins of the contemporary economic use of knowledge lie in the systematization of craft practices during the early years of the industrial revolution by workers and manufacturers alike (Landes, 1969). Knowledge was transmitted through the generations by an apprenticeship system (Pacey, 1976) and transformed through improvements in manufacturing practices into capital (Marx, 1973). The emergence of the professional engineering role during the eighteenth century (Artzt, 1966) and the incorporation of engineering training in the university during the nineteenth century was a step toward transforming craft practices into formal academic disciplines (Layton, 1966). The development of academic engineering research during the early twentieth century (Wildes & Lindgren, 1985) and the application of scientific principles to engineering practices expanded the process of transforming knowledge into economic goods (Noble, 1977). Although many of these activities took place in industrial laboratories (Reich, 1985), an important part of engineering research and training was incorporated into the university, an institution that was avowedly noneconomic in its values and norms. The recent industrial involvement of academic scientists, who had largely separated themselves from engineering in the late nineteenth century through development of an ideology and practice of basic research (Hounshell, 1980), was a further step in the drawing of knowledge producers into economic activities.

The transformation of scientific knowledge into economic activity is a fundamental social innovation, even as the worldwide spread of such activities presages a common form of economic development superseding traditional models of capitalism and socialism. The first step in the capitalization of science is to secure knowledge as private property. By its nature, knowledge is evanescent and temporary, because it is always in principle and practice replaceable by new knowledge. Thus property in knowledge with potential economic value must be captured quickly to secure value from it. Indeed, traditional means of securing the economic value of knowledge such as patents and copyrights make the temporal quality of knowledge an essential feature of their terms by specifying limited time periods of applicability. Through these mechanisms, a particular technique for producing antibodies that can fight cancer cells can take on the status of "property." Intellectual property is not only owned but, as such, carries all the exploitation rights ownership normally confers: It can be invested; it can be exchanged wholly or partly for other goods, services, or money; and it can be used to prevent other, similar ideas from trespassing on its intellectual domain, through its being granted patent protection.

The patent system is based on a presumption of individual contribution to technological innovation. The system was designed to reward and encourage inventors by extending to them a temporary monopoly on the economic rewards of their inventions in exchange for making their designs public. Initiated in 1790 in the United States, by the 1930s the patent system was under attack on the grounds that its original purpose had been subverted. It was argued that large corporations were exploiting the system to accumulate patents from individuals so as to exercise monopoly power over an industry, blocking new entrants from access to the marketplace and discouraging individual inventors from developing new ideas in areas of technology dominated by such companies (B. Stern, 1956). At the time, the patent system's effect on innovation could be judged relatively simply in terms of the large corporation versus individual inventor dichotomy. Recently, new actors have emerged on the intellectual property scene, potentially disrupting corporate control over intellectual property rights—changing the role of the patent system in innovation once again.

There was a marked increase in European and U.S. universities of patenting activity during the 1980s: The U.S. National Science Board's *Science and Engineering Indicators* (1989) notes that "U.S. universities received 2.0 percent of patents awarded to US inventors in 1988, more than double their 0.9 percent share in 1978. The Massachusetts Institute of Technology produces the highest number of patents each year, averaging 85. As Schuler (1991) has noted,

> When these figures are compared to the hundreds of patents issued to private industry, universities are not major holders of patents. On the other hand, many of these patents are basic patents that could dominate a new industry or technology. (p. 2)

Patenting activity has been encouraged, moreover, through the support of national and international science policy programs, such as the Stevensen-Wydler and Bayh-Dole acts in the U.S. national context, which, starting in 1980, made universities responsible for patenting and commercializing the intellectual property resulting from federally sponsored research, and the

European Commission's international VALUE program launched in mid-1991, the aim of which is to assist in the evaluation of the patentability of research, to help in the submission of patent claims, and, thereby, to promote the "valorization" of R&D.

The second step in the capitalization of knowledge is to accrue value from knowledge that has been secured. This typically takes place through marketing and licensing activities, based on the patenting and copyrighting mechanisms for securing property in knowledge. Although occasional copyrights and patents can be worth huge amounts (e.g., the patents for vitamin B-12), most are modest in value and there is skepticism over whether universities and government research centers can earn significant income through these devices: The French national research agency, CNRS, has recently disposed of a large number of patents that have generated little or no income. Nevertheless, patenting and licensing offices have become commonplace in research universities in the United States and elsewhere, often as a result of government policies to encourage the commercialization of research carried out with public support (Etzkowitz & Peters, 1991). As patenting and the wider commercialization of science develop, important changes are taking place in the character of science. In one area, biotechnology, particularly noted for its commercialization, MacKenzie et al. (1990) have declared that "there is a realignment drive by economic forces taking place between free information and proprietary information within [biotechnology], and that this process has significant implications for scientific practices . . . [involving] changes in the political economy of science and technology" (p. 6).

Particularly striking is the level of market influence among public science labs that regard basic science as their primary mission. In the United States 60% of these laboratories reported strong market-oriented research activities (Crow & Bozeman, 1991). Similar patterns can be found elsewhere; in the United Kingdom, for example, all the five government-funded research councils that support research in public science laboratories have devoted an increasing proportion of their grants to strategic and applied research (Healey, 1993).

The third stage in the capitalization of knowledge is to renew and increase its value. At present, traditional mechanisms for support of science for the sake of creating new knowledge as a cultural value often fall short despite their positive unanticipated effects (Mansfield, 1991). New approaches, typically involving coordination among industrial, academic, and governmental actors, have taken the form of a science-based industrial policy in many countries. The growing role of government in linking public sector science and industry, long a formal feature of state socialist societies, has become a real element of mixed economies. More traditional capitalist countries such as the United States and United Kingdom and the newly emerging postsocialist societies of eastern Europe often resist such a direct role for government on ideological grounds (Etzkowitz, 1992). Nevertheless, a broad shift in the relationship among public sector research institutions, industry, and government, in one form or another, can be identified worldwide.

This shift toward a more commercial orientation within public science reflects too the gradual growth in funding it has received from industrial sources over the past decade. This is true, for example, throughout all OECD (Organisation for Economic Cooperation and Development) countries. Although the average percentage contribution from industry to fund public sector research is quite small—4%-5%—in some countries (such as Germany, Sweden, and Norway) the figure can be as high as 12%-15%. Moreover, while the average proportion of industrial funding of public research is low, it is quite high for the principal academic centers in any one country. Harvard estimates, for example, that within the next decade it should receive approximately 25% of its research income from industrial sources; in Italy it is the prestigious academic institutions of Milan and Turin that attract most industrial support. Elsewhere, industry's closest linkages are with government research rather than with the less strong academic laboratories (which is also true of Australia, Greece, Portugal, and New Zealand, for example).

Linkage also varies within and between countries in terms of the disciplines most closely associated with the commercialization and exploitation of scientific knowledge. In the United States, for example, Krimsky et al. (1991) have shown that academic-corporate ties in biotechnology are notably high, confirming earlier findings of Blumenthal et al. (1986). In Japan, in contrast, linkage is much more focused on engineering and new-generation semiconductor technologies. Orsenigo (1989) has summarized the variety of links between academia and industry, which are all concerned with the exploitation of intellectual property in one form or another (see Table 21.1). Some of these links have existed for many years, such as consultancy work as well as short-term contract research that can be traced back to the mid- and late nineteenth century in the emergent chemical industry in Europe and the United States (Beer, 1959). But there are as well more recent and, some have argued, more contentious forms, such as for-profit university companies based on an alternative means of property creation: the stock market.

PROPERTY RIGHTS IN KNOWLEDGE

Intellectual property is commonly defined in legal terms as concerning statutory rights of ownership, such as in patents, registered designs, or copyright. There is a wider sense of intellectual property that covers defensible

TABLE 21.1 Main Types of Industry-University Relationships

Short-term	A. Consultancy and research by individual professors
	B. Industrial procurement of services:
	1. Education and training
	2. Testing
	3. Targeted contract research: problem solving
	4. Nontargeted technology transfers: diffusion oriented
	5. Patents
Medium-term	C. Corporate contributions:
	1. Fellowships
	2. Targeted contract research: design and engineering, development, applied research
	3. Nontargeted contract research: pre-competitive research
	D. Cooperative research:
	1. Joint research programs
	2. R&D consortia
	3. Joint R&D laboratories
Long-term	E. Privately funded research centers:
	1. Multicorporate
	2. Single funder
	F. Long-term research contracts: basic, fundamental, pre-competitive research
	G. University-controlled companies to exploit research
	H. Private companies that secure patent rights for resale

but nonstatutory claims over knowledge. Sociologists of science have shown how important eponymy and other forms of reputational credit are in determining intellectual rights, and thereby the distribution of rewards in the social institution of science (Merton, 1973b). Latour and Woolgar (1979) have examined in great detail the accretion of credit and, more important, the way intellectual capital is invested so as to enjoy the greatest return on one's scientific practice. As they note: "In this respect, scientists' behaviour is remarkably similar to that of an investor of capital. An accumulation of credibility is perquisite to investment. The greater this stockpile, the more able the investor to reap substantial returns and thus add further to his growing capital" (Latour & Woolgar, 1979, p. 197).

This individualistic quest has organizational consequences. The quest for credit—"the rewards and awards which symbolize peers' recognition of past scientific accomplishments" (Latour & Woolgar, 1979, p. 198)—is also a driving force in the formation of a research group. To aid in the pursuit of "credit" through paper production, scientists accumulate resources (funds, laboratory space, and assistants) to enable them to increase their rate of production. Scientists are able to maintain or even increase their access to resources insofar as they are able to sustain their credit rating. If a scientist's credit rating drops too far, access to resources will dissipate. The objective of the successful scientist, the speeding up of the "credibility cycle," has the consequence of building a larger organizational structure as additional recognition is used to gain more lab space, assistants, and equipment. Thus the building of a research group or quasi-firm is the organizational correlate of the social construction of scientific facts from the output of inscription machineries (Etzkowitz, 1992). The establishment of a business firm based on scientific research is often an attempt to create a self-generating mechanism for research support from the commercialization of research, reducing dependence on external funding sources (Etzkowitz, 1984).

This use of the term *capital* to describe scientific credibility is, of course, an analogy with economic capital. But this can and has been questioned; for example, Rip (1988) has suggested that

> the scientific entrepreneur is more similar to political, than to economic entrepreneurs. When scientists mobilize resources, as well as when they present their products to audiences, they have to justify their actions and their products. . . . Their success does not depend on a market taking up their products, but on their justifications being accepted. In a comparable way, the production of authoritative decisions in politics cannot proceed in a purely technical way, but depends critically on justifications and their acceptance. (p. 63)

What is, however, perhaps most noteworthy about today's science is that many claims to "credit" that previously would have been recognized only eponymously—as in "Boyle's law" or "Einstein's theory of relativity"—are recognized as belonging to a certain scientist, or team of scientists, because of the patent they hold on it—such as the Cohen-Boyer patent on DNA cloning techniques. Gaining "credibility" in science is increasingly tied to the ability to generate exploitable knowledge, making scientists more akin to "economic" entrepreneurs.

Disputes over ownership of intellectual property have, of course, a history as long as science itself, and are reflected in the many ways scientists have tried to ensure that they rather than other scientists receive the credit for new developments in a field. Secrecy or only partial disclosure of scientific information is commonplace, as is, according to Wade and Broad (1984), its more disreputable form, fraud. When secrecy is discovered, there can be acrimonious dispute among competitor scientists, with accusations and counter-accusations about "proper" scientific conduct, as was seen, for example, in the events surrounding the story of "pulsars" (see Woolgar, 1988) or the controversy in 1990 over cold fusion research. However, when associated with potential material rewards, disputes can be especially hard fought, as was true over the 1980 patent granted to Stanford University on the Cohen-Boyer genetic technique. In this case, the patent application had to be amended twice, while those colleagues who had been associated with the work found that, much to their annoyance, they were not to be included on the patent. A similar dispute between American and French scientists over patent rights to the discovery of the AIDS virus was apparently settled at the head of state level between Presidents Mitterand and Reagan, only to be reopened when the U.S. side was charged with unethical behavior (Haritos & Glassman, 1991).

Secrecy, partial disclosure, defensive patenting, and other forms of protecting one's claim to intellectual property clearly challenge two of the central norms of the social institution of science. "Disinterestedness" and "communality" have been said to be two of the key features of the scientific ethos (Merton, 1942). Through conforming to these—and to the norms of universalism and organized skepticism—scientists secure both objective knowledge and recognition for their contributions toward it. Other sociologists have challenged Merton's account of the scientific ethos, suggesting instead that it is an occupational ideology legitimating the relative independence of science from public scrutiny and accountability (Stehr, 1976), and find that, within science, behavior is not only a lot more variable but also at times quite counter to the Mertonian code (Mitroff, 1974a). Merton, himself, allowed for variability in normative behavior through his concept of scientists' ambivalence between norms and counternorms (Merton, 1976). Nevertheless, he stops well short of the hypothesis that a new normative structure of science has arisen, integrating "capitalization" and "eponymization" into a consistent ethos, and reflecting the transformation of science from a relatively minor institution encapsulated from social influence to a major institution that influences and is influenced by other social spheres (Etzkowitz, 1992).

In any event, whether a scientific code of conduct is a normative ethos or an occupational ideology, or both, the conception of a "disinterested" pursuit of knowledge becomes more difficult to sustain when intellectual property considerations enter into both problem choice and dissemination of results (Angier, 1988). At the same time, controversies over commercialization of academic research suggest that the values embodied in Mertonian norms persist in some fashion despite coming into conflict—else why the disputes? This inconsistency reflects a transitional period in which scientists, having taken on the multiple roles of investigator and entrepreneur, have not yet fully integrated them into a coherent normative code, occupational identity, and professional ideology. Nevertheless, a process of reinterpretation is under way. In the words of a molecular biologist in the early 1980s, reflecting a new awareness of his dual position as academic scientist and founder of a firm, "I can do good science and make money" (Etzkowitz, 1983).

Beyond the self-awareness of individual scientists stands an organizational structure that encourages an openness to new thinking. Neils Reimers of the Office of Technology Licensing at Stanford persuaded first Cohen and then Boyer to patent their work on DNA cloning, after some hesitation on Cohen's part. As Cohen (1982) himself was to comment later:

> My initial reaction to Reimers' proposal was to question whether basic research of this type could or should be patented and to point out that our work had been dependent on a number of earlier discoveries by others. . . . Reimers insisted that no invention is made in a vacuum and that inventions are always dependent on prior work by others. (p. 215)

The involvement of science in the creation of property is now institutionalized in the university as well as in government and industry. Intellectual property has become as important as the more common forms of material property; indeed, much material property could be neither created nor secured without the intellectual property on which it depends. Only a few decades ago, when nuclear magnetic resonance technology was invented at the University of California, Berkeley, scientists were content to receive the reputational rewards while allowing the General Electric Corporation to capture the financial profits. This episode is now viewed as a cautionary tale about what must not happen to future intellectual property developed in the university. Under current conditions of stringency, the university is determined to make money from the discoveries of its faculty. New science-based companies—such as in biotechnology—depend on the intellectual property of their innovations to sustain their corporate image and, more important, their capital base to pay for staff, buildings, and equipment.

CONTESTED KNOWLEDGE

Intellectual property is always contestable, and intellectual property rights over an area of research, a technique, a process, or product are often contested in and outside of courts. As with most other types of proprietary claims, those with the cultural, economic, and political resources will be most likely to secure their claim against competitors. Those who occupy this contested terrain include not only scientists, corporations, and academic institutions but also pressure groups, politicians, and whole countries, such as peripheral Third World states who challenge the claims of the more powerful. Political and ethical matters, reflecting the global economic divide between haves and have-nots, are therefore central to the debate over intellectual property rights. With the lessening of the distinction between science and technology, issues of intellectual property rights have entered the scientific workplace, becoming part of the relationship between research collaborators, mentors, and disciples. These developments pervade relationships among industry, university, and government in a number of countries across political divisions between the North and South, East and West, and take increasing prominence in trade relations among nations.

In the international arena, countries differ in the stance they have taken toward intellectual property rights depending upon their level of industrialization and intensity of research capabilities. For example, Americans and Japanese tend to differ on the importance to be attached to the original invention and its associated intellectual property rights versus the steps subsequently taken to improve a product. With its highly developed research capabilities and its difficulties in translating inventions into marketable products, the United States has adopted a strategy of attempting to increase its return from the sale of intellectual property. Thus the United States often takes the lead among industrialized nations in arguing for increased protection of intellectual property rights. Japan, on the other hand, has had great success in basing its industrial success on incremental improvements to existing technologies. Thus Japanese companies are skeptical about increased claims for payments based on intellectual property rights to an invention, believing that most of the value added has come later. They are especially resentful of claims from firms that have withdrawn from development of a technology. On the other hand, Americans feel "ripped off" when a technology originally "made in the U.S.A.," such as the VCR, turns into a major Japanese industry.

Historically, technology-rich countries favor increased protection while technology-poor countries favor less (David, 1992). In the nineteenth century this division existed between different parts of Europe. At present it is a division between developed and developing countries. As developing countries advance technologically to the status of newly industrialized countries, they begin to raise their intellectual property protection standards, presumably because they have crossed the divide from consumers to producers of new technology. However, developed countries argue that it is in the interest of less developed countries to strongly protect intellectual property rights as well. Nevertheless, this argument has gained only limited acceptance. Its most receptive audience is scientists and engineers in developing countries who might expect to develop new technology themselves and thus stand to benefit from expanded protection.

Cultural and economic differences among countries with respect to intellectual property are paralleled by those of large and small firms. Large corporations are interested in higher levels of intellectual property protection. They have the resources to obtain patent rights and defend them, whereas, for small firms, intellectual property rights issues can become a heavy financial burden, deflecting financial and managerial resources away from developing and marketing new technologies, although there is always the hope that they too can transform an intellectual property right into a dominant share of a market. Large firms tend to view intellectual property rights as a technique to maintain control over the technologies on which their companies are based. Employees are typically required to sign the rights to inventions over to their employers. Employees who leave a large firm to start a company based on technology that they worked on for their previous employer often find themselves subject to lawsuits. Because such litigation is typically quite expensive, the large firm has an advantage—depleting the smaller firm's resources for research and development. As in the case of Texas Instruments versus Compaq Computer, the dispute is often resolved by a payment to the large firm with the small firm allowed to continue its development of the technology in question. "Noting that Texas Instruments now makes more money from patent licensing than from operations, the smaller companies say its lawsuits amount to bullying" (L. Fisher, 1992). As between the United States and Japan, much of the disagreement here is over the value of the original invention in comparison to the subsequent effort that went into its development.

This difference in perspective and material interest can also be seen in the tense yet symbiotic relationship between large and small software firms using a particular language. In a typical instance, a large firm angrily threatened a lawsuit charging infringement on its copyright (R. Stern, 1988). Small companies who were selling supplementary programs with special features for the large company's product had established a group to develop common standards for the product's programming language. The large firm believed that it could extend its copyright entitlement to the original product into a right to control extensions of its use. The small firms sought to provide a common framework to encourage additional innovation and new products not under the control of the owners of the original product. Apro-innovation strategy would allow the large firm to recover a modest stream of licensing revenues from the small companies in recognition of its development of its original invention. The large company's attempts to inhibit the activities of the small companies failed to recognize that the purpose of the intellectual property system is not simply to protect revenues but to encourage the dissemination of knowledge and thereby enhance the possibility of further innovation. Additional debate has arisen over the question of whether to fit new technologies into the framework of existing intellectual property regimes or to develop special so-called sui generis laws to protect each major new technology (Armstrong, 1992; Frischtak, 1991).

Whether the proposal is to extend stringent protections worldwide or to tailor new protections for semiconductor chips or software, the emphasis of most legal approaches to intellectual property is to enhance the revenue of the inventor and/or the owner of the invention. An alternative approach to the intellectual property system is to view it as a potential source of future innovation rather than as merely a protector of existing inventions. In this view, intellectual property rights are but one element of a system of innovation in which an invention is viewed as a relatively minor part of the long-term process of developing a product and expanding an industry.

By deemphasizing the role of invention, an alternative ground can be sought for the intellectual property system, emphasizing its original purpose of encouraging innovation. This purpose is expected to be achieved, by increasing the flow of information, through the use of software databases and electronic communication systems rather than through the traditional incentive of temporary legal monopolies. For example, Japan is engaged in an effort to reformulate its patent system into an electronic database not only to increase efficiency in filing, storage, and retrieval of information but also to employ the database itself as a factor in the production of inventions. Patent archives have traditionally played this role, because a search of the patent literature has been an early resort of the perspicacious inventor. Nevertheless, the ability to electronically search and recombine existing elements of technologies into new patterns is expected to increase the velocity of this process and has raised, to a new level, the role of electronic media in the dissemination and protection of knowledge. It has also added an international dimension because much of the intellectual property available to be examined electronically will have originated in other countries.

The second issue concerns the growing complexity of the information system itself. LaFollette (1990b) has noted that there is a continual growth in the forms of media—especially electronic—through which scientific and technical information (STI) is disseminated, shared, accessed, or used as a resource. This makes control and regulation of intellectual property more difficult. As she says:

> Increasing the electronic interconnections has enhanced the sharing of information and enhanced creative collaborations in R&D, but it may have also endangered the heart of the intellectual property system, by making the boundaries of its legal and economic compartments—that is, "ownership" and "authorship"—even harder to define and protect. (LaFollette, 1990b, p. 135)

Moreover, the determination of just who does own scientific information takes on growing political importance as international exchange of STI is itself facilitated by electronic communication systems. Determining rights has become a national, not merely an individual or organizational, issue as "leakage" of the research effort goes abroad. In the past there has been an attempt to tighten up the dissemination of materials and information—particularly during the cold war period (Shattuck, 1991). The institution of secrecy, however, generates its own dynamic to overcome it. The contradictory and unexpected consequences from the use of export controls over sensitive technologies were a case in point. Not unexpectedly, competitors with similar technologies and less stringent controls gained some market advantages. However, countries subject to such controls often obtained forbidden devices and reverse engineered them to learn how to begin their own production. When political conditions changed, countries that developed their own versions of a controlled technology became fertile ground for the original manufacturers, given the knowledge base that was put in place through local efforts to thwart the purpose of the controls. The reverse engineering of Digital Equipment computers by Hungarian computer developers and that company's subsequent ease of entry into the Hungarian marketplace at the end of the cold war suggests the utility of spreading a knowledge base rather than excessively restricting it.

Of course, there will continue to be political reasons for restricting access to military technologies such as nuclear weapons. Nevertheless, restrictive policies have proved difficult to enforce, as in the case of Iraq's gaining access to sensitive nuclear technologies. Moreover, the cross-national development of research programs by several companies and the location of an individual company's research laboratories in several countries further serve to increase the porosity of national intellectual property boundaries. Typically, however, greater control has been associated more with public research programs than private ones; by implication, the more public sector science is commercialized, and particularly as overseas multinational companies participate in research centers in the United States, United Kingdom, Germany, France, or elsewhere, the less easy it will be to police and implement national policies on intellectual property rights.

In the global context, the dominance of core First World states in the areas of science and technology has important implications for developing Third World states and their ability to control local scientific and technological information for their own benefit. There has been much concern, for example, about the loss of control over the germ plasm and genetic material associated with indigenous seed in less developed countries. As multinational companies have moved into agricultural biotechnology, germ plasm has become targeted as a key resource for future profitable innovation

(Busch, Lacy, Burkhardt, & Lacy, 1991; Walgate, 1990). Biotechnology companies are clearly in a much stronger position to defend their claim that innovation—and thus patentable knowledge—with regard to seed development occurs through the developments in their high-tech labs. Yet as Mooney (1989) says:

> The argument that intellectual property is only recognizable when performed in laboratories with white lab coats is fundamentally a racist view of scientific development. . . . Farmers, gardeners and herbalists use much inventive genius and more as they continually modify and develop new plant, animal and microbial products and processes. (p. 33)

Where the property rights over seeds is secured on behalf of multinationals, these small-scale cultivators are likely to find their costs increase and ultimately to have their livelihood under threat. The long-term result is that many may become marginalized through having to sell more land to pay for increasing production costs (Juma, 1989). Moreover, the rapid and growing power of First World organizations over global intellectual property is indicated by the fact that less than 1% of world patents are now held by Third World nationals. Apart from the power imbalance here, this extension of ownership rights into areas such as traditional agriculture, living organisms, and the genetic bases that constitute them raises important ethical questions, as exemplified by the more recent debates surrounding the human genome project (Stewart, 1991).

We see from the foregoing discussion that the changing institutional, technological, and geopolitical circumstances associated with the growing commercialization of science make the control and regulation of intellectual property increasingly problematic. It is likely that the balance needed between public and private gain—which in many ways lies at the heart of the intellectual property debate—can only be resolved in the future via new structures of accountability built into both public and private sectors.

FACTORS ENCOURAGING COMMERCIALIZATION

Why, we now ask, has science increasingly come to be regarded as a "commodity," a form of intellectual property that can be commercialized? The STS literature suggests that there are five key socioeconomic processes at work that have stimulated the commercial exploitation of science.

The first factor is the growing "epistemic" character of scientific and technological knowledge (Rip, 1991) in the process of production, especially the appearance in recent years of pervasive enabling technologies that underpin a wide range of industrial sectors without being unique to any one. Information, communication, and bioscience technologies are especially good examples of these new generic forms of knowledge upon which a growing number of sectors must rely to remain competitive.

Related to this, and thereby to the growing commercialization of science, is the gradual breakdown of the distinction between basic and applied research. Thus, as Howells (1990) notes,

> In the field of high-temperature superconductors basic research is being undertaken with a view to future applications right from the beginning. Equally in some areas of biotechnology substantial progress in applications will not be possible without breakthroughs in the understanding of basic biology and biochemistry. (p. 273)

Hence it is of increasing importance for corporations to link with research centers to gain access to both basic and generic technologies through which future innovations will be secured. Academic-industry linkage has become especially important for research-intensive corporations. As the differentiation and complexity of the knowledge bases of new technologies grow, they create what Stankiewicz (1991) has called emerging "technological fields" (such as cybernetics, control engineering, computer sciences, and genetic engineering), which create new "communication systems" that link "sets of knowledge producers" across different organizational and technical areas.

Second, as the cognitive character of science and technology changes in the way just described, we see the emergence of a new division of labor that facilitates much greater linkage between what have been separate institutions. Disco et al. (1990) argue that, as these changes in the cognitive character of science and technology occur, "they allow a division of labour through which universities can participate in technological innovation" (p. 2). Universities have become an important site for so-called precompetitive research, often jointly supported by government and industry to supply the new strategic technologies of the future. Such work focuses on the broad generic principles and models that can be adapted and applied to specific technical systems—such as car design, furniture making, civil engineering, and so on. In this sense, technological fields are separate from actual industrial sectors. As Stankiewicz (1991) says,

> There is a partial decoupling of the technological fields from the concrete technological systems. Overlapping with the "communities" of practitioners [those immediately associated with a particular industrial sector] there emerge the communities of technologists developing and sharing knowledge relevant to many

different communities of practitioners. This is enhanced by the development of new basic technologies such as laser engineering, protein engineering etc. (p. 17)

Thus the institutional barriers between public and private sectors are broken down as the cognitive linkage between the two grows.

The third factor encouraging interorganizational linkage concerns the restructuring of capitalist production over the period dating from the early 1970s during which capitalist economies experienced major difficulties in their ability to sustain rates of profits. This period saw the shift from a "Fordist" production system based on large-scale mass production methods, geared toward uniform products for a homogeneous mass market, to the "flexible specialization" of the "post-Fordist" present. Corporations today maximize flexibility and keep costs down by splitting off parts of the production process once undertaken in-house to specialist subcontractor or subsidiary institutions whose services and skills can be bought quickly in response to market or technological shifts. An important implication of this has been closer commercial links with universities. As the OECD (1990) has remarked,

> The prolonged economic structural crisis and the beginning of a recovery have influenced the strategy of enterprises: in particular the contracting-out of many services has been intensified; in many cases this has led them to move out specialized work which can be done with better scientific and financial conditions in other laboratories (university laboratories, private or public contract research establishments) as well as advanced training in specific areas. (p. 10)

As corporations have sought to respond in this way to capital's own crisis, academic institutions in western Europe, the United States, and, more recently, in eastern Europe and the former Soviet Union have experienced their own crisis caused by the relative decline in state support. There has been, and this is our fourth argument, an erosion of the grants economy within which academia has traditionally operated and the appearance of a more competitive exchange economy that has required institutions to identify and build upon their specific areas of competence. Moreover, in the mid-1980s most western European and U.S. universities enjoyed for the first time, or expanded, the right to hold and exploit patents (Reams, 1986). Typically, this has meant a growing willingness to link with as well as establish commercial enterprises. Thus simultaneous financial and structural changes in public and private sectors have given further momentum to the interorganizational relations that cognitive changes in technological systems had already begun to facilitate.

Finally, a number of commentaries have stressed the regional developmental role that academia plays today, much more extensively than in the past. That is to say, government in a number of countries has sought to use the growing relationship between academia and industry along with the commercialization of public sector knowledge as a substitute for a state-driven industrial policy for economic development. This is especially true of the United Kingdom and the United States, where national planning for science and technology and economic development was regarded throughout the 1980s as politically unacceptable to center-right administrations. Rather than tackling industrial recession through a supply-side economic policy that devotes attention to improving industrial infrastructure, these administrations tried instead to engender a general environment that they believed would be conducive to technology transfer. In this context, by changing the requirements for receipt of public support, such as requiring steps to be taken to commercialize research results, universities were used as surrogate agents for industrial policy programs that governments were unwilling to undertake more directly (Etzkowitz, 1990a).

Science parks, a variant of industrial parks, to encourage the location of high-tech industrial concerns have been one notable feature of this process. Stanford (U.S.), Cambridge (U.S.), and Cambridge (U.K.) are regarded as the most successful examples of local technology poles around which regional development has occurred. In imitation of these successes, many other parks have been established around the world and are seen as a crucial mechanism for technology transfer and new employment. In some areas, such as the Basilicata region of Italy, the infrastructure that generally makes for success of such a park is lacking. Thus the developmental potential of these parks can be exaggerated, as Monck et al. (1988) have shown. On the other hand, their success can be underestimated by expecting results prematurely (Wheeler, 1993). For example, the park established by Rensellear Polytechnic Institute in Troy, New York, in the early 1980s, judged by some to have failed in its early years, was considered by the end of the decade to have succeeded in encouraging and retaining a significant number of firms.

ENTREPRENEURIAL SCIENCE IN THE UNIVERSITY

University-industry relationships capitalize knowledge through three similar procedures:

- The licensing to corporations of research results on exclusive or nonexclusive terms by administrators

- The establishment of university offices to market intellectual property to industry
- The exchange of ideas for equity in a firm, on the part of professors

It is no longer unusual for a university administration to negotiate with government agencies to organize an industrial park, for a group of faculty to bring together the companies in an industry to establish a research center, or for individual faculty members to participate in the formation of a firm to commercialize their research. Indeed, these activities are often encouraged by government even as they are sometimes a matter of dispute within the academy. The result of these individual and collective actions is that the university is playing a larger role in regional economic development and that the economic function of academic research, at least in the sciences, is becoming more explicit. Within the university, for example, this means that administrative offices are encouraging academic scientists to be aware of the commercial implications of their research and, of course, many scientists already have this awareness and need no encouragement. Although these organizational changes and the emergence of entrepreneurial behaviors among the professorate have become widespread during the past decade, they should properly be viewed as the transformation of an anomaly into a paradigm. At MIT the infusion of new science-based technology developed in the university into industry was viewed as an appropriate academic function as early as the mid-nineteenth century.

Entrepreneurial behavior in the university originated as part of the institutionalization of engineering as an academic discipline. The importance of maintaining a connection between what is taught in the university and what is practiced in industry affected how the academic engineering role was formed and legitimated. Much of the content of this model was created at MIT and then transferred to other institutions.

Having a dual role as a school with land grant origins that had evolved into a private institution, MIT transferred the land grant patent model to the private university sector. Because of their close contact with industry, MIT faculty and administrators were aware very early on that a patent by itself was worth very little. What counted was the subsequent development and marketing of the invention. Otherwise a patent was simply a piece of paper that could be counted as another academic accomplishment, much as articles on a curriculum vitae. Development and marketing could take place either through arrangement with existing companies or through the formation of a new firm. In working out these issues through a series of organizational experiments, MIT professors and administrators created a model for transferring technology from academia to the business world.

As the first and most developed of entrepreneurial universities, MIT provided—and still provides—a model for other institutions seeking to commercialize their activities. In Europe, for example, there are some notable examples of universities that have geared themselves, during the late 1970s and throughout the 1980s, toward the sort of organizational structure found at MIT, such as Salford University in the United Kingdom and the Technical University of Berlin in Germany. This is not to suggest they copied MIT or saw it as a blueprint for their own establishments. Rather, once the push toward the commercialization of science had become more generalized, the organizational logic of the MIT experience became very compelling.

The activities at MIT might have remained as a development peculiar to a unique institution were it not also for the parallel development of the faculty-formed firm, an explicit role for the university in shaping regional economic development strategy, the interdisciplinary research center, and the invention of the venture capital firm (Etzkowitz, 1988, 1993). During the postwar era, these innovations were transferred to Stanford University, where the model for the university-initiated firm had been independently invented just before the war and where a scheme for land development grew into the concept of the science park.

During World War II and the postwar era, government sponsorship of research eclipsed all other sources, as state support for frontier science grew rapidly during the 1950s and 1960s and industrial support for academia declined in relative terms. State support for science, within all NATO and generally all OECD countries, has grown over the past three decades, apart from a period during the 1970s when funding was hit by the then structural instability of most industrial economies. As with any statistical series, comparisons of spending on science across countries have to be treated with great care as they may not be measures of the same phenomenon: Statistical methods vary from country to country, the boundaries of R&D are regarded differently, and the relative buying power of state funding (allied to fiscal policy) means that the domestic and international "terms of trade" in R&D are very uneven.

While the level of cash support from industry is important in stimulating qualitative changes in academia, it is only one factor. The model of research is also crucial. Universities are undergoing a change from basic to applied to industrial product—a one-way flow linear model, in which they have been (since World War II) traditionally involved only in the first two steps—to a multidirectional spiral model in which the linear model is incorporated along with at least two other approaches:

where the solution of an industrial problem leads to basic research questions (e.g., the need for improved switching devices for telephony led to basic research in solid state physics in the 1940s at Bell Labs, U.S.), and

where the solution to a technological problem generates basic research questions and leads to a product (e.g., the ENIAC computer project led to Von Neumann's theorizing about the nature of computing machines and the UNIVAC computer system).

The latter two approaches—the first from a theoretically oriented industrial lab, the second from an interdisciplinary military-sponsored research project at a university—are currently being incorporated into academic research through the establishment of interdisciplinary research centers throughout most OECD countries, national programs such as the U.S. National Science Foundation's Industry-University and Science and Engineering Centers, and other state-supported research institutes. Moreover, the feedback cycle between basic and applied science is increasingly apparent within corporate research; for example, the Dutch firm Philips has developed a model of innovation based on evidence showing that maturing technologies need to be opened up to further innovation through a return to basic research in the company.

The commercialization of academia has grown through the internal activities of institutions as well as through corporate investment and participation in research activities. Many higher education and research institutes in industrialized countries have established agencies designed to exploit their own intellectual property. Rather than being a secondary consideration, research or training courses are often judged first in terms of the commercial value they might have for their institution. Technology transfer organizations have also appeared either as developments out of existing provisions or as entirely new organizational structures, perhaps peripheral to and semi-autonomous from the parent institution.

Some early studies argued that academic entrepreneurs were most likely to be those who had a predisposition toward commercializing their research (Peters & Fusfeld, 1982). In one of the more detailed examinations published thus far on the topic, Seashore Louis et al. (1989) argue that individual characteristics (such as motivational predisposition) "provide weak and unsystematic predictions of the forms of entrepreneurship" that are said to relate to innovation. Much more has to do with a combination of local institutional factors that provide the catalyst for successful commercial academic activity (see also Stankiewicz, 1986).

Academic science has also been transformed into a more organized mode, closer to the format of industrial labs (Etzkowitz, 1989). The model of the professor supervising a few Ph.D. students in one-on-one meetings has largely been relegated to the humanities in larger research-intensive universities. Instead, the so-called individual investigator is the leader of a research group comprising some or all of the categories of undergraduate and graduate students, postdocs, research associates, and technicians. In some countries the professor has become the fund-raiser, personnel manager, publicity agent, and research director for a team of researchers. These groups operate as firmlike entities or "quasi-firms" within the university, lacking only a direct profit motive to make them a firm (Etzkowitz, 1991). When professors move outside of the university to form a firm, their academic experience as entrepreneurs often stands them in good stead. They negotiate with venture capitalists instead of research agency program managers and hire employees instead of recruiting students and postdocs. This is in contrast to an earlier generation of technical entrepreneurs and firm founders who were viewed as bringing only technical expertise to a new firm, lacking the managerial skills and entrepreneurial background of many contemporary academic scientists. Indeed the holder of a professorship at a major university has even been held to be the equivalent of a CEO of a medium-sized corporation, in managerial experience. In turn, some organizers of large academic research centers have already had the experience of founding a private business firm and so they bring their new organizational experience back to the university (Etzkowitz, in press).

How far should we be concerned with these important trends with regard to the commercialization of academic research as a radical shift in the organizational structures of the two sectors and, in particular, in the specific role that academia plays within the wider society and economy? As Peters (1987) notes,

> Concern with economic issues, industrial competitiveness and technology transfer in the context of funding constraints is conducive to a period of organizational and intellectual experimentation within the university. It may bring about structures or an expansion that represent a significant departure from the past. (p. 178)

A key question, however, is whether these new structures will become encapsulated within only part of the university or whether the translation of intellectual property into economic enterprises will become an academic mission on par with teaching and research.

One view with respect to this argues that these developments are no more than an extension of earlier patterns and that the overall functional role of the two sectors and the nature of their relationship has not dramatically altered. The second position suggests that these quantitative and structural changes herald the appearance of a new type of academic institution, one that is oriented much more directly to playing a role on behalf of the state as an agency of economic development. Indeed, we may be witnessing a second academic revolution, following the first revolution when academic institutions took on the principal research role in the earlier part of this century (Etzkowitz, 1989). That first development was closely though not exclusively related to the state's need for academic research to contribute to the development of agricultural, medical, and military programs. Today's revolution ties in with the state's need to stimulate economic growth in the absence of formal industrial policies for doing so. Even where such policies exist, academic-industry relations are now given a very prominent, indeed central, place in the many policy proposals that have appeared over the past decade. It could be argued, then, that we are seeing the beginnings of a new "social contract" between academia and society replacing the postwar alignment of science and the military (Bush, 1945). This "contract" stipulates that large-scale government support for academic research will be sustained only so long as the research plays a key role in the new economy.

How can we determine whether the shifts we have noted are fundamental or cosmetic? One way is to chart the emergence of highly innovative organizational structures within academia and subsequently between academia and industry. For example, one important development has been the steady though stochastic growth of hybrid transinstitutional structures that combine academic and industrial R&D activities (Webster, 1990). These reflect the shift toward what Orsenigo (1989) has called "quasi-integrated" institutions, or what Stankiewicz (1986) terms "intermediate peripheral" institutions (such as Enterprise Forum at MIT, the Electronics Group at the University of Lund, and the Fraunhofer Gesellschaft in Germany). Commenting on these developments, Wasser (1990) believes that the university "may well be heading towards a transformation so radical as to become a qualitatively different structure" (p. 121). Indeed, "the present rapid and radical move to a university adaptive in a major fashion to economic development, to an entrepreneurial university," suggests that many institutions would no longer fit "the time-honoured definition of a university." Clearly, this analysis would refer primarily to those universities that are predominantly science based. More generally, we might say that the traditional distinctions between public and private sector R&D are becoming blurred, especially in bioscience, pharmaceuticals, and information technology.

However, as Etzkowitz (1983, 1989) has suggested elsewhere, the shift toward an "entrepreneurial university" need not imply that its members will have to acknowledge that they no longer operate in an academic culture, precisely because the norms of the academic culture have themselves been transformed over the past decade. The "traditional" pursuit of certified knowledge has been combined with and reinterpreted as compatible with commercially oriented research: "Among scientists, one of the most deeply held values is the extension of knowledge. The incorporation of this value into a compatible relationship with the capitalization of knowledge constitutes a normative change in science" (Etzkowitz, 1989, p. 27).

CONCLUSION: BEYOND CAPITALISM AND SOCIALISM

Intellectual property was identified as a formative element in the modern origins of science in the seventeenth century (Merton, 1933). In the late twentieth century, the confluence between science and the economy, mediated by the intellectual property system, has become increasingly central to the development of both spheres (Mansfield, 1992). Indeed, the commercialization of science, as this chapter has shown, may involve a normative shift in attitudes toward intellectual property so that the ethical presumptions of science themselves get redefined by the social actors involved. Even scientists who believe that direct involvement in commercialization is improper because it might compromise the openness of research are pushed in that direction. Krimsky notes that half of the U.S. scientists in his survey had strong ties with biotechnology firms but were also members of the NSF's peer review list for biomedical sciences. In these circumstances, "it is difficult to prevent people who are so inclined from pilfering ideas while they serve as peer reviewers." Ironically, then, "this may lead some scientists to seek commercial funds for their ideas rather than risk having them stolen through the peer-review process" (Krimsky et al., 1991, p. 285). The norm of "capitalization" has displaced "disinterestedness" as adherents, agnostics, and opponents of the legitimization of intellectual property regimes in the university all fulfill its requirements through a variety of available modes, ranging from filling out an intellectual property disclosure form to organizing a firm.

Clearly, commercial activity can be regulated according to conflict of interest principles but the implementation of such rules, separating academic from commercial activities, becomes even more problematic as the university itself becomes the entrepreneur. What was formerly an "outside" activity such as consulting when undertaken by a professor as a sideline of direct academic functions of teaching and research becomes an academic task itself when the mission of the university is expanded to incorporate economic development. The generation of intellectual property is making the university into a more central social institution in the "Millsian" sense (Mills, 1957), perhaps in the coming century displacing the military as a "core" sector in a revised tripartite schema of academy-industry-government. Moreover, as the number of interested parties involved directly or indirectly in the production of scientific information and research flows continues to grow, determinations of proper and rightful ownership become ever more uncertain (Zuckerman, 1988a). As intellectual property rights in knowledge are perceived to be ever more valuable, ownership rights that were only recently transferred from the federal government to the universities in the United States have become the subject of proposals to repatriate some of the proceeds to support the research enterprise as a whole ("Stanford Spin-Offs," 1993).

A broader context is also emerging for academic-industry relations. We are witnessing a subtle shift from a bipolar interaction between universities and industry to a multipolar interaction in which governmental authorities at several levels (international, national, and regional) are significant actors even within capitalist countries. In former socialist countries, changes within civil society are transforming the relations among actors from different sectors —academic, industry, and government—to make interaction among the sectors a real rather than a formal requirement. Indeed, in all of these societies, the sectors can no longer be seen as separate spheres because the legitimating sector or the political sphere is now engaged in promoting economic activities based on knowledge even in so-called capitalist societies. Accordingly, universities are creating their own industrial sector through spin-off firms and companies are producing knowledge and engaging in training in formats that are increasingly universitylike.

The direction of economic growth across countries suggests that industrial practice is becoming increasingly knowledge based at all levels of economic development. Among other things, this requires improved levels of education for populations in order for them to participate in the economy. Disparities of knowledge among sectors of a population have become indicators of class and predictors of levels of economic reward in all societies (Reich, 1991). The United States and the United Kingdom exhibit similar internal tendencies, with their parallel records of deindustrializing older cities like Detroit and Liverpool and newly emerging university-based high-technology towns such as Boulder, Colorado, Ann Arbor, Michigan, and Cambridge, England. Oxford, England, and Cambridge, Massachusetts, are, respectively, in the early and later stages of a dual track, losing older industrial firms and gaining new knowledge-based industries, with some recycling of buildings and reemployment of workers.

Beyond these specific trends, the larger significance of science-based economic development is that it supersedes a working class based on industrial labor and traditional forms of socialism. Indeed, knowledge as the basis of industrial development also supersedes traditional concepts of capital and capitalism. Capital and labor, traditionally organized, are insufficient to support economic growth in the coming era. The increasing role of knowledge in the economy makes irrelevant the old debate over the relative importance of capital and labor as factors of production. As a result, the following policy questions have come increasingly to the fore in all societies:

- On what grounds should knowledge production be publicly supported: cultural, economic, military, health, environmental?
- If economic, what role should government play in selecting economically relevant areas of knowledge—and, by implication, future industrial areas—for support?
- Under what conditions should publicly supported knowledge be privatized or publicized?

These issues of intellectual property protection and dissemination have moved to the forefront of the public agenda in diverse countries and regions, whether they are ranked high or low on scales of industrialization and research intensity (Dougherty & Etzkowitz, in press; Etzkowitz, Balazs, Healey, Stankiewics, & Webster, 1992). Important as recent developments in biotechnology, computers, and material science have been, they may merely foreshadow future forms of knowledge-based industrialization in such fields as nanotechnology, superconductivity, and artificial intelligence. The transcendence of traditional forms of socialism and capitalism through the universalization of science cuts across the traditional geographic divisions of North and South, East and West, overlaying a new common form of economic and social development upon the divergent political philosophies heretofore associated with these areas.

22

Scientific Knowledge, Controversy, and Public Decision Making

BRIAN MARTIN
EVELLEEN RICHARDS

THE central and increasingly contentious role of science and technology in modern society has given rise to a plethora of scientific and public controversies over scientific and technical issues. Such controversies often have profound social, political, and economic implications, and more and more often they feature public disagreements among scientific, technical, or medical experts. Whether the confrontation occurs over the control of AIDS, about the proposed introduction of the "abortion pill," about whether "cold fusion" exists, over the location of an airport, or over the implications of the "greenhouse effect," experts become involved. And many of them become involved not just as consultants or providers of expertise but as overt and committed defenders or opponents of one side or the other, as active participants in the debate.

AUTHORS' NOTE: We thank Pam Scott and an anonymous reviewer for helpful comments. A few passages from this chapter on fluoridation and on vitamin C and cancer are taken from earlier writings by the authors.

Disputes between experts provoke major difficulties for decision making and policy implementation in the case of such public confrontations, which, more often than not, are vociferous, protracted, rancorous, and unresolved. Traditionally, the neutral, disinterested, and objective expert has been promoted—not least by scientists themselves—as the rational and authoritative arbiter of public disputes over scientific or technical issues. But this old ideal of the appeal to facts and their interpretation by accredited experts has been eroded by the increasingly obvious limitations of experts and expert knowledge in resolving issues of public controversy. There is now a widespread public perception that experts can and do disagree, that they are not infallible by virtue of their specialist access to some rigorous scientific methodology that can guarantee their "objectivity," and that their purportedly "disinterested" advice may be influenced by professional, economic, or political considerations. Along with the well-documented decline of public trust in the infallibility and neutrality of expertise has come a growing demand for greater public participation in scientific and technical decision making and policy formulation.

For all of these reasons, as well as for their intrinsic interest and drama, scientific and technical controversies are the focus of an abundant and growing literature by social scientists and historians. These analysts have studied conflicts in and around science for insights into the science policy process; to learn more about the various roles of scientists and nonscientists in policymaking; to identify the ways in which the public might participate in decision making; to understand how controversies arise, are contained within the scientific community or expand into the public domain, are brought to a close, or why they persist; or to analyze the social construction and negotiation of scientific knowledge claims by disputing scientists.

Depending upon their purpose and point of view, researchers have developed a variety of approaches to controversy analysis. We have picked out four distinctive approaches, which we label positivist, group politics, constructivist, and social structural. We have selected these particular approaches because they cover a range of commonly used methods and illustrate a diversity of strengths and pitfalls of controversy analysis. In the following section, we describe each approach in turn, pointing out its advantages and limitations and illustrating it with characteristic accounts of the controversies over fluoridation and over vitamin C and cancer. Our aim is to illustrate and explain the approaches, not necessarily to judge or recommend them.

In the third section, we take this examination further by comparing the four selected approaches in a number of areas, such as epistemology, the focus of analysis, and the partisanship of the analyst. This examination also serves to highlight some of the assumptions made by analysts in undertaking their studies.

In the final section, we make the point that actual studies seldom fall simply into one of our four standard approaches. Indeed, these four approaches may be conceived of as "ideal types" in the sociological sense. Although actual studies may not fit one of these types precisely, ideal types are useful in helping to impose some conceptual order on the diversity of controversy studies. In conclusion, returning to our case studies of the fluoridation and the vitamin C and cancer controversies, we present an argument both for a plurality of approaches and for their integration.

The Fluoridation Debate

The question of whether fluoride should be added to public water supplies to prevent tooth decay has been perhaps the most vociferously contested public health issue in recent decades in the Western world, mobilizing enormous passions and requiring continual involvement by government bodies. The controversy involves scientific issues, such as the assessment of the effectiveness of fluoride in reducing tooth decay and the evaluation of alleged health risks, such as skeletal fluorosis, allergic and intolerance reactions, and genetic effects including cancer. It also involves ethical and political issues, such as whether a chemical should be added to the water supply to treat individuals, and the question of who should make decisions about fluoridation (Martin, 1991).

The Vitamin C and Cancer Dispute

The vitamin C and cancer controversy centers on the claim by Linus Pauling (Nobel laureate and world-famous advocate of peace and vitamin C) and Ewan Cameron (a Scottish surgeon) that megadoses of vitamin C (10 grams or more per day) can control or palliate cancer. The dispute has continued for more than 20 years amidst mounting charges of "bias," "fraud," and "misrepresentation" and has been punctuated by running battles over publication and funding, by personal attacks on the scientific and ethical credibility of the disputants, and by media and political interventions. It has become particularly intense over the claims and counterclaims surrounding the two negative randomized controlled clinical trials of vitamin C carried out by leading cancer specialists at the Mayo Clinic in 1979 and 1985 (Richards, 1988, 1991).

FOUR APPROACHES TO STUDYING CONTROVERSIES

The Positivist Approach

The essence of the positivist approach is that the social scientist accepts the orthodox scientific view and proceeds to analyze the issue from that stand-point. If the dominant scientists say that fluoridation is safe and effective, that vitamin C has no effect on cancer, that there are no hidden variables in quantum theory, that Velikovsky's ideas are discredited, or that continents drift, that is taken as the starting point for the social scientist.

Sometimes the scientific evidence is incomplete or contradictory. In these cases, scientific debate is legitimate. Once the uncertainties are resolved, though, only a few maverick scientists can be expected to hold out against the persuasive power of the evidence.

But even when, in positivist terms, the scientific issues are straightforward, controversy may persist. The problem then becomes one of explaining why there is a controversy at all. This usually means examining the critics of the orthodox view. Why do the critics persist in the face of the evidence? Who are the critics and what do they have to gain from persisting in their views? How do they relate to wider forces, such as corporations, governments, or groups of "true believers"? This approach is a "sociology of error": those who are wrong are analyzed to find out why.

For example:

> Fluoridation, the addition of fluoride to public water supplies to reduce tooth decay in children, has been scientifically proven to be effective and completely safe. Nevertheless, in the face of the evidence, there has been a continuing degree of citizen opposition to fluoridation since its inception. It is the task of social scientists to examine the reasons for this opposition—which may be explained by alienation, demography, or confusion—and perhaps also to draw lessons on how best to promote this proven health measure.

> Linus Pauling's claim that megadoses of vitamin C can cure cancer has been scientifically refuted by two randomized double-blind controlled clinical trials carried out by leading cancer researchers at one of the world's foremost medical research centers, the Mayo Clinic, and officially endorsed by the National Cancer Institute. In spite of the conclusively negative results of these objective and definitive trials, Pauling

> continues publicly and most unethically to promote his claims. It is the task of the social scientist to investigate the psychological, cultural, and social foundations of the popular adherence to such scientifically disproven treatments.

This approach is based on the separation of science from social science. Nature is assumed to hold a unique truth and the current state of scientific knowledge is assumed to be the best available approximation to that truth. There is no need to examine why scientists believe what they believe, because there are assumed to be no social factors intervening between nature and scientific truth. Those who disagree with these revelations of nature are treated differently. It is assumed that there must be some social explanation for their behavior. The familiar social science tools are brought to bear: analysis of individual psychology, belief systems, social roles, vested interest groups, and the like. There are dozens of studies of the fluoridation controversy that fit this model (Martin, 1989).

A continuing controversy is considered actually to be two controversies: a cognitive controversy (a controversy over knowledge) and a social controversy (a controversy over nonscientific issues) (Engelhardt & Caplan, 1987). The cognitive controversy can be settled by the supposedly tried-and-true scientific method, whereas the social controversy may persist indefinitely.

Sometimes there is a genuine cognitive controversy. Different scientists appear to have valid reasons for different beliefs about nature. In most cases, this does not persist once various objective tests are made, such as definitive experiments and repeated replications. But in the meantime, the role of the social analyst is not to second-guess the scientists but to examine the role of social factors in the social controversy (e.g., Mazur, 1973, 1981).

One implication of the positivist approach is that the social analyst becomes a de facto (and sometimes overt) supporter of scientific orthodoxy and, often, the causes associated with it. The social analysis is made of those holding out against the dominant view, and this tends to reduce the legitimacy of the critics, because their beliefs and behavior are explained in terms of psychology, sociology, or politics.

A limitation of the positivist approach lies in its dependence on scientists for determining what should be studied. If the orthodox view changes, then a new social analysis is required of any new opposition. Another limitation is that social scientists are precluded from studying social factors in the cognitive realm, especially in the formation and maintenance of the orthodox position. But these are "limitations" only from the point of view of those who reject some of the assumptions underlying the positivist approach.

The Group Politics Approach

This approach to scientific controversy concentrates on the activities of various groups, such as government bodies, corporations, citizens' organizations, and expert panels. Essentially, the controversy is dealt with as any other form of politics in the pluralist interpretation of liberal democracy: a process of conflict and compromise involving various groups contending in a political marketplace (Nelkin, 1971, 1975, 1979; also, Boffey, 1975; Dickson, 1984; Greenberg, 1967; Primack & von Hippel, 1974).

> Since the early 1950s, fluoridation has been backed by the U.S. Public Health Service, the American Dental Association, and other key professional bodies. In opposition have been numerous local citizens' groups, openly backed by a minority of scientists and dentists. The proponents have used their cognitive authority and their connections in the community to get fluoridation accepted and adopted, while the opponents have mobilized local supporters using claims of hazard and appeals against compulsory medication.

> Since 1972 Linus Pauling and Ewan Cameron have been locked in conflict with professional cancer researchers, orthodox nutritionists, and the National Cancer Institute over the interpretations and reinterpretations of a number of clinical trials of the effectiveness of vitamin C megadoses as a treatment for cancer. Pauling and Cameron have recruited support from the holistic health movement and the health food industry, from megavitamin therapists, and from the network of organizations promoting freedom of choice in cancer therapies to oppose the cancer research establishment.

Depending on the controversy and the analyst, different groups may be the focus of attention. In cases of nuclear power plants and chemical waste dumps, it is typically governments and corporations versus community groups. Also typically, mainstream scientists and engineers support the governments and corporations, with a few maverick scientists supporting the community groups.

There are a number of theoretical frameworks for proceeding with a group politics study. A commonly used one is resource mobilization, in which the focus is on how different groups are able to mobilize and use a range of "resources," including money, political power, supporters, status, belief systems, and scientific authority (Jenkins, 1983).

In group politics studies, scientific knowledge becomes a tool that can be and is used by the contending groups. When science and scientists are drawn into the dispute, this is characterized as the "politicization of expertise." The underlying assumption seems to be that science and scientists are normally neutral and apolitical, unless they are tainted by the political arena. This assumption is compatible with the positivist approach. In fact, using the positivist division of a controversy into scientific controversy and social controversy, the group politics approach is essentially the study of the social controversy, with passing attention to the scientific issues.

For this reason, the group politics approach seems best suited for those controversies where obvious contending groups are central to the dynamics of the dispute and where the state of scientific knowledge allows a number of interpretations. The study of contending groups is far less useful in probing disputes over knowledge that are largely restricted within the scientific community, such as theories of the origin of the universe or of superconductivity. That is why group politics is used almost exclusively where public policy issues are at stake.

The group politics approach works well when a dispute is active, namely, when there is overt controversy. But when the controversy fades from public view and there is little open contention—the usual state of "noncontroversial" science, or Kuhn's "normal" science—the group politics approach has little leverage to offer insight. This is not a limitation for studying controversies, but it suggests that group politics is not well suited for studying science, which is publicly controversial only occasionally.

The Sociology of Scientific Knowledge (SSK) or Constructivist Approach

For SSK analysts, scientific controversies are especially valuable sites for carrying out research into the nature of scientific knowledge claims. In the first place, they provide the sociologist with a set of ready-made alternative accounts of the natural world. They therefore suggest that these accounts are not directly given by nature but may be approached as the products of social processes and negotiations that mediate scientists' accounts of the natural world. Controversies have the further advantage that these social processes, which ordinarily are not visible to outsiders, are confronted and made overt by the contending disputants.

This approach to controversy analysis challenges the conventional positivist approach whereby scientific knowledge claims (being presumed uncontaminated by social and political influences) remain unscrutinized by the analyst, and social explanations are selectively applied to the side without authoritative scientific backing. The SSK program differs from this traditional approach in two major ways. First, the social analysis is applied to scientific knowledge claims as well as to wider social dynamics. Second, both sides in the controversy are examined using the same repertoire of conceptual tools. This principle of symmetry is the most important principle in the "strong program" in the sociology of scientific knowledge (Bloor, 1976; see also Barnes, 1974, 1977, 1982b; Mulkay, 1979) wherein the analyst is required to treat the conflicting claims of the disputants symmetrically or impartially. The sociologist or historian must attempt to explain adherence to all beliefs about the natural world, whether they are perceived to be true or false, rational or irrational, successful or failed, in an equivalent or symmetrical way. The same types of causes or determinants of beliefs—they may be psychological, economic, political, or historical as well as social—should be applied to both sides. No set of beliefs or their advocates may be privileged over another.

Unresolved controversies are particularly rewarding sites for carrying out empirical SSK research. They allow the analyst to study science that is still in the making. Retrospective judgments about the truth or falsity of the conflicting interpretations of nature may be avoided and the principle of symmetry is directly applied. By following the course of the controversy through to closure, the analyst is able to recover the sociological factors that explain how some beliefs become true and others false.

Beginning with Harry Collins's pioneering study of the dispute over the existence of high fluxes of gravitational radiation (Collins, 1975, 1981a, 1985), SSK analysts have accumulated an impressive array of empirical studies of scientific controversies that have compelled attention to their central programmatic claim that scientific knowledge is socially created or constructed (Collins & Pinch, 1979; Pickering, 1984; Pinch, 1986). In particular, their studies have contradicted the standard view that disputes over "facts" and their interpretation can be resolved by the impersonal or "objective" rules of experimental procedure. According to their revised view of scientific knowledge, where closure of a controversy has been achieved, it has resulted not from rigorous testing but from the pressures and constraints exerted by the adjudicating community. These pressures and constraints include not only the accepted knowledge of the community (the elements of its paradigm) but also the vested interests and social objectives that they embody. Together they shape the processes by which knowledge claims are accepted or rejected by the adjudicating community. Thus, within the terms of this "constructivist" approach, the "truth" or "falsity" of scientific claims is considered as deriving from the interpretations, actions, and practices of scientists rather than as residing in nature.

The key scientific studies underpinning fluoridation were trials comparing tooth decay in fluoridated and unfluoridated cities, showing reduced decay in children's teeth in the former. A few scientist critics pointed to methodological flaws in the trials, but these criticisms were ignored or rejected by the proponents of fluoridation, whose research and interpretations held sway in the field. Critics also raised questions about health hazards from fluoride, but these were also rejected. The proponents of fluoridation held key posts in dental journals and professional bodies and the claims of the critics were successfully ignored or dismissed.

The way in which the Mayo Clinic researchers designed and carried out their trials of vitamin C was determined by their own theoretical and clinical perspectives on how an anticancer drug should work. They achieved their negative results by disregarding the theoretical framework and associated clinical and evaluative practices of their alternative opponents. The dispute was closed, not by disproving Pauling's specific claims but by social and political means: by denying Pauling a professional platform for his criticisms and by blocking future trials of vitamin C on the grounds that it had been objectively tested and found wanting.

While the SSK approach has successfully opened up the content of disputed scientific knowledge to sociological analysis, the strong program's central theses of symmetry and impartiality require both the epistemological and the social neutrality of the analyst. This methodological prescription prohibits any evaluative or judgmental role for the controversy analyst. In practical terms, though, an insistence on treating two sides to a debate symmetrically gives more credibility to opponents of orthodoxy than would otherwise be the case, and thus provides de facto support for the opponents. SSK practitioners, in pursuing their aim of fine-grained sociological analysis of disputed knowledge claims, have focused almost exclusively on micro-level action and interaction between groups and actors within the scientific community. Their characteristic avoidance of the roles of professional and social power and broader structural influences in the constitution of scientific knowledge may be viewed as a serious limitation.

The Social Structural Approach

This approach uses concepts of social structure, such as class, the state, and patriarchy, to analyze society and to provide insights into controversial issues. Social structures are patterned sets of relations between people and groups. For example, in Marxist analysis, social class is determined by the relationship between groups and the means of production: The ruling class is made up of the owners of farms and factories and the proletariat is made up of those who sell their labor power. Use of social structures puts the focus on regular sets of social relations, in contrast to the group politics approach, which puts the focus on the activities of autonomous groups.

In Marxist analyses, which are carried out using categories such as class, capitalism, and the state, scientific controversy would be explained as a feature or outgrowth of class struggle or system contradictions. For example, the rise of the environmental movement and related disputes might be traced to attempts by the bourgeoisie to protect its class privileges once the working class began to acquire material affluence (Enzensberger, 1974). But a Marxist analysis could also be more sympathetic to environmentalists, for example, by analyzing the role of industry in causing pollution (Crenson, 1971) or the role of capitalism in transforming agricultural practices to require monocultures, artificial fertilizers, and pesticides.

Feminist analyses are carried out using categories such as gender and patriarchy. The controversy over reproductive technology, for instance, can be analyzed in terms of male doctors (or a patriarchal medical establishment) pursuing the latest stage of medicine's control over women's bodies (Corea, 1985; Spallone & Steinberg, 1987).

Another key structure is the state, which can be defined sociologically as a community based on a monopoly over the use of legitimate violence within a territory, and more conveniently thought of as government, the military, the legal system, and related bodies. Controversies over nuclear power can be analyzed as conflicts between state interests in a technology that grows out of and reinforces its control (only in the United States do private corporations play a major independent role in nuclear developments) versus citizen opponents.

Yet other controversies can be tackled using the concept of profession. The fluoridation controversy centrally involves the dental profession; the vitamin C and cancer controversy involves the medical profession; and yet other controversies involve the legal profession.

> Fluoridation was an attractive proposition for dental elites because it was a "scientific" measure in the service of dental health and because it did not challenge powerful groups implicated in tooth decay, especially the sugary food industry. Once the proponents of fluoridation captured key positions in the dental and medical establishment, they were able to dictate research and assessments of it. Research that showed the advantages of fluoridation was funded and published. Critics, whether scientists or nonscientists, were labeled cranks and ignored or attacked using the power of the dental profession.

> The cancer establishment is composed of researchers, the medical profession, and the drug and other industries. The standard treatments of

> surgery, radiation, and chemotherapy all depend on and reinforce this establishment. Vitamin C—a cheap, unpatentable substance that can be administered by patients themselves—is a threat to the cancer establishment. This powerful coalition of institutional and professional interests has brought about a political closure of the dispute by refusing Pauling a professional platform for his criticisms, denying him funding, and blocking any future trials of vitamin C.

Although structural analysis is commonly identified with well-known critiques of society, the examination of structures is not automatically critical of the status quo. For example, macroeconomics of the neoclassical variety proceeds by analyzing the market at a structural level rather than the decisions of individual buyers and sellers.

Some structural analysts stick to one framework, such as a Marxist class analysis or a feminist analysis of patriarchy. This may work well for some controversies. For others, a more eclectic approach, bringing together critiques from different traditions, may provide more insights, though perhaps at the risk of theoretical complexity or confusion.

One of the hazards of structural analysis is when the structures are assumed to take on a reality and a solidity that removes the prospect for struggle and change. This is the familiar problem of reification (hardening) of the categories. Another problem is that many of the categories in common use—such as class or patriarchy—appear to be too blunt to provide much insight about the dynamics of disputes at the local level. It is probably for this reason that there is no coherent body of social structural analyses of controversies, unlike each of the other approaches.

COMPARING THE FOUR APPROACHES TO CONTROVERSIES

We chose the four "ideal types" outlined above to explore some of the common approaches taken and also to emphasize that there is more than one way to study a controversy. Each approach has its peculiar strengths and limitations. We do not suggest that there is a best general approach, or even a best approach to a particular controversy. Rather, what is "best" depends on the purposes of those who produce and use the analysis.

To give further insight into different approaches to controversies, in this section we classify these four ideal types according to a number of criteria. This highlights the possibility of developing yet other approaches by varying one or more of the assumptions.

TABLE 22.1 Treatment of Scientific Knowledge

Approach	*Treatment of Scientific Knowledge*
Positivist	Positivist
Group politics	Unspecified (usually positivist)
SSK	Relativist
Social structural	Unspecified (usually positivist)

TABLE 22.2 Focus of Analysis

Approach	*Focus of Analysis*
Positivist	Inside the scientific community (on which outside events impinge)
Group politics	Outside the scientific community
SSK	Inside the scientific community
Social structural	Outside the scientific community

Epistemology

Table 22.1 classifies each approach according to its assumption about scientific knowledge. The availability of a "choice" between positivism and relativism reflects the continuing development of different approaches to the study of science. The positivist approach reflects the long tradition of positivism in the social sciences, which has its strongest following in the United States. The relativist sociology of scientific knowledge has been promoted by a group of researchers most closely identified with the British-based journal *Social Studies of Science*.

The group politics and social structural approaches do not include conceptual tools to examine scientific knowledge, and they are compatible with either a positivist or a relativist analysis. Most commonly, their authors seem to hold to positivism.

Focus of Analysis

Group politics and social structural approaches almost always focus on controversies and events outside the scientific community: government policy, public statements, social movements, class struggle, and so on. When scientists are mentioned, it is usually their role in public events, rather than their dealings at the laboratory bench, that is the focus of attention. Both these approaches have been used almost exclusively to deal with controversies with a major public dimension.

TABLE 22.3 Conceptual Tools

Approach	*Conceptual Tools*
Positivist	Actors
Group politics	Actors
SSK	Actors
Social structural	Social structures

By contrast, SSK analysts commonly deal with controversies that are largely restricted to disciplinary communities. Collins's (1975, 1981a, 1985) study of disputes between physicists investigating gravity waves is typical here.

The positivist approach in principle focuses on debates over knowledge within the scientific community. But in those cases where there is a social controversy (as well as a scientific controversy), an analysis of the social dynamics is necessary to understand why, for example, the controversy as a whole is not closed even when the scientific issues are no longer in dispute.

Conceptual Tools

Actors in Table 22.3 refers to people, groups of people, or organizations. Looking at what such "actors" do in society is the dominant form of analysis in studies of controversies, no doubt because controversies necessarily involve open confrontations between individuals and groups. The category "actor" may even be extended to include "nonhuman actors" such as scallops, door closers, and other technologies, an extension adopted by actor-network theorists such as Michel Callon (1986b) and Bruno Latour (Johnson, 1988).

Actor-oriented analyses do not always do so well when confronted with issues over which there is no controversy. The exception is the positivist approach, which works fine when there is no controversy; scientists have simply agreed about the facts and their interpretation. The group politics and SSK approaches attribute lack of controversy to the successful efforts of groups to gain cognitive or social authority. But they are hard pressed to explain why these efforts have been successful. Structural concepts such as hegemony and patriarchy are more useful here.

Closure

Why does a controversy end? The process by which a dispute ends or is resolved is called "closure."

TABLE 22.4 Closure of Debate

Approach	*Main Reason for Closure of Debate*
Positivist	Superior knowledge (for closing the scientific controversy)
Group politics	Superior political/economic/social resources
SSK	Superior persuasiveness or networking ability in the micropolitics of the scientific community; superior knowledge/politics
Social structural	Hegemony of dominant social structure

Each of the four approaches explains the closure of controversy in a distinctively different way. For positivists, closure of the scientific dispute is straightforward: In the absence of outside pressures, the scientifically correct side, as determined by rational analysis and investigation of the facts, will be readily acknowledged within the scientific community. However, where there are outside pressures, such as political or economic interests, they may impinge upon and prevent or overturn this result. These social pressures may keep alive a (social) controversy when the scientific issues have been decided, or close down a controversy when the scientific issues remain to be decided, or override the scientific consensus on the issues. Because social processes are seen to interrupt or distort the "proper" resolution of scientific controversies, the study of closure has been a preoccupation for positivist analysts (Engelhardt & Caplan, 1987).

SSK analysts also pay special attention to closure, but for a different reason. Their attention is directed to all social processes in knowledge production, and closure is a final stage in the certification of knowledge in the contentious course of controversies, and hence a revealing test of SSK analysis. Because the focus of attention is usually within the scientific community, the analysis of closure usually focuses on processes of successful persuasion at the level of scientists, research groups, and peer networks.

For group politics researchers, closure is the result of success by one group or the other in the political marketplace of contending interest groups. From this perspective, closure is not of special theoretical interest, because it is the struggle that is the center of attention.

For similar reasons, closure is not a central concern in structural analyses. Social structures do provide an effective way of explaining closure, namely, the dominance of the structure itself (or, in other words, the dominance of a particular pattern of social relations). Gramsci's term here is *hegemony.* This is especially useful in explaining the nonexistence of certain controversies. For example, the relative lack of controversy over automobile safety may be attributed to the dominance of automobile interests (Otake, 1982).

TABLE 22.5 Partisanship

Approach	*Partisanship of the Analyst*
Positivist	Assumed or open
Group politics	?
SSK	Denied, covert, de facto
Social structural	Dependent on the choice of structures

Partisanship of the Analyst

The activists in any controversy—such as the proponents of fluoridation, or Linus Pauling and Ewan Cameron—can be called partisans. If the social analyst of the controversy supports one side or the other, she also can be called a partisan.

The classic positivist ideal is that researchers—whether scientists or social scientists—should be neutral, nonpartisan students and commentators on the issue under study. In practice, this is seldom the case in the study of controversies, which almost invariably arouse the passions of analysts as well as the participants.

The positivist approach assumes there is an objective scientific truth that, sooner or later, will be revealed to support one side in the dispute. Because the social analyst accepts the judgment of mainstream science, there is inevitably a partisanship in favor of this judgment. Most often, this is not openly stated. Rather, it is a de facto partisanship that manifests itself by unquestioning acceptance of the position of the dominant scientists and a social analysis that undermines the legitimacy of contrary positions. Sometimes, though, positivists come out in the open with their commitments to the side of truth, as in the case of many sociological studies of fluoridation.

Group politics studies do not have to pass judgment on the scientific evidence but simply examine the jockeying for power that often uses the evidence as a resource. Analysts can support one side or the other by the direction and tenor of their study. There seems to be no easy generalization about partisanship.

SSK analysts vigorously claim that they are nonpartisan. After all, they use the principle of symmetry to study both sides with the same conceptual tools. However, by looking for social explanations for knowledge claims on both sides, SSK analysts tend to more severely undermine the side with greater cognitive authority. This is a predictable pattern of de facto partisanship (Scott, Richards, & Martin, 1990).

The social structural approach, like the group politics approach, has no strong pattern of partisanship. It is the choice of structures and the direction and tenor of the analysis that leads to support for one side or the other. A

TABLE 22.6 Decision-Making Procedures

Approach	*Preferred Decision-Making Procedures*
Positivist	Experts know best: use politics to help them win; science court
Group politics	Science hearings panel; citizen voice through a pluralist politics
SSK	(Not an issue: social analysis only)
Social structural	Alternative social structures (in which the controversy will not arise)

Marxist analysis of disputes over occupational health, by emphasizing the power of capital, is likely to support the claims of workers and unions. A feminist analysis of disputes over reproductive technology, by emphasizing the links between patriarchy and medicine, is likely to support the claims of women patients and critics.

Social scientists, like the scientists they study, increase the status of their work by presenting it as objective. As a result, analysts seldom discuss their own partisanship in the same pages as their analysis. Yet partisanship is a crucial issue for understanding both the strengths and the limitations of an analysis. It deserves much more attention.

Decision Making

There is no necessary reason that a social analyst who is studying a controversy should also have views about how the controversy should be resolved, namely, how decisions should be made about the issue. Yet, in practice, different approaches to studying controversies are commonly associated with characteristic attitudes of decision making.

Positivist analysts usually are quite committed to the triumph of scientific truth in any controversy. They assume that superior knowledge will win out within the scientific component of any controversy, but also that an accompanying social controversy may delay or reverse the victory of truth. Therefore they sometimes advocate intervening in the social controversy to make sure that correct science wins out. In the fluoridation controversy, for example, a number of social scientists explicitly advised profluoridation campaigners on how best to proceed.

In the positivist model, the expert scientists know best. What is called "politics" is generally a contamination of the pure world of science. The science court is one proposal to put the experts in proper decision-making roles. It uses the familiar legal adversary system to sort out technical disagreements, which are assumed to be separable from social issues.

The group politics approach addresses the conflicting claims and tactics of various interest groups within a pluralist model of society. This is often accompanied by a pluralist model of resolving conflict: Let the adversaries engage each other in some marketplace of ideas and opinions, with policy being made by some responsive and responsible arbiter such as government.

Group politics analyses typically (but not always) assume that scientific issues can be separated from social issues. In other words, the framework of group politics is used to handle the social controversy while, like the positivist approach, scientists are expected to sort out the strictly technical issues. However, unlike the positivists, group politics analysts are more likely to see the controversy as unresolvable unless both the technical and the social issues are addressed. The science hearings panel is one option. It includes both technical experts and laypeople who hear evidence and reach a decision on the issue as a whole (L. A. Cole, 1986).

More generally, group politics analysts implicitly look to the pluralist political arena—government agencies, business, unions, consumer groups, scientists, media—to provide a sufficiently diverse range of inputs into government so that a decision serving the general interest can be made. In this model, a robust system of representative democracy plus a vigorous "marketplace of ideas" is the best guarantee of reaching a satisfactory resolution to controversies (Goggin, 1986; Petersen, 1984).

Structural analysts tend to have a very different view of the world. They see the "marketplace of ideas" as inherently unequal, because the power of the different players varies enormously and the entire "playing field" is biased through structural inequality. For example, the debate over pesticides can never be fair as long as powerful chemical companies are pitted against a few volunteer citizen groups. A Marxist would say that resolving the pesticide dispute authentically must await the development of a socialist society.[1] In a society without capitalism, the true needs of the people could be recognized and a balanced assessment of pesticides be made. Similarly, feminists might argue that debates about reproductive technology will remain biased so long as male domination persists in medicine, government, and the family.

The question remains about what to do now, in present society. The usual approach is to support the claims of those groups challenging social structures considered oppressive. By this line of thinking, Marxists, for example, would support worker and community activists against chemical companies in disputes over pesticides.

The sociology-of-scientific-knowledge approach clearly states that it is in the business of analysis only, not prescription. SSK sets itself up in the model of natural science, trying to provide an objective or "naturalistic" account of the social factors in disputes over knowledge. It presents itself as a way of studying (social) reality, not for judging it or changing it.

Actually, even for the other three approaches, only a few social analysts are explicit about their favored means of decision making about controversies. Most of them simply analyze the controversy without saying what should be done. Nevertheless, it is possible to infer preferences on decision making from many accounts.

INTEGRATED APPROACHES AND THEIR IMPLICATIONS FOR PUBLIC DECISION MAKING

Many, indeed perhaps most, controversy studies do not fit neatly into any one of the above approaches but may draw upon the analytical tools of two or more. Controversy analysts have also argued the advantages of comparative analyses that examine and compare different or related disputes. These may be chosen for the opportunities they provide of exploring the differences between disputes that are confined to the scientific community and those that become issues of public debate, or of making cross-cultural comparisons, such as the analysis by Frances McCrea and Gerald Markle of the estrogen replacement disputes in the United States and the United Kingdom (Markle & Petersen, 1981; McCrea & Markle, 1984; Petersen & Markle, 1981).

In this final section, we wish to present the advantages of integrated approaches to controversy analysis. Such approaches have the special value of critically engaging both with the "inside" disputed scientific or technical knowledge and with the "outside" politics of competing interest groups and social structures, of integrating the investigation of both science and politics. They are also, we argue, crucial to the application of controversy analysis to realistic policymaking.

There are at least two ways to develop an "integrated" approach. One is to examine the controversy from several different perspectives, using each perspective both to illuminate a different facet of the issue and to throw light on the other perspectives. The following brief sketch of the fluoridation controversy illustrates such a "multifaceted" approach.

> The fluoridation debate can be analyzed at a number of different but interlinked levels. First, there is a technical debate over scientific claims and counterclaims about the benefits and risks. Second, there is a psychological dimension; as a result of the active role of leading partisans and of vehement debate itself, the arguments over benefits, risks, ethics, and decision making are polarized into two diametrically opposed, coherent wholes. Third, there is a struggle over credibility, using techniques such as endorsements and verbal and written attacks on the critics.

> Fourth, the power of the dental profession has been used to suppress opponents of fluoridation by blocking research funds, denying publication, and deregistering dentists, among other techniques. Fifth, the decision to promote fluoridation as a principal means to reduce tooth decay can be interpreted as a "path of least resistance" for the dental profession in the context of the power of corporations, especially the sugary food industries. In summary, fluoridation can be understood as a power struggle at a number of levels, from the details of scientific data to the organization of society, each of which throws light on the issue and provides a corrective to reliance on any single perspective.

This multiperspective account of the fluoridation controversy does not presume that there is a single "best" way to explain the issue.

A second way to offer an "integrated" approach is to combine several approaches into a single perspective that uses a range of conceptual tools. This type of explanation is illustrated by the following brief sketch of the vitamin C and cancer controversy.

> The history of the vitamin C and cancer controversy is best understood as a political struggle concerning control over the determination and evaluation of cancer treatments. By means of his personal prestige, his well-developed political and institutional skills and connections, and his alignment with holism and the health food lobby, Pauling succeeded not only in promoting vitamin C into a leading alternative treatment for cancer but also in organizing it onto the orthodox medical agenda. He thus brought it into competition with conventional cancer treatments and forced two professional evaluations of the Pauling-Cameron experimental claims via the professionally endorsed methodology. Both professionally conducted trials were problematic: They did not disprove the specific claims made by Pauling and Cameron. Nevertheless, through the assertion of its cognitive authority, backed by claims of objectivity and professional disinterest, and constituted by a powerful amalgam of institutional and professional interests, orthodox medicine appears to be in the process of foreclosing any future trials of vitamin C, thereby bringing about a political closure of the debate in its favor.

A comparison of these explanations of the fluoridation and vitamin C and cancer disputes with the single-approach explanations offered above demonstrates the significantly greater insight and explanatory power of an integrated approach to controversy analysis. In both cases, the inadequacies of the positivist approach are manifest once the disputed knowledge claims are subjected to sociological analysis. The scientifically adjudged efficacy and safety of fluoride and the inefficacy of vitamin C are shown to be cognitively underdetermined and are rendered problematic. When the SSK approach is linked with the group politics approach, the connections between these problematic judgments and the professional interests and wider social concerns of the adjudicating communities are displayed. The incorporation of structural analysis introduces the essential power dimension and explains why these adjudicating dental and medical professionals and their socially and economically powerful allies were able to exert their authority in the face of the problematic nature of their judgments, to marginalize or silence their opponents, and to close the disputes in their favor.

Integrated approaches are thus able to provide a more comprehensive and coherent understanding of scientific and technical disputes. They give political bite to the SSK analysis, without losing its primary advantage of opening up the "black box" of scientific and technical knowledge to sociological scrutiny. An integrated approach can be fine-grained enough to permit analysis of the nuances of controversy dynamics, of the complex and shifting negotiations and interactions of actors and groups, while engaging with the power relations of contending groups and larger social structures. It can be a sociologically rich and flexible approach that offers new insights for controversy analysts and decision makers.

Perhaps the most significant aspect of integrated approaches lies in their potential for enhancing opportunities for public participation in decision making. An approach to controversy analysis that does not privilege scientists and their knowledge, but integrates them as "partisan participants" into the wider political debate, necessarily democratizes the debate. It opens up the decision-making process and it permits an acknowledged and more prominent role for nonexperts, for the public at large, in the processes of scientific and technical assessment and decision making. Arie Rip has argued that controversies provide societies with an informal means of technology assessment that is often superior to any of the institutionalized methods of assessing the risks and benefits of new technologies (Rip, 1987). These institutional methods of assessment invariably are based on the acceptance of the misleading separation of the scientific from the social aspects of assessment. They function to protect the authority of scientific and technical experts and to exclude or disadvantage the public. An integrated approach exposes and undercuts the artificial separation of scientific and technical knowledge from its political contexts, from the social distribution of power. It presents a more realistic understanding of scientific and decision-making processes, and it offers a means of finding more effective, more democratic strategies for coping with the challenge of making informed, socially based decisions about contentious science and technologies.

NOTE

1. It should be noted that many Western Marxists see socialism as a more free and democratic system than state socialism of the traditional Soviet type.

PART VII
SCIENCE, TECHNOLOGY, AND THE STATE

Postwar studies of science policy and politics presumed an institutional separateness between science, technology, and the state that permitted each to be construed as an independent actor equipped with its own interests, resources, normative commitments, and internal political dynamics. Researchers operating with the familiar taxonomy of "policy for science" and "science in policy" rarely questioned the nature of the boundary between the modern state and one of its most powerful institutional supports. Recent work in science studies has rendered this picture highly problematic by calling attention to the multiple, mutually constitutive, and mutually reinforcing relationships in which science and technology engage with the state. In current STS scholarship, it is almost unthinkable that "knowledge" and "power" should be treated as autonomous black boxes that can interact without being transformed by the very processes of their interaction.

The conception in the 1990s is of a system of "coproduction," in which scientific and political order are simultaneously created and recreated so as to sustain each other through complex rituals of interdependence. This view of a continually shifting yet self-perpetuating linkage between science and the state promises to revitalize many traditional areas of STS research (funding policies, technical controversies, public participation, science advice, social regulation) as well as to open up new areas of inquiry (law and science, comparative policy, international trade, and global change, to name but a few).

Ironically, however, STS research on the politics of science and technology has been among the slowest to respond to the winds of constructivism set free in the 1970s by developments in the sociology of scientific knowledge (SSK). The turn to the micro level in SSK was perhaps partly to blame; richly detailed observational studies of scientific practice sometimes went hand-in-hand with a drastically oversimplified view of the world outside the laboratory, a world where ill-defined "interests" substituted for more deeply explored conceptions of state, society, politics, and culture. Equally, however, the traditionalists in both political science and "science, technology, and society" proved reluctant to take on board the far-reaching implications of a constructivist approach that permanently problematized the science-state boundary. Too often, it seemed that the scholarly literature on knowledge and power remained captive to a naive realism, resistant to less orthodox modes of interpretive scrutiny.

The chapters in this part of the *Handbook* illustrate both the extent to which constructivist ideas have now begun to penetrate the stubborn enclaves of science policy and politics and the immensity of the work that still lies ahead. The part begins with a chapter by Susan Cozzens and Edward Woodhouse that surveys the implications of the social constructionist assumption for the study of science and politics. Tracking science and technology (or "technoscience") from research funding through expert legitimation to appropriation by industry, Cozzens and Woodhouse show how the production of knowledge is at every step bound up with politics. An important theme in the chapter is that the authority of science is ultimately political, resting as it does on the ability to hide assumptions and worldviews behind a veneer of objectivity that is taken for granted by billions of people in their everyday lives. Because of this pervasive impact of technoscience, the authors conclude with a plea that the STS research agenda be expanded to include more work on the construction of scientific authority outside the purview of the formal institutions of science and the state.

Further elaborating on the constructivist framework, Bruce Bimber and David Guston show how STS research has undermined four classical assumptions about the "specialness" of science—the claims of epistemological, Platonic, sociological, and economic exceptionalism. Applying these findings to two case studies in U.S. science policy (the creation of the Office of Technology Assessment and the congressional investigation of scientific misconduct), Bimber and Guston argue that American political institutions have long understood the constructed character of scientific authority. Indeed, American politics seems at times an engine designed for the express purpose of deconstructing science. The challenge for STS scholars, in these authors' view, is not to enlighten an already skeptical political hierarchy about the limits of positivism but to explore in detail why the myth of scientific exceptionalism nevertheless retains such power to shape politics. The chapter in this respect defines another fruitful locus for empirical STS research.

Aant Elzinga and Andrew Jamison bring the lens of social studies of science to bear on a review of science and technology policy from the interwar period to the present. Central to their chapter is a demonstration of the ways in which earlier analyses and periodizations of science policy have reflected the adherence of STS scholars to four different policy cultures—bureaucratic, academic, entrepreneurial, and civic. Eschewing any single cultural framing, the authors trace the impact of all four cultures in their own decade-by-decade review of science and technology policy from Pearl Harbor to the present. A case study of science policymaking by the Organization for Economic Cooperation and Development (OECD) provides a connecting thread through the chapter as well as empirical grounding for the authors' assertion that science policy today can only be understood as a dynamic interplay among actors and interests who are involved in constructing a broader political sociology of knowledge.

Wim Smit's chapter analyzes the contribution of STS research to studies of military R&D, the policy context that above all others has defined the postwar interactions of science and technology with the state. Smit's major concerns are with the literature on military-academic relations and the exploitation of science and technology for military purposes. On the former topic, he finds that STS scholars have contributed largely descriptive studies that have failed to illuminate in full the political role of science and science advice or the social responsibility of scientists. With respect to technology, a few important case studies, such as those of missile guidance systems and early laser development, have used constructivist approaches to explore in detail the intricate interconnections between technology and its social context. But broader areas of investigation, like weapons innovation and the conversion and integration of civilian and military technologies, remain in Smit's view surprisingly untouched by SSK or modern social network approaches to the study of technology.

The constructivist strain in science policy studies plays a still less explicit role in the literature on science and technology in less developed countries as reviewed in the chapter by Wesley Shrum and Yehouda Shenhav. Given the diversity of theoretical and disciplinary commitments reflected in this literature, it is perhaps surprising that the authors are able to abstract any generalizations of value to STS. Yet, Shrum and Shenhav find that work in this area has indeed converged toward the shared conclusion that science and technology should be viewed in terms of interactions between context-specific knowledge and globally distributed social interests. Advances in this literature have hitherto come about less as a result of theoretical or epistemological speculation and more in consequence of localized concerns with the doing of technology. This focus has led to particular constellations of interest in the transfer, adaptation, and use of technology as well as in incremental technical change. The authors suggest that social network models may provide a framework for better integrating these studies in the coming decades.

Concluding the part, Vittorio Ancarani's chapter on globalization seeks to understand both how science and technology, and their volatility, are shaping the global environment and what role the STS literature can play in illuminating this question. The chapter exemplifies the distance that remains to be bridged between classical studies in the politics of science and technology and the newer approaches in social studies of science. For the most part, the work that Ancarani reviews maintains the theoretical model of science as an independent force, capable of transforming state-to-state relationships, although in ways that are as yet poorly understood and inadequately systematized. As in other chapters in this part, the review is equally informative with regard to what has already been done and what still has to be attempted. For STS researchers, some of the critical challenges ahead include explicating the phenomenon of globalization itself as well as refining the current largely determinist framework for addressing science, technology, and international relations.

Together, these six chapters make a compelling case for reintegrating the sociological strain in STS with the theoretical and methodological approaches explored in fields as diverse as development studies, international relations, and conventional science policy studies. STS can only gain in the next several years from a more sustained colloquy between micro- and macroinvestigations of the many-sided relationship between science, technology, and the state.

23

□

Science, Government, and the Politics of Knowledge

SUSAN E. COZZENS
EDWARD J. WOODHOUSE

WRITING about the political resources of American science in the inaugural volume of *Social Studies of Science* some two decades ago, Yaron Ezrahi felt obliged to open with an apology that his topic could be considered "somewhat perverse if not entirely heretical" because the traditional ethos of science emphasized "a complete separation between science and politics" (Ezrahi, 1971, p. 117). Since that time, the separation has broken down, in both scholarship and ordinary knowledge. The daily news is filled with stories of university pork barreling, research fraud, and controversies over climate warming and ozone depletion, cancer-causing chemicals, high-fat diets, and interventions in human genetics, to name just a few examples. By this indicator, the public image of scientists has declined: From guardians of the common good producing objective knowledge, scientists are now perceived as hired brains of special interests and lobbyists for their own. There is now no doubt, if there ever was, that scientists are intimately involved in politics.

While the public image of science has taken this political turn, science and technology studies have also moved toward a deeper understanding of science as a political phenomenon. The old understanding assumed that good science produced truth and that truth-producers deserved a special role in politics. The new understanding treats scientific knowledge as a negotiated product of human inquiry, formed not only via interaction among scientists but also by research patrons and regulatory adversaries. If all scientific knowledge is negotiated, then its content depends in crucial ways on how negotiating authority is distributed. Government becomes the key mediating institution where social actors participate, with varying degrees of influence and in a variety of structures, in shaping, interpreting, and using scientific knowledge claims. We examine those mutual influences of scientific knowledge and government authority in this chapter.

Such a holistic picture is somewhat unusual because the literature on science and government is highly fragmented. Government documents that touch on particular aspects of science appear in many policy arenas; and in the scholarly world, diverse disciplines and specialties pursue their distinctive sets of concerns, rarely framing or answering questions in terms that cut across fields. In their activities and writings, both scholars and government officials refrain from examining the overall relationship between science and government, and thus public understanding is also thwarted. Ezrahi's plaint remains apt; the political significance of science is not deeply understood.

STS has been evolving elements of a larger view, with longer time horizons and a broader institutional purview than science policy discussions generally use. In the *democratic* tradition of the field (implied in its original and still most descriptive name, "science, technology, and society"), many STS scholars pay special attention to less powerful, usually nongovernmental actors. Most STS scholars also assume that scientific knowledge is not the passive product of nature but an actively negotiated, social product of human inquiry; this is the *social constructionist* assumption. Latent within existing STS scholarship, then, is a considerable challenge to conventional ways of thinking about science politics. This chapter illustrates that challenge in several forms.

The first section takes up the claim that research knowledge is a product of politics. The authority structure of the funding system—who participates in it, in what network of power relationships—is a dominating influence. Because most research is supported with government funding, distributed through agencies established and maintained through political negotiation, the balance of knowledge among fields is a political product. Moreover, the assumptions and worldviews of science are shaped by expectations conveyed through the funding system and by the access it allows to various social groups. Industry, bureaucracy, and the organized public all play roles, which we discuss in this section.

The second section examines another major facet of the structure of authority, that is, the role of expertise in policymaking. In theory, there is no necessary contradiction between government's use of expertise and its responsiveness to public concerns. In actuality, conflicts abound. Professional scientists and their allies often win substantial autonomy to promote knowledge and resource claims in ways that advantage them at the expense of other equally legitimate social interests. Among other issues considered in this section is the extent to which environmental, consumer, and other public interest organizations serve as countervailing forces to the knowledge/power alliance.

Finally, science stands in a special relationship to industry, which is itself a source—perhaps the major source—of negotiating power in the modern state. The third section attempts to situate the actions of science elites, their allies, and their potential or actual adversaries within the structural setting of market-oriented societies.[1] When governments leave the development and distribution of technologies to private corporations, publicly funded research serves primarily as an inducement to the private sector to perform this function, and only secondarily as a form of public choice of knowledge or technology. Given the current structure of influence around science, research is much more likely to be pulled in directions chosen by industry than to be pushed toward democratically chosen ends, no matter how open the priority-setting process in government becomes. Thus democratic control of science depends ultimately on democratic control of technology.

We conclude with the implications of our analysis for the STS research agenda.

POLITICS AND THE SHAPING OF KNOWLEDGE

In what respects is it useful to think of science as a political phenomenon? Despite Don K. Price's (1967) effort to introduce the concept of the "scientific estate" as one of a few major forces in modern political life, mainstream political science still views the subject schizophrenically. Scientific claims are generally accepted as unproblematic truth; but, when controversy arises, science policymaking is perceived as being so much like any other kind of political activity as to be unworthy of special attention. The discipline treats the politics of the sciences as if they were about as worthy of study as the politics of Albania.[2]

The STS perspective counters that view, in particular through two branches of the sociology of scientific knowledge. *Interest theory* traces how the concerns of various actors are embodied in knowledge (Barnes, 1977; Bloor, 1976), and *social constructionism* demonstrates how actors attribute objectivity or fact status to the resulting knowledge through social processes

(Knorr Cetina & Mulkay, 1983; Latour & Woolgar, 1979). Negotiations between researchers and other political actors infuse scientific knowledge with the assumptions and worldviews of both scientists and their sponsors. As these assumptions are carried along with research knowledge into technological and professional practice, they are taken for granted as facts in the lives of millions of people.

Resource Coalitions

From its origins, the science-government relation has depended on coalitions between scientists and agency officials. As laboratory heads fly from meeting to meeting, they succeed in garnering resources only to the extent that funding officials find it in their interests to confer funds and authority on one would-be scientific elite rather than another. Elite scientists' networks have both an "inside," which looks like "pure science," and an "outside," where these other actors appear (Latour, 1987; see also Knorr Cetina, 1982, on trans-epistemic networks; Shrum, 1985b, on technical systems; Callon, 1986a, on actor networks). By "following the actors," we see interactive adjustments among scientists and government officials leading to the very formulation of scientific knowledge.

What power do scientists have to negotiate with government officials? Daniel Kevles's *The Physicists* (1978) points to several factors. Physics grew within the emerging research universities of the United States in the late nineteenth century, supported by the wealth of a growing educated upper class desiring prestige. Early on, physicists developed two attitudes that still appear among scientists in government: *best-science elitism*, an unrelenting lifestyle of invidious comparison, reinforced by competition for prizes and other forms of recognition, and *political elitism*, the notion that by virtue of their knowledge scientists should have a strong say in political matters.

The elite status and connections of physicists were important to the efforts of Vannevar Bush and other physicists to persuade both president and Congress in the late 1940s to adopt the principle of government support for basic science, but another element was critical. In the atomic bomb, the physicists had produced a source of tremendous political power for the United States (Rhodes, 1986). Other research communities had similarly, if less spectacularly, proved their problem-solving abilities to particular agency officials and were kept on after the war (Stine, 1986). In time, the scientific community thereby came to occupy a privileged position not unlike that of the business sector. By providing the trained personnel and base of technical knowledge that helped accomplish elites' purposes, scientists and their institutions won reciprocal resources not available to many other social interests.

The institutional arrangements set up in the late 1940s were varied, expressing different levels of dependence between agencies and researchers. (For an earlier period, see Dupree, 1964, 1990.) In the Public Health Service, scientists pushed aside the lifetime public health officers, whose claims to best-science elitism were weak, and established their own set of research programs in the National Institutes of Health (Strickland, 1989). The health research entrepreneurs adopted a mechanism that had been mandated for the National Cancer Institute in the 1930s as a defense against the old guard: Peer review appeared at NIH at this time. While broad-scale judgments about health research priorities continue to be made through the politics of Congress and pressure groups, peer review insulated "best-science" research from those politics at the level of selecting individual grants. Lay representatives sit on the institute councils that make the final decisions on grants, but the councils are strictly constrained by peer panel ratings of proposals.[3]

A somewhat different situation arose in oceanography, where "engineers, physicists, and biologists found themselves functioning as marine scientists during the war, and used funds after the war to continue this direction in their careers. Institutions of oceanography were expanded and became major research facilities in which these scientists would work" (Mukerji, 1989, p. 45). But oceanographers gained less institutional control than their biomedical colleagues. Funded by the Office of Naval Research (see Sapolsky, 1990), oceanography became a field that survives on government contracts, with researchers scrambling for resources within a priority scheme they do not control. Mukerji describes them as a "reserve labor force," with skills that prove useful in various unexpected capacities and that contribute to "military preparedness." Their relative lack of power is symbolized by their limited role, especially compared with researchers at NIH, in choosing among competing proposals; ONR program officers have wide latitude in project selection (Cozzens, 1987). Other fields developed different forms of interdependence with the agencies that fostered them (see Tatarewicz, 1990, and Mack, 1990, on NASA programs; Bromberg, 1982, and Kay, 1991-1992, on fusion research).

The third major interest group involved in shaping scientific knowledge is big business. The coalitions scientists make within government have depended on its political willingness. (See also the third section of this chapter.) The formation of the National Science Foundation is often described as a triumph of pure research (England, 1983), but major industrial firms cast the deciding, if informal, vote between the two options under discussion: an applied research agency that would make inventions widely accessible to small business through licensing (Senator Harley Kilgore's proposal) or a basic research agency with a more exclusive patenting policy (a proposal supported by the

National Association of Manufacturers) (Kleinman, 1993, p. 158). For decades, the representatives of large, science-intensive firms actively maintained their support for the latter option through their participation in discussions of research policy. Then through the 1980s industrial interest in government-funded research moved from passive support to active involvement (see Dickson, 1984, for a detailed account). In agriculture, for instance, after a long series of legislative and judicial victories securing property rights in plant forms, private seed companies have attempted to eliminate the century-old commitment to plant breeding in public research facilities (Kloppenberg, 1988).

Through government, not just interest groups but also interested individuals have played large roles in determining what science is done and for what purposes. For example, in the United States, the personal concerns of several wealthy and politically powerful figures, including Senator Edward M. Kennedy, redirected major portions of cancer research in the 1970s (Rettig, 1977; Strickland, 1972). Admiral Rickover's obsession with the nuclear submarine led eventually to a disastrous choice of nuclear technology (Morone & Woodhouse, 1989). The focus that STS maintains on actors' interests, perspectives, and interactions opens up the question of choice: If Rickover and Kennedy reshaped technoscience based on their conceptions of the public good, why not the rest of us? If the Navy can create a field for its needs, can the environmental or women's health movements do the same?

One arena of science-government relations where ordinary people get some of what they want is biomedical politics, even if the result may have escalated medical costs to levels no one wants. The enormous growth in funding for biomedical research over the 1950s and 1960s is attributable not only to alliances between research entrepreneurs and sympathetic senators but also to popular support for medical research. That support is reinforced by specific lobby groups formed by families of victims of various diseases (see, e.g., P. Fox, 1989). In the complex negotiations that characterize biomedical politics, the professionals sometimes prevail. For example,

> a rancorous dispute between the International Guild for Infant Survival and the National Institute for Child Health and Human Development . . . surfaced in three Congressional hearings. . . . Spokesmen for the Guild . . . urged Congress to recognize the social significance of SIDS [sudden infant death syndrome] by funding a large, vigorous, and focused research effort on the mysterious killer. Spokesmen for [the agency] . . . opposed targeting SIDS this way. (Hufbauer, 1986, p. 61)

But similar resistance to the idea of a "war on cancer" (which was conceived outside the research community) did not deter elected officials from allocating money (Rettig, 1977; Studer & Chubin, 1980). Shilts (1987) describes a third pattern, the long battle by gay groups to increase research attention to

AIDS. Likewise, in its research thrust on Alzheimer's disease, the National Institute on Aging at first encouraged the support of family groups that favored basic research over services but eventually built service research into its agenda as well (P. Fox, 1989).

Inside Scientific Knowledge

The process just described shapes the balance of research efforts across fields; scientists bargain for their research opportunities through government, with selected other groups that have tried and succeeded in entering those negotiations. According to the sociology of scientific knowledge, however, the negotiation process has further results. Because research itself reflects the assumptions and worldviews of those who participate in its creation, the worldviews of officials, industry, and a segment of the public are carried into scientific knowledge and then carried along with it into many areas of practice. What begins as someone's choice ends up perceived as fact by someone else.

This last point has been most powerfully advanced by the feminist critiques of scientific knowledge, particularly in biology. Feminist analysts in history and philosophy of science, as well as within biology itself, have examined the assumptions built into modern biological knowledge, produced by a male-dominated research community in a gendered culture (e.g., Harding, 1991; Hubbard, 1990; Keller, 1985). Longino (1989) uses these critiques to argue that the masculinity of biological knowledge is not mere bias, which can be removed by doing "good science," but the inevitable expression of worldview in the content of scientific knowledge itself. In areas like research on the brain, the assumption of sexual differences in brain function, while scientifically controversial, nonetheless ends up being taken for granted in social expectations about women. Likewise, in obstetric technology, technologically mediated pregnancies constitute the experience of parenthood for both men and women (Davis-Floyd, 1992; Layne, 1992; Mitford, 1992). The seemingly esoteric details of research results are shown, in this analysis, to be intimately linked to the character of everyday life.

Thus political action in relation to health research takes on new significance. In the United States, a coalition of women in the Senate and House of Representatives, in alliance with the women's health movement, has put pressure on the National Institutes of Health to pay more attention to women's health issues and include more women in studies of general health problems (Palca, 1991; Pinn, 1992; Veggeberg, 1992). The feminist analysis shows that, by raising awareness of women in biomedical research, this action not only changes women's life chances by generating more knowledge about their bodies but can also eventually change the cultural understanding of sex and gender. The social constructionist element of the STS perspective thus links the specific politics of research to the general politics of everyday life in a way that is far from apparent in the standard vocabulary of science policy.

The feminist case illustrates the general point that there are cultural and political ramifications of all the choices researchers make about their questions and assumptions. Those choices are being made across the government research system. Biotechnology (Wright, 1990, 1991) and computer science (Peters & Etzkowitz, 1988) have both been greatly stimulated by military interests. What assumptions do these bodies of knowledge now incorporate as a result? Why is the current allocation of environmental research spending dominated by the atmospheric sciences (Mandula & Blockstein, 1992) rather than, for instance, by the interests of minority communities who live with toxic wastes (Bryant & Mohai, 1992; Bullard, 1990)? It is only when we open the content of scientific knowledge to political inspection, as the social constructionist view does, that these issues emerge.

In short, government-funded science is "political" not only in Easton's (1958) sense of authoritatively allocating public resources but also in the sense of exerting symbolic authority by directly or indirectly shaping the ways people think about who they are and what is real and important (Edelman, 1985).

THE PROBLEM OF EXPERTISE

If the shaping and interpretation of scientific knowledge depend in vital ways on who participates in the negotiations, then the problems posed by professional expertise loom even larger than was heretofore understood by critics of technocracy (see Fischer, 1990; Winner, 1977). How do the insights from science studies modify our thinking about scientific expertise as a source of authority, that is, of socially legitimate power?

Claims to Authority

Philosophers once claimed the right to rule; scientists have been more modest. Only in rare areas of governance do scientists claim the right to rule as Platonic guardians, on the basis of both instrumental knowledge and the ability to judge the common good. Basic research policy is the notable example. Here the argument that the extension of knowledge, in whatever direction, serves humanity implicitly justifies leaving the specific choice of research projects to technical experts (Chubin & Hackett, 1990). The argument functions as little more than a generic ideology, however, within the varied institutional arrangements described in the last section, where both broader and narrower goals are frequently negotiated among researchers, their supporting agencies, elected officials, and ultimately public opinion.

Outside research policy, members of the scientific establishment have long assumed that they should have substantial influence over a range of government decisions by virtue of their claims to specialized knowledge. To be most helpful as advisers, however, they argue that they need to be separated from "politics." (See Brooks & Cooper, 1987; Golden, 1980, 1988.) Indeed, calls for a more politically committed advisory network have generally created low resonance in elite political circles.

This separatist position has proved convenient for agency officials and other politicians seeking safe havens in turbulent political environments (Brickman, Jasanoff, & Ilgen, 1985). In regulatory policy, for instance, officials are wont to draw sharp lines between "science" and "politics" in ways that serve their purposes but shift over time (see Rushefsky, 1986). In other areas, coalitions of bureaucrats, staffers, and political specialists raise the technical content of their policy areas in ways that enhance their authority and keep public discussion at a minimum. In weapons policy, for example, a narrow set of participants, including the technical community, defense contractors, congressional committees, and military management, frame issues in terms that screen most people out of the discussion. Even public interest groups specifically organized to lobby for slowing the arms race, whose staff members may have significant levels of technical knowledge themselves, have trouble penetrating the protective shield of expertise (Cohn, 1987; Greenwood, 1990; Hamlett, 1990b). Such areas form pockets of technocracy within systems of representative democracy that are otherwise more permeable.

The social constructionist perspective undermines all these claims to authority by demonstrating that there is in principle no way to separate science from values in any policy area, that any line drawn is artificial, temporary, and convenient to the purposes of the person or group drawing the line. The point is not merely that political uses of science are inevitable; rather, it is not even possible to think about what the science is apart from its various constructions. Because it is impossible to separate instrumental from moral judgments, and because no one has been able to show that any group of guardians is systematically better at making moral judgments than is the public as a whole, social theorists generally reject claims to guardianship (Beitz, 1990; Dahl, 1985, 1989). How, then, are the contributions of scientific experts to be integrated with others' judgments so as to construct science-intensive policy that is as fair, wise, and democratic as possible?

Accountable Expertise

Many elected officials, citizens, and scientists express a desire for scientific expertise to enter the regulatory arena as a neutral, mediating force, contributing to good or even "correct" decisions. They seem to want regulation to be a pocket of technocracy, working toward publicly defined ends. But the pocket is regularly ripped open by conflicting interests, as shown both in the daily news and in STS controversy studies ranging from nuclear power to vitamin C (e.g., Nelkin, 1992; Richards, 1991; Scott, Richards, & Martin, 1990). Scientific uncertainty coupled with the necessity of balancing incommensurable factors (e.g., cost versus safety) in policymaking means that "far from being an almost mechanical process safely relegated to technicians, the setting of health, safety, and environmental standards is in reality a microcosm in which conflicting epistemologies, regulatory philosophies, national traditions, social values, and professional attitudes are faithfully reflected" (Majone, 1984, p. 15).

Because of these factors, Collingridge and Reeve (1986) posit that science *cannot* be of use to policy. Debate over scientific knowledge claims will always be either overcritical (partisan and divergent) or undercritical (determined from the outset by elites' dictates). Others advance a less deterministic position: Scientific knowledge claims can sometimes contribute to the resolution of controversies, according to Brickman et al. (1985), depending on whether political structures and processes construct a workable context. Thus policymaking in the United States for nuclear power and for nuclear waste fits the Collingridge scenario, and some observers see a more general tendency toward gridlock in the American system with its multiple veto points (e.g., Hamlett, 1992). But European political processes tend to place "considerably lower demands and strains upon the role of scientific evidence . . . [where] both 'experts' and partisan interests are typically represented in a single deliberative forum . . . [and] scientific uncertainties can be papered over in the drive for a political compromise" among the most powerful social groups concerned with an issue (Brickman, 1984, p. 110).

Even in the United States, policy obviously does evolve. Although proposed legislation to reduce acid rain was stalemated throughout the 1980s, for example, sufficient dissatisfaction built up to produce the Clean Air Act amendments of 1990. Scientific agreement was not the key; the stalemate was broken by conventional politics—compromise among legislators from states producing and burning high-sulfur coal and those suffering the resulting acid rain (Bryner, 1993). Technocracy in this case was constrained by the fact that differing political constituencies construed evidence differently and by the fact that expert advice was rejected altogether to the extent that it conflicted with the bedrock political priority of saving coal miners' jobs.

What role does science play, then? Jasanoff (1990a) argues that the "political function of good science" is to certify that an "agency's scientific approach is balanced . . . and that its conclusions are sufficiently supported by the evidence" (p. 241). This can help make the scientific aspects of regulatory actions much less vulnerable to deconstruction and gridlock. This usually is not much of a problem in parliamentary systems because potential dissidents have little access to administrative rule making and rarely can mount effective challenges in court (Nelkin & Pollak, 1979). In the more open American system, both the Environmental Protection Agency and the Food and Drug Administration have had nontrivial problems in the past in making their judgments stick. Both agencies, Jasanoff found, have fared better since they learned to frame negotiations across the science-politics boundary in ways that (a) prevent scientific judgments insensitive to political realities and (b) head off administrative behaviors that cannot win scientific legitimacy. If carried out deftly, advisory committees are left with a domain that looks like "science" and regulators have a domain that looks like politics or administration.

Do such flexible negotiations produce relatively *democratic* judgments, however? This was not much of an issue when the standard myth of objective science prevailed. Scientists' public statements often "portrayed science as a method, rational and repeatable, that provided true understanding of realities that could be accurately described in terms of measurable formulae and a common, objective language . . . that provided a verifiable basis for action" (Fries, 1984, p. 336). Scientific "facts" were believed to place tight limits on the range of outcomes any thoughtful person would seek, so democratic scrutiny was not very important. Under a more interest-based and constructivist understanding of science, how can science-oriented policymaking be made responsive to a broad range of public concerns?

The classic answer in Western political thought is that such concerns must be effectively represented in the negotiations. One of the ways this can happen is through a

> pluralistic and conflicted advisory system . . . [where] clusters of experts and supporting constituencies and interests [come] into conflict with one another, . . . [allowing decision makers to obtain] leverage in dealing with other experts and their constituencies by co-opting them, by citing opposing authority against them, by enlisting broader constituencies through them in situations of conflict. (Burns, 1978, p. 411)

There is no substitute, in this view, for partisanship: If multiple, overlapping, and partially competing sets of interests champion those scientific knowledge claims perceived as useful for their purposes, then the resulting negotiated definitions of problems and policy options stand a reasonable chance of melding expertise with responsiveness to a broad range of social interests. This is the potential intelligence of democracy (Lindblom, 1965; Lindsay, 1943).

In contemporary political systems, however, there is nowhere near an equal competition among social interests, with some being professionally organized, well funded, and possessed of the necessary access and expertise to engage in reasoned persuasion on complex policy issues. In particular, industry groups such as the Chemical Manufacturers Association generally outmatch their would-be critics (Lindblom, 1977; Schlozman & Tierney, 1986). Partly redressing the imbalance is so-called public interest science, as mobilized by the European Green parties or by the Union of Concerned Scientists. Such organizations speak for upper-middle-class concerns far more than for the working class or underclass, however (Bullard, 1993). And even highly expert public interest groups (e.g., the Natural Resources Defense Council, Environmental Defense Fund) can only compete effectively in highly selected arenas—on a few high-profile pesticides and industrial chemicals, for example—not on the full range of 60,000 chemicals now in commerce, not even on the several hundred new chemicals introduced each year (Morone & Woodhouse, 1986).

The controversies that do occur to some degree help establish expectations that carry over to other cases, of course, and by no means do business elites always win; decision processes and outcomes vary across political systems, among policy areas, and over time (Vogel, 1986; G. Wilson, 1990). State elites play strong roles, sometimes dominant ones, in France, Japan, and other nations with prestigious public sectors (Brickman et al., 1985). And national workers' organizations in Israel, Norway, and other heavily unionized nations are influential in particular arenas such as workplace health and safety (but tend not to be active across the full range of science-intensive policymaking).

In principle, then, expertise potentially can be rendered both useful and accountable. How often this actually is approximated in practice is questionable, however. In relatively delimited arenas of regulatory politics—especially where highly visible, bellwether issues are hotly contested—science-laden policy may be about as democratically responsive as other types of policy. But is that saying very much? Contemporary political systems are characterized by a privileged position for business, substantial economic and political inequality, a stunted competition of ideas, and electoral-governmental institutions designed as if to insulate state elites from genuine accountability (Lindblom & Woodhouse, 1993).

Within these substantial constraints, negotiations such as those described by Jasanoff may achieve locally sensible answers integrating scientific and political considerations. And, in any given circumstance, elite scientists, elected officials, bureaucrats, and even business executives may speak for some larger sets of public interests. How often, how effectively, and for which outsiders' needs are subjects not well researched. But it is pretty clear that science policy participants do not often step outside their normal frame of reference and raise probing questions that challenge existing power relations. And STS research has not yet done well at conceptually handling the "liaison between an increasingly elitist, structured science and the particular interests of large social groups devoted to short-term goals of executive management, military gamesmanship, biomedical servicing, and corporate preservation" (Remington, 1988, pp. 65-66). Arguably, constructivist studies of science politics may need to supplement microbehavioral research on governmental negotiations with studies raising broader questions about what Restivo (1986) calls the "elitist and professionalist values of the scientific, academic, legal, military, political, and business communities" (p. 79). Ensuing sections consider how STS might attend to more of the voices and interests not presently participating influentially in science policy negotiations.

The Proknowledge Movements

One naive notion that runs through no small portion of the early STS controversy studies is the hope that citizen participation will somehow hold the experts accountable. STS quickly learned, however, that participation without decision power is meaningless. Nelkin and Pollak (1979) point out that much of what passes for "participation" in current governance can just as well be understood as attempts by the powerful to co-opt the public. Laird found this phenomenon in the Carter administration's energy policy procedures (Laird, 1993), and it has long characterized nuclear waste policy (Walker et al., 1983). Dickson (1984) makes a similar point:

> The promise of technology assessment was that it would ensure a better balance of the costs and benefits of scientific and technological progress by allowing for more democratic participation in the selection of technical choices. Genuinely opening up the channels for such participation, however, would have required a substantial shift in control over decision-making away from private into public channels. This . . . was a step that neither the scientific, the corporate, nor the political establishment was willing to take. (p. 259)

A major barrier to effective citizen participation in government decision making is the subtle denigration of nonprofessional knowledge in the educational system and throughout government interactions. The social constructionist arguments have shown that scientific knowledge has no privileged claim to truth and has thus placed all knowledges, in theory, on a common epistemological footing. A number of contemporary social movements are putting that insight into practice by challenging the authority of professional knowledge. The STS perspective helps to highlight the common goals of these disparate movements in relation to knowledge.

A common view among scientists is that the public is neither interested nor competent in the governmental matters scientists deal with. (See Prewitt, 1982.) According to this argument, a deficit of knowledge disqualifies citizens from participating in science-intensive areas such as regulatory policy.[4] Perhaps the single clearest accomplishment of interdisciplinary STS research has been to shred the "deficit of knowledge" argument. Dozens of studies of scientific controversies have tracked the involvement of citizens in issues they perceive as direct threats to their everyday lives (see Frankena & Frankena, 1988; Nelkin, 1992; Petersen, 1984; Piller, 1991). In adversary contexts, citizens not infrequently seek and acquire considerable technical knowledge when they need it, as did the Love Canal Homeowner's Association in studying the health effects associated with the movement of toxic substances through underground waterways in their neighborhood (Levine, 1982). For many years it stood as the only research on health effects in the neighborhood that used a local control group, yet New York State Health Department officials (before they even saw the results) dismissed the study as "information collected by housewives that is useless" (Levine, 1982, p. 93; see P. Brown, 1987, for a similar case). The example leads to the second point to emerge from studies of local controversies, which is that local knowledge is often more accurate or complete, even by conventional scientific standards, than the knowledge imported by "experts" to a local situation. For example, in the aftermath of Chernobyl, predictions based on average figures for environmental contamination were not credible to Cumbrian farmers, who knew in detail the variability of their own microenvironments (Wynne, 1991a, p. 115).

As social constructionists, STS researchers are able to identify the positive contributions of citizen knowledge rather than dismissing it as imperfect science. Professional science and citizen knowledge are often complementary, in different ways in different situations. When one looks back at the science-government relationship from this viewpoint, it appears that there are a number of areas where citizens care enough about the issues to build their own knowledge bases with which to challenge professional knowledge. The battlegrounds of the politics of knowledge are not necessarily fields of science but the issues raised in government by a range of contemporary social movements.

These movements seldom begin from knowledge goals. Instead, they develop them in the process of trying to solve some other problem. In the process of neighborhood environmental protest, for example, citizens may realize that the observations that spurred them to action—illness, visible pollutants, foul smells—are being discounted in favor of contradictory "scientific" evidence that claims safety. The political issue then becomes knowledge itself—whose will count, that is—as protest groups fight to have their own experience counted as real, true, and valuable. The women's health movement likewise revalues women's understanding of their own bodies and deprofessionalizes the health care setting (see Boston Women's Health Book Collective, 1984). The wider alternative health movement, as well as the deep ecology movement, have developed similar critiques of the limitations of professional knowledge and the political implications of science. The debates over abortion, creationism, and animal rights also illustrate that various American publics are willing to retrieve the definition of key aspects of their lives and cultures from the experts, when enough is seen to be at stake (see Jamison & Lunch, 1992; Jasper & Nelkin, 1992; Nelkin, 1982).

Because of the intimate connection to issues that arise from everyday life, these movements may be misunderstood as operating entirely in the politics of technology rather than science. Love Canal was primarily a controversy over a waste site, not over research. Women's clinics are primarily about health and healing, not about the authority of professional science. But in fact, the authority of science is inseparable from its applications when seen through the eyes of these citizens. Sensing the common elements in these disparate movements, leaders of the scientific establishment sometimes call them "antiscience." But in the social constructionist view, it is more accurate to label the common elements "proknowledge," because each movement seeks in part to revalue forms of knowledge that professional science has excluded rather than to devalue scientific knowledge itself. A range of citizen actions across many spheres of government, from courts through regulatory agencies to research sponsors, all share the implicit goal of reestablishing the legitimacy of knowledges other than professional ones. In the conventional politics of research, controversies and events that seem, from the viewpoint of the research community, like either threats to academic freedom or petty nuisances are reinterpreted, in the STS view, as part of the politics of knowledge.

The combined actions of the proknowledge social movements come closer to the ideal of participatory democracy envisioned by Mill and Dewey than the alternatives: technical guardianship or democracy by opinion poll. But it is important to remember that they, too, fall short of that ideal. Social movements only enter government negotiations when they have mobilized enough resources to get their issues on the agenda. Those members of society with the fewest resources are therefore the least likely to be mobilized—

though perhaps most likely to be affected by the outcomes of the negotiations (Jasanoff, 1993c; M. Reich, 1992).

SCIENCE AND BUSINESS

In the preceding sections, we have discussed the politics of research funding, the problems of expertise in policymaking, and movements that challenge the authority of professional knowledge. Moving out from authority relationships within government to broader patterns of power in society, we complete the analysis with an examination of the relations of science, business, and government. The STS literature does not for the most part examine the structure of this relationship. STS controversy studies and other studies of the regulatory process emphasize business as the adversary of citizen groups, but industry scarcely appears in the accounts of science provided by the sociology of scientific knowledge. Business historians and historians of technology describe the interpenetration of science and industry (see, e.g., Chandler, 1977; Hounshell, 1983; M. R. Smith, 1977) but seldom point to implications for the current situation. All these treatments underestimate the role of business in the politics of knowledge. The business-government relationship, we argue here, is in fact the most influential power relationship running through the politics of science. Too much attention to the details of negotiation within and between government agencies and laboratories can distract STS scholars from recognizing this essential structural situation.

We noted earlier that big business has played a distinctive role in the politics of funding for basic research. It gave the nod to the formation of a basic rather than an applied research agency as the flagship of the postwar research system (Kleinman, 1993), and industry executives participated directly and indirectly in basic research policy discussions over the years (B. L. Smith, 1991). As physicist Charles Schwartz found in examining the corporate affiliations of members of the President's Science Advisory Committee, at least 69% of the academic scientists supposedly giving "independent" advice to PSAC "had strong ties to private industry" (Schwartz, 1975). While billed as university faculty members, these "pure academics" sat on boards of directors or held significant consultantships with private sector businesses. Examples included Harvey Brooks, Raytheon; Philip Handler, Squibb-Beech Nut; and Jerome Wiesner, Celanese and others. By the 1980s industry was taking an even more active role in funding and privatizing scientific knowledge (Lewontin, 1992a).

None of this is surprising from a Marxist perspective, given that the state's function is seen as providing conditions for the accumulation of capital, including setting the legal framework for commercial exploitation of scientific knowledge while taxpayers fund exploratory research (Dickson, 1984; Kloppenberg, 1988). Non-Marxist political economists interpret the picture in a similar way, but with a twist: Individual business firms cannot be assured of capturing for themselves the profits that result from their expenditures on basic research and generic R&D and will therefore tend to invest less in this activity than the amount that many economists believe would be "optimal" for society as a whole. It is therefore appropriate that government pays for basic research and some early R&D as well as for development expenses in military and other government procurements (Mowery & Rosenberg, 1989; Nelson, 1959; Rosenberg, 1982).

Under either perspective, basic research is one of the inducements governments offer to businesses to keep a market-oriented economy "growing" and "competitive." This is no trivial function. Because the well-being of ordinary people is dependent on business sector investment, job creation, product innovation, and more (Lindblom, 1977), neither liberal democracies nor socialist states can survive without taking the needs of industry seriously. Thus, when governments take the advice of business leaders on what research activities will benefit their firms, the stakes are high, and business ends up with a major role in shaping the research agenda. This rationale has been part of the justification for government-sponsored research since the Bush report (1945) and has been spreading more widely recently (G. E. Brown, 1992), although neither party in the relationship necessarily operates with a sophisticated view of the investment-innovation relationship.

For liberal democrats, the privileged position of business has disturbing aspects because business cannot be trusted to take the broad public interest into account. Governments may need to make concessions, but they "have not in the past carefully distinguished between two types of privilege: those that directly assure profitability and those that give the corporation autonomy to pursue profits with little constraint" (Lindblom, 1977, p. 350). Government is thus inducing business to do *something*, but businesses decide the all-important details of what actually gets done. For participatory democrats, the situation is dangerous for a simpler reason—a key step in technological choice, the selection of which technologies to pursue, is privatized and left out of public decision processes. Moreover, intellectual property rules ensure that knowledge produced to serve business remains inaccessible to public scrutiny and use.

But does this problem have a solution, short of abandoning the market? Some members of the scientific elite have responded to the trend toward privatization by renewing calls for scientific autonomy, in the belief that scientists can play an independent, nonpartisan role in creating and applying public knowledge. STS scholars, however, tend to treat such proposals as naive (see, e.g., Cozzens, 1989; LaFollette, 1989; Restivo, 1986). A related response is to call on scientists and their institutions to reinvigorate their commitments to public responsibility (Busch, Lacy, Burkhardt, & Lacy, 1991). Discussions of codes of ethics, which for decades were liveliest in the engineering professions, caught fire in scientific associations in the 1980s regarding "conflict of interest," "public accountability," and ethics.[5] The branch of STS concerned with ethics and values studies has been active in these debates (see Hollander & Steneck, 1990). Allied with the advocates of public responsibility are STS scholars who argue for greater democratic conrol of the government research agenda, through such routes as greater public say in the governing boards of research agencies (e.g., Dickson, 1984).

While all these proposals deserve consideration, they are not structural responses to the privatization of scientific knowledge for they treat a symptom rather than the disease. If businesses need a technical knowledge base that is not being provided by public research, they will go where they can get it or hire researchers and produce it themselves. Science can be steered by the needs of business, but business cannot be steered through the provision of public funds for science. Thus we arrive at the ironic conclusion that democratic control of science depends in large part on democratic control of technology. The latter topic is beyond the scope of our chapter, but a few relevant proposals in the STS literature might be mentioned. Hamlett (1992) advocates federal chartering of corporations, so that government can require representatives of the consumer and environmental movements on corporate boards. Such participation could put into operation Morone and Woodhouse's (1989) proposal for wider and earlier debate as one of five steps necessary for intelligent democratic control of technology. The constructive technology assessment movement in Europe also embraces the goals of broader participation at earlier stages of the introduction of technology (Vig, 1992; see also Schwarz & Thompson, 1990). In relation to the politics embedded in artifacts, Winner (1986b) likewise argues for a new politics of design, one that would deliberately incorporate worker and consumer needs into the early stages of R&D. By affecting the direction of technological development, such measures would eventually change the agenda for government-sponsored research as well.

CONCLUSION

Seen in a broad context, then, science-government relations mediate a number of better-understood power relationships between state and society. They allow the translation of various organized social interests into scientific knowledge, with further reification into technologies, and they form the arenas of struggle over authoritative professional knowledge. STS perspectives reveal how deeply these dynamics reach into the content of scientific knowledge and how broadly they spread into the everyday lives of citizens. A major accomplishment of STS is to show that scientific practice is inherently political, because scientists help define a large part of what is taken for granted by billions of people—a type of influence that in some respects is the ultimate form of authority. As Schmandt and Katz (1986) put it, science and technology "continue to change our system of governance, the nature of issues policymakers deal with, the role of government in society, the responsibilities of the citizens, and the meaning of such central concepts as liberty, political accountability, and democracy" (p. 40).

There are aspects of this picture, however, that STS research has neglected. To date, for example, the field has overemphasized the *institutional* politics of science. If scientific knowledge claims ramify throughout technological societies, helping to shape and perhaps misshape the ways people think about themselves and their worlds, STS needs to devote more attention to the noninstitutional politics of science (Hess, 1992). That is, a portion of our research and teaching energies ought to be directed away from explicitly governmental and scientific institutions, following some feminist scholarship in examining the cultural politics of science in everyday life. If there is no sensible way to demarcate the original formulation of scientific knowledge claims from their subsequent interpretation within the scientific "community" (Latour, 1987), then it may be equally suspect for science studies scholars to concentrate on what happens either within science or within government—rather than devoting at least equal attention to pursuing what happens to shape (and fail to shape) scientific ideas in the larger contestations of knowledge markets.

Second, if there is no substitute for effective partisanship for achieving wiser and fairer outcomes from science-government relations, then STS research and teaching ought to focus even more than it now does on illuminating the barriers to wider sharing of political authority over technoscience. One illustrative research task would be more careful studies of the dissident scientists who provide countervailing expertise for labor, consumer, or other nonbusiness social forces (Messing, 1990). Another is the analysis of the dynamics of ordinary people's confrontations with experts (and failures to confront them) as part of the politics of knowledge: What issues are raised, what gets on the agenda of the controversy and what is left off, and under what circumstances do people succeed and fail in their efforts to make scientific expertise serve as a helpmate to ordinary knowledge?

Finally, scholars working on the politics of science need to pay more attention to the fact that modern science is inextricably bound up with business, advantaged by the connection, and partly responsible for the perversities emerging from corporate shapings of R&D-generated artifacts. Among other research tasks will be documenting the rhetorical-ideological moves made by scientists and their allies to avoid sharing the blame, improving our conceptual analysis of the science-business-government relationship, and analyzing much better than we have the pathways by which scientists and their institutions respond to their structural connection to business.

Implied but not explicitly stated in the foregoing is one general consideration: Ought STS to devote more effort to the study of the structural mobilization of bias, that is, to the issues that do *not* become controversial? Arguably, these reveal the deeper essence of political life, for

> in a sense, what we ordinarily describe as democratic politics . . . is the surface manifestation, representing superficial conflicts. Prior to (such superficial) politics, beneath it, enveloping it, conditioning it, is the underlying consensus on policy that usually exists in the society among a predominant portion of the politically active members. (Dahl, 1956, p. 132)

All of the dominant strands of science politics research—social constructionist, controversy studies, and governmentally oriented—emulate the news media in focusing on action, on what *is* or was happening. Of perhaps equal import are nondecisions, kept off the public agenda through identifiable structural processes and problem frameworks (Bachrach & Botwinick, 1992; Crenson, 1971), and uncontroversies—issues that do not emerge into public discussion because something in the situation, either a structural or a cultural force, suppresses latent disagreement.

Among other areas of non-decision making and uncontroversy that deserve enhanced scrutiny are the domination of research policy by scientific elites and their allies and the domination of corporate technological choice by corporate executives, scientists, technologists, and affluent consumers. The social democracies of western Europe, which have experimented in both these areas but now are retreating, form a natural comparative base for examining experience in the United States and Japan, which have experimented hardly at all. But no technological society has gone very far in attempting to make scientific institutions and scientific practitioners democratically accountable at the same time as effective in their tasks, and social theorists have barely begun to attempt to conceive how this could be accomplished.

In sum, both popular and scholarly discourse has moved considerably from the highly circumscribed view of science politics articulated in the longtime standard in the field, Don Price's (1967) *The Scientific Estate*, which naively claimed that "the notion of democracy, or ultimate rule by votes of the people, is simply irrelevant to science. For science is mainly concerned with the discovery of truths that are not affected by what the scientist thinks or hopes; its issues cannot be decided by votes" (p. 172). Nevertheless, in some respects, the agenda Ezrahi articulated in 1971 for developing a deep understanding of science as a political phenomenon remains to be fulfilled. For, although many particulars about the politicization of science now have become widely known, neither social thought nor social action has entirely faced up to the fact that, because science is sociopolitically constructed and constructing, knowledgeable societies need far more sophisticated processes for steering science democratically.

NOTES

1. Our case, of course, is the United States, which could be called a semidemocracy. As Kneen's (1984) study of Soviet scientists and the state makes apparent, however, the political-economic role of science is broadly similar across technological societies with very different polities and economies.

2. The phrase comes from Sheila Jasanoff.

3. The debate over where to draw the line of discretion is still open, however. For example, during the deliberations over the NIH 1992 negotiations, Representative David R. Obey (D-Wis) objected to the earmarking of $175 million specifically for research on cancer and Alzheimer's disease because of what he perceived as a precedent of politicians deciding research priorities: "I am a passionate supporter of NIH, but it's dangerous having dollars allocated on the basis of politics." In contrast, the conference committee agreed to "urge, in the strongest way, that the National Cancer Institute make breast, prostate, ovarian and cervical cancer their top priorities" (*Congressional Quarterly Weekly Report*, October 27, 1991, p. 3028).

4. This attitude is expressed, for example, in the chapter on public attitudes toward science and technology in *Science and Engineering Indicators* (National Science Board, 1989). See also Miller (1983a).

5. For a sample of discussion in this area, see the *Professional Ethics Report* published by the Program in Scientific Freedom, Responsibility and Law at the American Association for the Advancement of Science.

24

□

Politics by the Same Means

Government and Science in the United States

BRUCE BIMBER

DAVID H. GUSTON

OVER the past two decades the sociology of scientific knowledge (SSK) has eroded the traditional idea that science is a unique or exceptional branch of human activity. As chronicled in this volume and elsewhere (Cozzens & Gieryn, 1990; Knorr Cetina & Mulkay, 1983), SSK has contributed to an appreciation of the socially conditioned and contingent nature of scientific knowledge and practice. SSK portrays the laboratory as a nexus of human agenda and technical pursuits. It has questioned the uniqueness of scientific norms (Mitroff, 1974b; Schmaus, 1983), the degree of institutionalization of those norms (Mulkay, 1975), their effectiveness in motivating scientists in their research (Latour & Woolgar, 1979), and the ability of such norms to prescribe action unambiguously (Mulkay, 1980). SSK has questioned the exclusiveness of logic and empiricism in guiding the production of scientific facts (e.g., H. Collins, 1981b) and illustrated other influences such as professional

AUTHORS' NOTE: The authors wish to express their appreciation for helpful comments and suggestions to Sheila Jasanoff, Daryl Chubin, David Hart, Eugene Skolnikoff, and an anonymous reviewer.

interest (Pickering, 1984) and gender (Keller, 1985). An important accomplishment of this new tradition has been to reveal the many ways in which science is neither exceptional nor immune from the forces that affect other human activities.

To date, the study of the relationship between science and the polity has not been well informed by this perspective, particularly in political science. Scholarship on science and government exhibits some of the most persistent perceptions of science as exceptional, typically placing "science policy" in a unique category distinguished by the special qualities of science itself. For example, with his claim that "there is no difference in kind between one country and another in the relation of science to power," Salomon (1973, p. xxi) suggests that none of culture, social arrangements, institutional designs, or historical traditions can alter the relationship between science and power. In the same vein, Brooks (1987) more recently affirms a statement by the National Science Board in the United States that there is "only one science, one set of standards for evaluating evidence" (p. 4). Analyses of science policy have relied in a conspicuous way on the traditional view that science enjoys a special status. The poor execution of science policy analysis (Averch, 1985) may be symptomatic of this failure of scholarship on science policy. Merton's "The Normative Structure of Science" (1973a, originally published in 1942) and Polanyi's "The Republic of Science" (1962) still cast long shadows onto contemporary perceptions of the nature of science and government.

Similarly, SSK has avoided policy questions directly, despite their implicit role in both the theory and the practice of STS. For example, when the anthropologist in Latour and Woolgar's account (1979, p. 70) is "sorely tempted" by the explanation of science and scientific facts as rendered by the "natives," it is a policy question that helps him regain his critical equilibrium. "However in the back of his mind there remains a nagging question, How can we account for the fact that in any one year, approximately one and a half million dollars is spent to enable twenty-five people to produce forty papers?" Latour and Woolgar, and the laboratory ethnographies their study spawned, neglected this crucial question. The anthropologist's query speaks directly to the nature of the science-government relationship, because it raises analytic questions about how science policy is made as well as normative questions about how science policy is legitimated.

In this chapter, we identify four specific claims about scientific exceptionalism that underpin much writing on science policy and science-government relations in the United States and elsewhere. We argue that these claims are deficient and that a perspective that is better informed by SSK can provide a more productive view of government and science. We believe our argument is applicable across political systems, regardless of their characteristics. In this chapter we support our argument on the basis of the U.S. case only, because we are limited in the extent of the material we can present, but we have constructed this case in such a way as to facilitate international comparisons.

Inquiring about the nature of science-government relations requires detailing episodes of politics. We have chosen two illustrations from the United States that focus on an institutional feature of government that distinguishes U.S. policymaking processes from most other democracies: the separation of powers. We find science conforming to the contours of U.S. political institutions. Our focus is the U.S. Congress, an independent legislature, and we consider two episodes involving explicit claims to exceptionalism in Congress by proponents of science. In both instances, these claims about science have only limited relevance to political outcomes. We conclude that the central tenets of SSK are borne out in the politics of science and that more explicit recognition of these tenets would clarify future study of science policy. We suggest that SSK should broaden its focus to include directly the role of political institutions in shaping the nature of science.

FOUR CLAIMS TO SCIENTIFIC EXCEPTIONALISM

The practice and analysis of science policy have typically relied on one or more of four types of claims about scientific exceptionalism or specialness: epistemological exceptionalism, Platonic exceptionalism, sociological exceptionalism, and economic exceptionalism. None of these categories is a pure type; indeed, their interrelations have been subject to considerable attention. Yet each is intellectually distinct in the sense that each has been used independently to support claims about the scientific enterprise. Each has also been subject to academic scrutiny by SSK. Whereas that scrutiny has been successful in the critical sense, producing a body of coherent accounts of the social construction of scientific knowledge, it is far from clear that this critique has been influential in the policy arena. These four claims to scientific exceptionalism still shape political dialogue. We delineate below the four claims and the reasons for their persistence, after which we examine their bearing in the two cases.

The epistemological specialness of science rests on the claim that science is a search for truth. This "truth" is believed to be public, testable, and universal rather than merely particular or parochial in nature. Implicit in this conjunction of truth and science is that science is good because the truth is good in its own right. It follows therefore that every political community that recognizes the goodness of truth should both accept advice from scientists, who bring truth to government, and support science as an enterprise. Don K. Price (1967, 1981) argues that the closer a practice is to truth, the more distant it is entitled to be from direct political control or oversight. In distinguishing between science and the various professions (engineering, medicine, law, and so on), Price suggests that science may be effectively insulated from the realm of politics by these professions and by the nature of administrative bureaucracy. Ezrahi (1990) calls this insulation of science the "autonomy of truth" and describes its significance in the political thought of such seminal American thinkers as Jefferson and Franklin. This line of argument sees science as a unique truth-seeking activity. As a result, where science and politics meet, science has been ceded unique authority not conferred upon other segments of civil society.

For Ezrahi (1980, 1990), the autonomy of the truth-seeking community of science is important to the political community not only for the independent importance of truth (and the instrumental importance of science-based technology) but also because the truth community exemplifies the possibility of public, instrumental, and meliorist action that legitimates liberal-democratic regimes. The truth community successfully mediates among its members conflicts similar to those typical of the political community. The truth community exhibits fair consensus formation and generally peaceful revolutionary change, both of which can serve as ideals for the political community. The political community therefore suffers when this paradigmatic community on which it models itself is assaulted from without or within.

The next claim to scientific exceptionalism is the Platonic, and it recalls the philosopher-king and the Golden Lie of Plato's Republic: Only certain people of gilded intellect and heritage are privileged with policymaking responsibility. A century ago U.S. Geological Survey chief John Wesley Powell told a special congressional committee on government science that "it is impossible to directly restrict or control . . . scientific operations by law" and that therefore "central bureaus engaged in research should be left free to prosecute such research in all its details without dictation from superior authority" (I. B. Cohen, 1980, pp. 23-26). In the modern world, the supposed complexity and esoteric nature of science policy renders it inaccessible to popular comprehension and thus uncontrollable by democratic decision making. Studies finding fewer than 7% of the American public scientifically literate (Miller, 1991) and British attitudes toward science incoherent and inconsistent (Ziman, 1991) reinforce the postulate that science is too complex for nonexperts to administer and too subtle for the letter of the law to govern. Political processes developed for the governance of "simpler" policy problems within the ken of citizens and politicians are in this view inappropriately applied to science policy, where the entry costs—specialized training and facility with an arcane language—are inordinately high.

If the conduct of science policy is beyond the ken of the public and its representatives, handsome delegations of authority to the scientific community are in order. Thus the relationship between the scientific community and the U.S. federal government is often described as "a 'trusteeship' or social contract that delegates much of the decision making on federal research choices to scientific experts" (Meredith, Nelson, & Teich, 1991, p. 2; also see Brooks, 1985; Cozzens & Woodhouse, this volume; Guston, 1991; Hamlett, 1990a; Office of Technology Assessment [OTA], 1991a; Prewitt, 1982; B. L. R. Smith, 1991). The social contract model is consonant with the belief that scientific issues need to be clearly separated from policy issues to prevent their contamination with lay participation.

The sociological specialness of science involves the claim that science has a unique normative order providing for self-governance. Merton (1973a) proposed that scientists have a particular set of institutionalized norms, cognitive and social, and that these reinforce good behavior among the community of scientists. Polanyi (1962) argued that science is a voluntary association of mutually accommodating individuals, related to a free economic market, and that this market's most efficient—perhaps only efficient—mode of operation is without political interference. Because science is self-regulating through marketlike interactions, and because overt political control tends to distort these interactions, science should then be self-governing. "You can kill or mutilate the advance of science, [but] you cannot shape it" (Polanyi, 1962, p. 62). Efficient self-regulation and self-direction are the keystones of this view. In the context of politics, these qualities obviate the need for the explicit political oversight, regulation, and enforcement needed in the governance of business, medicine, transportation, and other sectors. The exceptionalism of the social community of science provides thematic content for some of the most familiar historical accounts of science (e.g., Daniels, 1967; Kevles, 1978). And, as Jasanoff (1987) writes, "Much of the authority of science in the twentieth century rests as well on its success in persuading decision-makers and the public that the Mertonian norms present an accurate picture of the way science really works" (p. 196).

The fourth category of argument involves economic exceptionalism. It relies on the claim that science is a uniquely productive investment of current resources for future gain. In the United States, supporters of a federal role in scientific research have been advancing such a claim since the nation's founding, privileging science by protecting intellectual monopolies under the U.S. Constitution (Article I, section 8). Vannevar Bush (1945) made economic promises central to his rationale for continued federal involvement in university research after World War II.

The rationale for federal science funding as well as priority setting in science were revisited during the turbulent 1960s (National Academy of Sciences [NAS], 1965). As U.S. economic and technological performance decayed through the mid-1980s despite an otherwise robust scientific enterprise, some analysts began to critically reexamine the science-technology connection (e.g., OTA, 1986; Shapley & Roy, 1985). The claim of economic specialness is current in the United States because of the contemporary fiscal climate and because of the overwhelming demand for increased economic competitiveness.

Recent calls for a redoubling of the U.S. academic research effort (Lederman, 1991) have cited at least one analyst who claims to have documented extraordinary social rates of return for investments in scientific research (Mansfield, 1991). One need look no further than the justifications that scientists offered for enormous expenditures on the superconducting supercollider, for example—in the presence of a crippling budget deficit—to see the conviction with which many are willing to argue for the economic specialness of science.

Claims of economic exceptionalism, as well as epistemological, Platonic, and sociological exceptionalism, have provided grounding for the defense of politically autonomous science in Western democracies. That these claims might have less merit than is traditionally thought has important implications for the conduct of scientific research as well as for the study of science policy. Examining the role of political institutions in influencing the making of science policy and the informing of policy by scientific or technical experts demonstrates the relevance of SSK to science policy and problems with the exceptional view of science. Because institutional arrangements vary across governmental systems, as well as within branches or departments of a government, the nature of the policy processes that take place within them varies as well. As political science has increasingly come to acknowledge, the state is not a neutral system for aggregating and reflecting social interests and demands, such as the desire of science for autonomy. On the contrary, the state is in many ways an independent agent that shapes how policy demands eventually result in policy outcomes and that even influences the expression of social interests and policy demands (Evans, Rueschmeyer, & Skocpol, 1985). The politics of science can be shaped as much by the dynamics of political institutions as by features of science itself.

The U.S. Congress is a good place to examine how important political institutions are in shaping science. Legislatures receive little attention in studies of science policy, which typically favor executive and administrative decision making over the ostensibly less orderly and less predictable processes of legislative policymaking. The executive focus is unfortunate, given that legislative processes reveal directly competing conceptions of how science should operate and of what utility science affords society. Legislatures also afford the benefit of being more open to scholarly scrutiny than administrative or executive hierarchies.

The divergent views of science that emerge in a legislature represent well-formed critiques of scientific exceptionalism. We have chosen two cases of science politics in Congress to illustrate the importance of institutional imperatives and the deficiency of an exceptional view of science politics. These cases represent examples of how political institutions can be assessed from a perspective informed by SSK. They also bear out the claim that routine politics replicate some of the findings of SSK. The cases also suggest an increasing awareness on the part of political actors of the importance of claims to economic exceptionalism and the decreasing relevance of epistemological, Platonic, and sociological exceptionalism.

The first case suggests that the competitive struggle between institutions in the U.S. system of separation of powers is an important part of the explanation for a well-known outcome commonly attributed instead only to the exceptionalism of science: the creation of the U.S. Office of Technology Assessment in 1972. The second case involves Congress's aggressive investigation of science a decade and a half later. It represents a blunt public confrontation between scientists' claims to special treatment and legislators' demand for political accountability in the context of institutional imperatives.

THE OFFICE OF TECHNOLOGY ASSESSMENT

The creation of the OTA was in many ways a case where claims of scientific exceptionalism were overtaken by more important institutional concerns within the government. The observer looking no further than the former would likely miss the significance of the latter. When members of Congress were considering establishing the OTA in the late 1960s and early 1970s, many claims were made, particularly by scientists, that Congress needed the new organization because of the special nature of science and science policy. These claims amounted to arguments for the Platonic and epistemological exceptionalism of science. In this view the OTA's mission was to provide expert technical advice to government officials for whom science is otherwise inaccessible. The agency was to bring the special power of science to the legislature.

Within the scientific community, there was a general feeling that Congress was uniquely unprepared to handle policy problems involving science and technology. Critics of the U.S. legislature explicitly advanced Platonic exceptionalism as the justification for creating a new agency, claiming that a closer relationship was needed between technical communities with special expertise and the decision makers on Capitol Hill, who had political authority but who did not understand science. Witnesses at hearings on the establishment of the OTA, who came from academia, the private sector, and especially scientific organizations, focused on the capacity of esoteric knowledge to improve the content of policy, tying the provision of better expert information in Congress to the future health of the democracy (U.S. House of Representatives, 1971; U.S. Senate, 1972).

The primary sponsor of the OTA, Representative Emilio Daddario (D-Conn), envisioned an agency that would provide neutral, competent judgment about important issues that legislators themselves did not have the training or background to understand fully. The implicit assumption behind this view was epistemological specialness—that the agency could provide Congress with access to truth claims it would otherwise forgo.

The National Academy of Sciences, in its well-known report on technology assessment, also viewed the problem in the context of the special problems of science and technology policy, advocating the creation of a new agency for assessing technology so that the entire system of national decision making could be improved (NAS, 1968). From this perspective, the OTA would be a unique liaison between the technical community—which has mastery over esoteric information—and political decision makers. The OTA was to be an organizational accommodation to the problem of the specialness of science; its effect would be better political decisions.

Another more important political dynamic was also at work in Congress during this period. For many legislators, the OTA was not just a mechanism for adapting to the exceptionalism of science and technology but part of the solution to quite another problem: the balance of power between branches. Many legislators were concerned with interbranch politics and control over the power to make policy in light of an aggressive executive branch whose power appeared to have been expanding for years (Sundquist, 1981). They acted on the OTA proposal not just because they saw it as a means to accommodate the epistemological or Platonic specialness of science but because they wanted a resource for battling the executive branch. The proposal for a technical agency was available at the right time to serve as the vehicle for pursuing these institutional goals. The OTA could serve a political purpose arising from the design of U.S. government as much as it could solve the problems generated by a special category of public policy. In the end, it was this institutional dynamic that carried the day.

When the OTA legislation was brought to the floor in the House and Senate in late 1972, monumental congressional reforms were under way. Legislators were reversing shifts in the balance of power in the favor of the executive branch after years of perceived abuses, most notably involving the conduct of the war in Vietnam and the impoundment of appropriated funds by President Nixon. The organizational and statutory changes of this period were the most strenuous attempts in Congress's history to strengthen its position in the system of separation of powers. Members passed the Congressional Budget and Impoundment Control Act of 1974, which created the Congressional Budget Office and restricted the president's capacity to impound appropriated funds. They passed the War Powers Resolution of 1973, staking a claim to exercise their constitutional war-making authority, and they increased the size and capabilities of the General Accounting Office. In the House, committee staff more than tripled during the 1970s (Smith & Deering, 1990). The current of reform was forceful, and it had little to do with science policy.

Part of the reason Congress's power had ebbed was that the executive branch had developed superior expertise and access to information of all kinds. This imbalance was inherent in the way that the two branches operated, but it had grown especially troublesome for Congress since the New Deal, with the expansion of the bureaucracy and the extended reach of the executive branch into more spheres of activity. Cold war politics also emphasized the presidential flavor of American government in foreign policy and defense, and it propelled the creation of the president's science advisory capacity (Killian, 1982). By the 1970s, Congress felt a spreading incapacity to assert itself effectively, in part because of its inferior expert resources.

Sundquist (1981) observes that "the clash with President Nixon heightened the demand for legislative self-sufficiency in information gathering and analysis to enable Congress to counter and combat the executive branch on something like equal terms" (p. 406). Congress recognized the need for information and analysis "independent of the traditional sources from the executive branch and interest groups" (Thurber, 1977, p. 101). Without such information, members were at a disadvantage politically when opposing bureaucrats or the president (Rieselbach, 1977). Members of Congress referred to the problem as a congressional "information gap" (Fuqua, 1972).

This gap was not the result of any special feature of science and technology; it existed across the policy spectrum. The legislature's historical disadvantage in expert resources is a well-known feature of the system of separation of powers (N. Carson, 1992; Caspar, 1981; J. H. Gibbons, 1988a; Gibbons & Gwin, 1985; Kingdon, 1981; Mezey, 1991; Pfiffner, 1991). It was against this background tension—heightened under the Nixon administration—that members of Congress accepted Daddario's proposal for a technology assessment agency as well as proposals for a new Congressional Budget Office, expansions of the General Accounting Office and Congressional Research Service, and across-the-board increases in staff sizes.

In debate over the creation of the OTA, at least a half-dozen proponents of the bill spoke of the need to lessen Congress's dependence on the executive branch as well as to improve members' analytic resources in an objective sense. Representative Olin Teague (1972) complained that, as the technical content of legislation increased during the 1950s and 1960s, congressional committees "had to depend more and more on experts from the executive branch or upon outside groups which have vested interests in the issues under consideration" (p. 3200). John Anderson (1972) reminded his colleagues of the legislature's difficult decisions about the SuperSonic Transport, recalling how "some members surely felt they were too dependent on the justifications of the executive branch and [suffered from] the lack of countervailing information of their own" (p. 3210). Representative Esch (1972) expressed his hope for the OTA in the following way: "I believe each branch [of government] should have access to its own group of scientific and technological experts—experts objective in their technical analyses, but loyal to the role and vested governmental interests of whichever branch they serve" (p. 3200).

Another legislator, John Wydler, commented several years later that in his view the agency had been created because legislators had grown suspicious of administration experts "whose bias was obvious." Wydler characterizes the testimony of administration officials before Congress by saying,

> Everything they were telling us was why the administration's program was the right one, that decisions that had been made were correct and so on. . . . [The] feeling was that what we needed was an independent group of scientific people in the Congress whom we could turn to as more or less our expert witnesses. . . . I really believe that was the motivating idea behind setting up the OTA. (U.S. House of Representatives, 1979, p. 59)

For these and other legislators, the OTA was not merely a new authority able to tackle previously inaccessible questions but instead was a political ally, an "expert witness" that could help Congress assert itself in the face of experts allied with the historically stronger executive branch. As the OTA's third director, Dr. John Gibbons (1988a), put it, "Congress created OTA in response to executive power" (p. 418).

It was not that the belief in scientific exceptionalism had no bearing on the creation of the OTA, for it was that belief that gave rise to the proposal for a new agency. But it took an institutional imperative for that proposal to be seized upon and passed into law in an environment where far more bills are introduced than passed. In the 92nd Congress, which created the OTA, over 20,000 bills were introduced, yet only 768 were approved (Stanley & Niemi, 1988). The perception of policy exceptionalism was not irrelevant but also was not sufficient to drive the creation of the agency. As is often the case in explaining political outcomes, there is no single "cause" of the creation of the OTA. The agency was the result of a confluence of institutional politics and claims about policy.

It does not follow from this account that an agency such as the OTA can only exist in a system like that of the United States. On the contrary, France, Germany, the Netherlands, Denmark, and the United Kingdom have assessment agencies that have been inspired by the OTA. But the reasons for OTA's creation in the United States during the 1970s, as well as the character of the work that it does, bear the distinctive marks of U.S. political institutions. Interbranch politics have continued to play an important part in the OTA's role in the legislative process since 1972—so much so that Gibbons (1988b) has remarked in a speech that the agency "spent much of the eighties countering executive branch activism." In this view, the OTA is "Congress's own little band that helps keep administration claims honest" (Hershey, 1989, p. F1).

The OTA is a manifestation of Congress's mistrust of executive branch experts, and this mistrust represents an implicit rejection of epistemological exceptionalism. Legislators reject the view that a valid technical "truth" can be identified and represented in the political process by scientists and engineers loyal to other institutional interests. And while the creation of the OTA was an acknowledgment of the high costs of entry into technical debate, the agency's existence does not necessarily support Platonic exceptionalism either. Many other policy areas also present esoteric problems requiring the knowledge of the expert. Tax law, not science policy, is often considered the most esoteric and complex subject of congressional politics. Legislators retain "experts" of all kinds, in the persons of lawyers, economists, accountants, and others. The establishment of the Congressional Research Service just 2 years before the OTA was created, and establishment of the Congressional Budget Office only 2 years after, testify to the need legislators felt in the early 1970s for experts of all stripes.

The OTA's functions in the legislature reflect a mix of problems: The challenge of science and technology policy, the institutional tensions inherent in U.S government, and, in the current period, jurisdictional politics within Congress as policy problems are overlaid on a highly developed system of standing committees and their myriad subcommittees. The OTA came into being because Daddario's proposal, couched in language of policy exceptionalism, could also serve a larger institutional purpose that was foremost on legislators' agenda at the time.

CONGRESSIONAL OVERSIGHT IN THE 1980S

In addition to being a consumer of technical advice from the OTA and other sources, the U.S. Congress funds and oversees the world's largest basic research enterprise. The nexus of forces—budgetary constraints, interest groups, jurisdictional rivalries, normative ideals—that intersect in Congress molds science in important ways. Some scholars have focused on influences of political institutions and actors on science (e.g., Jasanoff, 1987), but the significance of institutional politics for science has not been well explored. This fact is striking, because in recent years the courts and other political institutions have been doing the job of science studies; they have, in effect, taken up the scholarly agenda of deconstructing science. The U.S. Congress has engaged in what could be called the deconstruction of scientific exceptionalism. That is, members of Congress have discovered for themselves that "the science made, the science bought, the science known, bears little resemblance to science in the making, science in the searching, science uncertain and unknown" (Latour, 1991c, p. 7).

Representative John Dingell (D-Mich), chairman of the House Committee on Energy and Commerce—which has jurisdiction over the National Institutes of Health (NIH) among other parts of the research bureaucracy—has been at the vanguard of this discovery. Since 1981 Congress has pushed the NIH to control fraud and abuse in its grant programs (Broad & Wade, 1982; Gold, 1993). Since 1988 Dingell has conducted investigations into fraud and misconduct in science and other issues pertaining to the administration of research—including indirect cost payments to universities and financial conflicts of interest among researchers. These investigations have caused much consternation in the scientific community, which foresees burdensome federal regulations (Abelson, 1991a) and the "Dingellization of science" (Weiss, 1991). Despite this defensive posture by the scientific community, Dingell's inquiries can be seen not as an attack on the scientific community per se but as motivated by a strict fiscal accountability that induces the deconstruction of scientific exceptionalism as "collateral damage."

With the largest and most aggressive investigative staff on Capitol Hill, Dingell attempted to show that the purse strings of congressional appropriations ultimately bind even scientists to the public interest. Dingell reminded the scientific community that his committee had been very supportive of NIH research in the face of the fiscal stringency of the 1980s but that he would "inquire into matters relative to the expenditure of those funds," for the budget "cannot afford . . . meaningless or fraudulent work" (U.S. House of Representatives, 1988). Dingell also charged scientists with a "cavalier attitude . . . [that] is a curious one for those in charge of public money, public property and with responsibilities to the taxpayers" (U.S. House of Representatives, 1990, p. 3).

In overseeing and controlling an esoteric enterprise like science, politicians use a metaphorical technique called "metonymy," managing the entire scientific enterprise by controlling only a small but comprehensible piece of it (Turner, 1990). Financial accountability is an especially powerful metonym. Invoking accountability, Dingell equated congressional control over science with the shattering of scientific exceptionalism: "The scientific community, which apparently has been treated as a sacred cow for far too long, acts like a victim of government persecution if questioned about accountability for Federal funds. If this attitude does not change," the chairman warned, "the public's support for Federal contributions to science may decline" (U.S House of Representatives, 1991, p. 2). By framing the issue as one of financial accountability versus scientific exceptionalism, Dingell does not have to be "antiscience" to make his point. Members of Congress of all political inclinations find utility in standing up for accountability and frugality against privilege and waste; Dingell himself is not the problem for science but is a reflection of a universal political principle (Mayhew, 1974).

Prominent cases of misconduct in federally funded science originally attracted Dingell's attention. Not only did it appear that the normative structure of science failed to deter misconduct, but it also appeared to Dingell that the norms either were not as well institutionalized as some would claim or that they were systematically being flouted during university investigations of fraud allegations. Dingell warned that universities, which "frame the issues" and have "the first crack at the data or lack thereof," have "engaged in what appears to be a cover up effort to protect the institutions and some of their senior scientists." "In some cases, the investigations of scientific fraud and misconduct are carried out by the very targets of the investigations, and in other cases, by only the interested parties themselves" (U.S. House of Representatives, 1988, pp. 2-3). Misconduct challenges the peer review system (Ben-Yehuda, 1985; Chubin, 1990; Chubin & Hackett, 1990), especially as Congress discovers that peer reviews and departmental committees led not to universalistic decisions based on a skeptical analysis of the empirical evidence but to conflicts of interest and conspiracies.

Dingell challenged the epistemological specialness of science by defending whistle-blowers and their roles in identifying and correcting error and misconduct. Through whistle-blower Margot O'Toole, Dingell discovered that, unless a scientist questions possible errors in the literature, the self-correcting mechanism of science cannot be engaged. O'Toole, a junior colleague of Nobel laureate David Baltimore, represented a challenge to epistemological specialness. Her plight revealed the dependence of epistemological authority on social interaction and on the establishment of a network of resources and allies, ranging from the content of test tubes to elected officials. "The truth will out," perhaps, but not without someone to construct the network—a Pasteur or an O'Toole (Latour, 1988).

A second version of the whistle-blower argument suggested that the marketplace of scientific ideas is not very egalitarian and that an uneven distribution of resources among scientists can lead to an unfair outcome even under fair rules of exchange. Margot O'Toole complained in testimony that, after she had approached Baltimore, a coauthor of the paper under scrutiny, with her concerns, Baltimore challenged her to publish them. According to O'Toole's testimony, Baltimore threatened her by promising that he too would publish and then their scientific colleagues would deliberate. The postdoctoral fellow O'Toole plausibly inferred that the peers would believe Baltimore rather than her.

Dingell also challenged the construction of privileged scientific truth. In a May 1989 hearing on the "Baltimore case," Dingell interrogated the authors of the paper in question about an image of a published autoradiograph. As was conventional practice for contributors to the journal *Cell*, the authors had published a composite image. Separate columns of the autoradiograph had been photographed and exposed individually, and subsequently reconstructed in a way that best displayed the authors' scientific claims. The journal did not require that composites be labeled as such. Forensics experts detailed from the Secret Service to the investigation described in testimony the method of composing the separate exposures into a single image representing a scientific claim. Dingell wondered, how can the scientists know which photographs and which exposures to select? Because scientists often assert that research, by definition, is an investigation into the unknown, and that so little is known about the results of research that it cannot be planned or managed, how can it be known which exposures to select? If exposures are selected based on the hypothesis, then is the so-called evidence not tautological? Or fraudulent? The question of data selection gets to the heart of science as a constructive endeavor (e.g., Knorr Cetina & Amann, 1990; Kuhn, 1962/1970; Latour & Woolgar, 1979; Medawar, 1964).

After the 1988 misconduct hearings, Dingell and a colleague drafted legislation to create an office of scientific integrity, responsible for investigating scientific misconduct. Meeting with vocal resistance among the scientific community and a cramped legislative calendar, the amendment was never introduced (Mervis, 1988). Yet the following spring, the U.S. Public Health Service, the NIH's bureaucratic parent, created an Office of Scientific Integrity and an Office of Scientific Integrity Review. The existence of these congressionally inspired offices reinforced the weakness of the sociological exceptionalism of science, because it institutionalized the recognition that the Mertonian normative structure is incomplete and that the market of ideas needs regulation.

Dingell's refusal to accept the premise that science is especially deserving of autonomy from congressional oversight and investigation brought a good deal of criticism. He retaliated with an attack on the defenders of the Platonic specialness of science, who believe "that the Congress is not capable [of] understanding science, or even raising questions about science." Dingell countered that "no one questions the ability of the Congress" where appropriations are sought for cold fusion, the Strategic Defense Initiative, or particle accelerators (U.S. House of Representatives, 1989, p. 2). According to Dingell, congressional scrutiny is symmetric; if members must know enough to appropriate for science, and if scientists come to Congress for appropriations, then members must know enough to oversee and question the activities these appropriations support, and scientists must bear the scrutiny.

CONCLUSION: ECONOMIC EXCEPTIONALISM AND THE FUTURE OF SCIENCE AND POLITICS

Should we worry about the public exposure of the inadequacies of scientific exceptionalism that are borne out by congressional oversight and by the creation of the OTA? The traditional view of science as a sector of privilege would see the exposure accompanying oversight activities as damaging to the ability of science to generate both truth and technology. It might also interpret the fact that the OTA is a servant of institutional interests in Congress as damaging in the same way. Don K. Price (1967) might suggest that the criticisms of science and its appropriation by a legislature would adversely affect the balance among the set of fundamental democratic institutions. By treating science in this manner, political institutions may be arrogating to themselves authority that would preferably be distributed in a more pluralistic fashion. That is, a pluralist might fear the domination of university pursuits by Congress because of the belief that universities represent a sector of civil society whose independence is necessary to prevent a centralization of power.

STS should go beyond the pluralist commentary of this sort. If Ezrahi (1980) is correct that science represents "the possibility of supporting the constructive enterprise of liberal-democratic politics by homogeneous epistemological standards" (p. 51), the public deconstruction of science could undermine democracy in more fundamental ways than even by disrupting the plural distribution of power. Indeed, Ezrahi (1990) links the declining authority of science with the declining legitimacy of liberal-democratic political institutions. Science no longer seems capable of setting unimpeachable standards of integrity, civility, consensus, empiricism, and instrumental action for politics. Although Ezrahi is hopeful that the United States can regenerate a new democracy that is not reliant on the stabilization of ideas by science, the nondemocratic alternative he sees is not merely creeping authoritarianism but fascism. The alternative for scholarship is to focus on the science-politics boundary, perhaps seeking forces that present a counterweight to the deconstruction of science.

The brief studies of the OTA and congressional oversight of science presented above show how interbranch rivalry and financial accountability—politics by the same means—can drive science policy. The studies also show how this kind of politics erodes claims to the epistemological, Platonic, and sociological exceptionalism of science. Remaining is economic specialness, which, given its attachment to the primary rationale for government support of basic research, is perhaps the most consequential category of exceptionalism for the future of government-science relations. For even if science is neither "truth" nor beyond the ken of laypersons, science could still be the best investment for future economic productivity that a government could make.

This economic question has reached new levels of contention. The problem of "science in the steady state" has been of international concern for several years now (Cozzens et al., 1990), as has the education of scientists and engineers and the role of the university in a globally competitive economy (Zinberg, 1991). In fact there seems to be an unusual consensus on fundamental issues of science policy internationally, based primarily on the borderless problems thought to be within its purview: energy and environment, AIDS, and sustainable and internationally competitive economic development (Freier, 1991). Growing attention to science and science policy internationally seems predicated on increasing awareness of the instrumentality of science for application over a large domain of political problems (Ronayne, 1984).

The dynamic steady state of science in the United States over the past decade has been characterized by generally increasing research budgets, but similarly increasing demands for resources. In reporting an informal survey of 250 academic scientists, Leon Lederman (1991) described steady state science as "beset by flagging morale, diminishing expectations, and constricting horizons" (p. 4). For Lederman, "science pays." It is "impossible to imagine modern society without the fruits of 400 years of scientific research." Lederman invoked the economic activity generated by the Apollo program and the recent research by Mansfield (1991) demonstrating a relatively high social rate of return on academic research to argue for the efficacy of scientific relief.

For the politician, the steady state provides a different problem. Representative George Brown (D-Calif), chairman of the House Science, Space and Technology Committee and a congressional friend of science for nearly 30 years, predicted that domestic programs starved through the Reagan-Bush years are "going to set their hungry eyes on" the $75 billion of federal R&D spending. Commenting directly on the Lederman report, Brown confided that he worried about the report's "lack of uniqueness." He noted that "one could easily document a similar level of despair among 250 Medicare recipients, 250 disabled veterans . . . or even among 250 Members of Congress." Brown (1991) asserted:

> If we are going to justify the privileged treatment of research and development by the Federal Government—and we will have to justify it, if we hope to sustain it—then we must present a case that is based not on the frustration and discomfort of individual worthy scientists. Rather, we must present a case rooted in the welfare of our nation. (p. 25)

Brown, accepting the inadequacy of the "linear model" of R&D, suggested that federal support for basic research is neither necessary nor sufficient for economic and social gain. His recognition of what academic students of science and technology have known for years (Mowery & Rosenberg, 1989; Pinch & Bijker, 1987; Shapley & Roy, 1985; Wise, 1986) is significant because of the role that this linear model has played in policymakers' view of the economics of science. The model is often understood to mean that research funding is the best point of intervention by the government in the R&D system. Not only does the model constrain the locus of federal intervention to basic research, but it constrains the mode of intervention merely to adding more money to achieve more output. The simplicity of the linear model makes science a tempting target of federal investment, and it reinforces the view that science is uniquely capable of producing economic returns from the expenditure of public funds. The adherence of policymakers to this model has led to problems in conceptualizing and implementing technology policies in the United States (Alic, Branscomb, Brooks, Carter, & Epstein, 1992; Cozzens, 1988; Sultan, 1988).

Representative Brown's view of science represents new terms for the making of science policy that are likely to gain wider acceptance. The treatment of economic returns from science under these terms is subtle. Politicians want the scientific community to succeed in "justify[ing] the privileged treatment of research and development." They acknowledge that science will no longer be *assumed* to produce economic returns and to deserve implicitly privileged status but believe that scientists can and should "explicitly identify the connections between the nature of our R&D effort, and our economic vitality and quality of life" (G. E. Brown, 1991, p. 29). In other words, science should drop its pretensions and assumptions of exceptionalism, even while building a political case for economic importance.

The new political-economic terms for political debate about science call for increased attention from scholars of SSK to science policy. As Latour (1987) points out, political institutions are "obligatory passage points" for scientists. They should increasingly become such passageways for SSK as well. Tools for understanding how political institutions operate therefore may become as important to the study of science and government as are the tools for understanding how scientific institutions function. The challenge is open not only to political science but also to an array of disciplines. Without attention to political phenomena, accounts of science in society cannot be complete (Cozzens & Gieryn, 1990).

As scholars encounter political institutions, they can ask why so much stability exists: Why, if the political process can undermine scientific claims to specialness, is science still elevated and esteemed? Is it the importance of economic exceptionalism? Is it some recognition of Ezrahi's claim that liberal democracies need science to propagate ideals of consensual, meliorist change? Is it a cruder need to have an ostensibly objective and autonomous group to whom tough problems are delegated and from whom legitimacy is drawn? How is it that political processes sufficiently capable of deconstructing science manage repeatedly to reconstruct it (e.g., Woodhouse, 1991)?

These are important questions to SSK for remaining in contact with the society it studies. Politicians have been immersed in the manipulation of scientific and technical activity with only a tiny amount of guidance or scrutiny from academics who study science. Scholars in our field are well situated to ask questions about the importance of economic exceptionalism in maintaining the privileged position of science. Can economic specialness be a unique point of consensus between scientists and politicians, because the former can draw from it enough resources and autonomy, and the latter can claim credit and effect from their activities? Or will the two groups conflict because scientists will ultimately resist the manipulations of politicians, who want to make science work for the economy?

Finally, scholars can examine an important effect of the economic exceptionalism of science—that, in reducing the importance of science to its economic value, it also tends to reduce society to market and economic relations. In the political reconstruction of science, where are the points of intervention for nonmarket forces for the broader values of science and society?

25

Changing Policy Agendas in Science and Technology

AANT ELZINGA
ANDREW JAMISON

STUDIES of science and technology (S&T) policy occupy a weak and rather fragmented position within the broader STS community. A number of approaches representing different disciplinary and professional identities compete with each other for political and academic attention. The fragmentation may be explained by the variation in national context in which science policy and its academic study are carried out, including differences in intellectual traditions as well as different patterns of institutionalization. Often the primary audiences for such studies are policymakers and scientists rather than a distinct community of peers.

For the purposes of this chapter, we would like to make a distinction between science policy and the politics of science. The latter has to do with the interaction between science and power, that is, the mobilization of science as a resource in international relations, the use of science by interest groups or social classes to increase their power and influence in society, and the exercise of social control over knowledge. Science policy by contrast can be defined as the "collective measures taken by a government in order, on the one hand, to encourage the development of scientific and technical research, and, on the other, to exploit the results of this research for general political objectives" (Salomon, 1977b, pp. 45-46).

As a distinct public policy area, science policy was established in the immediate aftermath of World War II, even though important steps toward its development had been taken in the interwar period within private foundations and in the Soviet Union. In the United States, Vannevar Bush's 1945 report, *Science: The Endless Frontier*, served as a touchstone for policy thinking for some 40 years. It is only recently that it has been explicitly revised (Lederman, 1991; see Shapley & Roy, 1985). In Europe the rival visions of John Bernal and Michael Polanyi regarding the possibility of a political steering of science played a similar role. Starting with these experiences, this chapter will give an overview of changing agendas in science and technology policy against the background of changing relationships in the interface of science with society over the past 50 years.

Our approach follows that of science journalist David Dickson in his book *The New Politics of Science* (1984), in which science policy is seen as the outcome of the dynamic interplay among actors representing what we will refer to as different policy cultures. As part of a younger generation of scholars who came of age in the 1960s with both the academic challenge to positivist epistemology (e.g., Kuhn, 1962/1970) and the political challenge to technocratic civilization, we are critical of the technological determinism and scientistic optimism that characterized much of the "first generation" literature on science policy. Our approach seeks to link together what Ina Spiegel-Rösing in her introductory article in the 1977 STS anthology identified as two distinct subdisciplines within STS, namely, social studies of science and the study of science (and technology) policy (Spiegel-Rösing, 1977). To paraphrase Imre Lakatos, we believe the analysis of science and technology policy without the self-reflection that comes from science studies is blind, just as science studies are naive if not informed by a science and technology policy perspective (Lakatos, 1974).

In each of the last three decades, doctrines developed by the Organisation for Economic Cooperation and Development (OECD) have left their imprint on the science policy approaches that governments have followed. Before the OECD was established in 1961, there was less uniformity in the making of science policy, as the different national elites tended to have their own competing policy agendas. The OECD has served as a forum where government ministers responsible for science in the leading Western capitalist industrial nations regularly meet to develop a common frame of reference. At particular points in time, documents from international and interdisciplinary OECD panels have diagnosed problems, pointed to new issues, and articulated the assumptions and principles that have guided the member countries in the formulation of their science policies.

In what follows, we will trace the interaction of science policy and the broader politics of science from the interwar period to the present. A focal point in our analysis will be the shifts in OECD doctrines, because these have tended to articulate changes in policy agendas. The presentation highlights a number of policy-related themes on which, we feel, STS scholarship should be brought to bear to a greater extent: the intermittent narrowing and widening of the science policy discourse in a cyclical process of depoliticization and repoliticization; the underlying social and political factors that lie behind shifts in the main policy orientations; the interplay of various policy cultures; and the problems of periodizing and explaining science policy trends and transformations. We want to draw attention to the various ways in which science policy is embedded in a broader political sociology of scientific knowledge.

FROM POLITICS TO POLICY

Science policy and the politics of science can be seen as interacting at several levels. Most obviously, but often forgotten in the literature, the very idea of science policy is an integral part of a political program on behalf of those in power—the political, industrial, and military establishments—to use knowledge to achieve their goals. At the same time, many changes in policy have also been propelled by public debates and movements around controversial developments in science and technology, from the atomic bomb to genetic engineering and global warming. Institutional reforms, as well as regulatory innovations, have often responded to pressure emanating from interest groups and critical intellectuals. On another level, popular writers and opinion leaders influence the conceptual frameworks within which policy matters are discussed (Carson, 1962; Commoner, 1971; F. Graham, 1970). At this level, the politics of science becomes a rhetorical struggle over the ways that science and technology are interpreted, the worldviews and associated metaphors that give rise to alternative visions for the organization of knowledge. The political domain provides a space for a broader cultural assessment of scientific and technological choices as well as for a more specific process of accounting the costs and benefits to various groups in society.

Originally, in the formative period of science policy between World War I and World War II, there was an explicit connection between policy and politics, in the sense that policy proposals were embedded in conscious political programs of technocratic management, on the one hand, and democratic populism, on the other. Bernal's vision of a socially responsible science in its strong version had linked science with the struggle to create a socialist society (Bernal, 1939):

> The paradox is that Bernalism, the product of revolutionary thinking in the 1930s, was in fact taken over by captains of industry and ministers of government in the postwar period. . . . By 1964 what had happened was that the strong Bernalist thesis had been peeled away leaving the weaker Bernalism of planning, programming, people, money and equipment for efficient growth. The stronger version of Bernalism seemed to be forgotten even by Bernal himself. It is no wonder then that "Bernalism" could serve as the rationale or theoretical legitimation for science policy doctrines in countries in *both* East and West. (Elzinga, 1988, p. 94)

Science policy, as it has grown more professionalized and differentiated into particular societal sectors, has increasingly become separated from its political and ideological roots. To be more precise, these have often been rendered invisible, as policymaking has been reduced to a technocratic instrument for rationalization and planning in an established framework whose basic assumptions go unquestioned (Aronowitz, 1988a). Thus normalized, science policy proceeds until economic recession or political crisis in the broader society prompts a review of the existing doctrinal "paradigm" and a shift to a new conception of science-society relationships. Currently, there are signs that S&T policy might be in the process of being repoliticized, as the assumptions of the cold war era and the guiding conceptions of the state are in the throes of reformulation, as is the political value and nature of basic research itself.

THE CULTURAL DIMENSIONS OF SCIENCE AND TECHNOLOGY POLICY

One way to analyze the interaction between politics and policy is by focusing on the various actors who are involved in the making of science and technology policy. At work are what might be thought of as four main "policy cultures," coexisting within each society, competing for resources and influence, and seeking to steer science and technology in particular directions.[1] These cultures, which stand out as representative of the dominant voices in the literature we have reviewed, represent different political and social interests and draw on different institutional bases and traditions for their positions. Each policy culture has its own perceptions of policy, including doctrinal assumptions, ideological preferences, and ideals of science, and each has a different set of relationships with the holders of political and economic power. One might also say that these are the main constituencies in the realm of science and technology policy.

First, there is a bureaucratic policy culture, which in many countries is largely dominated by the military, based in the state administration with its agencies, committees, councils, and advisory bodies, concerned primarily with effective administration, coordination, planning, and organization. Here science is of interest primarily for what might be termed its social uses; the concern is with science for policy and indeed with making public policy scientific (Jasanoff, 1990a; B. L. R. Smith, 1990). In the second place, the academic culture, based among scientific practitioners themselves, is more concerned with policy for science and with preserving what are seen as traditional academic values of autonomy, integrity, objectivity, and control over funding and organization (Polanyi, 1958; Shils, 1968a; Wittrock & Elzinga, 1985). Third, there is an economic culture related to business and management, based in industrial firms, and focusing its attention on the technological uses of science. At work here is an entrepreneurial spirit or ethos that seeks to transform scientific results into successful innovations to be diffused in the commercial marketplace (Dosi, Freeman, Nelson, Siverbrerg, & Soete, 1988; Etzkowitz & Webster, this volume; Gibbons & Wittrock, 1985). Finally, we can refer to a civic culture, which in its most dynamic form is based in popular, social movements, such as environmentalism and feminism, and whose concerns are more with the social consequences and implications of science than with its production and application. The civic culture articulates its positions through public interest organizations as well as through campaigns and movements, and its influence is obviously determined by the relative strength of the civil society in a country's overall political culture (Almond & Verba, 1965; Blume et al., 1987; Nowotny & Rose, 1979). While the dominant cultures tend to draw science and technology policy into a "technocratic" direction, the civic culture stands for what has been called a "democratic strategy" for S&T policy (Dickson, 1984).

Although most countries have followed a similar overall pattern of policymaking, there are distinct "national styles" of science and technology policy, which reflect more general differences in policymaking and governmental regulation (Hilpert, 1991; Vogel, 1986). In the special case of S&T policy, national variations are dependent on the relative strengths and modes of interaction among the aforementioned policy cultures, on the one hand, and the more formalized country-specific institutional arrangements for production of knowledge, on the other. In this regard, one can distinguish between those countries, such as France, Japan, and Sweden, where the state has actively intervened in economic affairs, and other countries, such as the United States and Great Britain, where the invisible hand of the marketplace has been given greater leeway and where state intervention has largely been limited to strategic and military initiatives. Another difference is in the degree of centralized authority and regional influence over priorities. Here the German federal system represents an interesting counterpoint to the French republican system.

Despite these national differences in style, one can discern a number of underlying processes that have led to an international convergence in terms of issues and approaches. Important among these are (a) the increasing prominence of science-based technologies in the industrial political economy; (b) methodological conformity in identifying future priorities; (c) the globalization of knowledge production and diffusion in both public and private domains; (d) rising costs of research technologies, new large-scale experimental facilities, and other infrastructural supports; and (e) international agenda setting and "orchestration" from above through intergovernmental bodies leading to conformity in issue perception and management.

Some analysts are inclined to attribute an almost autonomous deterministic logic to the development of science and technology, giving rise to the same problems and issues in different countries. In the science of science movement of the 1950s and 1960s, which was especially influential in the former Soviet Union, some analysts even spoke of a Scientific-Technological Revolution (the STR theory) transforming communist and capitalist countries alike with a new science-based automated mode of production (Fleron, 1977; Richta et al., 1967). At roughly the same time in the West, such writers as Daniel Bell and John Kenneth Galbraith were propounding a theory of convergence, according to which all countries were moving to one and the same "postindustrial" formation (Bell, 1974; Galbraith, 1967). Our own perception is of a cyclical, socially constructed process of interaction between science and technology, on the one hand, and cultural critique or response, on the other. The "technological imperative" is in this view continually countered by political and social actors with their own visions and policy agendas (Baark & Jamison, 1986).

PERIODIZATIONS OF S&T POLICY

The different cultural orientations among S&T analysts have led to different approaches to the periodization of science policy: a bureaucratic focus on organizational orientation leading to one kind of analysis, an economic focus on R&D expenditures and profitability leading to another, and an academic focus on policy doctrines leading to yet a third way of analyzing recent history. A fourth perspective, emanating from the civic culture and focusing on the interrelations between science, technology, and social movements, opposes the "elitist" assumptions in all three of the foregoing approaches. While there is a certain agreement over key events and turning points, authors differ as to which issues are considered most important and deserve to be highlighted.

Harvey Brooks, one of the leading figures of American academically based science policy, distinguishes three distinct epochs: the cold war period from 1945 to 1965, the period dominated by social priorities from 1965 to 1978, and the period of innovation policy from 1978 to the late 1980s (Brooks, 1986, pp. 128-136). In each period, Brooks sees science policy primarily as a process of institutional coordination and management, and his periodization reflects to a considerable extent his own involvement in the process as a longtime adviser, at arm's length, to both legislative and executive branches of government.

Christopher Freeman, for many years director of the Science Policy Research Unit (SPRU) at the University of Sussex, presents a different framework for analysis. For Freeman, science policy is seen through the eyes of the economist, and his focus is on the role of research in industrial innovation. In his terminology, the 1940s and 1950s represent a period of "supply-side" research economics, when government effort was devoted to expanding the resource base for industrial innovation, in particular by investments in basic research and higher education. This corresponds to the period of postwar economic expansion, when a new "techno-economic paradigm" was taking shape, based on innovations in electronics, petrochemicals, and atomic energy (Freeman, 1987). As this postwar wave of innovation reached its peak in the early 1960s, with apparently diminishing rates of return on investment, science policy, according to Freeman, entered a period of "demand-side" economics. With a turning point around 1965, market concerns came to dominate science policy, as expenditures on basic research were questioned and "cost-consciousness and cost-effectiveness, as well as techniques of project evaluation, received increasing emphasis both in private industry and in government" (Freeman, 1988, p. 115). In the 1980s Freeman sees a third period, characterized by a combination of demand- and supply-side research economics aimed at fostering a perceived new techno-economic paradigm based on microelectronics, biotechnology, and new industrial materials. Science policy thus becomes innovation policy, an ingredient in a new kind of governmental framework for industrial innovation.

For Jean-Jacques Salomon, representing a political, or bureaucratic, approach with a continental European slant, science policy is seen in relation to the ideas or doctrines that dominate particular periods. To a large extent, science policy becomes a matter of international relations and political philosophy. Salomon's periodization thus seeks to trace the emergence and evolution of science policy as a specific realm of political life, with each period characterized in terms of strategic behavior and tactical accommodation to central doctrinal disputes (Salomon, 1977b). For Salomon, these periods are characterized in large measure by the dominant idea of the relationship between "science and power" (see Salomon, 1990).

Analysts of social movements, in seeing postwar history from the perspective of the civic culture, adopt a very different kind of periodization (see Jamison et al., 1990). The 1950s and early 1960s are seen as a time of "awakening," when the detrimental social and environmental consequences of industrial development were articulated and made visible. The late 1960s and early 1970s were a time of organization, when new forms of activism emerged and stimulated alternative thinking and calls for institutional reform. With the oil crisis of the mid-1970s, there was a further politicization around specific issues, especially nuclear power, which ebbed out toward the end of the decade. From this perspective, the 1980s represented a resurgence of the technocratic optimism of the immediate postwar era and a dispersion of activism into professional public interest organizations and environmental "think tanks," such as the World Resources Institute. Much of the "cognitive praxis" of the new social movements was transformed into new academic programs of environmental, peace, and women's studies (see Eyerman & Jamison, 1991). Our own periodization will try to integrate the various approaches while focusing on the relations of science policy to the broader politics of science and technology.

THE MAKING OF S&T POLICY

As an explicit and self-conscious activity, science and technology policy is a product of various initiatives taken between World Wars I and II (Ronayne, 1984). But these initiatives built on at least three centuries of preparation, during which the relevant actors had slowly developed their characteristic identities and institutional networks and science and technology had gradually moved from the margins into the center of political, social, and economic life. Science and technology had begun to take on something of their modern character in the course of the seventeenth-century scientific revolution. In the transition of European societies from feudalism to industrialism—or capitalism, if one so prefers—the modern scientist emerged as a kind of synthesis of the medieval scholar and the traditional artisan, with precursors among the artist-engineers of the Renaissance (Zilsel, 1942).

The scientific academies of the seventeenth century (the Academia del Cimento in Italy, the Royal Society in England, the Academie des Sciences in France) provided some of the first organized social spaces anywhere in the world for carrying out scientific research and communicating scientific results. It was in the social contract between the craftsman-scholar and the royal monarch that one finds the groundwork for the institutionalization of science as a specific endeavor, separate from religion, rhetoric, and politics (Mendelsohn, 1977; van den Daele, 1977). With the seventeenth-century scientific revolution, science came to be identified with an experimental practice, mediated by technical instruments; what remained separated in other parts of the world, divided into the separate realms of scholarly endeavor, on the one hand, and practical learning-by-doing, on the other, was combined in Europe in an academic scientific praxis (Ezrahi, 1990; Jamison, 1989; Shapin & Schaffer, 1985).

With the coming of the political and industrial revolutions of the late eighteenth century, science entered the universities and what had been until then a relatively marginal societal activity came to be transformed into a profession (Mendelsohn, 1964). The links with technology and industrial development were intensified during the nineteenth century, in new types of scientific universities, industrial research laboratories, and technological colleges. At the same time, scientists throughout the industrial world were creating associations for the advancement of science to further their own professional interests and acting, in various ways, as a lobby group to raise their social status and political influence. Engineers followed suit, professionalizing their training, expanding their activities throughout the state apparatus and the expanding corporate system, and forming academies of their own to sprout, as it were, a new, more practical branch within the academic culture. By the time of World War I, scientists and engineers had become key actors in the making of the modern industrial state, and there were those among them who dreamed of a technocratic society, run by the newfangled scientific experts in the name of instrumental or technological rationality (Noble, 1977).

In almost all industrial countries, responsibility for science and technology was largely in the hands of the private sector, which in the first decades of the twentieth century developed what might be called a corporate science policy through industrial research laboratories, university-industry linkages, and, perhaps most important, corporate foundations. The Carnegie, Ford, and Rockefeller foundations—to name only the most important ones—grew into significant actors in the interwar period, setting policy agendas in the physical and biological sciences as well as in certain areas of social science. Such foundation support was crucial in the consolidation of the "research university" and in the supersession of older and more traditional values in academic life (Geiger, 1986). The foundations also encouraged interdisciplinarity and the growth of international contacts among scientists by financing a range of international exchange programs in both research and education (see Abir-Am, 1982).

It was only in the Soviet Union that a more activist governmental science policy activity developed in the interwar years. In the words of Loren Graham (1967), "No previous government in history was so openly and energetically in favor of science. The revolutionary leaders of the Soviet government saw the natural sciences as the answer to both the spiritual and physical problems of Russia" (pp. 32-33). The Soviet government sought to integrate scientific research into what became an elaborate system of centralized economic planning, and the Academy of Sciences was reorganized to be better able to serve the interests of the new socialist society.

In 1931 a high-level delegation of Soviet officials and scientists attended an international congress for the history of science in London, and they presented the Soviet approach to science and technology as an alternative to the rather low status of science and technology in the capitalist countries, which were then in the throes of economic depression (Bukharin et al, 1931; see L. Graham, 1985). The facilities and support supplied to scientists in the Soviet Union were contrasted to the unemployment so rampant in the West, and the appeal of the Soviet "model" of state steering and control extended well beyond the card-carrying Communists to include many other scientists and intellectuals. In Britain, a highly "visible college" of leading scientists—Lancelot Hogben, Hyman Levy, J. B. S. Haldane, Joseph Needham, and John Bernal—agitated for more active state involvement in science policy and a more political role for scientists to play in solving social problems, and in 1939 Bernal published a kind of manifesto, *The Social Function of Science*, which brought together a range of historical, political, and economic arguments for increasing public support for science and technology (Werskey, 1978).

It was by virtue of these different initiatives that science and technology became, by the eve of World War II, an explicit area of political debate and public policy discussion. The economic culture had developed a number of institutions to finance and plan scientific research; the Soviet state had experimented with planning and science policy steering instruments; and activist scientists had tried to awaken their fellow members of the academic culture to the need for a greater social involvement in scientific research. But it would be the war itself that would put these different sources into the center of political attention.

FROM PEARL HARBOR TO SPUTNIK: THE 1940s AND 1950s

The experiences with military research in World War II, especially the establishment of large-scale, intricately planned, and multidisciplinary projects in operations research, electronics, radar, and atomic energy, marked a decisive turning point in the history of science and technology policy (Salomon, 1973). A kind of hybridization took place throughout the industrial world in the interface between representatives of the military or bureaucratic culture, on the one hand, and the scientific or academic culture, on the other (Weingart, 1982). A new breed of experts came into being, bridging the values and norm systems of the state and the academy, creating a new vocabulary and a new kind of social role as science and technology politicians. With experience in foundations or academic administration, these "new men," as C. P. Snow called them in one of his early postwar novels (Snow, 1954), recruited the personnel to staff the big science projects, lobbied and negotiated with military and government officials on behalf of the scientists, organized and managed the laboratories and project facilities, and, after the war, became the spokesmen for the political and professional ambitions of the academic culture (Greenberg, 1967). Some became expert critics, forming organizations to protest the arms race and thus making connections with activists within the civic culture, links that would grow in importance in the course of the 1960s (Blume, 1974). Others became members of the new policymaking bodies set up to finance and coordinate science and technology, the councils and commissions for various civil R&D programs, as well as the vastly expanded military research agencies. At the same time, the military connection contributed strongly to the shaping of scientific agendas, particularly in physics, the paragon discipline of the time (Forman, 1987; Hoch, 1988; Mendelsohn, 1990).

The immediate postwar period marked the acceptance by all industrial nations of an active state involvement in scientific and technological research. Science councils were created in most countries for basic research and engineering as well as for the main sectoral research areas: defense, health, agriculture, and (atomic) energy. Science councils were not entirely new—the U.S. National Institutes of Health, for instance, had been founded in 1930, and military research agencies had started even earlier—but they were given substantially increased budgets after the war, and thus a greater influence on distribution of national resources and the setting of national priorities. Also important in many countries were the national laboratories and research institutes that were created for military research, atomic energy, and agriculture. In the social sciences, the foundations played a significant role in this period in setting policy agendas; in area studies, sociology, survey research, and other fields, foundation support was essential in creating new departments and new subdisciplines within the university system. As such, the academic culture entered into more active alliances with the policy arms of both the bureaucratic and the economic cultures. And yet the academic culture managed to retain, in many countries, decisive control over the new funding and coordinating agencies. The achievements of the war effort—especially the atomic bomb—led to a "victory for elitism" and a strong belief on the part of state and corporate officials in leaving it to the scientists themselves to determine their own priorities (Kevles, 1978; Rose & Rose, 1969).

The result was a temporary defeat for a more socially conscious, populist view of science. The institutional norms of science, enunciated by Robert Merton in 1942 in the United States, corresponded to the tenets of the Society for Freedom in Science, which Michael Polanyi founded in 1941 as a counterpoint to Bernal's socialist planning doctrines. Both sought to (re)define the conditions for a free and open "scientific community" against the totalitarian movements of the right and the left. In 1945 Vannevar Bush, who had been one of the key science policymakers during the war, produced his report to President Roosevelt in which he argued for maximum autonomy for scientists in relation to political, economic, and social interests (Bush, 1945). Instead of controlling science, the public was encouraged to make better use of it and to learn more about it: A wave of popularization and public information largely replaced the activism that had been so widespread before the war.

Indeed, it is possible to characterize the immediate postwar period as a time of scientistic hegemony and see the new science and technology policy landscape as dominated by the voices of the academic culture defending academic autonomy and scientific freedom (Gilpin & Wright, 1964). By the end of the 1940s, however, the iron curtain had fallen with the dropping of the Soviet atomic bomb and the victory of the Communists in China; particularly in the United States, the anticommunist crusade took a heavy toll on the hard-won freedoms of the academic culture. The price of state patronage became a high one in terms of loyalty oaths and ideological conformity, but through most of the 1950s the main authority in the new relationships between science and the state continued to be exercised by scientists. Even censorship and political control were largely an internal affair of the academic culture (Schrecker, 1986).

It is, however, worth remembering that, between 1950 and 1957, when the Soviets launched their first Sputnik, the National Science Foundation in the United States remained a relatively small organization. The major part of R&D funding went through mission-oriented agencies such as the Office of Naval Research, the Atomic Energy Commission, and the National Institutes of Health. The relative share of the R&D budget devoted to basic research was still rather small compared with the military, health, and energy programs (Greenberg, 1967). But even within those programs, the emphasis was on science rather than technology; the main actors in the science policy discourse saw science as "pushing" or leading technology onward. Little attention was paid, either by policymakers or by academic researchers, to the interaction of science and technology or what later came to be called industrial innovation.

FROM SPUTNIK TO VIETNAM: THE 1960s

Sputnik, the Soviet space satellite, led to dramatic changes in the realm of science and technology policy, as the alliance between government and the universities took on new dimensions (Lakoff, 1977). A new wave of mobilization of scientists and engineers ensued in many countries, but it was particularly accentuated in the United States, where R&D budgets expanded by an average of 15% per year until the mid-1960s. The bulk of this spending went into military and space efforts, the two sectors that dominated the American R&D profile during that decade. This expansion, however, did not alter the fundamental structure of federal science support, which involved a division of labor between mission-oriented and basic research, a model adopted in most other Western industrial states. In this model the research community was given the prerogative of setting the internal goals of basic research and guaranteeing quality control by peer review processes, while the criteria underlying mission orientation were set by applied funding agencies.

The first OECD report, *Science and the Policies of Governments* (the so-called Piagnol report, 1963), formulated the distinction between "policy for science" and "science for policy" as well as the categories for calculating the flows of funds to various types of activities. These categories—basic research, applied research, technological development, or, simply, development (altogether abbreviated as R&D)—were later elaborated into a methodology for R&D statistics at an OECD workshop held in Frascati, Italy. The report also offered a set of recommendations for the member governments to follow in supporting scientific and technical research and in establishing scientific advisory bodies to government. And, perhaps most important, the document transformed a political ambition or vision into a strategic policy doctrine: namely, the idea that science, together with higher education, should be seen as a productive factor on par with labor and capital in the pursuit of economic growth. In the words of the report, "Expenditures on both education and research represent long-term investment in economic growth" (OECD, 1963, p. 30). This tenet of supporting science for the purpose of stimulating growth measured in terms of GNP may be called the first OECD science policy doctrine—OECD 1.

The role of scientists as experts expanded also in many civilian realms of politics and public decision making (D. K. Price, 1967). In the United States a special assistant to the president for science and technology was created by Congress in 1962 (Golden, 1988). In Sweden during the same year, a Science Advisory Council was set up, having direct access to the prime minister's office, while in Canada a Science Council was created with its own professional research policy staff at arm's length from government (it was abolished in 1992).[2]

The 1960s also saw the emergence of university-based units for science policy studies in many Western countries; for example, SPRU in Britain and the RPP (Research Policy Program) at Lund in Sweden were both founded in 1966. In eastern Europe and the former Soviet Union, similar developments took place under the auspices of a "science of science" movement (Goldsmith & MacKay, 1963). It was also during this period that the historian Derek de Solla Price developed his studies of what he saw to be the exponential growth of numbers of scientists and their publications, laying the statistical groundwork for the thesis that pointed to a transition from little science to Big Science, from a science of patronage and individual craftsmen to research based on teamwork, project funding, capital-intensive and politically steered efforts (D. J. S. Price, 1963). Internationally, within the United Nations a number of initiatives were also taken.[3]

Parallel to these developments, specialized journals devoted to science policy and R&D management began to appear (*Minerva, Research Policy, Impact of Science on Society, Technology Review*), while the journal *Science* in 1961 began a special news and comment section headed by Daniel Greenberg. The discussion concerning priorities in science came out into the open in the pages of *Minerva*, which featured articles written by prominent administrators, policymakers, and philosophers of science. The main issue in the so-called *Minerva* debate was how to choose between different fields of science, and how best to let scientific lobbies exert pressure on behalf of their own fields (Shils, 1968a). Alvin Weinberg introduced the idea of research investment as an insurance policy to guarantee a country a fund of basic technological development as well as a set of general criteria by which decisions might be made to select research for socially relevant areas (Weinberg, 1967).

Weinberg's criteria for scientific choice articulated two sets of principles. The first pertains to externalist concerns, that is, broad social and technological merit as well as potential impact on neighboring scientific fields. The other set of principles were internalist, for example: Was the particular topic ripe for exploitation, was the subject original or already overworked, was the problem fruitful for the development of the field of research? This provided a checklist for criteria that research councils eventually elaborated into more sophisticated guidelines.

It is interesting to note how, in periods with strong external pressures of relevance and accountability, there may be an epistemic drift of criteria whereby external assessments of societal relevance are pushed to the foreground while expert evaluations on the basis of internal criteria (peer review) may be pushed into the background (D. K. Price, 1979; Elzinga, 1985).

The bureaucratic culture's skepticism about the idea that investments in science lead directly to economic growth was the main reason for the launching of Project Hindsight, commissioned in the mid-1960s by the U.S. Defense Department. The study consisted of an inventory of military innovations to assess the relative contributions of science and technology to their development. The idea of the innovation chain, which had been an important early concept in science policy, was tested, and an attempt was made to trace innovations back to their roots in basic research (Sherwin & Isenson, 1967).

These investigations made a strong impression on industrialists, reinforcing their economistic ethos and strengthening their conviction that scientific expenditures needed to be justified directly in economic rather than scientific terms. It could no longer be taken for granted that investment in science contributed directly to economic growth; it seemed to be possible—indeed, rather common—for economic growth to happen without science. Rather than a science push, it seemed more relevant to refer to a market pull in explaining innovation.

The counteroffensive from the academic policy culture came in the form of a project called TRACES. Project Hindsight had served to cast doubt on the conventional wisdom—from Bernal to Bush and Brooks—which had claimed that generous investment in basic science was a prerequisite for maintaining technological supremacy. TRACES provided a methodological critique of Hindsight and tried to give evidence in favor of a science-push model, thereby showing the importance of basic science to industrial innovation (IIT Research Institute, 1968). The project, supported by the National Science Foundation, looked not only at the initial stages but at all the steps of a successful innovation and showed that basic research did play an important role in almost all of the projects investigated. There were similar investigations in other countries, which also sought to show that basic science was an economically important activity (see Irvine & Martin, 1984). At this point, the academic policy culture assumed as rather unproblematic the category of "basic research." Twenty years later, under external relevance and accountability pressure, and because of symbiotic linkages with new technologies, it became clear that this construction has been made from an internalist and positivist conception of science. Then the term *strategic research* began to be used as a more adequate category from the policy point of view.

The controversy over these studies was, in large measure, due to the difference in the views of the innovation chain that were held by different types of funding agencies. The controversy drew attention to the fact that the relationship between science and policy was socially or institutionally mediated through funding structures and interactive mechanisms at the science-industry-government interface and that the terms of the relationship helped determine the results of the innovation process, or at least the efficiency of their "use." These studies were carried out at a time when the United States had achieved technological supremacy by investing in military, nuclear, space, and related Big Science projects. But 1967 marked the beginning of a stagnation of federal R&D support, setting off a decade-long decline. Although the policy studies and debates mentioned above played a role, more important was the ideological tension that developed between the academic (and civic) policy cultures, on the one hand, and the bureaucratic-economic culture, on the other, primarily in relation to the war in Vietnam. Within this context, science and technology criticism was developed, putting the spotlight not only on the use of science in the Vietnam war but also on its role in pollution of the environment and other negative effects on society. It was out of this period of questioning and conflict among the main policy cultures that a new kind of accommodation would be put in place in the early 1970s.

THE 1970s: THE PERIOD OF SOCIAL RELEVANCE

The rise of what Theodore Roszak in the late 1960s called a "dissenting academy" was an important factor in the reticence of governments and industries to replenish basic research resources, and, for a time, tension even developed into animosity and the deterioration of communication. In the United States, the Office of Science and Technology Policy was dismantled, and the position of the presidential science policy adviser was seriously weakened (B. L. R. Smith, 1990). At the same time, a wave of criticism rolled over the academy, and new social movements based on feminism and environmentalism were forming their organizational identities throughout the industrial world. In this context, a number of new mission-oriented programs were initiated—the $100 million "war on cancer" (Studer & Chubin, 1980), the NSF program on research applied to national needs (RANN), and the establishment of the Office of Technology Assessment (OTA) in the Congress in 1972. With these measures—and similar ones like them in other countries—a new era of social accountability was ushered in. On the one hand, it signaled an alliance between the bureaucratic and economic policy cultures; on the other, at least in some places, there was an opening of the science and technology policy agendas to some of the concerns that were being voiced in the academic and civic cultures.

In 1971 an OECD panel chaired by Harvey Brooks produced the report *Science, Growth and Society: A New Perspective*. The Brooks report can be seen as a response, on the part of the member governments, to the period of questioning and social upheaval that marked the late 1960s. The report emphasized the need for greater societal control over applied research and a broadening of the domain of science policy to include the entire range of governmental policy sectors. A unified science was subdivided into separate component parts of distinct sectorial programs, and a new set of concepts—mission orientation, technology policy, social relevance—were brought to the forefront of the policy discourse. As such, OECD 2 represented a technocratic doctrine of social engineering that was shared as the conventional wisdom throughout the industrialized world.

Behind the shift to the socially responsive technocratic doctrine of the 1970s was a complex combination of factors. In the words of Harvey Brooks (1986):

> The epoch that began with euphoria about the capacity of science to solve social problems soon gave way to disillusionment and ended in what almost looked like a revolt against science, or at least against big science. . . . Very soon, however, science and rationality began to be viewed as the source of the problem rather than the basis of its solution, and the social problems became increasingly talked about as the secondary effects of progress in science and technology. (p. 130)

The social movements that arose in the 1960s—the antiwar, student, and environmental movements—led to what might be termed a repoliticization of science (see Nowotny & Rose, 1979; Rose & Rose, 1976a). The 1970s brought politics back into discussions of science and technology policy. On the one hand, there were a number of new organizational initiatives; new environmental groups emerged in almost all industrial countries, and after the U.N. Conference on the Environment in 1972 in Stockholm, environmental issues began to spread more actively to developing countries as well. The environmental movement became an active pressure group in many countries, developing a kind of counterexpertise and calling for greater public participation in science and technology decision making (Cotgrove, 1982; Skoie, 1979). In several countries, left-wing scientists developed "radical science" organizations, whose members played an important role in the public debates about military research and later laboratory experimentation with recombinant DNA or genetic engineering (Arditti, Brennan, & Cavrak, 1979; Balogh, 1991; W. Fuller, 1971).

Also important as new actors in the civic culture were the women's liberation groups, which raised issues of gender bias and focused on particular areas of medical technology, usually having to do with reproduction and birth control. As the 1970s progressed, women's studies emerged as a new area of research, and women's groups focused on discrimination and sexual differentiation in both scientific theories as well as in scientific institutions (Rothschild, 1983). As was to be the case with environmentalism, the feminist challenge grew into a substantial social movement in the 1970s and, in many countries, succeeded in placing new issues onto the national science and technology policy agendas. At the end of the 1970s, with a new wave of escalation in the nuclear arms race, the peace movement received a new lease on life, and mass demonstrations, as well as research and educational programs, developed throughout western Europe and the United States to urge "conversion" of military R&D to civilian uses (Thompson & Smith, 1980). Conversion projects and groups developed around many of the larger defense research establishments, and in Britain, workers' committees at Lucas Aerospace Corporation developed an alternative production plan, proposing that the firm produce a range of socially useful products rather than the weaponry that dominated the production list. The call for "alternative production" that spread from British to Scandinavian and German workers during the 1970s was of course a reflection of the less expansive economic situation, but it also represented a new activism within the labor movements of many countries, in part encouraged by the rise of the new social movements (Elliott & Elliott, 1976; Wainwright & Elliott, 1982).

The response to all of this "democratic" activity on the part of the other S&T policy cultures was to broaden substantially the realm of S&T policy to include many more social sectors as well as to establish new mechanisms for assessing the social and environmental impacts of technological development (Norman, 1979; OECD, 1981). Almost all industrial countries created environmental protection agencies, or departments, within which one or another form of environmental impact assessment was institutionalized. Environmental legislation was also reformed; and the environmental sector became an important area for science and technology policy initiatives through new R&D programs, committees, advisory boards, and so on (B. L. R. Smith, 1990). With the OPEC oil price increases of 1974—the so-called oil crisis—the environmental debate came increasingly to focus attention on energy and, in particular, the pros and cons of nuclear energy; this led some countries to develop some of the most ambitious nonmilitary R&D programs. Energy conservation, alternative energy, and energy systems studies became institutionalized in a host of new agencies and councils and, in many European countries, nuclear energy became a dominant political issue, bringing down governments and stimulating the formation of new political parties (see Rüdig, 1991).

The public debates about nuclear energy and genetic engineering, as well as the consequences of the computer "revolution," contributed to an interest in technology assessment. An Office of Technology Assessment was created at the U.S. Congress in 1972, and many other countries experimented with new bodies to carry out technology assessment and/or future studies as part of the science and technology policy activity (Wynne, 1975; for later assessments of the assessors, see Jamison & Baark, 1990). During the 1980s some of these TA activities would benefit from an interaction with STS scholars, particularly those involved in social constructivist approaches to technological development (Schot, 1992).

There were several large-scale, government-sponsored information campaigns as well as innovative attempts at the local level to increase the participation of the general public in S&T policy decisions. Policymaking was thus opened to representatives from the civic culture, under the general principle that people who were affected by science and technology should also have a voice in its operation. In Cambridge, Massachusetts, the city council established its own review board to regulate recombinant DNA experiments (Krimsky, 1982), and in London, the city council created its own research activity as both an environmental and an economic measure. This was also the period when science, technology, and society (STS) courses and programs were created at both technological universities and liberal arts colleges. The study of scientific controversies emerged at this time as an important area of attention (Mazur, 1981; Nelkin, 1979). For politicians and bureaucrats, the concern was with constructing mechanisms for resolving controversies, or achieving closure, which became a topic of STS scholarship (see Engelhart & Caplan, 1987).

One result of these initiatives was to increase the "external" influence over the academic culture, which eventually led to a kind of backlash. By the early 1980s, some were arguing that the call for social relevance had gone too far and that the increased public participation in S&T policy had taken authority away from the experts (Jasanoff, 1990a). More important, conservative governments came to power in both the United States and Great Britain, which saw their main task as strengthening the corporate or economic culture and cutting down on state expenditures. Part of this shift in emphasis was due to the growing economic and technological challenge from Japan and the so-called newly industrializing countries (NICs) of Korea, Taiwan, Singapore, and Hong Kong; apart from the failure to keep up with these countries' rates of growth, there was also an inability of the bureaucratic and civic cultures to develop functional alternatives and devise appropriate techniques to manage and organize their science and technology priorities. There was not sufficient quality control in many of the new sectorial R&D programs; nor did the economic payoff seem sufficiently high to motivate further increases and expansion.

THE 1980s: THE POLICY OF ORCHESTRATION

In the new context the OECD (1981) produced a new document, *Science and Technology Policy for the 1980s*, which served to guide the member countries in formulating their policy responses to what was increasingly seen as a Japanese challenge. The essence of OECD 3 was to stimulate the growth of new technologies by means of active industrial policy and encourage a more intimate relationship, and active partnership, between universities and industry. The commercial orientation given to science policy in the 1980s, with its emphasis on industrial innovation and technological forecasting, was something that was common to all of the industrialized countries, even though there were substantial differences in the mechanisms that were adopted.

The Japanese success in many branches of industry, especially in the important electronics industry, focused attention on the characteristic Japanese R&D policy approach with its systematic use of technology forecasting, or foresight, and its strong economic, or industrial, orientation (Freeman, 1987). In Japan and the east Asian NICs, science policy was incorporated into industrial policy, and the state played a more active, supportive role in relation to the export firms than was the case in most Western countries. Stimulated by the challenge of the Japanese miracle, most Western countries began to experiment with research foresight (Irvine & Martin, 1984; and, for a later comprehensive review of developments in eight countries, Martin & Irvine, 1989). There was also a strong move to develop so-called national programs to support the advanced, generic, or base technologies of microelectronics, biotechnology, and industrial materials while encouraging new linkages at the interface between universities and industries (Etzkowitz & Webster, this volume). Thus the values of the economic culture came to dominate. Third World countries followed suit, such as China, where science and technology policy reforms were strongly tied to the economic reform process that began in the late 1970s (Baark, 1991; Simon & Goldman, 1989).

Foresight became one of the central new policy methodologies. In the new committees and programs for advanced technology that were created in most countries, the state agencies tried to follow the Japanese consensual model of decision making (Martin & Irvine, 1989). The idea was to bring various actors together, chosen on the basis of their specialized knowledge, to visualize possible futures and select particular technological options that then became inputs in policy. Foresight thus involves a strong element of social construction, whereby governments or agencies connected to government seek to shape consensus among representatives from the academic, economic, and bureaucratic policy cultures. A similar process is conducted within industry under the rubric of "issue management." The introduction of such foresight methodologies in most Western industrial countries at the beginning of the 1980s suggests that there was a process of mutual accommodation taking place, which coincided with a weakening of the influence of the civic culture. The social movements declined in strength in most countries in the first half of the 1980s through processes of incorporation and marginalization, the more moderate demands being accepted and institutionalized, while the radical ones were in one way or another banished from the political agenda (see Vig & Kraft, 1984). In many countries, decisions were taken either to cut back on nuclear energy or to develop other sources of energy; at the same time, the corporate sector was developing a set of technological "solutions" to some of the environmental problems that had been so central to the public debates.

The project grant system that had developed in the 1970s had given more influence to the procedural managerial staff, which strengthened the alliance between the bureaucratic and economic policy cultures. Within academia, there emerged a critique of the increase in legal and administrative procedures and a call for less bureaucratic arrangements. Don Price (1979), an American science policy expert, suggested less direct yet more ambitious government efforts to guide university research:

> The most effective (though perhaps the most difficult) first step toward a long-range and integrated policy of governmental support for science and for universities might well be to work for a more disciplined system of responsibility within the Congress as a whole, and a more competent corps of generalist administrators (by no means excluding scientists) who could give the planning system of government more integrity and continuity. (p. 91)

This was essentially asking for the new approach of orchestration policy that began to make its mark in the 1980s. This formed part of what has been called a new social contract for science, replacing the older doctrine of "relative autonomy" that had been dominant throughout the postwar era and articulated so well in Vannevar Bush's report. This new contract is characterized by a stronger integration of academic science with the state as well as with the private sector, while at the same time emphasizing the importance of basic research (Kevles, 1990). Just as Vannevar Bush's report introduced a set of concepts that became ground-rule assumptions for the more technocratically oriented science policy studies literature, so also the "new contract" brings with it a set of new concepts that STS scholars appropriate, either critically or noncritically (Bimber & Guston, this volume).

With the shift in the external framework, there has emerged a new sensitivity in science policy studies to the internal cognitive dynamics of S&T. As new directions of research more obviously were established by factors previously considered "external" to the inquiry process itself, a number of concepts were developed to try to make sense of this. Pioneering work in this direction earlier, in the 1970s, had been that of the now disbanded Starnberg group in Germany that introduced "finalization theory." According to this theory, the inquiry process at certain times is more closed and at other times more open to external orientation, depending on various phases in an internal cognitive dynamics of science. During periods of strong theoretical development, it was postulated, disciplines will be relatively immune to external steering, while in pre- and postparadigmatic phases they will be more open; in the latter case it is because the discipline in question is held to be mature, an assumption that has been strongly criticized (Schäfer, 1983).

It has also been pointed out how in recent times there has evolved a social and cognitive differentiation into disciplinary communities and hybrid research communities, where the latter are policy generated on the basis of a consciously articulated social mandate and display reputational systems and criteria for the assessment of research results somewhat different than the traditional career patterns and criteria found in academic disciplines. The distinction is important for the analysis of differences between disciplinary and problem-oriented science, a consideration that is relevant for science policy. Essential here is what may be termed the *cognitive turn* in science policy studies (Rip, 1981), an approach further developed in some recent studies (Elzinga, 1985; Remington, 1988).

Significant in this cognitive turn are two further concepts: "collectivization" and "strategic research." The former refers to the process whereby Big Science involves researchers not only in teamwork but also in a process of integration over functional boundaries between science and decision making or production in society at large (Ziman, 1987). The latter concept has been defined by Martin and Irvine as "basic research carried out with the expectation that it will produce a broad base of knowledge likely to form the background to the solution of recognized current or future practical problems" (Irvine & Martin, 1984, p. 4).

The shift in the contract between science and society prompted a number of historical studies reviewing the institutional history of science and earlier experiences of university-industry relationships (Thackray, Sturchio, Bud, & Carroll, 1985). There was also a renewed interest in the history of industrial research (Dennis, 1987) as well as a number of policy-driven comparisons of so-called national systems of innovation (Nelson, 1984).

CURRENT TRENDS AND RESPONSES

In the 1990s we see a greater awareness of each country's particular characteristics as governments increasingly try to harmonize their practices and policies in OECD member countries in activities where potential sources of friction (subsidies, standards, and patent systems, for example) may be minimized. Science and technology policy, we are told by an OECD spokesman, "has to be related to both national contexts and to global change. Governments must plan their action more carefully in the social and institutional setting of their own countries and simultaneously make better use of science and technology policies to help solve the problems emerging in a rapidly changing world" (Aubert, 1992, p. 4; see OECD, 1992). *Globalization* and *indigenization* have become the new key words of the present decade. Intriguingly, this is at a time when "think globally, act locally" has become a popular slogan in nongovernmental organizations (J. Clark, 1991). The advent of new technologies and basic targeted research calls for policy analysis that is more sensitive to the cognitive and cultural dynamics of science. Differences in the sociocultural settings of science is an area where STS scholars drawing on the newer sociology of science are well equipped to provide important insights relevant to policymakers and managers of R&D programs.

A major motive force behind the new social contract of science-society relations has been the changing nature of global economic competition. The perceived threat of a Japanese challenge has led many countries to try to improve their own industrial competitiveness, and rivalry has developed between the United States and Europe. In the United States the "star wars" program, or Strategic Defense Initiative, helped serve this purpose (Elzinga, 1990), while the European response has been through the EUREKA program for coordinating research efforts on a continental scale. The latter is now part of a more general transnational policy framework for science and technology that operates at four distinct levels of activity: basic research, higher education, technological development, and standardization of regulation and assessment of technological impacts, particularly in relation to environmental protection. It is becoming increasingly common to refer to a European science and technology policy and, among STS analysts, to identify a special mission for Europe in the years to come. In the words of Sharp and Walker at SPRU, "The current conjuncture of events offers we believe an opportunity for Europe to take the lead in three important areas of policy—in gearing technology towards sustainability; in organising a 'Marshall Plan' for eastern Europe; and the setting up of a new international competition authority" (Freeman, Sharp, & Walker, 1991, p. 396). Altogether this motivates an increase in targeted projects and strategic research, requiring greater financial commitments from member countries and thus increasing the risk of political tensions over S&T policies (*New Scientist*, March 14, 1992). At the same time, the efforts to orchestrate a process of new institutional reforms tend to effect a more intimate integration of academic and economic policy cultures, while the civic culture articulates a critique of a sectorization of S&T policy at a higher—that is, European—level, locking individual countries into a technocratic mode of development steered by the new bureaucracies concentrated in Brussels.

The intensification of economic competition through high technology can be expected to lead to more conflictual relations among the Western countries, and perhaps especially between the three blocks of North America, Europe, and east Asia. The previous era of superpower animosity is being replaced by a more tripolar, or even multipolar, world in which regional traditions, as well as ethnic and religious identities, can come to play a more important role in S&T policy agendas.

Superimposed on this "regionalization" of S&T policies are several other global trends that should be noted. One is the centrality of biotechnology, with biology replacing physics as the core discipline in technoscience. Second, there are the new environmental problems that are having a significant impact on research agendas. Also, with the end of the cold war and the transformation of eastern Europe, the old East-West conflict has given way to a sharpening of the contradiction between the industrialized countries of the "north" and the developing countries of the "south," with major repercussions in the realm of S&T. We shall conclude this chapter by returning to the four policy cultures and briefly identify their responses to the aforementioned trends.

Increased commercialization and the spread of commercial values in the academic culture have evoked debate over several issues, among them fraud and corruption, scientific quality and integrity, as well as research ethics more generally. The concern about the weakening of internal peer review criteria in project selection prompted by the SDI has in some other sectors broadened to controversies over the validity of scientific results. In the defense sector the notion of spin-off is being replaced by a new catchphrase, *dual use,* while the conversion of military programs to civilian use continues to be hotly debated. In this process, the growing awareness of the difference between the bureaucratic and academic cultures will be significant in the planning and design of future institutions and policy programs. Policy analysts will have to increase their sensitivity to differences in cultural value systems and institutional arrangements.

It is clear that military investments in R&D no longer provide a guarantee of technological leadership in the international arena, in terms of neither weapons systems nor trade (Brooks & Branscomb, 1989). As a recent editorial in *Science* noted,

> Bringing profitable products to market requires a different set of motivations, organization, and culture than does the creation and dissemination of knowledge

> . . . . Consortiums and centers that include participation of several companies may be effective in basic science and in addressing generic issues in the development of new technologies, but they are said to be unlikely to have a significant role in the process of innovation. Companies are less and less interested in participating in industrial affiliates' programs or in consortiums. (Abelson, 1991b, p. 9)

The new entrepreneurial ethos that developed, perhaps most rapidly and controversially in the field of biotechnology, has collided with many of the traditional values of the academic culture (Kenney, 1986). Biotechnology has also brought to a head a new set of issues centering on questions of intellectual property, commodification, and genetic diversity and has inspired the emergence of the new field of bioethics. Particularly significant during the latter half of the 1980s were the policy questions associated with the attempt to map the human genome, one of the largest R&D efforts ever undertaken. Both the commercialization of biotechnology, as well as the organizational aspects of the human genome project, have given new substance to the older distinction between big and little science and point to the fact that biology is increasingly replacing physics as the privileged scientific area in the corridors of power and is the main source of imagery in the broader public discourse over science and technology policies and institutions.

The civic culture has been at a disadvantage when it comes to defining the terms of reference for the policy discourse. In the second half of the 1980s, however, as environmental issues received growing concern, a new, more professional environmentalism became an increasingly important factor in several countries and, perhaps especially, in international forums (Caldwell, 1990; J. McCormick, 1989). The pressure from environmental interest organizations, such as the World Wildlife Fund and the International Union for the Conservation of Nature (IUCN), led to a spate of initiatives. The new environmental problems—global warming, destruction of the tropical rain forests, biodiversity, and so on—led to a number of intergovernmental negotiations, signaling what one participant has called the coming of "ozone diplomacy" to the world of international relations (Benedick, 1991). At the present time, it seems fair to say that the commercial emphasis has slowed somewhat, as new environmental and distributive issues seek to place themselves more centrally onto the policy agenda. Much of this can be seen as a reflection of new alliances that have formed between critical academics, particularly feminists and environmentalists, with ever more professionalized "movement" organizations. In eastern Europe too, concern with pollution and public health was one of the driving forces behind the anticommunist movements, and the nuclear accident at Chernobyl in 1986 was, in many ways, the beginning of the end of the communist bureaucracies.

The emergence of postcolonial voices and movements in developing countries can also be expected to have a major impact on the attempts to repoliticize science policy in the industrialized world. The critical movements, as well as the critical intellectuals in developing countries, have in the course of the 1980s drawn special attention to the plight of women in developing countries. In the words of Vandana Shiva (1988), an STS researcher who has become one of the more influential environmentalists in India:

> Women producing survival are showing us that nature is the very basis and matrix of economic life. . . . They are challenging concepts of waste, rubbish and dispensability as the modern west has defined them. . . . They have the knowledge and experience to extricate us from the ecological cul-de-sac that the western masculine mind has maneuvred us into. (p. 224)

What is clear, at this point, is that the process of agenda reformulation can be expected to lead to new institutional initiatives, new doctrines and concepts about science and society, and new relations among the sciences. The difficult problems of reconciling cultural traditions with technological development, for example, can lead to a new role for the humanities; and the task of mobilizing knowledge resources in an increasingly competitive world can also inspire new educational efforts as well as new philosophical and theoretical syntheses. As the world changes, so too will the world of science and technology policy studies.

NOTES

1. The contribution of Erik Baark has been important in developing the concept of policy cultures (see Baark, 1991).

2. Similar arrangements were developed in other countries, frequently influenced by the discussions within the OECD, which started to publish national science policy reports based on visits by expert "reviewers." Parliamentary groups with an interest in science policy emerged, and in the U.S. Congress hearings were initiated in 1964 on science and technology. Congress also stimulated the National Academy of Sciences (NAC) to set up a committee on science and public policy (COSPUP) in 1963.

3. A conference in 1963 led to the creation of the Advisory Committee on the Application of Science and Technology to Development, and a series of ministerial conferences took place from 1965 onward. As a result of the connection between UNESCO and the IAEA, Professor Abdus Salam founded the International Centre for Theoretical Physics at Trieste in 1964. UNESCO also started its own unit for science policy led by the Belgian Yves d'Hemptine, who was instrumental in launching the International Comparative Study on the Performance of Research Units (ICSOPRU), a positivist quantitative survey of R&D productivity carried on in a large number of countries through the 1970s.

26

□

Science, Technology, and the Military

Relations in Transition

WIM A. SMIT

IN his advice to the president immediately after World War II, Vannevar Bush (1945) stated: "There must be more—and more adequate—military research in peacetime. It is essential that the civilian scientists continue in peacetime some portion of those contributions to national security which they have made so effectively during the war" (p. 6). His advice stood in sharp contrast to Thomas Alvin Edison's suggestion, many years before, during World War I, to the Navy, that it should bring into the war effort at least *one* physicist in case it became necessary to "calculate something" (Gilpin, 1962, p. 10).

In the highly organized and concentrated effort during World War II, a great number of scientists, mainly under the auspices of the newly established Office of Scientific Research and Development, had contributed to the development of a variety of new technologies (atomic bomb, radar, proximity fuse, but also penicillin). The decisive contribution of scientists to these war efforts implied a dramatic turn in the role for science and technology in future military affairs. It was on this experience that Bush could build his advice.

AUTHOR'S NOTE: I want to acknowledge the very valuable comments and suggestions by Arie Rip and Judith Reppy on previous versions of this chapter.

For the first time in history, military R&D became a large-scale institutionalized process in peacetime, indeed on a scale not seen before, legitimized as well as fueled by the climate of the cold war. In the decades following the war, weapons were replaced in a rapid process of "planned obsolescence." The R&D was carried out in national laboratories, defense industry, laboratories of the military services, and at the universities to a varying degree in different countries.

It has been estimated that annually some $100 billion is spent on military R&D and that the number of scientists and engineers involved is about 600,000 (Sen & Deger, 1990). In a United Nations study in 1981, it was estimated that at least 20% to 25% of all R&D was devoted to military purposes and that the same percentage of all scientists and engineers are engaged in military R&D (quoted in Thee, 1988). Most of these expenditures—for instance, in the United States, about 80%—are in the category of development and engineering.[1] Military R&D budgets have been most substantial for the United States, the former Soviet Union, France, and the United Kingdom. Less, but still substantial, is the portion in Germany, whereas Japan's expenditures on military R&D have always been low, though they have been increasing in recent years (to $750 million in 1990; see SIPRI Yearbook, 1991, p. 161).

This situation has led to the following two concerns about the role and impact of military R&D:

- The impact of military R&D and defense relations on academic research (mainly a concern in the United States) and on the course of scientific and technological developments more generally
- The exploitation of science and technology for military purposes and its impact both on national and on international security (which do not necessarily coincide)

Since the appearance of Sapolsky's chapter "Science, Technology and Military Policy" in the 1977 STS handbook, a number of far-reaching developments in quite distinct areas have occurred, most of them of the past few years, that place these traditional concerns in a new perspective.

First, the (institutional) role of the universities within the whole spectrum of R&D activities has changed in the past decennia and is still changing. This change is to be understood against the background of the pervasive role of R&D in modern society. While it has been generally recognized that the position of science and technology in society has fundamentally changed since World War II, these changes may now be considered as part of a transition phase toward a structurally new role in society of both the R&D system and the universities. This change may have consequences for the organization and character of not only civil but also military R&D.

Second, the radical political, military, economic, and social changes in eastern Europe and the former Soviet Union, accompanied by new relations between "East" and "West," have made it necessary to rethink the role of the military. This applies to the whole spectrum of doctrines, strategies, weapon systems, and military technology. In particular, old notions of arms race dynamics and the role of military R&D are coming under scrutiny.

Third, the relation between civilian and military technology is coming to be viewed from a new perspective. Traditional questions of spin-off and spin-in, are, for several reasons, now rephrased and analyzed with a somewhat different focus. First, civilian technology and R&D have acquired a "strategic" meaning; national security is now often seen as having both a military and a civil industrial/technological pillar. Second, the relation between civil and military technology has become a policy issue: In the "West," the issue of the integration of civilian and military technology is now on the agenda, whereas in the former Soviet Union and middle and eastern European countries, the question of conversion of military industry toward civilian activities is most pressing.

Fourth, various new approaches in STS studies have provided insight into developmental processes of science and technology in interaction with their social context. "Opening the black box" of technology, "social shaping" of technology, the "seamless web" of ST&S, removing interdisciplinary boundaries of sociology, philosophy, history, and (more recently) economics, as well as network approaches in the study of both developmental processes of science and technology and their impact on society have all contributed to these insights and will continue to do so. Studies from these perspectives on military technology and military R&D, however, have only recently begun (Elzen, Enserink, & Smit, 1990; Gummett & Reppy, 1990; MacKenzie, 1990a).

Fifth, in the past, many weapon impact assessments, both by governmental and by nongovernmental organizations, have been made within the framework of arms control endeavors. If they had influence, it was on the procurement of weapons, and not on the direction of military R&D and the military technological innovation process, basically because no coupling, institutional or otherwise, existed between the assessments and the innovation process. Since about 1990 a shift of attention toward influencing the earlier stages of the weapon innovation process can be noted (J. Dean, 1992; Greenwood, 1990; Hamlett, 1990b; Lakoff & Bruvold, 1990; Smit, 1989, 1991; Woodhouse, 1990).

The two traditional concerns will be taken as the starting point for organizing this chapter. The concerns themselves will be articulated, and the changes in focus will be discussed. The main tasks will be to discuss how far STS studies have provided relevant insights for approaching these concerns and to formulate a number of themes that might benefit from being addressed by future STS studies.

We will start with discussing the theme of military-university relations and then turn to the second theme, the exploitation of science and technology for military purposes.

THE UNIVERSITIES AND THE MILITARY

Two Types of Links: Advisers and Academic Research Funding

The involvement of the military with the universities varies greatly between different countries, and also over time. In the United States it has been relatively strong since World War II. Controversies that at times arose over these links have also been most pronounced in that country. The focus of this section therefore will be on the United States.

One main military-university link is faculty advisorships to the DoD, military laboratories, or the defense industry. This became institutionalized practice in particular after World War II (Gilpin, 1962; York & Greb, 1982).

Well known is the JASON group, consisting of excellent scientists, forming a "second generation" advisory pool for the military, often advising through "summer studies" (Kevles, 1978, p. 402).

No figures are available for the number of scientists that are serving as advisers to the military, but of importance is that often outstanding scientists are involved. In the United States, according to DeLauer (1989, p. 133), many university faculty members play an important role in the definition and direction of military R&D through such advisory bodies as the Defense Science Board, the Army Science Board, the Air Force Scientific Advisory Board, and the Naval Research Advisory Committee.

The other important link is military funding of science and engineering R&D at the universities, in the form of either research grants or funding of special research institutes. Such federally funded research and development centres (FFRDCs) are part of or administered by the university (see, for two case studies, Dennis, 1992). In 1966 eight such military-supported FFRDCs existed in the United States, but by 1975 five had been phased out or decoupled from the universities mainly, but not only, due to student protests. The Charles Stark Draper Laboratory, for instance, was separated from MIT, and the Stanford Research Institute decoupled from Stanford University (Dickson, 1984, p. 121; Gerjuoy & Baranger, 1989, p. 74).

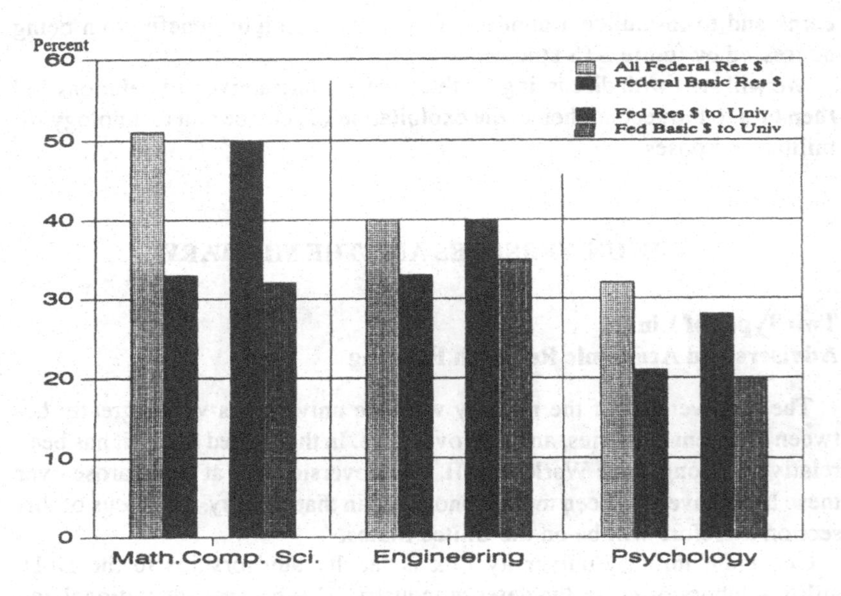

Figure 26.1. Leading Fields of DoD Research Support, Basic and Applied: 1986-1987

SOURCE: CRS, prepared using NSF data.

Some of the military-funded research, mainly in the categories of engineering and applied research, rather than basic research, is classified. It is mainly, but not exclusively, carried out in the FFRDCs. Publication of nonclassified research might, however, be subject to the sponsor's consent.

What is the size of military research at the universities in the United States? DoD funding of university research has fluctuated considerably over the past four decades. It increased in the first period after World War II and doubled (in constant [FY 1987] dollars) between 1960 and 1964 to a maximum of $1.063 billion, then decreased to a minimum of $0.451 billion in 1975, after which there was another increase, by a factor of 2.7, between 1975 and 1986 (Dickson, 1984, pp. 123-134; Gerjuoy & Baranger, 1989, p. 61; Kistiakowski, 1989, p. 143). Knezo (1990, pp. 24-25) cites NSF data showing that, averaged over 1986-1987, DoD support constituted 8% of total academic R&D funds received from all sources—public and private—and 32% of academic engineering R&D received from all sources. This support is very unevenly distributed over various fields and academic institutions. Heavily dependent on DoD funding are, for instance, electrical engineering (51%), aeronautical/space engineering (42%), and, to a lesser degree, mathematics

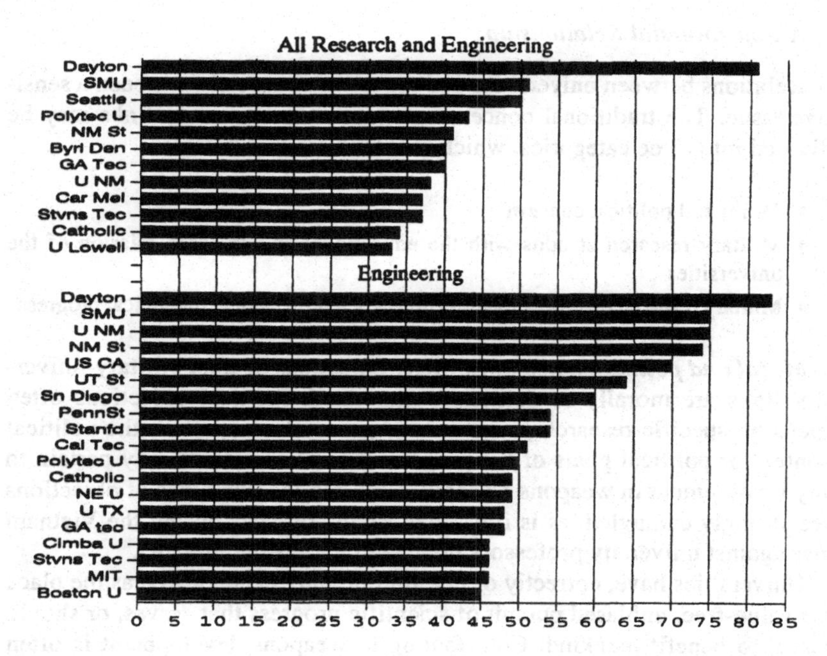

Figure 26.2. DoD Funding as a Percentage of All Funds Universities Received: 1986-1987

SOURCE: CRS, prepared using NSF data.

(15%) and computer sciences (14%). Less dependent is, for instance, psychology (6%), but DoD still supplied 26% of all federal research funding of academic psychology. A number of universities are to a considerable degree —typically between 30% and 55%—dependent on DoD funding for research and engineering. Most extreme in this respect is Dayton University, with a dependence of more than 80% (Gerjuoy & Baranger, 1989, p. 69; Knezo, 1990, p. 26). (See Figures 26.1, 26.2, and 26.3.)

No systematic data are available of academic R&D dependency on military funding in other countries. In the United Kingdom defense-funded research at the universities amounted to 10 million British pounds in 1984 (Gummett, 1988) and has increased to £19 million in 1990, which is still not more than about 1% of all governmental spending on military R&D (compared with 3.6% in the United States; see Wilson, 1989b, p. 10) and about 2% of government-funded R&D at the universities (U.K. Parliamentary Office, 1991).

A Controversial Relationship

Relations between universities and the military have always been a sensitive issue. The traditional concerns over military-university links may be divided into three categories, which will be discussed below:

- Moral and political concern
- Military research at odds with the educational and research mission of the universities
- Military research distorting the direction of scientific and technological progress

Moral and political objections. Many objections against military-university links are morally or politically motivated. Political objections often focus on specific research programs and generally depend on the political context or political goals of the program. Moral objections may pertain to any involvement in weapons research, but often the two types of objections are strongly entangled, as is illustrated by the protests during the Vietnam war against university professors involved in advising the DoD.

Universities have, correctly or not, traditionally been viewed as the place for value-free, unbiased pursuit of scientific progress that serves, or should serve, to benefit mankind. Contributing to weapons development is often considered to be at odds with this ideal.[2]

The fierceness of controversies over military-university links depends much on the (political) context. Thus contributions by scientists to the widely supported war efforts during World War II were seen as a duty rather than a debatable activity. And in the years after World War II, Vannevar Bush's advice for a continuation of the contribution by civilian scientists to military R&D gained broad support. It led to a substantial military funding of basic research at the universities (though its utility for defense was later doubted, as succinctly phrased by the Secretary of Defense Charles E. Wilson, in the mid-1950s: "Basic research is when you don't know what you are doing"; Kevles, 1978, p. 383). The deep controversy in the United States over the Vietnam war during the 1960s and early 1970s implied a vehement conflict over the various links between universities and the military (Dickson, 1984; Kevles, 1978). By contrast, the late 1970s and early 1980s again showed an increase in funding for defense-related research at the universities, which was welcomed, not in the least because of diminishing funding from other sources. The more focused SDI program, however, amplified the existing concerns of many scientists with the (uncontrolled) technological arms race (that received a new impetus with the ending of détente at the end of the 1970s) and created a lot of opposition to university participation (Kaysen, 1989, p. 37).

STS studies cannot resolve moral or political conflicts over university-military links, but some issues might, however, be clarified by STS studies. Indeed, the involvement of the military with the universities has been a subject of STS studies. Often these studies originated from a concern about the influence of the military on the character of the university as an institution of free research and education or about the use of science for destructive purposes. The aim of such studies often was to enhance consciousness and responsibility of faculty as to the nature and use of their research. These studies are to a great extent descriptive (compare Nelkin, 1972). Very few analytical STS studies exist that, specifically for military research and advisory tasks, analyze the relation between science and politics, the influence of faculty scientific advisory work, and the question of social responsibility of scientists. One early example is Gilpin's (1962) study of the role of American scientists in nuclear weapons policy (see also Leitenberg, 1973). On the other hand, many analytical studies exist on the nature of (scientific) controversies and expert advisory roles in other areas[3] that have certain relevance for the debate on military-university links.

Issues that played a central role in these debates are *objectivity* and *neutrality* of science and whether science is *value-free*. As the president of the World Federation of Scientific Workers, physicist Eric Burhop, concluded in 1974 in an exchange of letters with scientists involved in JASON activities during the Vietnam war:

> One suspects that many drifted into the work because it was technically interesting, challenging and lucrative and was satisfying for their ego. It is this so-called "value-free" attitude that has aroused suspicion about scientists among a section of the lay public, and especially among the youth. (p. 22)

Years before, in 1952, the physicist Condon had already warned:

> If in the years to come science and the scientists are closely identified in the public mind as the wizardry and wizards who have made all the fantastic new weapons of mass destruction . . . a horror and revulsion of war may, in that illogical and irrational way that so many things go in politics, be extended to science and the scientists. (Kevles, 1978, p. 369)

The position implied by Condon—that science merely provides "facts," that facts and politics can be neatly separated, and that scientists cannot be held responsible for the way scientific results are used—would be hard to defend in practice.

A variety of STS studies have illustrated that, even to the extent that scientific knowledge can be considered objective (in the sense of intersubjective validity or consensus), it is not neutral with respect to (its use in) politics. This is not only because, for instance, in advice, selections have to be made as to what is relevant scientific knowledge but also because "translations" must be made into policy terms or areas outside the scientific discipline for the knowledge to be useful. This entails a positioning with respect to both specific cognitive biases ("problem definition") and the various social groups involved in a controversy. York (1971), for instance, has in the case of the controversy over the antiballistic missile (ABM) system, neatly demonstrated that the question, "Will it work?" may have different meanings, dependent on the problem definition. It is often only within the framework of a particular problem definition that an acceptable separation of relevant "facts" and politics is possible.

Military research and advisory roles therefore are never neutral; they always contribute and are of relevance to one problem definition rather than another. Clarification of the different problem definitions involved in controversies over military research and advisory roles may therefore contribute to (though not resolve) the moral and political debate on military-university links. The important role of problem definitions also makes clear why acceptance of such links may change (as was illustrated above) and is dependent on the broader political context.

The picture of science that has emerged from STS studies, as a highly organized endeavor, also has implications for the issue of social responsibility. Currently the scientist's responsibility has been highly individualized. The scientist is not only held personally responsible for participating in (or not) or advising (or not) on military matters, but it is also generally assumed that he or she should be free to do so if he or she chooses, without, for instance, interference from the university administration. However, as science has become a highly organized endeavor, then "social responsibility" has also become an issue that requires organized activity too, which cannot be left merely to the "helpless" individual scientists. Therefore, without detracting from individual responsibility, it should be recognized that as an issue of social responsibility—military research at the universities—cannot be addressed solely by referring to the scientist's individual responsibility.

Divergent missions of the military and the universities. The second type of concern over relations between universities and defense agencies concerns their different missions and cultures—the military's main mission is to provide an adequate national defense; that of the universities is the education of students and carrying out research for the advancement of science and technology. Since World War II, advanced, science-based military technology is considered to be indispensable for defense, and the pursuit of technological superiority has become central in modern military thinking and defense policy. Military funding of university research is seen as a means not only to have access to the forefront of S&T developments but also to push advanced research into directions of interest to the military, though it should be added that there was not always agreement within the defense establishment on the utility of military support of basic research at the universities (DeLauer, 1989, p. 136). Moreover, the defense establishment and the major defense suppliers employ some 13% of all U.S. scientists and engineers (Kaysen, 1989, p. 30). Thus, as a major user of trained scientists and engineers, the defense establishment has an interest in the quality and type of research (e.g., relevant to weapons work; Kistiakowski, 1989, p. 145) carried out at the universities, just like other governmental agencies and civilian industry.

For the universities, openness, free communication, and criticism are considered to be of essence for their educational mission and as vital for the advancement of science. Thus we can see what unites universities and the military and also what divides them. To the university faculty and administration, military funding meant additional resources for research, availability of advanced equipment, and often access to knowledge of advanced developments in scientific and technological areas not available elsewhere (Kistiakowski, 1989, p. 145). However, the perceived incompatibility of classification with the universities' mission was for most of the universities a reason for banning almost all, or at least closely regulating, classified research (Wilson, 1989a, p. 51). In the 1960s most of the DoD-funded FFRDCs, which were often involved in classified research, were split off from the universities. The latest clash between the universities and DoD was in the mid-1980s, when DoD-funded research at the universities had again increased substantially, partly in relation to the SDI program. DoD's striving for strict control of the flow of sensitive scientific and technological information not only included restriction of publication on DoD-funded (unclassified) research but also prior reporting of discoveries in certain fields even if not funded by the government, the exclusion of foreign scientists from scientific conferences, and the exclusion of foreign students from "sensitive" research at the universities. The strong opposition from the universities and scientific professional organizations caused the DoD to retreat: It was agreed that military research conducted by the universities should be unclassified unless specifically excepted at the initiation of a project, and that unrestricted dissemination of research results in conferences and publications was permitted (DeLauer, 1989, pp. 138-139; Kaysen, 1989, p. 37; Wilson, 1989a, pp. 55 ff.).

Military influence on the development of science. Frequently the suggestion has been made that military funding has significantly shaped scientific development by stimulating the growth of specific fields and addressing particular research questions, thus causing a skewed development of science.

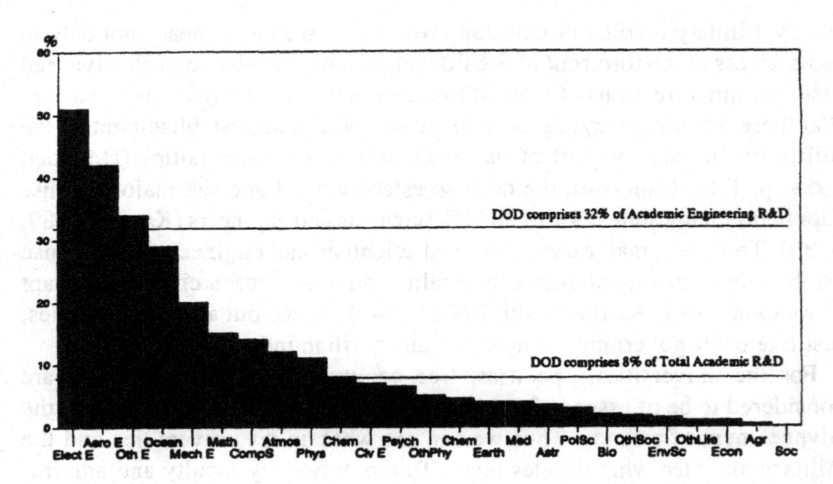

Figure 26.3. Importance of DoD to Academic R&D Fields: 1986-1987
SOURCE: NSF (1989).
NOTE: Based on 1986-1987 average R&D.

Rather than as a normative question, this theme should be addressed as an analytical-empirical question of the social shaping of science. (For what would be an unskewed development? Even research directions being dictated by the state of knowledge in a scientific discipline involves value judgments.) This issue differs from the other question, asked much less often but interesting from an STS point of view, which concerns whether military funding has also shaped the concepts and theories of science. It is this type of question that has been addressed by a number of social studies of science of the last two decades, though not with regard to military influences. These studies are part of the "strong program" in social studies of science or have been designated as "social constructivist" approaches (Pinch & Bijker, 1987).

Whereas systematic data exist on the degree that various academic fields in the United States have been funded by the DoD (see Figure 26.3 and Table 26.1), there is very little systematic evidence on the degree to which military funding has influenced the emergence and growth of scientific fields. Several examples of such an influence, however, have been suggested. Thus Gerjuoy and Baranger (1989) conclude that, due to the "merit review" process, "support by the mission orientated DoD agency is significantly but by no means overwhelmingly skewing research directions and subfield growth in all the quantitative sciences" (p. 71). Their main argument is the degree of dependency of the various fields on military funding. More specifically, Vera Kistiakowski (1989) mentions a disproportionate growth in materials science within the physical sciences, as well as in cryptology within mathematics,

TABLE 26.1 Analysis of DoD Support for University Research

	DoD Support for University Research (Millions of 1987 dollars)		*Percentage Share of Federal Support for University Research, Fiscal Year 1986*		*Percentage Share of Federally Supported Graduate Students, Fall 1986*	
	Fiscal Year 1976	*Fiscal Year 1986*	*DoD*	*NSF*	*DoD*	*NSF*
Physical sciences[a]	60.0	101.0	13.0	37.9	12.0	38.8
astronomy	0.6	6.3	9.0	39.6		
chemistry	15.3	44.6	16.3	38.9		
physics	41.0	49.2	11.8	37.8		
Mathematics	18.6	36.2	33.3	48.6	43.0	36.0
Computer science	23.7	118.2	67.8	22.3	55.8	30.0
Engineering[b]	112.8	284.2	45.0	28.7	33.1	25.2
electrical engineering	37.4	93.9	65.6	28.6		
mechanical engineering	15.6	48.1	56.3	27.5		
metallurgy and materials engineering	25.5	66.4	48.3	28.9		
Environmental sciences	61.7	92.4	24.5	55.2	15.0	44.8
Total (all fields)[c]	369.2	730.7	12.6	17.7	15.3	21.0

SOURCE: For information on support for university research: National Science Foundation (n.d., 1987). For percentage share of federally supported graduate students: NSF (1988).
a. Includes fields "not elsewhere classified" in addition to listed fields
b. Includes all engineering fields.
c. Includes all fields funded, in addition to listed fields.

whereas Forman (1987, pp. 200 ff.), in a detailed study, argues that military support pushed physics into quantum electronics and solid state physics to the detriment of other issues and caused an emphasis on applied rather than pure science. DeLauer (1989, p. 133) and also Flamm (1988) point at the DoD as the sponsor of major computer development and computing application programs since World War II, including advanced academic research efforts sponsored by DARPA on computer time sharing, packet switching, and the fifth generation computer project (which the U.S. computer industry, but not the Japanese, failed to exploit, according to Delauer).

It seems trivial that substantial funding has provided the possibilities for research activities in specific fields. The point is whether the distribution of research funding over fields or subfields has been mainly determined by scientists themselves or by the military funding organizations. Or, if scientists had a great say in the awarding of research funds, to what extent was their advice rooted in questions raised by the state of the art in a particular field or by the external mission? Would scientists have made different proposals or have given different advice when the research money had been provided by, say, the National Science Foundation?

From this viewpoint, further investigation of military and civil funding of, in particular, basic research (budget category 6.1) would be of interest. One should note that military support for basic research has changed over time, being the largest and not very selective in the first decade after World War II (after which it decreased and emphasis on engineering and technology development increased). Also of interest is whether the Mansfield amendment of 1969 (stemming from congressional concern about the relevance of basic research funded by the DoD), which stipulated that DoD-funded research projects should have "a direct and apparent relationship to a specific military function" (later softened by Congress into "potential relationship to a military function"), had any influence. Although intended to reduce the DoD's role in the shaping of basic research, one could, according to Kaysen (1989), hardly imagine "a more effective way of assuring that DoD's enormous financial sources would increasingly skew research to defense priorities" (p. 27). This claim, however, has not been substantiated by Kaysen.

In addition to DoD funding of research, participation of scientists in advisory committees might have influenced their research interests, according to the mechanism formulated by York as follows:

> Participation in these committees led to a remarkable and important synergism. I, and other laboratory leaders—both non-profit and industrial— . . . developed an understanding of the likely characteristics of future military requirements . . . we also acquired a first-hand working knowledge of the doctrines and strategies underlying our development and procurement plans . . . which each of us used to steer the course of the programs for which we were responsible." (Quoted in Kevles, 1988, p. 467)

More intricate is the question of whether a possible influence of the military on the course of science at the universities in the United States also had an effect on science development on a global scale. Here not only quantitative but above all qualitative contributions might be decisive. Also relevant is to what extent research elsewhere has been triggered or stimulated by U.S. DoD-funded research, thus exerting influence in an indirect—that is, "bandwagon"—way.

The normative question is whether one should designate developments influenced by DoD as "skewed." The underlying presupposition seems to be that science should be developed for its own sake and that therefore the course of development should be determined merely by an "internal" dynamic. However, there are no unambiguous (internal) criteria to decide whether research questions from one field are more interesting than those in another field. In practice, "external" criteria play, either implicitly or explicitly, a role in the decision-making process.

Another normative concern was about a shift at the universities from basic research to applied research due to DoD funding, thus hampering the advancement of science. Again it is not clear whether there has been such an effect. In the next section, moreover, we will discuss the normative base of this concern in relation to the changing role of the universities.

A different, and even more difficult, question is whether military funding has not only influenced the emergence and growth of specific (sub)fields but whether it also had an impact on the theories and basic concepts in these fields. The principal possibility of such a shaping has been shown for a variety of scientific fields and social contexts[4] (Latour, 1987; Stemerding, 1991). Very few claims that this has happened have been made for military influence, but one is from Forman (1987) for what he calls "a preoccupation with population redistribution, with modification of the normal or equilibrium populations of the various energy states of physical systems" (pp. 222 ff.) in nuclear, solid state, atomic, and molecular physics. He considers this a typical offspring of wartime and postwar military research interests and the result of a shift in epistemic goals toward "a kind of instrumentalist physics of virtuoso manipulations and *tours de force*, in which refined or gargantuan technique bears away the palm. It is just such physics as the military funding agencies would have wished." Not everybody, however, shares this interpretation (Gerjuoy & Baranger, 1989).[5] Very few studies within the "strong program" tradition have been devoted to a possible influence of the military on the nature of scientific theories. Subfields within computer science, which are substantially funded by the DoD, such as artificial intelligence and neural networks, because of the specific application context of their development, might be worthwhile domains for the study of such influences.

Implications of the Changing Role of Science, Technology, and the Universities in Society

One of the concerns cited above was about a shift from basic research to applications. The assumption here, as in the U.S. DoD budgetary categories 6.1-6.6, is that research processes can be divided into steps that make up a linear trajectory from basic research to technological application (Long & Reppy, 1984). For the development of many science-based technologies, there is no such linearity and R&D processes often show more of the characteristics of an Echternachter procession with much feedback (Böhme, van den Daele, et al., 1978; E. Layton, 1977, pp. 208-213; Pinch & Bijker,

1987; Stankiewicz, 1990). This pattern is well illustrated by the development of various lasers in which military research efforts were dominant (Seidel, 1987). Moreover, the production of scientific and of technological knowledge is often not separate activities and is often carried out in the same institution, with the same people involved (Galison, 1988; Stankiewicz, 1990). A general claim that the universities should stick to basic research is thus a claim for an artificial category, possibly excluding many interesting questions. In many cases a distinction between research and development, or between basic and applied research, will be deemed to fail because of lack of operational definitions and because often they differ more in intent than in content (L. Young, 1989).

Also, the evidently changing role of the universities in society—their mission—is relevant to this issue. Etzkowitz (1990) has pointed out that, after the first academic revolution in the United States at the end of the nineteenth century, in which research was added to the then-dominant task of teaching, a "second academic revolution" is now taking place. The change is in the adoption of an economic development function, which implies strong linkages with industrial companies, local governments, and other organizations. As in the first academic revolution, the traditional functions remain important, but a further extension of the tasks of the universities results. As a consequence, more emphasis is now put on what is called "strategic research"—research of strategic importance to societal issues—the economy, health, environment, defense, energy, and so forth. Thus, when science is pursued not merely for its own sake, the question of how much effort should be devoted to what is called basic research and how much to more application-oriented research becomes one of strategy rather than of principle. The adoption of this "service" task actually implies still another basic, new task for the universities: to manage and integrate these various tasks (education, advancement of science and technology, and research as a service to society) within the same institution. One aspect of this is that science has become a commodity (Gibbons & Wittrock, 1985) not only to industry but also to the universities. Thus the issue of secrecy is connected not only with military research but also more and more with patent-sensitive research and research in cooperation with industry. The way to deal with the issue of secrecy and science as commodity is also one of the new tasks of the university. Here, ironically, the need for openness in military research may be more urgent than for industrial research, not for "internal" reasons of scientific progress but because of the importance of transparency in military affairs as a "confidence building measure" in the new international security environment (Charter of Paris for a New Europe, 1990/1991; Smit, 1991).

This new framework for discussing the traditional concern over the "shift" at the universities from basic to applied research also puts in a different light the normative question of "skewing" through links with the military. Still, one main task for the universities would be to organize so that an independent and critical role of the universities is preserved in an environment of parochial and partisan interests connected with military funding. Actually one argument for links with the military given in the past was the need for an independent view, different perspectives, and a civilian influence on military R&D.

I do not suggest that this new perspective on university-military linkages would resolve all of the traditional moral and political issues regarding military research. One might even question whether military-funded research at the universities is necessary from a national defense perspective, especially as other advanced countries sustain such links to a much lesser degree. A comparative study of the United States and other countries might feed the discussion of the necessity of such links. Military R&D in the universities may be of more importance to universities as institutions than to the military, as most of the military R&D is carried out in defense industry or dedicated military laboratories and governmental institutes.

SCIENCE AND TECHNOLOGY FOR DEFENSE

The declared purpose of weapon innovation is enhancing national security (often broadly interpreted, including intervention and power projection). However, many have argued that in the past the steady increase in U.S. military power (expressed in terms of the quantitative and qualitative growth of weapons systems) actually resulted in a rapid decrease of national security (e.g., York, 1971, p. 228). Many studies, often carried out by scientists who had become concerned about the escalating arms race, have dealt with the impact of new weapons systems and new military technologies on national and international security. They were fed into the public discussion with the aim of containing the arms race and reaching international arms control agreements or preventing their erosion. Such impact analyses were begun in the second half of the 1960s on antiballistic missile systems (Garwin & Bethe, 1968), multiwarhead missiles (MIRVs), and strategic missile systems more generally (York, 1973).

A huge literature assessing the impact of emerging weapons and military technologies has appeared since. Most deal with nuclear weapons developments and the proliferation of nuclear technology as well as with the militarization of space. Other areas of concern have been chemical and biological weapons. Much less attention has been paid to conventional armament. In the 1980s both the United States and NATO emphasized the importance of a stronger conventional defense, which was believed to be feasible then due to new "emerging technologies," including sensor and guidance technologies, C^3I technologies (such as real-time data processing), electronic warfare, and a variety of new missiles and munitions. The emphasis on high technology in the conventional weapons area, in its turn, triggered a great variety of studies (pro and con) by academic and other defense analysts, not only in the United States but also in Europe. These included analyses of the technical feasibility of the proposed systems, the affordability of the acquisition of sufficient numbers of the ever more costly weapons, and the associated consequences for national defense. Since the early 1980s, the U.S. Office of Technology Assessment has been allowed by Congress to make assessments on defense technologies as well and has done so both on the SDI ballistic missile defense program and conventional armaments.

Making assessments of weapons systems is one thing, influencing procurement decisions is quite a different matter—not to speak of influencing the weapon innovation process itself. It is to this issue that we turn in the next section.

Past Analyses of Arms Race Dynamics

Many studies of the past that analyzed the technological arms race were inspired by the need to bring it under political control, the meaning of which, however, is not trivial and needs some elaboration.

In national politics there is often no consensus on the kind of armament that is desirable or necessary. Those who say that politics is not in control may actually mean that developments are not in accordance with their political preferences, whereas those who are quite content with current developments might be inclined to say that politics is in control. This is not quite satisfactory. But to say that politics is in control because actual weapon innovations are the outcome of the political process—which includes defense contractors' lobbying, interservice rivalry, bureaucratic politics, arguments over ideology, strategic concepts, and so forth, as Greenwood (1990) suggests—is not satisfactory either. Rather than speaking of control, one should ask whether it would be possible to influence the innovation process in a systematic way or to steer it according to some guiding principle (Smit, 1989). This implies that the basic issue of "control," even for those who are content with current developments, concerns whether it would be possible to change the course of these developments if it was desired.

A plethora of studies have appeared on what has been called the technological arms race (see, e.g., Gleditsch & Njolstad, 1990). Many of them deal with what President Eisenhower in his farewell address called the military-industrial complex, later extended to include the bureaucracy as well. These studies belong to what has been called the bureaucratic-politics school or domestic structure model (Buzan, 1987, chap. 7) in contrast to the action-reaction models (Buzan, 1987, chap. 6), which focus on interstate interactions as an explanation of the dynamics of the arms race. A third approach, the technological imperative model (Buzan, 1987, chap. 8), sees technological change as an independent factor in the arms race, causing an unavoidable advance in military technology, if only for the links with civil technological progress.[6] By contrast, Ellis (1987), in his social history on the machine gun, has shown the intricate interwovenness of weapon innovation with social, military, cultural, and political factors. To some extent, these approaches might be considered complementary, focusing on different elements in a complex pattern of weapon development and procurement. For instance, the "reaction" behavior in the interstate model might be translated into the "legitimation" process of domestically driven weapon developments.

Many of these studies are of a descriptive nature; some are more analytical and try to identify important determinants in the arms race (Allison & Morris, 1975; Brooks, 1975; Kaldor, 1983). Studies that relate military innovation, including military technology, to institutional and organizational military factors and to the role of civilians, are provided by Rosen (1991) and Demchak (1991). Weapon innovation in the former Soviet Union has been studied by Holloway (1983), Alexander (1982), Evangelista (1988), and MacKenzie (1990a). There are quite a number of case studies on specific weapons systems as well as empirical studies on the structure of defense industries (Ball & Leitenberg, 1983; Gansler, 1980; Kolodziej, 1987) and arms procurement processes (Cowen, 1986; Long & Reppy, 1980). Hardly any of these studies focus on the question of how the direction of the weapon innovation process and the course of military R&D might be influenced. One task for future STS studies, in my view, would be to combine insights from this great variety of studies for better understanding of weapon innovation processes. A first step in this direction has been carried out by Evangelista (1988), who performed an interesting comparative study on the weapon innovation process (of more radical innovations) in the United States and the former Soviet Union. He has distinguished a number of stages, each characterized by a specific type of social, political, and technological activity of the particular actors involved at that stage. One drawback of this model is that it is strongly linear. A network approach, in which the number of various actors involved may grow and shrink, allows for the more "oscillating" nature of the innovation process that one can often discern (Enserink, Smit, & Elzen, 1990).

In the past few years attention to the role of military R&D in what traditionally has been called the arms race has increased (Albrecht, 1990; J. Dean, 1992; Greenwood, 1990; Hamlett, 1990b; Lakoff & Bruvold, 1990; Smit, 1991; Smit, Grin, & Voronkov, 1992; Thee, 1986; United Nations,

1987; Woodhouse, 1990). Though these studies provide us with some interesting analyses, they are still far from giving clues to steering the weapon innovation process. As a research question in STS studies, it has not been addressed specifically until recently, and recent studies (Elzen et al., 1990; Enserink et al., 1990; Smit, 1991) are still in an embryonic stage.

Below, three issues will be discussed that are of interest both for the societally important question of steering of the weapon innovation process and for further developing the field of STS studies. The focus will be on the state of the art and on issues and approaches for future STS research:

- The relation between science and military technology
- Regulating military technological developments (regulation studies)
- Relations between civilian and military technology

Science and Military Technology

Many modern technological developments are rooted in or closely connected with scientific developments. What can be said about the nature of this relation?

First, one should recognize that there are no sharp definitions of the concepts of science and technology. Often the two concepts have been reified into two "things" that can have a mutual relation. Science, then, is understood as the body of knowledge widely accepted by the scientific community, which can be found mainly in scientific journals and books. Technology in its turn is understood to encompass both hardware and knowledge of the industrial arts (E. Layton, 1977, p. 199). One way of relating science and technology, then, might be to look for the similarities and differences between the two types of knowledge, for example, as to their nature and structure. A different way would be to ask how scientific knowledge is transformed into technology. The first is a static relationship; the latter, a dynamic one, which actually refers to the activities of scientists and technologists.

The more static interpretation underlays the Hindsight and TRACES studies. Hindsight, an initiative of the DoD to evaluate the usefulness of DoD-supported basic research to military technology, concluded that, of all considered critical events that had made weapons systems possible, only 0.3% fell into the category of basic research. TRACES, initiated by the NSF as a reaction to Hindsight, took a longer time perspective of about 45 years. It came to the different conclusion that 34% of the critical events were non-mission-oriented research, 38% were mission oriented, and 26% were classified as development (for a critical discussion, see M. Gibbons, 1984; E. Layton, 1977). In both studies, basic research and technological development were seen as separate activities, carried out by different persons and organizations. The central issue in both studies actually was to see whether technology is indeed a form of "applied science."

The point is not so much whether these concepts of science and technology and their mutual relation are right or wrong (if one could say so). Basic is that these images figure in our world and as such influence the activities of those who hold these images and who have to deal with funding or worry about the impact of developments in science (such as decision makers and advisers). These images are constructs of the participants involved in the development of S&T. The role that these images—that is, the actor-constructed distinctions between what science is and what technology is—play, for instance, in science and technology policy would be an interesting subject for further STS studies, not only for military but also for civilian technology.

Only a few detailed case studies exist on the role of science in the area of military technology. One is the extensive study on missile guidance technology by MacKenzie (1990a), which shows a complex relationship.

More studies exist on the relation between science (or basic research) and civilian technology. Pavitt (1991) has summarized the evidence from a number of economics-oriented innovation studies that have addressed the question of how far science contributes to technological developments. The intensity of direct transfers of knowledge from basic science to application varies widely among sectors of economic activity as well as among scientific fields. In chemicals and drugs firms, strong links exist to basic research in biology, while links of electronics firms are also intense but to more applied research in physics. The impact of basic research on technology also varies widely over different types of technology. To result in viable technology, knowledge from basic research has to be combined with knowledge from other sources, including design and engineering. The impact is not only through direct knowledge transfers but also through access to skills, methods, and instruments. Moreover, knowledge transfers are not primarily through literature but are mainly person embodied, involving personal contacts, movements, and participation in national and international networks.

The latter mechanism indeed shows the need for a more dynamic approach to the relationship between science and technology, focusing on the type of activities needed to link science and technology and on the conditions favorable to that end. Examples are actor- and sociotechnical-network approaches and social constructivist analyses, which emphasize the heterogeneous character of technology development. Central in these approaches is also the notion that there is no such thing as "technology" external to "society" and no "society" external to "technology" but only actions and actors connected in networks, constituting a "seamless web" (see MacKenzie, 1990a, chap. 8). It also implies that civilian technological applications often do not follow directly from science developed in the context of military technology (little "spin-offs"), and vice versa. The transformations, either from scientific insights to technology or from one particular technological application to another one, require the dedicated linking of actors, implying that in many cases developing science becomes an integral part of developing technology (see, e.g., the development of high-energy lasers, which will be discussed below, where a distinction between basic research and development is hard to make). This is of particular importance for the issues of the integration of civilian and military technology and of conversion, to be discussed in the last section.

Regulating and Steering the Weapon Innovation Process

Until recently, the issue of controlling military R&D was raised within the framework of the arms race between NATO and WTO. Analytical efforts were aimed at diagnosing rather than curing the problem and no way out was found. The ending of the cold war and the East-West confrontation in Europe, along with a substantial conventional disarmament agreement (CFE treaty), implies a profound change in the international security environment (even if ethnic and nationalist conflicts now surface). In addition, the prospect of U.N. peacekeeping or enforcing actions has increased after the U.N.-supported war of allied forces against Iraq when it annexed Kuwait in August 1990.

These changes call for changes in military strategy, doctrine, and postures, a process that has already started. The politically important issue then is to arrive at such changes that will cause international security to be enhanced and the process of demilitarization of international relations to be supported. Moreover, they should fit within a new international order in which the United Nations is to play an enhanced role in regulating international conflicts.

What are the implications for military technology and weapon innovation? One thing is already clear: Military budgets and forces will be substantially reduced both in the United States and in all European countries. The arms race in the traditional sense is over: The Russian economy has collapsed; its military industry is in disarray and is largely to be converted to civilian tasks. So far, however, the military R&D budgets of NATO countries have hardly been reduced and will decrease probably at a much slower pace than the overall military expenditures. The paradigm of "technological superiority" is still very vivid. The basic question here is whether such a vigorous effort for military technological innovation is still desirable (if it ever was). Not only its size but also its direction are to be under scrutiny. This implies the necessity of a new type of (impact) assessment studies and again the question of regulating the innovation process.

Whereas in the past arms control negotiations have dealt mainly with existing and not with possible future weapons (but see Smit, 1991, for a brief review of arms control treaties that have tried to deal with controlling "development"), it now seems compelling to also look into the earlier stages of the weapon innovation and acquisition process, including military R&D, and to bring these in alignment with the broader goals and requirements of a new international security system.

Such a shift of attention to these earlier stages, including questions of the necessity and possibility of steering military R&D, can already be noted (J. Dean, 1992; Greenwood, 1990; Hamlett, 1990b; Lakoff & Bruvold, 1990; Smit, 1989, 1991; Smit et al., 1992; Woodhouse, 1990). Future STS studies might contribute to the issue of regulating the weapon innovation process by analyzing the structure and dynamics of this process and subsequently making suggestions for some regulatory regime.

Weapon innovation and its associated military R&D differ in one respect from nearly all other technologies, in that there is virtually only one customer of the end product—that is, government. (Some civilian industries, such as nuclear power and telecommunication, however, show considerable similarities in market structure—monopolies or oligopolies coupled with one or at most a few dominant purchasers. They are also highly regulated and are markedly different than the competitive consumer goods markets. See Gummett, 1990.) Actually only a specific set of actors constitutes the actor network of military technological developments, including the defense industry, the military, the Ministry of Defense (MoD), and Parliament. The MoD, as the sole buyer on the monopsonistic armament market, has a crucial position. In addition, the MoD is heavily involved in the whole R&D process by providing, or refunding to industry, much of the necessary funds. Yet the MoD, in its turn, is dependent on the other actors, like the defense industry and military laboratories, which provide the technological options from which the MoD may choose. STS studies of this interlocked behavior seem a promising approach for making progress on the issue of regulatory regimes. Additional phenomena that also should be taken into account are the increasing international cooperation and amalgamation of the defense industry, the integration of civilian and military technology, and the constraining role of international agreements (see Smit et al., 1992).

In the past, few proposals were made on *how* to control military R&D, except for the suggestions by York (1976) and Brooks (1975) to reduce the budgets. But even budget reductions, because of their general character, would not exclude the development of, for instance, destabilizing weapons. Smit (1991) has discussed the arguments raised against the possibility of controlling military R&D and has suggested an alternative approach of "decentralized regulation." In this connection, he has discussed the role of "guiding principles" in a regulatory regime for military technological innovation. (See also Enserink et al., 1990.)

Callon, Larédo, Rabeharisoa, Gonrad, and Leray (in press) have developed a techno-economic network approach for directing research, in which interventions are accompanied by a continuous monitoring of their effects while the evolution of the contexts in which they have been inserted are attended to. It would be of interest to try and translate this approach to military R&D.

Conversion and the Integration of Civilian and Military Technology

A third issue that has drawn considerable interest due to the ending of the cold war is the conversion and the integration of civilian and military technology. Actually, the concerns in Western countries and in the states of the former Soviet Union are quite distinct.

To the latter states, the issue of conversion is of predominant importance because of the disproportional part of industry and technological development devoted to the military. In Russia the conversion process and a transition from centralized and bureaucratic production and decision structures toward more decentralized ones have to go hand in hand. This renders useless whatever insights might exist on technological change in a relatively stable societal environment.

In Western countries the relation between military and civilian technology is relevant to the policy issue of how and to what extent military and civilian technological development can and should be integrated. The main issues under debate (particularly in the United States and the United Kingdom but also in France) are as follows:

- How do we maintain a sound defense industry in view of decreasing weapon procurement and consequently less orders for the defense industry?
- How do we afford a sufficient defense capability in view of the ever increasing costs of advanced weaponry?
- How do we afford a military R&D capability for the (still undisputed) need of continuing weapon innovation and strive for technological superiority?
- How do we avoid the situation where (public) military R&D expenditures are made at the cost of technological innovation in the civilian sector, hampering international competitiveness?

Many reports, in particular in the United States and the United Kingdom, deal with these issues (Advisory Council on Science and Technology, 1989;

Defense Science Board, 1988; Gansler, 1989; Office of Technology Assessment, 1988a, 1989, 1991b).

Generally, these problems are felt as less pressing by countries with relatively smaller military and larger civilian R&D budgets. Regarding the issue of international competitiveness, the general opinion in the United States and United Kingdom is that countries like Germany and Japan are in a much better position.

Many of these questions are of a science and technology policy nature. They are also related to the two themes discussed before, that is, the science and military technology relation and the regulatory issue. For instance, how is a technology policy aimed at the integration of civilian and *military* technology to be shaped so that it does not interfere with attempts at steering military technological developments in less destabilizing directions.

STS studies may contribute to the question of the possibility of conversion and the integration of civilian and military technology by providing insight into the similarities and differences between civilian and military technology, as to both their nature and the specific social contexts and networks in which they are developed. Indeed, a central question is in what respect and to which degree civilian and military technology and R&D are distinct. Funding sources, possible applications, institutional context, and intentions, in themselves or together, are of relevance for designating R&D as civilian or military, but they are not always decisive on this issue. Military-funded basic research may have no military applications, even if the intention is there. On the other hand, civilian-funded research may lead to inventions with military applications.

One answer to the question of whether military and civilian technology is intrinsically different is that the distinction is an institutional rather than an intrinsic one (Gummett, 1991, p. 27) or, phrased more strongly, that there are no "civilian" and "military" technologies but only civilian and military markets. Whereas the institutional element is rightly emphasized, the main drawback of this statement is that it actually tries to draw a sharp division line between technology and its social context. It thus creates an image of technology as a "thing" that, so to speak, can exist independently of its institutional setting and is in itself neutral as to the civil-military distinction: The labeling of the technology as either military or civilian is then a mere projection of the nature of the institutional setting in which it functions or is being developed, without the technology itself showing "military" or "civilian" characteristics. Several case studies, however, illustrate how intricately interwoven the characteristics of technology may be with the social context in which it is being developed or in which it functions. MacKenzie (1990a) has made a very detailed study on the development of missile guidance technologies in relation to their social context and on the technological choices made for improving accuracy. He has shown how different emphases in requirements for missile accuracy and for civilian (and military) air navigation resulted in alternative forms of technological change: the former focusing on accuracy; the latter, on reliability, producibility, and economy.

Another example is provided by early laser development in the 1960s. The military, right from the start, showed great interest in this technology and vigorously supported its development. Seidel (1987) has shown how, through the early successes of the use of ruby and glass lasers in target illumination and range finding, and the failure of other researchers to employ gases as a lasing medium in such applications, the military services became wedded to a technological path that favored solid state laser types, neglecting the use of other possible lasing materials—and a particular "technological paradigm" emerged within this institutional military setting. In the subsequent search for high-power lasers, much resistance had to be overcome before a "technological paradigm" shift to gas laser types, in particular the gas dynamic laser, could occur. Moreover, some of the high-power laser projects (e.g., processing technologies for specific laser glasses) supported by the military within the "old" paradigm appeared to be of little use for civilian applications.

A third example is nuclear reactor technology, where the characteristics of the technology have been prescribed by military rather than civilian needs. Present nuclear reactor technology has been inherited from the 1950s and 1960s, when R&D was dominated by military interests: plutonium production and the development of a navy propulsion reactor (see, e.g., Hewlett & Holl, 1989, chap. 7; G. Thompson, 1984). The choice for a light water reactor was a direct result of the latter. Also the availability and even oversupply of enriched uranium from the gas-diffusion plants, built originally for the production of highly enriched uranium in the nuclear weapons program, played a role in the choice of this reactor type. If nuclear safety, rather than military needs, had been the primary concern, then quite different reactor types might have been developed for civilian use (Lilienthal, 1980, p. 99).

The three examples show that the characteristics of technologies and artifacts are tightly bound to the institutional setting in which they were developed. The nature of a technology can be better understood by not isolating it from this "social" context.[7] This is one of the important insights provided by modern social studies of technology (Hughes, 1983; Pinch & Bijker, 1987). It is true for the developmental process ("technology in the making") as well as for "technology in use." The technology, and the composition and nature of the social network that supports it, mutually influence each other. This is closely related to Winner's (1986b) observation that "artifacts do have politics."

A number of historical studies have addressed the question of the *relation* between civilian and military technology. Some of them have evidently shown that, in a number of cases in the past, the military has successfully guided technological developments in specific directions that also have penetrated the civilian sector. Thus M. R. Smith (1985a) has shown that the manufacturing method based on "uniformity" and "standardization" that emerged in the United States in the nineteenth century was more or less imposed (though not without difficulties and setbacks) by the Army Ordnance Department's wish for interchangeable parts. Noble (1985) has investigated, from a more normative perspective, three technical changes in which the military played a crucial role—namely, interchangeable parts manufacture, containerization, and numerical control—arguing that different or additional developments would have been preferable from a different value system. These cases make also clear that a distinction between civilian and military technology cannot be made by just trying to provide sharp definitions. As usual, it is better to start with an interesting problem and to proceed from there than to start with abstract definitions.

Studies going back to the nineteenth or even early twentieth century, though interesting from a historical perspective, may not always be relevant for modern R&D and current technological innovation processes. Systematic studies on the interrelation between *current* military and civilian technological developments have been begun only recently (see Gummett & Reppy, 1988). They point out that it might be useful to distinguish not only between different levels of technology, such as generic technologies and materials, components, and systems (Walker, Graham, & Harbor, 1988), but also between products and manufacturing or processing technologies (Alic, Branscomb, Brooks, Carter, & Epstein, 1992).

A variety of studies have dealt with conversion but few with the conversion of military R&D. Most of them have focused on conversion of production plants from military to civilian purposes, economic issues, or substituting jobs in the defense industry for jobs in the civilian sector (Dumas, 1982; for a comprehensive review, see Southwood, 1991). They have drawn little attention from the establishment. But because of the political and economic reforms in the eastern European countries and former Soviet Union, and the substantial fall in arms procurement, there is rising interest in conversion and many conferences on the issue have been organized. Still, little attention is being paid to the conversion of military R&D.[8]

Systematic studies on the issue of the integration of civilian and military technology are emerging too, often under the heading of "dual-use" technologies (e.g., Alic et al., 1992). To what extent such an integration will be possible depends on the divergence in technological transformations toward distinct applications. The closer to the "end application," the more divergent these transformations will likely be. This is certainly true for the differences between weapons and civilian goods, but it also applies to such technologies as civilian and military radar systems; though they have many basics in common, different requirements of the end product lead to different technological transformations in the development process. Thus the better prospects are in integration at the level of components, such as electronic data processing systems. One such example is the militarized version of a new model VAX computer produced by Raytheon Corporation under license from the Digital Equipment Corporation (DEC). It not only provided the military with state-of-the-art technology of the advanced civilian computer sector but also permitted the military to use the vast body of VAX-compatible software (Alic et al., 1992, p. 73). One prerequisite for integration is for the military to refrain from too exotic specifications of components and modules and to adapt more to civilian technology standards than is customary nowadays (which does not necessarily imply less performance or reliability; it might even result in better products; Gansler, 1988).

In the United States there have been a number of efforts in the last decade to integrate civilian and military R&D and production through the involvement of the Department of Defense. Examples are (Gansler, 1988) the high speed integrated circuits (VHSIC) and the Semiconductor Manufacturing Technology (SEMATECH) program; the development of advanced manufacturing technologies (e.g., "flexible manufacturing" and computer-integrated manufacturing); investments by the DoD in superconductivity to "ensure use of superconductivity technologies in military systems" and to "develop the required processing and manufacturing capabilities"—obviously a dual-purpose investment; and the Strategic Computing Program aimed at advanced machine intelligence, among other things, by a new generation of computers, parallel computing, and advances in artificial intelligence (Åkersten, 1987).

From an STS perspective, these attempts might serve well as case studies on the relation between civilian and military technological developments in a particular social and organizational context. A focus on the integration of civilian and military technology may also recast the often confusing discussion of "spin-off" (and "spin-in"). They may illuminate questions, as suggested by Gummett (1990), on whether market conditions matter more in directing of developments than the distinction between civilian and military activities. For instance, as mentioned before, considerable similarities in market structure (like special relationships between state agencies and companies) exist for the defense industry and some civil industries, such as nuclear power and telecommunication. The influence of (the structure of) organizational relationships on technological developments might be fruitfully analyzed by network approaches, or in terms of the interaction between strategic games, as both leave room for actor input into these developmental processes. Additional questions that, according to Gummett, could be raised relate to the impact of an increasing internationalization (collaboration) in the defense industry, the shift from mechanical engineering to electronics as the core technology in the defense sector (breaking down traditional industrial boundaries), and the process of diffusion of technology.

Sapolsky concluded his chapter, "Science, Technology and Military Policy," in the 1977 STS handbook with the observation that "the implications of military research and development run through the very fabric of science policy, and we must therefore claim for research in this area a much higher priority and a more integrative approach than hitherto" (p. 462). In spite of recent improvements (see Gummett, 1990), it is still true that defense science and technology policy are still remarkably understudied in view of the amount of money spent on them. This chapter has shown that, whereas the perspective and concerns may have changed since 1977, current issues and concerns still require much more insight into the processes of military technological development and its relation to civilian technology, to which STS studies may contribute.

NOTES

1. Though in the actual innovation process the distinction between research and development has become diffuse in many areas, it is customary to distinguish six categories in the U.S. DoD budget for Research, Development, Test, and Evaluation (RDT&E, which the NSF calls Defense R&D): 6.1 is research or basic research; 6.2 consists of exploratory development, also called applied research, intended to explore the use of a concept in an application or device with military relevance; 6.3 is advanced development; 6.4 is engineering development; 6.5 consists of management and support of R&D; and 6.6 is systems development. Activity 6.3a includes examination of alternative concepts through the advanced technology demonstration phase, and 6.3b represents advanced development.

According to DoD, the term *technology base* refers to the categories 6.1 and 6.2. The phrase *science and technology* generally refers to 6.1, 6.2, and 6.3a (Knezo, 1990, p. 1).

The percentage of basic research (6.1) in the RDT&E budget varied over the years between 2% and 5%; that of S&T (6.1 + 6.2 + 6.3a), roughly between 15% and 25% (Knezo, 1990, pp. 6-7).

2. See Edgerton (1992) for British scientists' and engineers' *public* reflections on science, technology, and war in the twentieth century.

3. See, for instance, many articles in *Social Studies of Science* during the past 20 years. See also Engelhardt and Caplan (1987).

4. In military technology, MacKenzie (1990a) has shown the "construction of technological facts" for accuracy claims of inertial guidance technology for missiles.

5. Forman refers to the analysis by Cini (1980) of a similar epistemic shift in theoretical particle physics in the latter 1950s, where a "dispersion philosophy" was remarkably rapidly established. Cini relates this cognitive commitment to the consultancies and advisory services of many U.S. elementary particle theorists to industry and the military.

6. It neglects, however, the investments, also institutional, needed for such a transfer from civil to military applications; see Reppy (1990) and the last section of this chapter.

7. The phenomenon of alternative shaping of technology can also be observed within the military domain. Trilling (1988) has shown that the different social and organizational contexts in the Soviet Union and the United States resulted in different styles and designs of fighter aircraft in these countries.

8. One exception is a Pugwash project on the issue, with a first meeting of the working group in Turin in 1991 and a second in London in 1993.

27

Science and Technology in Less Developed Countries

WESLEY SHRUM
YEHOUDA SHENHAV

It is not unusual for scientists to eat their subjects.

Thomas Bass, *Camping with the Prince, and Other Tales of Science in Africa*

THE literature on science and technology in less developed countries (LDCs) is immense and interdisciplinary, but not predominantly academic in character. It would be easy to damn much of it for lack of systematicity, methodological sophistication, and theoretical grounding. A good deal more of it could be condemned for its polemics, idealistic models, and naive assumptions. What must be kept in view, however, is that many of its contributors do not have

---

AUTHORS' NOTE: Our thanks go to Carl Bankston and D. Stephen Voss, who provided superb bibliographic assistance, and Xu Chao, who compiled the information on S&T studies in China. Frederick Buttel's reading was especially important in clarifying many of the issues involving the Green Revolution. For their comments and corrections, we appreciate the time and effort of Amiya Kumar Bagchi, Aant Elzinga, John Forje, Wu Yan-Fu, Jacques Gaillard, Esther Hicks, Chen Wen Hua, Isabel Licha, Gerardo Otero, Radhikka Ramasubban, Francisco Sagasti, Thomas Schott, Simon Schwartzman, and Dhirendra Sharma.

as their exclusive (or even primary) aim the enhancement of understanding. Instead, they focus on the betterment of that portion of humankind living in less developed countries or the advancement of organizational goals such as profitability. The work discussed below is much broader than the STS studies currently practiced in many institutions though it falls within this purview. The tasks of synthesizing this literature and theory construction lie in the future.[1]

In the first part we review perspectives on the relationship between science and economic development, the institutionalization of Westernized science throughout the world, and the research output of LDCs. In the second part we review research on technological development, including technology transfer, state intervention and regulation, technology generation, social effects of technological change, and appropriate technology arguments.[2]

The remarkable diversity of disciplinary perspectives, theoretical interests, and empirical studies precludes any simple summary of received or current wisdom. It is striking, however, that many authors have come to the conclusion—whether or not they state it in these terms—that *science and technology should be viewed in terms of context-specific forms of knowledge and practice that interact with a set of globally distributed social interests.* Such a conceptualization is preferable to those based on grand theory (whether "of" science, technology, or development) and suggests that social network approaches will be useful for capturing the interlocking set of individual and organizational interactions that propel world technoscience.

SCIENCE IN LESS DEVELOPED COUNTRIES

All theories of development imply some subtheory or account of the role of science and technology (Agnew, 1982). At one extreme, it is argued that lack of innovation and resistance to change are the main causes of underdevelopment. At the other, imported scientific ideologies and technological artifacts from industrialized countries are said to generate debilitating dependencies.[3] In all this, competing characterizations of social actors play a role. Are they passive recipients of Western aid? Are they hidebound traditionalists, reluctant to alter inefficient farming and production methods? Are they rational actors who reject or subvert inappropriate technologies imposed by elites? Are they stooges of multinationals in the capitalist core?[4]

It is possible to report that many studies, though certainly not all, now go beyond accounts of S&T in less developed countries in terms of individual properties or their simple aggregation. In place of this older "atomistic" view, *relational* concepts have become more important. Here the principal focus is on relations (also called "links" or "ties") between social actors, whether individual persons, organizations, or nations (Shrum & Mullins, 1988), and the effects of these relations in promoting or constraining forms of action. Macrotheories of development typically focus on the relations between nations or among nations, elites, and multinational corporations (MNCs).

Because national wealth and scientific effort are unquestionably associated, researchers have asked, Why and with what consequences do LDCs engage in various kinds of research? Does investment in scientific research, as well as investment in technology and engineering development, contribute to economic growth? Three theoretical perspectives have been brought to bear on this question: modernization, dependency, and institutional.

Modernization

According to neoclassical economic approaches to development (Pavitt, 1973; Williams, 1967) and sociological modernization theories, technology is a prime motor of social change among nation-states at the same level of modernization.[5] In extreme versions of the modernization view (and especially their characterizations by critics), the main causes of development are internal to a country while the configuration of external relationships is of little significance.

But scientific knowledge and the transfer of technology were thought to be a special case. Modernization theorists believed science was strongly linked to technology and, like education, improved the ability of a country to promote growth through more efficient use of its resources. Technology transfer, aid that promoted educational institutions, and other forms of scientific and technical assistance were encouraged as part of an appropriate developmental process.

This relationship between science and development was thought to be especially crucial for LDCs that sought to reach a developmental takeoff point (Dedijer, 1963; Sussex Group, 1971). First, an economic infrastructure, including both labor and capital, is required to absorb scientific knowledge and enable its use. Second, scientific knowledge must be relevant and applicable to economic endeavor.

But LDCs often lack a sound economic infrastructure and the question of relevance is still hotly debated. One of the most surprising findings that emerged from empirical work in the 1970s was the negligible influence of the expansion of higher education on economic growth (Meyer, Hannan, Rubinson, & Thomas, 1979). In sharp contrast, both primary and secondary educational expansion are known to have strong positive effects on national economic development (Benavot, Cha, Kamens, Meyer, & Wong, 1991; Rubinson & Ralph, 1984). School graduates at the secondary level reach the economic labor force (the production and service sectors), whereas higher education channels graduates into occupations that produce no tangible goods (e.g., lawyers, civil servants).[6]

Implicit in accounts of the relationship between science and economic growth is an underlying premise about the unidirectional influence of science on technology. The conventional belief that technology is deeply rooted in scientific knowledge has undergone substantial revision. While such influence relationships from science to technology do occur, this model does not generally reflect a modal process of research and innovation (Drori, in press; Shrum, 1986). Science penetrates the technological realm through a complex process consisting of several components but they do not occur in any determinate order. Often technological developments influence science.

Dependency

Dependency theory was the first major contribution to social science that originated in the LDCs themselves (Hettne, 1983). If modernization theory minimizes the influence of external relations on internal processes, dependency theory (and its close relation, "world system" theory) argues that external relations create obstacles for development because the direction of economic growth is conditioned by forces outside the country.[7] Position within an international network of relations is a central determinant of economic development. Dependency theorists claim that less developed economies are malintegrated into the international system *because of* their dependence on a small number of exchange relationships and because these relationships operate, through various mechanisms, to benefit interests in developed capitalist economies.

For example, investments by foreign businesses affect the division of labor and the economic structure in LDCs. Such economic interventions distort internal processes, increasing income inequality, suppressing political democracy, promoting the rise of core-linked economic sectors, and stunting economic development (Delacroix & Ragin, 1981; Evans & Timberlake, 1980; Semyonov & Lewin-Epstein, 1986).

Western science is viewed as another mechanism of domination, not just by producing the technological means for the subjugation of the masses (in some accounts) but also as an ideological force and an inappropriate developmental model. The creation and maintenance of scientific institutions not only absorb personnel and capital but constitute an irrelevant ideological diversion for countries without the resources or connections to pursue Western, specialty-oriented science.

Researchers in LDCs are linked to the "scientific core" in industrialized countries. Knowledge is produced in collaboration with foreign colleagues and research centers without taking LDC needs into account. Investments in fundamental research, and consequently in research productivity, are funded not only by domestic sources but also by foreign aid.[8] This enhances growth in the tertiary sector and increases the size of universities but results in misallocation of national resources at the expense of the productive sectors, interferes with the societal division of labor, and retards economic growth.

Cross-national empirical studies have consistently shown positive associations between economic wealth and research productivity. Such correlations may reflect no more than the general correspondence of a wide range of characteristics of industrialized societies and themselves imply no causal force. It might even be that the association exists because only wealthy countries can support the luxury of prestigious "fundamental" research. A major and still unresolved issue for LDCs is the extent to which investment in science by the state stimulates or retards economic development.

The relationship between scientific and economic status may be different for different levels of development. Shenhav and Kamens (1991) employed a longitudinal analysis (1973-1980), distinguishing between indigenous and Western scientific knowledge. For a sample of 73 LDCs, the latter has no relationship with economic performance and even a mild negative association with economic performance in the poorest countries. For the developed nations, on the other hand, scientific knowledge was found to be associated with economic development. Though LDCs invest more effort in applied research, they may also be less capable of converting theoretical knowledge into technological applications.

Institutional Theory

Whether or not Western science is useful or detrimental to LDCs, there remains the question of why countries are committed to its promotion. Institutional theory is concerned with the determinants of isomorphism, or the adoption of structurally similar forms throughout the world.[9] In general, the institutionalization of isomorphic science in LDCs is produced by a belief in the universality of science and its necessity for modernization. Scientists, elites, and policymakers in both developed and less developed countries share this orientation. Hence adoption of Western organizational forms serves as a legitimating device to other states and international agencies.

Through mimetic processes by which successful existing systems serve as models (DiMaggio & Powell, 1983), scientific institutions and beliefs are prescribed and diffused as a key component of the modern world system. Such processes are the main focus of institutional theory, according to which reality is socially constructed to create "truths" that acquire rulelike status.

Practices, processes, and national policies are adopted, transformed, and reproduced not necessarily because their technical superiority has been demonstrated but owing to participants' beliefs in the efficacy of certain ways of doing things. Science, like education, is one of the most significant institutions that provide interpretations, cultural meanings, and instrumental leverages throughout the world. Enormous environmental support allows actors and organizations that are accepted as "scientific" to define the meaning of rational activity and encourages worldwide diffusion.[10]

Scientists and scientific communities are principal carriers of the ideology of universalism, promoting the idea of a worldwide universal and context-free system of science. Relative consensus exists on the nature and aims of science because the most prominent scientists in LDCs who rise to administrative and policy positions are trained in industrialized countries (Gaillard, 1991). LDC scientists are linked initially to international scientific networks (often through educational experiences) and remain connected through cross-national collaborations, the exchange of scholars, binational funding, and international meetings.

The maintenance of these linkages tends to work against external interventions and goal-directed science, which conflict with the idea that the invisible hand of theoretical need should regulate the advancement of knowledge. Choudhuri's (1985) perceptive comparison of graduate work in India and the United States examines the consequences of peripheral location on the research practices and orientation of scientists, concluding that attempts to emulate "great scientists" can even be the cause of actual scientific "failure" rather than an agent of socialization into productive research activities.

By adopting Westernized science and Western organizational forms, LDCs help to promote comparability and compatibility but not solutions to local problems (Turnbull, 1989). Empirically, adoption of Western models does not imply that institutions developed in LDCs are simply mirrors of their Western counterparts, as Eisemon (1980, 1982) and Schwartzman (1991) show.

Institutional theory deemphasizes the power and interests of social actors in the institutional field, yet the processes it describes are often fueled by mechanisms of power and domination that are best explained by the dependency approach. The "taken for granted" can be used as a vehicle in the service of interests. Dependency theory is not at odds with institutional theory but is complementary, pointing to alternative network mechanisms as driving forces. The former emphasizes that establishing relations with powerful actors induces dependency in the absence of alternative sources. Institutional theory hypothesizes that actors in structurally equivalent positions (i.e., with similar sets of relations to other actors in the system) should behave similarly with respect to models provided by prestigious ("successful") positions.

In sum, whether one believes that most international ties induce development or dependency, it is a mistake to place such a large theoretical burden on any one type of linkage, external or internal. Though the causal importance of connections with MNCs and industrialized countries is now widely appreciated, it is preferable to recognize that developmental paths are historically contingent, without seeking a universal model of development.[11]

The Distribution of World Science

The late Mike Moravcsik (1928-1989) was a tireless promoter of "science development" for the benefits, both material and spiritual, that would accompany the advancement of science in developing countries.[12] But he remained harsh in the mid-1980s in his judgment of Third World science. An essay with John Ziman (1985) described a reality behind the "facade of science" in the fictitious land of "Paradisia": "no more than the fragments of a scientific community, disorganized, disunited, of limited professional competence, poverty stricken, intellectually isolated, and directed toward largely romantic goals—or no goals at all" (p. 701).

To what extent has the *practice* of modern research spread throughout the world? What level of scientific and technological capacity characterizes nations? The principal way of assessing capabilities has been via indicators such as publications, citations, and the number and distribution of technical personnel.[13]

High levels of concentration characterize the production of world science. Although the absolute quantity of world science and technology produced by LDCs has gradually increased—regardless of the measure used—many observers would agree with some portion of Moravcsik and Ziman's judgment. None of the LDCs (including India, China, and Brazil, the top producers) has the kind of active and integrated scientific community characteristic of OECD countries (Arunachalam, 1992; Gaillard, 1991).[14] Measures of inequality in the production of scientific publications are even larger than those in other social spheres (Frame, Narin, & Carpenter, 1977).[15]

This high level of inequality is illustrated by the fact that well over three quarters of world scientific output is produced by 10 countries. All are highly industrialized, apart from India, the leading LDC producer of science. The dominance of OECD and eastern European countries in producing scientific research is overwhelming. Together, the two regions contribute 94% of the indexed scientific literature.[16] Between 1981 and 1985, LDCs produced 5.8% of the world's scientific output (Braun, Glaenzel, & Schubert, 1988).[17]

There are, however, important limitations to these studies, particularly the focus on periodical output. Because the peer reviewed journal is the preferred Western form of disseminating research findings, researchers have often used the *Science Citation Index* to analyze differences in scientific productivity between countries and disciplines.[18] Work published in internationally oriented journals is distinct, in terms of both peer review procedures and impact, from that in national and local journals. Such databases cover only a small proportion of the total world technical literature, neglecting non-English sources and most of the periodicals in LDCs.

LDCs may only account for a fraction of journal coverage in the major indexes,[19] but the world share of LDCs in certain fields (such as soil sciences and agriculture) is larger (Arvanitas & Chatelin, 1988). Many researchers publish a great deal in domestic journals, even if they publish internationally (Davis & Eisemon, 1989; Lomnitz, Rees, & Cameo, 1987; Velho, 1986). While the LDC contribution to the mainstream scientific literature is small, there is no single database from which reliable estimates of the total contribution to the published literature can be made.

Though the developed countries occupy a position of centrality, linguistic, educational, and political factors affect the specific degree of influence one country has over another. Schott (1988) uncovered six structurally equivalent regions in the worldwide influence network. De Bruin, Braam, and Moed (1991), in an analysis of Gulf State coauthorships, show how closely scientific collaboration reflects political and past colonial alliances.

Even more important than *where* work is published is *what kind of* work is done. Twenty years ago the Sussex Group estimated that 98% of world R&D expenditures were by Western countries while only 1% was associated with problems directly related to the developing world.

Wherever resources originate, problem selection in LDCs is a core issue, both sociologically and politically. It is quite difficult to assess the degree to which published research is oriented toward local problems and needs.[20] The ideology of a universal science, with all its Western ethnocentrism, has consequences for the integration of the world scientific community and for science policy recommendations. Advocates of a world scientific system argue that smallness and provincialism impede the evolution of a scientific critical mass in a country and hinder communication with competent colleagues. If scientists must belong to the international system, scientific performance in LDCs should be associated with the adoption of foreign institutional forms.

This view, while it is readily explicable in terms of institutional theory, may have undesirable consequences for local economic performance by creating a gap between research effort and national economic needs. Local economies may exert less pressure on researchers when they are integrated into an international scientific community that promotes specialty-oriented science. State-of-the-art research directed toward a discourse of theoretical knowledge may be preferred to research that deals with local health problems, energy needs, and food production.

The rationalized belief in a context-free, specialty-oriented science, reflected in adherence to the values of universalism, is increasingly seen as quaint in contemporary science and technology studies.[21] But in the context of the Third World, it may actually be harmful.

THE PROCESS OF TECHNOLOGICAL DEVELOPMENT

The majority of research done in the context of LDCs relates more directly to technology than to science. Technology has more explicit relevance for dependency arguments because it includes the development and improvement of industrial processes, transfer or invention of artifacts, improvement of crops and food production, and shaping of social institutions. Just as the literature above points to problematic relationships between scientific and economic sectors, the presence or absence of ties is a key theme in research on technology.

Technological development in LDCs can refer to a wide spectrum of changes, their determinants, and their consequences. In this section we review current work on (a) the transfer of technology across international boundaries, (b) the process of state regulation, (c) the generation and adaptation of technology within LDCs, (d) the effects of technical change, and (e) arguments concerning Appropriate Technology.

Technology Transfer

Relations between countries are almost always *relations between organizations*, either public or private. "Technology transfer" examines several aspects of these relations.[22] It denotes movement of artifacts and/or knowledge. Products and processes developed in other countries are shifted across the boundaries of LDCs. It has often been observed that most technology transfer by MNCs does not take place for the benefit of the recipient country.

Technology transfer is clearly not a process that characterizes a "stage" in the development of the Third World. It occurs constantly in the First World as well. Nor does it only involve relations between industrialized and less industrialized countries. Headrick's (1988) historical overview of technology transfer between Britain and its colonies reveals no simple "advantage" gained from technological imports or exports. The same technology serving as a basis for the influence of colonizers sets in motion processes of social change undermining their power.

In theory, technology transfer is closely related to the *diffusion of innovations*, yet linkages between the two bodies of literature have not been close.

A bibliometric study of the subject showed that the two fields developed independently in the late 1960s and early 1970s (Cottrill, Rogers, & Mills, 1989). This had not changed by the end of the 1980s and the decreasing willingness to see technological diffusion as a process of adoption without adaptation suggests it may remain so.[23]

Neoclassical treatments of technology transfer, largely *economic* in character, stress such factors as (a) the complexity of the product or production techniques being transferred, (b) transfer environment in sender and recipient countries, (c) absorptive capacities of the recipient firm, and (d) transfer capability/profit-maximizing strategy of the donor firm (Baranson, 1970). Mansfield (1975) discusses forms of transfer, modeling transfer costs as a function of the experience of both source and recipient firm, the number of years the technology has been in existence, and the number of firms that have already applied the technology. In a study of 37 chemical, semiconductor, and pharmaceutical innovations, it was shown that the rate of technology transfer across international boundaries is increasing more rapidly than in the past as a result of the growing influence of multinationals (Mansfield et al., 1983).

Short-term strategies of private enterprises do not necessarily or even ordinarily promote "development" in its variety of meanings except in the trivial sense that some industrial or business activity takes place within an LDC. The notion that LDCs are free to choose autonomously from different technological alternatives or that entrepreneurial activity will always enhance growth and development in recipient countries is anachronistic. The relationship between supplier and receiver often involves a conflict of interests, a claim emphasized by dependency arguments. For instance, Kaplinsky's (1988) study of a large, export-oriented pineapple processing factory in Kenya shows in detail how the distribution of gains favors the MNC rather than the local area.

However, there are numerous criteria for successful technology transfer, reflecting the interests of a variety of constituencies. They include operational ones such as return-on-investment or growth in sales, second-order consequences like international competitiveness (export growth), or the development of indigenous capabilities (personnel training, ability to innovate) as well as infrequently measured but still critical social effects such as unemployment and inequality. Much of the literature on technology transfer consists of case studies that pay little systematic attention to these variable dimensions of transfer.

The *type of dyadic relationship* between the supplier and recipient may be the most significant single factor in predicting the consequences of technology transfer. As Derakshani (1984) argues, the location of control (ownership, managerial), personal interaction, initial supplier involvement, and stability of the relationship are crucial to the motivation of the supplier. Greater control, interaction, involvement, and stability are associated with supplier willingness to invest in transfer. Because the prices paid for foreign technology are not fixed, but instead depend on bargaining, this process may be a fruitful area for microsociological research on the pricing of technology. Helleiner (1988) discusses bargaining strategies in a preliminary way.

Technology embodied in artifacts can simply be shipped between countries (in exchange for currency or influence in the politics of the recipient) but the transfer of *manufacturing* technology takes place on a scale from direct investment (a wholly owned subsidiary), to independent licensees, through the intermediate arrangement of a joint venture.

Licensing agreements imply relatively distant relationships (fewer interactions) between source and recipient. Some argue that licensing (as well as trademarks and patents) is a form of technological domination owing to the high degree of technological concentration (Vayrynen, 1978) but it is also thought to assist in import-substitution industrialization. Although licensing may be enough for technology transfer to developed countries, LDCs often need more sustained relationships than are implied by the rights to use proprietary information (Marton, 1986).[24]

For the same reason, studies of *"turnkey" operations* (Akpakpam, 1986; Al-Ali, 1991; Beaumont, Dingle, & Reithinger, 1981) are generally critical of these agreements. The sale of manufacturing hardware, while relatively simple, fails to promote technological mastery in the recipient country owing to constraints on the nature of the content of interactions involved.

Economic arguments regarding the appropriability of innovations suggest that, for high technology, MNCs find it more efficient to *transfer within the firm*, and that many of their characteristics evolve to protect innovations from latecomers and copiers (Magee, 1981). Although part of the variation in transfer decisions is explained by the policies of host countries, the *organizational structure* of the MNC plays an important role. William Davidson's (1983) study of 57 U.S.-based multinationals and 954 new products shows that structures that centralize learning benefits are more efficient at transfer than those that distribute them (such as global product divisions).

It was generally assumed, particularly toward the beginning of the period we consider here, that the role of MNCs and establishment of "internal" or "intraorganizational" transfer mechanisms has eclipsed patents and licensing (Michalet, 1979) and has had the effect of centralizing R&D in the home country, with limited and local R&D labs of affiliates. We return to this point after considering state policies and regulatory mechanisms.

Received wisdom regarding the *R&D activities of multinationals* suggests a variety of negative effects, centering on the generation of dependence in recipients: (a) Much technology transfer consists of internal technical trade between MNCs and their subsidiaries; (b) within MNCs there is a lack of correspondence between book price and the real price of any internal technology transfer; (c) MNCs devote more resources to innovating new products than new production processes; and (d) R&D activities of MNCs are centralized in parent companies while R&D activities of subsidiaries are tightly controlled; and (e) the R&D burden is unevenly distributed, to the detriment of the subsidiary, as a result of the product cycle.[25]

Microstudies have emphasized the *characteristics of people* and the way training influences "malleability" of personnel or levels of understanding that can lead to adaptability (Holland, 1976; Holsinger & Theisen, 1977). MNCs are generally more interested in the talent pool, general education, and perceived personality traits of workers (Chen, 1981; Kollard, 1990) than the ability of managers to attend to "ethnomethodological cues," which are equally important (Washington, 1984).

Although the literature stresses the role of MNCs because of their dominant role in technology transfer, there has been some attention to "non-MNCs" (Alam & Langrish, 1981; Meissner, 1988), the role of regional and international organizations (Amarasuriya, 1987; Bailey, Cycon, & Morris, 1986; Del Campo, 1989; Haas, 1980; McCullock, 1981; Patel, 1984; Sussman, 1987; Weiss, 1985), and universities (Glyde & Sa-yakanit, 1985; Seitz, 1982; Shayo, 1986).

The *transfer of military technology* involves a unique set of issues and has been studied by a different community of scholars. Military transfers may increase the likelihood of armed conflict (McDonald & Tamrowski, 1987). They can, at minimum, consist simply of hardware and spare parts for counterregime coalitions or can involve a great deal of research as well (Varas & Bustamente, 1983). The recent discovery of the massive and covert Iraqi attempt to develop a nuclear weapons capability with long-term assistance from a variety of state and private sponsors heralds an emerging area of specialization: the reconstruction of technical systems. With the decline of competition between the United States and the former Soviet Union for military sponsorship of LDCs, many issues of technology transfer will concern ballistic missile capabilities and their technological features (Mahnken & Hoyt, 1990).

STATE INTERVENTION AND REGULATION

No theoretical argument is needed to underscore the fact that local LDC firms are at a disadvantage in their relationships with MNCs. State action (including industrial, trade, and R&D policy) constitutes third party intervention in these relationships and is deemed crucial to the promotion of science and technology for development. In the 1970s these policies focused on indigenous S&T institutionalization. Throughout the 1980s "self-reliant" policies, though never completely rejected, were rethought and redefined. Arguments that suggested the closure of economies by overt state action fell to more moderate positions that did not entirely reject the experiences of industrialized countries as relevant to development (Macioti, 1978).

Increasingly it is recognized that state organizations compete with other institutions in LDCs and that, apart from the few socialist countries remaining, they are often too weak to implement unilateral change (Migdal, 1988). The state may, with sufficient funds, set up national research institutions, promoting technology directly. Authoritarian regimes may actively seek to suppress research—a relatively easy task where resources are scarce (Puryear, 1982). Most of the literature deals with indirect mechanisms of state intervention such as changing the mode of association between foreign suppliers and local operations (full ownership to contractual relations), altering the costs of the transfer (balance of payments, restraints on local firms), and the content of the "technology package" (the information, services, rights, and restraints; Contractor, 1983). A massive early study of technology showed that links between research and production in LDCs are weak or nonexistent (Sagasti, 1978). The most important conclusion was that policies on foreign investments, credit and interest rates, patent and trade regulations, imports and exports, project analysis criteria, market protection, and social inequity are more influential in determining the direction of technological change than R&D policies per se.[26]

Although openness to trade is often simply argued as a corollary of nonrestrictive trade policies in general or formal economic models (Teece, 1981; Wang, 1990), some challenge the basis of LDC importation policies and argue they are misguided (e.g., Contractor, 1983). It is difficult if not impossible for governments to effectively regulate the mode, cost, and content of technology imports in the context of incompatible national objectives and competing constituencies.

Science and technology *policy planning in socialist countries* has been a major topic of discussion, especially historical accounts of policy shifts, organizational arrangements, and case studies of specific areas (Jamison & Baark, 1991; Saich, 1989; Simon, 1986; Vien, 1979; Volti, 1982). China's experience during and following the Cultural Revolution is an instructive but still little understood episode (R. Baum, 1982; Chai, 1981; G. Dean, 1972; Fangyi, 1987; Mendelssohn, 1976). On the one hand, a general policy of promoting science and technology (exemplified in Deng Xiao Peng's slogan "Science and Technology Are the First Productive Force") through the system of government ministries was designed to decentralize funding decisions by allowing regional and local bodies to make decisions and disburse funds (Blanpied, 1984). On the other hand, the networks of interpersonal relations (*guanxi*) so critical for support and resources in centralized economies tended to *re*centralize and undermine local control, as program directors who had been passed over used their direct ties with the ministries to gain resources (W. Fischer, 1984).

India's technology policies, which became increasingly restrictive from the mid-1960s after relatively open policies in the 1950s, have been subjected to intensive study, not only because it is the largest LDC for which information is readily available but because of its active social scientific community (Ahmad, 1981).[27] By the mid-1960s India had a collection of state-sponsored agencies whose activities were not directly related to the production system and an industrial system heavily dependent on foreign technology. During Indira Gandhi's regime, it witnessed the creation of new agencies for research and administration, changes in relative allocations among agencies, restrictive and selective policies toward foreign investments and collaborations as well as efforts to link the S&T system with other production sectors of the economy through an S&T plan (Natarajan, 1987). In spite of political infighting and bureaucratic conflicts, the development of research and technological capabilities in India has been considerable.

A study of the Indian scientific instruments industry from 1947 to 1963 found that restrictions on imports aided the growth of a domestic industry, but it required a decade before production could take the place of imports (Clark & Parthasarathi, 1982). Alam (1988) examined foreign collaborations approved between 1977 and 1983. Interviews with 211 technology-importing firms and government officials revealed a huge demand for technology, and restrictive policies had a limited effect in promoting indigenous development. A study of 42 British exporting firms found that, owing to the diversification of suppliers, restrictive Indian import policies were having a negative effect on the modernity of technologies (Bell & Scott-Kemmis, 1988). These authors argue that restrictions enlarged the gaps between potential, planned, and actual content of technologies transferred. The effects of the new Indian policies of the late 1980s—simplified licensing procedures, increased competition, and greater technology importation—are yet to be evaluated (Bhagavan, 1988).

Detailed studies of *how* science and technology policy is made in LDCs are still rare. Hill (1986) notes the general acceptance of S&T indicators for planning but a lack of influence of science and technology planners owing to an absence of ties to central decision makers. However, there are many case studies of the policies and performance of specific governments, for example, Sri Lanka (Ramanathan, 1988), a series of UNESCO studies, microelectronics in Brazil (Erber, 1985; Evans, 1986; Hobday, 1985; Langer, 1989; Westman, 1985), and petrochemicals in Nigeria (Turner, 1977). Such studies of policy do not thoroughly consider the role of elites and constituencies in policy formation but others tackle this problem (Adler, 1988; R. S. Anderson, 1983; Botelho, 1990; Morehouse, 1976; Ranis, 1990; Sharma, 1983).

In the midst of much legitimate concern over the "carriers" of self-reliance and its likelihood in the light of current international economic and political relations (Ernst, 1981; Forje, 1986), a number of instances arose of countries that seem to be success stories, defeating the predictions of those who thought underdevelopment a permanent state. Government involvement in science and technology policy is evident in studies of the "Newly Industrializing Countries" (South Korea, Hong Kong, Taiwan, Singapore, Mexico, Brazil). Following the lead of Japan, the idea that industrial development would *precede* technical development led to government investment in industry and in the education of workers. The importation of technology from North America and Europe was not seen as a threat but instead as an opportunity for assimilation, improvement, and reverse engineering (Chiang, 1989; Shishido, 1983; see Vogel, 1991, for a readable introduction to the "four dragons").

Some of the best studies use the Korean example to illustrate state involvement in a selective manner, particularly in negotiating the terms imposed on foreign technology suppliers (Enos & Park, 1984).[28] Arnold's (1988) comparison of Korea and Taiwan shows the dual motivation of this involvement in both upgrading industrial capacity and strengthening military capability. Moreover, different levels of centralization in science policy seem to be effective,[29] though the precise mechanisms are debated (Jacobsson, 1985). It should be noted that, in the past two decades, as Korea has industrialized, private expenditures on research have increased from 12.6% to over two thirds, while funds for public research labs have decreased from 84% to less than one fourth (Lee, Lee, & Bae, 1986).

Based on the relative success of these "upper-tier" LDCs, James's (1988) conclusions reflect the views of many: States should be selective in supporting specific areas for R&D, shift funds to applied projects, and avoid projects with international prestige but limited local usefulness. That is, restrictive policies should be targeted to specific ends rather than guided by a general philosophy. The initiation of production may require intervention, while production for export may require liberalization. Regulations must be specific to technologies rather than general (Marton, 1986).

Technology Generation

The most important development since the mid-1970s is the enlargement of research interests beyond the neoclassical question of choice of technique (the intensiveness of labor and capital in the production process). Just as there has been a general recognition in science and technology studies that researchers must examine the local conditions of production and the tacit character of much knowledge, a consensus emerged that technological capabilities in LDCs must be examined in more sophisticated and differentiated ways.

Purchasing, plant operations, duplication of existing technologies, and innovation (Desai, 1988) are all forms of knowledge that require detailed studies. In explaining the development of the NICs, the process of innovation is an important component, but innovating new products and processes is not crucial or even necessary for industrialization if a country acquires other capabilities (Pack & Westphal, 1986). As Fransman (1985/1986) effectively argues, relevant capabilities can be gained even in the search for new products. Processes and deep forms of technological knowledge are not necessarily preferable.

By the late 1970s technologies were not simply "adopted" but "adapted" to the local environment (Teitel, 1977). It was realized that significant processes of technical change were occurring within some LDCs and certain industrial sectors (Katz, 1987; Lall, 1987). For example, Girvan and Marcelle's (1990) study of a Jamaican company attributes its success to active strategies of developing relationships with suppliers of raw materials and in-plant experimentation. Teubal (1984) attempts to measure technological learning in firms and the extent to which it is embodied in exports. The drive for self-sufficiency can result in poor productivity; foreign and local technological elements must be combined. Dahlman, Ross-Larson, and Westphal (1987) provide a well-documented statement of this viewpoint. Commitment to local technological improvement is a critical factor (Bowonder & Mijake, 1988).

What is the relationship between research performed in LDCs and imported technology? Some studies show local R&D is insufficient for international competitive purposes (Agarwhal, 1985). Wionczek (1983) examines pharmaceuticals in Mexico and shows the expected relationship between imports and low domestic R&D, while Evans (1979, pp. 172-194) summarizes activities in Brazil, including a number of taken-for-granted justifications used by MNC managers for their lack of interest in local R&D.

Yet a number of studies challenge this "either/or" view. Fairchild and Sosin (1986) show Latin American firms competing successfully with MNCs through their own R&D activity. Blumenthal (1979) shows that industries/countries that import greater amounts of technology also spend more on R&D. The complementarity between technology imports and domestic R&D was confirmed for the private sector and "low-tech" firms in India by Siddarthan (1988). Katrak's (1989) study of Indian firms shows imports increased the chances a firm would begin R&D and that firms that spend more on imports also spend more on R&D.[30] A related study showed that firms that imported technology through licensing tended to complement this with more of their own R&D (Kumar, 1987).

A number of investigations parallel studies of communication, productivity, or value systems in developed countries. Ebadi and Dilts (1986) establish an association between frequency of communication and performance for a sample of 49 research groups in Afghanistan. Singh and Krishnaiah (1989) found work climate in R&D units was related to effectiveness for a large sample of units in five countries (including Egypt, Argentina, India, and Korea) and concluded that climate was more affected by the sociocultural setting than by institutional locus. Using the same data, Nagpaul and Krishnaiah (1988) found that external linkages to users and to researchers were related to effectiveness. Suttmeier's (1985) study of fraud in Chinese science is Mertonian in orientation, with an interesting twist on the notion of normative violation. A particularly fascinating example is Blecher and White's (1979) account of the effects of the Cultural Revolution on the organization and operations of a 248-person technical unit in western China, based on the encyclopedic recollections of a single participant.

Social Effects of Technological Change

At the beginning of the 1980s, Hebe Vessuri (1980) called for an understanding of the process of technical change in Latin American agriculture in terms of the establishment of relations among actors. To a limited extent this has occurred. There are many case studies of technical change and warnings about the negative impacts of new technologies (Muga, 1987; Nilsen, 1979). The largest body of work on the effects of technical change concerns the consequences of the Green Revolution—that is, developments in agricultural production technology—on productivity, employment, inequality, and landownership[31] as well as health, the environment, and social unrest (Anthony, 1988; Goody, 1980). More recently, the new biotechnologies have been a focus of interest.[32]

Studies of the transfer of Western agricultural practices (modern varieties of seeds, fertilizers, and pesticides, along with mechanized production) tend to fall into distinct camps, depending on whether the authors support or oppose these developments. Theories of the relationship between technology and employment suggest that new agricultural technologies may increase yields but require capital investments that increase local inequalities and dependency on suppliers.[33] Pearse (1980) provides the basic review of U.N. studies indicating that where inequalities exist, Green Revolution strategies result in the persistence and creation of poverty in rural areas.[34]

Another group of authors feels the negative consequences of the Green Revolution have been exaggerated.[35] Forsyth, Norman, McBain, and Solomon

(1980) attempted to show that the adoption of labor-intensive technologies does not reduce unemployment, while Gang and Gangopadhyay (1987) argue that they may actually create long-run unemployment. Bayri and Furtan (1989) examined the Turkish case, finding that high-yield crops only displaced labor because wheat is generally less labor intensive than the crops it replaced and that real wages fell owing to population growth. Blyn (1983) seeks to demonstrate the positive effects of tractorization in India.

Other studies find little impact at all. Herdt's (1987) study of Philippine rice farmers generally shows little change in the real incomes of rice farmers or laborers over the period from 1965 to 1982, while Diwan and Kallianpur (1985) found new fertilizers had a minor impact on grain production. Leaf's (1983) study of a village in the Punjab from 1965 to 1978 found real gains, economically and ecologically, including an increase in equality. Of interest, and perhaps telling, Zarkovic's (1987) study of the effects of technological innovation on agriculture over a 20-year period showed that the agricultural labor force increased in the Punjab but decreased in Haryana, suggesting that one cannot generalize about the Green Revolution as so many have done. Although labor-saving technology is usually not considered desirable, even this does not completely generalize (for Saudi Arabia, see Looney, 1988).

The comprehensive recent volume by Michael Lipton with Richard Longhurst (1989) seeks to solve the "mystery" of how modern seed varieties "work" but fail to alleviate poverty. Modern varieties *do* reach small farmers, reduce risk, raise employment, and decrease food prices. But because the poor are increasingly landless workers or near-landless farm laborers, these benefits are readily diluted or diverted. In general, the negative impacts of the Green Revolution have been greatest when modern technologies are introduced under conditions of high inequality (Buttel & Raynolds, 1989).

Clearly, much remains to be done in analyzing the social effects of new agricultural technologies in terms of their interactions with political systems and the increasing interest in gender. Sugarcane harvesting in Cuba has been used to suggest that socialist countries can adopt mechanized techniques with fewer social dislocations than capitalist countries (for the comparison with Jamaica, Edquist, 1985; for the Dominican Republic, Clemens & de Groot, 1988). The effects of technical change on women show that modernization has done little to free rural women from traditional roles (Ahmed, 1986; Stamp, 1989). In western Africa, von Braun (1988) found that technological change led to improved nutrition, but that productivity improvements, rather than improving the lot of women, simply led to an inflow of males into crop production.[36]

One must conclude that (a) different disciplinary perspectives and research traditions lack integration and (b) studies of the social impacts of technology in LDCs have yielded considerable information on particular types of situations, but there are too few comparative studies to allow a systematic assessment of social effects. The debate over Appropriate Technology helps to encapsulate many of the most fundamental practical and theoretical issues in the area.

Appropriate Technology and Technology Assessment

Throughout discussions of technology choice and the effects of technological change, the concepts of *Appropriate Technology* (AT) and the prior term *Intermediate Technology* (IT) as well as diverse but related offshoots such as *technological blending* (Bhalla & James, 1986) and *optimal technology* (Rao & Dubowl, 1984) are pervasive. The idea of "appropriateness" has been applied to everything from bicycle manufacturing (Onn, 1980) to management (Leonard, 1987) to information technology (Davies, 1985). The point of most discussions is that technology should be designed and assessed, adopted and adapted, with some concept of basic needs in mind (Yapa, 1982).[37]

Originating in the 1970s with the work of E. F. Schumacher, the idea that some of the negative consequences of capital-intensive technologies imported from highly industrialized countries could be reduced or prevented by the adoption of smaller scale, labor intensive, less mechanized technologies, was transformed into a kind of social movement (e.g., Dunn, 1978).[38]

The classic economic work on the subject, Frances Stewart's *Technology and Underdevelopment* (1977), lays out the basic assumptions. Investment in technology by LDCs is associated with dualistic development, benefits accruing primarily to the modernized sector, and growing unemployment. The technologies themselves are not to blame but poor selections of technology conditioned by the environment in industrialized countries where they are developed and hence ill-suited to the vastly different conditions of LDCs. This selection is influenced by the alliance of interests between the developed countries and the advanced technology sector in receiving countries. Stewart argues that choice of technique should be defined to include all the different ways in which basic needs may be met. Appropriate Technology is likely to be older technology from advanced countries, traditional technology from the Third World, or recent technology that has been designed with local conditions in mind.

Perkins's (1983) study of 10 industries in Tanzania exemplifies the problem. Although most of the output came from small-scale production practices, the state purchased more capital-intensive, less efficient technologies, owing in part to budgeting procedures. Ahiakpor (1989) surveyed 297 Ghanian manufacturing firms of five basic types: foreign owned, private, mixed foreign-private, state owned, and mixed state-foreign. Based on import dependency and the highest capital-labor ratio, he concludes that mixed state-foreign firms select the least "appropriate" technologies, suggesting to some the collusion of local and foreign elites supported by the state. A large number of empirical studies of Appropriate Technology are annotated in Ghosh and Morrison (1984).

Critics of AT have not been wanting (DeGregori, 1985; Ndonko & Anyang, 1981). It is ironic that the Appropriate Technology movement, *which* drew much of its appeal from its acknowledgment that technology is not context-free, was quickly criticized for neglecting the social and political context in which ATs were to be introduced (Howes, 1979).[39]

Some kind of assessment of technology would seem to be implied in the idea of Appropriate Technology, but assessments are often abstract and theoretical rather than systematic and empirical. Often they are simply of no interest to participants.[40]

Elzinga's (1981) discussion of the ideological and methodological presuppositions built into the evaluation process for development aid applies just as well to technology assessment. There are built-in limitations to the ability of LDCs to undertake technology assessment. Randolph and Koppel (1982) studied technology assessment in seven Asian countries, finding that activities had already peaked and that most countries were simply interested in accelerated adoption. The United Nations and the World Bank have been active in this area (programs and projects are reviewed in Chatel, 1979, and Weiss, 1985). Numerous proposals for various assessment and decision-making methods are now in existence (Flores-Moya, Evenson, & Yujirotayami, 1978; Hetman, 1977; Salmen, 1987; Sharif & Sundararajan, 1984; Thoburn, 1977; Trak & MacKenzie, 1980).

From the perspective of LDCs, factors that should be important in decision making include the question of whether the technology is at the right level of sophistication for the country in question and whether it offers the best value in the long run, rather than its initial cost. Ahmad's pragmatic and thoughtful discussion of the issue identifies the main issues of evaluation for developing countries in terms of cost, quality, scale, degree of sophistication, risk of failure, and environmental risks. Along with many authors, Ahmad (1989) feels that the AT movement has romanticized the problem and that frequent conflicts of interest prevent realistic assessment.[41]

CONCLUSION

This review has focused on published material in English on science and technology in less developed countries from 1976 through 1992. In this sense it is just as limited as the productivity studies discussed in the first part. Science and technology studies in the LDCs themselves have grown voluminous over this period.

In China, to take only one example, a field known as the Dialectics of Nature began with the translation of Engel's work in the 1930s. Populated mainly by scientists, the field was dominated by philosophical issues, Soviet influences, and a focus on pre-twentieth century materials until its disappearance during the Cultural Revolution. During the past decade, new professional associations and doctoral programs have been established. The Dialectics of Nature is now an umbrella for more than 3,000 professionals who study the history, philosophy, and sociology of science and technology as well as science policy. Marxism remains the dominant perspective, but the extent to which it provides a genuine organizing framework has decreased. Works by Merton, Kuhn, Popper, and Price are now widely read as a new generation of students searches the Western literature for alternative critical perspectives.[42]

Given our cultural focus, as well as the splendid interdisciplinarity of the field, it seems facile to speak of common understandings. Descriptively we identified (a) a new focus on the incremental technical change that characterizes most LDC activities, (b) a recognition that it is unproductive to distinguish sharply between the generation of new technology and the modifications necessary to new conditions, (c) a new emphasis on the transfer of technology that is tacit rather than explicit and codified, and (d) an awareness that the ability to produce basic science is not strongly associated with the adaptation and use of technology. The reorientation has not occurred, as the story goes in some areas of S&T studies, as a result of theoretical or epistemological considerations. It is more readily told that people wondering *how technology was done* in LDCs—rather than speculating about why it was not up to world standards—began to examine local conditions for the production of knowledge (see Fransman, 1985/1986, pp. 580-610; Rosenberg & Frischtak, 1985, pp. vi-xvii).

It has been argued, for instance, that the nature and dispersion of agricultural R&D in Africa are largely responsible for its irrelevance (Lipton, 1988). But researchers in LDCs have been, if anything, *more* sensitive to the process of technological adaptation, or "ethnoscience," in informal field and community-based research settings (Hainsworth, 1982; Herrera, 1981).[43] An important focus of research will remain the beliefs and practices of small groups of specialized S&T actors, but the generation of knowledge need not be studied in research laboratories and should certainly not be confined to the activities of scientists and engineers.

Nor will micro-level processes be sufficient to account for variation in the development of science and technology. As organizational theorists argue, the configuration of relationships within and among firms, national laboratories, and universities provides an important context for decision making and resource allocation. An understanding of technology transfer requires a sophisticated knowledge of the causes, varieties, and consequences of interorganizational relationships (Plucknett, Smith, & Ogediz, 1990; Shrum & Wuthnow, 1988). At the macro level, the question resolves itself into a debate over whether the nation or the organization is more important to competitive rivalry (Dore, 1989; Shrum & Bankston, 1994).

Social network models offer an opportunity for integrating micro and macro approaches through their focus on social actors—both individual and organizational—and a conceptualization that involves both the presence and the absence of relations within a social system.[44] A low density of ties between researchers and users, especially when combined with ties to Western research centers, can translate into inappropriate technologies or irrelevant research (Baark, 1987; Crane, 1977; Vessuri, 1986). Scientific "centers" in developing countries may even be less relevant, in the sense of responding to national policy, than regional institutions (Jimenez, Campos, & Escalante, 1991). Owing to the fact that organizations are differentiated entities and develop internally competing interests, the mechanism of accumulative advantage is different at the micro and macro levels.

The world technical community is now a reality, but one characterized by high levels of differentiation and inequality. The task of the next 15 years is to examine it without national bias.

NOTES

1. One of our reviewers suggested that the constructivist emphasis in STS programs during the 1980s was responsible for a decline in internationally oriented STS studies. Too, the 1970s' flow of funds into social scientific studies of development declined substantially during the 1980s. But we are aware of no decline in research or interest in S&T by writers across the broader range of disciplines.

For an introduction to both social constructivism and theories of development, see Yearley's *Science, Technology, and Social Change* (1988). For a masterful review of developmental economics in the late 1970s and early 1980s, see Fransman (1985/1986). Shrum, Bankston, and Voss (1994) provide references and annotations for the entire period covered in this chapter, from 1976 through 1992.

2. Notable omissions from the variety of topics considered under these broad headings are personnel issues such as the "brain drain" and the specialized information and library technology literature. We also exclude the large body of research on educational systems and university-industry relations. David Hess's (1991) work provides a case study of Brazilian Spiritism and an introduction to another neglected area, the shifting boundary between science and the occult.

3. For a Muslim view, see the works of Sardar (1977) or Zahlan (1980).

4. A surprise, in light of all that is said about the orientations of people in LDCs, studies of *attitudes* toward science and technological change are relatively scarce (but see Ghazanfar, 1980; Gomezgil, 1975; Malik, 1982, as well as Pattnaik, 1989, for science and religious beliefs).

5. See Portes (1976) and Badham (1984) for general reviews of these approaches.

6. Recently, however, Ramirez and Lee (1990) decomposed total tertiary enrollments into tertiary science and tertiary nonscience enrollments, finding that between 1960 and 1980 economic development was positively influenced by the relative number of tertiary science enrollments.

7. Although we do not properly review the distinctions here, both are "relational" approaches to development. While world system theory is a "holistic" network approach, treating the relations among positions within the world system of relations, dependency is an "actor-centered" approach that focuses on the array of (dyadic) relations influencing particular nations. The several versions of the dependency argument incorporate science and technology in different ways and some are critical of the "classical" dependency view that connections with highly industrialized countries prevent development. For example, Evans's (1979) account of dependent development in Brazil emphasizes the technological dependency that occurs in the context of a "triple alliance" between elite local capital, international capital, and the state.

8. Although it is difficult to calculate, dependency of LDCs on research funds from core countries is significant. Up to two thirds of LDC research budgets come from foreign sources while foreign aid pays for about 40% of agricultural research (Gaillard, 1991, pp. 142-144).

9. By "institutional" or "neo-institutional" theory, we refer to the Stanford school associated with John Meyer, as distinct from Ben-David's work on the institutionalization of science in Europe, which is Mertonian in emphasis. See Zucker (1987) or W. R. Scott (1987) for reviews.

10. Deborah Fitzgerald (1986) demonstrates that foreign aid of the Rockefeller Foundation in Mexico was guided by the ethnocentric strategy of increasing crop productivity through those crops, farmers, and agricultural students that most closely resembled their American counterparts. International organizations often enforce these mimetic processes.

11. See Evans and Stephens's (1988) account of the "new comparative historical political economy."

12. The reader may consult Moravcsik (1975) for a review of materials before the period covered by this chapter.

13. Morita-Lou (1985) and Arvanitis and Gaillard (1992) both provide excellent collections of work on S&T indicators for LDCs.

14. Moreover, as Schott (1991, p. 456) emphasizes, because population growth is more rapid in LDCs, their per capita share of world publications is decreasing.

15. For example, the Gini coefficient for inequality in scientific productivity in 1973 was 0.91, compared with 0.85 for economic inequality (measured by GNP), 0.75 for the distribution of national population, and 0.74 for the distribution of national land area (see also Hustopecky & Vlachy, 1978).

16. When the research contributions of Israel and South Africa are removed, non-Western countries contributed only 4.6% of the volume of world science (Frame et al., 1977, p. 506).

17. Garfield (1983) found that North America, Europe, and the former Soviet Union contributed 94% of world scientific production.

18. The database, compiled by the Institute for Scientific Information (ISI), includes bibliometric data for 150 countries and 106 scientific fields.

19. See Sen and Lakshmi (1992) for the Indian case.

20. Even this may not be relevant to the question of research emphasis, if, as Dhirendra Sharma estimates, countries such as India devote up to 60%-70% of their research funding to military, nuclear, and space science—fields in which results are largely unpublished.

21. See, for example, Goonatilake (1988) and compare Rabkin (1986).

22. Often the term *international technology transfer* (ITT) is used in the context of cross-national transfer to distinguish it from intranational processes.

23. Our assessment is shared by Reddy and Zhao (1990) in their excellent review of international technology transfer.

24. The relationship between licensing and dependency was examined by Mytelka (1978). A survey of general and production managers in Costa Rica whose firms were producing under license indicated problems with licensing (Grynspan, 1982). For investment licensing, see Bhatt (1979).

25. This shortened version is from an OECD paper by Michalet (quoted in Wionczek, 1976).

26. But see Pack and Westphal (1986), who argue that technology policies are more important than trade policies.

27. India is the best documented case for most problems in LDC science and technology (Krishna, 1992; Rahman, 1981; Visvanathan, 1985).

28. See also Lee (1988) and Choi (1988).

29. Taiwan's National Science Council is somewhat weaker than Korea's Ministry of Science and Technology but both have been relatively successful in promoting science and technology.

30. Imports of technology help promote in-house R&D, but the effect is limited. Larger firms have proportionately lower R&D spending.

31. This remains an area of much controversy. Alauddin and Tisdell (1989) review the effects of the Green Revolution in Bangladesh, showing a concentration of landownership along with a reduction in variability of food-grain production and yields. Chadney (1984) documents disparities in India, while Jaireth (1988) examines class differences in tubewell use.

32. Lawrence Busch, William Lacy, Jeffrey Burkhardt, and Laura Lacy provide an introduction and analysis, with some attention to LDC issues (Busch et al., 1991, pp. 169-190; also Buttel, Kenney, & Kloppenburg, 1985; Kenney, 1983; Rigg, 1989). See the debate between Buttel (1989, 1991) and Otero (1991) on the issue of whether biotechnology is "revolutionary" or "substitutionist."

33. R. V. Burke's (1979) study shows that larger Mexican farms benefit disproportionately from new technology. Quiroga's (1984) study of irrigation systems in El Salvador concludes that these require not only management but mechanisms to ensure that benefits actually accrue to peasant cultivators. Walker and Kshirsagar's (1985) study of the introduction of threshing machines shows primary benefits are to the owners of capital who buy and rent the machines. See also Shaw (1984).

34. See also Griffin (1974), Dahlberg (1979), and Byres (1981) for important Green Revolution critiques.

35. Y. Hayami and V. Ruttan (1971) should be consulted for their theory of induced innovation, suggesting economic forces induce technical change, which explains variations in agricultural productivity (for a history of the theory and a critique, see Koppel & Oasa, 1987). Induced innovation theory stresses the distinctively public nature of much agricultural R&D. A decentralized network of research centers are funded by developed countries specifically to promote innovation in crop science and its adaptation to local conditions (Ruttan, 1989). The Lipton and Longhurst study (1989) discussed below originated as an impact assessment of the work of these centers.

36. Technologies of control and communication have been explored by Buchner (1988), who examines the diffusion of the television and telephone cross-nationally with reference to their different possibilities for centralized control. Ogan (1988) examines the culturally specific uses of the videocassette recorder in Turkey. Straubhaar's (1989) study looks at the development of television in Brazil in the context of the transition from military to civilian rule.

37. Peter Heller (1985) treats most of the basic concepts of transfer while supplying 21 case studies focusing on its myriad effects. The volume is excellent for teaching purposes. For a

polemical and amusing introduction (anti-AT in orientation), Rybczinski's (1991) *Paper Heroes* has been reprinted. See also Inkster (1989) for terminology and Long (1978) for a related concept of basic needs.

38. Ghosh (1984) is a good reference source on this subject. McRobie (1979), Ovitt (1989), Smil (1976), and Riskin (1978) offer a number of interesting examples of successful AT.

39. See the informative 1987 debate between Frances Stewart and Richard Eckaus.

40. As one AT proponent said to the first author, when asked how he knew that a particular technology worked: "Well, someone may have that kind of information but we don't know. Our group just tries to disseminate it."

41. Few go as far as Hamelink (1988) in arguing that lack of responsibility in technological choice should be criminalized.

42. Gong Yuzhi, "Chinese History of Dialectics of Nature" (Parts I-IV) in *Studies in Dialectics of Nature* (1991, Nos. 1-4), and Huang Shunji, "Dialectics of Nature in New China" in *Studies in Dialectics of Nature* (1991, No. 1; Xu Chao, translator).

43. Biggs and Clay (1981), for example, examine the efficiency of formal and informal R&D in agriculture, focusing on biological and environmental characteristics that shape the process of innovation by means of case studies of farmers engaged in a continuous process of innovation. Paul Richards's (1985) *Indigenous Agricultural Revolution* provides an excellent introduction to this important line of work.

44. See the debate over the conceptualization of international science and technology between Schott (1993) and Shrum and Bankston (1994).

28

□

Globalizing the World

Science and Technology in International Relations

VITTORIO ANCARANI

THE introduction of advanced technologies to a global economy during the last decade has radically altered the relationships among states while enhancing the presence and effectiveness of multinational enterprises. As Stopford and Strange (1991) have said, "Growing economic interdependence—a much abused word—now means that competition between states and firms for market shares and investment opportunities have become much fiercer, far more intense" (p. 1). The recent slowdown in the world economy is revealing itself to be a stronger and worsening factor in the rivalry between states and firms. Trading blocs and parochial protectionism are likely to be included in the policy packages drawn up by most industrialized countries.

This chapter aims to describe the extent to which S&T can be seen as primary causes of new trends and events in international economic relations. A lot of attention has been focused on the ways in which S&T may be shaping a new global environment and how the volatility of S&T—that is, faster S&T change and increasing rates at which S&T move across state borders—may

AUTHOR'S NOTE: My thanks go to Jerry Markle, James Petersen, and an anonymous referee for comments on an earlier draft, and to Sheila Jasanoff, whose careful editing and helpful suggestions contributed very much to the final form of the chapter.

make the policy capabilities and effectiveness of governments increasingly vulnerable and sensitive. The first part of the chapter summarizes certain aspects of the alleged "globalization" of S&T and investigates the extent to which the reciprocal attitude of governments and firms is modified by interdependence and complexity. The second part focuses attention on S&T volatility and tries to analyze the origins of the present strain and potential conflicts among the so-called Triad countries (Europe, Japan, and United States) and their imminent new competitors. Though each of these points has been debated to some degree in the specialized literature of social studies of science, as well more extensively in the literature of international political economy and international relations, there is still little understanding of the impact of S&T on the present state of international affairs.

One reason for this may be the relatively underdeveloped state of theory building on this topic. In particular, there has been remarkably little contact to date between scholarship in international relations or political economy and the advances in science and technology studies that are reviewed in other sections of this volume. As a result, basic analytic concepts, including "globalization" and science and technology themselves, have been taken for granted and have not been problematized to a far greater degree than in more traditional areas of S&T research, such as laboratory studies or controversies.

The first two sections of this chapter therefore, largely drawing on existing literature, attempt to conceptualize how and why S&T have been considered a major factor enabling globalization as well as what globalization has meant in the context of the international political economy (R. Worthington, 1993). In the third and fourth sections, attention is focused on the notion of S&T volatility and the tensions and conflicts it is causing governments and multinational corporations. In the fifth section, some aspects of U.S.-Japanese S&T diplomacy are briefly reviewed. The case has been selected to show how previously conflict-free scientific exchange agreements have become contested and have thus gained prominence on the political agendas of nations.

THE IMPACT OF S&T ON INTERNATIONAL RELATIONS

Science and technology can be seen today as transnational activities entailing a worldwide diffusion of scientific personnel and activities.[1] Although this feature of S&T is not new, it is generally seen to have dramatically increased since World War II in terms of contacts, flows of people, information, and collaboration across state borders (Schott, 1991). The international dimension of science also includes more sophisticated relationships, such as joint research programs involving institutions and individuals across countries, and internationally shared research facilities, such as giant optical telescopes, space launching facilities, and macroaccelerators in the field of particle physics. The growth of "big science" and large-scale technological projects (Shrum, 1985a)—because of their size, complexity, high risk, and billion-dollar costs—is fostering international cooperation in scientific research (Granger, 1979; Rycroft, 1983; Wallerstein, 1984).

Science, and especially its technological outcomes, has stimulated other, more extended processes of globalization. By making communication easier—that is, by increasing information flows and shortening the time span needed to make information available—S&T have been widely recognized in the policy literature as key factors in causing greater international interdependence. Several observers[2] have suggested that S&T impacts have contributed some positive elements to the present international scene, though others remain more skeptical (Skolnikoff, 1993). The world—as a 1988 U.N. report perhaps naively observes—appears to be on the threshold of a new technological era thanks to high technology. Information technology has already been incorporated into the production processes of almost all sectors of the economies of industrial countries, while an increasing number of Third World countries have been able to benefit from technology transfer and to incorporate it into their own industrial structures. Globalization, however, is not a unilinear and symmetrical phenomenon forcing every sector of the international community to evolve toward a global community with no internal divisions. Such an essentially deterministic view—we may speak of a "benevolent evolutionism"—which is often present in the early theories of globalization,[3] is no longer found in recent analyses of international relations, at least overtly. Even though the massive use of science and technology in specific sectors such as the organization of the economy and finance, as well as in communication and transportation, appears to have accelerated the rhythm at which the world is shrinking, globalization has made neither nation-states and the national interest[4] nor the economic and political conflicts and confrontations for scarce resources obsolete, as the U.S.-Japanese case and the Gulf War have shown. However, some new institutional factors of globalization, such as the emergence of great multinational corporations (MNCs) and other transnational actors on the world scene, have affected world politics by causing a shift in the environment of international relations. Several analysts recognize that the growing role of MNCs (Ostry, 1990), and more generally the unavoidable presence of nongovernmental actors (J. N. Rosenau, 1992), represent potentially significant changes. One possibility for change is that these new actors, by vigorously challenging the authority of the nation-states, may compel governments to adjust their policies.

The conclusions of the various authors, however, do not always go in the same direction. Stopford and Strange (1991) show how global structural changes have altered the relationships among states and MNCs and focus on the mutual interdependence of states and firms throughout the world. As they have astutely observed: "Governments have come to recognize their increased dependence on the scarce resources controlled by firms" (p. 1). Contrasting with these views, authors working on new forms of politics, such as international environmental protection, have drawn attention to the rise of nongovernmental as well as nonbusiness actors who may be playing an equally important, though less easily recognized, part in the globalization process. Particular attention has also been devoted to networks of scientists and other providers of specialized knowledge (Haas, 1990, 1992; Schott, 1993) and, more recently, to organizations of citizen groups and transnational social movements (Jasanoff, 1994; M. Reich, 1994). Largely basing his analysis on the impact of technology-driven dynamics on societies, and on world politics, J. N. Rosenau (1990) has stressed the emergence of a multicentric world consisting of nonstate actors (he prefers to call them "sovereignty-free actors": MNCs, ethnic groups, subnational governments, transnational societies, international organizations, and other collectivities), which, he suggests, are fostering a profound transformation in world politics.

As a net result, even if national governments and international institutions (Haas, Keohane, & Levy, 1993) remain important actors in world politics, international relations and in particular the economic relations of the coming decades will be shaped not by governments or international institutions alone but by their *interaction* with global corporations and other transnational organizations. However, centering attention on globalization has not changed the core issue addressed in most scholarly writing in international relations. The race to gain control of resources and improve ranking is still seen as primary. But now governments have to achieve their aims, and ensure domestic welfare and security, while at the same time adjusting to the side effects of a decline in their authority and sovereignty that the new global actors are likely to bring about. The role of governments should not be underestimated. States are responsible for guaranteeing strategic resources such as science and technology, because of their high volatility,[5] through setting up costly infrastructure and selected policies. Paradoxical as it may be, governments have faced the growth of global interdependence by increasing their intervention. There is evidence of feverish activity being undertaken by governments as well as by global corporations to improve science and technology policies so as to capture their benefits. The search for competitive advantages is not only a source of conflict and distress among the Triad countries, but it is a difficult issue to tackle analytically and has to date received relatively little attention in the social studies of science (one exception being R. Worthington, 1993).

On a more general level, by enhancing and increasing the effectiveness of an endless list of activities (economic, financial, and so on), and by floating across countries, S&T are probably eroding the ability of government to master and determine policy in domestic as well international issues. Growing interdependence means that increasingly a variety of issues cannot be addressed, let alone solved, on a national basis. They require a wider approach and deserve ample international agreements, effective control and management activities, and huge allocations of resources. In other words, an ever-growing variety of policy problems deserve international coordination. International organizations seem quite obviously to be the appropriate ones to manage problems that stem from growing interdependence. To express it another way, we need to expand their role, perhaps by transferring power and authority from nation-states to international institutions. In international affairs, however, more than in any other field, what is needed is rarely what occurs.

Just by looking at the state of the art of international affairs after the collapse of bipolarism, analysts of international relations cannot easily understand the success or nonsuccess of several international institutions. While several multilateral institutions—such as the IMF (International Monetary Fund), the World Bank, and some organizations of the U.N. family—appear to be quite robust and able to adapt to the new conditions, and to accept newcomers from the ex-communist bloc (Caporaso, 1992; Haas et al., 1993; Ruggie, 1992), others, especially those related to key sectors of the economy and trade—for example, the GATT (General Agreement for Trade and Tariffs)—have found it increasingly difficult to draw up a new agenda. Trading blocs and new threats of protectionist retaliation are often adopted as comfortable policies to ward off uncomfortable economic slowdown and new challenges.

S&T IN INTERNATIONAL RELATIONS STUDIES: SHIFTING PERSPECTIVES

The relationship between international affairs and science and technology has been largely recognized. However, as Skolnikoff (1993) points out: "The most common approach to studying the interaction of science and technology with international affairs views the relationship in the context of specific policy areas, typically in relation to pressing policy concern" (p. 5). The lack of a more comprehensive approach has made, for most investigators, the issue a controversial one, especially when patterns of change and the consequences for international politics are a major concern.

Up to the 1970s, most studies dealing with the international aspects of S&T focused on the impact that S&T were having on the bipolar structure of power relations (see also Smit, this volume). The emergence of nuclear weapons was reputed not only to have changed the nature of warfare but to have made the threat of war a "virtually terrorizing effect by creating expectations of a danger going far beyond anything previously experienced" (Frei, 1982, p. 134). With the Cuban-U.S. missile crisis in 1963, nuclear deterrence reached its limit. Scholars of world politics began to question whether nuclear weapons were a realistic option in a crisis situation. Some analysts viewed nuclear deterrence in terms of a paradox and threat more than as a military solution (Allison, 1971; Mitroff, 1983). Others linked nuclear deterrence to game theory (Rapoport, 1974; Schelling, 1960) or to a strategy based on prevention rather than the conduct of war (Frei, 1982, p. 135). Still other political analysts concluded that bipolarism and mutual deterrence had developed the premises of a "security regime" and "nuclear learning" (J. S. Nye, 1987).

In the second half of the 1970s, some analysts began to think that the impact of S&T on world affairs was not limited to the nuclear challenge. S&T were singled out as powerful factors generating increasing interdependence and complexity in the world system (Keohane & Nye, 1977; Skolnikoff, 1977). Attention was shifted from a military-security issue to a broader range of topics and to a policy-centered analysis of the consequences of S&T on governments' ability to cope with the globalization of political, economic, and social life. The growth of international transactions and interactions dependent on new forms of transportation and communication, the emergence of new problems stemming from the impact of global technologies, the depletion of environmental resources and the permanent OPEC threat of an energy crisis, the population growth in the Third World countries, the emergence of new governmental and nongovernmental protagonists, and the increased "feasibility"[6] of actors and their capability to affect the daily life of others—all were factors to be included in an analysis of the interdependence and complexity of what many had come to regard as a "world society." In international studies, *interdependence* and *complexity* became two key words without which the fast-evolving network of relations in the world could not be understood. They functioned in the lexicon of these years as main features of the coming postindustrial society, just as S&T functioned as factors accounting for their exponential growth (Inkeles, 1975). S&T were also reputed to instigate, in a direct or indirect way, the new impetus of social and economic development toward an idealized postmaterialist or communal society (Bell, 1974).

With the oil shocks of 1973-1975 and the publication of the reports of the Club of Rome (1971 and 1975), there developed a new and more profound preoccupation with what Keohane and Nye (1977) called "conditions of extended interdependence" and J. N. Rosenau (1980) called "transnationalization of world affairs." In the 1970s and 1980s, world public opinion and academic circles had a new anxiety. During the ideological and military confrontation of the cold war, public opinion had not been concerned very much with S&T's role in daily life; for example, in the scholarly literature of science and technology studies, the topic was mostly relegated to historical work on technology. The Chernobyl disaster is only one example of events that changed all this. The peaceful application of hard science and hard technology to daily life seems no less threatening and dangerous than the superpowers' war games. Among the controversial issues facing scholars today are (a) whether or not the human race is able to counter the unwanted effects of S&T advances and to learn from them in a constructive way (Jasanoff, 1994); (b) how ready governments and democratic political bodies are to cope with the negative impacts of S&T; (c) how effective international organizations (the United Nations and others) are at facing the global problems arising from the use of technologies such as nuclear power, and what this means in terms of the resources and authority granted to such institutions (Haas et al., 1993); and, finally, (d) how well equipped the international community is to control and to detect the spread of biological weapons (Wright, 1990) or nuclear technologies in highly developed as well as in Third World countries (Greenwood, 1990; Lakoff & Bruvold, 1990; E. Lewis, 1990).

In the 1980s Japan and the "Newly Industrializing Countries" (NICs) began their commercial invasion of the rich markets of the Western world, and a new era of globalization was perceived in North America and in Europe. New technologies—often freely imported from the United States—began to be applied to "flexible" and high-tech production combined with low manpower costs. The erosion of the long-standing North American primacy in advanced technology, and the decline of U.S. international competitiveness in favor of Japan's emerging economic power, were perceived in the West more as a threat than a competitive challenge. The growth of the so-called dual-use technologies (Gummet, 1989; Gummet & Reppy, 1988; Wright & Ketcham, 1990), the increased role of private companies in military research and production, and especially the growing use of civilian technology in military products (Gansler, 1989) have called into question the theoretically tenuous distinctions between civilian and military strength and between economic and military threat. As a consequence, the traditional issue of security is now seen to extend far beyond traditional military analysis and to include economic as well as ecological factors. The contradictions, uncertainties, and instabilities inherent in these developments were ably, if controversially, surveyed by the World Commission on Environment and Development (1987) in its report, *Our Common Future*.

Shifts in public opinion and in academic and political circles have driven such topics to the top of the political and academic agenda. At the political level, questions have been raised about the tenability of the postwar free-trade regime, while the question has been raised concerning whether international organizations—such as the GATT—are capable of ensuring fairness and reciprocity among their partners. The ability of national governments and state institutions to manage policy on a variety of major subjects, such as S&T, is considered to be under strain. Finally, in a more restricted way, the power of governments to control technology transfers and their benefits has been raised as a problem of national interest.

The shift in academic research, though less conflict ridden, is no less pronounced. Changing interest is reflected in the growing body of work in the international political economy.[7] However, the basic emphasis on the power and influence of the state is not declining in the mainstream literature of the discipline. Aspects that previously were associated exclusively with military subjects, such as regional alliances and "strategic policies," now appear to be of greatest importance for the economic future of nations. From the viewpoint of the nation-state, Robert Gilpin (1987) has accurately described the dilemma of globalization as that of obtaining the benefits of international interdependence while preserving those of national freedom of action in both internal and external affairs. "The clash between the integrating forces of the world economy and the centrifugal forces of the sovereign states has become one of the critical issues in contemporary international relations" (Gilpin, 1987, p. 380).

Analysts focusing on globalization have frequently argued that one of the major consequences of such processes is the relative decline of national sovereignty. As mentioned earlier, the formation of transnational NGO communities is a part of this phenomenon. Yet it would be misleading to claim there has been a clear drift toward a broad restructuring of the international political system. Whether these changes simply add new dimensions to the international scene, or whether they foster major transformations, is impossible to say. However, at present the most likely outcome seems to be one of persistent conflicting patterns and diverging trends.

VOLATILITY OF S&T AND THE NEW INTERNATIONAL KNOWLEDGE-BASED ECONOMY

Technology appears to play a central role in the economic competition among the three main trade blocs on the world scene today: the United States, Japan, and the European Community. After World War II, developments in transportation and communication have provided the infrastructure for rapid expansion in international markets[8] and permitted MNCs to "act at a distance" (Latour, 1987, p. 223), that is, to build a global network of organizational ties and activities. Through the 1950s and 1960s there was impressive growth in foreign direct investment (FDI) by American corporations and, in the following decades, this policy was increasingly being emulated by European and later by Japanese companies. In the 1980s the FDIs grew more than four times as fast as GNP in OECD economies. Nevertheless, the flows of foreign direct investments have been concentrated mainly in the major economic areas.[9] More recently, there was a wave of joint international ventures and collaboration, often including technology transfers and also cooperation in R&D, creating a new dynamism and a new level of interaction in the world economy.

As internationalization processes have expanded in terms of world trade, cross-country investments, transnational joint ventures, and cooperation, several basic changes have also occurred in the global competitive and technological environment. While indicators in the 1980s showed a more even distribution of economic and technological capabilities among the largest industrial economies, the relative erosion of the unique postwar economic scientific and technological supremacy of the United States has stimulated growing competition.[10] American economic primacy, as R. Nelson (1990) has argued, was based on two separate factors: long-standing leadership in mass production and, after World War II, dominance in high-technology industries. The latter stems from the U.S. government's and private industry's massive investment in research and development. According to R. Nelson, as the dramatic postwar increase in R&D and education was perhaps the key to understanding U.S. primacy, so the subsequent slowdown may also explain its erosion. In the mid-1960s, after the Japanese and European industrial infrastructure had been rebuilt, figures indicating the percentage of the workforce trained in science and engineering and the percentage of GNP devoted to R&D began to show that the gap between the most industrialized countries was narrowing dramatically (R. Nelson, 1990, pp. 123-127). Even if the international preeminence of the United States still persists in a number of advanced R&D sectors, that country no longer monopolizes key technologies.

In effect, the scenario of international competition has radically changed in the last decade and is expected to change even more radically in the 1990s. What emerges is a new form of industrial competition largely dependent on technology and R&D-intensive work, often promoted through government intervention and occurring in imperfect markets. The critical competitive challenge offered by Japanese firms and the Japanese government is the most typical case. These new competitive patterns, especially evident in technology-intensive industries, have been analytically highlighted by the so-called New Trade Theory, which has questioned the customary noninterventionist view of international trade theory and has called for a more active trade and industrial policy.[11]

As research and development, and technological innovation, became fundamental elements of the industrial policy of some highly industrialized countries, and a major source of industrial competitive advantages between firms and nations, technology-related issues fostered, and are still fostering, friction and conflicts among governments. A focus of major concern has been scientific and technological volatility. The quickening pace of technological change and the expanding international diffusion of technology emphasize how technological advantages can easily be lost. So, fears of technological "leakage" are growing.

A substantial literature on the economics of technological change and technological innovation has begun to highlight the links between R&D, viewed as a system of search processes and organization, and their translation into commercial innovation. Consequently, major shifts in the research system affecting technological volatility have been discussed. As R. Nelson (1990) argues, "Technology is proving a resource much more accessible and less respecting of firms' and nations' boundaries than before" (p. 126). Mobility is reinforced by the *scientification* of technology—that is, the seamless incorporation of advanced scientific knowledge and techniques into technological research and innovation and the overlapping and converging patterns of communication between and within scientific and technological communities.[12]

A further related aspect of S&T mobility is the *internationalization of R&D systems* through a multifaceted and continuously evolving expansion of interorganizational research linkages, often involving R&D centers, universities, and national laboratories situated in different countries (Howells, 1990a, 1990b). The growth of international as well as domestic R&D cooperation has its roots in financial necessity—reflecting the rising costs and risks involved in industrial research and product development as well as in cognitive factors. To compete in highly dynamic R&D-intensive industries, firms are pushed to quickly develop expertise in a broad range of technologies, engineering specializations, and scientific disciplines that may not be available in-house or through national ventures (Mowery & Rosenberg, 1989, pp. 213-214; R. Nelson, 1990, pp. 128-129). Major companies are therefore extending their multinational operations either by expanding intracorporate research or by developing R&D linkages and cooperative ventures with other firms, university laboratories, and research centers.

The mounting activity in business circles and governments devoted to S&T policies reveals the extent to which S&T mobility is becoming a primary concern in drawing up the policy agenda. Just as financial volatility is becoming a matter of concern for the domestic economy (Frieden, 1991), because of the almost instantaneous ability of financial assets to cross countries in search of more profitable opportunities, similarly S&T volatility appears likely to produce competition between firms and nations trying to create favorable conditions to capture or attract their benefits. Beyond mere mobility, S&T volatility is expanded by the activities of public or private agents aimed at changing their flows and therefore the associated distributional effects and benefits. Competition between firms and nations in creating more suitable conditions for S&T is visible in R&D policy packages and in other technology-related policies.

More generally, S&T volatility is perhaps the major factor explaining the growing involvement of governments in industrialized and industrializing nations in policies to promote innovation and technological advance (Mowery & Rosenberg, 1993, p. 274). Indeed, S&T volatility could be singled out as a major factor in relaunching the role of state agencies in policing and managing long-terms investments and plans (Ikenberry, Lake, & Mastanduno, 1988). Evidence of this is seen in the policy packages planned by several highly industrialized as well as the Newly Industrializing Countries (the latter are engaged in technological "catch-up") in the sectors of communication, scientific and technical education, R&D facilities, as well as in policies to encourage and finance the growth of research consortia and to use and diffuse the results of research.[13]

Further, governments today are being asked to improve their capability to address and to manage difficult and unpredictable issues such as international scientific and technological cooperation and joint ventures because of the strategic and long-term consequences that they may have on the domestic economy and on implications for bilateral and multilateral relations. The perceived volatility of S&T also accounts for several "new issues" on the policy agenda. For example, this is the case for intellectual property rights and trade in services, which occupied a prominent role in the multilateral trade negotiations that took place within the GATT framework in the early 1990s.

Inside the Triad, concerns about competitiveness, fears of losing technological advantage, and difficulties in tapping benefits from indigenous R&D have created acute strains and potential for overreactions, giving rise to periodic demands for technological protectionism and even the introduction of restrictions on international scientific communication and collaboration.

PERCEPTUAL CHANGES IN INTERNATIONAL S&T POLICY

Attitudes toward international science and technology policy have changed very quickly in the last few years thanks to the prominence of trade issues on the political agenda. This change is particularly evident if one looks at the North American debate. In the postwar era, American international science policy was characterized by an open attitude but little political prominence. Traditionally, within the Western alliance, the American emphasis was mainly on science as an open international cultural activity. As one analyst argues: "Science was recognized as a valuable mechanism for encouraging greater cultural cooperation between nations and for greasing the wheels for political, economic, and military cooperation—but not necessarily seen as an integral part of such cooperation" (Dickson, 1984, p. 174).

This period can be labeled the hegemonic phase of American S&T policy. S&T policy with regard to Europe and Japan in the postwar era fit in well with this pattern. Unchallenged primacy in scientific and technological matters, together with political leadership within the Western world, made it possible for the United Stases to ignore the usual search for direct short-term reciprocity. This hegemonic posture greatly fostered cooperation among the advanced industrialized countries (Gilpin, 1987). As Nau (1975) put it:

> The presence of a decisively larger and more enthusiastic partner may be an important precondition for initiating successful international cooperation. This partner is not only willing to accept extra costs of cooperation but also places the highest values on the purpose of cooperation. . . . Studies show that, as states acquire increased capabilities in a particular sector, they exhibit a declining propensity to seek concrete gains in task performance through international cooperation. . . . They do so apparently for political rather than technical reasons. (p. 651)

Except for nuclear technology, S&T transfer had scarce political prominence inside the U.S. administration (B. L. R. Smith, 1991, pp. 68-70). This open-door attitude was shared by all government agencies and kept them from recognizing the long-term threat of the flow of technologies out of the United States because of government policy and MNCs operating inside Western countries.[14]

From the beginning of the 1970s, this climate changed to what can be labeled a posthegemonic phase in international S&T relations. Japan, even more than Europe, was able to benefit from the postwar open-door attitude toward S&T transfer. Indeed, it was Japan's success in overcoming the technological gap and challenging American leadership in high-tech industries that raised the level of U.S. public awareness of the international implications of technology transfer. By the mid-1980s the American public began to perceive economic competitors such as Japan as a threat to national economic security (Zinberg, 1993). In this context, technological issues have entered a completely new phase and taken on great political relevance.

The 1986 U.S.-Japanese dispute on trade in semiconductors perhaps marked the major turning point. Several different factors contributed to making this dispute a dangerous and greatly symbolic high-technology trade conflict. Semiconductors were an American invention at the core of the technological revolution of the postwar era and an American-born industry. The sector was also widely viewed as being of "strategic" economic and military importance. The dispute strengthened the support for restrictive trade legislation in Congress and encouraged a more skeptical opinion concerning international trade, stimulating a growing debate on industrial policy.

The challenge to the politically more sensitive U.S. high-tech industries extended the debate on foreign access to all areas of American science, including the traditional openness of the American university system. The use that foreign companies can make of U.S. university research (developed with federal funds) and the easy access to training and educational programs have been widely questioned. The thesis that attributes the alleged loss of technological primacy and market shares in high-tech industries primarily to "unfair access" to national research sparked off a mood of technological protectionism with strong arguments for limited access to technology and research, restrictions, and more appropriate pricing of high-tech exports as well as more active measures to prevent technological espionage.[15]

Even if this thesis, despite its political appeals, was questionable at best —it does not take into consideration all the factors that are involved in the exploitation of scientific knowledge and new technologies or the practical limitations on controlling the flow of scientific and technical information between countries—it did, however, foster a more activist technology policy in United States.[16]

The EC and Japan, as new global competitors, while largely benefiting from the "cultural" openness of U.S. science, showed that they could benefit from it with the support of new interventionist S&T policies. As the sources of globalization partly lie in the inner nature of the evolution of technology and its strong links with basic science, the organizational and financial capabilities of states, or clusters of states as in the EC, acted as powerful tools in enabling nations to take quick and full advantage of new knowledge, wherever developed (Keller, 1990, p. 1370). An active role for state policy has been particularly stressed by some analysts (Porter, 1990). The new highly competitive technological environment seems to call for a strategic state as a learning organization that is able to identify and sustain potential winners on the innovation and R&D fronts (Paquet, 1990). According to Rip, by adopting intelligent policies, governments can compensate firms for the time-consuming evolution of domestic processes (Rip, 1989, pp. 15-16).

As governments "have become catalysts and major actors in the process of socialization of risk for domestic and international firms" in international competition (Schmiegelow & Schmiegelow, 1990, p. 329), the intrusive presence of governments and state agencies has politicized international relations in S&T to a greater degree. Unfortunately, the consequences of the increasing convergence of major industrialized countries in following aggressive technology policies to support the competitive efforts of domestic firms has not yet received the attention it deserves. However, the effects of all these developments will certainly widely affect the future of world affairs and of international multilateral negotiations in the foreseeable future.

THE UNITED STATES AND JAPAN: MANAGING S&T FLOWS

As S&T are increasingly a central issue in international economic relations, attempts at cooperation are becoming increasingly difficult to manage. Whether or not global competitors are successful in steering S&T issues without hindering the positive effects inherent in the globalization process, will have significant implications in shaping the new international economic order and institutions, particularly inside the Triad. What are the factors that could enable international competitors to reach workable agreements in S&T issues? Could their expectations converge toward cooperative frameworks?

The negotiation of the U.S.-Japanese Science and Technology Agreement (S&TA), signed at the June 1988 economic summit in Toronto, illustrates the complications that trade issues are introducing into international scientific and technological relations. Formerly of concern only within the scientific community, the signing of the agreement involved trade policymakers and reached the top level of the international policy agenda (Mowery & Rosenberg, 1993). U.S.-Japanese negotiations on scientific and technological cooperation are an example of the complexity introduced by the increasing international flows of S&T and of the overlapping technical, economic, and political issues generated by the new role of science as an integral part of international economic policy. When such complex and interlocking issues are at stake, even when the parties are searching for a point of convergence, values, beliefs, perceptions, attitudes, subjective orientations, as well as uncertainties about reliable information and the like can make it very hard to draw up a mutually acceptable arrangement (Haggard & Simmons, 1987).

In this case, U.S. officials were convinced that Japan had taken unfair advantage of the openness of the American science system and thus was not really committed to working for change. Distrust between the two parties was widespread during the negotiations, as was the fear of each that the other was deliberately engaging in delaying tactics. Therefore the bargaining process typically ran into stalemates, and repeatedly there was fear that the entire negotiations would fail.

Even earlier experience in bilateral talks such as the previously mentioned negotiations on semiconductors, ending in the 1986 U.S.-Japanese agreement, proved to have a negative influence on the negotiations. The continuing difficulties of the American semiconductor industry and continuous U.S. criticism of Japanese industrial and trade policies seemed to account for the agreement's lack of effectiveness and its quick erosion. Further, any point of conflict that occurred in bilateral high-tech relations appeared to threaten the future of the whole bargaining process. An example was the Toshiba affair—involving the Japanese corporation's illegal transfer of militarily sensitive information to the Soviet Union—for it poisoned the atmosphere throughout the negotiation process.

Because international negotiations on S&T issues involve so many actors, address so many problems, and touch on so many complex issues, the arrangements to be worked out require careful delimiting and boundary drawing by the parties; even then it is almost impossible to foresee all the potential outcomes, let alone the problems of implementation (Ancarani, 1990). Choices inevitably cut across a broad area of issues and contexts, and complicated causal links among the different problems, as well as a variety of unexpected results, are typical factors that cause further complication.

For example, in the U.S.-Japanese S&TA negotiations, there were no clearly defined issues but wide-ranging topics with complex and crosscutting ramifications. The most contentious issues included the following:

Exchange of scientists. The main problem was the disparity between the large number of Japanese students and researchers studying at or visiting American universities and laboratories and the small number of Americans going in the opposite direction.

Funding of basic research. At stake was the disparity in basic research funding between the two countries. The U.S. complaint was that Japan was not adequately supporting the funding of basic science.

Exchange of information. Intellectual property rights were a particularly contentious issue, which often delayed the negotiations.

The problems arising under each of these headings are multifaceted, difficult to assess, and even more difficult to deal with. Consider the topic of exchange: How can the players clarify their self-interests on this issue? Every response is clearly indeterminate. *Nature* commented on the subject as follow:

> If a researcher takes time off from his regular job to work at a laboratory elsewhere, who in chauvinistic terms can be judged to gain? The country at which the researcher chooses to spend his time, or that to which he eventually returns? The simplest answer is that there is no simple answer, and that the balance of advantages will depend on the circumstances. ("Needless Quarrel," 1988, p. 375)

The parties may agree in principle that the imbalance in the flow of scientists must be reduced. Yet solutions are complicated by striking structural differences in the funding patterns and major scientific institutions in the two countries. In Japan roughly 80% of research is funded by industry, whereas the federal government funds about half of all research in the United States. Accordingly, a government-to-government agreement can prove widely ineffective as a means to redress the imbalance of scientific exchanges or to deal with the issue of mutual scientific access between the two countries. American government and university laboratories are open to Japanese researchers, while the largely corporate-funded Japanese research centers still remain closed to foreign access (Sun, 1987).

Other complicating factors have a cultural origin. For example, making formal provisions for American researchers to be invited to Japanese universities and laboratories may not produce the intended effects. In fact, in 1988 Japan began to set up a variety of schemes through which more than 100 long-term fellowships a year were offered to American scientists and engineers, but many of these went begging (Sun, 1989). Americans evidently were not attracted to doing research in Japan. Some doubted that the Japanese were at the cutting edge of research; others suspected that many laboratories lacked a stimulating atmosphere or distrusted the differences in research culture (Traweek, 1992), while others were also influenced by the language barrier and the cramped housing provisions.

One conclusion is that the traditional quick-fix solutions that can easily be identified by state negotiators are founded on an unrealistically simplistic conception of the problem. When the objective is to remove barriers to S&T mobility, no one can forecast whether the results stemming from a given agreement will in fact work well; complex monitoring systems are required to review, step by step, the implementation of the agreement. In any case, no guarantee of success is assured.

CONCLUDING REMARKS

An effort has been made in this chapter to understand why the global diffusion of S&T has not rendered obsolete either the conflicts and competition in the international arena between states and firms or the role of the nation-states. By focusing on the fierce competition that is developing in the Triad area, the chapter has mostly concentrated attention on concerns regarding possible breaking points in the coalition between the United States, Japan, and the EC, which has governed the economic expansion of the postwar world.

After analyzing how S&T facilitated the entrance of new actors into the environment of international relations, the chapter focused attention on the way S&T are also making world politics more dynamic, even if more conflictual. The notions of mobility and volatility were introduced to explain the extent to which S&T not only are generic goods, floating seemingly freely from one country to another, but require long-term investments and informed policies to capture their potential economic benefits. As S&T volatility embodies an apparently unavoidable propensity in science and technology to search for appropriate settings, it also seems to be causing fiercer competition and difficulties inside international institutions (the GATT is an example) and among OECD governments. Nevertheless, over the longer term, S&T volatility is likely to have a benign effect on the policy capabilities of state actors, which, by setting up new policy packages and more effective implementation strategies, are expected to improve the absorption and use of scientific and technological research. One consequence of focusing attention on absorption strategies and appropriate adjustments policies could be to soften and contain pressures toward S&T protectionist postures. As Mowery and Rosenberg (1989, p. 292) suggest, government efforts to stimulate adoption and use of S&T are preferable to a policy of restricting the international flow of basic scientific and technological information and research.

At present, however, intellectual property rights, technology transfer, unfair advantages stemming from national industrial policies, market access, dumping and other unfair trade practices, as well as the participation of foreign-based companies in government-funded projects are all factors that create never-ending disputes inside the Triad area. Attempts to shift from bilateral to multilateral negotiations to create a better international climate and to facilitate more effective action to prevent moves toward protectionist postures in S&T international relations have up to now failed to bring the expected results. However, some moves have been made, an example being the OECD meeting of science ministers in October 1987, where a proposal for a general framework of common principles for international cooperation in science and technology was put forward. The proposed guidelines stressed the need for an equitable contribution from all countries to the world's basic research and reciprocity in scientific exchanges (Dickson, 1987).

There is increasing evidence that the present international regulation of S&T is coming up against many difficulties in the search for concrete, appropriate institutional settings and adequate procedures to deal with the problems that arise in the multilateral arena. A long-term flexible institutional arrangement, such as the GATT, designed to evolve over time (through "rounds" of negotiations) and to give single national S&T systems the opportunity of choosing the most appropriate domestic strategies while satisfying international rules and requirements, has to be considered but does not yet seem on the point of being achieved. States and firms, for *their* part, appear to be operating with a vastly oversimplified understanding of the interconnected and culturally contextualized nature of S&T. For researchers in science and technology studies, this means that the globalization of S&T offers much room for investigation and theoretical clarification.

NOTES

1. Obviously this does not mean that research and other scientific and technical activities are evenly distributed. In terms of worldwide distribution, R&D expenditure is highly concentrated, with over 65% in the Triad—the United States, Japan, and the European Community—and around 20% in the former Soviet Union and eastern European countries (Freeman & Hagedoorn, 1992). In terms of international links, Eastern bloc countries until recently were isolated from the international scientific community. International links among scientists can suddenly be interrupted by changes in international relations, as has been documented in the case of Iran by De Bruin, Braam, and Moed (1991).

2. Perhaps the first scholar in international relations to recognize a positive impact of technology on the texture of world politics was Merriam (1945).

3. A benign scenario of the ongoing process of globalization can be found in Burton (1972).

4. After a period of criticism (see J. N. Rosenau, 1968), the old-fashioned notion of the national interest is being reviewed and revalued by some authors; see Krasner (1978) and, more recently, Ikenberry et al. (1988).

5. For the notion of volatility, see the third section of this chapter.

6. The notion of "feasibility" has been shaped in a general and philosophical way by the German philosopher K. O. Apel (1973). Apel stresses the asymmetry between the technological empowerment of human action and the scarce moral capability of human agents to assume responsibility for the macroeffects of their actions.

7. The emergence of international political economy as a special field of study that cuts across studies of international relations and the world economy is relatively recent. For a review of this comparatively new area of study, see B. J. Cohen (1990) and Gayle, Denamark, and Stiles (1991-1992).

8. From 1970 on the ratio of world exports to world gross domestic product dramatically increased and reached nearly 20% in 1987 (Litan & Suchman, 1990).

9. According to Howells and Wood (1991, p. 32), the G-5 nations accounted for around 75% of world FDI outflows.

10. The closing of the technological gap among the major industrialized countries can be indicated by a variety of input and output parameters of R&D. Input variables—such as R&D expenditure as a percentage of GNP and the number of scientists and engineers as a percentage of total workforce—and output variables—such as patents and shares of global markets in high-tech products—are commonly used. For an overview of such indicators in major industrialized countries, see the National Science Board (1989).

11. This new approach to trade policy is due to James A. Brander and Paul R. Krugman, among others (see Krugman, 1986).

12. The multiple linkages between scientific and technological networks or systems have been discussed in several subfields of science and technology studies, and also the increasingly international character of these networks. For a survey of sociological perspectives on the science-technology relationship, see Bijker, Hughes, and Pinch (1987). Another approach has sought to measure the transfer of scientific ideas into technology by counting the number of times the scientific literature has been cited in patent applications (Carpenter, Cooper, & Narin, 1980). One study indicated that in technological areas regarded as "science intensive," such as in some biotechnology areas, the citation time lag from scientific literature to patents is the same as within the scientific literature dealing with the same topics (Narin & Noma, 1985). The publishing habits and communication networks of technology experts and engineers are becoming increasingly transnational and similar to those of science. R. Nelson (1990) notes that the "scientification" of technology has stimulated an increase in the quantity of technology-relevant knowledge that is published in journals or discussed at national and international meetings.

13. European countries and Japan have supported their own firms in securing predominant positions in the global market for high-technology products. The active government support, which has contributed to the success of these national corporate ventures, is apparent not only in the industrial policies but also in science and technology policies including the creation of science parks.

14. An interesting example is the case of the European semiconductor industry. An analyst points out that "a new generation of European managers and professionals trained within U.S. multinational subsidiaries during the 1960s and 1970s . . ., together with a larger number of less well known engineers, managers and technicians, will add significantly to Europe's overall prospects in SCs" (Hobday, 1989, p. 182).

15. One well-known case was the decision by W. Grahm—White House science adviser during the second Reagan administration—to exclude noncitizens from a meeting on superconductivity in July 1987; another involves the criticisms raised about university-industry liaison programs in American universities that sell access to their research programs to foreign corporations. Public attention was raised in 1989 when a congressional subcommittee scheduled a hearing on the MIT Industrial Liaison Program to consider whether foreign corporations were benefiting unfairly from federally funded research (M. Crawford, 1989). For a brief discussion on the subject, see also Deutch (1991, p. 491).

16. By the mid-1980s new technology policy and new legislation were introduced in the United States. These initiatives reduced the diverging patterns of technology policy between the United States and its major competitors. Two examples worth mentioning here are the National Cooperative Research Act (1984), supported by the Reagan administration, which reduced antitrust legislation for interfirm cooperation in precompetitive research. According to Mowery and Rosenberg (1993, pp. 59-60), from 1984 to 1988, 111 cooperative ventures were registered under the terms of this act. Another example of policy initiative are agencies like the National Center for Manufacturing Science (NCMS), started in 1989, and the Defense Advances Research Projects Agency (DARPA), a research program in high-definition television, started in 1990. For an extended review of American science and technology policy, see Branscomb (1993).

References

Abbott, Andrew. (1988). *The system of professions.* Chicago: University of Chicago Press.

Abbott, G. F. (1990). American culture and its effects on engineering education. *IEEE Communications Magazine, 28*(12), 36-38.

Abelson, Philip H. (1991a, February 8). Federal impediments to scientific research. *Science, 251,* 605.

Abelson, Philip H. (1991b, April 5). Industrial interactions with universities. *Science*, p. 9.

Abir-Am, Pnina. (1982). The discourse of physical power and biological knowledge in the 1930s: A reappraisal of the Rockfeller Foundation's "policy" in molecular biology. *Social Studies of Science, 12,* 341-382.

Abrams, Philip. (1980). History, sociology, historical sociology. *Past and Present, 87,* 3-16.

Ackerman, Ron. (1985). *Data, instruments and theory: A dialectical approach to understanding science.* Princeton, NJ: Princeton University Press.

Adam, Ian, & Tiflin, Helen. (Eds.). (1991). *Past the last post: Theorising post-colonialism and post-modernism.* Hemel Hempstead, U.K.: Harvester.

Adler, Emanuel. (1988). State institutions, ideology and autonomous technological development: Computers and nuclear energy in Argentina and Brazil. *Latin American Research Review, 23,* 59-90.

Advisory Council on Science and Technology. (1989). *Defence R&D: A national resource.* London: Her Majesty's Stationery Office.

Agarwal, Anil, & Narain, Sunita. (1992). *Global warming in an unequal world: A case of environmental colonialism.* Delhi, India: Centre for Science and Environment.

Agarwhal, Suraj M. (1985). Electronics in India: Past strategies and future possibilities. *World Development, 13,* 273-292.

Agassi, Joseph. (1966). The confusion between science and technology in the standard philosophies of science. *Technology and Culture, 7,* 348-366.

Agnew, John A. (1982). Technology transfer and theories of development. *Journal of Asian and African Studies, 17,* 16-31.

Ahern, Nancy, & Scott, Elizabeth. (1981). *Career outcomes in a matched sample of men and women Ph.D.s.* Washington, DC: National Academy Press.

Ahiakpor, James. (1989). Do firms choose inappropriate technology in LDCs? *Economic Development and Cultural Change, 37,* 557-571.

Ahlstrom, G. (1982). *Engineers and industrial growth*. London: Croom Helm.

Ahmad, Aqueil. (1981). Sociologically oriented studies of science in India. *International Journal of Contemporary Sociology, 18*, 135-165.

Ahmad, Aqueil. (1989). Evaluating appropriate technology for development: Before and after. *Evaluation Review, 13*, 310-309.

Ahmed, Iftikar. (1986). Technology, production linkages and women's employment in South Asia. *International Labour Review, 126*, 21-40.

Åkersten, S. I. (1987). The strategic computing program. In A. M. Dinn (Ed.), *Arms and artificial intelligence* (pp. 87-99) (SIPRI). Oxford: Oxford University Press.

Akpakpam, Edet B. (1986). Acquisition of foreign technology: A case study of modern brewing in Nigeria. *Development and Change, 17*, 659-676.

Al-Ali, Salahaldeen. (1991). Technology dependence in developing countries: A case study of Kuwait. *Technology in Society, 13*, 267-278.

Alam, Ghayur. (1988). India's technology policy: Its influence on technology imports and technology development. In A. Desai (Ed.), *Technology absorption by Indian industry* (pp. 136-156). New Delhi, India: Wiley Eastern.

Alam, Ghayur, & Langrish, John. (1981). Non-multinational firms and transfer of technology to less developed countries. *World Development, 9*, 383-388.

Alauddin, Mohammad, & Tisdell, Clem. (1989). Poverty, resource distribution and security: The impact of new agricultural technology in rural Bangladesh. *Journal of Development Studies, 25*, 550-570.

Alberger, P. L., & Carter, V. L. (Eds.). (1981). *Communicating university research*. Washington, DC: Council for Advancement and Support of Education.

Alberts, Bruce, et al. (1990). *Molecular biology of the cell*. New York: Garland.

Albrecht, U. (1990). The role of military R&D in arms build-ups. In N. P. Gleditsch & O. Njolstad (Eds.), *Arms races: Technological and political dynamics* (pp. 87-104). London: Sage/Oslo: International Peace Research Institute.

Albu, A. (1980). British attitudes to engineering education: A historical perspective. In K. Pavitt (Ed.), *Technical innovation and British economic performance* (pp. 67-87). London: Macmillan.

Alexander, A. J. (1982). *Soviet science and weapons acquisition*. Santa Monica, CA: RAND Corporation.

Alic, John A., Branscomb, Lewis M., Brooks, Harvey, Carter, Ashton B., & Epstein, Gerald L. (1992). *Beyond spinoff: Military and commercial technologies in a changing world*. Cambridge, MA: Harvard Business School.

Allison, Graham T. (1971). *Essence of decision: Explaining the Cuban missile crisis*. Boston: Little, Brown.

Allison, Graham T., & Morris, F. A. (1975). Armaments and arms control: Exploring the determinants of military weapons. *Daedalus, 104*(3), 99-129.

Allport, P. (1991, October 5). Still searching for the Holy Grail. *New Scientist, 132*, 51-52.

Almond, Gabriel, & Verba, Sidney. (1965). *The civic culture*. Boston: Little, Brown.

Althusser, Louis. (1974). *La philosophie spontanée des savants*. Paris: Maspero.

Amann, Klaus. (1990). *Natürliche Expertise und künstliche Intelligenz: eine mikrosoziologische Untersuchung von Naturwissenschaftlern*. Unpublished doctoral dissertation, University of Bielefeld.

Amann, Klaus, & Knorr Cetina, Karin. (1988a). The fixation of (visual) evidence. [Special issue: Representation in Scientific Practice; M. Lynch & S. Woolgar, Eds.] *Human Studies*, 133-169.

Amann, Klaus, & Knorr Cetina, Karin. (1988b). Thinking through talk: An ethnographic study of a molecular biology laboratory. In Lowell Hargens, R. A. Jones, & Andrew Pickering

(Eds.), *Knowledge and society: Studies in the sociology of science past and present*. Greenwich, CT: JAI.

Amann, Klaus, & Knorr Cetina, Karin. (1989). Thinking through talk: An ethnographic study of a molecular biology laboratory. In R. A. Jones, Lowel Hargens, & Andrew Pickering (Eds.), *Knowledge and society: Studies in the sociology of science past and present* (Vol. 8, pp. 3-26). Greenwich, CT: JAI.

Amarasuriya, Nimala R. (1987). Development through information networks in the Asia-Pacific region. *Information Development, 3*, 87-94.

American Society for Engineering Education. (1955). Report of the Committee on Evaluation of Engineering Education. *Proceedings of the American Society for Engineering Education, 63*, 37.

Amor, A. J., Icamina, P. M., & Laing, M. (1987). *Science writing in Asia: The craft and the issues*. Manila: Press Foundation of Asia.

Ancarani, V. (1990). Regime theories and the management of the international relations of science & technology. *Reseax: Revue interdisciplinaire de philosophie morale et politique, 58*, 191-204.

Anderson, J. (1972, February 8). [Statement from the 92nd Congress, 2nd Session]. *Congressional Record*, p. 3210.

Anderson, R. L. (1970). Rhetoric and science journalism. *Quarterly Journal of Speech, 56*(4), 358-368.

Anderson, R. S. (1983). Cultivating science as cultural policy: A contrast of agricultural and nuclear science in India. *Pacific Affairs, 56*, 38-50.

Angier, Natalie. (1988). *Natural obsessions: The search for the oncogene*. Boston: Houghton Mifflin.

Anthony, Constance G. (1988). *Mechanization and maize: Agriculture and the politics of technology transfer in East Africa*. New York: Columbia University Press.

Apel, Karl Otto. (1973). *Transformation der philosophie*. Frankfurt am Main: Suhrkamp.

Arce, A., & Long, N. (1987). The dynamics of knowledge interfaces between Mexican agricultural bureaucrats and peasants: A case-study from Jalisco. *Boletin de Esudios Latinoamericanos y del Caribe, 43*, 5-30.

Arditti, R., Brennan, P., & Cavrak, S. (Eds.). (1979). *Science and liberation*. Boston: South End.

Armstrong, J. (1992). *Trends in global science and technology and what they mean for intellectual property systems*. Paper presented at the conference, "Global Dimensions of Intellectual Property Rights in Science and Technology," National Research Council, Washington, DC.

Armytage, W. H. G. (1965). *The rise of the technocrats*. London: Routledge & Kegan Paul.

Arnold, L. (1992). *Windscale 1957: Anatomy of a nuclear accident*. Dublin: Gill and Macmillan.

Arnold, W. (1988). Science and technology development in Taiwan and South Korea. *Asian Survey, 28*, 437-450.

Aronowitz, Stanley. (1988a). *Science as power*. Minneapolis: University of Minnesota Press.

Aronowitz, Stanley. (1988b). The science of sociology and the sociology of science. In Stanley Aronowitz, *Science as power: Discourse and ideology in modern society* (pp. 272-300). Minneapolis: University of Minnesota Press/London: Macmillan.

Artzt, Frederick. (1966). *The development of technical education in France, 1500-1850*. Cleveland: Society for the History of Technology.

Arunachalam, Subbiah. (1992). Peripherality in science: What should be done to help peripheral science get assimilated into mainstream science. In R. Arvanitis & J. Gaillard (Eds.), *Science indicators for developing countries* (pp. 67-76). Paris: ORSTOM.

Arvanitis, Rigas, & Chatelin, Y. (1988). National scientific strategies in tropical soil sciences. *Social Studies of Science, 18*, 113-146.

Arvanitis, Rigas, & Gaillard, Jacques. (Eds.). (1992). *Science indicators for developing countries*. Paris: ORSTOM.

Ascher, Maria, & Ascher, Robert. (1972). Numbers and relations from ancient Andean quipus. *Archives for the History of the Exact Sciences, 8*, 288-320.

Ascher, Maria, & Ascher, Robert. (1981). *Code of the quipu: A study in media, mathematics and culture*. Ann Arbor: University of Michigan Press.

Ashmore, Malcolm. (1989). *The reflexive thesis: Wrighting sociology of scientific knowledge*. Chicago: University of Chicago Press.

Ashmore, Malcolm, Mulkay, Michael, & Pinch, Trevor. (1989). *Health and efficiency: A sociology of health economics*. Milton Keynes, U.K.: Open University Press.

Ashmore, Malcolm, Myers, Greg, & Potter, Jonathan. (1995). Discourse, rhetoric, reflexivity: Seven days in the library. In Sheila Jasanoff, Gerald E. Markle, James C. Petersen & Trevor Pinch (Eds.), *Handbook of science and technology studies*. Thousand Oaks: Sage.

Astin, Helen, & Davis, Diane. (1985). Research productivity across the career- and life-cycle. In M. F. Fox (Ed.), *Scholarly writing and publishing: Issues, problems, and solutions*. Boulder, CO: Westview.

Atkinson, Paul. (1990). *The ethnographic imagination: The textual construction of reality*. London: Routledge.

Atkinson, Paul, & Delamont, Sara. (1977). Mock-ups and cock-ups: The stage management of guided discovery instruction. In P. Woods & Martin Hammersley (Eds.), *School experience*. London: Croom Helm.

Attewell, Paul. (1987). Big brother and the sweatshop: Computer surveillance in the automated office. *Sociological Theory, 5*, 87-99.

Aubert, J. E. (1992). What evolution for science and technology policies? *The OECD Observer, 174*, 4-6.

Auster, C. J. (1981). The changing role of women in the work force: The case of women engineers. In M. M. Trescott (Ed.), *Final report to the Rockefeller Foundation on "Refuting the image: A history of women engineers in the United States, 1850-1975."* Chicago: University of Illinois.

Averch, Harvey A. (1985). *A strategic analysis of science and technology policy*. Baltimore: Johns Hopkins University.

Axelrod, Robert. (1984). *The evolution of cooperation*. New York: Basic Books.

Baark, Eric. (1987). Commercialized technology transfer in China 1981-86: The impact of science and technology policy reforms. *China Quarterly, 111*, 390-406.

Baark, Eric. (1991). Fragmented innovation: China's science and technology policy reforms in retrospect. In *China's dilemmas in the 1990s: The problems of reforms, modernization and interdependence* (Joint Economic Committee). Washington, DC: Government Printing Office.

Baark, Eric, & Jamison, Andrew. (1986). The technology and culture problematique. In Eric Baark & Andrew Jamison (Eds.), *Technological development in China, India and Japan*. London: Macmillan.

Baber, Z. (1992). Sociology of scientific knowledge: Lost in the reflexive funhouse? *Theory and Society, 21*, 105-121.

Bachelard, Gaston. (1934). *Le Nouvel Esprit Scientifique*. Paris: PUF.

Bachrach, Peter, & Botwinick, Aryeh. (1992). *Power and empowerment*. Philadelphia: Temple University Press.

Bader, R. (1990, Spring). How science news sections influence newspaper science coverage: A case study. *Journalism Quarterly, 67*(1), 88-96.

Badham, Richard. (1984). The sociology of industrial and post-industrial societies. *Current Sociology, 32*, 1-136.

Bailes, Kendall. (1978). *Technology and society under Lenin and Stalin: Origins of the Soviet technical intelligentsia, 1917-1941*. Princeton, NJ: Princeton University Press.

Bailey, Conner, Cycon, Dean, & Morris, Michael. (1986). Fisheries development in the Third World: The role of international agencies. *World Development, 14*, 1269-1275.

Bailey, F. (1968). A peasant view of the bad life. In T. Shanin (Ed.), *Peasants and peasant society* (pp. 83-104). Harmondsworth, U.K.: Penguin.

Baily, Martin. (1991). Great expectations: PCs and productivity. In C. Dunlop & Robert Kling (Eds.), *Computerization and controversy* (pp. 111-117). New York: Academic Press.

Bailyn, L. (1980). *Living with technology: Issues at mid-career*. Cambridge: MIT Press.

Bailyn, L. (1985). Autonomy in the industrial R&D lab. *Human Resource Management, 24*, 129-146.

Bailyn, L., & Lynch, J. T. (1983). Engineering as a life-long career: Its meaning, its satisfactions, its difficulties. *Journal of Occupational Behavior, 4*, 263-283.

Bakx, K. (1991). The "eclipse" of folk medicine in Western society. *Sociology of Health and Illness, 13*(1), 20-38.

Ball, N., & Leitenberg, M. (1983). *The structure of the defense industry*. London: Croom Helm.

Balogh, Brian. (1991). *Chain reaction*. Cambridge: Cambridge University Press.

Baran, B. (1987). The technological transformation of white-collar work: A case study of the insurance industry. In H. Hartmann, R. Kraut, & L. Tilly (Eds.), *Computer chips and paper clips: Technology and women's employment* (Vol. 1, pp. 25-62). Washington, DC: National Academy Press.

Baranson, Jack. (1970). Technology transfer through the international firm. *American Economic Review, 60*, 435-440.

Barber, Bernard. (1962). *Science and social order*. New York: Collier.

Barber, Bernard. (1990). *Social studies of science*. New Brunswick, NJ: Transaction.

Barber, Bernard, & Hirsch, Walter. (Eds.). (1962). *The sociology of science*. New York: Free Press/London: Collier-Macmillan.

Barber, C. L. (1962). Some measurable characteristics of modern scientific prose. In *Contributions to English syntax and phonology*. Stockholm: Almquist and Wiksell. (Reprinted in *Episodes in ESP*, J. Swales, Ed., 1985, Oxford: Pergamon)

Barbezat, Debra A. (1987). Salary differentials by sex in the academic labor market. *Journal of Human Resources, 22*, 423-428.

Barnes, Barry. (1971). Making out in industrial research. *Science Studies, 1*, 157-175.

Barnes, Barry. (Ed.). (1972). *Sociology of science: Selected readings*. Harmondsworth, Middlesex, U.K.: Penguin.

Barnes, Barry. (1974). *Scientific knowledge and sociological theory*. London: Routledge & Kegan Paul.

Barnes, Barry. (1977). *Interests and the growth of knowledge*. London: Routledge & Kegan Paul.

Barnes, Barry. (1982a). The science-technology relationship: A model and a query. *Social Studies of Science, 12*, 166-171.

Barnes, Barry. (1982b). *T. S. Kuhn and social science*. New York: Columbia University Press.

Barnes, Barry. (1985). *About science*. Oxford: Blackwell.

Barnes, S. B., & Dolby, R. G. A. (1970). The scientific ethos: A deviant viewpoint. *Archives of European Sociology, 11*, 3-25.

Barnes, Barry, & Edge, David. (Eds.). (1982). *Science in context: Readings in the sociology of science*. Milton Keynes, U.K.: Open University Press/Cambridge: MIT Press.

Barnes, Barry, & Shapin, Steven. (Eds.). (1979). *Natural order: Historical studies in scientific culture*. London: Sage.

Barnett, C. (1986). *The audit of war: The illusion and reality of Britain as a great nation*. London: Macmillan.

Barnhart, B. (1989). The Department of Energy Human Genome Initiative. *Genomics, 5*, 657-660.

Barozzi, Anna, & Toschi, Vittoria. (1989). A cross-section of women engineers in Italy. *European Journal of Engineering Education, 14*, 381-388.

Barton, L., & Walker, S. (Eds.). (1983). *Gender, class and education.* Lewes, U.K.: Falmer.

Basalla, George. (1976). Pop science: The depiction of science in popular culture. In Gerald Holton & W. Blanpied (Eds.), *Science and its public* (pp. 261-278). Dordrecht, the Netherlands: Reidel.

Basalla, George. (1988). *The evolution of technology.* Cambridge: Cambridge University Press.

Bass, Thomas A. (1990). *Camping with the prince, and other tales of science in Africa.* New York: Houghton Mifflin.

Bastide, Françoise. (1990). The iconograpy of scientific texts: Principles of analysis. In Michael Lynch & Steve Woolgar (Eds.), *Representation in scientific practice* (pp. 187-229). Cambridge: MIT Press. (Original work published 1985)

Bateson, Gregory. (1980). *Mind and nature.* New York: Bantam.

The battle of the floods: Holland in February 1953, with a preface by Juliana, Queen of the Netherlands. (1953). Amsterdam: Netherlands Booksellers and Publishers Association, for the Benefit of the Netherlands Flood Relief Fund.

Bauer, Henry H. (1990). Barriers against interdisciplinarity: Implications for studies of science, technology and society (STS). *Science, Technology, & Human Values, 15*, 105-119.

Bauer, Henry H. (1992). *Scientific literacy and the myth of the scientific method.* Urbana: University of Illinois Press.

Bauer, Martin. (1992, August). *Mapping variety in public understanding of science.* Paper presented to the 4S/EASST Joint Meeting, Gothenburg, Sweden (also mimeo, London Science Museum).

Bauer, Martin, & Durant, John. (1992). *British public perceptions of astrology: An approach from the sociology of knowledge.* Paper presented to the annual meeting of the AAAS, Chicago.

Baum, E. (1990, December). Recruiting and graduating women: The underrepresented student. *IEEE Communications Magazine*, pp. 47-50.

Baum, Richard. (1982). Science and culture in contemporary China: The roots of retarded modernization. *Asian Survey, 22*, 1166-1186.

Baum, Robert J. (1980). *Ethics and engineering curricula.* New York: Hastings Center.

Baum, Robert J., & Flores, Albert W. (Eds.). (1980). *Ethical problems in engineering.* Troy, NY: Center for the Study of the Human Dimensions of Science and Technology.

Bayer, Alan, & Astin, Helen. (1975). Sex differentials in the academic reward system. *Science, 188*, 796-802.

Baynes, K., & Pugh, F. (1981). *The art of the engineer.* Woodstock, NY: Overlook.

Bayri, Tulay Y., & Furtan, W. Hartley. (1989). The impact of new wheat technology on income distribution: A Green Revolution case study—Turkey, 1960-1983. *Economic Development and Cultural Change, 38*, 113-128.

Bazerman, Charles. (1988). *Shaping written knowledge: The genre and activity of the experimental article in science.* Madison: University of Wisconsin Press.

Bazerman, Charles. (1989). Introduction. *Science, Technology, & Human Values, 14*, 3-6.

Bazerman, Charles. (1991). How natural philosophers can cooperate: The literary technology of coordinated investigation in Joseph Priestley's *History and present state of electricity* (1767). In C. Bazerman & J. Paradis (Eds.), *Textual dynamics of the professions* (pp. 13-44). Madison: University of Wisconsin Press.

Bazerman, Charles, & Paradis, J. (Eds.). (1991). *Textual dynamics of the professions.* Madison: University of Wisconsin Press.

Beaumont, C., Dingle, J., & Reithinger, A. (1981). Technology transfer and applications. *R&D Management, 11,* 149-156.

Beck, Ulrich. (1992). *Risk society: Towards a new modernity.* London: Sage. (Translation of 1986 German version, *Risikogesellsschaft: auf dem Weges einem andere Moderne*, Frankfurt: Suhrkampf)

Becker, Gary. (1984). Pietism and science: A critique of Robert K. Merton's hypothesis. *The American Journal of Sociology, 89,* 1065-1090.

Becker, Howard. (1982). *Art worlds.* Berkeley: University of California Press.

Beechey, V. (1988). Rethinking the definition of work. In J. Jenson et al. (Eds.), *Feminization of the labour force* (pp. 45-62). Cambridge: Polity.

Beer, Gillian, & Martins, Herminio. (1990). Introduction. *History of the Human Sciences, 3,* 163-175.

Beer, John. (1959). *The emergence of the German die industry.* Urbana: University of Illinois Press.

Beitz, Charles R. (1990). *Political equality: An essay in democratic theory.* Princeton, NJ: Princeton University Press.

Bell, Daniel. (1974). *Coming of post-industrial society.* New York: Basic Books.

Bell, Martin, & Scott-Kemmis, Don. (1988). Technology import policy: Have the problems changed? In A. Desir (Ed.), *Technology absorption in Indian industry* (pp. 30-70). New Delhi, India: Wiley Eastern.

Belt, Henk van den, & Rip, Arie. (1987). The Nelson-Winter-Dosi model and synthetic dye chemistry. In Wiebe E. Bijker, Thomas P. Hughes, & Trevor J. Pinch (Eds.), *The social construction of technological systems: New directions in the sociology and history of technology* (pp. 135-158). Cambridge: MIT Press.

Benavot, A., Cha, Yun-Kyung, Kamens, David, Meyer, John, & Wong, Suk-Ying. (1991). Knowledge for the masses: World models and national curricula, 1920-1986. *American Sociological Review, 56,* 85-100.

Ben-David, Joseph. (1968). *Fundamental research and the universities.* Paris: OECD.

Ben-David, Joseph. (1971). *The scientist's role in society: A comparative study.* Englewood Cliffs, NJ: Prentice-Hall.

Ben-David, Joseph. (1991). *Scientific growth: Essays on the social organization and ethos of science* (G. Freudenthal, Ed.). Berkeley: University of California Press.

Benedick, Richard. (1991). *Ozone diplomacy: New directions in safeguarding the planet.* Cambridge, MA: Harvard University Press.

Beniger, James R. (1986). *The control revolution: Technological and economic origins of the information society.* Cambridge, MA: Harvard University Press.

Benveniste, G. (1972). *The politics of expertise.* Berkeley, CA: Glendessary.

Ben-Yehuda, N. (1985). *Deviance and moral boundaries.* Chicago: University of Chicago.

Berg, H. M., & Ferber, Marianne. (1983). Women and women graduate students: Who succeeds and why. *Journal of Higher Education, 54,* 629-648.

Berger, Peter, & Luckmann, Thomas. (1967). *The social construction of reality.* Garden City, NY: Anchor/London: Allen Lane.

Berlin, B., & Kay, P. (1969). *Basic colour terms: Their universality and evolution.* Berkeley: University of California Press.

Berman, M. (1974). "Hegemony" and the amateur tradition in British science. *Journal of Social History, 8,* 30-50.

Berman, M. (1984). *The reenchantment of the world.* New York: Bantam.

Bernal, John D. (1939). *The social function of science.* London: Routledge.

Bernal, John D. (1954). *Science in history.* London: C. A. Watts.

Bernal, John D. (1965, October 21). Voluntary underemployment [Letter]. *New Scientist,* p. 215.

Berner, Boel. (1992). Professional or wage worker? Engineers and economic transformation in Sweden. In C. Smith & P. Meiksins (Eds.), *Engineering class politics*. London: Verso.

Bezilla, M. (1981). *Engineering education at Penn State: A century in the land grant tradition*. University Park: Pennsylvania State University Press.

Bhagavan, M. R. (1988). India's industrial and technological policies into the late 1980s. *Journal of Contemporary Asia, 18*, 220-233.

Bhalla, A. S., & James, D. D. (1986). Technological blending: Frontier technology in traditional economic sectors. *Journal of Economic Issues, 20*, 453-462.

Bhatt, V. V. (1979). Indigenous technology and investment licensing: The case of the Snaraj Tractor. *Journal of Development Studies, 15*, 320-330.

Biggs, Stephen D., & Clay, Edward J. (1981). Sources of innovation in agricultural technology. *World Development, 9*, 321-336.

Bijker, Wiebe E. (1987). The social construction of Bakelite: Towards a theory of invention. In Wiebe E. Bijker, Thomas P. Hughes, & Trevor J. Pinch (Eds.), *The social construction of technological systems: New directions in the sociology and history of technology* (pp. 159-187). Cambridge: MIT Press.

Bijker, Wiebe E. (1992). The social construction of fluorescent lighting—or how an artefact was invented in its diffusion stage. In Wiebe Bijker & John Law (Eds.), *Shaping technology/building society*. Cambridge: MIT Press.

Bijker, Wiebe E., Hughes, Thomas P., & Pinch, Trevor J. (1987). *The social construction of technological systems: New directions in the sociology and history of technology*. Cambridge: MIT Press.

Bijker, Wiebe E., & Law, John. (Eds.). (1992). *Shaping technology/building society*. Cambridge: MIT Press.

Bijker, Wiebe E., & Pinch, Trevor J. (1987). The social construction of facts and artefacts: Or how the sociology of science and the sociology of technology might benefit each other. In Wiebe E. Bijker, Thomas P. Hughes, & Trevor J. Pinch (Eds.), *The social construction of technological systems: New directions in the sociology and history of technology* (pp. 17-50). Cambridge: MIT Press.

Billig, Michael. (1991). *Repopulating the pages of social psychology*. Manuscript submitted for publication, Loughborough University.

Billington, D. P. (1979). *Robert Maillart's bridges: The art of engineering*. Princeton, NJ: Princeton University Press.

Billington, D. P. (1983). *The tower and the bridge: The new art of structural engineering*. New York: Basic Books.

Bimber, B. (1990). Karl Marx and the three faces of technological determination. *Social Studies of Science, 20*, 333-351.

Birke, Lynda, & Silverton, J. (Eds.). (1984). *More than the parts: Biology and politics*. London: Pluto.

Blackburn, P., Coombs, Rod, & Green, Kenneth. (1985). *Technology, economic growth and the labour process*. London: Macmillan.

Blanpied, William A. (1984). Balancing central planning with institutional autonomy: Notes on a visit to the People's Republic of China. *Science, Technology, & Human Values, 9*(2), 67-72.

Blaut, J. M. (1979). Some principles of ethnogeography. In S. Gale & G. Olsson (Eds.), *Philosophy in geography* (pp. 1-7). Dordrecht, the Netherlands: Reidel.

Blaxter, M. (1983). The causes of disease: Women talking. *Social Science and Medicine, 17*(2), 59-69.

Blecher, Marc J., & White, Gordon. (1979). Micropolitics in contemporary China: A technical unit during and after the Cultural Revolution. *International Journal of Politics, 9*, 1-135.

Bleier, Ruth. (1984). *Science and gender: A critique of biology and its themes on women*. New York: Pergamon.
Block, E. (1985). T. H. Huxley's rhetoric and the mind-matter debate 1868-1874. *Prose Studies, 8*, 19-39.
Bloomfield, Brian. (1986). Capturing expertise by rule induction. *Knowledge Engineering Review, 1*, 30-36.
Bloomfield, Brian. (1987). The culture of artificial intelligence. In Brian Bloomfield (Ed.), *The question of artificial intelligence* (pp. 59-105). London: Croom Helm.
Bloomfield, Brian. (1989). On speaking about computing. *Sociology, 23*, 409-426.
Bloor, David. (1976). *Knowledge and social imagery*. London: Routledge & Kegan Paul.
Bloor, David. (1991). *Knowledge and social imagery* (2nd ed.). Chicago: Chicago University Press.
Bloor, David. (1992). Left and right Wittgensteinians. In A. Pickering (Ed.), *Science as practice and culture*. Chicago: Chicago University Press.
Blume, Stuart. (1974). *Toward a political sociology of science*. New York: Free Press.
Blume, Stuart. (1992). *Insight and industry: On the dynamics of technological change in medicine*. Cambridge: MIT Press.
Blume, Stuart S., & Sinclair, Ruth. (1973). Chemists in British universities: A study of the reward system in science. *American Sociological Review, 38*, 126-138.
Blume, Stuart, et al. (Eds.). (1987). *The social direction of the public sciences* (Sociology of the Sciences Yearbook). Dordrecht, the Netherlands: Reidel.
Blumenthal, Tuvia. (1979). A note on the relationship between domestic research and development and imports of technology. *Economic Development and Cultural Change, 27*, 303-306.
Blyn, George. (1983). The Green Revolution revisited. *Economic Development and Cultural Change, 31*, 705-726.
Bodmer, W. F., et al. (1985). *The public understanding of science*. London: Royal Society.
Boffey, Phillip M. (1975). *The brain bank of America: An inquiry into the politics of science*. New York: McGraw-Hill.
Boguslaw, Robert. (1965). *The new utopians*. Englewood Cliffs, NJ: Prentice Hall.
Böhme, G., van den Daele, Wolfgang, Hohfeld, R., Krohn, Wolfgang, Schäfer, W., & Spengler, T. (1978). *Die gesellschaftliche Orientierung des Wissenschaftlichen Fortschritts*. Frankfurt am Main: Suhrkamp Verlag.
Böhme, G., van den Daele, W., Hohfeld, R. Krohn, W. Schäfer, W. & Spengler, T. (1978). The "scientification" of technology. In Wolfgang Krohn et al. (Eds.), *The dynamics of science and technology* (pp. 219-228). Dordrecht, the Netherlands: Reidel.
Boltanski, Luc. (1987). *The making of a class: Cadres in French society*. Cambridge: Cambridge University Press.
Boltanski, Luc, & Maldidier, P. (1970). Carriere scientifique, moral scientifique et vulgarisation. *Social Science Information, 9*, 99-118.
Boltanski, Luc, & Thévenot, Laurent. (1991). *De la justification: Les économies de la grandeur.* Paris: Gallimard.
Booker, Peter J. (1979). *A history of engineering drawing*. London: Northgate.
Boorstin, D. (1978). *The republic of technology*. New York: Harper & Row.
Borning, Alan. (1987). Computer system reliability and nuclear war. *Communications of the ACM, 30*(2), 112-131.
Bose, C., Bereano, P., & Malloy, M. (1984). Household technology and the social construction of housework. *Technology and Culture, 25*, 53-82.
Bostian, L. R. (1983). How active, passive and nominal styles affect readability of science writing. *Journalism Quarterly, 60*, 635-640.

Boston Women's Health Book Collective. (1984). *The new our bodies, ourselves*. New York: Simon & Schuster.

Botelho, Antonio Jose Junqueira. (1990). The professionalization of Brazilian scientists, the Brazilian Society for the Progress of Science (SBPC) and the state, 1958-60. *Social Studies of Science, 20*, 473-502.

Bourdieu, Pierre. (1971). Le marché des biens symboliques. *L'Année Sociologique, 22*, 49-126.

Bourdieu, Pierre. (1975a). La spécificité du champ scientifique et les conditions sociales du progrès de la raison. *Sociologie et Sociétés, 7*(1).

Bourdieu, Pierre. (1975b). The specificity of the scientific field and the social conditions of the progress of reason. *Social Science Information, 14*, 19-47.

Bowden, Lord. (1965a, September 30). Expectations for science: 1: To the limits of growth. *New Scientist*, pp. 849-853.

Bowden, Lord. (1965b, October 7). Expectations for science: 2: The administrator becomes important. *New Scientist*, pp. 48-52.

Bowden, Lord. (1965c, November 11). Growth rate for science [Letter]. *New Scientist*, p. 438.

Bowes, J. E., Stamm, K. R., Jackson, K. M., & Moore, J. (1978, May). *Communication of technical information to lay audiences*. Seattle: University of Washington, School of Communications.

Bowker, Geof, & Latour, Bruno. (1987). A blooming discipline short of discipline: (Social) studies of science in France. *Social Studies of Science, 17*(4), 715-747.

Bowonder, B., & Mijake, T. (1988). Measuring innovativeness of an industry: An analysis of the electronics industry in India, Japan and Korea. *Science and Public Policy, 15*, 279-303.

Boyer, P. (1985). *By the bomb's early light: American thought and culture at the dawn of the atomic age*. New York: Pantheon.

Bracken, Paul. (1984). *The command and control of nuclear forces*. New Haven, CT: Yale University Press.

Bramwell, A. (1989). *Ecology in the 20th century: A history*. New Haven, CT: Yale University Press.

Brancher, D. (Ed.). (1980). *The environment in engineering education*. Paris: UNESCO.

Brannigan, Augustine. (1981). *The social basis of scientific discoveries*. Cambridge: Cambridge University Press.

Branscomb, Lewis M. (Ed.). (1993). *Empowering technology: Implementing a U.S. strategy*. Cambridge: MIT Press.

Braudel, F. (1977). *Afterthoughts on material civilization and capitalism*. Baltimore: Johns Hopkins University Press.

Braun, T., Glaenzel, W., & Schubert, A. (1988). The newest version of the facts and figures on publication output and relative citation impact of 100 countries, 1981-85. *Scientometrics, 13*, 181-188.

Braverman, Harry. (1974). *Labor and monopoly capital: The degradation of work in the twentieth century*. New York: Monthly Review Press.

Brickman, Ronald. (1984). Science and the politics of toxic chemical regulation: U.S. and European contrasts. *Science, Technology, & Human Values, 9*(1), 107-111.

Brickmann, Ronald, Jasanoff, Sheila, & Ilgen, Thomas. (1985). *Controlling chemicals: The politics of regulation in Europe and the United States*. Ithaca, NY: Cornell University Press.

Bright, James R. (1964). *Research, development, and technological development*. Homewood, IL: Irwin.

Brighton Women and Science Group. (1980). *Alice through the microscope*. Brighton, U.K.: Author.

British Association for the Advancement of Science. (1976). *Science and the media: Report of a study group*. London: British Association for the Advancement of Science.

British Medical Association. (1987). *Living with risk*. Chichester: Wiley.

Broad, William, & Wade, Nicholas. (1982). *Betrayers of the truth*. New York: Simon & Schuster.

Brodkey, L. (1987). *Academic writing as social practice.* Philadelphia: Temple University Press.

Bromberg, Joan. (1982). *Fusion: Science, politics, and the invention of a new energy source.* Cambridge: MIT Press.

Bromberg, Joan L. (1986). Engineering knowledge in the laser field. *Technology and Culture, 27,* 798-818.

Brooks, Harvey. (1975). The military innovation system and the qualitative arms race. *Daedalus, 104*(3), 75-97.

Brooks, Harvey. (1985). *Current science and technology policy issues.* Washington, DC: George Washington University.

Brooks, Harvey. (1986). National science policy and technological innovation. In R. Landau & N. Rosenberg (Eds.), *The positive sum strategy: Harnessing technology for economic growth.* Washington, DC: National Academy of Sciences.

Brooks, Harvey. (1987). Introduction and overview. In Harvey Brooks & Chester L. Cooper (Eds.), *Science for public policy* (pp. 1-10). New York: Pergamon.

Brooks, Harvey, & Branscomb, Lewis. (1989, August-September). [Interview]. *Technology Review,* pp. 55-64.

Brooks, Harvey, & Cooper, Chester. (Eds.). (1987). *Science for public policy.* New York: Pergamon.

Brown, George W., interviewed by Richard R. Mertz (1973, March 15). Smithsonian Computer Oral History, AC NMAH #196. Archive Center, National Museum of American History, Washington, DC.

Brown, George E., Jr. (1991). A perspective on the federal role in science and technology. In M. O. Meredith, S. D. Nelson, & A. H. Teich (Eds.), *Science and technology yearbook, 1991* (pp. 23-30). Washington, DC: American Association for the Advancement of Science.

Brown, George E., Jr. (1992, October 9). Rational science, irrational reality: A congressional perspective on basic research and society. *Science, 258,* 200-201.

Brown, M., & May, J. (1989). *The Greenpeace story.* London: Dorling Kindersley.

Brown, Martin (Ed.). (1971). *The social responsibility of the scientist.* New York: Free Press/London: Collier-Macmillan.

Brown, Michael, & Lyon, Katherine A. (1992). Holes in the ozone layer. In D. Nelkin (Ed.), *Controversy.* Newbury Park, CA: Sage.

Brown, Phil. (1987). Popular epidemiology: Community response to toxic waste-induced disease in Woburn, Massachusetts. *Science, Technology, & Human Values, 12,* 78-85.

Brown, Phil, & Mikkelski, Edwin. (1990). *No safe place: Toxic waste, leukemia and community action.* Berkeley: University of California Press.

Brown, R. H. (1977). *A poetic for sociology.* Cambridge: Cambridge University Press.

Brueckner, Leslie, & Borrus, Michael. (n.d.). The commercial impacts of military spending: The VHSIC case. In P. N. Edwards & R. Gordon (Eds.), *Strategic computing: Defense research and high technology.* Unpublished manuscript.

Bruer, John T. (1984). Women in science: Toward equitable participation. *Science, Technology, & Human Values, 9*(3), 3-7.

Brush, Stephen. (1974, March 22). Should the history of science be rated X? *Science, 183,* 1164-1172.

Brush, Stephen. (1991). Women in science and engineering. *American Scientist, 79,* 404-419.

Bryant, Bunyan, & Mohai, Paul. (Eds.). (1992). *Race and the incidence of environmental hazards: A time for discourse.* Boulder, CO: Westview.

Bryant, Lynwood. (1976). The development of the diesel engine. *Technology and Culture, 17,* 432-446.

Bryner, Gary. (1993). *Blue skies, green politics: The Clean Air Act of 1990.* Washington, DC: Congressional Quarterly Press.

Brzezinski, Zbigniew. (1970). America and the technetronic age. In *Between two ages: America's role in the technetronic era*. New York: Viking.

Bucciarelli, L. L. (1988). Engineering design process. In F. Dubinskas (Ed.), *Making time: Ethnographies of high-technology organizations* (pp. 92-122). Philadelphia: Temple University Press.

Buchanan, R. Angus. (1983). Gentlemen engineers: The making of a profession. *Victorian Studies, 26*, 407-429.

Buchanan, R. Angus. (1985a). Institutional proliferation in the British engineering profession, 1847-1914. *Economic History Review, 38*, 42-60.

Buchanan, R. Angus. (1985b). The rise of scientific engineering in Britain. *British Journal of the History of Science, 18*, 218-233.

Buchanan, R. Angus. (1986). The diaspora of British engineering. *Technology and Culture, 27*, 501-524.

Buchanan, R. Angus. (1988). Engineers and government in nineteenth-century Britain. In R. MacLeod (Ed.), *Government and expertise* (pp. 41-58). Cambridge: Cambridge University Press.

Buchanan, R. Angus. (1989). *The engineers: A history of the engineering profession in Britain, 1750-1914*. London: Jessica Kingsley.

Buchanan, R. Angus. (1991). Theory and narrative in the history of technology. *Technology and Culture, 32*, 365-376.

Buchner, Bradley Jay. (1988). Social control and diffusion of modern telecommunications technologies: A cross-national study. *American Sociological Review, 53*, 446-453.

Bud, Robert. (1988). The myth and the machine: Seeing science through museum eyes. In G. Fyfe & J. Law (Eds.), *Picturing power: Visual depiction and social relations* (pp. 138-164). London: Routledge & Kegan Paul.

Bukharin, Nicholas, et al. (1931). *Science at the cross roads*. London: Cass. (Reprint 1971)

Bullard, Robert D. (1990). *Dumping in Dixie: Race, class, and environmental quality*. Boulder, CO: Westview.

Bullard, Robert D. (1993). *Confronting environmental racism: Voices from the grass roots*. Boston: South End.

Bullard, Robert, Bryant, B., & Mohai, P. (Eds.). (1992). *Race and the incidence of environmental hazards*. Boulder, CO: Westview.

Bunders, J. (1987). The practical management of scientists' actions: The influence of patterns of knowledge development in biology on cooperations between university biologists and non-scientists. In S. Blume et al. (Eds.), *The social direction of the public sciences* (pp. 39-72). Dordrecht, the Netherlands: Reidel.

Bunge, Mario. (1966). Technology as applied science. *Technology and Culture, 7*, 329-347.

Burchfield, J. D. (1975). *Lord Kelvin and the age of the earth*. London: Macmillan.

Burke, John G. (Ed.). (1966). *The new technology and human values*. Belmont, CA: Wadsworth.

Burke, Robert V. (1979). Green Revolution technologies and farm class in Mexico. *Economic Development and Cultural Change, 28*, 135-154.

Burkett, W. D. (1986). *News reporting: Science, medicine, and high technology*. Ames: Iowa State University Press.

Burnham, J. (1987). *How superstition won and science lost: Popularization of science and health in America*. New Brunswick, NJ: Rutgers University Press.

Burhop, E. H. S. (1974). The social responsibility of the scientist: A report of correspondence with members of the Jason Committee of the Institute for Defence Analysis. *Scientific World, 28*(1), 20-23.

Burns, James MacGregor. (1978). *Leadership*. New York: Harper & Row.

Burton, J. (1972). *World society*. Cambridge: Cambridge University Press.

Busch, Lawrence, Lacy, William B., Burkhardt, Jeffrey, & Lacy, Laura R. (1991). *Plants, power, and profit: Social, economic, and ethical consequences of the new biotechnologies.* Cambridge: Blackwell.

Bush, Vannevar. (1945). *Science: The endless frontier* (Charter document for the U.S. National Science Foundation). Washington, DC: Government Printing Office. (Reprinted 1960/1980)

Buttel, Frederick. (1989). How epoch marking are high technologies? The case of biotechnology. *Sociological Forum, 4,* 247-260.

Buttel, Frederick. (1991). Beyond deference and demystification in the sociology of science and technology. *Sociological Forum, 6,* 567-577.

Buttel, Frederick, & Raynolds, Laura. (1989). Population growth, agrarian structure, food production, and food distribution in the Third World. In *Food and natural resources* (pp. 325-362). New York: Academic Press.

Buttel, Frederick H., Kenney, Martin, & Kloppenburg, Jack, Jr. (1985). From Green Revolution to biorevolution: Some observations on the changing technological bases of economic transformations in the Third World. *Economic Development and Cultural Change, 34,* 31-56.

Button, Graham. (1990). Going up a blind alley: Conflating conversation analysis and computational modelling. In P. Luff, D. Frohlich, & G. N. Gilbert (Eds.), *Computers and conversation.* London: Academic Press.

Buzan, B. (1987). *An introduction to strategic studies: Military technology and international relations.* Houndmills, U.K.: Macmillan.

Byatt, I. C. R., & Cohen, A. V. (1969). *An attempt to quantify the economic benefits of scientific research* (Science Policy Studies No. 4). London: Her Majesty's Stationery Office.

Byres, T. J. (1981). The new technology, class formation, and class action in the Indian countryside. *Journal of Peasant Studies, 8,* 405-454.

Caldecott, L., & Leland, S. (Eds.). (1983). *Reclaim the Earth.* London: Women's Press.

Caldwell, L. (1990). *Between two worlds: Science, the environmental movement and policy choice.* Cambridge: Cambridge University Press.

Callon, Michel. (1980a). The state and technical innovation: A case study of the electrical vehicle in France. *Research Policy, 9,* 358-376.

Callon, Michel. (1980b). Struggles and negotiations to decide what is problematic and what is not: The socio-logics of translation. In Karin Knorr, Roger Krohn, & Richard Whitley (Eds.), *The social process of scientific investigation* (pp. 197-219). Dordrecht, the Netherlands: Reidel.

Callon, Michel. (1981). Pour une sociologie des controverses technologiques. *Fundamenta Scientiae, 2,* 381-399.

Callon, Michel. (1986a). The sociology of an actor-network: The case of the electric vehicle. In Michel Callon, John Law, & Arie Rip (Eds.), *Mapping the dynamics of science and technology* (pp. 19-34). Basingstoke, U.K.: Macmillan.

Callon, Michel. (1986b). Some elements of a sociology of translation: Domestication of the scallops and the fishermen of St Brieux Bay. In John Law (Ed.), *Power, action and belief: A new sociology of knowledge?* (Sociological Review Monograph, pp. 196-229). London: Routledge & Kegan Paul.

Callon, Michel. (1987). Society in the making: The study of technology as a tool for sociological analysis. In Wiebe E. Bijker, Thomas P. Hughes, & Trevor J. Pinch (Eds.), *The social construction of technological systems: New directions in the sociology and history of technology.* Cambridge: MIT Press.

Callon, Michel (Ed.). (1989). *La science et ses réseaux: Genèse et circulation des faits scientifiques* (Anthropologie des sciences et des techniques). Paris: La Découverte.

Callon, Michel. (1991). Techno-economic networks and irreversibility. In J. Law (Ed.), *A sociology of monsters: Essays on power, technology and domination* (Sociological Review Monograph, pp. 132-164). London: Routledge & Kegan Paul.

Callon, Michel. (1992). Variety and irreversibility in networks of technique conception and adoption. In D. Foray & C. Freeman (Eds.), *Technology and the wealth of nations*. London: Frances Printer.

Callon, Michel, Larédo, P., Rabeharisoa, V., Gonard, T., & Leray, T. (in press). The management and evaluation of technological programs and the dynamics of techno-economic networks: The case of AFME. *Research Policy.*

Callon, Michel, & Latour, Bruno. (1992). Don't throw the baby out with the Bath school! A reply to Collins and Yearley. In Andrew Pickering (Ed.), *Science as practice and culture* (pp. 343-368). Chicago: University of Chicago Press.

Callon, Michel, & Law, John. (1982). On interests and their transformation: Enrollment and counter-enrollment. *Social Studies of Science, 12*(4), 615-625.

Callon, Michel, Law, John, & Rip, Arie. (Eds.). (1986). *Mapping the dynamics of science and technology: Sociology of science in the real world.* London: Macmillan.

Cambrosio, Alberto. (1988). Going monoclonal: Art, science, and magic in the day-to-day use of hybridoma technology. *Social Problems, 35*, 244-260.

Cambrosio, Alberto, & Keating, Peter. (1988, August). "Going monoclonal": Art, science and magic in the day-to-day use of hybridoma technology. *Social Problems, 35*, 244-260.

Cambrosio, Alberto, & Keating, Peter. (1991). *A matter of facts: Constituting novel entities in immunology.* Paper presented at the annual meeting of the American Anthropological Association, Chicago.

Cambrosio, Alberto, & Keating, Peter. (1992). A matter of FACS: Constituting novel entities in immunology. *Medical Anthropology Quarterly, 6*, 362-384.

Cambrosio, Alberto, Keating, Peter, & MacKenzie, M. (1990). Scientific practice in the courtroom: The construction of sociotechnical identities in a biotechnology patent dispute. *Social Problems, 37*, 301-319.

Cambrosio, Alberto, Limoges, Camille, & Pronovost, Denyse. (1990). Representing biotechnology: An ethnography of Quebec science policy. *Social Studies of Science, 20*, 195-227.

Cameron, Susan Wilson. (1978). *Women faculty in academia: Sponsorship, informal networks, and scholarly success.* Unpublished doctoral dissertation, University of Michigan.

Campbell, Donald. (1974). Evolutionary epistemology. In Paul A. Schlipp (Ed.), *The philosophy of Karl Popper* (Vol. 141, pp. 413-463). La Salle, IL: Open Court.

Campbell, Donald T. (1988). *Methodology and epistemology for social sciences: Selected papers.* Chicago: University of Chicago Press.

Campbell, J. A. (1987). Charles Darwin: Rhetorician of science. In J. Nelson, A. Megill, & D. McCloskey (Eds.), *The rhetoric of the human sciences: Language and argument in scholarship and public affairs* (pp. 69-86). Madison: University of Wisconsin Press.

Canguilhem, Georges. (1988). *Ideology and rationality in the histories of the life sciences.* Cambridge: MIT Press.

Cantley, Mark. (1988). Biotech safety regulations and public attitudes in the EEC. In *World Biotech Report.* London.

Cantor, Geoffrey. (1987). Weighing light: The role of metaphor in eighteenth-century optical discourse. In A. E. Benjamin, Geoffrey N. Cantor, & John R. R. Christie (Eds.), *The figural and the literal: Problems of language in the history of science and philosophy, 1630-1800* (pp. 124-146). Manchester: Manchester University Press.

Caporaso, J. A. (1992). International relations theory and multilateralism: The search for foundations. *International Organization, 3*, 599-632.

Cardwell, Donald S. L. (1957). *The organisation of science in England.* London: Heinemann.

Cardwell, Donald S. L. (1971). *From Watt to Clausius: The rise of thermodynamics in the early industrial age*. Ithaca, NY: Cornell University Press.

Carlson, W. Bernard. (1988). Academic entrepreneurship and engineering education: Dugald C. Jackson and the MIT-GE Cooperative Engineering Course, 1907-1932. *Technology and Culture, 29*(3), 536-567.

Carnap, Rudolf. (1936). Testability and meaning. *Philosophy of Science, 3*, 419-471

Carnap, Rudolf. (1937) Testability and meaning. *Philosophy of Science, 4*, 1-40.

Carnap, Rudolf. (1955). Testability and meaning. In H. Feigl & M. Brodbeck (Eds.), *Readings in the philosophy of science*. New York.

Carpenter, M. P., Cooper, M., & Narin, F. (1980). Linkage between basic research literature and patents. *Research Management, 23*, 30-35.

Carroll, James. (1971, February 19). Participatory technology. *Science, 171*, 647-653.

Carr-Saunders, A. M., & Wilson, P. A. (1933). *The professions*. Oxford: Clarendon.

Carson, N. (1992). Process, prescience, and pragmatism: The Office of Technology Assessment. In C. W. Weiss (Ed.), *Organizations for policy analysis: Helping government think* (pp. 236-251). Newbury Park, CA: Sage.

Carson, Rachel. (1962). *Silent spring*. Harmondsworth, U.K.: Penguin.

Cartwright, Nancy. (1983). *How the laws of physics lie*. Oxford: Clarendon.

Caspar, B. (1981). The rhetoric and reality of congressional technology assessment. In T. J. Kuehn & A. L. Porter (Eds.), *Science, technology, and national policy* (pp. 327-345). Ithaca, NY: Cornell University Press.

Casper, Barry, & Wellstone, Paul. (1978). The science court on trial in Minnesota. *The Hastings Center Report, 8*(4), 5-7. (Reprinted in Barnes & Edge, Eds., 1982, *Science in context: Readings in the sociology of science*, pp. 282-289, Milton Keynes, U.K.: Open University Press)

Casper, Barry M., & Wellstone, Paul D. (1981). *Powerline: The first battle of America's energy war*. Amherst: University of Massachusetts Press.

Catalyst. (1992). *Women in engineering: An untapped resource*. New York: Author.

Caudill, E. (1987, Winter). A content analysis of press views of Darwin's evolution theory. *Journalism Quarterly, 64*, 782-786, 946.

Caudill, E. (1989a). *Darwinism in the press: The evolution of an idea*. Hillsdale, NJ: Lawrence Erlbaum.

Caudill, E. (1989b). The roots of bias: An empiricist press and coverage of the Scopes trial. *Journalism Monographs* (114).

Chadney, James G. (1984). The economic implications of the new technologies in Punjab. *Eastern Anthropologist, 37*, 227-237.

Chai, T. R. (1981). Chinese academy of sciences in the Cultural Revolution: A test of the red and expert concept. *Journal of Politics, 43*, 1215-1229.

Chandler, Alfred D., Jr. (1977). *Visible hand: The managerial revolution in American business*. Cambridge, MA: Belknap.

Channell, David F. (1982). The harmony of theory and practice: The engineering science of W. J. M. Rankine. *Technology and Culture, 23*, 39-52.

Channell, David F. (1984). The distinction between engineering science and natural science: W. J. M. Rankine. In *Technology and science: Important distinctions for liberal arts colleges* (pp. 52-59). Davidson, NC: Davidson College.

Channell, David F. (1986). *William John Macquorn Rankine*. Edinburgh: Scotland's Cultural Heritage.

Channell, David F. (1988). Engineering science as theory and practice. *Technology and Culture, 29*, 98-103.

Channell, David F. (1989). *The history of engineering science: An annotated bibliography*. New York: Garland.

Charlesworth, Max, Farrall, Lyndsay, Stokes, Terry, & Turnbull, David. (1989). *Life among the scientists: An anthropological study of an Australian scientific community.* Melbourne: Oxford University Press.

Charlton, T. M. (1982). *A history of theory of structures in the nineteenth century*. New York: Cambridge.

Charter of Paris. (1991). In SIPRI Yearbook, *World armaments and disarmament* (pp. 603-610). Oxford: Oxford University Press. (Original date of document 1990)

Chartier, Roger. (1987). *The cultural uses of print in early modern France.* Princeton, NJ: Princeton University Press.

Chartier, Roger. (1988). *Cultural history.* Ithaca, NY: Cornell University Press.

Charvolin, Florian, Limoges, Camille, & Cambrosio, Alberto. (1991, November). *Discourse and policy: An ethnography of the constitution of Canadian bio-safety regulations.* Paper presented at the 4S Meeting, Cambridge, MA.

Chatel, B. H. (1979). Technology assessment and developing countries. *Technological Forecasting and Social Charge, 35*, 339-350.

Chen, Edward. (1981). The role of MNCs in the production and transfer of technology in host countries. *Development and Change, 12*, 579-599.

Chiang, Jong-tsong. (1989). Technology and alliance strategies for follower countries. *Technological Forecasting and Social Change, 35*, 339-350.

Child, John, et al. (1983). A price to pay? Professionalism and work organization in Britain and West Germany. *Sociology, 17*, 63-78.

Chodorow, Nancy. (1974). Family structure and feminine personality. In M. C. Rosaldo & L. Lamphere (Eds.), *Woman, culture, and society*. Stanford, CA: Stanford University Press.

Choi. (1988). Science policy mechanisms and technology development in the developing countries. *Technological Forecasting and Social Change, 33*, 279-292.

Choudhuri, A. R. (1985). Practising Western science outside the West: Personal observation on the Indian scene. *Social Studies of Science, 15*, 475-505.

Chubin, Daryl. (1974). Sociological manpower and womanpower: Sex differences in career patterns of two cohorts of American doctorate scientists. *The American Sociologist, 9*, 83-92.

Chubin, Daryl. (1992, Fall). The elusive second "S" in "STS": Who's zoomin' who? *Technoscience*, pp. 12-13.

Chubin, Daryl, & Hackett, Edward J. (1990). *Peerless science: Peer review and U.S. science policy.* Albany: State University of New York Press.

Chubin, Daryl E. (1990). Scientific malpractice and the contemporary politics of knowledge. In Susan E. Cozzens & Thomas F. Gieryn (Eds.), *Theories of science in society* (pp. 144-263). Bloomington: Indiana University Press.

Churchill, Frederick. (1979). Sex and the single organism. *Studies in History of Biology, 3*, 139-177.

Cicourel, Aaron. (1974). *Cognitive sociology*. New York: Free Press.

Cini, M. (1980). The history and ideology of dispersion relations: The pattern of internal and external factors in a paradigm shift. *Fundamenta Scientiae, 1*, 157-172.

Clark, J. (1991). *Democratizing development: The role of voluntary organizations*. London: Earthscan.

Clark, Norman, & Juma, C. (1987). *Long-run economics: An evolutionary approach to economic growth.* London: Frances Pinter.

Clark, Norman, & Parthasarathi, Ashok. (1982). Science based industrialization in a developing country: The case of the Indian scientific instruments industry 1947-1968. *Modern Asian Studies, 16*, 657-682.

Clark, T., & Westrum, Ron. (1987). Paradigms and ferrets. *Social Studies of Science, 17*, 3-33.

Clarke, Adele. (1987). Research materials and reproductive science in the United States, 1910-1940. In G. L. Geison (Ed.), *Physiology in the American context, 1850-1940*. Bethesda, MD: American Physiological Society.

Clarke, Adele. (1990a). A social worlds research adventure: The case of reproductive science. In S. Cozzens & T. Gieryn (Eds.), *Theories of science in society*. Bloomington: Indiana University Press.

Clarke, Adele. (1990b). Controversy and the development of reproductive sciences. *Social Problems, 37*, 18-37.

Clarke, Adele, & Gerson, Elihu. (1990). Symbolic interactionism in social studies of science. In H. Becker & M. McCall (Eds.), *Symbolic interaction and cultural studies* (pp. 179-214). Chicago: University of Chicago Press.

Clarke, Lee. (1992). The wreck of the Exxon Valdez. In Dorothy Nelkin (Ed.), *Controversy*. Newbury Park, CA: Sage.

Clemens, E. S. (1986). Of asteroids and dinosaurs: The role of the press in the shaping of scientific debate. *Social Studies of Science, 16*(3), 421-456.

Clemens, Harrie, & de Groot, Jan P. (1988). Agrarian labor market and technology under different regimes: A comparison of Cuba and the Dominican Republic. *Latin American Perspectives, 15*, 6-36.

Clifford, James. (1988). *The predicament of culture: Twentieth-century ethnography, literature and art*. Cambridge, MA: Harvard University Press.

Clifford, James, & Marcus, George E. (Eds.). (1986). *Writing culture: The poetics and politics of ethnography*. Berkeley: University of California Press.

Coase, R. (1937). The nature of the firm. *Economica, 4*, 386-405.

Cockburn, Cynthia. (1983). *Brothers: Male dominance and technological change*. London: Pluto.

Cockburn, Cynthia. (1985). *Machinery of dominance: Women, men and technical know-how*. London: Pluto.

Cohen, A. V., & Ivens, L. N. (1967). *The sophistication factor in science expenditure* (Science Policy Studies No. 1). London: Her Majesty's Stationery Office.

Cohen, B. J. (1990). The political economy of international trade. *International Organization, 2*, 261-281.

Cohen, I. Bernard. (Ed.). (1980). Testimony before the Joint Commission to consider the present organizations of the Signal Service, Geological Survey, Coast and Geodetic Survey and the Hydrographic Office of the Navy Department (Senate Miscellaneous Document 82, 49th Congress, 1st session, reprint from 1886). In *Three centuries of science in America*. New York: Arno.

Cohen, Stephen S., & Zysman, John. (1987). *Manufacturing matters: The myth of the post-industrial economy*. New York: Basic Books.

Cohn, Carol. (1987). Sex and death in the rational world of defense intellectuals. *Signs: Journal of Women in Culture and Society, 12*(4), 687-718.

Cohn, V. (1989). *News & numbers: A guide to reporting statistical claims and controversies in health and related fields*. Ames: Iowa State University Press.

Cole, B. J. (1975). Trends in science and conflict coverage in four metropolitan newspapers. *Journalism Quarterly, 52*, 465-471.

Cole, Jonathan. (1973). *Social stratification in science* (S. Cole, Ed.). Chicago: University of Chicago Press.

Cole, Jonathan. (1979). *Fair science: Women in the scientific community*. New York: Free Press.

Cole, Jonathan, & Cole, Stephen. (1973). *Social stratification in science*. Chicago: University of Chicago Press.

Cole, Jonathan, & Zuckerman, Harriet. (1984). The productivity puzzle: Persistence and change in patterns of publication among men and women scientists. In P. Maehr & M. W. Steinkamp (Eds.), *Women in science*. Greenwich, CT: JAI.

Cole, Jonathan, & Zuckerman, Harriet. (1987). Marriage, motherhood, and research performance in science. *Scientific American, 255*(2), 119-125.

Cole, Leonard A. (1986). Resolving science controversies: From science court to science hearings panel. In M. Goggin (Ed.), *Governing science and technology in a democracy* (pp. 244-261). Knoxville: University of Tennessee Press.

Cole, Stephen. (1991). *Social influences on the growth of knowledge*. Cambridge, MA: Harvard University Press.

Cole, Stephen. (1992). *Making science: Between nature and society*. Cambridge, MA: Harvard University Press.

Coler, Myron A. (Ed.). (1963). *Essays on creativity in the sciences*. New York: New York University Press.

Collingridge, David. (1980). *The social control of technology*. London: Frances Pinter.

Collingridge, David, & Reeve, Colin. (1986). *Science speaks to power: The role of experts in policymaking*. New York: St. Martin's Press.

Collins, A., & Gentner, D. (1987). How people construct mental models. In D. Holland & N. Quinn (Eds.), *Cultural models in language and thought* (pp. 243-265). Cambridge: Cambridge University Press.

Collins, H. M. (1974). The TEA set: Tacit knowledge and scientific networks. *Science Studies, 4*, 165-186.

Collins, H. M. (1975). The seven sexes: A study in the sociology of a phenomenon, or the replication of experiments in physics. *Sociology, 9*, 205-224.

Collins, H. M. (Ed.). (1981a). Knowledge and controversy: Studies in modern natural science. [Special issue] *Social Studies of Science, 11*(1).

Collins, H. M. (1981b). Stages in the empirical programme of relativism. *Social Studies of Science, 11*, 3-10.

Collins, H. M. (1983). The sociology of scientific knowledge: Studies of contemporary science. *Annual Review of Sociology, 2*, 265-285.

Collins, H. M. (1985). *Changing order: Replication and induction in scientific practice*. London: Sage.

Collins, H. M. (1987). Certainty and the public understanding of science: Science on television. *Social Studies of Science, 17*(4), 689-713.

Collins, H. M. (1988). Public experiments and displays of virtuosity: The core-set revisited. *Social Studies of Science, 18*(4), 725-748.

Collins, H. M. (1989). Computers and the sociology of scientific knowledge. *Social Studies of Science, 19*, 613-624.

Collins, H. M. (1990). *Artificial experts: Social knowledge and intelligent machines*. Cambridge: MIT Press. (Editions du Seuil, French language, 1992)

Collins, H. M. (1992a). *Changing order: Replication and induction in scientific practice* (rev. ed.). Chicago: University of Chicago Press.

Collins, H. M. (1992b, January 10). The good, the bad and unquoted. *Times Higher Education Supplement*, p. 15.

Collins, H. M. (in press-a). Knowing and growing: Building an expert system for semi-conductor crystal growers. In Steve Woolgar & Fergus Murray (Eds.), *Social perspectives on software*. Cambridge: MIT Press.

Collins, H. M. (in press-b). Skill and the Turing test. In Graham Button (Ed.), *A sociology of new technology*. London: Routledge.

Collins, H. M., de Vries, Gerard H., & Bijker, Wiebe E. (in press). *The grammar of skill.*

Collins, H. M., Green, Rodney H., & Draper, Robert C. (1985). Where's the expertise: Expert systems as a medium of knowledge transfer. In M. J. Merry (Ed.), *Expert systems 85* (pp. 323-334). Cambridge: Cambridge University Press.

Collins, H. M., & Pinch, Trevor. (1979). The construction of the paranormal, nothing unscientific is happening. In R. Wallis (Ed.), *On the margins of science: The social construction of rejected knowledge* (Sociological Review Monograph). Keele, U.K.: University of Keele.

Collins, H. M., & Pinch, Trevor. (1982a). The construction of the paranormal: Nothing unscientific is happening. In R. Wallis (Ed.), *On the margins of science: The social construction of rejected knowledge* (Sociological Review Monographs 27). Keele, U.K.: University of Keele.

Collins, H. M., & Pinch, Trevor. (1982b). *Frames of meaning: The social construction of extraordinary science*. London: Routledge & Kegan Paul.

Collins, H. M., & Pinch, Trevor. (1993). *The golem: What everyone should know about science*. Cambridge: Cambridge University Press.

Collins, H. M., & Shapin, Steven. (1983). The historical role of the experiment. In F. Bevilacqua & P. J. Kennedy (Eds.), *Using history of physics in innovatory physics education* (pp. 282-292). Pavia: Centro Studi per la Didattica della Facolta di Scienze Matematiche, Fisiche e Naturali, Universita die Pavia, and the International Commission on Physics Education.

Collins, H. M., & Shapin, Steven A. (1989). Experiment, science teaching and the new history and sociology of science. In Michael Shortland & Andrew Warwick (Eds.), *Teaching the history of science*. London: Blackwell.

Collins, H. M., & Yearley, Stephen. (1992a). Epistemological chicken. In A. Pickering (Ed.), *Science as practice and culture* (pp. 301-326). Chicago: University of Chicago Press.

Collins, H. M., & Yearley, Stephen. (1992b). Journey into space. In A. Pickering (Ed.), *Science as practice and culture* (pp. 369-389). Chicago: University of Chicago Press.

Collins, Randall. (1975). *Conflict sociology*. New York: Academic Press.

Collins, Randall. (1988). *Theoretical sociology*. New York: Harcourt Brace Jovanovich.

Collins, Randall. (1989). Toward a theory of intellectual change: The social causes of philosophies. *Science, Technology, & Human Values, 14*, 107-140.

Collins, Randall, & Restivo, Sal. (1983). Robber barons and politicians in mathematics: A conflict model of science. *The Canadian Journal of Sociology, 8*, 199-227.

Commission on Professionals in Science and Technology. (1964-1991). *Manpower comments*. Washington, DC: Commission on Professionals in Science and Technology.

Commission on Professionals in Science and Technology. (1989). *Professional women and minorities: A manpower data resource service* (8th ed.). Washington, DC: Author.

Commoner, Barry. (1971). *The closing circle*. New York: Knopf.

Conant, James Bryant. (1947). *On understanding science*. New Haven, CT: Yale University Press.

Conant, James Bryant. (Ed.). (1948). *Harvard case histories in experimental science* (2 vols.). Cambridge, MA: Harvard University Press.

Conklin, H. (1964). Ethnogenealogical method. In W. H. Goodenough (Ed.), *Explorations in cultural anthropology*. New York: McGraw-Hill.

Conrad, J. (Ed.). (1980). *Society, technology and risk assessment*. New York: Academic Press.

Constant, Edward W. (1980). *The origins of the turbojet revolution*. Baltimore: Johns Hopkins University Press.

Constant, Edward W. (1983). Scientific theory and technological testability: Science, dynamometers, and water turbines in the 19th century. *Technology and Culture, 24*, 183-198.

Contractor, Farok J. (1983). Technology importation policies in developing countries: Some implications of recent theoretical and empirical evidence. *Journal of Developing Areas, 17*, 499-520.

Cook-Degan, Robert M. (1991). The human genome project: The formation of federal policies in the United States, 1986-1990. In K. E. Hanna (Ed.), *Biomedical politics* (pp. 99-168). Washington, DC: National Academy Press.

Coombs, Roderick, Savioti, P., & Walsh, Vivienne. (1987). *Economics and technological change*. London: Macmillan Education.

Cooter, Roger. (1984). *The cultural meaning of popular science: Phrenology and the organization of consent in nineteenth century Britain*. Cambridge: Cambridge University Press.

COPUS (n.d.). *COPUS looks forward: The next five years*. London: Royal Society Committee for the Public Understanding of Science (Chair, Sir Walter Bodmer).

Corea, G. (1985). *The mother machine: Reproductive technologies from artificial insemination to artificial wombs*. New York: Harper & Row.

Corea, G., et al. (1985). *Man-made women: How new reproductive technologies affect women*. London: Hutchinson.

Cotgrove, Stephen. (1982). *Catastrophe or cornucopia: The environment, politics and the future*. Chichester, U.K.: John Wiley.

Cotgrove, Stephen, & Box, Stephen. (1970). *Science, industry and society*. London: George Allen and Unwin.

Cotkin, G. (1984). The socialist popularization of science in America, 1901 to the First World War. *History of Education Quarterly, 24*(2), 201-214.

Cottrill, Charlotte A., Rogers, Everett M., & Mills, Tamsy. (1989). Co-citation analysis of the scientific literature of innovation research traditions: Diffusion of innovations and technology transfer. *Knowledge: Creation, Diffusion, Utilization, 11*, 181-208.

Couchman, P., & Fink-Jensen, K. (1990). *Public attitudes to genetic engineering in New Zealand*. Christchurch, N.Z.: Department of Scientific and Industrial Research.

Council for Science and Society. (1977). *The acceptability of risks*. London: Barry Rose/CSS.

Cowan, Ruth S. (1976). The "industrial revolution" in the home: Household technology and social change in the twentieth century. *Technology and Culture, 17*, 1-23. (Reprinted as The industrial revolution in the home. In D. MacKenzie & J. Wajcman, Eds., 1985, *The social shaping of technology: How the refrigerator got its hum*, Milton Keynes, U.K.: Open University Press)

Cowan, Ruth S. (1979). From Virginia Dare to Virginia Slims: Women and technology in American life. *Technology and Culture, 20*, 51-63.

Cowan, Ruth S. (1983). *More work for mother: The ironies of household technology from the open hearth to the microwave*. New York: Basic Books.

Cowen, R. (1986). *Defense procurement in the Federal Republic of Germany: Politics and organization*. Boulder, CO: Westview.

Cozzens, Susan E. (1987). Expert review in research evaluation. *Science and Public Policy*.

Cozzens, Susan E. (1988). Derek Price and the paradigm of science policy. *Science, Technology, & Human Values, 13*(3-4), 361-372.

Cozzens, Susan E. (1989). Autonomy and power in science. In Susan E. Cozzens & T. F. Gieryn (Eds.), *Theories of science in society*. Bloomington: Indiana University Press.

Cozzens, Susan E. (1990). Autonomy and power in science. In Susan E. Cozzens & T. Gieryn (Eds.), *Theories of science in society* (pp. 164-184). Bloomington: Indiana University Press.

Cozzens, Susan E. (1992, March 13). *Women in the politics of science*. Paper presented to a Science Policy Support Group Forum, Women in Science, London.

Cozzens, Susan E., & Gieryn, Thomas. (Eds.). (1990). *Theories of science in society*. Bloomington: Indiana University Press.

Cozzens, Susan E., Healey, P., Rip, Arie, & Ziman, John. (Eds.). (1990). *The research system in transition*. Boston: Kluwer Academic.

Cramer, Jacqueline. (1987). *Mission orientation in ecology: The case of Dutch fresh-water ecology*. Amsterdam: Rodopi.

Cramer, Jacqueline, Eyerman, R., & Jamison, Andrew. (1987). The knowledge interests of the environmental movement and their potential for influencing the development of science. In S. Blume et al. (Eds.), *The social direction of the public sciences* (pp. 89-115). Dordrecht, the Netherlands: Reidel.

Cramer, Jacqueline, & Hagendijk, Rob. (1985). Dutch fresh-water ecology: The links between national and international scientific research. *Minerva, 23*, 43-61.

Cramer, Jacqueline, & van den Daele, W. (1985). Is ecology an "alternative" natural science? *Synthese, 65*, 347-375.

Crane, Diana. (1972). *Invisible colleges*. Chicago: University of Chicago Press.

Crane, Diana. (1977). Technological innovation in developing countries: A review of the literature. *Research Policy, 6*, 374-395.

Crary, J., & Kwinter, S. (Eds.). (1992). *Zone 6: Incorporations*. New York: Urzone.

Crawford, M. (1989, June 9). MIT industry links draw congressional attention. *Science*, p. 1136.

Crawford, Stephen. (1989). *Technical workers in an advanced society: The work, career and politics of French engineers*. Cambridge: Cambridge University Press.

Crawford, Stephen. (1991). Changing technology and national career structures: The work and politics of French engineers. *Science, Technology, & Human Values, 16*, 173-194.

Crenson, Matthew A. (1971). *The un-politics of air pollution: A study of non-decisionmaking in the cities*. Baltimore: Johns Hopkins University Press.

Critchley, O. H. (1988). The enigma of the engineer: Hero of the industrial revolution—Mere henchman in the age of science. *IEEE Proceedings, 135*(A-5), 253-260.

Crompton, Rosemary, & Jones, G. (1984). *White-collar proletariat: Deskilling and gender in clerical work*. London: Macmillan.

Cronholm, M., & Sandell, R. (1981). Scientific information: A review of research. *Journal of Communication, 31*(2).

Crown, Patricia L., & Judge, W. James. (1991). *Chaco and Hohokam: Prehistoric regional systems in the American Southwest*. Sante Fe: School of American Research Press.

Crowther, J. G. (1970). *Fifty years with science*. London: Barrie & Jenkins.

Cutcliffe, Stephen H. (1989a). The emergence of STS as an academic field. *Research in Philosophy and Technology, 9*, 287-301.

Cutcliffe, Stephen H. (1989b). Science, technology, and society: An interdisciplinary academic field. *National Forum, 69*(2), 22-25.

Cutcliffe, Stephen H. (1990). The STS curriculum: What have we learned in twenty years? *Science, Technology, & Human Values, 15*(3), 360-372.

Daedalus. (1990). Risk. [Special issue] *Daedalus, 119*(4).

Daey Ouwens, C., Hoogstraten, P. van, Jelsma, J., Prakke, F., & Rip, A. (1987). *Constructive technology assessment*. The Hague, the Netherlands: Nota.

Dahl, Robert A. (1956). *A preface to democratic theory*. Chicago: University of Chicago Press.

Dahl, Robert A. (1985). *Controlling nuclear weapons: Democracy versus guardianship*. Syracuse, NY: Syracuse University Press.

Dahl, Robert A. (1989). *Democracy and its critics*. New Haven, CT: Yale University Press.

Dahlberg, Kenneth A. (1979). *Beyond the Green Revolution: The ecology and politics of global agricultural development*. New York: Plenum.

Dahlman, Carl J., Ross-Larson, Bruce, & Westphal, Larry. (1987). Managing technological development: Lessons from the newly industrializing countries. *World Development, 15*, 759-775.

Danhof, Clarence. (1968). *Government contracting and technological change*. Washington, DC: Brookings Institution.

Daniels, G. H. (1967, June 30). The pure-science ideal and democratic culture. *Science, 156*, 1699-1706.

Darnton, Robert. (1984). Philosophers trim the tree of knowledge: The epistemological strategy of the *Encyclopédie*. In R. Darnton, *The great cat massacre* (pp. 191-213). New York: Basic Books.

Daston, Lorraine, & Galison, Peter. (in press). *The image of objectivity*.

Daston, Lorraine, & Otte, M. (Eds.). (1991). Style in science. *Science in Context, 4*(2).

David, Paul A., Mowery, D. C., & Steinmueller, W. E. (1992). Analysing the economic payoffs from basic research. *Economics of Innovation and New Technology, 2*, 73-90.

Davidson, Donald. (1984). *Truth and interpretation*. Oxford: Oxford University Press.

Davidson, William. (1983). Structure and performance in international technology transfer. *Journal of Management Studies, 20*, 453-465.

Davies, D. M. (1985). Appropriate information technology. *International Library Review, 17*, 247-258.

Davis, B. D., et al. (1990, July 27). The human genome project and other initiatives. *Science, 249*, 342-343.

Davis, C. H., & Eisemon, T. O. (1989). Mainstream and non-mainstream scientific literature in four peripheral Asian scientific communities. *Scientometrics, 15*, 215-239.

Davis, M. (1991). Thinking like an engineer: The place of a code of ethics in the practice of a profession. *Philosophy and Public Affairs, 20*, 150-167.

Davis, Natalie Zemon. (1975). *Society and culture in early modern France*. Stanford, CA: Stanford University Press.

Davis-Floyd, Robbie E. (1992). *Birth as an American rite of passage*. Berkeley: University of California Press.

Day, C. R. (1978). The making of mechanical engineers in France: The Ecoles d'Arts et Metiers, 1803-1914. *French Historical Studies, 10*, 443.

Day, C. R. (1987). *Education for the industrial world: The Ecoles d'Arts et Metiers and the rise of French industrial engineering*. Cambridge: MIT Press.

Dean, G. (1972). Science, technology, and development: China as a "case study." *The China Quarterly, 51*, 520-534.

Dean, J. (1992). Constraining technological innovation in weapons. In H. G. Brauch, H. van der Graaf, J. Grin, & Wim A. Smit (Eds.), *Controlling destabilizing R&D and the export of dual use technology: Lessons from the past and challenges for the 1990s*. Amsterdam: Free University Press.

Dear, Peter. (1985). Totius in Verba: Rhetoric and authority in the early Royal Society. *Isis, 76*, 145-161.

Dear, Peter. (1987). Jesuit mathematical science and the reconstruction of experience in the early 17th century. *Studies in the History and Philosophy of Science, 18*, 133-175.

Dear, Peter. (Ed.). (1991). *The literary structure of scientific argument: Historical studies*. Philadelphia: University of Pennsylvania Press.

Deaux, K., & Emswiller, T. (1974). Explanations of successful performance in sex-linked tasks. *Journal of Personality and Social Psychology, 22*, 80-85.

De Bruin, Renger E., Braam, Robert R., & Moed, Henk F. (1991). Bibliometric lines in the sand. *Nature, 349*, 559-562.

Dedijer, S. (1963). Underdeveloped science in underdeveloped countries. *Minerva, 2*, 61-81.

Deem, R. (Ed.). (1980). *Schooling for women's work*. London: Routledge & Kegan Paul.

Defense Science Board. (1988, October). *The defense industrial and technology base* (Summer study). Washington, DC: Department of Defense, Office of the Under Secretary of Defense for Acquisition.

DeGregori, Thomas R. (1985). *A theory of technology: Continuity and change in human development.* Ames: Iowa State University Press.

Delacroix, J., & Ragin, C. C. (1981). Structural blockage: A cross national study of economic dependency, state efficacy and underdevelopment. *American Journal of Sociology, 84,* 1311-1147.

Delamont, Sara. (1987). Three blind spots. *Social Studies of Science, 17,* 163-170.

DeLauer, R. D. (1989). The good of it and its problems. [Special issue: Universities and the Military] *The Annals of the American Academy of Political and Social Science, 502,* 130-140.

Del Campo, Enrique Martin. (1989). Technology and the world economy: The case of the American hemisphere. *Technological Forecasting and Social Change, 35,* 351-364.

Deleuze, Giles, & Guattari, Felix. (1987). *A thousand plateaus: Capitalism and schizophrenia.* Minneapolis: Minneapolis University Press.

Demchak, C. C. (1991). *Military organizations, complex machines: Modernization in the U.S. armed services.* Ithaca, NY: Cornell University Press.

Dennis, Michael A. (1987). Accounting for research: New histories of corporate laboratories and the social history of American science. *Social Studies of Science, 17,* 479-518.

Dennis, Michael A. (1992). *"Our first line of defense": University laboratories and the making of the postwar American state* (Mimeo). (Submitted for publication in *Isis*)

Derakshani, Shidan. (1984). Factors affecting success in international transfers of technology: A synthesis and a test of a new contingency model. *Developing Economies, 22,* 27-46.

Dertouzos, Michael L., et al. (1991). Communications, computers, and networks: Special issue. *Scientific American, 265*(3), 62-164.

Desai, Ashok V. (1988). Technological performance in Indian industry: The influence of market structures and policies. In *Technology and absorption in Indian industry* (pp. 1-29) New Delhi: Wiley.

Desmond, Adrian. (1987). Artisan resistance and evolution in Britain, 1819-1848. *Osiris, 3*(2nd Series), 77-110.

Devlin, Keith. (1991, December 12). A yen for teamwork. *Guardian,* p. 31.

Diamond, Stanley. (1974). *In search of the primitive: A critique of civilization.* New Brunswick, NJ: Transaction.

Dibella, S. M., Ferri, A. J., & Padderud, A. B. (1991). Scientists' reasons for consenting to mass media interviews: A national survey. *Journalism Quarterly, 68*(4), 740-749.

Dickens, P. (1992). *Society and nature: Towards a green social theory.* New York: Harvester Wheatsheaf.

Dickson, David. (1979). Science and political hegemony in the 17th century. *Radical Science Journal, 8,* 7-37.

Dickson, David. (1984). *The new politics of science.* New York: Pantheon.

Dickson, David. (1987, November 6). OECD to set rules for international science. *Science,* p. 743.

Dickson, David. (1988). *The new politics of science* (2nd ed.). Chicago: University of Chicago Press.

Dietz, T., Stern, P., & Rycroft, R. (1989). Definitions of conflict and the legitimation of resources: The case of environmental risk. *Sociological Forum, 4,* 47-69.

Dijksterhuis, Eduard Jan. (1961). *The mechanization of the world picture.* Oxford: Clarendon.

Dijkstra, Edsger. (1989). On the cruelty of really teaching computing science (with replies by others). *Communications of the ACM, 32*(12), 1398-1414.

DiMaggio, P. J., & Powell, W. W. (1983). The iron cage revisited: Institutional isomorphism and collective rationality in organizational fields. *American Sociological Review, 48*, 147-160.

Dimento, Joseph. (1977). Citizen environmental litigation and administrative process. *Duke Law Journal, 22*, 409-452.

Dinnerstein, Dorothy. (1977). *The mermaid and the minataur*. New York: Harper & Row.

Disco, Cornelius. (1990). *Made in Delft: Professional engineering in the Netherlands, 1880-1940*. Amsterdam: University of Amsterdam.

DiSessa, A. (1985). *Intuition as knowledge, final report, laboratory for computer science*. Cambridge: MIT Press.

Divall, Colin. (1990). A measure of agreement: Employers and engineering studies in the universities of England and Wales, 1897-1939. *Social Studies of Science, 20*, 65-112.

Divall, Colin. (1991). Fundamental science versus design: Employers and engineering studies in British universities, 1935-1976. *Minerva, 29*, 166-194.

Diwan, Romesh, & Kallianpur, Renu. (1985). Biological technology and land productivity: Fertilizers and food production in India. *World Development, 13*, 627-638.

Dix, L. S. (1987a). *Women: Their underrepresentation and career differentials in science and engineering*. Washington, DC: NAS Press.

Dix, L. S. (1987b). *Minorities: Their underrepresentation and career differentials in science and engineering*. Washington, DC: NAS Press.

Dobson, A. (1990). *Green political thought*. London: Unwin Hyman.

Donohue, G. A., Tichenor, P. J., & Olien, C. N. (1973). Mass media functions, knowledge, and social control. *Journalism Quarterly, 50*, 652-659.

Donovan, Arthur. (1986). Thinking about technology. *Technology and Culture, 27*, 674-679.

Doorman, S. J. (1989). *Images of science: Scientific practice and the public*. Aldershot, U.K.: Gower.

Doran, C. (1989). Grasping reflexivity. *Social Studies of Science, 19*, 755-759.

Dore, Ronald. (1989). Technology in a world of national frontiers. *World Development, 17*, 1665-1676.

Dorf, R. C. (1974). *Technology and society*. San Francisco: Boyd and Fraser.

Dornan, C. (1988). The "problem" of science and the media: A few seminal texts in their context, 1956-1965. *Journal of Communication Inquiry, 12*(2), 53-70.

Dornan, C. (1990). Some problems in conceptualizing the issue of "science and the media." *Critical Studies in Mass Communication, 7*, 48-71.

Dosi, Giovant. (1984). *Technical change and industrial transformation: The theory and an application to the semiconductor industry*. London: Macmillan.

Dosi, Giovant, Freeman, Chrie, Nelson, Richard, Siverberg, G., & Soete, L. (Eds.). (1988). *Technical change and economic theory*. London: Frances Pinter.

Dougherty, K., & Etzkowitz, Henry. (1993). *The hidden industrial policy: Science and technology policy at the state level*. Paper presented at the American Sociological Association Annual Meetings.

Douglas, Mary. (1966). *Purity and danger*. New York: Praeger.

Douglas, Mary. (1970). *Natural symbols: Explorations in cosmology*. Harmondsworth, U.K.: Penguin.

Douglas, Mary. (1975). Deciphering a meal. In M. Douglas, *Implicit meanings*. London: Routledge & Kegan Paul.

Douglas, Mary. (1982). *Essays in the sociology of perception*. London: Routledge & Kegan Paul.

Douglas, Mary. (1986a). *How institutions think*. Syracuse, NY: Syracuse University Press.

Douglas, Mary. (1986b). *Risk acceptability according to the social sciences*. London: Routledge & Kegan Paul.

Douglas, Mary, & Wildavsky, Aaron. (1984). *Risk and culture*. Berkeley: University of California Press.

Douglas, Susan. (1987). *Inventing American broadcasting, 1899-1922*. Baltimore: Johns Hopkins University Press.

Deutch, John. (1991, August 2). The foreign policy of U.S. universities. *Science*, p. 492.

Downey, Gary L. (1986). Ideology and the clamshell identity. *Cultural Anthropology, 33*, 35-37.

Downey, Gary L. (1992a). CAD/CAM saves the nation? Toward an anthropology of technology. *Knowledge and Society, 9*, 143-168.

Downey, Gary L. (1992b). Human agency in CAD/CAM technology. *Anthropology Today, 8*, 2-6.

Downey, Gary L. (1992c). Steering technology through computer-aided design. In Arie Rip, Thomas Misa, & Johan Schot (Eds.), *Managing technology in society: New forms for the control of technology*. Under review at Cambridge University Press.

Downey, Gary L. (in press-a). *CAD/CAM culture: An excavation of cyborg practices.*

Downey, Gary L. (in press-b). Training engineers as boundary subjects. *Science as Culture.*

Downey, Gary L., Donovan, Arthur, & Elliott, T. J. (1989). The invisible engineer: How engineering ceased to be a problem in science and technology studies. *Knowledge and Society: Studies in the Sociology of Science Past and Present, 8*, 189-216.

Downey, Gary L., Hegg, S., & Lucena, J. (in press). *Weeded out: Critical reflection in engineering education.* Unpublished manuscript.

Drake, Karl, & Wildavsky, Aaron. (1990). Theories of risk perception: Who fears what and why? *Daedalus, 119*(4), 41-60.

Dresselhaus, Mildred S. (1984). Responsibilities of women faculty in engineering schools. In V. B. Hass & C. C. Perrucci (Eds.), *Women in scientific and engineering professions* (pp. 128-136). Ann Arbor: University of Michigan Press.

Dreyfus, Hubert. (1972). *What computers can't do.* New York: Harper & Row. (2nd ed., 1979)

Drori, Gili S. (in press). The relationship between science, technology and the economy in lesser developed countries. *Social Studies of Science, 23*.

Dubeck, L., Mosher, S., & Boss, J. (1988). *Science in cinema: Teaching science fact through science fiction films*. New York: Teacher's College Press.

Dubinskas, F. (Ed.). (1988). *Making time: Ethnographic studies of high-technology organization.* Philadelphia: Temple University Press.

Dulong, R., & Ackermann, W. Yian (1971). Popularisation of science for adults. *Social Science Information, 11*(1), 113-148.

Dumas, L. J. (Ed.). (1982). *The political economy of arms reduction: Reversing economic decay.* Boulder, CO: Westview.

Dunlop, Charles, & Kling, Rob. (1991). *Computerization and controversy.* New York: Academic Press.

Dunn, P. D. (1978). *Appropriate technology: Technology with a human face*. London: Macmillan.

Dunwoody, S. (1980, Winter). The science writing inner club: A communication link between science and the lay public. *Science, Technology, & Human Values, 5*, 14-22.

Dunwoody, S. (1982, December). A question of accuracy. *IEEE Transactions on Professional Communication, PC-25*, 196-199.

Dunwoody, S., & Long, M. (Compilers). (1991). *Annotated bibliography of research on mass media science communication.* Madison: University of Wisconsin, Center for Environmental Communications and Education Studies.

Dunwoody, S., & Scott, B. T. (1982). Scientists as mass media sources. *Journalism Quarterly, 59*, 52-59.

Dupree, A. Hunter. (1964). *Science in the federal government: A history of policies and activities to 1940.* New York: Harper & Row.

Dupree, A. Hunter. (1990). Science policy in the United States: The legacy of John Quincy Adams. *Minerva, 28*, 259-271.

Durant, John. (Ed.). (1992). Media coverage of Chernobyl. [Special issue] *Public Understanding of Science, 1*(3).

Durant, John R., Evans, G. A., & Thomas, G. P. (1989). The public understanding of science. *Nature, 340*, 11-14.

Durant, John R., Evans, G. A., & Thomas, G. P. (1992). Public understanding of science in Britain: The role of medicine in the popular representation of science. *Public Understanding of Science, 1*(3), 161-182.

Durant, John R., Miller, J. D., Tchernia, J., & van Deelen, W. (1991). *Europeans, science and technology*. Paper presented to the annual meeting of the AAAS, Washington, DC.

Durbin, Paul T. (Ed.). (various years). *Research in philosophy and technology* (various vols.). London: JAI.

Durkheim, Émile. (1938). *The rules of sociological method.* New York: Free Press. (Original work published 1895)

Durkheim, Émile. (1961). *The elementary forms of the religious life*. New York: Collier.

Duster, Troy. (1990). *Backdoor to eugenics*. New York: Routledge & Kegan Paul.

Easlea, Brian. (1980). *Witch hunting, magic, and the new philosophy*. Brighton, U.K.: Harvester.

Easton, David. (1958). *The political system*. Glencoe, IL: Free Press.

Ebadi, Y. M., & Dilts, D. A. (1986). The relation between research and development project performance and technical communication in a developing country: Afghanistan. *Management Science, 32*, 822-830.

Eckaus, Richard S. (1987). Appropriate technology: The movement has only a few clothes on. *Issues in Science and Technology, 3*(2), 62-71.

Eco, Umberto. (1983). *The name of the rose* (W. Weaver, Trans.). London: Martin Secker & Warburg.

Edelman, Murray. (1985). *The symbolic uses of politics*. Urbana: University of Illinois Press.

Edelstein, M. (1988). *Contaminated communities*. Boulder, CO: Westview.

Edge, David. (1974). Moral education and the study of science. In G. Collier, J. Wilson, & P. Tomlinson (Eds.), *Values and moral development in higher education* (pp. 147-159). London: Croom Helm.

Edge, David. (1975). On the purity of science. In W. R. Niblett (Ed.), *The sciences, the humanities, and the technological threat* (pp. 42-64). London: University of London Press.

Edge, David. (1977). Why I am not a co-citationist. *4S Newsletter, 2*(3), 13-19.

Edge, David. (1979). Quantitative measures of communication in science: A critical review. *History of Science, 17*, 102-134.

Edge, David. (1985). Dominant scientific methodological views: Alternatives and their implications. In B. Musschenga & D. Gosling (Eds.), *Science education and ethical values* (pp. 1-9). Geneva: WCC Publications/Washington, DC: Georgetown University Press.

Edge, David. (1988a). Twenty years of science studies in Edinburgh. In E. Mayer (Ed.), *Ordnung, rationalisierung, kontrolle* (Symposium an der Technischen Hochschule Darmstadt vom 7. bis 9. Mai 1987, pp. 17-29), Darmstadt. *THD-schriftenreihe wissenschaft und technik, 42*.

Edge, David. (1988b). Review. *British Journal of Educational Studies, 36*(1), 76-77.

Edge, David. (1990). Competition in modern science. In T. Frängsmyr (Ed.), *Solomon's house revisited: The organization & institutionalization of science* (pp. 208-232). Canton, MA: Science History Publications.

Edgerton, David. (1992, August 12-15). *British scientists and engineers and the relations of science, technology and war.* Paper presented at the 4S-EASST joint Conference, Gothenborg.

Edgerton, David. (1993). Tilting at paper tigers. *British Journal for the History of Science, 26*, 67-75.

Edmondson, R. (1984). *Rhetoric in sociology.* London: Macmillan.

Edquist, Charles. (1985). *Capitalism, socialism and technology: A comparative study of Cuba and Jamaica.* London: Zed.

Edwards, D., & Mercer, N. (1987). *Common knowledge: The development of understanding in the classroom.* London: Routledge.

Edwards, Paul. (1991). Paper presented to the Society for Social Studies of Science Annual Meeting, Cambridge, MA.

Edwards, Paul N. (1987). A history of computers in weapons systems. In D. Bellin & G. Chapman (Eds.), *Computers in battle* (pp. 45-60). New York: Harcourt.

Edwards, Paul N. (1989). The closed world: Systems discourse, military policy, and post-WWII US historical consciousness. In L. Levidow (Ed.), *Cyborg worlds: The military information society* (pp. 135-158). London: Free Association Books.

Edwards, Paul N. (1990). The army and the microworld: Computers and the militarized politics of gender. *Signs, 16*(1), 102-127.

Edwards, Paul N. (in press). *The closed world: Computers and the politics of discourse in Cold War America.* Cambridge: MIT Press.

Eichler, Margrit. (1986). The relationship between sexist, non-sexist, woman-centered and feminist research. *Studies in Communication, 3*, 37-74.

Einsiedel, E. F. (1992, January). Framing science and technology in the Canadian press. *Public Understanding of Science, 1*(1), 89-101.

Eisemon, Thomas Owen. (1980). Scientists in Africa. *Bulletin of Atomic Scientists, 36*, 17-22.

Eisemon, Thomas Owen. (1982). *The science profession in the Third World.* New York: Praeger.

Eisenstein, Elizabeth. (1979). *The printing press as an agent of change.* Cambridge: Cambridge University Press.

Eisenstein, H. (1984). *Contemporary feminist thought.* London: Allen and Unwin.

Eliade, Mircea. (1972). *Shaminism: Archaic techniques of ecstasy.* Princeton, NJ: Princeton University Press.

Elliott, David, & Elliott, R. (1976). *The control of technology.* London: Wykeham.

Ellis, Jacques. (1987). *The social history of the machine gun.* London: Cresset Library. (Reprint of 1975 ed.)

Ellul, Jacques. (1964). *The technological society.* New York: Knopf. (Published in French, *La Technique ou l'enjeu du siècle.* Paris: Librairie Armand Colin, 1954)

Elster, Jon. (1983). *Explaining technical change: A case study in the philosophy of science.* Cambridge: Cambridge University Press.

Elzen, Boelie, Enserink, B., & Smit, Wim A. (1990). Weapon innovation: Networks and guiding principles. *Science and Public Policy, 17*(3), 171-193.

Elzinga, Aant. (1980). "Science studies" in Sweden. *Social Studies of Science, 10*(2), 181-214.

Elzinga, Aant. (1981). *Evaluating the evaluation game: On the methodology of project evaluation with special reference to development cooperation.* Stockholm: Swedish Agency for Research Cooperation (SAREC) R1.

Elzinga, Aant. (1985). Research, bureaucracy and the drift of epistemic criteria. In B. Wittrock & A. Elzinga (Eds.), *The university research system.* Stockholm: Almqvist & Wiksell.

Elzinga, Aant. (1988). Bernalism, Comintern and the science of science: Critical science movements then and now. In J. Annerstedt & A. Jamison (Eds.), *From research policy to social intelligence.* London: Macmillan.

Elzinga, Aant. (1990, August). Large scale military funding induces culture clash. *Space Policy*, pp. 187-194.

Engelhardt, H. Tristram, Jr., & Caplan, Arthur L. (Eds.). (1987). *Scientific controversies: Case studies in the resolution and closure of disputes in science and technology*. Cambridge: Cambridge University Press.

England, J. Merton. (1983). *A patron for pure science: The National Science Foundation's formative years, 1945-57*. Washington, DC: National Science Foundation.

Enos, J. L., & Park, W. H. (1984). *The adoption and diffusion of imported technology: The case of Korea*. London: Croom Helm.

Enserink, B., Smit, Wim A., & Elzen, Boelie. (1990). Assessments and the B-1 bomber network. *Project Appraisal, 5*(4), 235-254.

Environmental Health Center. (1990). *Chemicals, the press, and the public: A journalist's guide to reporting on chemicals in the community*. Washington, DC: National Safety Council, Environmental Health Center.

Enzensberger, Hans-Magnus. (1974). A critique of political ecology. *New Left Review, 84*, 3-31.

Erber, Fabio Stefano. (1985). The development of the "electronics complex" and government policies in Brazil. *World Development, 13*, 293-309.

Ernst, D. (1981). Technology policy for self-reliance: Some major issues. *International Social Science Journal, 33*, 466-480.

Esch, M. (1972, February 8). [Statement from the 92nd Congress, 2nd Session]. *Congressional Record*, p. 3200.

Etzkowitz, Henry. (1983). Entrepreneurial scientists and entrepreneurial universities in American academic science. *Minerva, 21*, 198-233.

Etzkowitz, Henry. (1988). The making of an entrepreneurial university: The traffic among MIT, industry and the military, 1860-1960. In E. Mendelsohn et al. (Eds.), *Science and the military*. Dordrecht, the Netherlands: Reidel.

Etzkowitz, Henry. (1989). Entrepreneurial science in the academy: A case of the transformation of norms. *Social Problems, 36*(1), 14-29.

Etzkowitz, Henry. (1990a). The capitalization of knowledge. *Theory and Society, 19*, 107-121.

Etzkowitz, Henry. (1990b). The second academic revolution: The role of the research university in economic development. In Susan E. Cozzens, Peter Healy, Arie Rip, & John Ziman (Eds.), *The R&D system in transition*. Dordrecht, the Netherlands: Kluwer.

Etzkowitz, Henry. (1991a). Inventions. In *Encyclopedia of sociology*. New York: Macmillan.

Etzkowitz, Henry. (1991b). Regional industrial and science policy in the United States. *Science and Technology Policy, 18*.

Etzkowitz, Henry. (1992a, Spring). Individual investigators and their research groups. *Minerva*.

Etzkowitz, Henry. (1992b). Redesigning Solomon's house: The university and the internationalization of science and business. In S. Sorlin et al. (Eds.), *Sociology of science yearbook*. Amsterdam: Kluwer.

Etzkowitz, Henry. (1992c, Fall). Capitalizing science in post-socialist eastern Europe. *Science, Knowledge and Technology*.

Etzkowitz, Henry. (1993a, February). The National Science Foundation and United States industrial and science policy. *Science and Technology Policy*.

Etzkowitz, Henry. (1993b). Enterprises from science: The origins of science-based regional economic development and the venture capital firm. *Minerva*.

Etzkowitz, Henry. (in press). *Entrepreneurial science: The second academic revolution*.

Etzkowitz, Henry, Balazs, K., Healey, Peter, Stankiewics, R., & Webster, Andrew. (in press). Beyond capitalism and socialism: The role of academic-industry-government relations in economic development. *Science and Public Policy*.

Etzkowitz, Henry, & Peters, L. (1991, Summer). Profit from knowledge: Organizational innovations and normative change in American universities. *Minerva*.

Evangelista, M. (1988). *Innovation and the arms race: How the United States and the Soviet Union develop military technologies*. Ithaca, NY: Cornell University Press.

Evans, Peter B. (1979). *Dependent development: The alliance of multinational, state and local capital in Brazil*. Princeton, NJ: Princeton University Press.

Evans, Peter B. (1986). State, capital, and the transformation of dependence: The Brazilian computer case. *World Development, 14*, 791-808.

Evans, Peter B., Rueschemeyer, D., & Skocpol, Theda. (1985). *Bringing the state back in*. Cambridge: Cambridge University Press.

Evans, Peter B., & Stephens, John D. (1988). Development in the world economy. In N. Smelser (Ed.), *Handbook of sociology* (pp. 739-773). Newbury Park, CA: Sage.

Evans, Peter B., & Timberlake, M. (1980). Dependence, inequality and the growth of the tertiary: A comparative analysis of less developed countries. *American Sociological Review, 48*, 421-428.

Evans, W. A., Krippendorf, M., Yoon, J. H., Posluszny, P., & Thomas, S. (1990). Science in the prestige and national tabloid presses. *Social Science Quarterly, 71*(1), 105-117.

Evered, D., & O'Connor, M. (Eds.). (1987). *Communicating science to the public*. Chichester, U.K.: Wiley.

Eyerman, R., & Jamison, Andrew. (1991). *Social movements: A cognitive approach*. Cambridge, MA: Polity.

Ezrahi, Yaron. (1971). The political resources of American science. *Science Studies, 1*, 117-133.

Ezrahi, Yaron. (1980). Science and the problem of authority in democracy. In T. Gieryn (Ed.), *Science and social structure* (pp. 43-60). New York: New York Academy of Sciences.

Ezrahi, Yaron. (1990). *The descent of Icarus: Science and the transformation of contemporary democracy*. Cambridge, MA: Harvard University Press.

Ezrahi, Yaron. (1991). *The descent of Icarus*. Cambridge, MA: Harvard University Press.

Fagen, M. D. (1978). *A history of engineering and science in the Bell system*. Murray Hill, NJ: Bell Telephone Laboratories.

Fahnestock, J. (1989). Arguing in different forums: The Bering Strait crossover controversy. *Science, Technology, & Human Values, 14*, 26-42.

Fairchild, Loretta G., & Sosin, K. (1986). Evaluating differences in technological activity between transnational and domestic firms in Latin America. *Journal of Development Studies, 22*, 697-708.

Fairclough, Sir John. (1992, October 23). Sizzling start for the white heat. *Times Higher Education Supplement*, p. 17.

Fangyi, Huang. (1987). China's introduction of foreign technology and external trade: Analysis and options. *Asian Survey, 27*, 577-594.

Farley, John. (1982). *Gametes and spores*. Baltimore, MD: Johns Hopkins University Press.

Farrall, Lyndsay. (1981). Knowledge and its preservation in oral cultures. In D. Denoon & R. Lacey (Eds.), *Oral traditions in Melanesia*. Port Moresby: Institute of Papua New Guinea Studies.

Faulkner, Wendy, & Arnold, Erik. (Eds.). (1985). *Smothered by invention: Technology in women's lives*. London: Pluto.

Fausto-Sterling, Anne. (1989). Life in the XY Corral. *Women's Studies International Forum, 12*(3), 319-331.

Favereau, Olivier. (in press). Règles, organisation et apprentissage collectif. In A. Orléan (Ed.), *Analyse économique des conventions*. Paris: Presses Universitaires de France.

Fayard, P. (1988). *La communication scientifique publique: De la vulgarisation a la mediatisation*. Lyon: Chronique sociale.

Feldberg, R., & Glenn, E. (1983). Technology and work degradation: Effects of office automation on women clerical workers. In J. Rothschild (Ed.), *Machina ex dea: Feminist perspectives on technology* (pp. 59-78). New York: Pergamon.

Feldt, Barbara. (1986). *The faculty cohort study: School of medicine.* Ann Arbor, MI: Office of Affirmative Action.

Ferber, Marianne, & Kordick, Betty. (1978). Sex differentials in the earnings of Ph.D.s. *Industrial and Labor Relations Review, 31*, 227-238.

Ferguson, E. (1992). *Engineering and the mind's eye.* Cambridge: MIT Press.

Ferguson, H. A. (1988). *Delta-Visie: Een terugblik op 40 jaar natte waterbouw in Zuidwest-Nederland.* 's-Gravenhage: Rijkswaterstaat.

Feyerabend, Paul. (1975). *Against method.* London: New Left Books.

Fidell, L. S. (1975). Empirical verification of sex discrimination in hiring practices in psychology. In R. K. Unger & F. L. Denmark (Eds.), *Woman: Dependent or independent variable?* New York: Psychological Dimensions.

Figert, Anne E. (1992). *Women and the ownership of PMS: The professional, gendered and scientific structuring of a psychiatric disorder.* Unpublished doctoral dissertation, Indiana University, Bloomington.

Finnegan, R., & Horton, Robin. (Eds.). (1973). *Modes of thought.* London: Faber.

Fischer, A. W., & McKenney, J. L. (1993). The development of the ERMA banking system: Lessons from history. *IEEE Annals of the History of Computing, 15*(1), 44-57.

Fischer, Claude. (1988). "Touch someone": The telephone industry discovers sociability. *Technology and Culture, 29*, 32-61.

Fischer, Frank. (1990). *Technocracy and the politics of expertise.* London: Sage.

Fischer, William A. (1984). Scientific and technological planning in the People's Republic of China. *Technological Forecasting and Social Change, 25*, 189-208.

Fisher, Donald. (1990). Boundary work and science: The relation between power and knowledge. In S. Cozzens & T. Gieryn (Eds.), *Theories of science in society* (pp. 98-119). Bloomington: Indiana University Press.

Fisher, Lawrence. (1992, February 22). Texas Instruments gets chip case ruling. *The New York Times*, p. 39.

Fiske, R. S. (1984). Volcanologists, journalists, and the concerned local public: A tale of two crises in the eastern Caribbean. In *Studies in geophysics: Explosive volcanism: Inception, evolution, and hazards.* Washington, DC: National Academy Press.

Fitchen, Jance M., Fessenden Raden, June, & Heath, Jennifer S. (1987). Providing risk information in communities: Factors influencing what is heard and accepted. *Science, Technology, & Human Values, 12*(3-4), 94-101.

Fitzgerald, Deborah. (1986). Exporting American agriculture: The Rockefeller Foundation in Mexico, 1943-1953. *Social Studies of Science, 16*, 457-483.

Flamm, Kenneth. (1987). *Targeting the computer: Government support and international competition.* Washington, DC: Brookings Institution.

Flamm, Kenneth. (1988). *Creating the computer: Government, industry and high technology.* Washington, DC: Brookings Institution.

Fleck, Jamie. (1987). Development and establishment of artificial intelligence. In B. Bloomfield (Ed.), *The question of artificial intelligence* (pp. 106-164). London: Croom Helm.

Fleck, Ludwik. (1935). *Genesis and development of a scientific fact.* Chicago: University of Chicago Press.

Fleron, F. (Ed.). (1977). *Technology and communist culture.* New York: Praeger.

Fletcher, Ronald. (1991). *Science, ideology and the media: The Cyril Burt scandal.* New Brunswick, NJ: Transaction.

Flores, A. W. (Ed.). (1989). *Ethics and risk management in engineering.* Boulder, CO: Westview.

Flores-Moya, Piedad, Evenson, Robert E., & Hyami, Yujiro. (1978). Social returns to rice research in the Philippines: Domestic benefits and foreign spillover. *Economic Development and Cultural Change, 26*, 591-608.

Florman, S. (1968). *Engineering and the liberal arts*. New York: McGraw-Hill.

Florman, S. (1976). *The existential pleasures of engineering*. New York: St. Martin's Press.

Florman, S. (1981). *Blaming technology: The irrational search for scapegoats*. New York: St. Martin's Press.

Florman, S. (1987). *The civilized engineer*. New York: St. Martin's Press.

Fores, Michael. (1988). Transformations and the myth of "engineering science": Magic in a white coat. *Technology and Culture, 29*, 62-81.

Forgan, S. (1989). The architecture of science and the idea of a university. *Studies in the History and Philosophy of Science, 20*, 405-434.

Forje, John W. (1986). Two decades of science and technology in Africa. *Science and Public Policy, 13*, 89-96.

Forman, Paul. (1987). Behind quantum electronics: National security as basis for physical research in the United States, 1940-1960. *Historical Studies in the Physical and Biological Sciences, 18*(1), 149-229.

Forrester, Jay W. (1946-1948). *Computation books*. Cambridge: MIT Archives.

Forrester, Jay W., Boyd, Hugh R., Everett, Robert E., & Fahnestock, Harris. (1948). *A plan for digital information handling equipment in the military* (Project DIC 6345). Cambridge: MIT Servomechanisms Laboratory.

Forsyth, David, Norman, J. C., McBain, S., & Solomon, Robert F. (1980). Technical rigidity and appropriate technology in less developed countries. *World Development, 9*, 371-398.

Forsythe, Diana. (1987, November 19-22). *Engineering knowledge: An anthropological study of an artificial intelligence laboratory*. Paper presented at the 12th Annual Meeting of the Society for Social Studies of Science, Worcester, MA.

Forsythe, Diana. (1993a). Engineering knowledge: The construction of knowledge in artificial intelligence. *Social Studies of Science, 23*, 445-477.

Forsythe, Diana. (1993b). The construction of work in artificial intelligence. *Science, Technology, & Human Values, 18*(4), 460-479.

Forsythe, Diana E., & Buchanan, B. G. (1988). An empirical study of knowledge elicitation: Some pitfalls and suggestions. [Special issue: Methods in Knowledge Engineering; P. E. Lehner & L. Adelman, Eds.] *IEEE Transactions on Systems, Man and Cybernetics*.

Forsythe, Diana, & Buchanan, B. G. (1989). Knowledge acquisition for expert systems: Some pitfalls and suggestions. *IEEE Transactions on Systems, Man and Cybernetics, 19*(3), 435-442.

Forsythe, Diana, & Buchanan, B. G. (1991). Non-technical problems in knowledge engineering: Suggestions for project managers. In J. Liebowitz (Ed.), *Proceedings of the World Congress of Expert Systems*. New York: Pergamon.

Foucault, Michel. (1970). *The order of things*. New York: Random House.

Foucault, Michel. (1975). *Surveiller et Punir.* Paris: Gallimard.

Foucault, Michel. (1980). *Power/knowledge: Selected interviews and other writings 1972-77* (C. Gordon, Ed.). New York: Pantheon.

Foucault, Michel. (1984). Space, knowledge, power. In P. Rabinow (Ed.), *The Foucault reader* (pp. 239-256). New York: Pantheon.

Fowler, R. (1977). *Linguistics and the novel*. London: Methuen.

Fox, D., & Lawrence, Christopher. (1988). *Photographing medicine: Images and power in Britain and America since 1840*. New York: Greenwood.

Fox, Mary Frank. (1981). Sex, salary, and achievement: Reward-dualism in academia. *Sociology of Education, 54*, 71-84.

Fox, Mary Frank. (1991). Gender, environmental milieu, and productivity in science. In H. Zuckerman, J. Cole, & J. Bruer (Eds.), *The outer circle: Women in the scientific community*. New York: Norton.

Fox, Mary Frank, & Faver, Catherine. (1985). Men, women, and publication productivity. *The Sociological Quarterly, 26*, 537-549.

Fox, Mary Frank, & Hesse-Biber, Sharlene. (1984). *Women at work*. Palo Alto, CA: Mayfield.

Fox, Patrick. (1989). From senility to Alzheimer's disease: The rise of the Alzheimer's disease movement. *Milbank Quarterly, 67*(1), 58-102.

Frake, C. (1962). The ethnographic study of cognitive systems. In T. Gladwin et al. (Eds.), *Anthropology and human behaviour*. Seattle: Anthropology Society of Washington.

Fraley, P. C. (1963). The education and training of science writers. *Journalism Quarterly, 40*, 323-328.

Frame, J. D., Narin, F., & Carpenter, M. P. (1977). The distribution of world science. *Social Studies of Science, 7*, 501-516.

Franke, Richard H. (1989). Technological revolution and productivity decline: The case of US banks. In T. Forester (Ed.), *Computers in the human context* (pp. 281-290). Cambridge: MIT Press.

Frankena, Frederick, & Frankena, Joann Koelin. (1988). *Citizen participation in environmental affairs, 1970-1986: A bibliography*. New York: AMS Press.

Franklin, H. Bruce. (1988). *War stars: The superweapon and the American imagination*. New York: Oxford University Press.

Fransman, Martin. (1985). Conceptualising technical change in the Third World in the 1980's: An interpretive survey. *Journal of Development Studies, 21*, 572-652. (Republished as *Technology and economic development*, Boulder, CO: Westview, 1986)

Frazier, Kendrick. (1986). *People of Chaco: A canyon and its culture*. New York: Norton.

Freeman, Christopher. (1969). *Measurement of output of research and research and experimental development: A review paper*. Paris: UNESCO.

Freeman, Christopher. (1977). Economics of research and development. In I. Spiegel-Rösing & D. Price. (Eds.), *Science, technology and society: A cross-disciplinary perspective* (pp. 223-275). London: Sage.

Freeman, Christopher. (1982). *The economics of industrial innovation*. London: Frances Pinter.

Freeman, Christopher. (1987). *Technology policy and economic performance: Lessons from Japan*. London: Frances Pinter.

Freeman, Christopher. (1988). Quantitative and qualitative factors in national policies for science and technology. In J. Annerstedt & A. Jamison (Eds.), *From research policy to social intelligence*. London: Macmillan.

Freeman, Christopher, & Hagedoorn, J. (1992). *Globalisation of technology* (Monitor-Fast paper). Brussels: Commission of the European Communities.

Freeman, Christopher, Sharp, Margaret, & Walker, W. (Eds.). (1991). *Technology and the future of Europe*. London: Frances Pinter.

Frei, D. (1982). *Risks of unintentional nuclear war*. Geneva: United Nations.

Freidson, Eliot. (1970). *Profession of medicine*. Chicago: University of Chicago Press.

Freier, S. (1991). Report on the International Forum on Government and Science. In W. T. Golden (Ed.), *Worldwide science and technology advice to the highest levels of government* (pp. 15-27). New York: Pergamon.

Frenkel, Karen A. (1990). Women & computing. *Communications of the ACM, 33*(11), 34-46.

Freudenburg, Nicholas. (1984). *Not in our backyards!* New York: Monthly Review Press.

Freudenburg, William. (1992). Heuristics, biases and not-so-general publics. In S. Krimsky & D. Golding (Eds.), *Social theories of risk* (pp. 229-250). New York: Praeger.

Freudenburg, William R. (1988, October 7). Perceived risk, real risk: Social science and the art of probabilistic risk assessment. *Science, 242*, 44-49.

Freudenthal, Gad. (1986). *Atom and individual in the age of Newton*. Dordrecht, the Netherlands: Reidel.

Freudenthal, Gad. (1990). Science studies in France: A sociological view. *Social Studies of Science, 20*(2), 353-369.

Frieden, J. A. (1991). Invested interests: The politics of national economic policies in a world of global finance. *International Organization, 4*, 425-451.

Friedman, A. J. (1987). The influence of pseudoscience, parascience and science fiction. In D. Evered & Maureen O'Connor (Eds.), *Communicating science to the public* (pp. 190-204). Chichester, U.K.: Wiley.

Friedman, S., & Friedman, K. (1988). *Reporting on the environment: A handbook for journalists*. Bangkok, Thailand: Asian Forum of Environmental Journalists.

Friedman, S. M. (1989). TMI: The media story that will not die. In L. M. Walters, L. Wilkins, & T. Walters (Eds.), *Bad tidings: Communication and catastrophe* (pp. 161-170). Hillsdale, NJ: Lawrence Erlbaum.

Friedman, S. M., Dunwoody, S., & Rogers, C. L. (Eds.). (1986). *Scientists and journalists: Reporting science as news*. New York: Free Press.

Friends of the Earth U.K. (FoE). (1990). *Annual report and accounts 1989/90*. London: Author.

Fries, Sylvia D. (1984). The ideology of science during the Nixon years: 1970-1976. *Social Studies of Science, 14*, 323-341.

Frischtak, C. (1991). *Harmonization vs. differentiation in IPR regimes*. Paper presented at the conference "Global Dimensions of Intellectual Property Rights in Science and Technology," National Research Council, Washington, DC.

Fuchs, Stephen. (1992). *The professional quest for truth: A social theory of science and knowledge*. Albany: State University of New York Press.

Fuhrman, E., & Oehler, Kay. (1986). Discourse analysis and reflexivity. *Social Studies of Science, 16*, 293-307.

Fujimura, Joan H. (1987). Constructing "do-able" problems in cancer research: Articulating alignment. *Social Studies of Science, 17*, 257-293.

Fujimura, Joan H. (1988). The molecular biological bandwagon in cancer research: Where social worlds meet. *Social Problems, 35*(3), 261-283.

Fujimura, Joan H. (1992a). Crafting science: Standardized packages, boundary objects and "translation." In A. Pickering (Ed.), *Science as practice and culture*. Chicago: University of Chicago Press.

Fujimura, Joan H. (1992b). Problem paths: A tool for dynamic analysis of situated scientific problem construction. In Andrew Pickering (Ed.), *Science as practice and culture*. Chicago: Chicago University Press.

Fujimura, Joan H. (n.d.). *A tool for dynamic analysis of situated scientific problem construction*. Manuscript submitted for publication.

Fuller, Steve. (1988). *Social epistemology*. Bloomington: University of Indiana Press.

Fuller, Steve. (1992). STS as a social movement: On the purpose of graduate programs. *Science, Technology & Society: Curriculum Newsletter of the Lehigh University STS Program & Technology Studies Resource Center, 91*, 1-5.

Fuller, W. (Ed.). (1971). *The social impact of modern biology*. London: Routledge & Kegan Paul.

Funtowicz, Silvio, & Ravetz, Jerome. (1990). *Global environmental science and the emergence of second-order science* (EUR 12803EN). Ispra, Italy: Joint Research Centre of the Commission of the European Communities.

Fuqua, Don. (1972, February 8). [Statement from the 92nd Congress, 2nd Session.] *Congressional Record*, p. 3209.

Fusfeld, H. (1986). *The research enterprise*. Cambridge, MA: Ballinger.

Fuss, Diana. (1989). *Essentially speaking: Feminism, nature and difference*. New York: Routledge & Kegan Paul.

Fyfe, Gordon, & Law, John (Eds.). (1988). *Picturing power: Visual depiction and social relations* (Sociological Review Monograph 35). London: Routledge & Kegan Paul.

Gabriel, Kathryn. (1991). *Roads to center place: A cultural atlas of Chaco Canyon and the Anasazi*. Boulder, CO: Johnson.

Gaddy, G. D., & Tanjong, E. (1986, Spring). Earthquake coverage by the Western press: Testing geographical bias in international news. *Journal of Communication, 36,* 105-112.

Gaillard, Jacques. (1991). *Scientists in the Third World*. Lexington: University Press of Kentucky.

Galbraith, John K. (1967). *The new industrial state*. London: Hamish Hamilton.

Galison, Peter. (1987). *How experiments end*. Chicago: University of Chicago Press.

Galison, Peter. (1988). Physics between war and peace. In E. Mendelsohn, M. R. Smith, & P. Weingart (Eds.), *Science, technology and the military* (Sociology of Sciences Yearbook, Vol. XII/1, pp. 47-86). Dordrecht, the Netherlands: Kluwer Academic.

Galison, Peter. (1989, October 20). *The trading zone: Coordination between experiment and theory in the modern laboratory*. Paper presented at the History of Science Colloquium, Princeton University.

Galison, Peter, & Assmus, Alexi. (1989). Artificial clouds, real particles. In David Gooding, Trevor Pinch, & Simon Schaffer (Eds.), *The uses of experiment* (pp. 225-274). Cambridge, MA: Cambridge University Press.

Gambetta, D. (Ed.). (1988). *Trust: The making and breaking of cooperative relations*. Oxford: Blackwell.

Gang, Ira N., & Gangopadhyay, Shubhashis. (1987). Employment, output and the choice of techniques: The trade-off revisited. *Journal of Development Economics, 25,* 321-327.

Gansler, J. S. (1980). *The defense industry*. Cambridge: MIT Press.

Gansler, J. S. (1988). The need—and opportunity—for greater integration of defence and civil technologies in the United States. In P. Gummett & J. Reppy (Eds.), *The relations between defence and civil technologies* (pp. 138-158). Dordrecht, the Netherlands: Kluwer Academic.

Gansler, J. S. (1989). *Affording defense*. Cambridge: MIT Press.

Ganz Brown, Carole, & Rushing, Francis. (1990). Intellectual property rights in the 1990s. In F. Rushing & C. Ganz (Eds.), *Intellectual property rights in science, technology and economic performance*. Boulder, CO: Westview.

Gardner, R. E. (1976). Women in engineering: The impact of attitudinal differences on educational institutions. *Engineering Education, 67,* 233-240.

Garfield, Eugene. (1983). Mapping science in the Third World. *Science and Public Policy, 10,* 112-127.

Garfinkel, Harold, Lynch, Michael, & Livingston, Eric. (1981). The work of discovering science constructed with material from the optically discovered pulsar. *Philosophy of the Social Science, 11,* 131-158.

Garson, Barbara. (1988). *The electronic sweatshop*. New York: Simon & Schuster.

Garvey, William D. (1979). *Communication: The essence of science—facilitating information exchange among librarians, scientists, engineers and students*. Oxford: Pergamon.

Garwin, Richard L., & Bethe, Hans A., (1968, March). Anti-ballistic missile systems. *Scientific American*, pp. 21-31.

Gastel, B. (1983). *Presenting science to the public*. Philadelphia: ISI Press.

Gaston, Jerry. (1973). *Originality and competition in science*. Chicago: University of Chicago Press.

Gaston, Jerry. (1978). *The reward system in British and American science*. New York: Wiley.

Gauthier, L. (1991). *Construction de représentations visuelles et organisation spatio-temporelle des ressources en astronomie*. Unpublished doctoral dissertation, University of Montreal.

Gayle, D. J., Denamark, R. A., & Stiles, K. W. (1991-1992). International political economy: Evolution and prospects. *International Studies Notes, 3*(1), 64-68.

Geertz, Clifford. (1973a). *The interpretation of culture*. New York: Basic Books.
Geertz, Clifford. (1973b). Thick description: Toward an interpretive theory of culture. In C. Geertz, *The interpretation of cultures: Selected essays* (pp. 3-32). New York: Basic Books.
Geertz, Clifford. (1983). *Local knowledge: Further essays in interpretative anthropology*. New York: Basic Books.
Geertz, Clifford. (1990, November 8). A lab of one's own. *The New York Review of Books, 37*, 19-23.
Geiger, R. (1986). *To advance knowledge: The growth of American research universities in the 20th century, 1900-1940*. Oxford: Oxford University Press.
Gellner, Ernst. (1964). *Thought and change*. Chicago: University of Chicago Press.
Gentner, D., & Gentner, D. R. (1983). Flowing waters or teeming crowds: Folk models of electricity. In D. Gentner & A. Stevens (Eds.), *Mental models* (pp. 99-129). Hillsdale, NJ: Lawrence Erlbaum.
Gentner, D., & Stevens, A. (Eds.). (1983). *Mental models*. Hillsdale, NJ: Lawrence Erlbaum.
Gerbner, G., Gross, L., Morgan, M., & Signorelli, N. (1981). Health and medicine on television. *New England Journal of Medicine, 303*(15), 901-904.
Gerjuoy, Edward, & Baranger, E. U. (1989). The physical sciences and mathematics. [Special issue: Universities and the Military] *The Annals of the American Academy of Political and Social Science, 502*, 58-81.
Gerson, Elihu. (1983). Scientific work and social worlds. *Knowledge, 4*, 357-377.
Gerver, Elisabeth. (1985). *Humanizing technology.* New York: Plenum.
Ghazanfar, S. M. (1980). Individual modernity in relation to economic-demographic characteristics: Some evidence from Pakistan. *Studies in Comparative International Development, 15*, 37-53.
Ghosh, Pradip. (Ed.). (1984). *Technology policy and development: A Third World perspective*. Westport, CT: Greenwood.
Ghosh, Pradip, & Morrison, Denton. (Eds.). (1984). *Appropriate technology in Third World development*. Westport, CT: Greenwood.
Gibbons, Ann. (1992, March 13). Key issue: Tenure. *Science, 255*, 1386.
Gibbons, John H. (1988a). Technology and law in the third century of the Constitution. In W. T. Golden (Ed.), *Science advice to the president, Congress and the judiciary*. New York: Pergamon.
Gibbons, John H. (1988b, April 28). *Technology policy.* Speech presented at the Massachusetts Institute of Technology, sponsored by the MIT Program in Science, Technology, and Society.
Gibbons, John H., & Gwin, H. L. (1986). Technology and governance. *Technology in Society, 7*, 333-352.
Gibbons, Michael. (1984). Is science industrially relevant? The interaction between science and technology. In M. Gibbons & P. Gummett (Eds.), *Science, technology and society today.* Manchester, U.K.: Manchester University Press.
Gibbons, Michael, & Wittrock, B. (Eds.). (1985). *Science as a commodity: Threats to the open community of scholars*. Harlow, Essex, U.K.: Longman.
Giddens, Anthony. (1977). *New rules of sociological method.* New York: Basic Books.
Giddens, Anthony. (1979). *Central problems in social theory.* Berkeley: University of California Press.
Giedion, S. (1948). *Mechanization takes command: A contribution to anonymous history*. Oxford: Oxford University Press.
Giere, Ron. (1988). *Explaining science: A cognitive approach*. Chicago: University of Chicago Press.

Gieryn, Thomas F. (1983). Boundary work and the demarcation of science from non-science: Strains and interests in professional ideologies of scientists. *American Sociological Review, 48,* 781-795.

Gieryn, Thomas F. (1992). The ballad of Pons and Fleischmann: Experiment and narrative in the (un)making of cold fusion. In E. McMullin (Ed.), *The social dimensions of science* (pp. 217-243). Notre Dame, IN: University of Notre Dame Press.

Gieryn, Thomas F., & Figert, Anne. (1986). Scientists protect their cognitive authority: The status degradation ceremony of Sir Cyril Burt. In G. Böhme & N. Stehr (Eds.), *The knowledge society* (pp. 67-86). Dordrecht, the Netherlands: Reidel.

Gieryn, Thomas F., & Figert, Anne. (1990). Ingredients for the theory of science in society: O-rings, ice water, c-clamp, Richard Feynman and the press. In Susan E. Cozzens & Thomas Gieryn (Eds.), *Theories of science in society* (pp. 67-97). Bloomington: Indiana University Press.

Gilbert, G. Nigel. (1991, November). *Artificial societies*. Inaugural lecture, University of Surrey.

Gilbert, G. Nigel, & Mulkay, Michael. (1984). *Opening Pandora's box: A sociological analysis of scientists' discourse*. Cambridge: University of Cambridge Press.

Gilbert, G. Nigel, & Woolgar, Steve. (1974). The quantitative study of science: An examination of the literature. *Science Studies, 4*(3), 279-294.

Gilbert, G. Nigel, & Wooffitt, Robin. (in press). Sociology in machines: Applying sociology to software design and software to sociology. In Steve Woolgar & Fergus Murray (Ed.), *Social perspectives on software*. Cambridge: MIT Press.

Gilbert, Sandra, & Gubar, Susan. (1979). *The madwoman in the attic*. New Haven, CT: Yale University Press.

Gilbert, W. (1991, January 10). Towards a paradigm shift in biology. *Nature, 349,* 99.

Gilfillan, S. C. (1935a). *The sociology of invention*. Cambridge: MIT Press. (Paperback ed., 1970)

Gilfillan, S. C. (1935b). *Inventing the ship*. Cambridge: MIT Press.

Gillespie, Brendan, Eva, Dave, & Johnston, Ron. (1979). Carcinogenic risk assessment in the United States and Great Britain: The case of Aldrin/Dieldrin. *Social Studies of Science, 18,* 265-301.

Gilligan, Carol. (1982). *In a different voice*. Cambridge, MA: Harvard University Press.

Gilpin, Robert. (1962). *American scientists and nuclear weapons policy*. Princeton, NJ: Princeton University Press.

Gilpin, Robert. (1987). *The political economy of international relations*. Princeton, NJ: Princeton University Press.

Gilpin, Robert, & Wright, Christopher. (Eds.). (1964). *Scientists and national policy-making*. New York: Columbia University Press.

Gingras, Yves. (1991). *Physics and the rise of scientific research in Canada*. Montreal: McGill-Queen's University Press.

Gingras, Yves. (in press). Following scientists through society? Yes, but at arm's length! In J. Z. Buchwald (Ed.), *The autonomy of experiment, the sovereignty of practice*. Chicago: University of Chicago Press.

Gingras, Yves, & Trépanier, Michel. (1993). Constructing a Tokamak: Political, economic and technical factors as constraints and resources. *Social Studies of Science, 23,* 5-36.

Ginzberg, Eli. (Ed.). (1964). *Technology and social change*. New York: Columbia University Press.

Girvan, N. P., & Marcelle, G. (1990). Overcoming technological dependency: The case of Electric Arc (Jamaica) Ltd., a small firm in a small developing country. *World Development, 18,* 91-108.

Gispen, C. W. R. (1988). German engineers and American social theory: Historical perspectives on professionalization. *Comparative Studies in Society and History, 30,* 550-574.

Gispen, C. W. R. (1990). *New profession, old order: Engineers and German society, 1815-1914*. Cambridge: Cambridge University Press.

Gladwin, T. (1970). *East is a big bird: Navigation and logic on Puluwat*. Cambridge, MA: Harvard University Press.

Gleditsch, N. P., & Njolstad, O. (Eds.). (1990). *Arms races: Technological and political dynamics*. London: Sage/Oslo: International Peace Research Institute.

Glenn, Marian, Monroe, Dorrie, & Lamont, Judith. (1993). Pathways to the podium: Women organizing women speaking. In D. Fort (Ed.), *A hand up: Women mentoring women in science*. Washington, DC: The Association for Women in Science.

Glover, Ian A., & Kelly, Michael P. (1987). *Engineers in Britain: A sociological study of the engineering dimension*. London: Allen and Unwin.

Glyde, Henry R., & Sa-yakanit, Virulh. (1985). Institution links: An example in science and technology. *Higher Education in Europe, 10*, 51-59.

Goggin, Malcolm. (Ed.). (1986). *Governing science and technology in a democracy*. Knoxville: University of Tennessee.

Gold, Barry. (1993). Congressional activities regarding misconduct and integrity in science. In *Responsible science: Ensuring the integrity of the research process* (Vol. 2). Washington, DC: National Academy Press.

Golden, William T. (Ed.). (1980). *Science advice to the president*. New York: Pergamon.

Golden, William T. (Ed.). (1988). *Science and technology advice to the president, Congress, and judiciary*. New York: Pergamon.

Goldsmith, Maurice. (1965, January 2). The Science of Science Foundation. *Nature, 205*, 10.

Goldsmith, Maurice, & MacKay, A. (Eds.). (1963). *The science of science*. Harmondsworth: Penguin.

Goldstein, J. H. (Ed.). (1986). *Reporting science: The case of aggression*. Hillsdale, NJ: Lawrence Erlbaum.

Goldstine, Herman. (1972). *The computer from Pascal to von Neumann*. Princeton, NJ: Princeton University Press.

Goldstrum, M. (1985). Popular political economy for the British working class reader in the nineteenth century. In T. Shinn & R. Whitley (Eds.), *Expository science* (pp. 259-273). Dordrecht, the Netherlands: Reidel.

Golinski, Jan V. (1987). Robert Boyle: Skepticism and authority in seventeenth-century chemical discourse. In A. Benjamin, G. Cantor, & J. R. R. Christie (Eds.), *The figural and the literal: Problems of language in the history of science and philosophy 1630-1800* (pp. 58-82). Manchester, U.K.: Manchester University Press.

Golinski, Jan V. (1990a). Language, discourse and science. In R. Olby, G. Cantor, J. R. R. Christie, & M. Hodge (Eds.), *Companion to the history of science* (pp. 110-123). London: Routledge & Kegan Paul.

Golinski, Jan V. (1990b). The theory of practice and the practice of theory: Sociological approaches in the history of science. *ISIS, 81*, 492-505.

Golinski, Jan V. (1992). *Science as public culture: Chemistry and enlightenment in Britain, 1760-1820*. Cambridge: Cambridge University Press.

Gomezgil, Maria Luisa Rodriguez Salade. (1975). Mexican adolescents' image of the scientist. *Social Studies of Science, 5*, 355-361.

Gooday, Graeme. (1991). "Nature" in the laboratory: Domestication and discipline with the microscope in Victorian life science. *British Journal History of Science, 24*, 307-341.

Gooday, Graeme. (1992, July). *Laboratory electricians at large: Electrotechnical translation in the late Victorian landscape*. Paper presented at the BSHS/CSHPS/HSS Anglo American Conference, Toronto, Canada.

Goodell, R. (1977). *The visible scientists*. Boston: Little, Brown.

Goodell, R. (1980, November/December). The gene craze. *Columbia Journalism Review,* pp. 41-45.

Goodell, R. (1986). How to kill a controversy: The case of recombinant DNA. In S. Dunwoody, S. M. Friedman, & C. Rogers (Eds.), *Scientists and journalists* (pp. 170-181). New York: Free Press.

Goodenough, W. H., & Thomas, S. D. (n.d.). Traditional navigation in the Western Pacific: A search for pattern. In *Expedition* (Vol. 30). Philadelphia: University of Pennsylvania, Museum of Archaeology and Anthropology.

Goodfield, J. (1981). *Reflections on science and the media.* Washington, DC: American Association for the Advancement of Science.

Gooding, David. (1989). History in the laboratory: Can we tell what really went on? In F. A. J. L. James (Ed.), *The development of the laboratory: Essays on the place of experiment in industrial civilisation.* London: Macmillan.

Gooding, David. (1990). *Experiment and the making of meaning.* Dordrecht, the Netherlands: Kluwer.

Gooding, David. (1992). Putting agency back into experiment. In A. Pickering (Ed.), *Science and practice and culture.* Chicago: Chicago University Press.

Gooding, David C., & Addis, T. R. (1990, September). Towards a dynamical representation of experimental procedures. In D. Gooding (Ed.), *Bath 3: Rediscovering skill* (pp. 61-68). Bath, U.K.: Science Studies Centre.

Gooding, David, Pinch, Trevor, & Schaffer, Simon. (Eds.). (1989). *The uses of experiment.* Cambridge: Cambridge University Press.

Goody, Jack. (1977). *The domestication of the savage mind.* Cambridge: Cambridge University Press.

Goody, Jack. (1980). Rice burning and Green Revolution in Northern Ghana. *Journal of Development Studies, 16,* 136-155.

Goonatilake, Susantha. (1988). Epistemology and ideology in science, technology and development. In A. Wad (Ed.), *Science, technology, and development* (pp. 93-114). Boulder, CO: Westview.

Gopnik, M. (1972). *Linguistic structures in scientific texts.* The Hague, the Netherlands: Mouton.

Gordon, D. R. (1982). *Rochester Institute of Technology: Industrial development and educational innovation in an American city.* New York: Edwin Mellen.

Gorman, Michael. (1989). Beyond strong programmes: How cognitive approaches can complement SSK. *Social Studies of Science, 19,* 643-653.

Gorman, Michael E., & Carlson, W. Bernard. (1990). Interpreting invention as a cognitive process: The case of Alexander Graham Bell, Thomas Edison, and the telephone. *Science Technology and Human Values, 15,* 131-164.

Gouldner, Alvin. (1979). *The future of intellectuals and the rise of the new class.* New York: Seabury.

Graham, F. (1970). *Since silent spring.* Boston: Houghton Mifflin.

Graham, Huggan. (1991). Decolonising the map: Post-colonialism, post-structuralism and the cartographic connection. In A. Ian & H. Tiffin (Eds.), *Past the last post: Theorising post-colonialism and post-modernism* (pp. 125-138). New York: Harvester Wheatsheaf.

Graham, L. (1985). The socio-political roots of Boris Hessen: Soviet Marxism and the history of science. *Social Studies of Science, 15,* 705-722.

Graham, Loren. (1979). Concerns about science. In G. Holton & R. Morison (Eds.), *Limits of scientific inquiry.* New York: Norton.

Graham, Loren. (1967). *The Soviet Academy of Sciences and the Communist party 1927-1932.* Princeton, NJ: Princeton University Press.

Granger, J. V. (1979). *Technology and international relations.* San Francisco: Freeman.

Granovetter, Marc S. (1973). The strength of weak ties. *American Journal of Sociology, 78,* 1360-1380.

Graubard, Stephen. (Ed.). (1983). Science literacy. [Special issue] *Daedalus, 112*(2).

Gray, Chris. (1989). The cyborg soldier: The US military and the post-modern warrior. In Les Levidow & Kevin Robins (Eds.), *Cyborg worlds: The military information society*. London: Free Association Books.

Gray, Chris Hables. (1991). *Computers as weapons and metaphors: The US military 1940-1990 and postmodern war.* Unpublished doctoral dissertation, University of California, Santa Cruz.

Grayson, L. P. (1977, December). A brief history of engineering education in the United States. *Engineering Education,* pp. 246-264.

Greenberg, Daniel S. (1967). *The politics of pure science*. New York: New American Library.

Greenfield, L. B., Holloway, E. L., & Remus, L. (1982). Women students in engineering: Are they so different from men? *Journal of College Student Personnel, 23,* 508-514.

Greenwood, Ted. (1990). Why military technology is difficult to restrain. *Science, Technology, & Human Values, 15,* 412-429.

Greimas, A. J., & Courtés, J. (1979). *Sémiotique. Dictionnaire raisonné de la théorie du langage*. Paris: Hachette.

Griffin, Keith. (1974). *The political economy of agrarian change*. Cambridge, MA: Harvard University Press.

Griffin, Susan. (1978). *Woman and nature*. New York: Harper & Row.

Griliches, Zvi. (1958). Research costs and social returns: Hybrid corn and related innovations. *Journal of Political Economy, 66,* 419-431.

Grimston, M. (1991, August). *Green science as pseudo science.* Paper presented at the annual meeting of the British Association for the Advancement of Science, Plymouth.

Gross, Alan G. (1990). *The rhetoric of science.* Cambridge, MA: Harvard University Press.

Grove, J. W. (1989). Nonsense and good sense about women in science. *Minerva, 27*(4), 535-546.

Gruber, William H., & Marquis, Donald G. (Eds.). (1969). *Factors in the transfer of technology.* Cambridge: MIT Press.

Grünbaum, Adolf. (1960). The Duhemian argument. *Philosophy of Science, 27*(1), 75-87.

Grünbaum, Adolf, & Salmon, W. (Eds.). (1988). *The limitations of deductivism.* Berkeley: University of California Press.

Grunig, J. E. (1980). Communication of scientific information to non-scientists. In B. Dervin & M. J. Voight (Eds.), *Progress in communication sciences* (Vol. 1, pp. 167-214). Norwood, NJ: Ablex.

Grynspan, D. (1982). Technology transfer patterns and industrialization in LDC's: A study of licensing in Costa Rica. *International Organization, 36,* 795-806.

Guagnini, Anna. (1988). Higher education and the engineering profession in Italy: The Scuole of Milan and Turin, 1859-1914. *Minerva, 26,* 512-548.

Guillierie, R., & Schoenfeld, A. C. (1979). *An annotated bibliography of environmental communication research and commentary.* Columbus, OH: ERIC/SMEAC Clearinghouse for Science, Mathematics, and Environmental Education.

Gummett, Philip. (1988). The government of military R&D in Britain. In E. Mendelsohn, M. R. Smith, & P. Weingart (Eds.), *Science, technology and the military* (Sociology of Sciences Yearbook, Vol. XII/1, pp. 481-506). Dordrecht, the Netherlands: Kluwer Academic.

Gummett, Philip. (1989, November). *What issues are raised for science & technology studies by defence science & technology policy?* Paper presented to the 4S/EASST meeting, Amsterdam.

Gummett, Philip. (1990). Issues for STS raised by defence science and technology policy. *Social Studies of Science, 20,* 541-558.

Gummett, Philip. (Ed.). (1991). *Future relations between defence and civil science and technology: A report for the [UK] Parliamentary Office of Science and Technology* (SPSG Review Paper No. 2). London: Science Policy Support Group.

Gummett, Philip, & Reppy, Judith. (Eds.). (1988). *The relations between defence and civil technologies*. Dordrecht, the Netherlands: Kluwer Academic.

Gummett, Philip, & Reppy, Judith. (1990). Military industrial networks and technical change in the new strategic environment. *Government and Opposition, 25*(3).

Gusfield, Joseph. (1976). The literary rhetoric of science. *American Sociological Review, 41*, 16-34.

Guston, David H. (1991, September). *Science and the social contract.* Paper presented at the annual meeting of the American Political Science Association, San Francisco.

Haas, E. B. (1980). Technological self-reliance for Latin America: The OAS contribution. *International Organization, 34*, 541-570.

Haas, Peter M. (1989). Do regimes matter? Epistemic communities and Mediterranean pollution control. *International Organization, 43*, 377-403.

Haas, Peter M. (1990). *Saving the Mediterranean: The politics of international environmental cooperation*. New York: Columbia University Press.

Haas, Peter M. (Ed.). (1992). Knowledge, power, and international policy coordination. [Special issue] *International Organization, 46*(1).

Haas, Peter M., Keohane, Robert O., & Levy, Marc A. (Eds.). (1993). *Institutions for the heart.* Cambridge: MIT Press.

Haber, Samuel. (1964). *Efficiency and uplift*. Chicago: University of Chicago Press.

Haberer, Joseph. (1969). *Politics and the community of science*. New York: Van Nostrand Reinhold.

Haberfeld, Yitchak, & Shenhav, Yehouda. (1990). Are women and blacks closing the gap? Salary discrimination in American science during the 1970s and 1980s. *Industrial and Labor Relations Review, 44*, 68-82.

Habermas, Jürgen. (1987). *Théorie de l'agir communicationnel, 2: Pour une critique de la raison fonctionaliste*. Paris: Fayard.

Hacker, Sally. (1981). The culture of engineering: Woman, workplace and machine. *Women's Studies International Quarterly, 4*, 341-353.

Hacker, Sally. (1989). *Pleasure, power and technology: Some tales of gender, engineering, and the cooperative workplace*. Boston: Unwin Hyman.

Hacker, Sally. (1990). *"Doing it the hard way": Investigations of gender and technology*. Boston: Unwin Hyman.

Hacking, Ian. (1983). *Representing and intervening: Introductory topics in the philosophy of natural science*. Cambridge: Cambridge University Press.

Hacking, Ian. (1992). The self-vindication of the laboratory sciences. In A. Pickering (Ed.), *Science as practice and culture*. Chicago: Chicago University Press.

Hafner, Katie, & Markoff, John. (1991). *Cyberpunk: Outlaws and hackers on the computer frontier.* New York: Simon & Schuster.

Hagedoorn, J. (1989). *The dynamic analysis of innovation and diffusion.* London: Frances Pinter.

Hagendijk, Rob. (1990). Structuration theory, constructivism, and scientific change. In S. Cozzens & T. Gieryn (Eds.), *Theories of science in society* (pp. 43-65). Bloomington: Indiana University Press.

Haggard, S., & Simmons, B. A. (1987). Theories of international regimes. *International Organization, 41*, 491-517.

Hagstrom, Warren O. (1966). *The scientific community.* New York: Basic Books.

Hainsworth, Geoffrey B. (1982). *Village level modernization in South East Asia.* Vancouver: University of British Columbia Press.

Halloran, M. (1986). The birth of molecular biology. *Rhetoric Review, 3*, 70-83.

Hamelink, Cees. (1988). *The technology gamble: A study of technology choice*. Norwood, NJ: Ablex.

Hamlett, Patrick W. (1990a). Dialogue on science and Congress. In *Science, technology and politics, 1990: A yearbook* (pp. 23-74). Ottowa, Canada: Odda Tala.

Hamlett, Patrick W. (Ed.). (1990b). Technology and the arms race. *Science, Technology, & Human Values, 15*, 461-473.

Hamlett, Patrick W. (1992). *Understanding technological politics: A decision-making approach*. Englewood Cliffs, NJ: Prentice Hall.

Hamlin, Chris. (1992). Reflexivity in technology studies: Toward a technology of technology. *Social Studies of Science, 22*, 511-544.

Hanmer, J. (1985). Transforming consciousness: Women and the new reproductive technologies. In G. Corea et al. (Eds.), *Man-made women: How new reproductive technologies affect women* (pp. 88-109). London: Hutchinson.

Hansen, A. (1990). Socio-political values underlying media coverage of the environment. *Media Development, 2*, 3-6.

Hansen, A. (Ed.). (1993). *The mass media and environmental issues*. Leicester, U.K.: Leicester University Press.

Hanson, N. R. (1965). *Patterns of discovery*. Cambridge: Cambridge University Press.

Haraway, Donna. (1985). A manifesto for cyborgs: Science, technology and socialist feminism in the 1980s. *Socialist Review, 15*, 65-108.

Haraway, Donna. (1989). *Primate visions: Gender, race, and nature in the world of modern science*. New York: Routledge & Kegan Paul.

Haraway, Donna. (1991a). *Simians, cyborgs, and women*. New York: Routledge & Kegan Paul.

Haraway, Donna. (1991b). Situated knowledges: The science question in feminism and the privilege of partial perspective. In *Simians, cyborgs and women: The reinvention of nature* (pp. 183-203). New York: Routledge & Kegan Paul.

Haraway, Donna. (1992). *Symians, cyborgs, and women: The reinvention of nature*. London: Free Association Books.

Harding, Jan. (1983). *Switched off: The science education of girls*. York, U.K.: Longman, for the Schools Council.

Harding, Sandra. (1986). *The science question in feminism*. New York: Cornell University Press.

Harding, Sandra. (1991). *Whose science? Whose knowledge?* New York: Cornell University Press.

Harding, Sandra, & Hintikka, M. (Eds.). (1983). *Discovering reality*. Dordrecht, the Netherlands: Reidel.

Harré, R. (1981). Rituals, rhetoric and social cognition. In J. Forgas (Ed.), *Social cognition*. London: Academic Press.

Harris, R. A. (1991). Rhetoric of science. *College English, 53*, 282-307.

Hart, H. L. H. (1955). Are there any natural rights? *Philosophical Review, 64*(2), 175-191.

Hartland, J. (in press). The use of intelligent machines for electrocardiograph interpretation. In G. Button (Ed.), *A sociology of new technology*. London: Routledge & Kegan Paul.

Hartmann, H. (1976). Capitalism, patriarchy and job segregation by sex. In M. Blaxall & B. Reagan (Eds.), *Women and the workplace* (pp. 137-169). Chicago: University of Chicago Press.

Hartmann, H., Kraut, R., & Tilly, L. (Eds.). (1986-1987). *Computer chips and paper clips: Technology and women's employment* (Vols. 1, 11). Washington, DC: National Academy Press.

Harvey, Bill. (1981). Plausibility and the evaluation of knowledge. *Social Science Studies of Science, 11*(1), 95-130.

Hass, V. B., & Perrucci, C. C. (Eds.). (1984). *Women in scientific and engineering professions*. Ann Arbor: University of Michigan Press.

Hayami, Yujiro, & Ruttan, Vernon W. (1971). *Agricultural development: An international perspective.* Baltimore: Johns Hopkins University Press.

Hayden, D. (1982). *The grand domestic revolution: A history of feminist designs for American homes, neighborhoods, and cities.* Cambridge: MIT Press.

Hayes, D. P. (1992, April 30). The growing inaccessibility of science. *Nature, 356,* 739.

Hayles, Katherine. (1990). *Chaos bound: Orderly disorder in contemporary literature and science.* Ithaca, NY: Cornell University Press.

Hays, S. P. (1987). *Beauty, health and permanence: Environmental politics in the United States 1955-1985.* Cambridge: Cambridge University Press.

Hazen, R. M., & Trefil, J. (1991). *Science matters: Achieving scientific literacy.* New York: Doubleday.

Headrick, Daniel R. (1988). *The tentacles of progress: Technology transfer in the age of imperialism, 1850-1940.* New York: Oxford University Press.

Hedges, A. (1991). *Attitudes towards energy conservation in the home.* London: U.K. Department of the Environment.

Heffner, Alan. (1980). Authorship recognition of subordinates in collaborative research. *Social Studies of Science, 9,* 377-384.

Heimer, Carol A. (1984). Organizational and individual control of career development in engineering project work. *Acta Sociologica, 4,* 283-310.

Hein, H. S. (1990). *The exploratorium: The museum as laboratory.* Washington, DC: Smithsonian Institution Press.

Helleiner, G. K. (1988). Transnational enterprises in the manufacturing sector of the less developed countries. In H. W. Singer et al. (Eds.), *Technology transfer by multinationals* (pp. 203-223). New Delhi, India: Ashish.

Heller, Peter B. (1985). *Technology transfer and human values: Concepts, applications, cases.* Lanham, MD: University Press of America.

Helman, C. (1978). "Feed a cold, starve a fever": Folk models of infection in an English suburban community, and their relation to medical treatment. *Culture, Medicine and Psychiatry, 2,* 107-137.

Helmreich, Robert, Spence, Janet, Beane, William, Lucker, G. William, & Matthews, Karen. (1980). Making it in academic psychology: Demographic and personality correlates of attainment. *Journal of Personality and Social Psychology, 39,* 896-908.

Henderson, Kathryn. (1991a). Flexible sketches and inflexible data bases. *Science, Technology, & Human Values, 16*(4), 448-473.

Henderson, Kathryn. (1991b). *On line and on paper: Visual representations, visual culture, and computer-graphics in design engineering.* Unpublished doctoral dissertation, University of California, San Diego.

Henderson, Kathryn. (1992, December). *The political career of a prototype.* Paper presented at the Visualization Workshop, Princeton University, Department of History.

Henig, Robin Marantz. (1979). Science for the people: Revolution's evolution. *BioScience, 29*(6), 341-344.

Hennen, L. (1992). *Technisierung des Alltags* (Studien zur Sozialwissenschaft, Band 104). Opladen: Westdeutscher Verlag.

Herbert, S. (1991). Charles Darwin as a prospective geological author. *British Journal of the History of Science, 24,* 159-192.

Herdt, Robert W. (1987). A retrospective view of technological and other changes in Philippine rice farming, 1965-1982. *Economic Development and Cultural Change, 35,* 329-350.

Herken, Gregg. (1983). *Counsels of war.* New York: Knopf.

Herrera, Amilcar O. (1981). The generation of technologies in rural areas. *World Development, 9,* 21-36.

Hershey, R. D., Jr. (1989, July 16). Capitol Hill's high-tech tutor. *The New York Times*, pp. F1-F2.

Hess, David. (1991). *Spirits and scientists: Ideology, spiritism, and Brazilian culture*. University Park: University of Pennsylvania Press.

Hess, David J. (1992). Introduction: The new ethnography and the anthropology of science and technology. In D. J. Hess & L. Layne (Eds.), *Knowledge and society: The anthropology of science and technology* (Vol. 9). Greenwich, CT: JAI.

Hesse, Mary. (1974). *The structure of scientific inference*. London: Macmillan.

Hessenbruch, Arne. (1992, July). *Balancing the lab and the non-lab*. Talk presented at the joint BSHS, HSS & CSHS meeting, Toronto.

Hetman, Francois. (1977). Social assessment of technology and some of its international aspects. *Technological Forecasting and Social Change, 11*, 303-314.

Hettne, Bjorn. (1983). The development of development theory. *Acta Sociologica, 26*, 247-266.

Hétu, C. (1989). Skills, knowledge and models: An ethnographic study of conversations among engineers and workers in the semiconductor industry. In M. de Montmollin & A. Hingel (Eds.), *Information technology, competence and employment*. New York: Wiley.

Hewlett, R. G., & Holl, J. M. (1989). *Atoms for peace and war, 1953-1961*. Berkeley: University of California Press.

Hicks, E. K., & Callebaut, W. (Eds.). (1989). *Evaluative proceedings 4S/EASST 1988* (SISWO publikatie no. 343). Amsterdam: SISWO.

Hilgartner, Stephen. (1990, August). The dominant view of popularization: Conceptual problems, political uses. *Social Studies of Science, 20*, 519-539.

Hilgartner, Stephen. (1992). Who speaks for science: Disputes among experts in the diet-cancer debate. In D. Nelkin (Ed.), *Controversy*. Newbury Park, CA: Sage.

Hilgartner, Stephen, & Brandt-Rauf, S. (1990). *Controlling access to scientific data: Boundary construction in molecular genetics*. Paper presented at the meeting of the Society for Social Studies of Science.

Hill, Stephen. (1986). From light to dark: Seeing development strategies through the eyes of S&T indicators. *Science and Public Policy, 13*, 254-275.

Hilpert, U. (Ed.). (1991). *State policies and techno-industrial innovation*. London: Routledge.

Hilton, R. H., & Sawyer, P. H. (1952). Technical determinism: The stirrup and the plough. *Past and Present, 24*, 90-100.

Hinkle, G., & Elliot, W. R. (1989). Science coverage in three newspapers and three supermarket tabloids. *Journalism Quarterly, 66*, 353-358.

Hirsch, Walter. (1968). *Scientists in American society*. New York: Random House.

Hirschauer, S. (1991). The manufacture of bodies in surgery. *Social Studies of Science, 21*(2), 279-319.

Hirschauer, S. (1992). *Die medizinische Konstruktion von Transsexualität*. Frankfurt: Suhrkamp.

Hobbs, J., & Moore, R. (Eds.). (1985). *Formal theories of the common-sense world*. Norwood, NJ: Ablex.

Hobday, Michael. (1985). The impact of microelectronics on developing countries: The case of Brazilian communications. *Development and Change, 16*, 313-340.

Hobday, Michael. (1989). The European semiconductor industry: Resurgence and rationalization. *Journal of Common Market Studies, 28*, 155-186.

Hoch, Paul. (1988). The crystallization of a strategic alliance: The American physics elite and the military in the 1940s. In E. Mendelsohn et al. (Eds.), *Science technology and the military* (Sociology of the Sciences Yearbook). Dordrecht, the Netherlands: Kluwer.

Hodges, Andrew. (1983). *Alan Turing: The enigma*. New York: Simon & Schuster.

Hoffmann, Erik P. (1979). Contemporary Soviet theories of scientific, technological and social change. *Social Studies of Science, 9*(1), 101-113.

Hohenemser, Cristoph, Kasperson, Roger, & Kates, Robert. (1977, April 1). The distrust of nuclear power. *Science, 196*, 25-34.

Holdren, John P. (1982, April). Energy hazards: What to measure, what to compare. *Technology Review*, pp. 32-38, 74-75.

Holland, D., & Quinn, N. (1987). *Cultural models in language and thought*. Cambridge: Cambridge University Press.

Holland, Susan S. (1976). Exchange of people among international companies: Problems and benefits. *Annals of the American Academy of Political and Social Science, 424*, 52-66.

Hollander, Rachelle, & Steneck, Nicholas. (1990). Science- and engineering-related ethics and values studies: Characteristics of an emerging field of research. *Science, Technology, & Human Values, 15*(1), 84-104.

Hollinger, David. (1983). The defense of democracy and Robert K. Merton's formulation of the scientific ethos. *Knowledge and Society, 4*, 1-15.

Hollis, Martin, & Lukes, Stephen. (1982). *Rationality and relativism*. Oxford: Blackwell.

Holloway, D. (1983). *The Soviet Union and the arms race*. New Haven, CT: Yale University Press.

Holmes, Frederic Lawrence. (1987). Scientific writing and scientific discovery. *Isis, 78*, 220-235.

Holmes, Frederic Lawrence. (1992). *Between biology and medicine: The formation of intermediary metabolism* (Berkeley Papers in the History of Science no. 24). Berkeley: University of California, Office for History of Science and Technology.

Holmstrom, Engin Inel, & Holmstrom, Robert. (1974). The plight of the woman graduate student. *American Educational Research Journal, 11*, 1-17.

Holsinger, Donald B., & Theisen, Gary L. (1977). Education, individual modernity, and national development: A critical appraisal. *Journal of Developing Areas, 11*, 315-334.

Holton, Gerald. (Ed.). (1965). *Science and culture*. Boston: Beacon.

Holton, Gerald. (1973). *Thematic origins of scientific thought: Kepler to Einstein*. Cambridge, MA: Harvard University Press.

Holton, Gerald. (1992). How to think about the "anti-science" phenomenon. *Public Understanding of Science, 1*(1), 103-128.

Holton, Gerald, & Blanpied, William A. (Eds.). (1976). *Science and its public: The changing relationship*. Dordrecht, the Netherlands: Reidel.

Holtzman, Neil A. (1989). *Proceed with caution: Predicting genetic risks in the recombinant DNA era*. Baltimore: Johns Hopkins University Press.

Hornig, S. (1990). Science stories: Risk, power, and perceived emphasis. *Journalism Quarterly, 67*(4), 767-776.

Horton, Robin. (1967). African traditional thought and Western science. *Africa, 37*, 1-2.

Horton, Robin, & Finnegan, R. (1973). *Modes of thought*. London: Faber and Faber.

Hounshell, David. (1980, February 8). Edison and the pure science ideal in America. *Science*, pp. 612-617.

Hounshell, David. (1983). *From the American system to mass production, 1800-1932: The development of manufacturing technology in the United States*. Baltimore: Johns Hopkins University Press.

Howells, J. (1990a). The location and organization of research and development: New horizons. *Research Policy, 19*, 133-146.

Howells, J. (1990b). The globalisation of research and development: A new era of change? *Science & Public Policy, 17*, 273-285.

Howells, J., & Wood, M. (1991). *The globalisation of technology* (Monitor-Fast FOP 274). Brussels: Commission of the European Communities.

Howes, Michael. (1979). Appropriate technology: A critical evaluation of the concept and the movement. *Development and Change, 10*, 115-124.

Hoyrup, Else. (1978). *Women and mathematics, science and engineering: A partially annotated bibliography . . . .* Roskilde, Denmark: Roskilde University Library.

Hoyrup, Else. (1987). *Women of science, technology and medicine: A bibliography*. Roskilde, Denmark: Roskilde University Library.

Hubbard, Ruth. (1990). *The politics of women's biology*. New Brunswick, NJ: Rutgers University Press.

Hubbard, Ruth, Henefin, Mary Sue, & Fried, Barbara. (Eds.). (1982). *Biological woman: The convenient myth*. Cambridge, MA: Schenkman.

Hubbard, Ruth, & Lowe, Marian. (Eds.). (1979). *Genes and gender*. Staten Island, NY: Gordian.

Hubbard, Ruth, & Wald, Elijah. (1993). *Exploding the gene myth*. Boston: Beacon.

Huber, Peter W. (1991). *Galileo's revenge: Junk science in the courtroom*. New York: Basic Books.

Huddleston, R. D. (1971). *The sentence in written English: A syntactic study based on the analysis of scientific texts*. Cambridge: Cambridge University Press.

Hudson, Liam. (1962). Converger/diverger. *Nature, 196*, 601.

Hudson, Liam. (1963). Converger/diverger. *Nature, 198*, 913.

Hudson, Liam. (1966). *Contrary imaginations: A psychological study of the English schoolboy*. London: Methuen.

Hufbauer, Karl. (1986). Federal funding and sudden infant death research, 1945-80. *Social Studies of Science, 16*(1), 61-78.

Hughes, Everett. (1971). *The sociological eye*. Chicago: Aldine.

Hughes, P. (1992). *Preparing for accidents at nuclear power stations: The role of emergency planning in a technological society*. Unpublished master of philosophy thesis, University of Lancaster.

Hughes, Thomas P. (1983). *Networks of power: Electrification in Western society, 1880-1930*. Baltimore: Johns Hopkins University Press.

Hughes, Thomas P. (1986). The seamless web: Technology, science, etcetera, etcetera. *Social Studies of Science, 16*, 281-292.

Hughes, Thomas P. (1987a). The evolution of large technological systems. In Wiebe E. Bijker, Thomas P. Hughes, & Trevor Pinch (Eds.), *The social construction of technological systems*. Cambridge: MIT Press.

Hughes, Thomas P. (1987b). The seamless web: Technology, science, etcetera, etcetera. *Social Studies of Science, 16*, 281-292.

Hughes, Thomas P. (1989a). *American genesis: A century of invention and technological enthusiasm*. Harmondsworth, U.K.: Penguin.

Hughes, Thomas P. (1989b). The evolution of large technological systems. In Wiebe E. Bijker, Thomas P. Hughes, & Trevor Pinch (Eds.), *The social construction of technological systems* (pp. 51-82). Cambridge: MIT Press.

Hughes, Thomas P., & Hughes, A. C. (1990). *Lewis Mumford: Public intellectual*. Oxford: Oxford University Press.

Hull, David. (1988). *Science as a process: An evolutionary account of the social and conceptual development of science*. Chicago: University of Chicago Press.

Hunt, Lynn. (Ed.). (1989). *The new cultural history*. Berkeley: University of California Press.

Hustopecky, J., & Vlachy, J. (1978). Identifying a set of inequality measures for science studies. *Scientometrics, 1*, 85-98.

Hutchins, Edwin. (1983). Understanding micronesian navigation. In D. Gentner & A. Stevens (Eds.), *Mental models* (pp. 191-225). Hillsdale, NJ: Lawrence Erlbaum.

Hutchins, Edwin. (1989). The technology of team navigation. In J. Galegher, R. Kraut, & C. Egido (Eds.), *Intellectual teamwork: Social and technical bases of cooperative work*. Hillsdale, NJ: Lawrence Erlbaum.

Hutton, S., & Lawrence, P. (1981). *German engineers: The anatomy of a profession*. Oxford: Clarendon.

IIT Research Institute. (1968). *Technology in retrospect and critical events in science (TRACES)*. Chicago: Author.

Ikenberry, G. J., Lake, D. A., & Mastanduno, M. (Eds.). (1988). *The state and American foreign economic policy*. Ithaca, NY: Cornell University Press.

Ilerbaig, Juan. (1992). The two STS subcultures and the sociological revolution. *Science, Technology & Society: Curriculum Newsletter of the Lehigh University STS Program & Technology Studies Resources Center, 90*, 1-6.

Inkeles, A. (1975). The emerging social structure of the world. *World Politics, 27*, 467-495.

Inkster, Ian. (1989). Appropriate technology, alternative technology, and the Chinese model: Terminology and analysis. *Annals of Science, 46*, 263-277.

Irvine, John, & Martin, Ben. (1984). *Foresight in science*. London: Frances Pinter.

Irwin, Alan. (1993). Acid pollution and public policy: The changing climate of environmental decision-making. In M. Radojevic & R. Harrison (Eds.), *Atmospheric acidity: Sources, consequences and abatement*. Amsterdam: Elsevier.

Irwin, Alan, Dale, A., & Smith, D. (in press). Science and Hell's Kitchen: The local understanding of hazard issues. In A. Irwin & B. Wynne (Eds.), *Misunderstanding science*. Cambridge: Cambridge University Press.

Irwin, Alan, & Wynne, Brian. (Eds.). (in press). *Misunderstanding science*. Cambridge: Cambridge University Press.

Ivins, W. M. (1973). *On the rationalization of sight*. New York: Plenum.

Jackson, T. (1991, November 15). Moonshine and magnanimity of the fusion fantasists. *The Guardian*, p. 31.

Jacky, Jonathan. (n.d.). Software engineers and hackers: Programming and military computing. In Paul N. Edwards & R. Gordon (Eds.), *Strategic computing: Defense research and high technology*. Unpublished manuscript.

Jacob, Margaret. (1988). *The cultural meaning of the scientific revolution*. Philadelphia: Temple University Press.

Jacob, Pierre. (1981). *De Vienne à Cambridge*. Paris: Gallimard.

Jacobi, D. (1985). References iconiques et modèles analogiques dans des discours de vulgarisation scientifique. *Information sur les Sciences Sociales/Social Science Information, 24*, 847-867.

Jacobi, D., & Schiele, B. (1989). Scientific imagery and popularized imagery: Differences and similarities in the photographic portraits of scientists. *Social Studies of Science, 19*, 731-753.

Jacobs, John F. (1983). SAGE overview. *Annals of the History of Computing, 5*(4), 323-329.

Jacobsson, Staffan. (1985). Technical change and industrial policy: The case of computer numerically controlled lathes in Argentina, Korea and Taiwan. *World Development, 13*, 353-370.

Jacobus, Mary, Keller, Evelyn F., & Shuttleworth, S. (Eds.). (1990). *Body/politics: Women and the discourses of science*. New York: Routledge.

Jagacinski, Carolyn M. (1987a). Engineering careers: Women in a male-dominated field. *Psychology of Women Quarterly, 11*, 97-110.

Jagacinski, Carolyn M. (1987b). Androgyny in a male-dominated field: The relationship of sex-typed traits to performance and satisfaction in engineering. *Sex Roles, 17*, 529-547.

Jagacinski, C. M., & LeBold, W. K. (1981). A comparison of men and women undergraduate and professional engineers. *Engineering Education, 72*, 213-220.

Jaireth, Jasveen. (1988). Class relations and technology use: A study of Tubewell utilization in Punjab (India). *Development and Change, 19*, 89-114.

James, Dilmus D. (1988). Accumulation and utilization of internal technological capabilities in the Third World. *Journal of Economic Issues, 22*, 339-353.
James, John. (1982). *Chartres: The masons who built a legend*. London: Routledge & Kegan Paul.
James, John. (1989). *The template-makers of the Paris basin*. Leura: West Grinstead.
Jamison, Andrew. (1988). Social movements and the politicization of science. In J. Annerstedt & A. Jamison (Eds.), *From research policy to social intelligence* (pp. 69-86). London: Macmillan.
Jamison, Andrew. (1989). Technology's theorists: Conceptions of innovation in relation to science and technology policy. *Technology and Culture, 30*(3), 505-533.
Jamison, Andrew, & Baark, Erik. (1990). Modes of biotechnology assessment in the USA, Japan and Denmark. *Technology Analysis and Strategic Management, 2*(2), 111-127.
Jamison, Andrew, & Baark, Erik. (1991). *Technological innovation and environmental concern: Contending policy models in China and Vietnam*. Lund, Sweden: Research Policy Institute.
Jamison, Andrew, et al. (1990). *The making of the new environmental consciousness: A comparative study of the environmental movements in Sweden, Denmark and the Netherlands*. Edinburgh: Edinburgh University Press.
Jamison, E. (1985). *Women of the world: A chartbook for developing regions*. Washington, DC: U.S. Department of Commerce, Bureau of the Census, Center for International Research.
Jamison, Wesley, & Lunch, William. (1992). The rights of animals, perceptions of science, and political activism: A demographic, attitudinal, and behavioral profile of American animal rights activists. *Science, Technology, & Human Values, 17*(4), 438-458.
Jansen, S. C. (1990). Is science a man? New feminist epistemologies and reconstructions of knowledge. *Theory and Society, 19*, 235-246.
Jasanoff, Sheila. (1986). *Risk management and political culture*. New York: Russell Sage.
Jasanoff, Sheila. (1987). Contested boundaries in policy-relevant science. *Social Studies of Science, 17*, 195-230.
Jasanoff, Sheila. (1990a). *The fifth branch: Science advisers as policymakers*. Cambridge, MA: Harvard University Press.
Jasanoff, Sheila. (1990b). American exceptionalism and the political acknowledgment of risk. *Daedalus, 119*, 61-81.
Jasanoff, Sheila. (1992a). Science, politics, and the renegotiation of expertise at EPA. *Osiris, 7*, 1-23.
Jasanoff, Sheila. (1992b). What judges should know about the sociology of science. *Jurimetrics, 32*, 345-359.
Jasanoff, Sheila. (1993). Bridging the two cultures of risk analysis. *Risk Analysis, 13*(2), 123-129.
Jasanoff, Sheila. (Ed.). (1994). *Learning from disaster: Risk management after Bhopal*. Philadelphia: University of Pennsylvania Press.
Jasper, James. (1990). *Nuclear politics*. Princeton, NJ: Princeton University Press.
Jasper, James M., & Nelkin, Dorothy. (1992). *The animal rights crusade: The growth of a moral protest*. New York: Free Press.
Jenkins, J. Craig. (1983). Resource mobilization theory and the study of social movements. *Annual Review of Sociology, 9*, 527-553.
Jerome, Fred. (1986, September). Gagging government scientists: A new government policy? *Technology Review*, pp. 25-35.
Jewkes, John, Sawers, David, & Sillerman, Richard. (1969). *The sources of invention*. London: Macmillan.
Jimenez, J., Campos, M. A., & Escalante, J. C. (1991). Distribution of scientific tasks between center and periphery in Mexico. *Social Science Information, 30*, 471-820.

Joerges, Bernard. (1988). *Technik im Alltag*. Frankfurt am Main: Suhrkamp Verlag.

Johnson, Deborah. G. (1989). The social/professional responsibility of engineers. *Annals of the New York Academy of Science, 577*, 106-114.

Johnson, Jim. (aka Latour, Bruno). (1988). Mixing humans and non-humans together: The sociology of a door-closer. *Social Problems, 35*, 298-310.

Johnson, J. K., Schatzberg, W., & Waite, R. A. (Eds.). (1987). *The relations of literature and science: An annotated bibliography of scholarship, 1880-1980*. New York: Modern Language Association of America.

Johnson, L. Z. (1957, Spring). Status and attitudes of science writers. *Journalism Quarterly, 34*, 247-251.

Jokisch, R. (1982). *Techniksoziologie*. Frankfurt am Main: Suhrkamp Verlag.

Jonsen, Albert R. (1991). American moralism and the origin of bioethics in the United States. *The Journal of Medicine and Philosophy, 16*, 113-130.

Jordan, Kathleen, & Lynch, Michael. (1992). The sociology of a genetic engineering technique: Ritual and rationality in the performance of the plasmid prep. In Adele Clarke & Joan Fujimara (Eds.), *The right tools for the job: At work in 20th century life sciences*. Princeton, NJ: Princeton University Press.

Jordanova, Ludmilla. (1989). *Sexual visions: Images of gender in science and medicine between the eighteenth and twentieth centuries*. New York: Harvester.

Joseph, L. E. (1991). *Gaia: The growth of an idea*. London: Arkana.

Joynson, Robert B. (1989). *The Burt affair*. New York: Routledge & Kegan Paul.

Judge, W. James. (1984). New light on Chaco Canyon. In D. G. Noble (Ed.), *New light on Chaco Canyon* (pp. 1-12). Santa Fe, NM: School of American Research Press.

Juma, C. (1989). *The gene hunters: Biotechnology and the scramble for seeds*. Princeton, NJ: Princeton University Press.

Jupp, A. (1989). *The provision of hazard information to the public*. Unpublished master of science thesis, University of Manchester.

Kachaunov, Stefan, & Simeonova, Kostadinka. (1979). Social studies of science in Bulgaria. *Social Studies of Science, 9*(1), 91-99.

Kaldor, M. (1983). *The baroque arsenal*. London: Sphere.

Kaplan, Barbara Beigun. (1991). STS, women's studies, and the transformation of the undergraduate curriculum. *Science, Technology & Society: Curriculum Newsletter of the Lehigh University STS Program & Technology Studies Resource Center, 86/87*, 1-11.

Kaplan, Norman. (Ed.). (1965). *Science and society*. Chicago: Rand McNally.

Kaplinsky, Raphael. (1988). Export-oriented growth: A large international firm in a small developing country. In H. W. Singer et al. (Eds.), *Technology transfer by multinationals* (pp. 466-487). New Delhi, India: Ashish.

Karp, H., & Restivo, Sal. (1974). Ecological factors in the emergence of modern science. In S. Restivo & C. K. Vanderpool (Eds.), *Comparative studies in science and society* (pp. 123-143). Columbus, OH: Charles E. Merrill.

Kashet, Eva, Robbins, Mary Louise, Lieve, Loretta, & Huang, Alice. (1974, February). Status of women microbiologists. *Science*, pp. 488-494.

Katrak, Homi. (1989). Imported technologies, and R&D in a newly industrialising country: The experience of Indian enterprises. *Journal of Development Economics, 31*, 123-140.

Katriel, T., & Sanders, R. E. (1989). The meta-communicative role of epigraphs in scientific text construction. In H. W. Simons (Ed.), *Rhetoric in the human sciences* (pp. 183-194). London: Sage.

Katz, Jorge M. (Ed.). (1987). *Technology generation in Latin American manufacturing industries*. New York: St. Martin's Press.

Kay, W. D. (1991-1992, Winter). The politics of fusion research. *Issues in Science and Technology*.

Kaysen, Carl. (1989). Can universities cooperate with the defense establishment? [Special issue: Universities and the Military] *The Annals of the American Academy of Political and Social Science, 502,* 29-39.

Keller, A. (1984). Has science created technology. *Minerva, 22,* 160-182.

Keller, Evelyn Fox. (1978). Gender and science. *Psychoanalysis and Contemporary Thought, 1*(3), 409-433.

Keller, Evelyn Fox. (1983). *A feeling for the organism: The life and work of Barbara McClintock.* San Francisco: Freeman.

Keller, Evelyn Fox. (1985). *Reflections on gender and science.* New Haven, CT: Yale University Press.

Keller, Evelyn Fox. (1992). *Secrets of life, secrets of death: Essays on language, gender, and science.* New York: Routledge & Kegan Paul.

Keller, K. H. (1990). Science and technology. *Foreign Affairs, 69,* 123-138.

Kellert, Stephen. (1993). *In the wake of chaos.* Chicago: University of Chicago Press.

Kelly, Alison. (1979). *Girls and science* (Monographs of the International Association for the Evaluation of Educational Achievement, No. 9). Stockholm: Almqvist & Wiksell International.

Kelly, Alison. (Ed.). (1987). *Science for girls?* Milton Keynes, U.K.: Open University Press.

Kelly-Godol, Joan. (1976). The social relations of the sexes: Methodological implications of women's history. *Signs, 1*(4), 809-823.

Kemp, Louis W. (1986). Aesthetes and engineers: The occupational ideology of highway design. *Technology and Culture, 27,* 759-797.

Kempton, W. (1987). Two theories of home heat control. In D. Holland & N. Quinn (Eds.), *Cultural models in language and thought* (pp. 222-242). Cambridge: Cambridge University Press.

Kempton, W. (1991). Public understanding of global warming. *Society and Natural Resources, 4,* 331-345. [Expanded version: *Global Environmental Change, 1*(3), 1991, pp. 183-208]

Kempton, W., & Montgomery, L. (1982). Folk quantification of energy. *Energy: The International Journal, 7,* 817-827.

Kennedy, M. D. (1987). Polish engineers' participation in the Solidarity movement. *Social Forces, 65,* 641-669.

Kenney, Martin. (1983). Is biotechnology a blessing for less developed nations? *Monthly Review, 34,* 10-19.

Kenney, Martin. (1986). *Biotechnology: The university-industrial complex.* New Haven, CT: Yale University Press.

Keohane, Robert O., & Nye, Joseph S. (1977). *Power and interdependence.* Boston: Little, Brown.

Kevles, Daniel J. (1978). *The physicists: The history of a scientific community in modern America.* New York: Knopf.

Kevles, Daniel J. (1985). *In the name of eugenics: Genetics and the uses of human heredity.* New York: Knopf.

Kevles, Daniel J. (1988). R&D and the arms race: An analytical look. In E. Mendelsohn, M. R. Smith, & P. Weingart (Eds.), *Science, technology and the military* (Sociology of Sciences Yearbook, Vol. XII/1, pp. 481-506). Dordrecht, the Netherlands: Kluwer Academic.

Kevles, Daniel J. (1990). Principles and politics in federal R&D policy, 1945-1990: An appreciation of the Bush report. Preface. In V. Bush, *Science: The endless frontier.* Washington, DC: NSF.

Kidd, J. S. (1988). The popularization of science: Some basic measurements. *Scientometrics, 14*(1-2), 127-142.

Killian, J. R. (1982). *Sputnik, scientists, and Eisenhower.* Cambridge: MIT Press.

Kingdon, John W. (1981). *Congressmen's voting decisions* (2nd ed.). New York: Harper & Row.

Kingsland, Sharon E. (1992, July). *An elusive science: Ecological enterprise in the southwestern United States.* Paper presented at the HSS/BSHS meetings, Toronto.

Kinmoth, Earl. (1986). Engineering education and its rewards in the United States and Japan. *Comparative Education Review, 30,* 396-416.

Kinsella, J. (1989). *Covering the plague: AIDS and the American media.* New Brunswick, NJ: Rutgers University Press.

Kirkup, G., & Keller, L. S. (Eds.). (1992). *Inventing women: Science, technology and gender.* Cambridge: Polity.

Kistiakowski, Vera. (1989). Military funding of university research. [Special issue: Universities and the Military] *The Annals of the American Academy of Political and Social Science, 502,* 141-154.

Kitzinger, C. (1988). *The social construction of lesbianism.* London: Sage.

Kjerulff, Kristen, & Blood, Milton. (1973). A comparison of communication patterns in male and female graduate students. *Journal of Higher Education, 44,* 623-632.

Klaidman, Stephen D. (1991). *Health in the headlines: The stories behind the stories.* New York: Oxford University Press.

Klein, R. (1985). What's "new" about the "new" reproductive technologies? In G. Corea et al. (Eds.), *Man-made women: How new reproductive technologies affect women* (pp. 64-73). London: Hutchinson.

Kleinman, Daniel. (1993). *Groping toward the endless frontier: The politics of postwar research policy in the United States.* Unpublished doctoral dissertation, University of Wisconsin, Madison.

Kline, Ronald. (1987). Science and engineering theory in the invention and development of the induction motor, 1880-1900. *Technology and Culture, 28,* 283-313.

Kling, Rob, & Iacono, Suzanne. (1984). Computing as an occasion for social control. *Journal of Social Issues, 40*(3), 77-96.

Kling, Rob, & Scacchi, Walt. (1982). The web of computing: Computing technology as social organization. *Advances in Computers, 21,* 3-85.

Kloppenberg, Jack Ralph, Jr. (1988). *First the seed: The political economy of plant biotechnology, 1492-2000.* Cambridge: Cambridge University Press.

Kneen, P. (1984). *Soviet scientists and the state.* London: Macmillan.

Knezo, G. J. (1990). *Defense basic research priorities: Funding and policy issues* (CSR Report for Congress). Washington, DC: Congressional Research Service.

Knight, N. (1986). The new light: X rays and medical futurism. In J. J. Corn (Ed.), *Imagining tomorrow* (pp. 10-34). Cambridge: MIT Press.

Knights, David, & Willmott, H. (Eds.). (1988). *New technology and the labour process.* London: Macmillan.

Knorr, Karin. (1977). Producing and reproducing knowledge: Descriptive or constructive? Toward a model of research production. *Social Science Information, 16,* 669-696.

Knorr, Karin. (1981). *The manufacture of knowledge: An essay on the constructivist and contextual nature of science.* Oxford: Pergamon. (rev. ed. 1984, *Die Fabrikation von Erkenntnis,* Frankfurt: Suhrkamp)

Knorr Cetina, Karin. (1982). Scientific communities of transepistemic arenas of research? A critique of quasi-economic models of science. *Social Studies of Science, 12*(1), 101-130.

Knorr Cetina, Karin. (1987). Toward a reconception of macrosocial order. In N. Fielding et al. (Eds.), *Macrotheory and microresearch.* London: Sage.

Knorr Cetina, Karin. (1988). The internal environment of knowledge claims: One aspect of the knowledge-society connection. *Argumentation, 2,* 369-389.

Knorr Cetina, Karin. (1991). Epistemic cultures: Forms of reason in science. *History of Political Economy, 23*, 105-122.
Knorr Cetina, Karin. (1992a). The couch, the cathedral, and the laboratory: On the relationship between experiment and laboratory in science. In A. Pickering (Ed.), *Science as practice and culture* (pp. 113-138). Chicago: University of Chicago Press.
Knorr Cetina, Karin. (1992b). *Liminal and referent epistemologies in contemporary science: An ethnography of the empirical in two sciences.* Paper presented at the Thursday Seminar, Princeton Institute for Advanced Study.
Knorr Cetina, Karin. (in press). *Epistemic cultures: How scientists make sense*. Chicago: Indiana University Press.
Knorr Cetina, Karin, & Amann, Klaus. (1990). Image dissection in natural scientific inquiry. *Science, Technology, & Human Values, 15*(3), 259-283.
Knorr Cetina, Karin, & Amann, Klaus. (1992). Konsensprozesse in der Wissenschaft. In H. Giegel (Ed.), *Kommunikation and Konsens in modern Gesellschaften*. Frankfurt: Suhrkamp.
Knorr Cetina, Karin, & Mulkay, Michael. (1983). *Science observed: Perspectives on the social study of science*. London: Sage.
Kollard, F. (1990). National cultures and technology-transfer: The influence of Mexican life-style on technology transfer. *International Journal of Intercultural Relations, 14*, 319-336.
Kolodziej, E. A. (1987). *Making and marketing arms: The French experience and its implications for the international system*. Princeton, NJ: Princeton University Press.
Koppel, Bruce, & Oasa, Edmund. (1987). Induced innovation theory and Asia's Green Revolution: A case study of an ideology of neutrality. *Development and Change, 18*, 29-67.
Kornhauser, William. (1962). *Scientists in industry: Conflict and accomodation*. Berkeley: University of California Press.
Kramarae, Chris. (Ed.). (1988). *Technology and women's voices*. New York: Routledge & Kegan Paul.
Kranakis, Eda. (1982). The French connection: Giffard's injector and the nature of heat. *Technology and Culture, 23*, 3-38.
Kranakis, Eda. (1989). Social determinants of engineering practice: A comparative view of France and America in the nineteenth century. *Social Studies of Science, 19*, 5-70.
Kranzberg, Melvin, & Pursell, Carroll W., Jr. (Eds.). (1967). *Technology in Western civilization* (2 vols.). New York: Oxford University Press.
Krasner, Stephen D. (1978). *Defending the national interest*. Princeton, NJ: Princeton University Press.
Kreps, David, & Wilson, Robert. (1982). Reputation and imperfect information. *Journal of Economic Theory, 27*, 253-279.
Krieghbaum, H. (1940). The background and training of science writers. *Journalism Quarterly, 17*, 15-18.
Krieghbaum, H. (Ed.). (1963). Science news [Special section]. *Journalism Quarterly, 40*, 291-338.
Krieghbaum, H. (1967). *Science and the mass media*. New York: New York University Press.
Krimsky, Sheldon. (1982). *Genetic alchemy*. Cambridge: MIT Press.
Krimsky, Sheldon. (1984). Epistemic considerations on the value of "folk-wisdom" in science and technology. *Policy Studies Review, 3*, 246-262.
Krimsky, Sheldon. (1991). *Biotechnics and society*. New York: Praeger.
Krimsky, Sheldon, et al. (1991). Academic corporate ties in biotechnology: A quantitative study. *Science, Technology, & Human Values, 16*(3), 275-287.
Krishna, Venni Venkata. (1992). The colonial "model" and the emergence of national science in India: 1876-1920. In P. Petitjean et al. (Eds.), *Science and empires: Historical studies about*

scientific development and European expansion (pp. 57-72). Dordrecht, the Netherlands: Kluwer.

Krohn, Wolfgang, Kuppers, G., & Nowotny, Helga. (Eds.). (1990). *Self-organization: Portrait of a scientific revolution* (Sociology of the Sciences Yearbook). Dordrecht, the Netherlands: Kluwer Academic.

Krugman, P. R. (1986). Introduction: New thinking about trade policy. In P. R. Krugman (Ed.), *Strategic trade policy and the new international economics*. Cambridge: MIT Press.

Krupat, A. (1992). *Ethnocriticism: Ethnography, history, literature*. Berkeley: University of California Press.

Kuhn, Thomas S. (1962). *The structure of scientific revolutions*. Chicago: University of Chicago Press. (rev. 2nd ed., 1970)

Kuhn, Thomas S. (1970). Logic of discovery or psychology of research? In I. Lakatos & A. Musgrave (Eds.), *Criticism and the growth of knowledge* (pp. 1-23). Cambridge: Cambridge University Press.

Kuhn, Thomas S. (1977a). *The essential tension*. Chicago: University of Chicago Press.

Kuhn, Thomas S. (1977b). The function of measurement in modern physical science. In T. S. Kuhn, *The essential tension: Selected studies in scientific tradition and change*. Chicago: University of Chicago Press.

Kuhn, Thomas S. (1992). *The trouble with the historical philosophy of science* (Robert & Maurine Rothschild Distinguished Lecture, November 19, 1991). Cambridge, MA: Harvard University, Department of the History of Science.

Kuklick, Henrika. (1980). Boundary maintenance in American sociology: Limitations to academic "professionalization." *Journal of the History of the Behavioral Sciences, 16*, 201-219.

Kumar, Nagesh. (1987). Technology imports and local research and development in Indian manufacturing. *Developing Economies, 25*, 220-233.

Kumazawa, M., & Yamada, Jomoko. (1989). Jobs and skills under the lifelong Nenko employment practice. In S. Wood (Ed.), *The transformation of work*. London: Unwin Hyman.

Kunda, G. (1992). *Engineering culture: Control and commitment in a hi-tech corporation*. Philadelphia: Temple University Press.

Kuznick, P. J. (1987). *Beyond the laboratory: Scientists as political activists in 1930s America*. Chicago: University of Chicago Press.

Kwa, C. (1987). Representations of nature mediating between ecology and science policy: The case of the International Biological Program. *Social Studies of Science, 17*, 413-442.

Kwa, C. (1989). *Mimicking nature: The development of systems ecology in the United States, 1950-1975*. Amsterdam: University of Amsterdam.

Lachmund, J. (1994). *Die Transformation der medizinischen Kultur*. Unpublished doctoral dissertation, University of Bielefeld, Faculty of Sociology.

Laetsch, W. M. (1987). A basis for better public understanding of science. In D. Evered & M. O'Connor (Eds.), *Communicating science to the public* (pp. 1-10). Chichester, U.K.: Wiley.

LaFollette, Marcel C. (1989). ["Panel on autonomy."] In S. Cozzens et al. (Eds.), *The research system in transition*. Dordrecht, the Netherlands: Kluwer.

LaFollette, Marcel C. (1990a). *Making science our own: Public images of science, 1910-1955*. Chicago: University of Chicago Press.

LaFollette, Marcel C. (1990b). US policy on intellectual property issues in R&D. In S. Cozzens et al. (Eds.), *The research system in transition*. Dordrecht, the Netherlands: Kluwer.

Laird, Frank N. (1993). Participatory analysis, democracy, and technological decision making. *Science, Technology, & Human Values, 18*(4), 341-361.

Lakatos, Imre. (1974). History of science and its rational reconstructions. In Yehuda Elkana (Ed.), *The interaction between science and philosophy*. Atlantic Highlands, NJ: Humanities.

Lakatos, Imre, & Musgrave, Alan. (Eds.). (1970). *Criticism and the growth of knowledge*. Cambridge: Cambridge University Press.

Lakoff, Sanford A. (Ed.). (1966). *Knowledge and power*. New York: Free Press/London: Collier-Macmillan.

Lakoff, Sanford A. (1977). Scientists, technologists and political power. In I. Spiegel-Rösing & D. Price (Eds.), *Science, technology and society: A cross-disciplinary perspective*. London: Sage.

Lakoff, Sanford A., & Bruvold, W. E. (1990). Controlling the qualitative arms race: The primacy of politics. *Science, Technology, & Human Values, 15*(4), 382-411.

Lall, Sanjaya. (1987). *Learning to industrialize: The acquisition of technological capability in India*. London: Macmillan.

Lambert, H., & Rose, Hilary. (in press). Disembodied knowledge? Making sense of medical science. In Alan Irwin & Brian Wynne (Eds.), *Misunderstanding science*. Cambridge: Cambridge University Press.

Lambourne, B., Shallis, M., & Shortland, Michael. (1990). *Close encounters? Science and science fiction*. Bristol, U.K.: Adam Hilger.

Landes, David S. (1969). *The unbound Prometheus*. Cambridge: Cambridge University Press.

Langer, Erick D. (1989). Generations of scientists and engineers: Origins of the computer industry in Brazil. *Latin American Research Review, 24*(2), 95-112.

Langley, P., Simon, H. A., Bradshaw, G. L., & Zytkow, J. M. (1987). *Scientific discovery: Computational explorations of the creative process*. Cambridge: MIT Press.

Lankford, John. (1981). Amateurs vs. professionals: The controversy over telescope size in late Victorian science. *Isis, 72*, 11-28.

La Porte, Todd R. (Ed.). (1989). *Social responses to large technical systems: Control or anticipation*. Dordrecht, the Netherlands: Kluwer.

Lapp, Ralph E. (1965). *The new priesthood*. New York: Harper & Row.

Laqueur, Tom. (1990). *Making sex*. Cambridge, MA: Harvard University Press.

Larson, Magali Sarfatti. (1977). *The rise of professionalism*. Berkeley: University of California Press.

Lash, Scott, & Friedman, J. (Eds.). (1992). *Modernity and identity*. Oxford: Blackwell.

Latour, Bruno. (1980). The three little dinosaurs or a sociologist's nightmare. *Fundamenta Scientiae, 1*, 79-85.

Latour, Bruno. (1981). Insiders and outsiders in the sociology of science: Or, how can we foster agnosticism? In R. A. Jones & H. Kuklick (Eds.), *Knowledge and society* (pp. 199-216). Greenwich, CT: JAI.

Latour, Bruno. (1983). Give me a laboratory and I will raise the world. In K. Knorr Cetina & M. Mulkay (Eds.), *Science observed* (pp. 141-170). Beverly Hills, CA: Sage.

Latour, Bruno. (1984). *Les Microbes: Guerre et Paix, suivi de Irréductions*. Paris: Editions A. M. Métailié.

Latour, Bruno. (1985, June). Les "vues" de l'esprit. In B. Latour & J. de Noblet (Eds.), *Les "vues" de l'esprit*. [Special issue] *Culture Technique, 14*. (Translated as Visualization and cognition, in Henriker Kuklick, Ed., *Knowledge and society*, Vol. 6, pp. 1-40, Greenwich, CT: JAI, 1986; also as Drawing things together, in Michael Lynch & Steve Woolgar, Eds., *Representation in scientific practice*, pp. 19-68, Cambridge: MIT Press, 1990)

Latour, Bruno. (1986). Visualisation and cognition: Thinking with eyes and hands. *Knowledge and Society: Studies in the Sociology of Culture Past and Present, 6*, 1-40.

Latour, Bruno. (1987). *Science in action: How to follow scientists and engineers through society*. Cambridge, MA: Harvard University Press.

Latour, Bruno. (1988). *The pasteurization of France* (followed by *Irreductions: A politico-scientific essay*). Cambridge, MA: Harvard University Press.

Latour, Bruno. (1989, April). *Do we really need the notion of ideology? A case to get rid of the notion by using Pasteur's historiography.* Paper presented at the Conference on Ideology in the Life Sciences, Harvard University.

Latour, Bruno. (1990). Postmodern? No, simply amodern! Steps towards an anthropology of science. *Studies in the History and Philosophy of Science, 21*(1), 145-171.

Latour, Bruno. (1991a). *Nous n'avons jamais été modernes: Essai d'anthropologie symétrique.* Paris: La Découverte.

Latour, Bruno. (1991b). Technology is society made durable. In J. Law (Ed.), *A sociology of monsters: Essays on power, technology and domination* (Sociological Review Monograph, pp. 103-130). London: Routledge & Kegan Paul.

Latour, Bruno. (1991c). The impact of science studies on political philosophy. *Science, Technology, & Human Values, 16*(1), 3-19.

Latour, Bruno. (1992). *ARAMIS, ou l'amour des techniques.* Paris: Éditions la découverte.

Latour, Bruno, & de Noblet, J. (Eds.). (1985). *Les "vues" de l'esprit.* [Special issue] *Culture Technique, 14.*

Latour, Bruno, Mauguin, Philippe, & Teil, Geneviève. (1992). A note on socio-technical graphs. *Social Studies of Science, 22*(1), 33-57.

Latour, Bruno, & Woolgar, Steve. (1979). *Laboratory life: The social construction of scientific facts.* Beverly Hills, CA: Sage.

Latour, Bruno, & Woolgar, Steve. (1986). *Laboratory life: The construction of scientific facts* (2nd ed.). Princeton, NJ: Princeton University Press.

Laudan, Larry. (1983). The demise of the demarcation problem. In R. Laudan (Ed.), *The demarcation between science and pseudo-science* (pp. 7-35). Blacksburg: Virginia Tech, Center for the Study of Science in Society.

Laudan, Larry. (1990). *Science and relativism: Some key controversies in the philosophy of science.* Chicago: University of Chicago Press.

Laudan, Rachel. (Ed.). (1984). *The nature of technological knowledge: Are models of scientific change relevant?* Dordrecht, the Netherlands: Reidel.

Lave, Jean. (1988). *Cognition in practice: Mind, mathematics and culture in everyday life.* Cambridge: Cambridge University Press.

Lave, Jean, Murtaugh, M., & de la Rocha, O. (1984). The dialectic of arithmetic in grocery shopping. In B. Rogoff & J. Lave (Eds.), *Everyday cognition: Its development in social context* (pp. 67-94). Cambridge, MA: Cambridge University Press.

Law, John. (1986a). Laboratories and texts. In M. Callon, J. Law, & A. Rip (Eds.), *Mapping the dynamics of science and technology.* London: Macmillan.

Law, John. (1986b). On the methods of long-distance control vessels navigation and the Portuguese route to India. In J. Law (Ed.), *Power, action and belief: A new sociology of knowledge?* (Sociological Review Monograph 38, pp. 234-263). Keele, U.K.: University of Keele.

Law, John. (Ed.). (1986c). *Power, action and belief: A new sociology of knowledge?* London: Routledge & Kegan Paul.

Law, John. (1987a). Technology and heterogeneous engineering: The case of Portuguese expansion. In W. Bijker, T. Hughes, & T. Pinch (Eds.), *The social construction of technological systems: New directions in the sociology and history of technology* (pp. 111-134). Cambridge: MIT Press.

Law, John. (1987b). The structure of sociotechnical engineering: A review of the new sociology of technology. *Sociological Review, 35,* 404-425.

Law, John. (1988). The anatomy of a sociotechnical struggle: The design of the TSR 2. In B. Elliot (Ed.), *Technology and social process* (pp. 44-69). Edinburgh: Edinburgh University Press.

Law, John. (1991a). Power, discretion and strategy. In J. Law (Ed.), *A sociology of monsters: Essays on power, technology and domination* (Sociological Review Monograph, pp. 165-191). London: Routledge & Kegan Paul.

Law, John. (1991b). Theory and narrative in the history of technology: Response. *Technology and Culture, 32,* 377-384.

Law, John. (Ed.). (1991c). *A sociology of monsters: Essays on power, technology and domination* (Sociological Review Monograph 38). London: Routledge & Kegan Paul.

Law, John. (1993). *Modernity, myth and materialism.* London: Blackwell.

Law, John, & Bijker, Wiebe E. (1992). Postscript: Technology, stability and social theory. In W. Bijker & J. Law, *Shaping technology/building society.* Cambridge: MIT Press.

Law, John, & Callon, Michel. (1988). Engineering and sociology in a military aircraft project: A network analysis of technological change. *Social Problems, 35,* 284-297.

Law, John, & Williams, R. (1982). Putting facts together: A study of scientific persuasion. *Social Studies of Science, 12,* 535-558.

Lawless, E. (1977). *Technology and social shock.* New Brunswick, NJ: Rutgers University Press.

Laymon, Ronald. (1989). Applying idealized scientific theories to engineering. *Synthese, 81,* 353-371.

Layne, Linda. (1992). Of fetuses and angels: Fragmentation and integration in narratives of pregnancy loss. In D. Hess & L. Layne (Eds.), *Knowledge and society: Vol. 9. The anthropology of science and technology* (pp. 29-59). Greenwich, CT: JAI.

Layton, D. (1973). *Science for the people.* London: Allen and Unwin.

Layton, D., Davy, A., & Jenkins, E. (1986). Science for specific social purposes. *Studies in Science Education, 13,* 17-40.

Layton, D., Jenkins, E., MacGill, S., & Davy, A. (1993). *Inarticulate science?* Driffield, U.K.: Studies in Education.

Layton, Edwin T. (1971). Mirror image twins: The communities of science and technology in 19th-century America. *Technology and Culture, 12,* 562-580.

Layton, Edwin T. (1974). Technology as knowledge. *Technology and Culture, 15,* 31-41.

Layton, Edwin T. (1976). American ideologies of sciences and engineering. *Technology and Culture, 17,* 688-700.

Layton, Edwin T. (1977). Conditions of technological development. In I. Spiegel-Rösing & D. Price (Eds.), *Science, technology and society: A cross-disciplinary perspective* (pp. 197-222). London: Sage.

Layton, Edwin T. (1984). Engineering and science as distinct activities. In *Technology and science: Important distinctions for liberal arts colleges* (pp. 6-13). Davidson, NC: Davidson College.

Layton, Edwin T. (1985). Engineering needs a loyal opposition: An essay review. *Business and Professional Ethics Journal, 2,* 57.

Layton, Edwin T. (1986). *The revolt of the engineers: Social responsibility and the American engineering profession.* Baltimore, MD: Johns Hopkins University Press.

Layton, Edwin T. (1987). Through the looking glass, or news from Lake Mirror Image. *Technology and Culture, 28,* 594-607.

Layton, Edwin T. (1988a). Science as a form of action: The role of the engineering sciences. *Technology and Culture, 29,* 82-97.

Layton, Edwin T. (1988b). The dimensional revolution: The new relations between theory and experiment in engineering in the age of Michelson. In S. Goldberg & R. H. Stuewer (Eds.), *The Michelson era in American science, 1870-1930* (pp. 23-39). New York: American Institute of Physics.

Leaf, Murray. (1983). The Green Revolution and cultural change in a Punjab village, 1965-1978. *Economic Development and Cultural Change, 31,* 227-270.

Leavis, F. R. (1963). *Two cultures? The significance of C. P. Snow.* New York: Pantheon.

Leder, P. (1990, October 5). Can the human genome project be saved from its critics . . . and itself? *Cell, 63,* 1-3.

Lederman, Leon. (1991). *Science: The end of the frontier?* Washington, DC: American Association for the Advancement of Science.

Lee, Chong-ouk. (1988). The role of the government and R&D infrastructure for technology development. *Technological Forecasting and Social Change, 33,* 33-54.

Lee, Jinjoo, Lee, Sangjin, & Bae, Zong-tae. (1986). The practice of R&D management: An empirical study of Korean firms. *R&D Management, 16,* 297-308.

Le Grand, H. E. (1986). Steady as a rock: Methodology and moving continents. In John A. Schuster & R. R. Yeo (Eds.), *The politics and rhetoric of scientific method* (pp. 259-297). Dordrecht, the Netherlands: Reidel.

Leitenberg, M. (1973). The dynamics of military technology today. *International Social Science Journal, 25*(3), 336-357.

Leith, P. (1987). Involvement, detachment and programming: The belief in PROLOG. In B. Bloomfield (Ed.), *The question of artificial intelligence* (pp. 220-257). London: Croom Helm.

Lekson, Stephen, Windes, Thomas C., Stein, John R., & Judge, W. James. (1988, July). The Chaco Canyon community. *Scientific American, 259,* 100-109.

Leonard, D. K. (1987). The political realities of African management. *World Development, 15,* 899-910.

Leonard, H. J. (1988). *Pollution and the struggle for the world product.* Cambridge: Cambridge University Press.

Leveson, Nancy. (1989). *Women in computer science.* Washington, DC: National Science Foundation.

Levin, Alexey. (1984). Soviet science studies: A dissident view. *Social Studies of Science, 14*(3), 451-467.

Levine, Adeline. (1982). *Love Canal: People, science and politics.* Boulder, CO: Westview.

Levy, Steven. (1984). *Hackers.* New York: Anchor.

Levy-Leblond, J. M. (1992). About misunderstandings about misunderstandings. *Public Understanding of Science, 1*(1), 17-22.

Lewenstein, Bruce V. (1992a). Cold fusion and hot history. *Osiris, 7*(2nd series), 135-163.

Lewenstein, Bruce V. (1992b, October). Industrial life insurance, public health campaigns, and public communication of science. *Public Understanding of Science, 1*(4), 347-366.

Lewenstein, Bruce V. (1992c, January). The meaning of "public understanding of science" in the United States after World War II. *Public Understanding of Science, 1*(1), 45-68.

Lewenstein, Bruce V. (Ed.). (1992d). *When science meets the public.* Washington, DC: American Association for the Advancement of Science.

Lewis, D. (1975). *We the navigators: The ancient art of landfinding in the Pacific.* Canberra: Australian National University Press.

Lewis, E. (1990). The qualitative arms race: Pluralism gone mad? *Science, Technology, & Human Values, 15*(4), 430-441.

Lewontin, Richard C. (1992a). *Biology as ideology: The doctrine of DNA.* New York: Harper-Perennial.

Lewontin, Richard C. (1992b, May 28). The dream of the human genome. *New York Review of Books,* pp. 31-40.

Leyten, A., & Smits, R. (1987). *The revival of technology assessment: The development of TA in five European countries and in the USA.* The Hague, the Netherlands: Ministry of Education and Science.

Lighthall, Frederick F. (1991). Launching the space shuttle Challenger: Disciplinary deficiencies in the analysis of engineering data. *IEEE Transactions on Engineering Management, 38,* 63-74.

Lilienthal, D. E. (1980). *Atomic energy: A new start.* New York: Harper & Row.

Lindblom, Charles E. (1965). *The intelligence of democracy.* New York: Free Press.

Lindblom, Charles E. (1977). *Politics and markets.* New York: Basic Books.

Lindblom, Charles E., & Woodhouse, Edward J. (1993). *The policy-making process* (3rd ed.). New York: Prentice Hall.

Lindsay, A. D. (1943). *The modern democratic state.* London: Oxford University Press.

Lipscombe, Barry. (1989). Expert systems and computer-controlled decision making in medicine. *AI and Society, 3,* 184-197.

Lipscombe, Barry. (1990). *Minds machines and medicine: An epistemological study of computer diagnosis.* Unpublished doctoral dissertation, Bath University.

Lipton, Michael. (1988). The place of agricultural research in the development of sub-Saharan Africa. *World Development, 16,* 1231-1257.

Lipton, Michael, with Longhurst, Richard. (1989). *New seeds and poor people.* Baltimore: Johns Hopkins University Press.

Litan, R. E., & Suchman, P. O. (1990, January 5). US trade policy at a crossroad. *Science, 247,* 33-38.

Livingston, Eric. (1986). *The ethnomethodological foundations of mathematics.* London: Routledge & Kegan Paul.

Locke, R. R. (1984). *The end of the practical man: Entrepreneurship and higher education in France, Germany and Great Britain, 1880-1940.* London: JAI.

Logan, R. A. (1991). Popularization vs. secularization: Media coverage of health. In L. Wilkins & P. Patterson (Eds.), *Risky business: Communicating issues of science, risk, and public policy* (pp. 43-59). New York: Greenwood.

Lomnitz, Larissa A., Rees, Martha A., & Cameo, Leon. (1987). Publication and referencing patterns in a Mexican research institute. *Social Studies of Science, 17,* 115-133.

Long, Franklin A. (1978). Basic needs strategy for development of technology in low income countries. *American Journal of Economics and Sociology, 37,* 261-270.

Long, Franklin A., & Reppy, Judith. (Eds.). (1980). *The genesis of new weapons.* New York: Pergamon.

Long, Franklin A., & Reppy, Judith. (1984). The decision process for US military R&D. In K. Tsipis & P. Janeway (Eds.), *Review of US military research and development* (pp. 4-19). Washington, DC: Pergamon-Brassey's.

Long, J. Scott. (1987). Discussion: Problems and prospects for research on sex differences. In L. Dix (Ed.), *Women: Their underrepresentation and career differentials in science and engineering.* Washington, DC: National Academy Press.

Long, J. Scott. (1990). The origins of sex differences in science. *Social Forces, 68,* 1297-1315.

Long, J. Scott. (1992, September). Measures of sex differences in scientific productivity. *Social Forces,* pp. 159-178.

Long, J. Scott, & McGinnis, Robert. (1985). The effects of the mentor on the academic career. *Scientometrics, 7,* 255-280.

Longino, Helen. (1989). *Science as social knowledge: Values and objectivity in scientific inquiry.* Princeton, NJ: Princeton University Press.

Longino, Helen, & Doell, Ruth. (1983). Body, bias, and behavior. *Signs, 9*(2), 206-227.

Looney, R. E. (1988). The impact of technology transfer on the structure of the Saudi Arabian labor force. *Journal of Economic Issues, 22,* 485-492.

Lotka, A. J. (1926). The frequency distribution of scientific productivity. *Journal of the Washington Academy of Sciences, 26,* 317.

Loughlin, J. (1993). The challenge of feminism to social studies of science. In T. Brante, Steve Fuller, & William Lynch (Eds.), *Controversial science* (pp. 3-20). Albany: State University of New York Press.

Low, Morris Fraser. (1989). The butterfly and the frigate: Social studies of science in Japan. *Social Studies of Science, 19*(2), 313-342.

Lowe, P., & Flynn, A. (1989). Environmental politics and policy in the 1980s. In J. Moran (Ed.), *The political geography of contemporary Britain* (pp. 255-279). London: Macmillan.

Lowrance, William M. (1976). *Of acceptable risk: Science and the determination of safety*. Los Altos, CA: William Kaufmann.

Lubek, I. (1976). Some tentative suggestions for analysing and neutralising the power structure in social psychology. In L. H. Strickland, F. E. Aboud, & K. J. Gergen (Eds.), *Social psychology in transition*. New York: Plenum.

Luff, P., Gilbert, G. Nigel, & Frohlich, D. M. (1990). *Computers and conversation*. London: Academic Press.

Lynch, Michael. (1982). Technical work and critical inquiry: Investigations in a scientific laboratory. *Social Studies of Science, 12*, 499-534.

Lynch, Michael. (1985a). *Art and artifact in laboratory science: A study of shop work and shop talk in a research laboratory*. London: Routledge & Kegan Paul.

Lynch, Michael. (1985b). Discipline and the material form of images: An analysis of scientific visibility. *Social Studies of Science, 15*, 37-66.

Lynch, Michael. (1988). The externalized retina: Selection and mathematization in the visual documentation of objects in the life sciences. *Human Studies, 11*(2/3), 201-234.

Lynch, Michael. (1990). The externalized retina: Selection and mathematization in the visual documentation of objects in the life sciences. In Michael Lynch & Steve Woolgar (Eds.), *Representation in scientific practice*. Cambridge: MIT Press.

Lynch, Michael. (1991). Laboratory space and the technological complex: An investigation of topical contextures. *Science in Context, 4*(1), 81-109.

Lynch, Michael. (1992). Extending Wittgenstein: The pivotal move from epistemology to sociology of science. In Andrew Pickering (Ed.), *Science as practice and culture*. Chicago: University of Chicago Press.

Lynch, Michael, & Edgerton, Samuel. (1988). Aesthetics and digital image processing: Representational craft in contemporary astronomy. *Sociological Review Monograph, 35*, 184-220.

Lynch, Michael, Livingstone, Eric, & Garfinkel, Harold. (1983). Temporal order in laboratory work. In Karin Knorr Cetina & Michael Mulkay (Eds.), *Science observed: Perspectives on the social study of science*. London: Sage.

Lynch, Michael, & Woolgar, Steve (Eds.). (1990). *Representation in scientific practice*. Cambridge: MIT Press.

Lyne, J., & Howe, H. (1986). "Punctuated equilibria": Rhetorical dynamics of a scientific controversy. *Quarterly Journal of Speech, 72*, 132-147.

Lyotard, J. F. (1984). *The postmodern condition: An inquiry into knowledge*. Manchester: Manchester University Press.

MacCormack, C., & Strathern, M. (Eds.). (1980). *Nature, culture and gender*. Cambridge: Cambridge University Press.

Macdonald, S., & Silverstone, R. (1992). Science on display: The representation of scientific controversy in museum exhibitions. *Public Understanding of Science, 1*(1), 69-87.

Macdonald-Ross, M. (1987). The role of science books for the public. In D. Evered & M. O'Connor (Eds.), *Communicating science to the public* (pp. 175-189). Chichester, U.K.: Wiley.

Machlup, Fritz. (1980). *Knowledge and knowledge production*. Princeton, NJ: Princeton University Press.

Macioti, Manfredo. (1978). Technology and development: The historical experience. *Impact of Science on Society, 28,* 93-108.
Mack, Pamela Etter. (1990). *Viewing the earth: The social construction of the Landsat satellite system.* Cambridge: MIT Press.
MacKay, Alan, & Goldsmith, Maurice. (Eds.). (1964, 1966). *The science of science.* London: Souvenir (1964)/Harmondsworth, Middlesex, U.K.: Pelican (rev. ed., 1966).
MacKenzie, Donald A. (1981). *Statistics in Britain: 1865-1930.* Edinburgh: Edinburgh University Press.
MacKenzie, Donald A. (1984). Marx and the machine. *Technology and Culture, 25,* 473-502.
MacKenzie, Donald. (1989). From Kwajalein to Armageddon? Testing and the social construction of missile accuracy. In David Gooding, Trevor Pinch, & Simon Schaffer (Eds.), *The uses of experiment* (pp. 409-436). Cambridge, MA: Cambridge University Press.
MacKenzie, Donald. (1990a). *Inventing accuracy: A historical sociology of nuclear missile guidance.* Cambridge: MIT Press.
MacKenzie, Donald. (1990b, September). *Economic and sociological explanation of technical change.* Paper presented at the Conference, "Firm Strategy and Technical Change: Micro Economics or Micro Sociology," Manchester.
MacKenzie, Donald. (1991, August). The fangs of the VIPER. *Nature, 352,* 467-468.
MacKenzie, Donald. (n.d.). Negotiating arithmetic, constructing proof: Information technology and the sociology of mathematics. Unpublished manuscript submitted to *Social Studies of Science.*
Mackenzie, Donald, & Milne, R. (1989, August 19). A fresh green tinge to Europe's research. *New Scientist,* pp. 23-25.
MacKenzie, Donald, & Spinardi, Graham. (1988). The shaping of nuclear weapon system technology: US fleet ballistic missile guidance and navigation: II: "Going for broke"—The path to Trident II. *Social Studies of Science, 18,* 581-624.
MacKenzie, Donald, & Wajcman, Judy. (Eds.). (1985). *The social shaping of technology: How the refrigerator got its hum.* Milton Keynes, U.K.: Open University Press.
Mackie, Marlene. (1977). Professional women's collegial relations and productivity. *Sociology and Social Research, 61,* 277-293.
Magee, Stephen P. (1981). The appropriability theory of the multinational corporation. *Annals of the American Academy of Political and Social Sciences, 458,* 123-135.
Mahnken, T. G., & Hoyt, T. D. (1990). The spread of missile technology to the Third World. *Comparative Strategy, 9,* 245-263.
Mahoney, Michael S. (1988). The history of computing in the history of technology. *Annals of the History of Computing, 10*(2), 113-125.
Majone, Giandomenico. (1984). Science and trans-science in standard setting. *Science, Technology, & Human Values, 9*(1), 15-22.
Malcolm, Shirley Mahaley, Hall, Paula Quick, & Brown, Janet Welsh. (1976, April). *The double bind: The price of being a minority woman in science* (AAAS Report No. 76-R-3). Washington, DC: American Association for the Advancement of Science.
Malik, Yogendra K. (1982). Attitudinal and political implications of diffusion of technology: The case of North Indian youth. *Journal of Asian and African Studies, 17,* 1-12.
Mallet, Serge. (1975). *New working class.* Nottingham, U.K.: Spokesman Books.
Mamdani, M. (1972). *The myth of population control: Family caste and class in an Indian village.* New York: Monthly Review Press.
Mandula, Barbara, & Blockstein, David. (1992). Federal funding for environmental research. Unpublished manuscript submitted to *Environmental Science and Technology.*

Manegold, Karl-Heinz. (1978). Technology academised: Education and training of the engineer in the nineteenth century. In Wolfgang Krohn & Edward Layton (Eds.), *The dynamics of science and technology* (pp. 137-158). Dordrecht, the Netherlands: Reidel.

Mansfield, Edwin. (1965). Rates of return from industrial research and development. *American Economic Review, 55*(2), 310-322.

Mansfield, Edwin. (1975). International technology transfer: Forms, resource requirements, and policies. *American Economic Review, 65*, 372-376.

Mansfield, Edwin. (1991). Academic research and industrial innovation. *Research Policy, 20*(1), 1-12.

Mansfield, Edwin. (1992). *Unauthorized use of intellectual property: Effects on investment, technology transfer, and innovation.* Paper presented at the conference, "Global Dimensions of Intellectual Property Rights in Science and Technology," National Research Council, Washington, DC.

Mansfield, Edwin, Romeo, Anthony, Schwartz, Mark, Teece, David, Wagner, Samuel, & Brach, Peter. (1983, March-April). New findings in technology transfer, productivity and development. *Research Management*, pp. 11-20.

Marcson, Simon. (1960). *The scientist in American industry.* New York: Harper.

Marcus, George, & Cushman, D. (1982). Ethnographies as texts. *Annual Review of Anthropology, 11*, 25-69.

Marcus, George E., & Fischer, Michael M. J. (1986). *Anthropology as cultural critique: An experimental moment in the human sciences.* Chicago: University of Chicago Press.

Markle, Gerald E., & Petersen, James C. (1980). *Politics, science and cancer: The Laetrile phenomena.* Boulder, CO: Westview.

Markle, Gerald E., & Petersen, James C. (1981). Controversies in science and technology: A protocol for comparative research. *Science, Technology, & Human Values, 6*, 25-30.

Markus, George. (1987). Why is there no hermeneutics of the natural sciences? Some preliminary theses. *Science in Context, 1*, 5-54.

Markusen, Ann, & Yudken, J. (1992). *Dismantling the cold war economy.* New York: Basic Books.

Marlier, E. (1992). Eurobarometer 35.1: Opinions of Europeans on biotechnology in 1991. In J. Durant (Ed.), *Biotechnology in public: A review of recent research* (pp. 52-108). London: Science Museum.

Marshack, Alexander. (1989). North American Indian Calendar Sticks: The evidence for a widely distributed tradition. In A. F. Aveni (Ed.), *World archaeoastronomy: Selected papers from the 2nd Oxford International Conference on Archaeoastronomy held at Merida, Yucatan, Mexico, 1986* (pp. 308-324). Cambridge: Cambridge University Press.

Martin, Brian. (1989). The sociology of the fluoridation controversy: A reexamination. *Sociological Quarterly, 30*, 59-76.

Martin, Brian. (1991). *Scientific knowledge in controversy: The social dynamics of the fluoridation debate.* Albany: State University of New York Press.

Martin, Ben, & Irvine, John. (1989). *Research foresight.* London: Frances Pinter.

Martin, Emily. (1989). *The woman in the body.* Milton Keynes, U.K.: Open University Press.

Martin, Emily. (1990). Science and women's bodies: Forms of anthropological knowledge. In M. Jacobus, E. F. Keller, & S. Shuttleworth (Eds.), *Body/politics: Women and the discourses of science* (pp. 69-82). New York: Routledge & Kegan Paul.

Martin, Emily. (1991). The egg and the sperm: How science has constructed a romance based on stereotypical male-female roles. *Signs, 16*(3), 485-501.

Martin, Michelle. (1991.) *"Hello Central?" Gender, technology, and culture in the formation of telephone systems.* Montreal: McGill-Queen's University Press.

Martin, M. W., & Schinzinger, R. (1989). *Ethics in engineering.* New York: McGraw-Hill.

Martin, S., & Tait, Joyce. (1992). Attitudes of selected groups in the UK to biotechnology. In J. Durant (Ed.), *Biotechnology in public: A review of recent research* (pp. 28-41). London: Science Museum.

Marton, Katherin. (1986). Technology transfer to developing countries via multinationals. *World Economy, 9*, 409-426.

Marvin, C. (1987). *When old technologies were new: Thinking about communications in the late nineteenth century*. New York: Oxford University Press.

Marwell, Gerald, Rosenfeld, Rachel, & Spilerman, Seymour. (1979, September 21). Geographic constraints on women's careers in academia. *Science, 205*, 1225-1231.

Marx, Karl. (1867). *Das Kapital: Kritik der politischen Ökonomie. Erster Band*. Hamburg: Verlag von Otto Meissner. (Modern republications are commonly based on the 4th ed., revised by F. Engels and published in 1890; 1976 English publication, *Capital: A Critique of Political Economy*, Harmondsworth, U.K.: Penguin)

Marx, Karl. (1956). *Economic and philosophic manuscripts of 1844*. Moscow: Foreign Languages Publishing House.

Marx, Karl. (1973). *Grundrisse: Foundations of the critique of political economy*. New York: Vintage.

Marx, Karl, & Engels, Friedrich. (1974). *The German ideology*. New York: International.

Marx, Leo. (1964). *The machine in the garden: Technology and the pastoral ideal in America*. New York: Oxford University Press.

Masterman, Margaret. (1970). The nature of a paradigm. In Imre Lakatos & Alan Musgrave (Eds.), *Criticism and the growth of knowledge* (pp. 59-89). Cambridge: Cambridge University Press.

Mathews, J. (1989a). *Tools of change: New technology and the democratisation of work*. Sydney: Pluto.

Mathews, J. (1989b). *Age of democracy: The politics of post-Fordism*. Oxford: Oxford University Press.

Mayhew, D. R. (1974). *Congress: The electoral connection*. New Haven, CT: Yale University Press.

Maynard-Moody, Steven. (1992). The fetal research dispute. In D. Nelkin (Ed.), *Controversy*. Newbury Park, CA: Sage.

Mayntz, Renate, & Hughes, Thomas P. (Eds.). (1988). *The development of large technical systems*. Frankfurt am Main: Campus Verlag.

Mazur, Allan. (1973). Disputes between experts. *Minerva, 11*, 243-262.

Mazur, Allan. (1977). Science courts. *Minerva, 15*, 1-14.

Mazur, Allan. (1981). *The dynamics of technical controversy*. Washington, DC: Communications Press.

McAfee, N. (1974). Brighter prospects for women in engineering. *Engineering Education, 64*, 23-25.

McCall, R. B., & Stocking, S. H. (1982, September). Between scientists and public: Communicating psychological research through the mass media. *American Psychologist, 37*, 985-995.

McCarthy, John, & Zald, M. (1973). *The trend of social movements in America: Professionalization and resource mobilization*. Morristown, NJ: General Learning Press.

McCloskey, M. (1983). Naive theories of motion. In D. Gentner & A. Stevens (Eds.), *Mental models* (pp. 299-324). Hillsdale, NJ: Lawrence Erlbaum.

McCloskey, Donald. (1985). *The rhetoric of economics*. Madison: University of Wisconsin Press/Brighton, U.K.: Wheatsheaf.

McCloskey, Donald. (1987). A strong programme in the rhetoric of science (review of H. M. Collins *Changing order*, 1985). *Journal of Economic Psychology, 8*, 128-133.

McCluskey, Stephen. (1980). Science, society, objectivity, and the Southwest. In J. B. Carlson & W. J. Judge (Eds.), *Astronomy and ceremony in the prehistoric Southwest* (Papers of the Maxwell Museum of Anthropology No. 2; pp. 205-217). Albuquerque: University of New Mexico.

McCluskey, Stephen. (1982). Historical archaeoastronomy: The Hopi example. In A. F. Aveni (Ed.), *Archaeoastronomy in the New World: Proceedings of an international conference at Oxford 1981* (pp. 31-59). Cambridge: Cambridge University Press.

McCormick, J. (1989). *The global environmental movement*. London: Belhaven.

McCormick, Kevin. (1988). Engineering education in Britain and Japan: Some reflections on the use of the "best practice" models in international comparison. *Sociology, 22*, 583-605.

McCormick, Kevin. (1989). *The development of engineering education and training in Britain and Japan* (mimeo). Sussex, U.K.: University of Sussex, School of Social Sciences.

McCormick, Kevin. (1992). Japanese engineers as corporate salary men. In C. Smith & P. Meiksins (Eds.), *Engineering class politics*. London: Verso.

McCrea, Frances B., & Markle, Gerald E. (1984). The estrogen replacement controversy in the USA and UK: Different answers to the same question? *Social Studies of Science 14*, 1-26.

McCullock, Rachel D. (1981). Technology transfer to developing countries: Implications of international regulation. *Annuals of the American Academy of Political and Social Science, 458*, 110-122.

McDermott, D. (1981). Artificial intelligence meets natural stupidity. In J. Haugeland (Ed.), *Mind design* (pp. 143-160). Cambridge: MIT Press.

McDonald, Ronald H., & Tamrowski, Nina. (1987). Technology and armed conflict in Central America. *Journal of Interamerican Studies and World Affairs, 29*, 93-108.

McGaw, Judith. (1982). Women and the history of American technology. *Signs: A Journal of Women in Culture and Society, 7*, 798-828.

McGrew, Anthony. (1990). The political dynamics of the new environmentalism. *Industrial Crisis Quarterly, 4*, 291-305.

McGucken, W. (1984). *Scientists, society, and state: The social relations of science movement in Great Britain, 1931-1947*. Columbus: Ohio State University Press.

McIlwee, Judith S., & Robinson, J. Gregg. (1992). *Women in engineering: Gender, power and workplace culture*. Albany: State University of New York Press.

McIntosh, R. (1976). Ecology since 1900. In B. Taylor & T. White (Eds.), *Issues and ideas in America* (pp. 353-372). Norman: University of Oklahoma Press.

McIntosh, R. (1985). *The background of ecology: Concept and theory*. Cambridge: Cambridge University Press.

McKechnie, R. (in press). Insiders and outsiders: Identifying experts on home ground. In A. Irwin & B. Wynne (Eds.), *Misunderstanding science*. Cambridge: Cambridge University Press.

McKinlay, Andrew, & Potter, Jonathon. (1987). Model discourse: Interpretative repertoires in scientists' conference talk. *Social Studies of Science, 17*, 443-463.

McKusick, V. A., & Ruddle, F. H. (1987). A new discipline, a new journal, a new name. *Genomics, 1*, 1-2.

McMahon, A. M. (1984). *The making of a profession: A century of electrical engineering in America*. New York: IEEE Press.

McMath, R. C. (1985). *Engineering the New South: Georgia Tech, 1885-1985*. Athens: University of Georgia Press.

McNeil, Maureen. (Ed.). (1987). *Gender and expertise*. London: Free Association Books.

McNeil, M., Varcoe, I., & Yearley, S. (Eds.). (1990). *The new reproductive technologies*. London: Macmillan.

McRobie, George. (1979). Intermediate technology: Small is successful. *Third World Quarterly, 1,* 71-86.
Mead, M., & Metraux, R. (1957, August 20). The image of the scientist among college students. *Science, 126,* 384-390.
Meadows, A. J. (1974). *Communication in science.* London: Butterworth.
Medawar, Peter. (1964, August 1). Is the scientific paper fraudulent? *The Saturday Review,* pp. 42-43.
Meehan, Richard L. (1981). *Getting sued and other tales of the engineering life.* Cambridge: MIT Press.
Megill, Allan. (Ed.). (1992, Spring). *Rethinking objectivity* [Special issue]. *Annals of Scholarship, 9*(1-3).
Mehan, H. (1979). *Learning lessons: Social organization in the classroom.* Cambridge, MA: Harvard University Press.
Meiksins, Peter F. (1982). Science in the labor process: Engineers as workers. In C. Derber (Ed.), *Professionals as workers* (pp. 121-140). Boston: G. K. Hall.
Meiksins, Peter F. (1986). Professionalism and conflict: The case of the American Association of Engineers. *Journal of Social History, 19,* 403-421.
Meiksins, Peter F. (1988). The "revolt of the engineers" reconsidered. *Technology and Culture, 29,* 219-246.
Meiksins, Peter, & Smith, Chris. (1991, August). *The organization of professional technical workers: A comparative analysis.* Paper presented at the American Sociological Association Meetings, Cincinnati, OH.
Meiksins, Peter F., & Smith, Chris. (in press). Why American engineers aren't unionized: A comparative perspective. *Theory and Society.*
Meiksins, P. F., & Watson, J. M. (1989). Professional autonomy and organizational constraint: The case of engineers. *The Sociological Quarterly, 30,* 561-585.
Meissner, Frank. (1988). *Technology transfer in the developing world: The case of the Chile Foundation.* New York: Praeger.
Meltsner, A. (1979). The communication of scientific information to the wider public: The case of seismology. *Minerva, 17,* 331-354.
Mendelsohn, Everett. (1964). The emergence of science as a profession in nineteenth century Europe. In K. Hill (Ed.), *The management of scientists.* Boston: Beacon.
Mendelsohn, Everett. (1977). The social production of scientific knowledge. In E. Mendelsohn et al. (Eds.), *The social production of scientific knowledge* (Sociology of the Sciences Yearbook). Dordrecht, the Netherlands: Reidel.
Mendelsohn, Everett. (1990). Science, technology and the military: Patterns of interaction. In J. Salomon (Ed.), *Science, war and peace.* Paris: Economica.
Mendelssohn, Kurt. (1976). *Science and Western domination.* London: Thames and Hudson.
Merchant, Carolyn. (1980). *The death of nature: Women, ecology and the scientific revolution.* New York: Harper & Row.
Meredith, M. O., Nelson, S. D., & Teich, Alan H. (Eds.). (1991). *AAAS science and technology yearbook, 1991.* Washington, DC: AAAS.
Merleau-Ponty, Maurice. (1945). *Phenomenologie de la perception.* Paris: Gallimard. (English trans., *Phenomenology of perception,* London: Routledge & Kegan Paul, 1962)
Merriam, C. E. (1945). *Systematic politics.* Chicago: Chicago University Press.
Merrifield, J. (1989). *Putting the scientists in their place: Participatory research in environmental and occupational health.* New Market, TN: Highlander Centre.
Merton, Robert K. (1933). *Science, technology and society in seventeenth century England.* Bruges, Belgium: St. Catherines.

Merton, Robert K. (1942). Science and technology in a democratic order. *Journal of Legal and Political Sociology, 1,* 15-26.

Merton, Robert K. (1968). The self-fulfilling prophecy. In *Social theory and social structure* (pp. 475-490). New York: Free Press. (Original work published 1948)

Merton, Robert K. (1970). *Science, technology and society in seventeenth-century England.* New York: Harper & Row. (Original work published 1938 in *Osiris*)

Merton, Robert K. (1973a). The normative structure of science. In R. K. Merton, *The sociology of science: Theoretical and empirical investigations.* Chicago: University of Chicago Press.

Merton, Robert K. (1973b). *The sociology of science: Theoretical and empirical investigations* (N. W. Storer, Ed.). Chicago: University of Chicago Press.

Merton, Robert K. (1973c). The Matthew effect in science. In *The sociology of science.* Chicago: University of Chicago Press.

Merton, Robert K. (1976). *Sociological ambivolence and other essays.* New York Free Press.

Merton, Robert K. (1977). The sociology of science: An episodic memoir. In R. Merton & J. Gaston (Eds.), *The sociology of science in Europe* (pp. 3-141). Carbondale: Southern Illinois University Press.

Merton, Robert K. (1984). The fallacy of the latest word: The case of "pietism and science." *The American Journal of Sociology, 89,* 91-1121.

Merton, Robert K., & Gieryn, Thomas F. (1978). Institutionalized altruism: The case of the professions. In M. S. Das & T. Lynn Smith (Eds.), *Sociocultural change since 1950* (pp. 309-344). New Delhi, India: Vikas.

Mervis, J. (1988, October 31). A threat to monitor science is quashed. *The Scientist,* p. 1.

Messing, Karen. (1990). Union-initiated research in genetic effects of workplace agents. *Genewatch, 6*(4-5), 8-14.

Mey, Marc De. (1982). *The cognitive paradigm.* Dordrecht, the Netherlands: Reidel.

Meyer, J., Hannan, M., Rubinson, R., & Thomas, G. (1979). National economic development, 1950-70: Social and political factors. In J. Meyer & M. Hannan (Eds.), *National development and the world system: Educational, economic, and political change, 1950-1970.* Chicago: University of Chicago Press.

Mezey, M. (1991). The legislature, the executive, and public policy: The futile quest for congressional power. In J. A. Thurber (Ed.), *Divided democracy: Cooperation and conflict between the president and Congress.* Washington, DC: Congressional Quarterly Press.

Michael, M. (1992). Lay discourses of science: Science-in-general, science-in-particular, and self. *Science, Technology, & Human Values, 17*(3), 313-333.

Michalet, Charles Albert. (1979). The international transfer of technology and the multinational enterprise. *Development and Change, 2,* 157-174.

Michie, D., & Johnston, R. (1985). *The creative computer: Machine intelligence and human knowledge.* Harmondsworth, U.K.: Pelican.

Mies, M. (1987). Why do we need all this? A call against genetic engineering and reproductive technology. In P. Spallone & D. Steinberg (Eds.), *Made to order: The myth of reproductive and genetic progress* (pp. 34-47). Oxford: Pergamon.

Migdal, Joel S. (1988). *Strong societies and weak states: State-society relations and state capabilities in the Third World.* Princeton, NJ: Princeton University Press.

Milkman, Ruth. (1987). *Gender at work: The dynamics of job segregation by sex during World War II.* Chicago: University of Illinois Press.

Millar, Robin. (1989). *Doing science: Images of science in science education.* Brighton, U.K.: Falmer.

Millard, Rodney J. (1988). *The master spirit of the age: Canadian engineers and the politics of professionalism 1887-1922.* London: University of Toronto Press.

Miller, C. (1989). The rhetoric of decision science, or Herbert A. Simon says. *Science, Technology, & Human Values, 14*, 43-46. (Summary of a longer article in H. W. Simons, Ed., *The rhetorical turn: Invention and persuasion in the conduct of inquiry*, Chicago, University of Chicago Press, 1990)
Miller, Jean Baker. (1974). *Toward a new psychology of women*. Boston: Beacon.
Miller, Jon D. (1983a). *The American people and science policy: The role of public attitudes in the policy process*. New York: Pergamon.
Miller, Jon D. (1983b). Scientific literacy: A conceptual and empirical review. *Daedalus, 112*(2), 29-48.
Miller, Jon D. (1987). *Scientific literacy in the United States: Communicating science to the public*. New York: Wiley.
Miller, Jon D. (1990, December). *The public understanding of science and technology in the United States*. Washington, DC: National Science Foundation.
Miller, Jon D. (1991). *The public understanding of science and technology in the US: Report to the US National Science Foundation*. DeKalb, IL: National Opinion Research Centre, University of Chicago.
Miller, Perry. (1962). *The New England mind*. Cambridge, MA: Harvard University Press.
Millett, Kate. (1970). *Sexual politics*. New York: Doubleday.
Mills, C. Wright. (1957). *The power elite*. New York: Oxford University Press.
Mirsky, E. M. (1972). Science studies in the USSR (history, problems, prospects). *Science Studies, 2*(3), 281-294.
Misa, Thomas. (1988). How machines make history, and how historians (and others) help them to do so. *Science, Technology, & Human Values, 13*, 308-331.
Mitcham, Carl. (1980). Philosophy of technology. In Paul T. Durbin (Ed.), *A guide to the culture of science, technology and medicine* (pp. 282-363). New York: Free Press.
Mitchell, Juliet. (1975). *Psychoanalysis and feminism*. New York: Vintage.
Mitford, Jessica. (1992). *The American way of birth*. New York: Dutton.
Mitman, Gregg. (1991). Donna Haraway: *Primate visions: Gender, race, and nature in the world of modern science. Isis, 82*(1), 163-165.
Mitroff, Ian I. (1974a). Norms and counter-norms in a select group of the Apollo moon scientists: A case study in the ambivalence of scientists. *American Sociological Review, 39*, 579-595.
Mitroff, Ian I. (1974b). *The subjective side of science: A philosophical inquiry into the psychology of the Apollo moon scientists*. Amsterdam: Elsevier.
Mitroff, J. I. (1983). *The 1980: Policymaking amid turbulence*. Lexington, MA: Lexington.
Moore, Sally Falk. (1985). *Power and property in Inca Peru*. Westport, CT: Greenwood.
Moravcsik, Michael. (1975). *Science development: The building of science in less-developed countries*. Bloomington, IN: International Development Center.
Moravcsik, Michael, & Ziman, John. (1985). Paradisia and Dominitia: Science and the developing world. *Foreign Affairs, 53*, 699-724.
Morehouse, Ward. (1976). Professional estates as political actors: The case of the Indian scientific community. *Philosophy and Social Action, 2*, 61-95.
Morgall, J. (1992). *Developing technology assessment*. Philadelphia: Temple University Press.
Morison, Elting. (1966). *Men, machines and modern times*. Cambridge: MIT Press.
Morita-Lou, Hiroko. (Ed.). (1985). *Science and technology indicators for development*. Boulder, CO: Westview.
Morone, Joseph G., & Woodhouse, Edward J. (1986). *Averting catastrophe: Strategies for regulating risky technologies*. Berkeley: University of California Press.
Morone, Joseph G., & Woodhouse, Edward J. (1989). *The demise of nuclear energy: Lessons for democratic control of technology*. New Haven, CT: Yale University Press.

Morse, Dean, & Warner, Aaron W. (Eds.). (1966). *Technological innovation and society*. New York: Columbia University Press.

Moscovici, S. (1984). The phenomenon of social representations. In R. M. Farr & S. Moscovici (Eds.), *Social representations* (pp. 3-70). Cambridge: Cambridge University Press.

Mowery, David C., & Rosenberg, Nathan. (1989). *Technology and the pursuit of economic growth*. Cambridge: Cambridge University Press.

Mowery, David C., & Rosenberg, Nathan. (1993). The U.S. national innovation system. In R. Nelson (Ed.), *National innovation systems: A comparative analysis*. Oxford: Oxford University Press.

Muga, D. A. (1987). The effect of technology on an indigenous people: The case of the Norwegian Sami. *Journal of Ethnic Studies, 14*, 1-24.

Mukerji, Chandra. (1989). *A fragile power: Scientists and the state*. Princeton, NJ: Princeton University Press.

Mulkay, Michael. (1969). Some aspects of cultural growth in the natural sciences. *Social Research, 36*, 22-52.

Mulkay, Michael. (1972). *The social process of innovation*. London: Macmillan.

Mulkay, Michael. (1975). Norms and ideology in science. *Social Science Information, 4-5*, 637-656.

Mulkay, Michael. (1976). Norms and ideology in science. *Social Science Information, 15*, 637-656.

Mulkay, Michael. (1979). *Science and the sociology of knowledge*. London: Allen and Unwin.

Mulkay, Michael. (1980). Interpretation and the use of rules: The case of the norms of science. In T. Gieryn (Ed.), *Science and social structure: A festschrift for Robert K. Merton* (pp. 111-125). New York: New York Academy of Sciences.

Mulkay, Michael. (1985). *The word and the world: Explorations in the form of sociological analysis*. London: Allen and Unwin.

Mulkay, Michael. (1989). Looking backward. *Science, Technology, & Human Values, 14*, 441-459.

Mulkay, Michael, Potter, J., & Yearley, Stephen. (1983). Why an analysis of scientific discourse is needed. In K. Knorr Cetina & M. Mulkay (Eds.), *Science observed: Perspectives on the social study of science* (pp. 171-203). London: Sage.

Mullins, Nicholas C. (1972). The development of a scientific specialty: The Phage Group and the origins of molecular biology. *Minerva, 10*(1), 51-82.

Mullins, Nicholas C. (1973). *Science: Some sociological perspectives*. Indianapolis: Bobbs-Merrill.

Mullins, Nicholas C. (1977). *Rhetorical resources in natural science papers*. Princeton, NJ: Princeton University, Institute for Advanced Study.

Mumford, Lewis. (1934). *Technics and civilization*. New York: Harcourt, Brace.

Mumford, Lewis. (1938). *The culture of cities*. New York: Harcourt Brace Jovanovich.

Mumford, Lewis. (1961). *The city in history: Its origins, its transformations, and its prospects*. New York: Harcourt, Brace.

Mumford, Lewis. (1964). Authoritarian and democratic technics. *Technology and Culture, 5*, 1-8.

Mumford, Lewis. (1967). *The myth of the machine: Vol. 1. Technics and human development*. New York: Harcourt Brace Jovanovich.

Mumford, Lewis. (1970). *The myth of the machine: Vol. 2. The pentagon of power*. New York: Harcourt, Brace.

Murrell, R. K. (1987). Telling it like it isn't: Representations of science in *Tomorrow's World*. *Theory, Culture, Society*, pp. 489-506.

Myers, Greg. (1985). Texts as knowledge claims: The social construction of two biologists' articles. *Social Studies of Science, 15*, 593-630.

Myers, Greg. (1990a). *Writing biology: Texts and the social construction of scientific knowledge.* Madison: University of Wisconsin Press.

Myers, Greg. (1990b). The double helix as an icon. *Science as Culture, 9,* 49-72.

Myers, Greg. (1992). Fictions from facts: The form and authority of the scientific dialogue. *History of Science, 30,* 221-247.

Mytelka, Lynn Krieger. (1978). Licensing and technology dependence in the Andean group. *World Development, 6,* 447-459.

Nagel, T. (1986). *The view from nowhere.* New York: Oxford University Press.

Nagpaul, P. S., & Krishnaiah, V. S. R. (1988). Dimensions of research planning: Comparative study of research units in six countries. *Scientometrics, 14,* 383-410.

Nandy, Ashis. (1988). *Science, hegemony and violence: A requiem for modernity.* Delhi, India: Oxford University Press.

Nanney, David. (1957). The role of cytoplasm in development. In W. D. McElroy & H. B. Glass (Eds.), *The chemical basis of heredity.* Baltimore: Johns Hopkins University Press.

Narin, F., & Noma, E. (1985). Is technology becoming science? *Scientometrics, 7,* 369-381.

Natarajan, R. (1987). Science, technology and Mrs. Ghandi. *Journal of Asian and African Studies, 22,* 232-249.

National Academy of Engineering. (1983). *International competition in advanced technology: Decisions for America.* Washington, DC: Author.

National Academy of Engineering. (1985). *Technological frontiers and foreign relations.* Washington, DC: Author.

National Academy of Engineering. (1986). *The positive sum strategy: Harnessing technology for economic growth.* Washington, DC: Author.

National Academy of Engineering. (1987a). *Technology and global industry: Companies and nations in the world economy.* Washington, DC: Author.

National Academy of Engineering. (1987b). *Technology and employment: Innovation and growth in the U.S. economy.* Washington, DC: Author.

National Academy of Engineering. (1989). *Education and employment of engineers: A research agenda for the 1990s.* Washington, DC: NAS Press.

National Academy of Sciences (NAS). (1965). *Basic research and national goals.* Washington, DC: National Academy Press.

National Academy of Sciences (NAS). (1968). *Technology: Processes of assessment and choice.* Washington, DC: National Academy Press.

National Academy of Sciences (NAS). (1988). *Engineering personnel data needs for the 1990s.* Washington, DC: NAS Press.

National Academy of Sciences, Institute of Medicine. (1991). *Biomedical politics.* Washington, DC: National Academy Press.

National Institutes of Health and Department of Energy. (1990). *Understanding our genetic inheritance, the U.S. human genome project: The first five years, FY 1991-1995* (DOE/ER-0452P). Washington, DC: Department of Health and Human Services and Department of Energy.

National Research Council, Committee on the Education and Employment of Women in Science and Engineering. (1979). *Climbing the academic ladder: Doctoral women scientists in academe.* Washington, DC: National Academy Press.

National Research Council, Committee on the Education and Employment of Women in Science and Engineering. (1983). *Climbing the ladder: An update on the status of doctoral women scientists and engineers.* Washington, DC: National Academy Press.

National Research Council. (1985a). *Engineering technology education.* Washington, DC: NAS Press.

National Research Council. (1985b). *Engineering graduate education and research*. Washington, DC: NAS Press.

National Research Council. (1985c). *Engineering in society*. Washington, DC: NAS Press.

National Research Council. (1985d). *Support organizations for the engineering community*. Washington, DC: NAS Press.

National Research Council. (1985e). *Engineering employment characteristics*. Washington, DC: NAS Press.

National Research Council. (1985f). *Continuing education of engineers*. Washington, DC: NAS Press.

National Research Council. (1985g). *Engineering education and practice in the United States*. Washington, DC: NAS Press.

National Research Council. (1986a). *Engineering undergraduate education*. Washington, DC: NAS Press.

National Research Council. (1986b). *Engineering infrastructure diagraming and modeling*. Washington, DC: NAS Press.

National Research Council, Committee on Mapping and Sequencing the Human Genome. (1988). *Mapping and sequencing the human genome*. Washington, DC: National Academy Press.

National Research Council. (1991). *Women in science and engineering: Increasing their numbers in the 1990s*. Washington, DC: National Academy Press.

National Science Board. (1989). *Science and engineering indicators: 1989*. Washington, DC: Government Printing Office.

National Science Board. (1991). *Science and engineering indicators: 1991*. Washington, DC: Government Printing Office.

National Science Foundation, Division of Science Resources Studies. (1987). *Federal funds for research and development: Fiscal years 1986, 1987 and 1988* (NSF Science Resources Series). Washington, DC: Government Printing Office.

National Science Foundation, Division of Science Resources Studies. (1988). *Academic science/engineering: Graduate enrollment and support* (NSF Science Resources Series). Washington, DC: Government Printing Office.

National Science Foundation. (1989, December 28). *Effect of downsized DOD budget on engineering* (memo from John White and F. Karl Willenbrick). Washington, DC: Author.

National Science Foundation. (1990). *Women and minorities in science and engineering* (NSF 90-301). Washington, DC: Author.

National Science Foundation. (1991a). Biennial Ph.D survey. *Science, 252,* 1113.

National Science Foundation. (1991b). Attitudes towards science and technology. In *Science and engineering indicators, 1991* (pp. 166-191). Washington, DC: Government Printing Office.

National Science Foundation, Division of Science Resources Studies. (n.d.). *Federal funds for research and development: Federal obligations for research to universities and colleges by agency and detailed field of science/engineering—fiscal years 1973-1987*. Washington, DC: Government Printing Office.

Nau, H. R. (1975). Collective responses to R&D problems in western Europe: 1955-1958. *International Organization, 3,* 617-653.

Ndonko, W. A., & Anyang, S. O. (1981). Concept of appropriate technology: An appraisal from the Third World. *Monthly Review, 32,* 35-43.

Needless quarrel over research. (1988, February 4). *Nature,* pp. 375-376.

Nelkin, Dorothy. (1971). *Nuclear power and its critics: The Cayuga Lake controversy.* Ithaca, NY: Cornell University Press.

Nelkin, Dorothy. (1972). *The university and military research: Moral politics at MIT.* Ithaca, NY: Cornell University Press.

Nelkin, Dorothy. (1975). The political impact of technical expertise. *Social Studies of Science, 5,* 35-54.

Nelkin, Dorothy. (1977a). Technology and public policy. In I. Spiegel-Rösing & Derek Price (Eds.), *Science, technology and society: A cross-disciplinary perspective* (pp. 393-442). Beverly Hills, CA: Sage.

Nelkin, Dorothy. (1977b). Thoughts on the proposed science court. *Harvard Newsletter on Science, Technology, & Human Values, 18,* 20-31.

Nelkin, Dorothy. (1977c). Scientists and professional responsibility: The experience of American ecologists. *Social Studies of Science, 7,* 75-95.

Nelkin, Dorothy. (Ed.). (1979). *Controversy: Politics of technical decisions.* Beverly Hills, CA: Sage. (2nd ed., 1984; 3rd ed., 1991)

Nelkin, Dorothy. (1982). *The creation controversy: Science or scripture in the schools.* New York: Norton.

Nelkin, Dorothy. (1984a). *The creation controversy.* New York: Norton.

Nelkin, Dorothy. (1984b). *Science as intellectual property.* Washington, DC: American Association for the Advancement of Science.

Nelkin, Dorothy. (Ed.). (1985). *The language of risk: Conflicting perspectives on occupational health.* Beverly Hills, CA: Sage.

Nelkin, Dorothy. (1987, January). Science, technology and public policy. [Entire issue] *Newsletter* (of the History of Science Society).

Nelkin, Dorothy. (1989). Communicating technological risk. *Annual Reviews of Public Health, 10,* 95-113.

Nelkin, Dorothy. (Ed.). (1992). *Controversies: Politics of technical decisions* (3rd ed.). Newbury Park, CA: Sage.

Nelkin, Dorothy. (1994). *Selling science: How the press covers science and technology* (2nd ed.). New York: Freeman.

Nelkin, Dorothy, & Pollak, Michael. (1979). Public participation in technological decisions: Reality or grand illusion? *Technology Review, 81*(8), 55-64.

Nelkin, Dorothy, & Tancredi, Laurence. (1994). *Dangerous diagnostics: The social power of biological information* (2nd ed.). Chicago: University of Chicago Press.

Nelson, J. S., & Megill, Alan. (1986). Rhetoric of inquiry: Projects and prospects. *Quarterly Journal of Speech, 72,* 20-37.

Nelson, J. S., Megill, Alan, & McCloskey, Don. (Eds.). (1987). *The rhetoric of the human sciences: Language and argument in scholarship and public affairs.* Madison: University of Wisconsin Press.

Nelson, Lynn. (1991). *Who knows? From Quine to a feminist empiricism.* Philadelphia: Temple University Press.

Nelson, Richard R. (1959). The simple economics of basic scientific research: A theoretical analysis. *Journal of Political Economy, 67,* 297-306.

Nelson, Richard R. (1984). *High-technology policies: A five nation comparison.* New York: American Enterprise Institute.

Nelson, Richard R. (1990). U.S. technological leadership: Where did it come from and where did it go? *Research Policy, 19,* 117-132.

Nelson, Richard R., Peck, Merton J., & Kalachek, Edward D. (Eds.). (1967). *Technology, economic growth and public policy.* Washington, DC: Brookings Institution.

Nelson, Richard R., & Winter, S. G. (1977). In search of a useful theory of innovation. *Research Policy, 6,* 36-76.

Nelson, Richard R., & Winter, S. G. (1982). *An evolutionary theory of economic change.* Cambridge, MA: Belknap Press of Harvard University Press.

Nelson, William R. (Ed.). (1968). *The politics of science.* New York: Oxford University Press.

Newby, Howard. (1992, January 17). Join forces in modern marriage. *Times Higher Education Supplement*, p. 20.

Nickles, Thomas. (1988). Reconstructing science: Discovery and experiment. In D. Batens & T. P. Van Bendegem (Eds.), *Theory and experiment*. Dordrecht, the Netherlands: Reidel.

Nietzsche, Friedrich. (1974). *The gay science*. New York: Vintage.

Nieva, Veronica, & Gutek, Barbara. (1980). Sex differences in evaluation. *Academy of Management Review, 5*, 267-276.

Nilsen, Svein Erik. (1979). The use of computer technology in some developing countries. *International Social Science Journal, 31*, 513-528.

Noble, David F. (1977). *America by design: Science, technology, and the rise of corporate capitalism*. Oxford: Oxford University Press.

Noble, David F. (1979). Social choice in machine design. In A. Zimbalist (Ed.), *Case studies on the labor process*. New York: Monthly Review Press.

Noble, David F. (1984). *Forces of production: A social history of industrial automation*. New York: Knopf.

Noble, David F. (1985). Command performance: A perspective on the social and economic consequences of military enterprise. In M. R. Smith (Ed.), *Military enterprise and technological change: Perspectives on the American experience*. Cambridge: MIT Press.

Noble, David F. (1992). *A world without women*. New York: Knopf.

Norman, C. (1979). *The god that limps*. New York: Norton.

Nowotny, Helga, & Rose, Hilary. (Eds.). (1979). *Counter-movements in the sciences* (Sociology of the Sciences Yearbook). Dordrecht, the Netherlands: Reidel.

Nuffield Foundation. (1975). *Case-studies in interdisciplinarity: 2: Science, technology and society*. London: Author.

Nye, Joseph S. (1987). Nuclear learning and U.S.-Soviet security regimes. *International Organization, 41*, 371-402.

Nye, Mary Jo. (1992). New views of old science. In D. Calhoun (Ed.), *1993 yearbook of science and the future* (pp. 220-240). Chicago: Encyclopedia Britannica.

Oakley, A. (1987). From walking wombs to test-tube babies. In M. Stanworth (Ed.), *Reproductive technologies: Gender, motherhood and medicine* (pp. 36-56). Cambridge: Polity.

O'Brien, James A. (1968). *The impact of computers on banking*. Boston: Bankers.

Ochs, K. H. (1992). The rise of American mining engineers: A case study of the Colorado School of Mines. *Technology and Culture, 33*(2), 278-301.

O'Connor, J. G. (1969). Growth of multiple authorship. *DRTC Seminar, 7*, 463-483.

Office of Technology Assessment (OTA), U.S. Congress. (1986). *Basic research as an investment: Can we measure the returns?* Washington, DC: Government Printing Office.

Office of Technology Assessment (OTA), U.S. Congress. (1987). *New developments in biotechnology* (Background paper, Public perceptions of biotechnology). Washington, DC: Government Printing Office.

Office of Technology Assessment (OTA), U.S. Congress. (1988a). *The defense technology base*. Washington, DC: Government Printing Office.

Office of Technology Assessment (OTA), U.S. Congress. (1988b). *Mapping our genes—The genome projects: How big, how fast?* (OTA-BA-373). Washington, DC: Government Printing Office.

Office of Technology Assessment (OTA), U.S. Congress. (1989). *Holding the edge: Maintaining the defense technology base*. Washington, DC: Government Printing Office.

Office of Technology Assessment (OTA), U.S. Congress. (1991a). *Federally funded research: Decisions for a decade*. Washington, DC: Government Printing Office.

Office of Technology Assessment (OTA), U.S. Congress. (1991b). *Redesigning defense: Planning the transition to the future U.S. defense industrial base*. Washington, DC: Government Printing Office.

Ogan, Christine. (1988). Media imperialism and the videocassette recorder: The case of Turkey. *Journal of Communication, 38*, 93-106.

Ogburn, W. F. (1945). *The social effects of aviation*. Boston: Houghton Mifflin.

Ogburn, W. F., with the assistance of Gilfillan, S. C. (1933). The influence of invention and discovery. In *Recent social trends in the United States* (Report of the President's Research Committee on Social Trends, pp. 122-166). New York: McGraw-Hill.

Ogburn, W. F., & Meyers Nimkoff, F. (1955). *Technology and the changing family*. Boston: Houghton Mifflin.

O'Keefe, M. (1970). The mass media as sources of medical information for doctors. *Journalism Quarterly, 47*, 95-100.

Olson, M., Hood, L., Cantor, C., & Botstein, D. (1989, September 29). A common language for physical mapping of the human genome. *Science, 245*, 1434-1435.

Onn, Fong Chan. (1980). Appropriate technology: An empirical study of bicycle manufacturing in Malaysia. *Developing Economies, 18*, 96-115.

Ophir, A., Shapin, Steven, & Schaffer, Simon (Eds.). (1991). *The place of knowledge: The spatial setting and its relation to the production of knowledge*. [Special issue] *Science in Context, 4*(1).

Organisation for Economic Cooperation and Development (OECD). (1963). *Science and the policies of governments*. Paris: Author.

Organisation for Economic Cooperation and Development (OECD). (1965). *The research and development effort*. Paris: Author.

Organisation for Economic Cooperation and Development (OECD). (1966a). *Government and allocation of resources to science*. Paris: Author.

Organisation for Economic Cooperation and Development (OECD). (1966b). *Fundamental research and the policies of governments*. Paris: Author.

Organisation for Economic Cooperation and Development (OECD). (1971). *Science, growth and society: A new perspective*. Paris: Author.

Organisation for Economic Cooperation and Development (OECD). (1981). *Science and technology policy for the 1980s*. Paris: Author.

Organisation for Economic Cooperation and Development (OECD). (1984). *Industry and university: New forms of co-operation and communication*. Paris: Author.

Organisation for Economic Cooperation and Development (OECD). (1990). *University enterprise relations in OECD member countries* (DSTI/SPR/89.37; 1st rev.). Paris: Author.

Organisation for Economic Cooperation and Development (OECD). (1992). *Science and technology policy outlook*. Paris: Author.

Orlans, Harold. (Ed.). (1968). *Science policy and the university*. Washington, DC: Brookings Institution.

Orsenigo, Luigi. (1989). *The emergence of biotechnology*. London: Frances Pinter.

Orth, Charles D., Bailey, Joseph C., & Wolek, Francis W. (Eds.). (1964). *Administering research and development*. New York: Irwin.

Ortner, Sherry B. (1974). Is female to male as nature is to culture? In M. Z. Rosaldo & L. Lamphere (Eds.), *Women, culture, and society*. Stanford, CA: Stanford University Press.

Ostry, S. (1990). *Governments and corporations in a shrinking world: Trade and innovation policies in the United States, Europe and Japan*. New York: Council on Foreign Relations.

Otake, Hideo. (1982). Corporate power in social conflict: Vehicle safety and Japanese motor manufacturers. *International Journal of the Sociology of Law, 10*, 75-103.

Otero, Gerardo. (1991). The coming revolution in biotechnology: A critique of Buttel. *Sociological Form, 6*, 551-565.

Ott, M. D. (1978a). Differences between men and women engineering students. *Journal of College Student Personnel, 19*, 552-557.

Ott, M. D. (1978b). Retention of men and women engineering students. *Research in Higher Education, 9*, 127-150.

Ott, M., & Reese, N. A. (Eds.). (1975). *Women in engineering: Beyond recruitment*. Ithaca, NY: Cornell University Press.

Otway, Harry. (1992). Public wisdom, expert fallibility: Towards a contextual theory of risk. In S. Krimsky & D. Golding (Eds.), *Social theories of risk* (pp. 215-228). New York: Praeger.

Otway, Harry, & Gow, H. (1990). *Communicating information to the public about major industrial hazards*. London: Elsevier.

Ovensen, N. K. (1980). *Advances in the continuing education of engineers*. Paris: UNESCO.

Ovitt, G. (1989). Appropriate technology: Development and social change. *Monthly Review, 40*, 22-32.

Owens, Larry. (1986). Vannevar Bush and the differential analyzer: The text and context of an early computer. *Technology and Culture, 27*, 63-95.

Pacey, Arnold. (1976). *The maze of ingenuity*. Cambridge: MIT Press.

Pack, Howard, & Westphal, Larry E. (1986). Industrial strategy and technological change: Theory versus reality. *Journal of Development Economics, 22*, 87-128.

Paine, R. (1992). "Chernobyl" reaches Norway: The accident, science and the threat to cultural knowledge. *Public Understanding of Science, 1*(3), 261-280.

Palca, Joseph. (1991, November 8). NIH unveils plan for women's health project. *Science, 254*, 792.

Palter, R. (1987). Saving Newton's text: Documents, readers, and the ways of the world. *Studies in the History and Philosophy of Science, 18*, 385-439.

Pancaldi, Giuliano. (1980). The history and social studies of science in Italy. *Social Studies of Science, 10*(3), 351-374.

Paquet, G. (1990). Internationalization of domestic firms and governments: Anamorphosis of a palaver. *Science & Public Policy, 17*, 327-332.

Parker, I., & Shotter, J. (Eds.). (1990). *Reconstructing social psychology*. London: Routledge & Kegan Paul.

Parsons, Talcott. (1954). The professions and social structure. In *Essays in sociological theory* (pp. 34-49). New York: Free Press. (Original work published 1939)

Patel, Surendra. (1984). Technology in UNCTAD: 1970 to 1984. *IDS Bulletin, 15*(3), 63-66.

Pattnaik, Binay Kumar. (1989). Scientific temper and religious beliefs. *Journal of Sociological Studies, 8*, 13-40.

Pauly, P. J. (1979, Fall). The world and all that is in it: The National Geographic Society, 1888-1918. *American Quarterly, 31*(4), 517-532.

Pavitt, Keith. (1973). Technology, international competition and economic growth: Some lessons and perspectives. *World Politics, 25*, 183-205.

Pavitt, Keith. (1991). What makes basic research economically useful? *Research Policy, 20*, 109-119.

Pearl, Amy, Pollack, Martha E., Riskin, Eve, Thomas, Becky, Wolf, Elizabeth, & Wu, Alice. (1990). Becoming a computer scientist: A report by the ACM Committee on the Status of Women in Computing Science. *Communications of the ACM, 33*(11), 48-57.

Pearse, Andrew. (1980). *Seeds of plenty, seeds of want: Social and economic implications of the Green Revolution*. Oxford: Clarendon.

Pelaez, E. (in press). Software: A very peculiar commodity. In S. Woolgar & F. Murray (Ed.), *Social perspectives on software*. Cambridge: MIT Press.

Pelz, Donald C., & Andrews, Frank M. (1966). *Scientists in organizations: Productive climates for research and development*. New York: Wiley.

Penick, James L., Jr., Pursell, Carroll W., Jr., Sherwood, Morgan B., & Swain, Donald C. (Eds.). (1965). *The politics of American science*. Cambridge: MIT Press.

Perkins, F. C. (1983). Technology choice, industrialisation and development experiences in Tanzania. *Journal of Development Studies, 19*, 213-243.

Perrucci, Carolyn C. (1970). Minority status and the pursuit of professional careers: Women in science and engineering. *Social Forces, 49*, 245-259.

Perry, Ruth, & Greber, Lisa. (1990). Women and computers: An introduction. *Signs, 16*(1), 74-101.

Pestre, Dominique. (1990). *Louis Neel: Le magnétisme et Grenoble* (Vol. Cahier d'Histoire du CNRS). Paris: CNRS.

Peters, Lois S. (1987). *Academic crossroads: The US experience*. Centre for Technology Policy, RPI.

Peters, Lois S., & Etzkowitz, Henry. (1988, November 16-19). *The institutionalization of academic computer science*. Paper presented at the joint conference of the Society for Social Studies of Science and the European Association for the Study of Science and Technology, Amsterdam.

Peters, Lois S., & Fusfeld, H. (1982). *University-industry research relationships* (NSF NSB82-2). Washington, DC: National Science Foundation.

Petersen, James C. (Ed.). (1984). *Citizen participation in science policy*. Amherst: University of Massachusetts Press.

Petersen, James C., & Markle, Gerald E. (1981). Expansion of conflict in cancer controversies. In L. Kriesberg (Ed.), *Research in social movements, conflicts and change* (Vol. 4, pp. 151-169). Greenwich, CT: JAI.

Petersen, Victor. (1991, September). NASA aims. *Scientific American*.

Petroski, Henry. (1985). *To engineer is human: The role of failure in successful design*. New York: St. Martin's Press.

Pettegrew, L. S., & Logan, R. (1987). The health care context. In C. R. Berger & S. H. Chaffee (Eds.), *Handbook of communication science* (pp. 657-710). Newbury Park, CA: Sage.

Pfeffer, N. (1987). Artificial insemination, in-vitro fertilization and the stigma of infertility. In M. Stanworth (Ed.), *Reproductive technologies: Gender, motherhood and medicine* (pp. 81-97). Cambridge: Polity.

Pfetsch, Frank R. (1979). The "finalization" debate in Germany: Some comments and explanations. *Social Studies of Science, 9*(1), 115-124.

Pfiffner, J. P. (1991). Divided government and the problems of government. In J. A. Thurber (Ed.), *Divided democracy: Cooperation and conflict between the president and Congress*. Washington, DC: Congressional Quarterly Press.

Pheterson, G. T., Kiesler, S. B., & Goldberg, P. A. (1971). Evaluation of the performance of women as a function of their sex, achievement, and personal history. *Journal of Personality and Social Psychology, 19*, 110-114.

Phillips, D. P., Kanter, E. J., Bednarczyk, B., & Tastad, P. L. (1991). Importance of the lay press in the transmission of medical knowledge to the scientific community. *New England Journal of Medicine, 325*, 1180-1183.

Pickering, Andrew. (1984). *Constructing quarks: A sociological history of particle physics*. Chicago: University of Chicago Press.

Pickering, Andrew. (1989). Living in the material world. In D. Gooding, Trevor Pinch, & Simon Schaffer (Eds.), *The uses of experiment*. Cambridge, MA: Cambridge University Press.

Pickering, Andrew. (1990). Knowledge, practice and mere construction. *Social Studies of Science, 20*, 682-729.

Pickering, Andrew. (1992a). From science as knowledge to science as practice. In A. Pickering (Ed.), *Science as practice and culture* (pp. 1-26). Chicago: University of Chicago Press.

Pickering, Andrew. (Ed.). (1992b). *Science as practice and culture.* Chicago: University of Chicago Press.

Pickering, Andrew. (in press). The mangle of practice. *AJS.*

Pickering, Andrew, & Stephanides, A. (1992). Constructing quaternions: On the analysis of conceptual practice. In A. Pickering (Ed.), *Science as practice and culture.* Chicago: University of Chicago Press.

Piller, Charles. (1991). *The fail-safe society: Community defiance and the end of American technological optimism.* New York: Basic Books.

Pinch, Trevor. (1982). Kuhn: The conservative and radical interpretations. *4S Newsletter, 7,* 10-25.

Pinch, Trevor. (1985). Towards an analysis of scientific observation: The externality and evidential significance of observation reports in physics. *Social Studies of Science, 15,* 167-187.

Pinch, Trevor. (1986). *Confronting nature: The sociology of neutrino detection.* Dordrecht, the Netherlands: Reidel.

Pinch, Trevor J. (1989). How do we treat technical uncertainty in systems failure? The case of the space shuttle *Challenger*. In La Porte (Ed.), *Social responses to large technical systems: Control or anticipation* (pp. 143-158). Dordrecht, the Netherlands: Kluwer.

Pinch, Trevor. (1993). "Turn, turn, and turn again: The Woolgar formula." *Science, Technology, & Human Values, 18,* 511-522.

Pinch, Trevor J., & Bijker, Wiebe E. (1984). The social construction of facts and artifacts, Or how the sociology of science and the sociology of technology might benefit each other. *Social Studies of Science, 14,* 399-441.

Pinch, Trevor, & Bijker, Wiebe E. (1987). The social construction of facts and artifacts, Or how the sociology of science and the sociology of technology might benefit each other. In W. E. Bijker, T. P. Hughes, & T. Pinch (Eds.), *The social construction of technological systems* (pp. 17-50). Cambridge: MIT Press.

Pinch, Trevor J., & Collins, Harry. (1992, December). *Inside knowledge: The phenomenology of surgical skill.* Paper presented at the Visualization Workshop, Princeton University, Department of History.

Pinch, Trevor, & Pinch, Trevor. (1988). Reservations about reflexivity and new literary forms or why let the devil have all the good tunes? In S. Woolgar (Ed.), *Knowledge and reflexivity* (pp. 178-197). London: Sage.

Pinn, Vivian. (1992, October 14). Women's health research: Prescribing change and addressing the issues. *Journal of the American Medical Association, 268*(14), 1921-1922.

Pion, G. M., & Lipsy, M. W. (1981). Public attitudes towards science and technology: What have the surveys told us? *Public Opinion Quarterly, 45,* 303-316.

Pitt, Joe. (n.d.). *Thinking about technology.* Boulder, CO: Westview.

Pletta, Dan H. (1984). *The engineering profession: Its heritage and its emerging public purpose.* Lanham, MD: University Press of America.

Plucknett, Donald, Smith, Nigel, & Ozgediz, Selcuk. (1990). *Networking in international agricultural research.* Ithaca, NY: Cornell University Press.

Polanyi, Michael. (1958). *Personal knowledge.* London: Routledge & Kegan Paul/Chicago: University of Chicago Press.

Polanyi, Michael. (1962). The republic of science: Its political and economic theory. *Minerva, 1,* 54-73.

Polanyi, Michael. (1967). *The tacit dimension.* London: Routledge & Kegan Paul.

Pomerantz, Anita. (1975). *Second assessments: A study of some features of agreements/disagreements*. Unpublished doctoral dissertation, University of California, Irvine, School of Social Sciences.

Poovey, Mary. (1989). *Uneven developments: The ideological work of gender in mid-Victorian England*. Chicago: University of Chicago Press.

Popper, Karl R. (1945). *The open society and its enemies*. London: Routledge & Kegan.

Popper, Karl R. (1959). *The logic of scientific discovery*. New York: Harper. (Original work published 1934)

Popper, Karl R. (1963). *Conjectures and refutations*. New York: Harper.

Popper, Karl R. (1970). Normal science and its dangers. In I. Lakatos & A. Musgrave (Eds.), *Criticism and the growth of knowledge* (pp. 51-58). Cambridge: Cambridge University Press.

Popper, Karl R. (1972). *Objective knowledge: An evolutionary approach*. Oxford: Clarendon.

Porritt, J. (1989). Green shoots, rotten roots. *BBC Wildlife, 7*, 352-353.

Porter, M. E. (1990). *The competitive advantage of nations*. New York: Free Press.

Portes, A. (1976). On the sociology of national development: Theories and issues. *American Journal of Sociology, 82*, 55-85.

Potter, Jonathon. (1984). Testability, flexibility: Kuhnian values in psychologists' discourse concerning theory choice. *Philosophy of the Social Sciences, 14*, 303-330.

Potter, Jonathon. (1988). Cutting cakes: A study of psychologists' social categorizations. *Philosophical Psychology, 1*, 17-33.

Potter, Jonathon, & Mulkay, Michael. (1985). Scientists' interview talk: Interviews as a technique for revealing participants' interpretative practices. In M. Brenner, J. Brown, & D. Canter (Eds.), *The research interview: Uses and approaches*. London: Academic Press.

Potter, Jonathon, Stringer, Peter, & Wetherell, Margaret. (1984). *Social texts and context*. London: Routledge & Kegan Paul.

Potter, Jonathon, & Wetherell, Margaret. (1987). *Discourse and social psychology: Beyond attitudes and behaviour*. London: Sage.

Potter, Jonathon, & Wetherell, Margaret. (in press). *Facts, rhetoric and reality*. London: Sage.

Potter, Jonathon, Wetherell, Margaret, & Chitty, Andrew. (1991). Quantification rhetoric: Cancer on television. *Discourse and Society, 2*, 333-365.

Powell, W., & DiMaggio, P. (1991). *The new institutionalism in organizational analysis*. Chicago: University of Chicago Press.

Prelli, L. J. (1989). *A rhetoric of science: Inventing scientific discourse*. Columbia: University of South Carolina Press.

Presser, Stanley. (1980). Collaboration and the quality of research. *Social Studies of Science, 10*, 95-101.

Prewitt, Kenneth. (1982). The public and science policy. *Science, Technology, & Human Values, 7*(39), 5-14.

Price, Derek J. de Solla. (1963). *Little science, big science*. New York: Columbia University Press.

Price, Derek J. de Solla. (1967). Networks of scientific papers. *Science, 149*, 510-515.

Price, Derek J. de Solla. (1971). Principles for projecting funding of academic science in the 1970s. *Science Studies, 1*(1), 85-99.

Price, Don K. (1967). *The scientific estate*. Cambridge, MA: Belknap Press of Harvard University Press.

Price, Don K. (1979). Endless frontier or bureaucratic morass? In G. Holton & R. Morrison (Eds.), *Limits of scientific inquiry*. New York: Norton.

Price, Don K. (1981). The spectrum from truth to power. In T. J. Kuehn & A. L. Porter (Eds.), *Science, technology, and national policy* (pp. 95-131). Ithaca, NY: Cornell University Press.

Price, F. (in press). Now you see it, now you don't: Mediating science and managing uncertainty in reproductive medicine. In A. Irwin & B. Wynne (Eds.), *Misunderstanding science*. Cambridge: Cambridge University Press.

Primack, Joel, & von Hippel, Frank. (1974). *Advice and dissent: Scientists in the political arena.* New York: Basic Books.

Proctor, Robert. (1991). *Value-free science?* Cambridge, MA: Harvard University Press.

Proctor, Robert. (in press). *The great cancer debates.*

Puryear, Jeffrey M. (1982). Higher education, development assistance, and repressive regimes. *Studies in Comparative International Development, 17*(2), 3-35.

Putnam, Hilary. (1978). *Meaning and the moral science*. London: Routledge & Kegan.

Quine, Willard V. O. (1953). Two dogmas of empiricism. In W. V. Quine (Ed.), *From a logical point of view*. Cambridge, MA: Harvard University Press.

Quine, Willard V. O. (1969). *Ontological relativity and other essays.* New York: Columbia University Press.

Quiroga, Eduardo R. (1984). Irrigation planning to transform subsistence agriculture: Lessons from El Salvador. *Human Ecology, 12*, 183-202.

Rabkin, Yakov. (1986). Cultural variations in scientific development. *Social Science Information, 25*, 967-989.

Radder, Hans. (1992). Normative reflections on constructivist approaches to science and technology. *Social Studies of Science, 22*(1), 141-173.

Radnitzky, G., & Bartley, W. W. (Eds.). (1987). *Evolutionary epistemology, rationality, and the sociology of knowledge*. La Salle, IL: Open Court.

Raelin, J. A. (1986). *The clash of cultures: Managers and professionals.* Boston: Harvard Business School Press.

Rahman, A. (1981). Interaction between science, technology, and society: Historical and comparative perspectives. *International Social Science Journal, 33*, 508-521.

Raina, D., Raza, G., Dutt, B., & Singh, S. (1992). *The social context of scientific attitudes: Representations of science at a congregation of religious pilgrims in Northern India* (mimeo). New Delhi, India: National Institute for Science, Technology and Development Studies.

Rakow, L. (1988). Women and the telephone: The gendering of a communications technology. In C. Kramarae (Ed.). *Technology and women's voices.* New York: Routledge & Kegan Paul.

Rakow, L. (1992). *Gender on the line: The telephone and community life*. Champaign: University of Illinois Press.

Ramanathan, K. (1988). Evaluating the national science and technology base: A case study on Sri Lanka. *Science and Public Policy, 15*, 304-320.

Ramirez, Francisco, & Lee, Molly. (1990). *Education, science, and development.* Paper presented at the Conference on Comparative Sudies of Education. Washington, DC: Department of Education.

Rammert, W. (1990). *Computerwelten—Alltagswelten: Wie verändert der Computer die soziale Wirklichkeit*. Opladen: Westdeutscher Verlag.

Randolph, Robert H., & Koppel, Bruce. (1982). Technology assessment in Asia: Status and prospects. *Technological Forecasting and Social Change, 22*, 363-384.

Ranis, Gustav. (1990). Science and technology policy: Lessons from Japan and the East Asian NICs. In R. Evenson & G. Ranis (Eds.), *Science and technology: Lessons for development* (pp. 157-178). Boulder, CO: Westview.

Rao, K. Naguraja, & Dubowl, Joel B. (1984). The allure of optimum technologies and the social realities of the developing world. *Bulletin of Science, Technology and Society, 4*, 345-355.

Rapoport, A. (Ed.). (1974). *Game theory as a theory of conflict resolution*. Dordrecht, the Netherlands: Reidel.

Rapp, F. (1981). *Analytical philosophy of technology* (Boston Series in the Philosophy of Science, Vol. 63). Dordrecht, the Netherlands: Reidel.

Raup, D. (1986). *The nemesis affair: A story of the death of dinosaurs and the ways of science.* New York: Norton.

Raven, D., Tijssen, L., & de Wolf, J. (Eds.). *Cognitive relativism and social science.* New Brunswick, NJ: Transaction.

Ravetz, A. (1965). Modern technology and an ancient occupation: Housework in present-day society. *Technology and Culture, 6,* 256-260.

Ravetz, Jerome R. (1971). *Scientific knowledge and its problems.* Oxford: Oxford University Press.

Rayner, Steve. (1991). A cultural perspective on the structure and implementation of global environmental agreements. *Evaluation Review, 15,* 75-102.

Raza, G., Dutt, B., Singh, S., & Wahid, A. (1991). *Prototype of the forms of scientific cognition: A survey of cultural attitudes to natural phenomena* (Report NISTADS 108[AV]91). New Delhi, India: National Institute for Science, Technology and Development Studies.

Reader, W. J. (1987). *A history of the Institution of Electrical Engineers 1871-1971.* London: Peter Perigrinus.

Reams, Bernard. (1986). *University-industry research partnerships: The major legal issues in research and development agreements.* Westport, CT: Quorum.

Reddy, N. Mohan, & Zhao, Liming. (1990). International technology transfer: A review. *Research Policy, 19,* 285-307.

Redmond, Kent C., & Smith, Thomas M. (1980). *Project Whirlwind: The history of a pioneer computer.* Boston: Digital.

Rees, Mina. (1982). The computing program of the Office of Naval Research, 1946-1953. *Annals of the History of Computing, 4*(2), 103-113.

Reich, Leonard S. (1983). Irving Langmuir and the pursuit of science and technology in the corporate environment. *Technology and Culture, 24,* 199-221.

Reich, Leonard S. (1985). *The making of American industrial research.* Cambridge: Cambridge University Press.

Reich, Michael. (1992). *Toxic politics: Responding to chemical disasters.* Ithaca, NY: Cornell University Press.

Reich, Michael. (1993). Toxic politics and pollution victims in the Third World. In S. Jasanoff (Ed.), *Learning from disaster* (pp. 180-203). Philadelphia: University of Pennsylvania Press.

Reich, Robert. (1991). *The work of nations: Preparing ourselves for 21st century capitalism.* New York: Knopf.

Reiter, Rayna Rapp. (Ed.). (1975). *Toward an anthropology of women.* New York: Monthly Review Press.

Remington, John A. (1988). Beyond big science in America: The binding of inquiry. *Social Studies of Science, 18,* 45-72.

Rényi, A. (1984). *A diary in information theory.* Chichester, U.K.: Ellis Horwood.

Reppy, Judith. (1990). The technological imperative in strategical thought. *Journal of Peace Research, 27*(1), 101-106.

RSGB (Research Surveys of Great Britain). (1988). *Public perceptions of biotechnology: Interpretive report.* London: Author.

Reskin, Barbara. (1976). Sex differences in status attainment in science: The case of post-doctoral fellowships. *American Sociological Review, 41,* 597-612.

Reskin, Barbara. (1977). Scientific productivity and the reward structure of science. *American Sociological Review, 42,* 491-504.

Reskin, Barbara. (1978a). Scientific productivity, sex, and location in the institution of science. *American Journal of Sociology, 83,* 1235-1243.

Reskin, Barbara. (1978b). Sex differentiation and the social organization of science. *Sociological Inquiry, 48*, 6-37.

Reskin, Barbara, & Roos, Patricia. (1990). *Job queues, gender queues.* Philadelphia: Temple University Press.

Restivo, Sal. (1983). *The social relations of physics, mysticism, and mathematics.* Dordrecht, the Netherlands: Kluwer/D. Reidel.

Restivo, Sal. (1986). Science, secrecy, and democracy. *Science, Technology, & Human Values, 11*, 79-84.

Restivo, Sal. (1990). The social roots of pure mathematics. In S. Cozzens & T. Gieryn (Eds.), *Theories of science in society* (pp. 120-143). Bloomington: Indiana University Press.

Restivo, Sal. (1992). Zen and the art of science studies. *Science, Technology, & Human Values, 17*, 402-406.

Restivo, Sal. (1994). [Review of self-organization: Portrait of a scientific revolution.] *Science, Technology, & Human Values, 19(1)*, 117-119.

Rettig, Richard A. (1977). *Cancer crusade: The story of the National Cancer Act of 1971.* Princeton, NJ: Princeton University Press.

Reynolds, Terry S. (1983). *75 years of progress: A history of the American Institute of Chemical Engineers.* New York: American Institute of Chemical Engineers.

Reynolds, Terry S. (1986). Defining professional boundaries: Chemical engineering in the early twentieth century. *Technology and Culture, 27*(4).

Reynolds, Terry S. (Ed.). (1991). *The engineer in America.* Chicago: University of Chicago Press.

Rhees, D. J. (1987). *The chemists' crusade: The rise of an industrial science in modern America, 1907-1922.* Unpublished doctoral dissertation, University of Pennsylvania.

Rhodes, Richard. (1986). *The making of the atomic bomb.* New York: Simon & Schuster.

Richards, Evelleen. (1988). The politics of therapeutic intervention: The vitamin C and cancer controversy. *Social Studies of Science, 18*, 653-701.

Richards, Evelleen. (1991). *Vitamin C and cancer: Medicine or politics?* London: Macmillan.

Richards, Paul. (1985). *Indigenous agricultural revolution: Ecology and food production in West Africa.* Boulder, CO: Westview.

Richta, R., et al. (1967). *Civilization at the crossroads: Social and human implications of the scientific and technological revolution.* Prague: Czechoslovak Academy of Sciences.

Rieselbach, L. N. (1977). *Congressional reform in the seventies.* Morristown, NJ: General Learning.

Rigg, Jonathan. (1989). The new rice technology and agrarian change: Guilt by association. *Progress in Human Geography, 13*, 374-400.

Rip, Arie. (1981). A cognitive approach to science policy. *Research Policy, 10*, 294-331.

Rip, Arie. (1987). Controversies as informal technology assessment. *Knowledge, 8*, 349-371.

Rip, Arie. (1988). Contextual transformation in contemporary science. In A. Jamison (Ed.), *Keeping science straight: A critical look at the assessment of science and technology.* Gothenburg, Sweden: University of Gothenburg, Department of Theory of Science.

Rip, Arie. (1989, June). *Quasi-evolutionary model of technological development and a cognitive approach to technology policy.* Unpublished paper, Center for Studies of Science, Technology, and Society, University of Twente.

Rip, Arie, & Boeker, Egbert. (1974). Scientists and social responsibility in the Netherlands. *Social Studies of Science, 5*(4), 457-484.

Riskin, Carl. (1978). Intermediate technology in China's rural industries. *World Development, 6*, 1297-1311.

Roberts, L. (1989, September 29). New game plan for genome mapping. *Science, 245*, 1438-1440.

Robertson, P. L. (1981). Employers and engineering education in Britain and the United States, 1890-1914. *Business History, 23,* 42-58.

Robinson, J. G., & McIlwee, J. S. (1992). *Women in engineering.* Albany: SUNY Press.

Rogers, Everett M. (1962). *The diffusion of innovations.* New York: Free Press/London: Collier-Macmillan.

Rogoff, B. (1990). *Apprenticeship in thinking: Cognitive development in social context.* New York: Oxford University Press.

Ronayne, Jarlath. (1978). Scientific research, science policy and social studies of science and technology in Australia. *Social Studies of Science, 8*(3), 361-384.

Ronayne, Jarlath. (1984). *Science in government.* Baltimore: Edward Arnold.

Rorty, Richard. (1979). *Philosophy and the mirror of nature.* Princeton, NJ: Princeton University Press.

Rorty, Richard. (1985). Habermas and Lyotard on postmodernity. In R. J. Bernstein (Ed.), *Habermas and modernity.* Cambridge: MIT Press.

Rosaldo, M. Z., & Lamphere, L. (Eds.). (1974). *Women, culture, and society.* Stanford, CA: Stanford University Press.

Rose, C. (1990). *The dirty man of Europe: The great British pollution scandal.* London: Simon & Schuster.

Rose, Hilary, & Rose, Steven. (1969). *Science and society.* Harmondsworth, Middlesex, U.K.: Penguin.

Rose, Hilary, & Rose, Steven. (Eds.). (1976a). *Ideology of/in the natural sciences: Vol. 1. The radicalisation of science.* London: Macmillan.

Rose, Hilary, & Rose, Steven. (Eds.). (1976b). *Ideology of/in the natural sciences: Vol. 2. The political economy of science.* London: Macmillan.

Rosen, B., & Jerdee, T. H. (1974). Influence of sex-role stereotypes on personnel decisions. *Journal of Applied Psychology, 59,* 9-14.

Rosen, S. P. (1991). *Winning the next war: Innovation and the modern military.* Ithaca, NY: Cornell University Press.

Rosenau, James N. (1968). The national interest. In *The international encyclopedia of the social sciences* (Vol. 11, pp. 34-40). New York: Crowell-Collier.

Rosenau, James N. (1980). *The study of global interdependence.* London: Frances Pinter.

Rosenau, James N. (1990). *Turbulence in world politics.* New York: Harvester Wheatsheaf.

Rosenau, James N. (1992). Normative challenges in a turbulent world. *Ethics & International Affairs, 6,* 1-19.

Rosenau, P. M. (1992). *Post-modernism and the social sciences.* Princeton, NJ: Princeton University Press.

Rosenberg, Nathan. (1976). *Perspectives on technology.* Cambridge: Cambridge University Press.

Rosenberg, Nathan. (1982). *Inside the black box: Technology and economics.* Cambridge: Cambridge University Press.

Rosenberg, Nathan. (1986). *Civilian spillovers from military R&D spending: The American experience since World War II.* Paper presented at the Conference on Technical Cooperation and International Competitiveness, Lucca, Italy.

Rosenberg, Nathan. (1990). Why do firms do basic research (with their own money)? *Research Policy, 19,* 165-174.

Rosenberg, Nathan, & Frischtak, Claudio. (Eds.). (1985). *International technology transfer: Concepts, measures and comparisons.* New York: Praeger.

Rosenberg, Nathan, & Vincenti, W. G. (1978). *The Britannia Bridge: The generation and diffusion of technological knowledge.* Cambridge: MIT Press.

Rosenfeld, Dave. (1984). Don't just reduce risk—transform it! *Radical Science Journal, 14,* 38-57.

Rosner, David, & Markowitz, Gerald. (1991). *Deadly dust: Silicosis and the politics of occupational disease in 20th century America.* Princeton, NJ: Princeton University Press.
Ross, Andrew. (1991). *Strange weather: Culture, science and technology in an age of limits.* London: Verso.
Rossi, Alice S. (1965). Women in science: Why so few? *Science, 148,* 1196-1202.
Rossi, Alice S. (1972). Barriers to the career choice of engineering, medicine, or science among American women. In J. M. Bardwick (Ed.), *Readings on the psychology of women* (pp. 72-82). New York: Harper & Row.
Rossiter, Margaret. (1982). *Women scientists in America: Struggles and strategies to 1940.* Baltimore, MD: Johns Hopkins University Press.
Roszak, Theodore. (1968). *The making of a counter culture.* New York: Doubleday.
Roszak, Theodore. (1974, Summer). The monster and the titan. *Daedalus,* p. 31.
Roszak, Theodore. (1986). *The cult of information.* New York: Pantheon.
Rothman, R. A. (1972). A dissenting view on the scientific ethos. *British Journal of Sociology, 23,* 102-108.
Rothschild, Joan. (Ed.). (1983). *Machina ex dea: Feminist perspectives on technology.* New York: Pergamon.
Rouse, Joseph. (1987). *Knowledge and power: Toward a political philosophy of science.* Ithaca, NY: Cornell University Press.
Rouse, Joseph. (1992). What are cultural studies of scientific knowledge? *Configurations, 1,* 1-22.
Rowan, K. E. (1988). A contemporary theory of explanatory writing. *Written Communication, 5,* 23-56.
Rowan, K. E. (1992). Strategies for enhancing comprehension of science. In B. V. Lewenstein (Ed.), *When science meets the public* (pp. 131-143). Washington, DC: American Association for the Advancement of Science.
Rowland, R. (1985). Motherhood, patriarchal power, alienation and the issue of "choice" in sex preselection. In G. Corea et al. (Eds.), *Man-made women: How new reproductive technologies affect women* (pp. 74-87). London: Hutchinson.
Royal Society. (1985). *The public understanding of science.* London: Author.
Royal Society. (1992). *Risk: Analysis, perception, management.* London: Author.
Rubenstein, E. (1973). Profiles in persistence. *IEEE Spectrum, 10,* 52-57, 60-64.
Rubin, Gayle. (1975). The traffic in women: Notes on the political economy of sex. In R. R. Reiter (Ed.), *Toward an anthropology of women.* New York: Monthly Review Press.
Rubinson, R., & Ralph, J. (1984). Technical change and the expansion of schooling in the United States, 1890-1970. *Sociology of Education, 57,* 144-152.
Rüdig, W. (1991). *Anti-nuclear movements: A global survey.* London: Longman.
Rudwick, Martin J. S. (1976). The emergence of a visual language for geological science, 1760-1840. *History of Science, 14,* 149-195.
Rudwick, Martin J. S. (1985). *The great Devonian controversy: The shaping of scientific knowledge among gentlemanly specialists.* Chicago: University of Chicago Press.
Ruggie, John G. (1992). Multilateralism: The anatomy of an institution. *International Organization, 3,* 561-598.
Rupke, N. (1987). Introduction. In N. Rupke (Ed.), *Vivisection in historical perspective* (pp. 1-18). London: Croom Helm.
Rushefsky, Mark E. (1986). *Making cancer policy.* Albany: State University of New York Press.
Russell, Nicholas. (1992, June 12). A boost blurring the edge. *Times Higher Education Supplement,* p. 17.

Ruttan, Vernon. (1989). The international agricultural research system. In J. L. Compton (Ed.), *The transformation of international agricultural research and development* (pp. 173-205). Boulder, CO: Lynne Rienner.

Ryan, M., & Dunwoody, S. L. (1975). Academic and professional training patterns of science writers. *Journalism Quarterly, 52,* 239-246, 290.

Rybczinski, Witold. (1991). *Paper heroes: Appropriate technology: Panacea or pipe dream.* Garden City, NY: Anchor/Doubleday.

Rycroft, R. W. (1983). International cooperation in science policy. The U.S. role in macroprojects. *Technology in Society, 5,* 51-68.

Rydell, R. W. (1984). *All the world's a fair: Visions of empire at American international expositions, 1876-1916.* Chicago: University of Chicago Press.

Rydell, R. W. (1985, December). The fan dance of science: American world's fairs in the Great Depression. *Isis, 76*(284), 525-542.

Sagasti, F. R. (1978). *Science and technology for development: Main comparative report of the science and technology policy instruments.* Ottawa, Canada: International Development Research Centre.

Saich, Tony. (1989). *China's science policy in the 80s.* Manchester, U.K.: Manchester University Press.

Said, Edward. (1978). *Orientalism.* New York: Pantheon.

Said, Edward. (1990). Figures, configurations transfigurations. *Race and Class, 32,* 1-16.

Salmen, Lawrence F. (1987). *Listen to the people: Participant-observer evaluation of development projects.* Oxford: Oxford University Press.

Salomon, Jean-Jacques. (1973). *Science and politics* (N. Lindsay, Trans.). Cambridge: MIT Press.

Salomon, Jean-Jacques. (1977a, October). Crisis of science, crisis of society. *Science and Public Policy,* pp. 414-433.

Salomon, Jean-Jacques. (1977b). Science policy studies and the development of science policy. In I. Spiegel-Rösing & D. Price (Eds.), *Science, technology and society: A cross-disciplinary perspective.* London: Sage.

Salomon, Jean-Jacques. (1990, June). *Science policy trends in industrially advanced countries.* Paper presented to the conference, "Science and Social Priorities: Perspectives of Science Policy for the 1990s," Prague.

Salter, Liora. (1988). *Mandated science: Science and scientists in the making of standards.* Dordrecht, the Netherlands: Kluwer Academic.

Samson, D., & Schiele, B. (1989). *L'evaluation museale publics et expositions: Bibliographie raisonnee.* Paris: Expo Media.

Samuelson, Pamela. (1987). Innovation and competition: Conflicts over intellectual property in new technologies. *Science, Technology, & Human Values, 11*(1), 6-21.

Sanders, Jo Schuchat, & Stone, Antonia. (1986). *The neuter computer: Computers for girls and boys.* New York: Neal-Schuman.

Sapolsky, Harvey M. (1977). Science, technology and military policy. In I. Spiegel-Rösing & D. Price (Eds.), *Science, technology and society: A cross-disciplinary perspective* (pp. 443-472). London: Sage.

Sapolsky, Harvey M. (1990). *Science & the navy: The history of the Office of Naval Research.* Princeton, NJ: Princeton University Press.

Sapp, Jan. (1986). Inside the cell: Genetic methodology and the case of the cytoplasm. In J. A. Schuster & R. R. Yeo (Eds.), *The politics and rhetoric of scientific method* (pp. 167-202). Dordrecht, the Netherlands: Reidel.

Sapp, Jan. (1987). *Beyond the gene.* Oxford: Oxford University Press.

Sardar, Ziauddin. (1977). *Science, technology, and development in the Muslin world*. London: Croom Helm.

Schäfer, W. (Ed.). (1983). *Finalization in science*. Dordrecht, the Netherlands: Reidel.

Schaffel, Kenneth. (1989). The US Air Force's philosophy of strategic defense: A historical overview. In S. J. Cimbala (Ed.), *Strategic air defense* (pp. 3-22). Wilmington, DE: Scholarly Resources.

Schaffer, Simon. (1983). Natural philosophy and public spectacle in the eighteenth century. *History of Science, 21*, 1-43.

Schaffer, Simon. (1986). Scientific discoveries and the end of natural knowledge. *Social Studies of Science, 16*, 387-420.

Schaffer, Simon. (1988). Astronomers mark time: Discipline and the personal equation. *Science in Context, 2*, 115-145.

Schaffer, Simon. (1989). Glass works, Newton's prisms and the uses of experiment. In D. Gooding, T. Pinch, & S. Schaffer (Eds.), *The uses of experiments: Studies in the natural sciences* (pp. 67-104). Cambridge: Cambridge University Press.

Schaffer, Simon. (1991). *Where experiments end: Table-top trial in victorian astronomy*. Unpublished manuscript, Cambridge.

Schatten, G., & Schatten, H. (1983). The energetic egg. *The Sciences, 23*(5), 28-34.

Schaub, J. H., & Pavlovic, K. (Ed.). (1983). *Engineering professionalism and ethics*. New York: Wiley.

Schelling, Thomas. (1960). *The strategy of conflict*. London: Oxford University Press.

Scherer, C. W., & Yarbrough, P. (1988, December 15-17). Media focus, personal dispositions, and activation of health risk reduction behavior: A longitudinal study. In *Proceedings of symposium on science communication: Environmental and health research*. Los Angeles: University of Southern California.

Schiebinger, Londa. (1989). *The mind has no sex? Women in the origins of modern science*. Cambridge, MA: Harvard University Press.

Schmandt, Jürgen, & Katz, James E. (1986). The scientific state: A theory with hypotheses. *Science, Technology, & Human Values, 11*(1), 40-52.

Schmaus, Warren. (1983). Fraud and the norms of science. *Science, Technology, & Human Values, 8*(4), 12-22.

Schmiegelow, H., & Schmiegelow, M. (1990). How Japan affects the international system. *International Organization, 44*, 553-588.

Schmookler, Jacob. (1966). *Invention and economic growth*. Cambridge, MA: Harvard University Press.

Schneider, I. (1987, August). The theory and practice of movie psychiatry. *American Journal of Psychiatry, 144*(8), 996-1002.

Schoeck, Helmut, & Wiggins, James W. (Eds.). (1960). *Scientism and values*. Princeton, NJ: D. Van Nostrand.

Scholzman, Kay Lehman, & Tierney, John T. (1986). *Organized interests and American democracy*. New York: Harper & Row.

Schön, Donald A. (1967). *Technology and change*. Oxford: Pergamon.

Schopman, P. (1987). Frames of artificial intelligence. In B. Bloomfield (Ed.), *The question of artificial intelligence* (pp. 165-219). London: Croom Helm.

Schot, J. (1992). Constructive technology assessment, opportunities for the control of technology: The case of clean technologies. *Science, Technology, & Human Values, 17*,36-56.

Schott, Thomas. (1988). International influence in science: Beyond center and periphery. *Social Science Research, 17*, 219-238.

Schott, Thomas. (1991). The world scientific community: Globality and globalisation. *Minerva, 29*, 440-462.

Schott, Thomas. (1993). World science: Globalization of institutions and participation. *Science, Technology, & Human Values, 18,* 196-208.

Schrecker, E. (1986). *No ivory tower: McCarthyism and the universities.* New York: Oxford University Press.

Schuler, Eugene. (1991). University patenting. In *SUNY research.* Albany: State University of New York Research Foundation.

Schumpeter, Joseph A. (1939). *Business cycles.* New York: McGraw-Hill.

Schumpeter, Joseph A. (1975). *Capitalism, socialism and democracy.* New York: Harper & Row (Original work published 1942)

Schumpeter, Joseph A. (1980). *The theory of economic development.* London: Oxford University Press. (Original work published 1934)

Schuster, John A., & Yeo, R. R. (Eds.). (1986). *The politics and rhetoric of scientific method.* Dordrecht, the Netherlands: Reidel.

Schwartz, Charles. (1975, June). *Public interest science: A critique* (Hearings on the National Science Policy and Organization Act of 1975, pp. 819-825). Washington, DC: 94th Congress, first session, U.S House of Representatives, Committee on Science and Technology.

Schwartzman, Simon. (1991). *A space for science: The development of the scientific community in Brazil.* University Park: Pennsylvania State University.

Schwarz, Michiel, & Thompson, Michael. (1990). *Divided we stand: Redefining politics, technology and social choice.* London: Harvester Wheatsheaf.

Science. (1987, April 17). Risk assessment [Special issue]. *Science, 236.*

Scott, James K. (1992). Exploring socio-technical analysis: Monsieur Latour is not joking! *Social Studies of Science, 22*(1), 33-57.

Scott, Pam, Richards, Evelleen, & Martin, Brian. (1990). Captives of controversy: The myth of the neutral social researcher in contemporary scientific controversies. *Science, Technology, & Human Values, 15,* 474-494.

Scott, R. L. (1976). On viewing rhetoric as epistemic: Ten years later. *Central States Speech Journal, 27,* 258-266.

Scott, W. Richard. (1987). The adolescence of institutional theory. *Administrative Science Quarterly, 32,* 493-511.

Scranton, Philip. (1991). Theory and narrative in the history of technology: Comment. *Technology and Culture, 32,* 385-393.

Searing, Susan E., & Apple, Rima D. (1987). *The history of women and science, health, and technology: A bibliographic guide to the professions and the disciplines.* University of Wisconsin System, Women's Studies.

Seashore Louis, K., et al. (1989). Entrepreneurs in academe: An explanation of behaviours among life scientists. *Administration Science Quarterly, 34,* 110-131.

Secord, James A. (1985). Newton in the nursery: Tom Telescope and the philosophy of tops and balls, 1761-1838. *History of Science, 23*(2), 127-151.

Seely, Bruce E. (1984). The scientific mystique in engineering: Highway research at the bureau of public roads, 1918-1940. *Technology and Culture, 25,* 798-831.

Seely, Bruce E. (1988). *Building the American highway system: Engineers as policy makers.* Philadelphia: Temple University Press.

Segal, L. (1987). *Is the future female? Troubled thoughts on contemporary feminism.* London: Virago.

Seidel, R. W. (1987). From glow to flow: A history of laser research and development. *Historical Studies in the Physical and Biological Sciences, 18*(pt. 1), 111-147.

Seitz, Frederick. (1982). The role of universities in the transnational interchange of science and technology for development. *Technology in Society, 4,* 33-40.

Semyonov, M., & Lewin-Epstein, N. (1986). Economic development, investment dependence and the rise of services in less developed nations. *Social Forces, 64*, 582-598.

Sen, B. K., & Lakshmi, V. V. (1992). Indian periodicals in the Science Citation Index. *Scientometrics, 23*, 291-318.

Sen, S., & Deger, S. (1990). The reorientation of military R&D for civilian purposes. In *Proceedings of the Fortieth Pugwash Conference on Science and World Affairs* (pp. 429-452). Egham, U.K.

Serrell, B. (Ed.). (1990). *What research says about learning in science museums*. Washington, DC: Association of Science-Technology Centers.

Serres, Michel. (1982). *Hermes: Literature, science, philosophy*. Baltimore: Johns Hopkins University Press.

SFS. (1982). *Care in society*. Oxford: Pergamon.

Shanker, S. G. (1987). AI at the crossroads. In B. Bloomfield (Ed.), *The question of artificial intelligence* (pp. 1-58). London: Croom Helm.

Shapin, Steven. (1974). The audience for science in eighteenth century Edinburgh. *History of Science, 12*, 95-121.

Shapin, Steven. (1979). The politics of observation: Cerebral anatomy and social interests in the Edinburgh phrenology disputes. In R. Wallis (Ed.), *On the margins of science: The social construction of rejected knowledge* (Sociological Review Monograph no. 27; pp. 139-178). London: Routledge & Kegan Paul.

Shapin, Steven. (1982). History of science and its sociological reconstructions. *History of Science, 20*, 157-211.

Shapin, Steven. (1984). Pump and circumstance: Robert Boyle's literary technology. *Social Studies of Science, 14*, 481-520.

Shapin, Steven A. (1988a). Following scientists around (review of Latour). *Social Studies of Science, 18*(3), 533-550.

Shapin, Steven. (1988b). The house of experiment in seventeenth-century England. *ISIS, 79*, 373-404.

Shapin, Steven. (1988c). Robert Boyle and mathematics: Reality, representation, and experimental practice. *Science in Context, 2*, 23-58.

Shapin, Steven. (1989). The invisible technician. *American Scientist, 77*, 553-563.

Shapin, Steven. (1990). Public understanding of science. In R. C. Olby et al. (Eds.), *Companion to the history of modern science*. London: Routledge & Kegan Paul.

Shapin, Steven. (1992a). Discipline and bounding: The history and sociology of science as seen through the externalism-internalism debate. *History of Science, 30*, 333-369.

Shapin, Steven A. (1992b). Why the public ought to understand science-in-the-making. *Public Understanding of Science, 1*(1), 27-30.

Shapin, Steven. (1993). Mertonian concessions. *Science, 259*, 839-841.

Shapin, Steven, & Barnes, Barry. (1977). Science, nature, and control: Interpreting mechanics institutes. *Social Studies of Science, 7*, 31-74.

Shapin, Steven, & Schaffer, Simon. (1985). *Leviathan and the air pump: Hobbes, Boyle and the experimental life*. Princeton, NJ: Princeton University Press.

Shapley, D., & Roy, R. (1985). *Lost at the frontier: U.S. science and technology policy adrift*. Philadelphia: ISI.

Sharf, B. F. (1986). Send in the clowns: The image of psychiatry during the Hinckley trial. *Journal of Communication, 36*(4), 80-93.

Sharif, M. Nawaz, & Sundararajan, V. (1984). Assessment of technological appropriateness: The case of Indonesian rural development. *Technological Forecasting and Social Change, 25*, 225-238.

Sharma, Dhirendra. (1983). *India's nuclear estate*. New Delhi, India: Lancer International.

Sharp, Andrew. (1963). *Ancient voyagers in Polynesia.* Sydney: Angus and Robertson.
Sharpe, R. (1988). *The cruel deception: The use of animals in medical research.* Wellingborough, U.K.: Thorsons.
Shattuck, J. (1991). National security information controls in the United States: Implications for international science and technology. In D. S. Zinberg (Ed.), *The changing university* (pp. 153-163). Dordrecht, the Netherlands: Kluwer.
Shaw, Anthony B. (1984). Impact of new technology on the Guyanese rice industry: Efficiency and equity considerations. *Journal of Developing Areas, 18,* 191-218.
Shaw, D. L., & Van Nevel, J. P. (1967). The informative value of medical science news. *Journalism Quarterly, 44*(3), 548.
Shayo, L. K. (1986). The transfer of science and technology between developed and developing countries through co-operation among institutions of higher learning. *Higher Education in Europe, 11,* 19-23.
Sheets-Pyenson, S. (1981). War and peace in natural history publishing: *The Naturalist's Library.* 1833-1843. *Isis, 72,* 50-72.
Sheets-Pyenson, S. (1985). Popular science periodicals in Paris and London: The emergence of a low scientific culture, 1820-1875. *Annals of Science, 42*(6), 549-572.
Shen, B. S. P. (1975). Science literacy and the public understanding of science. In S. Day (Ed.), *Communication of scientific information* (pp. 44-52). Basel: Karger.
Shenhav, Yehouda, & Kamens, David. (1991). The cost of institutional isomorphism: Science in less developed countries. *Social Studies of Science, 21,* 527-545.
Shepherd, R. G. (1979). Science news of controversy: The case of marijuana. *Journalism Monographs, 62.*
Shepherd, R. G. (1981). Selectivity of sources. *Journal of Communication, 31,* 129-137.
Sherwin, C., & Isenson, R. (1967, June 23). Project hindsight. *Science,* p. 1571.
Shils, Edward. (Ed.). (1968a). *Criteria for scientific development: Public policy and national goals.* Cambridge: MIT Press.
Shils, Edward. (1968b, June). The profession of science. *Advancement of Science,* pp. 469-480.
Shils, Edward. (1974). The public understanding of science. *Minerva, 12,* 153-158.
Shilts, Randy. (1987). *And the band played on: Politics, people, and the AIDS epidemic.* New York: Penguin.
Shinn, Terry. (1980a). Des Corps de l'Etat au Secteur Industriel: Genese de la Profession d'Ingenieur, 1750-1920. In R. Fox & G. Weisz (Eds.), *The organization of science and technology in France, 1808-1914.* Cambridge: Cambridge University Press.
Shinn, Terry. (1980b). The genesis of French industrial research, 1880-1940. *Social Science Information, 19,* 607-640.
Shinn, Terry. (1980c). *Savoir Scientifique et Pouvoir Sociale: L'Ecole Polytechnique, 1794-1914.* Paris: Presses de la Fondation Nationale des Sciences Politiques.
Shinn, Terry. (1984). Reactionary technologists: The struggle over the Ecole Polytechnique, 1880-1914. *Minerva, 22,* 329-345.
Shinn, Terry, & Paul, H. W. (1981-1982). The structure and state of science in France. *Contemporary French Civilization, 6,* 153-193.
Shinn, Terry, & Whitley, Richard. (Eds.). (1985). *Expository science: Forms and functions of popularisation.* Dordrecht, the Netherlands: Reidel.
Shishido, Toshio. (1983). Japanese industrial development and policies for science and technology. *Science, 219,* 259-264.
Shiva, Vandana. (1989). *Staying alive: Women, ecology and development.* London: Zed.
Shortland, Michael. (1987, October). Screen memories: Towards a history of psychiatry and psychoanalysis in the movies. *British Journal of the History of Science, 20,* 421-452.

Shortland, Michael. (1988). Wonder stories: Science and culture in alienland. *Science as Culture, 5.*

Shortland, Michael. (1989). *Medicine and film: A checklist, survey, and research resource.* Oxford: Wellcome Unit for the History of Medicine.

Shortland, Michael, & Gregory, J. (1991). *Communicating science.* Chichester, U.K.: Wiley.

Shove, Elizabeth. (1992). *Professional environments of building research.* Paper presented to the British Sociological Association Annual Conference, University of Kent.

Showalter, Elaine. (1970). *A literature of their own.* Princeton, NJ: Princeton University Press.

Shrum, Wesley. (1985a). *Networking of innovations in technological systems.* West Lafayette, IN: Purdue University Press.

Shrum, Wesley. (1985b). *Organized technology: Networks and innovation in technical systems.* West Lafayette, IN: Purdue University Press.

Shrum, Wesley. (1986). Are "science" and "technology" necessary? The utility of some old concepts in contemporary studies of the research process. *Sociological Inquiry, 56,* 324-340.

Shrum, Wesley, & Bankston, Carl. (1994). Organizational and geopolitical approaches to international science and technology networks. *Knowledge and Policy.*

Shrum, Wesley, Bankston, Carl, & Voss, D. Stephen. (1993). *Science and technology in less developed countries: An annotated bibliography, 1976-1992.* Metuchen, NJ: Scarecrow.

Shrum, Wesley, & Mullins, Nicholas. (1988). Network analysis in the study of science and technology. In A. van Raan (Ed.), *Handbook of quantitative studies in science and technology* (pp. 107-143). Amsterdam: Elsevier Science.

Shrum, Wesley, & Wuthnow, Robert. (1988). Reputational status of organizations in technical systems. *American Journal of Sociology, 93,* 882-912.

Siddarthan, N. S. (1988). In house R&D, imported technology, and firm size: Lessons from Indian experience. *Developing Economies, 26,* 212-221.

Silverstone, Roger. (1985). *Framing science: The making of a BBC documentary.* London: British Film Institute Books.

Silverstone, Roger, & Hirsch, E. (Eds.). (1992). *Consuming technologies.* London: Routledge & Kegan Paul.

Simmel, George. (1900). *Philosophes des Geldes.* Leipzig: Duncker and Humblot.

Simon, D., & Goldman, M. (Eds.). (1989). *Science and technology in post-Mao China.* Cambridge, MA: Harvard University Press.

Simon, Denis Fred. (1986). The challenge of modernizing industrial technology in China: Implications for Sino-U.S. relations. *Asian Survey, 26,* 420-439.

Simon, Herbert. (1991). Comment on the symposium, "Computer Discovery and the Sociology of Scientific Knowledge." *Social Studies of Science, 21,* 143-148.

Simons, Herb. (Ed.). (1989). *Rhetoric in the human sciences.* London: Sage.

Simons, Herb. (Ed.). (1990). *The rhetorical turn: Invention and persuasion in the conduct of inquiry.* Chicago: University of Chicago Press.

Sinclair, B. (1986). Local history and national culture: Notions on engineering professionalism in America. *Technology and Culture, 27,* 683-693.

Singh, Prithpal, & Krishnaiah, V. S. R. (1989). Analysis of work climate perceptions and performance in R&D units. *Scientometrics, 17,* 333-352.

Singleton, V., & Michael, M. (1993). Actor-networks and ambivalence in cancer screening. *Social Studies of Science, 23,* 227-264.

SIPRI Yearbook. (1991). *World armaments and disarmament.* Oxford: Oxford University Press.

Sismondo, Sergio. (1993). Some social constructions. *Social Studies of Science, 23*(3), 515-553.

Siwolop, S. (1981). Readership and coverage of science and technology in newspapers and magazines. In P. L. Alberger & V. L. Carter (Eds.), *Communicating university research* (pp. 197-205). Washington, DC: Council for the Advancement and Support of Education.

Skoie, H. (Ed.). (1979). *Scientific expertise and the public*. Oslo, Norway: Institute for Studies in Research and Higher Education.

Skolimowski, Henry. (1966). The structure of thinking in technology. *Technology and Culture, 7*, 371-383.

Skolnikoff, Eugene B. (1967). *Science, technology and American foreign policy*. Cambridge: MIT Press.

Skolnikoff, Eugene B. (1977). Science, technology and the international system. In I. Spiegel-Rösing & D. Price (Eds.), *Science, technology and society: A cross-disciplinary perspective* (pp. 507-533). Newbury Park, CA: Sage.

Skolnikoff, Eugene B. (1993). *The elusive transformation: Science, technology, and the evolution of international relations*. Princeton, NJ: Princeton University Press.

Sladovich, H. E. (1991). *Engineering as a social enterprise*. Washington, DC: National Academy Press.

Slezak, Peter. (1989). Scientific discovery by computer as empirical reputation of the strong program. *Social Studies of Science, 19*, 563-600.

Slinn, Judy. (1989). *Engineers in power: 75 years of the EPEA*. London: Lawrence and Wishart.

Smelser, Neil, & Content, R. (1980). *The changing academic market: General trends and a Berkeley case study*. Berkeley: University of California Press.

Smil, Vaclav. (1976). Intermediate energy technology in China. *World Development, 4*, 429-438.

Smit, Wim A. (1989). Defense technology assessment and the control of emerging technologies. In M. ter Borg & W. A. Smit (Eds.), *Non-provocative defence as a principle of arms reduction* (pp. 61-76). Amsterdam: Free University Press.

Smit, Wim A. (1991). Steering the process of military technological innovation. *Defense Analysis, 7*(4), 401-415.

Smit, Wim A., Grin, John, & Voronkov, Lev. (Eds.). (1992). *Military technological innovation and stability in a changing world: Politically assessing and influencing weapon innovation and military research and development*. Amsterdam: VU-Free University Press.

Smith, Alice Kimball. (1965). *A peril and a hope*. Chicago: University of Chicago Press.

Smith, Barbara. (Ed.). (1983). *Home girls: A black feminist anthology*. New York: Kitchen Table, Women of Color Press.

Smith, Bruce L. R. (1991). *American science policy since World War II*. Washington, DC: Brookings Institution.

Smith, Cecil O. (1990). The longest run: Public engineers and planning in France. *American History Review, 95*, 657-692.

Smith, Chris. (1987). *Technical workers: Class, labour and trade unionism*. London: Macmillan.

Smith, Chris. (1991). *How are engineers formed? Professionals, nation and class politics: Review article*. Birmingham: Aston University Business School.

Smith, Chris, & Meiksins, Peter F. (1992). *Engineering class politics*. London: Verso.

Smith, Dorothy. (1974). Women's perspective as a radical critique of sociology. *Sociological Inquiry, 44*, 7-13.

Smith, Keith. (1991). *Economic returns to R&D: Methods, results and challenges* (SPSG Review Paper No. 3). London: Science Policy Support Group.

Smith, Merritt Roe. (1977). *Harper's Ferry Armory and the new technology: The challenge of change*. Ithaca, NY: Cornell University Press.

Smith, Merritt Roe. (1985a). Army ordnance and the "American system" of manufacturing, 1815-1861. In M. R. Smith (Ed.), *Military enterprise and technological change: Perspectives on the American experience*. Cambridge: MIT Press.

Smith, Merritt Roe. (1985b). *Military enterprise and technological change*. Cambridge: MIT Press.

Smith, Robert W., & Tatarewicz, Joseph N. (1985). Replacing a technology: The large space telescope and CCDS. *Proceedings of the IEEE, 73*(7), 1221-1235.

Smith, Steve. (1989). Information technology in banks: Taylorization or human-centered systems? In T. Forester (Ed.), *Computers in the human context* (pp. 377-390). Cambridge: MIT Press.

Smith, S. S., & Deering, C. J. (1990). *1990 committees in Congress* (2nd ed.). Washington, DC: Congressional Quarterly.

Snow, C. P. (1954). *The new men*. London: Macmillan.

Snow, C. P. (1959). *The two cultures and the scientific revolution*. Cambridge: Cambridge University Press.

Snow, C. P. (1971). *Public affairs*. London: Macmillan.

Söderqvist, T. (1986). *The ecologists: From merry naturalists to saviours of the nation*. Stockholm: Almqvist and Wiksell.

Sohn-Rethel, A. (1978). *Intellectual and manual labor.* London: Macmillan.

Sonnert, Gerhart. (1990, August). *Careers of women and men postdoctoral fellows in the sciences.* Paper presented at meetings of the American Sociological Association.

Sorensen, Knut H., & Berg, Anne-Jorunn. (1987). Genderization of technology among Norwegian engineering students. *Acta Sociologica, 2*, 151-171.

Southwood, P. (1991). *Disarming military industries: Turning an outbreak of peace into an enduring legacy.* Houndmills, U.K.: Macmillan.

Soyland, A. J. (1991). Analyzing therapeutic and professional discourse. In *Therapeutic and everyday discourse as behavior change: Towards a microanalysis in psychotherapy process research.* Norwood, NJ: Ablex.

Spallone, Patricia, & Steinberg, Deborah Lynn. (Eds.). (1987). *Made to order: The myth of reproductive and genetic progress.* Oxford: Pergamon.

Spengler, Oswald. (1926). *The decline of the West* (Vol. 1). New York: International.

Spiegel-Rösing, Ina S. (1973). Science policy studies in a political context: Conceptual and institutional development of science policy studies in the German Democratic Republic. *Science Studies, 3*(4), 393-413.

Spiegel-Rösing, Ina S. (1977). The study of science, technology and society (SSTS): Current trends and future challenges. In I. Spiegel-Rösing & D. Price (Eds.), *Science, technology and society: A cross-disciplinary perspective*. London: Sage.

Spiegel-Rösing, Ina S., & Price, Derek de Solla. (1977). *Science, technology and society: A cross-disciplinary perspective*. London: Sage.

Squires, Arthur M. (1986). *The tender ship: Governmental management of technological change.* Boston: Birkhauser.

Stabile, Donald. (1984). *Prophets of order: The rise of the new class, technocracy and socialism in America.* Boston: South End.

Stamp, Patricia. (1989). *Technology, gender, and power in Africa.* Ottawa, Canada: International Development Research Centre.

Stanford spin-offs bring back congressional investigators. (1993). *Technology Access Report, 6*(6), 1.

Stankiewicz, R. (1986). *Academics and entrepreneurs: Developing university-industrial relations*. London: Frances Pinter.

Stankiewicz, R. (1990). Basic technologies and the innovation process. In J. Sigurdson (Ed.), *Measuring the dynamics of technological change* (pp. 13-38). London: Frances Pinter.

Stankiewicz, R. (1991). Technology as an autonomous socio-cognitive system. In H. Grupp (Ed.), *Technology at the cross-roads between science and innovation.* New York: Springer.

Stanley, A. (1992). *Mothers and daughters of invention: Notes for a revised history of technology.* Metuchen, NJ: Scarecrow.

Stanley, H. W., & Niemi, R. G. (1988). *Vital statistics on American politics*. Washington, DC: Congressional Quarterly Press.

Stanworth, Michelle. (Ed.). (1987). *Reproductive technologies: Gender, motherhood and medicine*. Cambridge: Polity.

Star, Susan Leigh. (1983). Simplification in scientific work: An example from neuroscience research. *Social Studies of Science, 13*, 205-228.

Star, Susan Leigh. (1985). Scientific work and uncertainty. *Social Studies of Science, 15*, 391-427.

Star, Susan Leigh. (1986). Triangulation clinical and basic research: British localizationists, 1870-1906. *History of Science, 24*, 29-48.

Star, Susan Leigh. (1988). The structure of ill-structured solutions: Boundary objects and heterogeneous distributed problem solving. In M. Huhns & L. Gasser (Eds.), *Distributed artificial intelligence* (pp. 37-54). Menlo Park, CA: Morgan Kauffman.

Star, Susan Leigh. (1989a). *Regions of mind: Brain research and the quest for scientific certainty*. Stanford, CA: Stanford University Press.

Star, Susan Leigh. (1989b). *Human beings as material for computer science, Or what computer science can't do*. Paper presented to meeting of Society for Social Studies of Science, Irvine, CA.

Star, Susan Leigh. (1990). Layered space, formal representations and long-distance control: The politics of information. *Fundamenta Scientiae, 10*, 125-155.

Star, Susan Leigh. (1991a). Invisible work and silenced dialogues in representing knowledge. In I. V. Eriksson, B. A. Kitchenham, & K. G. Tijdens (Eds.), *Women work and computerization: Understanding and overcoming bias at work and education* (pp. 81-92). Amsterdam: North Holland.

Star, Susan Leigh. (1991b). Power, technologies, and the phenomenology of conventions: On being allergic to onions. In J. Law (Ed.), *A sociology of monsters*. London: Routledge.

Star, Susan Leigh. (in press). The sociology of the invisible: The primacy of work in the writings of Anselm Strauss. In D. Maines (Ed.), *Social organization and social processes: Essays in honor of Anselm Strauss*. New York: Aldine de Gruyter.

Star, Susan Leigh, & Griesemer, James. (1989). Institutional ecology, "translations" and boundary objects: Amateurs and professionals in Berkeley's Museum of Vertebrate Zoology, 1907-1939. *Social Studies of Science, 19*, 387-420.

Starr, Paul. (1982). *The social transformation of American medicine*. New York: Basic Books.

Staudenmaier, John M. (1985). *Technology storytellers: Reweaving the human fabric*. Cambridge: MIT Press.

Stehr, Nico. (1976). The ethos of science revisited: Social and cognitive norms. *Sociological Inquiry, 48*(3-4), 173-196.

Stein, D. (1985). *Ada: A life and legacy*. Cambridge: MIT Press.

Stemerding, Dirk. (1991). *Plants, animals and formulae: Natural history in the light of Latour's Science in Action and Foucault's The Order of Things*. Unpublished doctoral dissertation, University of Twente, Enschede.

Stepan, Nancy. (1986). Race and gender: The role of analogy in science. *Isis, 77*, 261-277.

Stern, Bernhard. (1956). *Historical sociology*. New York: Citadel.

Stern, Richard. (1988, February). Reflections on Dirty Harry and dBase. *IEEE Micro*, pp. 70-72.

Stevenson, R., Jr. (1980). *Corporations and information*. Baltimore: Johns Hopkins Press.

Stewart, A. (1991). Ethical and social implications of the human genome project. *Science and Public Policy, 18*, 123-129.

Stewart, Frances. (1977). *Technology and underdevelopment*. Boulder, CO: Westview.

Stewart, Frances. (1987). The case for appropriate technology. *Issues in Science and Technology, 3*(4), 101-109.

Stine, Jeffrey. (1986). *A history of science policy in the United States, 1940-1985*. Washington, DC: Government Printing Office.

Stocking, George W., Jr. (Ed.). (1983). *History of anthropology: Vol. 1. Observers observed: Essays on ethnographic fieldwork*. Madison: University of Wisconsin Press.

Stone, C. D. (1988). *Should trees have standing? Toward legal rights for natural objects*. Portola Valley, CA: Tioga.

Stopford, J., & Strange, S. (1991). *Rival states, rival firms: Competition for world market shares*. Cambridge: Cambridge University Press.

Storer, Norman. (1966). *The social system of science*. New York: Holt, Rinehart & Winston.

Storer, Norman. (1973). *The sociology of science*. Chicago: University of Chicago Press.

Strasser, S. (1982). *Never done: A history of American housework*. New York: Pantheon.

Strathern, M., & Franklin, S. (1993). *Kinship and the new genetic technologies: An assessment of existing research* (Report to the Commission of the European Communities, DG-XII, Human Genome Analysis Programme). Brussels: Commission of the European Communities.

Straubhaar, Joseph. (1989). Television and video in the transition from military to civilian rule in Brazil. *Latin American Research Review, 24*, 140-154.

Strauss, Anselm. (1978). A social worlds perspective. *Studies in Symbolic Interaction, 1*, 119-128.

Strauss, Anselm. (1982). Social worlds and legitimation processes. *Studies in Symbolic Interaction, 4*, 171-190.

Strauss, K. (1988). Engineering ideology. *IEEE Proceedings, 135*(A-5), 261-265.

Strickland, Stephen P. (1972). *Politics, science, and dread disease: A short history of United States medical research policy*. Cambridge, MA: Harvard University Press.

Strickland, Stephen P. (1989). *The story of the NIH grants program*. Lanham, MD: University Press of America.

Stringer, Peter. (1985). You decide what your title is to be and [read] write to that title. In D. Bannister (Ed.), *Issues and approaches in personal construct theory*. London: Academic Press.

Studer, Kenneth E., & Chubin, Daryl E. (1980). *The cancer mission: Social contexts of biomedical research*. Beverly Hills, CA: Sage.

Sturtevant, W. (1964). Studies in ethnoscience. *American Anthropologist, 66*, 99-131.

Suchman, Lucy. (1987). *Plans and situated actions: The problem of human-machine communication*. Cambridge: Cambridge University Press.

Sultan, P. E. (1988). Passage on the rope bridge between science and technology: Tales of valor with virtue and vanity with vertigo. *Technological Forecasting and Social Change, 34*(3), 213-230.

Sun, M. (1987). Strains in U.S.-Japan exchanges. *Science, 237*, 476-478.

Sun, M. (1989, March 24). Japan lays out welcome-mat for U.S. scientists. *Science*, pp. 1546-1547.

Sundquist, J. (1981). *The decline and resurgence of Congress*. Washington, DC: Brookings Institution.

Survey Research Center. (1958). *The public impact of science in the mass media*. Ann Arbor: University of Michigan, Survey Research Center.

Survey Research Center. (1959). *Satellites, science, and the public*. Ann Arbor: University of Michigan, Institute for Social Research, Survey Research Center.

Sussex Group. (1971). Science in underdeveloped countries. *Minerva, 9*, 101-121.

Susskind, Laurence, & Weinstein, Alan. (1980). Towards a theory of environmental dispute resolution. *Environmental Affairs, 9*, 311-356.

Sussman, Gerald. (1987). Banking on telecommunications: The World Bank in Philippines. *Journal of Communication, 37*, 90-105.

Suttmeier, Richard P. (1985). Corruption in science: The Chinese case. *Science, Technology, & Human Values, 10*, 49-61.

Sutton, C. (1992). *Words, science and learning*. Milton Keynes, U.K.: Open University Press.

Tannen, Deborah. (1990). *You just don't understand: Women and men in conversation*. New York: Morrow.

Tatarewicz, Joseph N. (1990). *Space technology and planetary astronomy*. Bloomington: Indiana University Press.

Tauber, A. I., & Sarkar, S. (1992). The human genome project: Has blind reductionism gone too far? *Perspectives in Biology and Medicine, 35*, 220-235.

Taylor, Calvin W., & Barron, Frank. (Eds.). (1963). *Scientific creativity*. New York: Wiley.

Teague, O. (1972, February 8). [Statement from the 92nd Congress, 2nd Session]. *Congressional Record*, p. 3200.

Technology Review. (1984). Women in technology [Special section]. *Technology Review, 87*(8), 29-52.

Teece, David J. (1981). The market for know-how and the efficient international transfer of technology. *Annals of the American Academy of Political and Social Science, 458*, 81-96.

Teich, Albert H., & Gold, Barry D. (1986). Education in science, engineering and public policy: A stocktaking. *Social Studies of Science, 16*(4), 685-704.

Teitel, Simon. (1977). On the concept of appropriate technology for less developed countries. *Technological Forecasting and Social Change, 11*, 349-370.

Tesler, Lawrence G. (1991, September). Networked computing in the 1990s. *Scientific American, 265*(3), 86-93.

Teubal, Morris. (1984). The role of technological learning in the exports of manufactured goods: The case of selected capital goods in Brazil. *World Development, 12*, 849-865.

Thackray, Arnold, Sturchio, J., Bud, Robert, & Carroll, Thomas. (1985). *Chemistry in America: Historical indicators*. Dordrecht, the Netherlands: Reidel.

Thagard, P. (1989). Welcome to the cognitive revolution. *Social Studies of Science, 19*, 653-657.

Thee, M. (1986). *Military technology, military strategy, and the arms race*. London: Croom Helm/New York: St. Martin's Press.

Thee, M. (1988). Science and technology for war and peace. *Bulletin of Peace Proposals, 19*(3), 261-292.

Theocharis, T., & Psimipoulis, M. (1987, October 15). Where science has gone wrong. *Nature, 329*, 595-598 (and ensuing correspondence in *330* [1987], 308, 689-690; *331* [1988], 129-131, 204, 384, 558; and the authors' reply, *332* [1988], 389).

Thill, G. (1972). *La Fête Scientifique*. Paris: Institut Catholique de Paris.

Thoburn, John T. (1977). Commodity prices and appropriate technology: Some lessons from tin mining. *Journal of Development Studies, 14*, 35-52.

Thomas, G., & Durant, John. (1987). Why should we promote the public understanding of science. In M. Shortland (Ed.), *Science literacy papers* (pp. 1-14). Oxford: University of Oxford Science Literacy Group.

Thomas, S. (1990). Science in public. [Special issue] *Critical Studies of Mass Communication, 7*(1).

Thompson, E., & Smith, D. (Eds.). (1980). *Protest and survive*. Harmondsworth, U.K.: Penguin.

Thompson, G. (1984). The genesis of nuclear power. In J. Tirman (Ed.), *The militarization of high technology* (pp. 63-75). Cambridge, MA: Ballinger.

Thompson, P. (1983). *The nature of work: An introduction to debates on the labour process*. London: Macmillan.

Thurber, J. A. (1977). Policy analysis on Capitol Hill: Issues facing the four analytic support agencies of Congress. *Policy Sciences Journal, 6*, 101-111.

Tichi, C. (1987). *Shifting gears: Technology, literature, culture in modernist America*. Chapel Hill: University of North Carolina Press.

Timoshenko, Stephen P. (1953). *History of strength of materials*. New York: McGraw-Hill.

Tobey, R. (1971). *The American ideology of national science*. Pittsburgh: University of Pittsburgh Press.

Todhunter, Isaac. (1960). *A history of the theory of elasticity and the strength of materials, from Galileo to the present*. London: Dover.

Torstendahl, Rolf. (1982a). Engineers in industry, 1850-1910: Professional men and new bureaucrats: A comparative perspective. In C. G. Bernhard, E. Crawford, & P. Sorbom (Eds.), *Science, technology and society in the time of Alfred Nobel* (pp. 253-270). Oxford: Pergamon.

Torstendahl, Rolf. (1982b). *Dispersion of engineers in a traditional society: Swedish technicians 1860-1940*. Uppsala, Sweden: Studia Historica Uppsaliensa.

Torstendahl, Rolf. (1985). Engineers in Sweden and Britain 1820-1914: Professionalization and bureaucratization in comparative perspective. In W. Conze & J. Kocka (Eds.), *Bildungsburgertum in 19e Jahrhundert*. Stuttgart: Klett-Cotta.

Touraine, Alain. (1980). *La Prophecie Anti-Nucleaire*. Paris: Edition du Seuil.

Trachtman, L. E. (1981). The public understanding of science effort: A critique. *Science, Technology, & Human Values, 6*(36), 10-15.

Trak, Ayse, & MacKenzie, Michael. (1980). Appropriate technology assessment: A note on policy considerations. *Technological Forecasting and Social Change, 17*, 329-388.

Travis, H. David L. (1981). Replicating replication? Aspects of the social construction of learning in planarian worms. *Social Studies of Science, 11*(1), 11-32.

Traweek, Sharon. (1988). *Beamtimes and lifetimes: The world of high energy physicists*. Cambridge, MA: Harvard University Press.

Traweek, Sharon. (1992). Border crossings: Narrative strategies in science studies and among physicists in Tsukuba Science City, Japan. In A. Pickering (Ed.), *Science as practice and culture*. Chicago: University of Chicago Press.

Trescott, Martha M. (Ed.). (1979a). *Dynamos and virgins revisited: Women and technological change in history*. Metuchen, NJ: Scarecrow.

Trescott, Martha M. (1979b). A history of women engineers in the United States, 1850-1975: A progress report. In *Proceedings of the Society of Women Engineers 1979 National Convention* (pp. 1-14). New York: Society of Women Engineers.

Trescott, Martha M. (1982). Women in the intellectual development of engineering: Studies in persistence and holism. In G. Kass-Simon & P. Farnes (Eds.), *Intellectual history of women in science*. Stanford, CA: Stanford University Press.

Trescott, Martha M. (1984). Women engineers in history: Profiles in holism and persistence. In V. B. Hass & C. C. Perrucci (Eds.), *Women in scientific and engineering professions* (pp. 181-204). Ann Arbor: University of Michigan Press.

Trilling, L. (1988). Styles of military technical development: Soviet and U.S. jet fighters—1945-1960. In E. Mendelsohn, M. R. Smith, & P. Weingart (Eds.), *Science, technology and the military* (Sociology of Sciences Yearbook, Vol. XII/1, pp. 155-185). Dordrecht, the Netherlands: Kluwer Academic.

Troitzsch, U., & Wohlauf, G. (1980). *Technik-Geschichte: Historische Beiträge unde neuere Ansätze*. Frankfurt am Main: Suhrkamp.

Tuana, Nancy. (Ed.). (1987). *Feminism and science*. Bloomington: Indiana University Press.

Tuckman, Howard. (1976). *Publication, teaching, and the academic reward structure*. Lexington, MA: Lexington.

Tuckman, Howard, & Hagemann, R. (1976). An analysis of the reward structure in two disciplines. *Journal of Higher Education, 47*, 447-464.

Tudor, Andrew. (1989). *Monsters and mad scientists: A cultural history of the horror movie*. Oxford: Blackwell.

Turing, Alan M. (1936). On computable numbers. *Proceedings of the London Mathematical Society, 42*(2), 230-265.

Turing, Alan M. (1950). Computing machinery and intelligence. *Mind, 59*(236), 433-460. (Reprinted in D. Hofstadter & D. Dennet, Eds., 1982, *The mind's I*, pp. 53-66, Harmondsworth: Penguin)

Turkle, Sherry. (1984). *The second self: Computers and the human spirit.* London: Granada.

Turkle, Sherry, & Papert, Seymour. (1990). Epistemological pluralism: Styles and voices within the computer culture. *Signs, 16*(1), 128-157.

Turnbull, David. (1989). The push for a malaria vaccine. *Social Studies of Science, 19*, 283-300.

Turnbull, David. (1991). *Mapping the world in the mind: An investigation of the unwritten knowledge of the Micronesian navigators.* Geelong, Australia: Deakin University Press.

Turnbull, David. (1993). The ad hoc collective work of building Gothic cathedrals with templates, string, and geometry. *Science, Technology, & Human Values, 18* (3), 315-340.

Turner, J., & Wynne, Brian. (1992). Risk communication: A literature review and some implications for biotechnology. In J. Durant (Ed.), *Biotechnology in public* (pp. 109-141). London: Science Museum.

Turner, Stephen P. (1990). Forms of patronage. In Susan E. Cozzens & Thomas F. Gieryn (Eds.), *Theories of science in society* (pp. 185-211). Bloomington: Indiana University Press.

Turner, Terisa. (1977). Two refineries: A comparative study of technology transfer to the Nigerian refining industry. *World Development, 5*, 235-256.

Turow, J. (1989). *Playing doctor: Television, storytelling and medical power*. New York: Oxford University Press.

Tybout, Richard A. (Ed.). (1965). *Economics of research and development*. Columbus: Ohio State University Press.

Tyler, Stephen. (1987). *The unspeakable*. Madison: University of Wisconsin Press.

UNESCO. (1988). *Engineering schools and endogenous technology development*. Paris: Author.

Unger, S. H. (1982). *Controlling technology: Ethics and the responsible engineer*. New York: Holt, Rinehart & Winston.

Unger, S. H. (1989). Engineering ethics and the question of whether to work on military projects. *Annals of the New York Academy of Science, 577*, 211-215.

U.K. Cabinet Office. (1990a). *Annual review of government funded research and development 1990*. London: Her Majesty's Stationery Office.

U.K. Cabinet Office. (1990b). *This common inheritance: Britain's environmental strategy.* London: Her Magesty's Stationery Office.

U.K. Cabinet Office, Advisory Council on Science and Technology. (1991). *Science and technology: Education and employment*. London: Her Majesty's Stationery Office.

U.K. Committee on Manpower Resources for Science and Technology. (1968). *The flow into employment of scientists, engineers and technologists (The Swann Report)* (cmnd 3760). London: Her Majesty's Stationery Office.

U.K. Parliamentary Office of Science and Technology. (1991). *Relationships between defence and civil science and technology.* London: Author.

United Nations. (1987). *The military use of R&D: A report pursuant to General Assembly Resolution 37/99J*. New York: U.N. Department of Disarmament Affairs.

U.S. Bureau of the Census. (1988). *Recent data on scientists and engineers in industrialized countries*. Washington, DC: U.S. Department of Commerce, Bureau of the Census, Center for International Research.

U.S. Department of Agriculture. (1979). *Popular reporting of agricultural science: Strategies for improvement* (Proceedings of the National Agricultural Science Information Conference). Ames: Iowa State University.

U.S. House of Representatives, Committee on Science and Technology, Subcommittee on Science, Research, and Technology. (1979, October 10). *Hearing* (96th Congress, 1st Session). Washington, DC: Government Printing Office.

U.S. House of Representatives, Committee on Science and Astronautics. (1971, August 16). *Establishing the Office of Technology Assessment and amending the National Science Foundation Act of 1950* (92nd Congress, 1st Session; House Report 92-469). Washington, DC: Government Printing Office.

U.S. House of Representatives, Committee on Energy and Commerce, Subcommittee on Oversight and Investigations. (1988, April 12). *Fraud in NIH grant programs* (100th Congress, 2nd Session). Washington, DC: Government Printing Office.

U.S. House of Representatives, Committee on Energy and Commerce, Subcommittee on Oversight and Investigations. (1989, May 4, 9). *Scientific fraud* (101st Congress, 1st Session). Washington, DC: Government Printing Office.

U.S. House of Representatives, Committee on Energy and Commerce, Subcommittee on Oversight and Investigations. (1990, April 30, May 14). *Scientific fraud* (101st Congress, 2nd Session). Washington, DC: Government Printing Office.

U.S. House of Representatives, Committee on Energy and Commerce, Subcommittee on Oversight and Investigations. (1991, March 6). *Scientific fraud* (102nd Congress, 1st Session). Washington, DC: Government Printing Office.

U.S. House of Representatives, Committee on Science and Technology, Subcommittee on Science, Research, and Technology. (1979, October 10). *Hearing* (96th Congress, 1st Session). Washington, DC: Government Printing Office.

U.S. National Science Foundation, Office of Economic & Manpower Studies, Division of Science Resources & Policy Studies. (1971). *A review of the relationship between research and development and economic growth/productivity*. Washington, DC: Author.

U.S. National Science Foundation. (1977). *Women and minorities in science and engineering* (NSF 77-304). Washington, DC: Author.

U.S. Senate, Committee on Rules and Administration, Subcommittee on Computer Services. (1972, March 2). *Technology assessment for the Congress* (92nd Congress, 2nd Session). Washington, DC: Government Printing Office.

Usher, A. P. (1954). *A history of mechanical inventions*. Cambridge, MA: Harvard University Press.

Valenstein, Elliot. (Ed.). (1980). *The psychosurgery debate*. New York: Freeman.

Valley, George E., Jr. (1985). How the SAGE development began. *Annals of the History of Computing, 7*(3), 196-226.

Van den Berghe, W. (1986). *Engineering manpower*. Paris: UNESCO.

van den Daele, W. (1977). The social construction of science: Institutionalisation and definition of positive science in the latter half of the seventeenth century. In E. Mendelsohn et al. (Eds.), *The social production of scientific knowledge* (Sociology of the Sciences Yearbook). Dordrecht, the Netherlands: Reidel.

Van Tassel, David D., & Hall, Michael G. (Eds.). (1966). *Science and society in the United States*. Homewood, IL: Dorsey.

Varas, Augusto, & Bustamente, Fernando. (1983). The effect of R&D transfer of military technology to the Third World. *International Social Science Journal, 35*, 141-161.

Vasantha, A. (1992). *Public perception of science: An exploratory study in a developing society* (mimeo). New Delhi, India: Nehru University.

Vaughan, D. K. (1990). The image of the engineer in the popular imagination, 1880-1980. *Bulletin of Science, Technology and Society, 10,* 301-304.

Vayrynen, Raimo. (1978). International patenting as a means of technological dominance. *International Social Science Journal, 30,* 315-338.

Veen, J. van. (1962). *Dredge drain reclaim: The art of a nation* (5th ed.). The Hague, the Netherlands: Martinus Nijhoff.

Vega, Garcilaso de la. (1961). *The Incas: The royal commentaries of the Inca* (M. Jolas, Trans.). New York: Avon.

Veggeberg, Scott. (1992, November 23). NIH women's health study takes a giant step forward. *The Scientist, 1.*

Velho, Lea. (1986). The meaning of citation in the context of a scientifically peripheral country. *Scientometrics, 9,* 71-89.

Vernant, Jean-Pierre. (1990). La formation de la pensée positive dans la Grèce archaïque. In J. Vernant & P. Vidal-Naquet (Eds.), *La Grèce ancienne* (pp. 196-228). Paris: Seuil.

Vessuri, Hebe M. C. (1980). Technological change as the social organization of agricultural production. *Current Anthropology, 21,* 315-327.

Vessuri, Hebe M. C. (1986). Universities, scientific research, and the national interest in Latin America. *Minerva, 24,* 1-36.

Vessuri, Hebe M. C. (1987). The social study of science in Latin America. *Social Studies of Science, 17*(3), 519-554.

Vetter, Betty M. (1981). Women scientists and engineers: Trends in participation. *Science, 214,* 1313-1321.

Vien, Nguyen Khac. (1979). The scientific and technical revolution in the socialist Republic of Vietnam. *Impact of Science on Society, 29,* 241-246.

Vig, Norman J. (1992). Parliamentary technology assessment in Europe: A comparative perspective. In *Science, technology, and politics: Policy analysis in Congress* (pp. 209-226). Boulder, CO: Westview.

Vig, Norman J., & Kraft, Michael. (Eds.). (1984). *Environmental policy in the 1980s.* Washington, DC: CQ Press.

Vincenti, Walter G. (1979). The air-propeller tests of W. F. Durand and E. P. Lesley: A case study in technological methodology. *Technology and Culture, 20,* 712-751.

Vincenti, Walter G. (1982). Control-volume analysis: A difference in thinking between engineering and physics. *Technology and Culture, 23,* 145-174.

Vincenti, Walter G. (1984). Technological knowledge without science: The innovation of flush riveting in American airplanes, ca. 1930-ca. 1950. *Technology and Culture, 25,* 540-576.

Vincenti, Walter G. (1986). The Davis wing and the problems of airfoil design: Uncertainty and growth in engineering knowledge. *Technology and Culture, 27,* 717-758.

Vincenti, Walter G. (1988). How did it become "obvious" that an airplane should be inherently stable? *American Heritage of Invention and Technology, 4,* 50-56.

Vincenti, Walter G. (1990). *What engineers know and how they know it: Analytical studies from aeronautical history.* Baltimore, MD: Johns Hopkins University Press.

Vinck, Dominique. (1991). *La Coordination du travail scientifique. Etude de deux formes spécifiques: le laboratoire et le réseau.* Unpublished doctoral dissertation, University of Paris.

Vinck, Dominique, Kahane, B., Larédo, Philippe, & Meyer, J. (1993). A network approach to studying research programmes: Mobilizing and coordinating public responses to HIV/AIDS. *Technology Analysis and Strategic Management, 5*(1), 39-54.

Visvanathan, Shiv. (1985). *Organizing for science: The making of an industrial research laboratory.* Delhi, India: Oxford University Press.

Vivian, R. Gwinn. (1974). Conservation and diversion: Water-control systems in the Anasazi Southwest. In T. E. Downing & M. Gibson (Eds.), *Irrigation's impact on society* (pp. 95-112). Tucson: University of Arizona Press.

Vogel, David. (1986). *National styles of regulation: Environmental policy in Great Britain and the United States*. Ithaca, NY: Cornell University Press.

Vogel, David. (1989). *Fluctuating fortunes: The political power of business in America*. New York: Basic Books.

Vogel, Ezra. (1991). *The four little dragons: The spread of industrialization in East Asia*. Cambridge, MA: Harvard University Press.

Volti, Rudi. (1982). *Technology, politics and society in China*. Boulder, CO: Westview.

Von Braun, Joachim. (1988). Effects of technological change in agriculture on food consumption and nutrition: Rice in a West African setting. *World Development, 16*, 1083-1098.

Vos, R. (1991). *Drugs looking for diseases: Innovative drug research and the development of the beta blockers and the calcium antagonists*. Dordrecht, the Netherlands: Kluwer Academic.

Vries, G. de. (1992). *Wittgenstein and the sociology of scientific knowledge: Consequences to a farewell to epistemology* (mimeo).

Wad, Atul. (Ed.). (1988). *Science, technology, and development*. Boulder, CO: Westview.

Waddell, C. (1989). Reasonableness versus rationality in the construction and justification of science policy decisions: The case of the Cambridge experimentation review board. *Science, Technology, & Human Values, 14*, 7-25.

Wade, Nicholas. (1981). *The Nobel duel: Two scientists' 21-year race to win the world's most coveted research price*. Garden City, NY: Anchor/Doubleday.

Wade, S., & Schramm, W. (1969). The mass media as sources of public affairs, science, and health knowledge. *Public Opinion Quarterly, 33*(2), 197-209.

Wainwright, H., & Elliott, D. (1982). *The Lucas plan: A new trade unionism in the making?* London: Allison and Busby.

Wajcman, Judy. (1991). *Feminism confronts technology*. University Park: Pennsylvania State University Press.

Wakeland, H. L. (1990). International education for engineers: A working model. *Annals AAPSS, 511*, 122-131.

Walentynowicz, Bohdan. (1975). The science of science in Poland: Present state and prospects of development. *Social Studies of Science, 5*(2), 213-222.

Walgate, R. (1990). *Miracle or menace? Biotechnology and the Third World*. London: Panos Institute.

Walker, Charles, et al. (1983). *Too hot to handle? Social and policy issues in the management of radioactive wastes*. New Haven, CT: Yale University Press.

Walker, T. (1988). Whose discourse? In S. Woolgar (Ed.), *Knowledge and reflexivity* (pp. 55-79). London: Sage.

Walker, Thomas S., & Kshirsagar, K. G. (1985). The village impact of machine-threshing and implications for technology development in the semi-arid tropics of peninsular India. *Journal of Development Studies, 21*, 215-231.

Walker, W., Graham, M., & Harbor, B. (1988). From components to integrated systems: Technological diversity and integration between military and civilian sectors. In P. Gummett & J. Reppy (Eds.), *The relations between defence and civil technologies* (pp. 17-37). Dordrecht, the Netherlands: Kluwer Academic.

Walkerdine, V. (1988). *The mastery of reason: Cognitive development and the production of rationality*. London: Routledge.

Wallerstein, M. B. (Ed.). (1984). *Scientific and technological cooperation among industrialized countries: The role of the United States*. Washington, DC: National Academy Press.

Wallis, Roy. (Ed.). (1979). *On the margins of science: The social construction of rejected knowledge* (Sociological Review Monograph). Keele, U.K.: University of Keele.

Wallis, Roy. (1985). Science and pseudo-science. *Social Science Information, 24,* 585-601.

Walters, L. M., Wilkins, L., & Walters, T. (Eds.). (1989). *Bad tidings: Communication and catastrophe*. Hillsdale, NJ: Lawrence Erlbaum.

Wang, Jian-Ye. (1990). Growth, technology transfer, and the long-run theory of international capital movements. *Journal of International Economics, 29,* 255-271.

Warner, Aaron W., Morse, Dean, & Eichner, Alfred S. (Eds.). (1965). *The impact of science on technology*. New York: Columbia University Press.

Washington, R. O. (1984). Designing a management information system in the Arab Republic of Egypt: A case study of factors influencing technology transfer in Third World countries. *Social Development Issues, 8,* 158-171.

Wasser, H. (1990, Spring). Changes in the European university. *Higher Education Quarterly*.

Watson, G., & Seiler, R. (Eds.). (1992). *Text in context*. London: Sage.

Watson, Helen. (1990). Investigating the social foundations of mathematics: Natural number in diverse forms of life. *Social Studies of Science, 20,* 283-312.

Watson, Helen, with the Yolngu Community at Yirrkala, & Chambers, D. W. (1989). *Singing the land, signing the land.* Geelong, Australia: Deakin University Press.

Watson, Helen, et al. (1989). Australian aboriginal maps. In D. Turbull (Ed.), *Maps are territories*. Geelong, Australia: Deakin University Press.

Watson, James D. (1968). *The double helix.* London: Weidenfeld & Nicolson/New York: Athaneum.

Watson, James D., & Jordan, E. (1989). The human genome program at the National Institutes of Health. *Genomics, 5,* 654-656.

Watson, James M., & Meiksins, Peter F. (1991). What do engineers want? Work values, job rewards, and job satisfaction. *Science, Technology, & Human Values, 16,* 140-172.

Way, P. O., & Jamison, E. (1986). *Scientists and engineers in industrialized countries*. Washington, DC: U.S. Department of Commerce, Bureau of the Census, Center for International Research.

Weart, Spencer R. (1988). *Nuclear fear: A history of images*. Cambridge, MA: Harvard University Press.

Weaver, Warren. (1955, December 30). Science and people. *Science, 122,* 1255-1259.

Weber, Max. (1947). *Theory of economic and social organization*. Glencoe: Free Press.

Weber, Max. (1948). Science as a vocation. In H. Gerth & C. Wright Mills (Eds.), *From Max Weber*. New York: Oxford University Press.

Weber, Max. (1958). *The Protestant ethic and the spirit of capitalism*. New York: Scribner. (Original work published 1904-1905)

Webster, A. (1990). Institutional stability: Engineering an environment for biotechnology. *Science and Public Policy, 17*(5).

Webster, Andrew, & Etzkowitz, Henry. (1991). *Academic-industry relations: The second academic revolution?* (SPSG Concept Paper No. 12). London: Science Policy Support Group

Webster, J. (1989). *Office automation: The labour process and women's work in Britain.* Hemel Hempstead, U.K.: Wheatsheaf.

Webster, S. (1982). Dialogue and fiction in ethnography. *Dialectical Anthropology, 7,* 91-114.

Weil, Vivian, & Snapper, J. D. (Eds.). (1989). *Owning scientific and technical information*. New Brunswick, NJ: Rutgers University Press.

Weimer, W. B. (1977). Science as a rhetorical transaction: Toward a nonjustificational conception of rhetoric. *Philosophy and Rhetoric, 10,* 1-29.

Weinberg, Alvin M. (1967). *Reflections on big science*. Cambridge: MIT Press.

Weinberg, Alvin M. (1972). Science and trans-science. *Minerva, 10,* 209-222.

Weinberg, Alvin M. (1989-1990). Engineering in an age of anxiety. *Issues in Science and Technology, 6*, 37-43.
Weinberg, Sandy. (1990). Expanding access to technology: Computer equity for women. In A. H. Teich (Ed.), *Technology and the future* (pp. 277-287). New York: St. Martin's Press.
Weingart, Peter. (1982). The scientific power elite—a chimera: The deinstitutionalization and politicization of science. In N. Elias et al. (Eds.), *Scientific establishments and hierarchies* (Sociology of the Sciences Yearbook). Dordrecht, the Netherlands: Reidel.
Weingart, Peter. (1989). *Technik als sozialer Prozess*. Frankfurt am Main: Suhrkamp Verlag.
Weiss, Charles, Jr. (1985). The World Bank's support for science and technology. *Science, 227*, 261-265.
Weiss, C. H., & Singer, E. (1988). *The reporting of social science in the mass media*. New York: Russell Sage.
Weiss, John Hubbel. (1982). *The making of technological man: The social origins of French engineers*. Cambridge: MIT Press.
Weiss, R. (1991, February 1). NIH: The price of neglect. *Science, 251*, 508-511.
Weizenbaum, Joseph. (1976). *Computer power and human reason: From judgment to calculation*. San Francisco: Freeman.
Wenzel, G. (1991). *Animal rights, human rights*. Toronto: University of Toronto Press.
Werskey, Gary. (1978). *The visible college: The collective biography of British scientific socialists of the 1930s*. New York: Holt, Rinehart & Winston.
Westman, John. (1985). Modern dependency: A "crucial case" study of Brazilian government policy in the minicomputer industry. *Studies in Comparative International Development, 201*(2), 25-47.
Westrum, Ron. (1977). Social intelligence about anomalies: The case of UFOs. *Social Studies of Science, 7*, 271-302.
Westrum, Ron. (1978). Social intelligence about anomalies: The case of meteorites. *Social Studies of Science, 8*, 461-493.
Westrum, Ron. (1991). *Technologies and society*. Belmont, CA: Wadsworth.
Whalley, Peter. (1984). Deskilling engineers? The labor process, labor markets, and labor segmentation. *Social Problems, 32*, 117-132.
Whalley, Peter. (1986). *The social production of technical work: The case of British engineers*. Albany: SUNY Press.
Whalley, Peter. (1987). Constructing an occupation: The case of British engineers. *Current Research on Occupations and Professions, 4*, 3-20.
Whalley, Peter. (1991). Negotiating the boundaries of engineering: Professionals, managers, and manual work. In *Research in the sociology of organizations* (Vol. 8, pp. 191-215). Greenwich, CT: JAI.
Whalley, Peter, & Crawford, S. (1984). Locating technical workers in the class structure. *Politics and Society, 13*, 235-248.
Wheeler, P. (1993). *Memorias*. Mexico City: Centro para la innovacion technologica, UNAM.
White, L., Jr. (1962). *Medieval technology and social change*. Oxford: Oxford University Press.
Whitley, Richard. (1984). *The intellectual and social organization of the sciences*. Oxford: Oxford University Press.
Wiener, M. (1985). *English culture and the decline of the industrial spirit 1850-1980*. Harmondsworth, Middlesex, U.K.: Penguin.
Wildes, K. L., & Lindgren, N. A. (1985). *A century of electrical engineering and computer science at MIT, 1882-1982*. Cambridge: MIT Press.
Wilensky, Harold. (1964). The professionalization of everyone? *American Journal of Sociology, 70*, 137-158.

Wilkins, L. (1987). *Shared vulnerability: The media and American perception of the Bhopal disaster.* New York: Greenwood.

Wilkins, L., & Patterson, P. (Eds.). (1991). *Risky business: Communicating issues of science, risk, and public policy.* New York: Greenwood.

Williams, Bruce R. (1967). *Technology, investment and growth.* London: Chapman & Hall.

Williams, R. (1990). *Notes on the underground: An essay on technology, society and the imagination.* Cambridge: MIT Press.

Wilson, Brian. (Ed.). (1977). *Rationality.* Oxford: Oxford University Press.

Wilson, David A. (1989a, March). Consequential controversies. [Special issue: Universities and the Military] *The Annals of the American Academy of Political and Social Science, 502,* 40-57.

Wilson, David A. (1989b, March). Preface. [Special issue: Universities and the Military] *The Annals of the American Academy of Political and Social Science, 502,* 9-14.

Wilson, Graham. (1990). *Business and politics: A comparative introduction* (2nd ed.). Chatham, NJ: Chatham House.

Wilson, Richard, & Crouch, E. A. C. (1987, April 17). Risk assessment and comparisons: An introduction. *Science, 236,* 267-270.

Winner, Langdon. (1977). *Autonomous technology: Technics-out-of-control as a theme in political thought.* Cambridge: MIT Press.

Winner, Langdon. (1986a). On the foundations of science and technology studies. *Science, Technology & Society: Curriculum Newsletter of the Lehigh University STS Program & Technology Studies Resource Center, 53,* 6-8.

Winner, Langdon. (1986b). *The whale and the reactor: A search for the limits in an age of high technology.* Chicago: University of Chicago Press.

Winner, Langdon. (1991). Upon opening the black box and finding it empty: Social constructivism and the philosophy of technology. In C. Pitt & E. Lugo (Eds.), *The technology of discovery and the discovery of technology: Proceedings of the 6th International Conference of the Society for Philosophy and Technology* (pp. 503-519). Blacksburg, VA: Society for Philosophy and Technology.

Winner, Langdon. (1993). Upon opening the black box and finding it empty: Social constructivism and the philosophy of technology. *Science, Technology, & Human Values, 18*(3), 362-378.

Winograd, Terry. (1991). Strategic computing research and the universities. In C. Dunlop & R. Kling (Eds.), *Computerization and controversy* (pp. 704-716). New York: Academic Press.

Winograd, Terry, & Flores, Fernando. (1986). *Understanding computers and cognition.* Reading, MA: Addison-Wesley.

Winterston, J. (1992). *Written on the body.* London: Jonathan Cape.

Wionczek, Miguel. (1976). Notes on technology transfer through multi-national enterprises in Latin America. *Development and Change, 7,* 135-155.

Wionczek, Miguel. (1983). Research and development in pharmaceuticals: Mexico. *World Development, 11,* 243-250.

Wise, G. (1985). Science and technology. *Osiris* (2nd ser.), *1,* 229-246.

Wise, G. (1986). Science and technology. In S. G. Kohlstedt & M. W. Rossiter (Eds.), *Historical perspectives on American science: Perspectives and prospects.* Baltimore: Johns Hopkins University Press.

Wise, Norton. (1988). Mediating machines. *Science in Context, 2-1,* 77-113.

Wise, Norton, & Smith, C. (1988). *Energy and empire.* Cambridge: Cambridge University Press.

Withey, S. B. (1959). Public opinion about science and scientists. *Public Opinion Quarterly, 23,* 382-388.

Wittgenstein, Ludwig. (1921). *Tractatus Logico-philosophicus.* London: Routledge & Kegan.

Wittgenstein, Ludwig. (1953). *Philosophical investigations.* Oxford: Blackwell.

Wittgenstein, Ludwig. (1976). *Philosophical investigations*. Oxford: Blackwell.

Wittrock, B., & Elzinga, Aant. (Eds.). (1985). *The university research system*. Stockholm, Sweden: Almqvist & Wiksell.

Wolpert, Lewis. (1992). *The unnatural nature of science: Why science does not make (common) sense*. London: Faber & Faber.

Wood, S. (Ed.). (1982). *The degradation of work: Skill, deskilling and the labour process*. London: Hutchinson.

Woodhouse, Edward J. (1990). Is large-scale military R&D defensible theoretically? *Science, Technology, & Human Values, 15*(4), 442-460.

Woodhouse, Edward J. (1991). The turn toward society? Social *re*construction of science. *Science, Technology, & Human Values, 16*(3), 390-404.

Woolgar, Steve. (1976). Writing an intellectual history of scientific development: The use of discovery accounts. *Social Studies of Science, 6*, 395-422.

Woolgar, Steve. (1980). Discovery: Logic and sequence in a scientific text. In K. Knorr, R. Krohn, & R. Whitley (Eds.), *The social process of scientific investigation* (pp. 239-268). Dordrecht, the Netherlands: Reidel.

Woolgar, Steve. (1981). Interests and explanation in the social study of science. *Social Studies of Science, 11*(3), 365-394.

Woolgar, Steve. (1983). Irony in the social studies of science. In K. Knorr Cetina & M. Mulkay (Eds.), *Science observed: Perspectives on the social study of science* (pp. 239-266). London: Sage.

Woolgar, Steve. (1985). Why not a sociology of machines? The case of sociology and artificial intelligence. *Sociology, 19*, 557-572.

Woolgar, Steve. (1986). Discourse and praxis. *Social Studies of Science, 16*, 309-318.

Woolgar, Steve. (Ed.). (1988a). *Knowledge and reflexivity: New frontiers in the sociology of knowledge*. London: Sage.

Woolgar, Steve. (1988b). *Science: The very idea*. London: Tavistock.

Woolgar, Steve. (1988c). Reflexivity is the ethnographer of the text. In S. Woolgar (Ed.), *Knowledge and reflexivity* (pp. 14-34). London: Sage.

Woolgar, Steve. (1989a). *A coffeehouse conversation on the possibility of mechanizing discovery and its sociological analysis. Social Studies of Science, 19*, 658-668.

Woolgar, Steve. (1989b). What is the analysis of scientific rhetoric for? A comment on the possible convergence between rhetorical analysis and social studies of science. *Science, Technology, & Human Values, 14*, 47-49.

Woolgar, Steve. (1991a). Beyond the citation debate: Towards a sociology of measurement technologies and their use in science policy. *Science and Public Policy, 18*, 319-326.

Woolgar, Steve. (1991b). The turn to technology in social studies of science. *Science, Technology, & Human Values, 16*, 20-50.

Woolgar, Steve. (in press). Configuring the user. In John Law (Ed.), *Technology, power and society* (Sociological Review Monograph 38).

Woolgar, Steve, & Ashmore, Malcolm. (1988). The next step: An introduction to the reflexive project. In Steve Woolgar (Ed.), *Knowledge and reflexivity*. London: Sage.

Woolgar, Steve, & Pawluch, D. (1985). Ontological gerry-mandering: The anatomy of social problems explanations. *Social Problems, 32*, 214-227.

Woolgar, Steve, & Russell, G. (in press). The social basis of computer viruses. In S. Woolgar & F. Murray (Eds.), *Social perspectives on software*. Cambridge: MIT Press.

World Commission on Environment and Development. (1987). *Our common future*. Oxford: Oxford University Press.

Worster, D. (1985). *Nature's economy: A history of ecological ideas*. Cambridge: Cambridge University Press.

Worthington, E. (1983). *The ecological century: A personal appraisal.* Oxford: Oxford University Press.

Worthington, Richard. (1993). Science and technology as a global system. *Science, Technology, & Human Values, 18*(2), 175-185.

Wright, B. (Ed.). (1987). *Women, work and technology.* Ann Arbor: University of Michigan Press.

Wright, Susan. (Ed.). (1990). *Preventing a biological arms race.* Cambridge: MIT Press.

Wright, Susan. (1991). Biowar treaty in danger. *Bulletin of the Atomic Scientists, 47,* 36-40.

Wriston, W. B. (1988). Technology and sovereignty. *Foreign Affairs, 67,* 63-75.

Wynne, Brian. (1975). The rhetoric of consensus politics: A critical review of technology assessment. *Research Policy, 4*(2), 108-158.

Wynne, Brian. (1979). Between orthodoxy and oblivion: The normalisation of deviance in science. In R. Wallis (Ed.), *On the margins of science: The social construction of rejected knowledge* (Sociological Review Monograph). Keele, U.K.: University of Keele.

Wynne, Brian. (1980). Technology, risk, and participation: The social treatment of uncertainty. In J. Conrad (Ed.), *Society, technology and risk* (pp. 83-107). London: Academic Press.

Wynne, Brian. (1982). *Rationality and ritual: The Windscale inquiry and nuclear decision in Britain* (Monograph No. 3). Chalfont St. Giles, U.K.: British Society for the History of Science.

Wynne, Brian. (1987). *Risk management and hazardous wastes: Implementation and the dialectics of credibility.* Berlin: Springer.

Wynne, Brian. (1989a). Establishing the rules of law. In R. Smith & B. Wynne (Eds.), *Expert evidence: Interpreting science in the law* (pp. 23-55). London: Routledge.

Wynne, Brian. (1989b). Sheep farming after Chernobyl: A case study in communicating scientific information. *Environment Magazine, 31*(2), 10-15, 33-39.

Wynne, Brian. (1989c). Frameworks of rationality in risk management: Towards the testing of naive sociology. In J. Brown (Ed.), *Environmental threats* (pp. 33-45). London: Belhaven.

Wynne, Brian. (1991a). Knowledges in context. *Science, Technology, & Human Values, 19,* 1-17.

Wynne, Brian. (1991b). Public perception and communication of risks: What do we know? *The Journal of NIH Research, 3,* 65-71.

Wynne, Brian. (1992a). Representing policy constructions and interests in SSK. *Social Studies of Science, 22*(3), 575-580.

Wynne, Brian. (1992b). Public understanding of science: New horizons or hall of mirrors? *Public Understanding of Science, 1,* 37-43.

Wynne, Brian. (1992c). Uncertainty and environmental learning: Reconceiving science and policy in the preventive paradigm. *Global Environmental Change, 2,* 137-154.

Wynne, Brian. (1992d). Misunderstood misunderstanding: Social identities and public uptake of science. *Public Understanding of Science, 1*(3), 281-304.

Wynne, Brian. (1992e). Risk and social learning: Reification to engagement. In S. Krimsky & D. Golding (Eds.), *Social theories of risk* (pp. 275-300). New York: Praeger.

Wynne, Brian. (1992f, October). *Research cultures, policy cultures and innovation in the UK: Environment and clean technology.* Paper presented to the Science Policy Support Group, London.

Wynne, Brian. (1993). Public uptake of science: A case for institutional reflexivity. *Public Understanding of Science, 2*(4), 321-337.

Wynne, Brian, McKechnie, R., & Michael, M. (1990). *Frameworks for understanding public interpretations of science and technology* (End of Award Report to ESRC). Lancaster, U.K.: Lancaster University, Centre for Science Studies and Science Policy.

Wynne, Brian, & Millar, Robin H. (1988). Public understanding of science: From contents to processes. *International Journal of Science Education.*

Yapa, Lakshman S. (1982). Innovation bias, appropriate technology and basic goods. *Journal of Asian and African Studies, 17*, 32-44.

Yearley, Steven. (1981). Textual persuasion: The role of social accounting in the construction of scientific arguments. *Philosophy of the Social Sciences, 11*, 409-435.

Yearley, Steven. (1988). *Science, technology, and social change*. London: Unwin Hyman.

Yearley, Steven. (1989). Bog standards: Science and conservation at a public inquiry. *Social Studies of Science, 19*, 421-438.

Yearley, Steven. (1992a). *The green case*. London: Routledge & Kegan Paul.

Yearley, Steven. (1992b). Skills, deals and impartiality: The sale of environmental consultancy skills and public perceptions of scientific neutrality. *Social Studies of Science, 22*, 435-453.

Yearley, Steven. (1992c). Green ambivalence about science: Legal-rational authority and the scientific legitimation of a social movement. *British Journal of Sociology, 43*, 511-532.

Yeo, R. R. (1986). Scientific method and the rhetoric of science in Britain, 1830-1917. In John A. Schuster & R. R. Yeo (Eds.), *The politics and rhetoric of scientific method* (pp. 259-297). Dordrecht, the Netherlands: Reidel.

York, H., & Greb, A. (1982). Scientists as advisers to governments. In J. Rotblat (Ed.), *Scientists, the arms race and disarmament: A Unesco/Pugwash symposium* (pp. 83-99). London: Taylor and Francis/Paris: UNESCO.

York, H. F. (1971). *Race to oblivion: A participant's view of the arms race*. New York: Simon & Schuster.

York, H. F. (Compiler). (1973). *Arms control* (Readings from *Scientific American*). San Francisco: Freeman.

York, H. F. (1976, November 29-30). Discussion regarding Dr. Yorks lecture. In *Technological innovation: A socio-political problem* (Proceedings of the symposium, "Control of Technological Innovation). Enschede: University of Twente.

Young, L. (1989). Electronics and computing. [Special issue: Universities and the Military] *The Annals of the American Academy of Political and Social Science, 502*, 82-93.

Young, Robert. (1986). *Darwin's metaphor.* Cambridge: Cambridge University Press.

Yoxen, Edward. (1985). Speaking out about competition: An essay on the double helix as popularisation. In T. Shinn & R. Whitley (Eds.), *Expository science* (pp. 163-181). Dordrecht, the Netherlands: Reidel.

Yoxen, Edward. (1986). *The gene business: Who should control biotechnology?* London: Free Association Books.

Yoxen, Edward. (1988). *Public concern and the steering of science* (SPSG Concept Paper 7). London: Science Policy Support Group.

Zahlan, A. B. (1980). *Science and science policy in the Arab world*. New York: St. Martin's Press.

Zarkovic, Milica. (1987). Effects of economic growth and technological innovation on the agricultural labor force in India. *Studies in Comparative International Development, 22*, 103-120.

Zehr, Stephen C. (1990). *Acid rain as a social, political and scientific controversy.* Unpublished doctoral dissertation, Indiana University, Bloomington.

Zeldenrust, S. (1985, October). *Strategic action in the laboratory: (Inter)organizational resources and constraints in industrial and university research.* Paper presented at the 19th Annual Meeting of the Society for Social Studies of Science, Rennselaer Polytechnic Institute, Troy, NY.

Zenzen, Michael, & Restivo, Sal. (1982). The mysterious morphology of immiscible liquids: A study of scientific practice. *Social Science Information, 21*(3), 447-473.

Zhongliang, Z., & Jiansheng, Z. (1991). *A survey of public science literacy in China* (mimeo).

Zilsel, E. (1942). The sociological roots of science. *American Journal of Sociology, 47*, 245-279.

Ziman, John. (1968). *Public knowledge: An essay concerning the social dimension of science*. Cambridge: Cambridge University Press.

Ziman, John. (1987). *Science in a steady state*. London: Science Policy Support Group.

Ziman, John. (1989). *Science in a "steady state"* (SPSG Concept Paper No. 8). London: Science Policy Support Group.

Ziman, John. (1991). Public understanding of science. *Science, Technology, & Human Values, 16*(1), 99-105.

Zimbalist, A. (Ed.). (1979). *Case studies in the labour process*. New York: Monthly Review Press.

Zimmerman, Jan. (Ed.). (1983). *The technological woman: Interfacing with tomorrow*. New York: Praeger.

Zinberg, Dorothy S. (Ed.). (1991). *The changing university*. Boston: Kluwer Academic.

Zinberg, Dorothy S. (1993). Putting people first: Education, jobs, and economic competitiveness. In L. M. Branscomb (Ed.), *Empowering technology: Implementing a U.S. strategy*. Cambridge: MIT Press.

Ziporyn, T. (1988). *Disease in the popular American press: The case of diphtheria, typhoid fever, and syphilis, 1870-1920*. New York: Greenwood.

Zuboff, Shoshana. (1988). *In the age of the smart machine: The future of work and power*. New York: Basic Books.

Zucker, Lynn G. (1987). Institutional theories of organization. *Annual Review of Sociology, 13*, 443-464.

Zuckerman, Harriet. (1977a). Deviant behavior and social control in science. In E. Sagarin (Ed.), *Deviance and social change* (pp. 87-138). Beverly Hills, CA: Sage.

Zuckerman, Harriet. (1977b). *Scientific elite: Nobel laureates in the United States*. New York: Free Press.

Zuckerman, Harriet. (1987). Persistence and change in the careers of men and women scientists and engineers. In L. Dix (Ed.), *Women: Their underrepresentation and career differentials in science and engineering*. Washington, DC: National Academy Press.

Zuckerman, Harriet. (1988a). Intellectual property and diverse rights of ownership in science. *Science Technology and Human Values, 13*, 7-16.

Zuckerman, Harriet. (1988b). The sociology of science. In N. Smelser (Ed.), *Handbook of sociology* (pp. 511-574). Newbury Park, CA: Sage.

Zuckerman, Harriet, & Cole, Jonathan. (1975). Women in American science. *Minerva, 13*, 82-102.

Zuckerman, Harriet, Cole, Jonathan, & Bruer, John. (1991). *The outer circle: Women in the scientific community*. New York: Norton.

Zuidema, R. Tom. (1977). The Inca calendar. In A. F. Aveni (Ed.), *Native American astronomy* (pp. 219-259). Austin: University of Texas Press.

Zuidema, R. Tom. (1982a). Bureaucracy and systematic knowledge in Andean civilization. In G. A. Collier, R. I. Rosaldo, & J. D. Wirth (Eds.), *The Inca and Aztec states: 1400-1800 anthropology and history* (pp. 419-458). New York: Academic Press.

Zuidema, R. Tom. (1982b). The sidereal lunar calendar of the Incas. In A. F. Aveni (Ed.), *Archaeoastronomy in the New World: Proceedings of an international conference at Oxford 1981* (pp. 59-107). Cambridge: Cambridge University Press.

Zussman, Robert. (1984). The middle levels: Engineers and the "working middle class." *Politics & Society, 13*, 217-237.

Zussman, Robert. (1985). *Mechanics of the middle class: Work and politics among American engineers*. Berkeley: University of California Press.

Index

About the Authors

Vittorio Ancarani has been trained in sociology and political science and is currently senior researcher at Turin University where he teaches sociology of science. He is the author of a book on the French epistemologist Gaston Bachelard (1981) and has written extensively on academic science including an edited book: *La scienza accademica nell'Italia post-unitaria* (1989). His main research interests center on the relations of science and technology with international politics.

Malcolm Ashmore is a Lecturer in Sociology at Loughborough University, United Kingdom. He is the author of *The Reflexive Thesis: Wrighting Sociology of Scientific Knowledge* (1989) and coauthor, with Michael Mulkay and Trevor Pinch, of *Health and Efficiency: A Sociology of Health Economics* (1989). His main research interests lie in the sociology of scientific knowledge and currently include an investigation of the social practice of debunking in science.

Wiebe E. Bijker is Professor of Technology and Society at the University of Limburg, Maastricht, the Netherlands. He was trained as an engineer in applied physics at Delft University of Technology and studied philosophy at the University of Groningen. His *Of Bicycles, Bakelites, and Bulbs: Toward a Theory of Socio-Technical Change* is forthcoming in 1994 as part of the MIT Press series Inside Technology.

Bruce Bimber is Assistant Professor of Political Science at the University of California, Santa Barbara, where he teaches courses on American political institutions and public policy. His research interests include the politics of expertise and the politics of coordination among policymaking institutions. Prior to joining the University of California, he worked as a social scientist at RAND in Washington, D.C., and at the RAND-sponsored Critical Technologies Institute, an advisory organization for the White House Office of Science and Technology Policy. He has also served as a Research Fellow at the Brookings Institution.

Gary Bowden is Associate Professor of Sociology and Co-director of the Environmental Studies Program at the University of New Brunswick. His research interests center on the sociology of natural resources, and he recently edited "Social Studies of the Environment," a special issue of the *Canadian Review of Sociology and Anthropology.*

Michel Callon is Professor of Sociology and Director of the Centre de Sociologie de l'Innovation at the Ecole Nationale Supérieure des Mines de Paris. He has published widely in the sociology of science and technology, the economics of research and development, the sociology of translation, and scientometrics. Together with Bruno Latour, John Law, and other colleagues, he contributed to the development of actor-network theory. He has pioneered a range of techniques for describing scientific and technical change, including the co-word method. He was directly involved in the definition and implementation of policy on research in France in the early 1980s. He is the coeditor of *Mapping the Dynamics of Science and Technology*.

H. M. Collins is Director of the Science Studies Centre at the University of Bath, where he holds a personal professorial chair in sociology. He is a Fellow of the Royal Society of Arts. Between 1991 and 1993 he was President of the International Society for Social Studies of Science. He has published over 60 papers and six books on two broad themes: the nature of science, explored through comparative case studies of physics and parapsychology, and the nature of human knowledge, investigated through the study of the prospects for artificial intelligence and our interactions with machines in general. His most recent books are *Changing Order* (1985, second edition, 1992), *Artificial Experts* (1990), and *The Golem: What Everyone Should Know About Science* (1993, with Trevor Pinch). He is currently working on a new theory of action and applying it to understanding the knowledge and skills of surgeons.

Susan E. Cozzens is Associate Professor and Director of Graduate Studies in the Department of Science and Technology Studies at Rensselaer Polytechnic Institute. Her current research is on democratic processes of scientific and technological choice and the social shaping of science, including case studies in Scandinavian science policy, biomedical politics, and the politics of research on women and minorities. She has been active in science policy analysis for a dozen years, specializing in evaluation of research programs and science and technology indicators. She is coeditor of *Theories of Science in Society, The Research System in Transition,* and *Invisible Connections* and author of *Social Control and Multiple Discovery in Science.* She has been a council member for the Society for Social Studies of Science and is past editor of *Science, Technology, & Human Values.*

Gary Lee Downey is Associate Professor of Science and Technology Studies and Cultural Anthropology at Virginia Polytechnic Institute and State University. He holds a B.S. degree in mechanical engineering from Lehigh University. Author of the forthcoming ethnography, *CAD/CAM Culture: An Excavation of Cyborg Practices,* and coeditor with Joseph Dumit and Sharon Traweek of *Cyborgs and Citadels: Anthropological Interventions in the Borderlands of Technoscience,* he is coauthoring with Shannon Hegg and Juan Lucena an ethnography that explores how the engineering curriculum constructs engineers as persons.

David Edge read Physics at Cambridge, with a Ph.D. in Radio Astronomy in 1959. He spent 6 years in London, as a Radio Talks Producer in the BBC Science Unit before, in 1966, becoming the first Director of the Science Studies Unit at Edinburgh University. He resigned from that post in 1989, and retired from the university, as Emeritus Reader, in 1992. In 1971 he was jointly responsible for launching the journal *Social Studies of Science*; he is now its editor. He was elected a Fellow of the AAAS in 1989 and of the Royal Society of Edinburgh in 1991. From 1985 to 1987, he was President of 4S and was awarded the 1993 Bernal Prize. From 1989 to 1993, he was Chairman of the UK Science Policy Support Group. He is the coauthor (with Michael Mulkay) of *Astronomy Transformed* (1976) and coeditor (with Barry Barnes) of *Science in Context* (1982).

Paul N. Edwards is Visting Assistant Professor in the Program in Science, Technology, and Society and the Department of Computer Science at Stanford University. From 1990 to 1992, he taught in the Department of Science and Technology Studies at Cornell University. He holds a Ph.D. in History of Consciousness from the University of California at Santa Cruz and has recently completed his first book, *The Closed World: Computers and the Politics of Discourse in Cold War America* (in press).

Aant Elzinga was born in Friesland, the Netherlands, and immigrated to Canada, obtaining his B.A. in theoretical physics at the University of Western

Ontario in 1960. He holds a M.Sc. in the history and philosophy of science from London and a Ph.D in theory of science from Gothenburg (1971). He has written widely in the history of science and more recently on the politics and epistemology of science. His interest in science policy was particularly cultivated during a time as Science Adviser to the Canadian government in the mid-1980s. His most recent book is *Changing Trends in Antarctic Research* (1993). He is President of the European Association for the Study of Science and Technology (EASST) and a member of the International Council for Science Policy Studies. He is Full Professor in the Theory of Science at the University of Gothenburg and Vice Dean of the Faculty of Humanities. He has served as an evaluator in several international panels, for Unesco, among others. He is coeditor (with Björn Wittrock) of *The University Research System* (1985).

Henry Etzkowitz is Associate Professor of Sociology at the State University of New York at Purchase. He is also Co-director of the International Study Group on Academic-Industry Relations, Science Policy Support Group, London, and a member of the steering committee of the Centro di Politica della Scienza e della Technologia, Fondazione Rosselli, Milan. He is also the founding chair of the Section on Science, Knowledge and Technology in the American Sociological Association. His research on academic-industry relations, women in science, and state science and technology policy is supported by the U.S. National Science Foundation. He is coauthor (with M. Richter) of *Science, Technology and Society* (in press). His next book is titled *Entrepreneurial Science: The Second Academic Revolution*. He is currently in residence as Visiting Scientist at the Computer Science Department of Columbia University.

Mary Frank Fox is Associate Professor of Sociology, School of History, Technology, and Society, Georgia Tech. Her work focuses upon women and men in occupations and organizations, especially academic and scientific ones. Her current research, supported by the National Science Foundation, is a study with H. Etzkowitz of women in doctoral education in science and engineering. Her work has been published in over 20 different scholarly journals and collections. She is Associate Editor of *Sex Roles* and member of the Council of the Sex and Gender Section of the American Sociological Association. She is past Publication Chair of 4S and Chair of the 4S Handbook Committee.

Thomas F. Gieryn is Associate Professor in the Department of Sociology at Indiana University, where he is also Director of Graduate Studies and Director of the Program on Scientific Dimensions of Society. His *Handbook* chapter is drawn from his book-in-progress, *Making Space for Science: Cultural Cartography Episodically Explored,* a study of pragmatic demarcations of science from non-science and the deployment of such maps in political and professionalizing projects. He is also investigating the stabilization of the boundaries of science as they assume concrete form in the design of new research buildings for biotechnology. In 1993 he taught the sociology of science at Nankai University in Tianjin, China, and in 1994 he was honored with the President's Award for Distinguished Teaching at Indiana University.

David H. Guston is Assistant Professor of Public Policy at the Eagleton Institute of Politics, in the Bloustein School of Planning and Public Policy at Rutgers, the State University of New Jersey. He received his Ph.D. in political science from the Massachusetts Institute of Technology and has been a Research Fellow and Adjunct Lecturer at the Center for Science and International Affairs in the Kennedy School of Government at Harvard University. He has also helped staff major studies at the National Academy of Sciences and the congressional Office of Technology Assessment. His primary research interest is the interaction of democratic institutions and scientific practice, and he has published articles on the congressional oversight of science, both historical and contemporary. His first book, *The Fragile Contract: University Science and the Federal Government* (1994, coedited with K. Keniston), examines current dynamics between the political and scientific communities.

Stephen Hilgartner, a sociologist, is Assistant Professor in the Center for the Study of Society and Medicine, Columbia University. His work has addressed the popularization of science, cultural authority in the diet-cancer debate, and the process through which social problems are collectively defined. He is currently conducting a long-term, prospective study of the effort to map and sequence the human genome.

Andrew Jamison was born in California and graduated from Harvard University in history and science in 1970 before moving to Sweden. He received a Ph.D. in theory of science at the University of Gothenburg in 1983 and is currently Associate Professor and Director of the Graduate Program in Science and Technology Policy at the Research Policy Institute, University of Lund. He has served as consultant to a number of U.N. agencies. His main interests are in the relations between science, technology, and culture and environmental politics. He has contributed to several anthologies on science policy in developing countries, most recently to *The Uncertain Quest: Science, Technology and Development* (1994). He is editor of *Keeping Science Straight* (1988); coeditor (with Erik Baark) of *Technological Development in China,*

India and Japan (1986), and coauthor (with Ron Eyerman) of *The Making of the New Environmental Consciousness* (1990), *Social Movements: A Cognitive Approach* (1991), and *Seeds of the Sixties* (1994).

Sheila Jasanoff, Professor of Science Policy and Law, is the founding chair of the Department of Science and Technology Studies at Cornell University. She holds an A.B. in mathematics from Harvard College, a Ph.D. in linguistics from Harvard University, and a J.D. from Harvard Law School. Her primary research interests are in the areas of risk management and environmental regulation, interactions between science, technology, and the law, and the implications of social studies of science for science policy. Her books and articles on these topics include *The Fifth Branch: Science Advisers as Policymakers* (1990) and *Learning from Disaster: Risk Management After Bhopal* (edited; University of Pennsylvania Press, 1994). She is currently completing a book on science and the courts. She has taught at Yale University (1990-1991) and Boston University School of Law (1993) and has been a Visiting Research Associate at the Centre for Socio-Legal Studies, Oxford University (1986). She is a Fellow of the American Association for the Advancement of Science and recipient (1992) of the Distinguished Achievement Award of the Society for Risk Analysis. She has served on the Council of the Society for Social Studies of Science as well as on numerous advisory committees and panels of the National Science Foundation, the National Academy of Sciences, and the Institute of Medicine; she is a past member of the AAAS-ABA National Conference of Lawyers and Scientists (1985-1991). She is an editorial adviser to several journals and has been a consultant to the Organization for Economic Cooperation and Development, the Office of Technology Assessment, the National Science Foundation, the Institute of Medicine, and Oak Ridge National Laboratory.

Evelyn Fox Keller received her Ph.D. in theoretical physics at Harvard University, worked for a number of years at the interface of physics and biology, and is now Professor of History and Philosophy of Science in the Program in Science, Technology and Society at MIT. She is perhaps best known as the author of *A Feeling for the Organism: The Life and Work of Barbara McClintock*; *Reflections on Gender and Science*; and, most recently, *Secrets of Life, Secrets of Death: Essays on Language, Gender and Science*. Her current research is on the history of developmental biology.

Karin Knorr Cetina is Professor of Sociology at the Faculty of Sociology, University of Bielefeld, Germany. She has published widely in social studies of science (constructivist approach, laboratory studies) and (micro) social theory. Her works include *The Manufacture of Knowledge* (1981), *Science*

Observed (edited with M. Mulkay, 1983), *Advances in Social Theory and Methodology* (edited with A. Cicourel), and *Epistemic Cultures* (in print), a book that is the result of the first cross-disciplinary study of scientific practice. She is currently writing a book on theoretical constructionism.

Bruce V. Lewenstein is Associate Professor in the Departments of Communication and Science & Technology Studies at Cornell University. After an undergraduate education at the University of Chicago that combined rhetoric and physical sciences, he worked for several years as a science writer. He then earned M.A. (1985) and Ph.D. (1987) degrees in the history and sociology of science at the University of Pennsylvania. His research focuses on the history of science communication, including the cold fusion saga. He is the author of many articles about the popularization of science and edited *When Science Meets the Public* (1992). He serves as Associate Editor of the journal *Public Understanding of Science*.

Juan C. Lucena is currently a doctoral student in STS at Virginia Polytechnic Institute & State University. He holds B.S. degrees in mechanical and aeronautical engineering and an M.S. degree in STS from Rensselaer. He has worked for the Division of Engineering Infrastructure Development at the National Science Foundation. He is writing his dissertation on the relationship between power and knowledge in engineering education. He is also coauthoring a book (along with Gary Downey and Shannon Hegg) that analyzes how the engineering curriculum constructs engineers as persons.

Gerald E. Markle, originally trained as a molecular biologist, is Professor of Sociology at Western Michigan University. His work has been published in journals such as *Social Studies of Science* and *Science, Technology, & Human Values* as well the *American Sociological Review* and *Social Problems.* He is coauthor of two books, *Minutes to Midnight: Nuclear Weapons Protest in America* (1989) and *Cigarettes: The Battle Over Smoking* (1983) and coeditor of *Politics, Science, and Cancer: The Laetrile Phenomenon* (1980). His most recent book is *The Gray Zone,* a sociological meditation on the Holocaust.

Brian Martin is Senior Lecturer in the Department of Science and Technology Studies, University of Wollongong, Australia. After completing his Ph.D. at Sydney University, he worked for 10 years as an applied mathematician before switching to social science. He is the author of several books and numerous articles in the sciences and social sciences. In recent years he has worked especially on nonviolent defense, a range of scientific controversies (including nuclear power, fluoridation, pesticides, and the origin of AIDS), suppression of intellectual dissent, critique of experts, and information in a free society.

Greg Myers is Lecturer in the Department of Linguistics and Modern English Language at Lancaster University, United Kingdom, where he teaches on the Culture and Communication degree. He has published *Writing Biology* (1990) and *Words in Ads* (1994) as well as several papers on the rhetorical analysis of academic and scientific writing. His current research interests include an examination of the role of texts in science policy.

Dorothy Nelkin is University Professor at New York University, teaching in the Department of Sociology and School of Law. She is a member of the National Academy of Sciences' Institute of Medicine. She has served on the Board of Directors of the AAAS and is now on the NIH National Advisory Council for the Human Genome Project and the Advisory Panel on Embryo Research. She has been a Guggenheim Fellow and a Visiting Scholar at the Russell Sage Foundation. Her books include *Controversy: Politics of Technical Decisions; Science as Intellectual Property; The Creation Controversy; Workers at Risk; Selling Science: How the Press Covers Science and Technology* (with Laurence Tancredi); *Dangerous Diagnostics: The Social Power of Biological Information;* and (with Susan Lindee) *Supergene—DNA in American Popular Culture* (forthcoming).

James C. Petersen, Professor of Sociology at Western Michigan University, has conducted research on science-related controversies, dissent by engineers, and public involvement in the science policy process. He is the editor of *Citizen Participation in Science Policy* (1984), coeditor of *Politics, Science, and Cancer: The Laetrile Phenomenon* (1980), and coauthor of *Whistleblowing: Ethical and Legal Issues in Expressing Dissent* (1986). His work has also been published in *Social Studies of Science* and *Science, Technology, & Human Values* and in edited books including, Daryl Chubin and Ellen Chu, *Science off the Pedestal* (1989); H. Tristram Engelhardt Jr. and Arthur Caplan, *Scientific Controversies* (1987); and Dorothy Nelkin, *Controversy: Politics of Technical Decisions* (1979 and 1984, first and second editions).

Trevor Pinch is Associate Professor in the Department of Science and Technology Studies, Cornell University. He has a degree in physics from Imperial College London University, an M.A. from Manchester University in the social organization of science and technology, and a Ph.D. in sociology from Bath University. He has been a researcher at the University of Bath and has taught in the Sociology Department of York University in England. His research interests are primarily in the sociology of science and technology, about which he has published many articles and several books. His most recent book (coauthored with Harry Collins) is *The Golem: What Everyone Should Know About Science* (1993/1994). He is currently writing a book on the rhetoric of street selling.

Jonathan Potter is Reader in Discourse Analysis at Loughborough University, United Kingdom. He is the author, with Margaret Wetherell, of *Discourse and Social Psychology* (1987) and *Mapping the Language of Racism* (1992); and, with Derek Edwards, *Discursive Psychology* (1993) as well as many articles on the analysis of scientific discourse. He is interested in the construction and use of factual discourse in science, social science, and the media. He is Associate Editor of *Theory and Psychology.*

Sal Restivo is Professor of Sociology and Science Studies in the Department of Science and Technology Studies at Rensselaer Polytechnic Institute in Troy, New York, and President (1993-1995) of the Society for Social Studies of Science. He is the author of *The Social Relations of Physics, Mysticism, and Mathematics* (1983), *The Sociological Worldview* (1991), *Mathematics in Society and History* (1992), and *Science, Society, and Values: Toward a Sociology of Objectivity* (1994) and coeditor of *Comparative Studies in Science and Society* (with C. K. Vanderpool, 1974) and of *Math Worlds: Philosophical and Social Studies of Mathematics and Mathematics Education* (with J. P. Van Bendegem and R. Fischer, 1993).

Evelleen Richards is Associate Professor in the Department of Science and Technology Studies at the University of Wollongong, Australia. She completed her Ph.D. at the University of New South Wales and is a social historian of science with a special interest in the social construction of biological and medical knowledge. She is the author of *Vitamin C and Cancer: Medicine or Politics?* (1991). She has also published widely in the areas of the social history of evolutionary biology, the politics of medicine and health, and gender and science. She is currently at work on a book, *Darwin and the Descent of Woman: The Making of Sexual Selection*, and a number of projects in the sociology of drug evaluation.

Yehouda Shenhav is Associate Professor of Sociology at Tel Aviv University. His Ph.D. is from Stanford University and he has taught at several universities, including Stanford University and the University of Wisconsin–Madison. He is currently studying the social, political, and epistemological roots of the science of organizations using 19th-century engineering magazines like *The American Machinist* and *The Engineering Magazine*.

Wesley Shrum has been a faculty member in the Department of Sociology at Louisiana State University since 1982. His main interest has been organizational networks in technical systems. Concurrent with the review essay in this volume, he has recently published an annotated bibliography on science and technology in less developed countries (with Carl Bankston and Stephen Voss, 1994). His current field study in Kenya, Ghana, and India examines research networks in sustainable agriculture and natural resource management. Other research interests include the role of critics in the performing arts, the relationship between crime and community security practices, and ritual disrobement at Mardi Gras.

Wim A. Smit is Associate Professor and Director of the Centre for Studies of Science Technology and Society, University of Twente. After receiving his Ph.D. in Physics (1973), he went into STS studies. He has published on such issues as Nuclear Proliferation, Assessment and Dynamics of Military Technological Developments, Assessments of Nuclear Technology, and Risk Assessments. He was coauthor of the SIPRI publication *Uranium Enrichment and Nuclear Weapon Proliferation* (1983). The books he has coedited include *Military Technological Innovation and Stability in a Changing World: Politically Assessing and Influencing Weapon Innovation and Military Research and Development* (1992), *Controlling the Development and Spread of Military Technology* (1992), and *Non-Provocative Defense as a Principle of Arms Reduction and Its Implications for Assessing Defense Technologies* (1989).

David Turnbull is a lecturer in the Social Studies of Science in the Arts Faculty, Deakin University, Geelong, Victoria, Australia. He has recently published *Maps Are Territories: Science Is an Atlas* (1993). His current research compares the construction of knowledge spaces in different scientific traditions.

Judy Wajcman is a Professor of Sociology at the University of New South Wales in Sydney, Australia, and is currently Principal Research Fellow of the Industrial Relations Research Unit at the University of Warwick in England. She has been active in the women's movement in both Britain and Australia since the early 1970s. Her previous research has focused particularly on feminist approaches to science and technology and on gender relations in the workplace. At Warwick she is currently researching the different experience of women and men senior managers, both at work and at home. She is the author of *Women in Control: Dilemmas of a Workers' Co-Operative* (1983), *Feminism Confronts Technology* (1991), and coeditor, with Donald MacKenzie, of *The Social Shaping of Technology* (1985).

Helen Watson-Verran is Senior Lecturer in the Department of History and Philosophy of Science, University of Melbourne, Australia. During 7 years at Obafemi Awolowo University in Nigeria in the early 1980s, she began work on understanding how disparate knowledge systems might be worked together. Since returning to Australia, she has worked with the Yolngu Aboriginal community in northern Australia articulating a framework and elaborating new ways of "mapping" the logic of Yolngu and Western knowledge.

Andrew Webster is Reader in Sociology at Anglia Polytechnic University and Director of the Science and Technology Studies Unit (SATSU). He is also Co-director of an International Study Group on Academic-Industry relations and coordinator of an EC-funded evaluation of trans-European innovation. His principal area of interest is in the sociological aspects of science and technology policy, with particular reference to the relationships between corporate and public sector research science. Projects have examined the sociological dynamics of long-term research collaborations between universities and corporations, and the emergence of a patenting culture in academia. One of his recent texts is *Science, Technology and Society* (1992).

Edward J. Woodhouse (Ph.D., political science, Yale, 1983) studies the obstacles to wise policymaking about science and technology. Believing that technoscience as now practiced is a partisan conflict with more losers than winners worldwide, he attempts to develop usable knowledge helpful to have-nots in their (fettered and halting) efforts to constrain the privileged position of business, reduce global economic inequality, and reorient socializing institutions and media to promote more vigorous inquiry about social goals and technological means. His books include *The Policy-Making Process* (1993, with Charles Lindblom), *The Demise of Nuclear Energy? Lessons for Democratic Control of Technology* (1989, with Joseph Morone), and *Averting Catastrophe: Strategies for Regulating Risky Technologies* (1986, with Joseph Morone). He now is writing a utopian novel while continuing research on why humans and political-economic institutions repeat their errors more than learning from them.

Brian Wynne is Professor of Science Studies and Director of the Centre for Science Studies and Science Policy at Lancaster University. He has an M.A. and Ph.D. in materials science from Cambridge University and M.Phil. in sociology of science from Edinburgh University. After sociological work in the history of the 20th century physics his research has focused on scientific and public knowledge in the fields of risk and technology politics. This has increasingly turned to critical analysis of the globalization of environmental issues and of the mutual reinforcement of forms of political order and dominant scientific cultures. His publications include: *Rationality and Ritual: The Windscale Inquiry and Nuclear Decision Making in Britain* (1982); *Risk Management and Hazardous Wastes* (1987); and *Misunderstanding Science* (forthcoming, edited with Alan Irwin).

Steven Yearley is Professor of Sociology at the Queen's University of Belfast; he was formerly Professor of Sociology at the University of Ulster, near Belfast; he was also formerly Reader in Sociology at the Queen's University of Belfast. He has worked extensively on the sociology of the green movement, on social studies of the earth sciences, and on the areas in which science studies and studies of environmentalism overlap. Representative publications include *The Green Case: A Sociology of Environmental Issues, Arguments, and Politics* (1992) and *Making Sense of Science* (1995).

Printed in the United States
By Bookmasters

Printed in the United States
By Bookmasters